Advances in Intelligent Systems and Computing

Volume 345

Series editor

Janusz Kacprzyk, Polish Academy of Sciences, Warsaw, Poland
e-mail: kacprzyk@ibspan.waw.pl

About this Series

The series "Advances in Intelligent Systems and Computing" contains publications on theory, applications, and design methods of Intelligent Systems and Intelligent Computing. Virtually all disciplines such as engineering, natural sciences, computer and information science, ICT, economics, business, e-commerce, environment, healthcare, life science are covered. The list of topics spans all the areas of modern intelligent systems and computing.

The publications within "Advances in Intelligent Systems and Computing" are primarily textbooks and proceedings of important conferences, symposia and congresses. They cover significant recent developments in the field, both of a foundational and applicable character. An important characteristic feature of the series is the short publication time and world-wide distribution. This permits a rapid and broad dissemination of research results.

Advisory Board

More information about this series at http://www.springer.com/series/11156

Jong-Hwan Kim · Weimin Yang
Jun Jo · Peter Sincak · Hyun Myung
Editors

Robot Intelligence Technology and Applications 3

Edition of the Selected Papers from the
3rd International Conference on Robot
Intelligence Technology and Applications

 Springer

Editors

Jong-Hwan Kim
Korea Advanced Institute of Science and
    Technology (KAIST)
Daejeon
Korea

Weimin Yang
Beijing University of Chemical
    Technology
Beijing
China

Jun Jo
Griffith University
Gold Coast
Australia

Peter Sincak
Technical University of Kosice
Kosice
Slovakia

Hyun Myung
KAIST
Daejeon
Korea

ISSN 2194-5357   ISSN 2194-5365   (electronic)
Advances in Intelligent Systems and Computing
ISBN 978-3-319-16840-1   ISBN 978-3-319-16841-8   (eBook)
DOI 10.1007/978-3-319-16841-8

Library of Congress Control Number: 2015934216

Springer Cham Heidelberg New York Dordrecht London

Printed on acid-free paper

Springer International Publishing AG Switzerland is part of Springer Science+Business Media
(www.springer.com)

Preface

The "intelligent" robots have different possibilities and abilities as humans. On one hand, the humans have more and unbeatable intellect vice versa robots, whose advantage is in natural connectivity with other technical systems. These facts results in the so called "networked robots" which are connected to different external sensor arrays, cameras or even other robots. The robots can extend their information source and overcome intelligence of humans. So in fact we live in an era when humans with highly developed intellect are being compared with multisource supported robots with ability to get pieces of information from cyberspace or any kind of local and remote sensors in real time manner.

This is the third edition that aims at serving the researchers and practitioners in related fields with a timely dissemination of the recent progress on robot intelligence technology and its applications, based on a collection of papers presented at the 3rd International Conference on Robot Intelligence Technology and Applications (RiTA), held in Beijing, China, November 6 - 8, 2014. For better readability, this edition has the total 74 papers grouped into 3 parts: Part I: Ambient, Behavioral, Cognitive, Collective, and Social Robot Intelligence, Part II: Computational Intelligence and Intelligent Design for Advanced Robotics, Part III: Applications of Robot Intelligence Technology, where individual parts, edited respectively by Peter Sincak, Hyun Myung, Jun Jo along with Weimin Yang and Jong-Hwan Kim, begin with a brief introduction written by the respective part editors.

Part I. Ambient, Behavioral, Cognitive, Collective, and Social Robot Intelligence

RITA conference is one of the few conferences which address a Robot Intelligence Theory and Applications. These perspectives of robots intelligence technology including ambient, behavioral, cognitive, collective, genetic and social intelligences are very important in creating an alternative to single self-learning robot. In the first sentence of this part I have written the word Intelligence in quotes so honestly believe that the notion "robot intelligence" is going to evolve and the artificial intelligence in many of their various forms is a key factor for creating robots as machines with different level of intelligence related to given task from humans or companionship with humans.

In this first part we can find 24 papers dealing with various partial contribution to Ambience, Cognition, Behavioral modeling, Social Robots and Collective Intelligence. I can divide the papers into four sections;

A. Robot navigations, localization, path planning and related problems in various environments

1) Directional drilling localization using graph SLAM and magnetic field in backward travel
2) Visual Odometry Algorithm Using an RGB-D Sensor and IMU in a Highly Dynamic Environment
3) Graph Structure-based Simultaneous Localization and Mapping with Iterative Closest Point Constraints in Uneven Outdoor Terrain
4) A Usability Study for Signal Strength based Localization
5) Path Mapping and Planning with Partially known Paths using Hierarchical State Machine for Service Robot
6) Dual Multiobjective Quantum-inspired Evolutionary Algorithm for a Sensor Arrangement in a 2D Environment
7) RRT*-Quick: A Motion Planning Algorithm with Faster Convergence Rate

B. Robot movement and related problems of control

1) Wave based tracking control of a flexible arm using lumped model
2) Stable Modifiable Walking Pattern Generator with a Vertical Foot Motion by Evolutionary Optimized Central Pattern Generator
3) Falling Prevention System from External Disturbances for Humanoid Robots
4) Modifiable Walking Pattern Generator on Unknown Uneven Terrain
5) A Four-legged Social Robot based on a Smartphone

C. Robot behaviors, agent system for complex tasks

1) Behavior Selection Method of Humanoid Robots to Perform Complex Tasks
2) A Research Platform for Flapping Wing Micro Air Vehicle Control Study
3) Genetic Network Programming with Fuzzy Reinforcement Learning for Multi-Behaviour Robot Control
4) Robust and Reliable Feature Extraction Training by Using Unsupervised Pre-training with Self-Organization Map
5) Self-organization in Groups of Intelligent Robots
6) Formation task in a group of quad rotors
7) Cooperative Control of UAVs Using a Single Master Subsystem for Multi-task Multi-target Operations
8) Applying Self Organization Maps to Cluster Analysis of Regional Industrial Structures

D. Human robot interaction and related topics

1) Research of a self-adaptive robot impedance control method for robot-environment interaction
2) Gaze Control Factors for Natural Human Robot Interaction from Scanpath Comparisons
3) The Affective Loop: A Tool for Autonomous and Adaptive Emotional Human-Robot Interaction
4) 3D Visibility Check in Webots for Human Perspective Taking in Human-Robot Interaction

Generally we can consider these papers as contributions to the selected issues of Robot Intelligence from Theory and Applications. This issues are extremely important since communities in Artificial Intelligence and Robotics tends to be isolated and we do need to prevent the re-invention of artificial intelligence in robotics community to help science, research and technology and on the other hand also practical application of artificial intelligence in Robotics can give a valuable feedbacks to artificial intelligence researchers to rethink the theory since applications are providing giving an important experience and data for the theory. The both communities need to be in a closer contact and benefit from each other and to share the obtained knowledge to for a positive future of the mankind.

Part II. Computational Intelligence and Intelligent Design for Advanced Robotics

This part consists of three segments of different topics which cover broad spectrum of topics related to robot intelligence; Computational Intelligence, Behavioral Intelligence, and Intelligent robot design.

The *Computational Intelligence* is a methodology involving computing that exhibits an ability to learn or to deal with new situations. It usually comprises of soft computing techniques such as evolutionary computation, neural networks, and fuzzy systems.

By representing a problem with a chromosome and genes, and describing fitness of this chromosome in the form of objective function, evolutionary computation effectively solves a solution using meta-heuristics inspired by genetics. The problems hard to be solved due to their inherent complexity, or the problems that do not have mathematical model that is necessary for classical optimization methods or hard computing techniques, can be candidates for the application of computational intelligence approach. The robots that use these problem solving capabilities can be regarded to have computational intelligence. The following papers present excellent examples of these approaches.

1) Obstacle Avoidance and Dribbling Strategy for Humanoid Soccer Robots
2) Path Planning of Nonholonomic Mobile Robot for Maximum Information Collection in Dynamic Environments
3) DNSGA: Dual Nondominate Sorting Genetic Algorithm

To achieve complex intelligent behavior of robots, computational intelligence techniques as well as various perception/recognition techniques can be used. The following papers treat various kinds of examples related to the topic of *Behavioral Intelligence*.

1) Practical Real-Time System for Object Counting based on Optical Flow
2) A Study on Motion Energy of Humanoid Robot in Different Walking Gaits or Postures
3) Dense 3D Mapping for indoor environment based on Kinect-style depth cameras
4) Exponential Backoff-Sampling RRT For Smart Carpet
5) The Mobile Robot SLAM Based on Depth and Visual Sensing in Structured Environment
6) Application Exploring of Ubiquitous Pressure Sensitive Matrix as Input Resource for Home-Service Robots
7) Arm Trajectory Generation based on RRT* for Humanoid Robot
8) Natural Motion Generation for Humanoid Robot Dancing Show Using Humanoid Robot : TOTO(Training Oriented Test Object robot)
9) A Colour Detection and Connected-Objects Separation Methodology for VEX Robotics

The above papers deal with behavioral intelligence applied to humanoid and mobile robots robots. They show advances in the localization, mapping, and path planning issues to implement behavioral intelligence.

The rest of the papers are dedicated to the *Design of Intelligent Robots* using intelligent perception and sensing capabilities. These robots can be also designed to be used as educational purposes as can be seen in the following papers:

1) Drawing Pressure Estimation Using Torque Feedback Control Model of A 4-DOF Robotic Arm
2) Evolving Honeycomb Pneumatic Finger in Bullet Physics Engine
3) Towards Coexistence of Human and Robot: How Ubiquitous Computing Can Contribute?
4) Electronic Artificial Skin for application in Pressure Sensor
5) An Intelligent Rover Design Integrated with Humanoid Robot for Alien Planet Exploration
6) Unified Minimalistic Modelling of Piezoelectric Stack Actuators for Engineering Applications
7) Conception of a Tendon-Sheath and Pneumatic System Driven Soft Rescue Robot
8) AUT-UofM Humanoid TeenSize Joint Team; A New Step Toward 2050's Humanoid League Long Term RoadMap
9) ROBO+EDU: Project and implementation of Educational Robotics in Brazilian public schools
10) Smartphone Controlled Robot Platform for Robot Soccer and Edutainment
11) AMiRoSoT: An autonomous, vision based, low cost robot soccer league
12) Model Checking of a Training System using NuSMV for Humanoid Robot Soccer
13) Some Effects of Culture, Gender and Time on Task of Student Teams Participating in the Botball Educational Robotics Program

Part III. Applications of Robot Intelligence Technology

There are so many aspects of life that robots may be applied to make everyday life safer and more convenient. Robots should be intelligent enough to go beyond simple preprogrammed reactions to environmental stimuli. Intelligent robots should be able to solve complicated problems, self-learn and perform actions that were not explicitly programmed, and thus be able to participate in complex interactions with humans or other robots. This part presents 25 novel papers that will introduce some of the many possible applications of robots with intelligent technologies. The intelligent technologies utilized in the papers are as follows:

A. Computer Algorithms and Computational Intelligence techniques

1) Coordinated Control of a New Pneumatic Gripper
2) Robust Camera Calibration for the MiroSot and the AndroSot Vision Systems using Artificial Neural Networks
3) Image Classification Using Convolutional Neural Networks With Multi-Stage Feature
4) Traversability Classification using Super-voxel Method in Unstructured Terrain
5) Soft Peristaltic Actuation for the Harvesting of Ovine Offal
6) The Design of ARM-based Control System of Unmanned Research Catamaran
7) The Design and Experiment of Unmanned Surveyed Catamaran

B. Computer Vision and Image Processing

1) Techniques for Designing an FPGA-Based Intelligent Camera for Robots
2) Robust Object Recognition Under Partial Occlusions Using an RGB-D Camera
3) Face Verification Across Pose via Look-Alike Ranked List Comparison
4) Landmark Tracking Using Unrectified Omnidirectional Image For Automated Guided Vehicle
5) Human Pose Estimation Algorithm for Low-Cost Computing Platform Using Depth Information Only
6) Dense Optical Flow Estimation with 3D Structure Tensor Models

C. Various Sensor Technologies/Devices other than cameras

1) Improvement of Dust Detection System using Infra-red Sensors
2) Assessment of the Effectiveness of Acupuncture on Facial Paralysis based on sEMG Decomposition
3) Human detection Algorithm Based on Edge Symmetry
4) Pose-Sequence-Based Graph Optimization Using Indoor Magnetic Field Measurements
5) An efficient ego-lane detection model to avoid false-positives detection of guardrails
6) Peak Detection with Pile-up Rejection using Multiple- Template Cross-correlation for MWD (Measurement While Drilling)

D. Wireless Technologies

1) Wireless Remote Control of Robot Dual Arms and Hands Using Inertial Measurement Units for Learning from Demonstration
2) Online learning-prediction Based diagnosis Decision Support System towards swallowing dysfunction in Rehabilitation Medicine

E. Cloud Computing, Simulation, Mechanical Structure and Solar Energy

1) Cloud-based Image Recognition for Robots
2) Rubio Simulator: A PBL Based Project of a Rubik Cube Solving Robot Simulator
3) Magnetorheological Damper Control in a Leg Prosthesis Mechanical
4) Solar-hydrogen energy: an effective combination of two alternative energy sources that can meet the energy requirements of mobile robots

We do hope that readers find the third edition of Robot Intelligence Technology and Applications 3, RiTA 3, stimulating, enjoyable and helpful for their further research.

November, 2014   Weimin Yang
Beijing   Jong-Hwan Kim
                                                General Chairs of RiTA 2014

                                                        Peter Sincak
                                                        Hyun Myung
                                                            Jun Jo
                                                Program Chairs of RiTA 2014

Contents

Part II: Computational Intelligence and Intelligent Design for Advanced Robotics

Part III: Applications for Robot Intelligence Technology

Part I
Ambient, Behavioral, Cognitive, Collective, and Social Intelligences

Part I

Ambient, Behavioral, Cognitive, Collective, and Social Intelligences

Directional Drilling Localization Using Graph SLAM and Magnetic Field in Backward Travel

Byeolteo Park and Hyun Myung

Urban Robotics Laboratory, KAIST (Korea Advanced Institute of Science and Technology),
Daejeon 305-701, Republic of Korea
{starteo,hmyung}@kaist.ac.kr

Abstract. Directional drilling is a method to bore toward a desired path. For control of directional drilling, the information about underground localization is important. However, in underground environment, GPS signal is unreachable and wireless beacon system is useless. Conventional researches focused on the methods based on IMU (Inertial Measurement Unit), but vibration of drilling and distortion of magnetic fields interfere with the IMU measurement. In this paper, a new underground localization algorithm for directional drilling using graph SLAM (simultaneous localization and mapping) and magnetic fields in backward travel is proposed. The proposed algorithm records magnetic fields in forward travel and optimizes the poses using the graph SLAM by matching magnetic fields in backward travel. The proposed algorithm is verified by simulations.

Keywords: graph SLAM, underground localization, magnetic field, directional drilling.

1 Introduction

Directional drilling is widely used for gathering unconventional resources such as tight gas and shale gas. The purpose of directional drilling is to bore toward a desired direction in underground [1]. This method is more advanced than conventional vertical or horizontal drilling. Directional drilling promotes the accessibility to complex locations. And it is possible to drill the large area through a well. Since directional drilling can have a curved path. Therefore directional drilling has the advantage that is the economic efficiency [2, 3].

RSS (rotary steerable system) is the latest directional drilling technology. RSS facilitates changing the direction of drill bit in real time [4]. For control of RSS to a desired direction, RSS must predict the current pose. Therefore the information about underground localization is important. However, in underground environment, GPS signal is unreachable and wireless beacon system is useless.

Conventional researches use IMU and describe about compensation of the noise of the sensors. There are methods to analyze the noise properties and to use external physical measurement. Methods to analyze the noise properties use neural network, wavelet transform, and low pass filter [5, 6]. However, these methods cannot com-

© Springer International Publishing Switzerland 2015
J.-H. Kim et al. (eds.), *Robot Intelligence Technology and Applications 3,*
Advances in Intelligent Systems and Computing 345, DOI: 10.1007/978-3-319-16841-8_1

pensate unsuspected noise. Methods to use external physical measurement employ gravity and geomagnetic field [7, 8]. Since vibration and distortion of magnetic field exist in drilling environment, these methods have limitation. Furthermore, a method using the penetration rate is introduced [9]. But this method is usable in a small curvature, because of linear system equations. Thus, existing methods cannot solve problems such as unknown environment, vibration, distortion of magnetic field, and a bent path.

To address these problems, we propose a novel underground localization algorithm for directional drilling. The proposed algorithm uses graph SLAM and magnetic field in backward travel. First, odometry of the drill system is predicted by the penetration length and the angular rate. Simultaneously, magnetic field in forward travel is recorded at each pose. Second, in backward travel, the proposed algorithm examines the similarity of magnetic field between forward and backward travels using normalized cross-correlation. The drill system has occasionally backward travel in order to change the drill bit or to case the hole. And, in backward travel, the drill system passes the location where the drill system passed in forward travel through the hole. As a result, the proposed algorithm can seek the same poses. Conventional methods need geomagnetic field. However, observed magnetic field is composed of geomagnetic field and local magnetic interference. Geomagnetic field is generated by the inner part of Earth. Local magnetic interference is generated by the metallic drill system and soils. Conventional methods need geomagnetic field. Since the proposed algorithm compare the stream of magnetic field, the result of the proposed algorithm is not affected. Finally, the proposed algorithm optimizes the poses using these results and graph SLAM. In this paper, the second section describes the proposed algorithm. And the proposed algorithm is validated by simulations in the third section. The forth section provides future work and conclusions.

2   Methods

2.1   Odometry

The odometry of the drill system is obtained by the penetration length and the angular rate. The penetration length is measured by wheel encoder at ground. And the angular rate is measured by the gyroscope which is installed nearby drill bit. Fig. 1 describes coordinate systems and locations of sensors. The origin of the body frame is the center of the gyroscope. Reference frame is located between encoders.

If the movement of the drill system is too small in a short time, the pose of the drill system is predicted by two parts which are the rotation and the translation. The transformation matrix of the drill system is represented as

$$T_b = T_r * T_t = \begin{pmatrix} & & & 0 \\ & R_b & & 0 \\ & & & 0 \\ 0 & 0 & 0 & 1 \end{pmatrix} \begin{pmatrix} 1 & 0 & 0 & 0 \\ 0 & 1 & 0 & 0 \\ 0 & 0 & 1 & D_p \\ 0 & 0 & 0 & 1 \end{pmatrix} \tag{1}$$

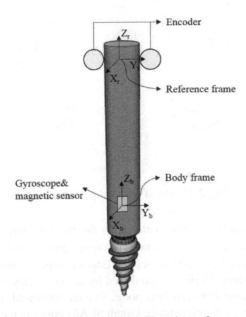

Fig. 1. Coordinate system and locations of sensors

where T_b is the transformation matrix of the drill system, T_r is the transformation matrix about the rotation, T_t is the transformation matrix about the translation, R_b is the rotation matrix of the drill system, and D_p is the penetration length. R_b is obtained by the gyroscope. And D_p is measured by the encoder.

2.2 Similarity of Magnetic Field

The proposed algorithm records the strength of magnetic field by each pose. Distortion of magnetic field reflects regional characteristics. When the drill system moves to backward direction, the proposed algorithm compares magnetic field between the recorded data in forward travel and the current data in backward travel for the similarity test. This process uses normalized cross-correlation. Normalized cross-correlation is widely used in computer vision researches for template matching. Normalized cross-correlation is robust to noise, signal scale change, and offset difference [10]. The result of normalized cross-correlation exhibits a value between -1 and 1. If the result of normalized cross-correlation is closed to 1, it means that two signal are similar. By discovering the result over a threshold, the proposed algorithm treats that two pose are the same. As a result, the noisy path is adjusted.

2.3 Pose Optimization

The proposed algorithm optimizes the poses using graph SLAM. Graph SLAM is the state-of-art SLAM algorithm and optimizes for the full path [11]. Fig. 2 describes

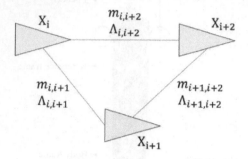

Fig. 2. Graphical model of graph SLAM

graphical model of graph SLAM, where X_i is the pose vector, $m_{i,j}$ is the relative pose between two nodes, and $\Lambda_{i,j}$ is and information matrix with covariance. Graph SLAM is comprised of edges and nodes [12]. Edges compose constraints. Nodes present poses of drill system. Nodes are predicted by the odometry. If the result of normalized cross-correlation between two nodes is over threshold, the proposed algorithm treats two nodes as identical node. Graph SLAM corrects the full path using the maximum likelihood estimation method, as follows:

$$x^* = \arg\min_x \frac{1}{2}\sum_{<i,j>\in\Gamma} r_{i,j}^T(x)\Lambda_{i,j}r_{i,j}(x) \qquad (2)$$

where $x = \{X_1, X_2, ...\}$, $r_{i,j}$ is the residual of the predicted and observed relative poses between two nodes, $\Lambda_{i,j}$ is the information matrix which has the covariance information, and Γ is the set of edges. In the present study, this method is used for optimization of the all poses. The proposed algorithm uses GTSAM (Georgia Tech smoothing and mapping) graph SLAM algorithm for computational efficiency [13].

3   Simulations

In order to verify the proposed algorithm, the simulations are performed. The total travelled length is 1000m in the simulations. The path contains a single bent path with curvature of 30°/100m. The distance between nodes is 5m. 3-axis angular rate and 3-axis magnetic field are generated by considering the curvature. Noise of data was modeled following a Gaussian distribution. The generated data is shown in Fig. 3.

In Fig. 3, the travelled distance represents the total travelled length regardless of the penetration direction. Fig.4 shows the results of the simulation.

In Fig. 4, while the odometry diverges, the results of the proposed algorithm is close to the ground truth. And Fig. 5 shows errors of the odometry and the proposed algorithm by each node.

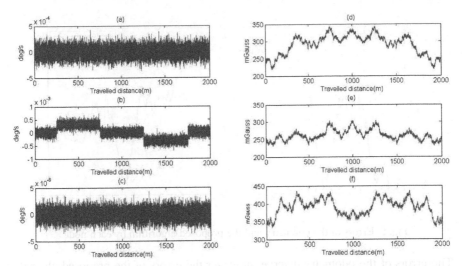

Fig. 3. Generated data of 3-axis angular rate and 3-axis magnetic field. (a) Angular rate of X-axis. (b) Angular rate of Y-axis. (c) Angular rate of Z-axis. (d) Magnetic field of X-axis. (e) Magnetic field of Y-axis. (f) Magnetic field of Z-axis.

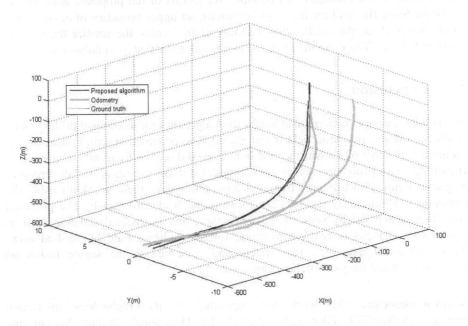

Fig. 4. Results of the odometry and the proposed algorithm. The result of the proposed algorithm is close to the ground truth.

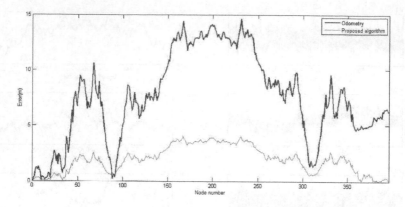

Fig. 5. Errors of the odometry and the proposed algorithm by each node

The errors of the odometry diverge, however the errors of the proposed algorithm are maintained with small value. RMSE (root mean square error) of the odometry is 7.5810m, and RMSE of the proposed algorithm is 1.933m. Furthermore, in the 10 tests, RMSE of the odometry is 7.6934m, and RMSE of the proposed algorithm is 1.8959m. Since the drill environment is various, an upper boundary of errors is not fixed. Nonetheless, the smaller the boundary of the errors, the smaller the interval between holes is. Therefore the proposed algorithm is verified by simulations.

4   Conclusion

In this paper, a novel underground localization algorithm for directional drilling using graph SLAM and magnetic field is introduced. The proposed algorithm uses the attribute that the drill bit revisits the hole in backward direction frequently. Conventional methods have problems such as the accumulated noise, vibration, and distortion of magnetic fields. However, the proposed algorithm solves these problems using the magnetic field similarity test between forward and backward travels. The proposed algorithm is verified by simulations. Since there are unexpected situation such as temperature changes in real environment, further studies should be verified using the experimental data of a real system and compared to the sensor fusion on accelerometer and magnetic field sensor.

Acknowledgments. This work was supported by the Technology Innovation Program (#10048079, Pilot Test Research for Directional Drilling System and #2011201030001D, Technical Development of Stable Drilling and Operation for Shale/Tight Gas Field) funded by the Ministry of Trade, Industry & Energy(MOTIE, Korea).

References

1. Park, B., Kim, J., Park, J., Shin, J., Myung, H.: Hybrid 4-pad Rotary Steerable System for Directional Drilling of Unconventional Resources. In: Proc. 10th Int. Conf. on Ubiquitous Robots and Ambient Intelligence (URAI), pp. 659–660 (2013)
2. Jurkov, A.S., Cloutier, J., Pecht, E., Mintchev, M.P.: Experimental Feasibility of the in-Drilling Alignment Method for Inertial Navigation in Measurement-While-Drilling. IEEE Trans. Instrum. Meas. 60, 1080–1090 (2011)
3. Moody, M., Jones, S., Leonard, P.: Development & Field-Testing of a Cost Effective Rotary Steerable System. In: Proc. Society of Petroleum Engineers (SPE) Annual Technical Conf. and Exhibition, SPE-90482 (2004)
4. Park, J., Kim, J., Park, B., Myung, H.: Design and Analysis of a New Hybrid Rotary Steerable System for Directional Drilling. In: Proc. 44th Int. Symp. on Robotics (ISR), pp. 1–3 (2013)
5. Zhang, Y., Wang, Y., Yang, T., Yin, R., Fang, J.: Dynamic Angular Velocity Modeling and Error Compensation of One-Fiber Fiber Optic Gyroscope (OFFOG) in the Whole Temperature Range. Meas. Sci. Technol. 23, 025101 (2012)
6. Yanbo, Z., Xiaofeng, S., Huihui, Z., Jun, Z.: Dynamical Measurement of Drill Tool Attitude at Stick-Slip and Continuous Rotation Mode in Vertical Drilling. In: Proc. 6th IEEE Conf. on Industrial Electronics and Applications (ICIEA), pp. 2698–2701 (2011)
7. Noureldin, A., Irvine-Halliday, D.: Measurement-While-Drilling Surveying of Highly Inclined and Horizontal Well Sections Utilizing Single-Axis Gyro Sensing System. Meas. Sci. Technol. 15, 2426 (2004)
8. Jian, K., BoXiong, W., ZhongXiang, H., Rui, W., Tao, L.: Study of Drill Measuring System Based on MEMS Accelerative and Magnetoresistive Sensor. In: Proc. IEEE 9th Int. Conf. on Electronic Measurement & Instruments (ICEMI), pp. 2112–2116 (2009)
9. Noureldin, A., Tabler, H., Irvine-Halliday, D., Mintchev, M.P.: A New Borehole Surveying Technique for Horizontal Drilling Processes using One Fiber Optic Gyroscope and Three Accelerometers. In: Proc. Int. Association of Drilling Contractors (IADC)/Society of Petroleum Engineers (SPE) Drilling Conf. and Exhibition, SPE-59198 (2000)
10. Zhao, F., Huang, Q., Gao, W.: Image Matching by Normalized Cross-Correlation. In: Proc. IEEE Int. Conf. on In Acoustics, Speech and Signal Processing (ICASSP), pp. 14–19 (2006)
11. Thrun, S., Leonard, J.: Simultaneous Localization and Mapping, pp. 871–889. Springer, Berlin (2008)
12. Ferandez-Madrigal, J.A., Blanco, C.J.L.: Simultaneous Localization and Mapping for Mobile Robots: Introduction and Methods, pp. 336–389. Information Science Reference, Hershey (2012)
13. Frank, D.: Factor Graphs and GTSAM: A Hands-on Introduction. Georgia Tech Technical Report, GT-RIM-CP&R-2012-002 (2012)

Visual Odometry Algorithm Using an RGB-D Sensor and IMU in a Highly Dynamic Environment

Deok-Hwa Kim, Seung-Beom Han, and Jong-Hwan Kim

Department of Electrical Engineering, KAIST
291 Daehak-ro, Yuseong-gu, Daejeon, 305-701, Republic of Korea
{dhkim,sbhan,johkim}@rit.kaist.ac.kr

Abstract. This paper proposes a robust visual odometry algorithm using a Kinect-style RGB-D sensor and inertial measurement unit (IMU) in a highly dynamic environment. Based on SURF (Speed Up Robust Features) descriptor, the proposed algorithm generates 3-D feature points incorporating depth information into RGB color information. By using an IMU, the generated 3-D feature points are rotated in order to have the same rigid body rotation component between two consecutive images. Before calculating the rigid body transformation matrix between the successive images from the RGB-D sensor, the generated 3-D feature points are filtered into dynamic or static feature points using motion vectors. Using the static feature points, the rigid body transformation matrix is finally computed by RANSAC (RANdom SAmple Consensus) algorithm. The experiments demonstrate that visual odometry is successfully obtained for a subject and a mobile robot by the proposed algorithm in a highly dynamic environment. The comparative study between proposed method and conventional visual odometry algorithm clearly show the reliability of the approach for computing visual odometry in a highly dynamic environment.

1 Introduction

Nowadays, many different kinds of robots have been used in various environments like home, museum, school, etc. To carry out tasks in such environments, autonomous navigation systems become more important [1, 2]. In the autonomous navigation systems, humanoid robots [3, 4] or aerial vehicles [5, 6, 7] might not directly use their encoder values for the odometry information. However, wheeled robots can use their encoder values for odometry information. For this reason, usage of the encoder sensors for the odometry information limits the scalability of platform to conduct various tasks. Because of this limitation, visual odometry, which is not much affected by platform types, has become more important than the classical encoder odometry. Actually, in robotics and computer vision, many techniques and algorithms have been developed for 3-D mapping and visual odometry using monocular cameras [8, 9], fish-eye vision sensors [10], stereo cameras [11] and RGB-D sensors [12, 13].

© Springer International Publishing Switzerland 2015
J.-H. Kim et al. (eds.), *Robot Intelligence Technology and Applications 3*,
Advances in Intelligent Systems and Computing 345, DOI: 10.1007/978-3-319-16841-8_2

Most of the 3-D mapping and visual odometry systems require spatial alignment of successive images. To solve the relationship between two consecutive images, various methods were proposed. Chen and Medioni proposed the method of point-to-plane registration [14], which has been widely used as a more accurate variant of the standard iterative closest point (ICP) algorithm. Challis organized the procedure for point-to-point registration based on singular value decomposition (SVD) [15, 16]. This point-to-point registration technique is useful to estimate transformation parameters using the matched image points obtained by feature detection algorithms. Based on rotation and scale invariant feature points, the spatial alignment between the two consecutive images is generally computed by the RANSAC algorithm [17]. This algorithm is usually used for estimating specific model parameters, like a line, plane, sphere, rotation, etc. To estimate the rigid body transformation, the RANSAC algorithm is much faster than the ICP. However, due to insufficient information, the RANSAC algorithm is more inaccurate than the ICP. Steinbrucker et al. proposed a visual odometry algorithm in an aspect of maximizing photoconsistency using the GPGPU parallelization technique [18].

Many researchers have conducted research to determine the ego-motion using various sensors with the methods of spatial alignment in the field of computer vision. Davision et al. presented MonoSLAM which can recover the 3-D trajectory of a monocular camera [8]. However, it has a drawback that initialization process is required to measure feature point depths or odometry. Based on this MonoSLAM research, Tardif et al. suggested monocular visual odometry using an omndiriectional camera [19]. This research conducted the decoupled estimation of the rotation and translation using epipolar geometry. Nister et al. proposed a simple visual odometry algorithm using a stereo camera through the preemptive RANSAC [20]. To solve the problem of an unstructureed outdoor navigation, Konolige et al. conducted research on outdoor mapping and navigation using a stereo camera [21]. This research integrated information from IMU and GPS devices. Using a Kinect-style RGB-D sensor, Newcombe et al. proposed a system, Kinectfusion for accurate real-time mapping of indoor scenes [22]. Through the ICP algorithm, Kinectfusion could successfully obtain the ego-motion of a RGB-D sensor, only using the depth information. Henry et al. proposed a RGB-D mapping method, which combines the ICP and RANSAC algorithm to calculate visual odometry [23]. Furthermore, they introduced optimization techniques [24, 25, 26] when loop-closure [27, 28] was detected.

Although the previous researches has been applied to estimating ego-motion and achieved successful results, they have not taken highly dynamic environments into account. In this paper, dynamic environments that cannot properly estimate the ego-motion using only vision sensors are defined as "highly dynamic environments". In real life, there are various moving objects such as people, animals, and vehicles. Those moving objects usually affect the calculation of visual odometry as a type of noise. Generally, the noise of moving objects can be eliminated by the RANSAC algorithm. However, in a highly dynamic environment, the RANSAC algorithm cannot eliminate the noise of moving objects when static feature points are overwhelmed by dynamic feature points. Therefore, it is difficult to compute accurately odometry information in such a highly dynamic environment.

In this paper, considering the highly dynamic environment, a robust visual odometry algorithm using a Kinect-style RGB-D sensor and inertial measurement unit (IMU) is proposed. Based on the SURF descriptor, 3-D feature points are generated by incorporating depth information into RGB color information. In particular, rotation components between the two consecutive images are compensated using an IMU in order to generate motion vectors of the feature points. Using the descriptor information of each feature point, the proposed algorithm makes the correspondence between the rotary-compensated feature points of the two successive images. Motion vectors of the matched feature points are generated by calculating translation vectors and velocities between them. Before calculating the rigid body transformation matrix between the successive images from the RGB-D sensor, the generated 3-D feature points are filtered into dynamic or static feature points using the motion vectors. Using the static feature points, the rigid body transformation matrix is finally computed by the RANSAC algorithm.

This paper is organized as follows. Section 2 proposes a robust visual odometry algorithm using an RGB-D sensor and IMU in a highly dynamic environment. In Section 3, experimental environment and scenarios are presented and the experimental results for a subject and a mobile robot in various environments are discussed. Finally, concluding remarks follow in Section 4.

2 Visual Odometry in a Highly Dynamic Environment

This section describes the proposed visual odometry algorithm for applying in a highly dynamic environment. A diagram of the overall algorithm is shown in Fig. 1. The proposed visual odometry algorithm has five different modules to estimate a rigid body transformation of the RGB-D sensor. Through the five different modules, such as the feature extractor, the rotation compensator, the motion vector generator, the motion vector filter, the sensor motion estimator and the moving object detector, the disturbance of dynamic movements is effectively removed for calculation of visual odometry.

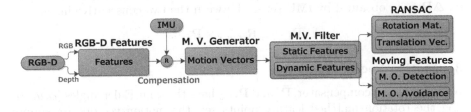

Fig. 1. A diagram of the proposed visual odometry algorithm

2.1 Feature Extraction

For extracting feature points, the proposed algorithm uses the SURF algorithm. The SURF algorithm is based on the sum of 2-D Haar wavelet responses and makes efficient use of integral images. The SURF feature point has scale-invariant and rotational-invariant features. To speed up the process of the SURF algorithm, graphic processing unit (GPU) is applied. After the extraction of feature points, the matching process of the feature points is performed between the two consecutive images by the Brute-force matcher. The Brute-force matcher matches feature points between the two images using descriptors of the feature points.

After the matching process of the feature points, the coordinate conversion from image plane to 3-D space is followed. The 3-D feature points are calculated from

$$
p_{i,j} = d(u_{i,j}, v_{i,j}) \begin{pmatrix} \frac{1}{f_u} & 0 & 0 \\ 0 & \frac{1}{f_v} & 0 \\ 0 & 0 & 1 \end{pmatrix} \begin{pmatrix} u_{i,j} - c_u \\ v_{i,j} - c_v \\ 1 \end{pmatrix} \tag{1}
$$

where $p_{i,j}$ is the 3-D feature point and $p_{i,j} \in \mathbf{P}_i$; i the number of images; j the index of the feature points; d the distance from a depth image; u, v the pixel coordination of the feature points; f_u, f_v the focal length of the sensor; c_u, c_v the principal point of the sensor. The camera parameters like $f_u, f_v, c_u,$ and c_v can be obtained from the chess-board calibration.

2.2 Rotation Compensator

In this paper, motion vectors of the feature points are used to remove the disturbance of dynamic movements. To generate the motion vectors, the rotary compensation process should be done first. Rotation for each frame is represented with Euler angles (roll-pitch-yaw). The rotation matrix at the i-th frame using Euler angles is as follows:

$$
\mathbf{R}_i = \mathbf{R}_{i,z}(\psi) \times \mathbf{R}_{i,y}(\theta) \times \mathbf{R}_{i,x}(\phi). \tag{2}
$$

An IMU sensor is also used to compensate the rotation between the consecutive images. The main reason for using the IMU sensor is to remove the effect of dynamic movements. The only vision-based odometry algorithms have the limitation in getting rid of the disturbance caused by moving objects in a highly dynamic environment. $\bar{\mathbf{P}}_i$ is the rotary-compensated $\mathbf{P}_i$ with the rotation matrix, $\triangle \mathbf{R}_{imu,i}$, obtained by IMU sensor between the two consecutive images:

$$
\bar{\mathbf{P}}_i = \triangle \mathbf{R}_{imu,i} \times \mathbf{P}_i. \tag{3}
$$

2.3 Motion Vector Generator

From the rotation compensator, $\bar{\mathbf{P}}_i$ and $\mathbf{P}_{i-1}$ have the same Euler angles (ϕ, θ, ψ). Using the rotation matched feature points set, the motion vectors are simply defined as a velocity of feature points as follows:

$$
m_{i,j} = \frac{p_{i-1,j} - \bar{p}_{i,j}}{t_i} \tag{4}
$$

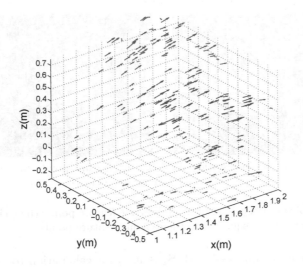

Fig. 2. Example of motion vectors of feature points for the x-axis moving RGB-D sensor (the arrows are motion vectors)

where $m_{i,j}$ is the motion vector of the j-th feature point at the i-th frame; t_i is the time interval between the two consecutive images. If there are no movements of the RGB-D sensor, the magnitude of each motion vector is near to zero. On the other hand, the RGB-D sensor moves to a certain direction, the motion vectors also show the same tendencies. In this paper, it is assumed that the motion vectors showing different tendencies with the RGB-D sensor are generated by moving objects or noise. Fig. 2 shows an example of motion vectors in 3-D space. In the figure, most of motion vectors have the same tendencies with the RGB-D sensor movement. However, the small number of motion vectors show the different tendencies with the sensor movement, which are caused by moving objects or noise.

2.4   Motion Vector Filter

To improve the estimation precision of the rigid body transformation matrix in a highly dynamic environment, the feature points set caused by moving objects should be removed. For this purpose, the following motion vector filter is applied to classify motion vectors as dynamic or static:

$$m_{ref,i} = v_{i-1} \tag{5}$$

$$\{p_{i-1,j}, p_{i,j}\} \in \mathbf{S}_i \quad : \|m_{i,j} - m_{ref,i}\| < \epsilon \tag{6}$$

$$\{p_{i-1,j}, p_{i,j}\} \in \mathbf{D}_i \quad : \|m_{i,j} - m_{ref,i}\| \geq \epsilon \tag{7}$$

where $\mathbf{S}_i$ is the static feature points set at the i-th frame; $\mathbf{D}_i$ is the dynamic feature points set; $m_{ref,i}$ is the reference velocity for classifying motion vectors. In this paper, $m_{ref,i}$ is defined as the velocity of the sensor at the $(i-1)$-th

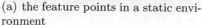

(a) the feature points in a static envi- (b) the feature points in a dynamic
ronment   environment

Fig. 3. Example of the classified static and dynamic feature points (blue circles are the static feature points; green squares are the dynamic feature points)

frame and ϵ is heuristically assigned. $\mathbf{S}_i$ is used for calculation to estimate the pose of the RGB-D sensor.

2.5   Estimating 3-D Rigid Body Transformations

To calculate visual odometry, the relationship between the two consecutive images should be clearly defined. This section deals with the determination method of parameters for the rigid body transformation. The relationship between feature points sets of the two consecutive images is represented as follows:

$$\mathbf{P}_{i-1} = {}^{i-1}\mathbf{R}_i \times \mathbf{P}_i + {}^{i-1}\mathbf{t}_i. \tag{8}$$

As the parameters for ${}^{i-1}\mathbf{R}_i$ or ${}^{i-1}\mathbf{t}_i$ are still unknown, the least squares method [15] is employed to approximately estimate ${}^{i-1}\mathbf{R}_i$ and ${}^{i-1}\mathbf{t}_i$ with the following least squares:

$$\frac{1}{n}\sum_{j=1}^{n}({}^{i-1}\mathbf{R}_i \times p_{i,j} + {}^{i-1}\mathbf{t}_i - p_{i-1,j})^{\mathbf{T}}({}^{i-1}\mathbf{R}_i \times p_{i,j} + {}^{i-1}\mathbf{t}_i - p_{i-1,j}) \tag{9}$$

The least squares method is a standard approach to solve the overdetermined systems in which there are more equations than unknown parameters. The rigid body transformation can be estimated when the least squares is minimized. To simplify the least squares, the following average of 3-D feature points is used:

$$\bar{p}_i = \frac{1}{n}\sum_{i=1}^{n}p_{i,j}. \tag{10}$$

From (9) and (10), a translation vector between the two consecutive images is approximately estimated as follows:

$$^{i-1}\mathbf{t}_i = \bar{p}_{i-1} - {}^{i-1}\mathbf{R}_i \times \bar{p}_i. \tag{11}$$

From (11), the least squares is simplified as follows:

$$\frac{1}{n}\sum_{j=1}^{n}(p_{i-1,j}^{'} - {}^{i-1}\mathbf{R}_i \times p_{i,j}^{'})^{\mathbf{T}}(p_{i-1,j}^{'} - {}^{i-1}\mathbf{R}_i \times p_{i,j}^{'}) \tag{12}$$

$$\frac{1}{n}\sum_{j=1}^{n}(p_{i-1,j}^{'\mathbf{T}}p_{i-1,j}^{'} + p_{i,j}^{'\mathbf{T}}p_{i,j}^{'} - 2p_{i-1,j}^{'\mathbf{T}} \times {}^{i-1}\mathbf{R}_i \times p_{i,j}^{'}). \tag{13}$$

where $p_{i,j}^{'} = p_{i,j} - \bar{p}_i$. Minimizing (13) is the same as maximizing the following equation:

$$\frac{1}{n}\sum_{j=1}^{n}(p_{i-1,j}^{'\mathbf{T}} \times {}^{i-1}\mathbf{R}_i \times p_{i,j}^{'}) = \mathrm{tr}\left\{{}^{i-1}\mathbf{R}_i^{\mathbf{T}} \times \frac{1}{n}\sum_{j=1}^{n}p_{i-1,j}^{'}p_{i,j}^{'\mathbf{T}}\right\}. \tag{14}$$

The above equation is represented by trace of the rotation matrix multiplied by correlation matrix of the 3-D feature points between the two consecutive images as follows:

$$\mathbf{C}_i = \frac{1}{n}\sum_{j=1}^{n}p_{i-1,j}^{'}p_{i,j}^{'\mathbf{T}} = \mathbf{U}_i\mathbf{W}_i\mathbf{V}_i^{\mathbf{T}}. \tag{15}$$

Through the singular value decomposition (SVD), the correlation is represented as $\mathbf{C}_i$. To maximize (14), the rotation matrix should satisfy the following condition:

$${}^{i-1}\mathbf{R}_i = \mathbf{U}_i\mathbf{V}_i^{\mathbf{T}}. \tag{16}$$

From (11) and (16), the rotation matrix and translation vector can be estimated by the least squares method. The rigid body transformation matrix is represented by the rotation matrix and translation vector as follows:

$${}^{i-1}\mathbf{T}_i = \left(\begin{array}{c|c} {}^{i-1}\mathbf{R}_i & {}^{i-1}t_i \\ \hline 0 & 1 \end{array}\right). \tag{17}$$

By accumulating the estimated rigid body transformation matrix every frame, the visual odometry of the RGB-D sensor is calculated.

2.6   RANSAC Algorithm

In this paper, the Euclidean error RANSAC algorithm (EE-RANSAC) is applied for estimation of the rotation matrix and the translation vector which come from sensor movements. Basically, the RANSAC algorithm is used for estimation of specific model parameters from noisy data sets. Through the RANSAC algorithm, the remaining noise, which is caused by inaccurate measurements, interference of light, incorrect camera parameters, etc., is eliminated to estimate the rigid body transformation of the sensor.

Algorithm 1. EE-RANSAC algorithm to estimate the parameters of the rigid body transformation at the i-th frame

---

$\mathbf{X}_i \Leftarrow \{$**Noisy Input Data Set at i-th frame**$\}$

$\mathbf{e}_{min} \Leftarrow \infty$

for $k = 0$ to max iterations **do**

    $\mathbf{Y}_{i,k} \Leftarrow$ RandomSampling$(\mathbf{X}_i)$

    $^{i-1}\mathbf{T}^*_{i,k} \Leftarrow$ ComputeModelParameter$(\mathbf{Y}_{i,k})$

    $\{\mathbf{e}_{i,k}, \mathbf{I}_{i,k}, \mathbf{O}_{i,k}\} \Leftarrow$ VerifyingModelParameter$(\mathbf{X}_i, {}^{i-1}\mathbf{T}^*_{i,k})$

    if $\mathbf{e}_{i,k} < \mathbf{e}_{min}$ **then**

        $\mathbf{e}_{min} \Leftarrow \mathbf{e}_{i,k}$

        $^{i-1}\mathbf{T}_i \Leftarrow$ ComputeModelParameter$(\mathbf{I}_{i,k})$

        if $\mathbf{e}_{min} < \varepsilon_0$ **then**

            return $^{i-1}\mathbf{T}_i$

        end if

    end if

end for

return $^{i-1}\mathbf{T}_i$

In Algorithm 1, the parameters of the rigid body transformation between the two successive images are recursively estimated by random sampling feature points set. To estimate the parameters at each iteration, the least squares method is used along with randomly sampled feature points. Using the estimated parameters, the fitness evaluation of the model is conducted by applying the estimated model to all feature points. Finally, using the inliers, the parameters of the rigid body transformation is once again estimated after model verification. Visual odometry is calculated by accumulating the rigid body transformation matrix at every frame as follows:

$$^{0}\mathbf{T}_n = {}^{0}\mathbf{T}_1 \times {}^{1}\mathbf{T}_2 \times {}^{2}\mathbf{T}_3 \ldots \times {}^{n-1}\mathbf{T}_n. \tag{18}$$

3   Experiment

The proposed visual odometry algorithm was tested using a differential wheeled mobile robot equipped with an RGB-D sensor, known as the Kinect sensor, as shown in Fig. 4. The experimental setup was composed of Gentoo OS, Intel i5 3.3GHz dual-core processor, NVIDIA GTX 560 GPU and 6GB RAM. The average computation time was 60.0 ms per frame at a 640 x 480 image resolution. Two kinds of experiments respectively in a static environment and a dynamic environment were conducted to verify the effectiveness of the proposed algorithm. A 9-DOF IMU was used to divide feature points into dynamic or static ones. The proposed visual odometry algorithm was tested in the Robot Intelligence Technology laboratory at KAIST. The RGB-D sensor was carried by the mobile robot controlled remotely by an experimenter. The EE-RANSAC visual odometry algorithm was used for comparison purpose. The EE-RANSAC algorithm computes the visual odometry recursively through minimizing the Euclidean error between the two consecutive images.

3.1 Visual Odometry Experiment in a Static Environment

As most of the visual odometry algorithms have been researched in a static environment, they do not need to filter dynamic feature points. Considering the real environment with moving objects, our proposed visual odometry algorithm employs a filtering step for the dynamic feature points. In a static environment, the additional filtering step does not much affect to the estimation of the rigid body transformation of the sensor because this step only deals with dynamic feature points. The experiment in a static environment demonstrated it. The experiment was carried out with a small-sized mobile robot, $7.5 \times 7.5 \times 7.5 \ cm^3$, equipped with an RGB-D sensor and an IMU sensor. The mobile robot travelled about 1.0 m with forward, turning right and forward movements.

Table 1. Visual odometry error result of forward, turning right and forward movements in a static environment

	EE RANSAC V.O.	Proposed V.O.
Transl. Error	0.0733 m	0.1043 m
Rot. Error	8.1610°	6.6571°

The experiment results in a static environment are shown in Fig. 5. For the comparison, the EE-RANSAC visual odometry was calculated in addition to the proposed algorithm. As a result, the both algorithms showed similar experimental results in a static environment. The ground truth of this experiment was obtained by measuring the robot path on a plaid ground. The final pose error of the EE-RANSAC visual odometry algorithm was $0.0733m \angle 8.1610°$ and the final pose error of the proposed algorithm was $0.1043m \angle 6.6571°$.

Fig. 4. Differential wheeled robot equipped with an RGB-D sensor and IMU sensor on the top

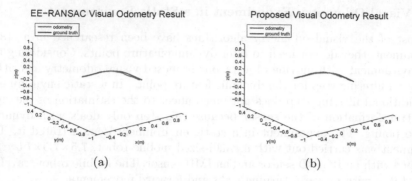

Fig. 5. Visual odometry results of forward, turning right and forward movements in a static environment. (a) Result from the EE-RANSAC algorithm. (b) Result from the proposed algorithm.

3.2   Visual Odometry Experiment in a Dynamic Environment

In real life, there are a lot of dynamic movements of cars, people, animals, etc., which affect the visual odometry calculation in case of having poor detected feature points. In the proposed algorithm, the disturbances from moving objects are effectively removed through the motion vector filtering step. To verify the effectiveness of the proposed algorithm, two kinds of experiments depending on the stopping and moving states of the mobile robot were conducted in a dynamic environment having the poor detected feature points. As was in the previous experiment in a static environment, the EE-RANSAC visual odometry algorithm was calculated in addition to the proposed algorithm for comparison purpose.

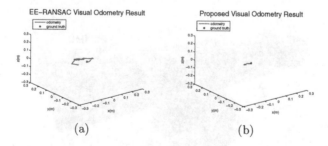

Fig. 6. Visual odometry results with the stopped mobile robot in a dynamic environment having the poor detected feature points. (a) Result from the EE-RANSAC algorithm. (b) Result from the proposed algorithm.

Firstly, the visual odometry experiment with the mobile robot stopping at an origin position was carried out in a dynamic environment. The dynamic environment included an experimenter walking around in the front of the RGB-D sensor. Ideally, the trajectory of the robot should be at an origin position

Table 2. Visual odometry error result with the stopped mobile robot in a dynamic environment having the poor detected feature points

	Transl. Error	Rot. Error
EE-RANSAC V.O.	0.0866 m $\pm$ 0.0143 m	4.8547° $\pm$ 0.8055°
Proposed V.O.	0.0431 m $\pm$ 0.0080 m	2.3239° $\pm$ 0.4241°

because there was no movements of the sensor. However, as Fig. 6 shows, it was not at the origin position because the experimenter made disturbances. The figure shows that the disturbances of the moving experimenter affected the EE-RANSAC visual odometry algorithm much more than the proposed visual odometry. The final pose error of the EE-RANSAC visual odometry algorithm was $0.0866m\angle 4.8547°$ and that of the proposed algorithm was $0.0431m\angle 2.3239°$.

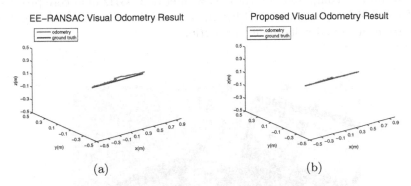

(a)   (b)

Fig. 7. Visual odometry results with the mobile robot of moving forward in a dynamic environment having the poor detected feature points. (a) Result from the EE-RANSAC algorithm. (b) Result from the proposed algorithm.

Table 3. Visual odometry error result with the mobile robot of moving forward in a dynamic environment having the poor detected feature points

	Transl. Error	Rot. Error
EE-RANSAC V.O.	0.0327 m $\pm$ 0.0058 m	1.3007° $\pm$ 0.2855°
Proposed V.O.	0.0254 m $\pm$ 0.0046 m	1.7516° $\pm$ 0.2566°

Secondly, the visual odometry experiment with the mobile robot moving forward in a dynamic environment was performed. To make a dynamic environment, the experimenter roamed around in the front of the robot as in the previous experiment. A ground truth of this experiment would be a straight line because the mobile robot moved forward about 0.9 m. As was in the previous experiment, the visual odometry was also influenced by experimenter roaming around in the front of the sensor. As Fig. 7 shows, in the situation having poor detected feature points, the proposed algorithm was less affected by the experimenter's movement than the EE-RANSAC visual odometry algorithm. A final pose error of the EE-RANSAC visual odometry algorithm was $0.0327m\angle 1.3007°$ and that of the proposed algorithm was $0.0254m\angle 1.7516°$.

3.3   Application for a Subject with a Blindfold

The proposed visual odometry algorithm can be used to a wearable navigation system for the visually impaired persons. Fig. 8 shows the mobile visual odometry system. For experiments, a subject with a blindfold wore the RGB-D sensor around his neck and carried a bag containing a laptop on his back. The experimental setup was composed of Gentoo OS, Intel i7 2.7GHz dual-core processor, NVIDIA GT 555M GPU and 8GB RAM. Average computation time was 100.0 ms per frame at a 640 x 480 image resolution.

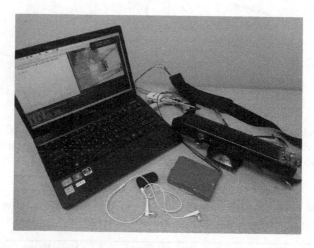

Fig. 8. The wearable mobile visual odometry system for the visually impaired persons

In a static environment, two kinds of experiments were conducted for a subject with a blindfold. The first experiment was to chase a 3-meter-square with such a system. A ground truth of this experiment would be the 3-meter-square. As Fig. 9(a) shows, the final pose error was $0.3188m\angle 19.5635°$. The final pose error of the mobile visual odometry system was slightly bigger than that of the previous experiments using the mobile robot because the sway motion of the subject was not considered. The second experiment was conducted for going up the stairs. A ground truth of this experiment was measured by the number of steps and by properties of the stairs. As Fig. 9(b) shows, the final pose error of this experiment was $0.3049m\angle 5.5460°$.

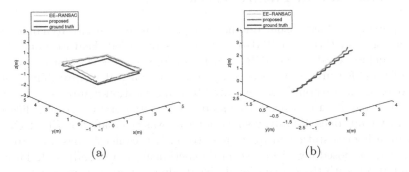

(a)   (b)

Fig. 9. Visual odometry results using the wearable mobile visual odometry system in a static environment. (a) Result of chasing the $3.0m \times 3.0m$ square (b) Result of going up the stairs.

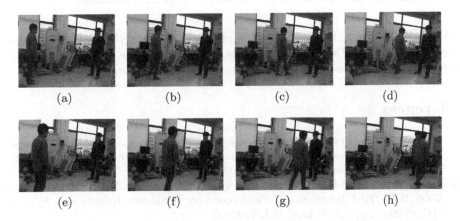

(a)   (b)   (c)   (d)

(e)   (f)   (g)   (h)

Fig. 10. An experiment result for a subject equipped with a blindfold to avoid a moving person

The proposed algorithm can be used for another purpose like a moving object detection in addition to visual odometry. Using the dynamic feature points after the motion vector filtering step, moving objects can be detected. For simplicity of the moving object detection, this paper assumes that a moving object is on a moving plane. To detect a plane from the dynamic feature points, this paper uses the RANSAC algorithm, which classifies the dynamic feature points into a plane or noise. If a moving object in front of the subject with a blindfold is detected, our proposed mobile system will notify the position of the moving object to the subject equipped with the system through bluetooth earphones. As Fig. 10 shows, a subject wearing a blindfold could avoid an approaching person with a random direction. This experiment demonstrated that the proposed visual odometry could be used in the other application such as collision avoidance as well as visual odometry.

4   Conclusion and Future Work

In this paper, a robust visual odometry algorithm in a highly dynamic environment using an IMU and RGB-D sensor was proposed. Feature points were obtained and matched by the SURF descriptor from successive color images. Rotation of these feature points was compensated by an IMU sensor. The motion vector of each feature point was computed by using rotation-compensated feature points. By the motion vector filtering step, the feature points were classified into static or dynamic ones. Finally, visual odometry was calculated by EE-RANSAC algorithm using the static feature points. To verify the effectiveness and applicability of the proposed algorithm, several experiments were conducted. Through the comparison with the conventional visual odometry algorithm, our proposed algorithm was proved to be more reliable for computing visual odometry in a highly dynamic environment. In addition, considering the visually impaired persons, the applicability of the proposed algorithm to the collision avoidance was verified through the experiment with a subject wearing a blindfold.

Acknowledgment. This work was supported by the Technology Innovation Program, 10045252, Development of robot task intelligence technology, funded by the Ministry of Trade, Industry & Energy (MOTIE, Korea).

References

[1] Mourikis, A.I., Trawny, N., Roumeliotis, S.I., Johnson, A.E., Matthies, L.: Vision-aided inertial navigation for precise planetary landing: Analysis and experiments. In: Robotics: Science and Systems (2007)

[2] Jeong, W., Lee, K.M.: Cv-slam: A new ceiling vision-based slam technique. In: 2005 IEEE/RSJ International Conference on Intelligent Robots and Systems (IROS 2005), pp. 3195–3200. IEEE (2005)

[3] Baglietto, M., Sgorbissa, A., Verda, D., Zaccaria, R.: Human navigation and mapping with a 6dof imu and a laser scanner. Robotics and Autonomous Systems 59(12), 1060–1069 (2011)

[4] Yoo, J.K., Kim, J.H.: Fuzzy integral-based gaze control architecture incorporated with modified-univector field-based navigation for humanoid robots. IEEE Transactions on Systems, Man, and Cybernetics, Part B: Cybernetics 42(1), 125–139 (2012)

[5] Grzonka, S., Grisetti, G., Burgard, W.: Towards a navigation system for autonomous indoor flying. In: IEEE International Conference on Robotics and Automation, ICRA 2009, pp. 2878–2883. IEEE (2009)

[6] Bachrach, A., Huang, A.S., Maturana, D., Henry, P., Krainin, M., Fox, D., Roy, N.: Visual navigation for micro air vehicles (2011)

[7] Huang, A.S., Bachrach, A., Henry, P., Krainin, M., Maturana, D., Fox, D., Roy, N.: Visual odometry and mapping for autonomous flight using an rgb-d camera. In: Int. Symposium on Robotics Research (ISRR), Flagstaff, Arizona, USA (2011)

[8] Davison, A.J., Reid, I.D., Molton, N.D., Stasse, O.: Monoslam: Real-time single camera slam. IEEE Transactions on Pattern Analysis and Machine Intelligence 29(6), 1052–1067 (2007)

[9] Kneip, L., Chli, M., Siegwart, R.: Robust real-time visual odometry with a single camera and an imu. In: BMVC, pp. 1–11 (2011)

[10] Han, S., Kim, J., Myung, H., et al.: Landmark-based particle localization algorithm for mobile robots with a fish-eye vision system. IEEE/ASME Transactions on Mechatronics PP(99), 1–12 (2012)

[11] Konolige, K., Agrawal, M., Solà, J.: Large-scale visual odometry for rough terrain. In: Kaneko, M., Nakamura, Y. (eds.) Robotics Research. STAR, vol. 66, pp. 201–212. Springer, Heidelberg (2010)

[12] Hu, G., Huang, S., Zhao, L., Alempijevic, A., Dissanayake, G.: A robust rgb-d slam algorithm. In: 2012 IEEE/RSJ International Conference on Intelligent Robots and Systems (IROS), pp. 1714–1719. IEEE (2012)

[13] Endres, F., Hess, J., Engelhard, N., Sturm, J., Cremers, D., Burgard, W.: An evaluation of the rgb-d slam system. In: IEEE International Conference on Robotics and Automation (ICRA), pp. 1691–1696. IEEE (2012)

[14] Yang, C., Medioni, G.: Object modelling by registration of multiple range images. Image and Vision Computing 10(3), 145–155 (1992)

[15] Challis, J.H.: A procedure for determining rigid body transformation parameters. Journal of Biomechanics 28(6), 733–737 (1995)

[16] Kim, D.-H., Kim, J.-H.: Image-based ICP algorithm for visual odometry using a RGB-D sensor in a dynamic environment. In: Kim, J.-H., Matson, E., Myung, H., Xu, P. (eds.) Robot Intelligence Technology and Applications. AISC, vol. 208, pp. 423–430. Springer, Heidelberg (2013)

[17] Nistér, D.: Preemptive ransac for live structure and motion estimation. Machine Vision and Applications 16(5), 321–329 (2005)

[18] Steinbrucker, F., Sturm, J., Cremers, D.: Real-time visual odometry from dense rgb-d images. In: 2011 IEEE International Conference on Computer Vision Workshops (ICCV Workshops), pp. 719–722. IEEE (2011)

[19] Tardif, J.P., Pavlidis, Y., Daniilidis, K.: Monocular visual odometry in urban environments using an omnidirectional camera. In: IEEE/RSJ International Conference on Intelligent Robots and Systems, IROS 2008, pp. 2531–2538. IEEE (2008)

[20] Nistér, D., Naroditsky, O., Bergen, J.: Visual odometry. In: Proceedings of the 2004 IEEE Computer Society Conference on Computer Vision and Pattern Recognition, CVPR 2004, vol. 1, p. I-652. IEEE (2004)

[21] Konolige, K., Agrawal, M., Bolles, R.C., Cowan, C., Fischler, M., Gerkey, B.: Outdoor mapping and navigation using stereo vision. In: Khatib, O., Kumar, V., Rus, D. (eds.) Experimental Robotics. STAR, vol. 39, pp. 179–190. Springer, Heidelberg (2008)

[22] Newcombe, R.A., Davison, A.J., Izadi, S., Kohli, P., Hilliges, O., Shotton, J., Molyneaux, D., Hodges, S., Kim, D., Fitzgibbon, A.: Kinectfusion: Real-time dense surface mapping and tracking. In: 2011 10th IEEE International Symposium on Mixed and Augmented Reality (ISMAR), pp. 127–136. IEEE (2011)

[23] Henry, P., Krainin, M., Herbst, E., Ren, X., Fox, D.: Rgb-d mapping: Using kinect-style depth cameras for dense 3d modeling of indoor environments. The International Journal of Robotics Research 31(5), 647–663 (2012)

[24] Kaess, M., Ranganathan, A., Dellaert, F.: isam: Incremental smoothing and mapping. IEEE Transactions on Robotics 24(6), 1365–1378 (2008)

[25] Kuemmerle, R., Grisetti, G., Strasdat, H., Konolige, K., Burgard, W.: g2o: A general framework for graph optimization. In: Proc. of the IEEE Int. Conf. on Robotics and Automation, ICRA (2011)

[26] Kaess, M., Johannsson, H., Roberts, R., Ila, V., Leonard, J., Dellaert, F.: isam2: Incremental smoothing and mapping with fluid relinearization and incremental variable reordering. In: 2011 IEEE International Conference on Robotics and Automation (ICRA), pp. 3281–3288. IEEE (2011)
[27] Ho, K.L., Newman, P.: Detecting loop closure with scene sequences. International Journal of Computer Vision 74(3), 261–286 (2007)
[28] Kim, D.H., Kim, J.H.: Visual loop-closure detection method using average feature descriptors. In: Kim, J.-H., Matson, E., Myung, H., Xu, P. (eds.) Robot Intelligence Technology and Applications 2. AISC, vol. 274, pp. 113–118. Springer, Heidelberg (2014)

Graph Structure-Based Simultaneous Localization and Mapping with Iterative Closest Point Constraints in Uneven Outdoor Terrain

Taekjun Oh[1], Hyongjin Kim[1], Donghwa Lee[1], Hyun Chul Roh[2], and Hyun Myung[1,2]

[1] Department of Civil and Environmental Engineering, KAIST, Daejeon, 305-701, Korea
[2] Robotics Program, KAIST, Daejeon, 305-701, Korea
{buljaga,hjkim86,leedonghwa,rohs_,hmyung}@kaist.ac.kr

Abstract. The purpose of this study is to propose a novel mobile robot localization method applicable to outdoor environments, such as an uneven terrain. In order to solve the robot localization problem, we exploit state of the art graph-based SLAM (Simultaneous Localization and Mapping) algorithm and ICP (Iterative Closest Point) algorithm considering the gyroscopic data as a constraint for a graph structure. We confirm our method by testing actual sensor data acquired from a vehicle in outdoor environments and show that our proposed method is improved and suitable for uneven terrain.

Keywords: SLAM, Graph Structure, ICP, Point Cloud, UGV.

1 Introduction

In recent years, the field robot for performing various tasks in a variety of environments has emerged with advancement of robot technologies whereas the industrial robot repeatedly executes the same work at a fixed location from the past. There are a lot of difficulties for robot automation, since the field robot works in various environments. One of the key issues in the automation of field robot is to solve the robot localization problem [1].

There are a wide range of sensors and algorithms for the robot localization problem. A GPS (Global Positioning System) and IMU (Inertial Measurement Unit) are generally used as positioning sensors and the VO (Visual Odometry) [2] algorithm using camera images and ICP (Iterative Closest Point) [3] algorithm using laser scanner data are employed for obtaining a robot's pose. However, in some situation like a non-line-of-sight, a significant position error occurs because GPS signal cannot be received directly from the satellite [4]. In case of IMU and VO algorithm, error is accumulated when used for a long period of time. Also ICP algorithm causes errors, if the algorithm is processed using data from an uneven surface. In this case, an alternative method is necessary and SLAM (Simultaneous Localization and Mapping) might be a solution to this problem [5-12].

In this paper, we propose a novel method for the robot localization exploiting the graph structure-based SLAM algorithm considering the constraint of graph structure

© Springer International Publishing Switzerland 2015
J.-H. Kim et al. (eds.), *Robot Intelligence Technology and Applications 3*,
Advances in Intelligent Systems and Computing 345, DOI: 10.1007/978-3-319-16841-8_3

from ICP algorithm for point cloud data matching with roll and pitch angles from the IMU and kinematics for a wheeled robot to reduce the robot pose error in exhaustive out door such as rough terrain.

The remainder of this paper is organized as follows. Section 2 explains our approach on how to minimize the robot pose error using graph-based SLAM and constraint from ICP algorithm. In Section 3, we then provide overall experimental environments and system and describe the experimental results with our approach to confirm our method. Detailed conclusion and future work are discussed in Section 4.

2   Graph SLAM with ICP Constraint

In this section we introduce a novel graph SLAM with ICP Constraint. First, we present the ICP algorithm considering the gyroscopic data for the graph structure constraints. Then we describe how to generate constraints of the graph structure. The overall localization algorithm is shown in Fig. 1.

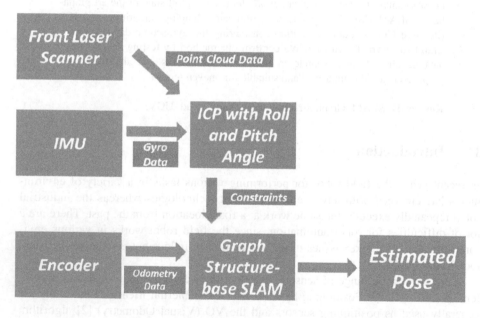

Fig. 1. The overview of the localization algorithm

2.1   ICP with Roll and Pitch Angle

The widely used two dimensional ICP is a method to match two dimensional point cloud data generated from a laser scanner. The two dimensional ICP is able to estimate pose relationship successfully in indoor environments. However the method induces the pose error in outdoor environments when the surface is not even with the laser scanner looking forward. The laser scanner fluctuates in outdoor, because the floor is uneven. Therefore, it is necessary to consider the roll and pitch angles of the laser scanner as shown in Fig. 2. We conducted the ICP algorithm using point cloud data generated from a 2D laser scanner and IMU instead of conducting a naïve ICP algorithm. If the laser scanner fluctuates on uneven road, the error of ICP is inevitable.

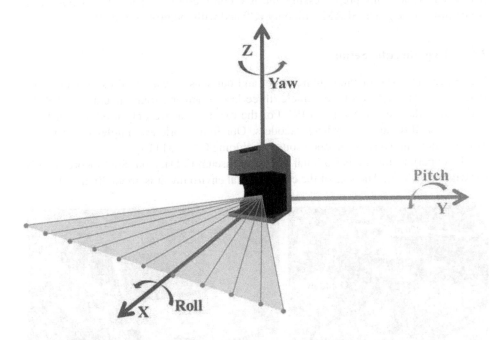

Fig. 2. The axis of the laser scanner

2.2   Constraints of Graph Structure

The constraints of a graph-structure are important in a graph-based SLAM, because it is possible to induce the pose error due to improper constraint. When the robot moves one meter by using the odometry obtained from encoder, we generate a node and form constraints of the graph structure and valid constraints can be generated from the ICP using the method of Section 2.1. Also we incorporate the wheeled robot kinematics. That is, if the rotation angle of the ICP result exceeds the range of wheeled robot kinematics or the translation of the ICP result moves rapidly to sideways along to the robot axis, it was considered an invalid constraint. In comparison with odometry, if the translation of the ICP result is too small or too large, it was also considered an invalid constraint.

3   Experiments

This section presents an experimental setup and the results of our method. For the experiments, we used our sensor system and tested experiments in outdoor. We compared ICP without any preprocessing and ICP considering roll and pitch angles. Also we examined the graph SLAM with only ICP and with the proposed method.

3.1   Experimental Setup

Fig. 3 presents experimental environment and our sensor system for experiment. The system is composed of a road vehicle, three laser scanners, four cameras, a GPS, an IMU, two wheel encoders and a PC. For the experiment, we only used a front laser scanner, an IMU and two wheel encoders. Our framework was implemented in the Matlab platform based on an open source including GTSAM [13].

The experimental site is National Science Museum in Daejeon, South Korea and is shown in Fig. 3 (a). The size of the experimental environment is about 300m x 440m.

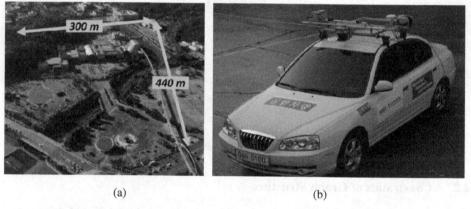

(a)   (b)

Fig. 3. Experimental setup (a) experimental environment, (b) sensor system

3.2   Results

Fig. 4 presents the results of ICP without any preprocessing and ICP with the preprocessing considering roll and pitch angles. The result of the ICP without a gyroscopic data is shown in Fig. 4 (a) and the result of the ICP performed in consideration of the roll and pitch angles is shown in Fig. 4 (b). The red dots indicate a model data and the green dots template data. The blue dots represent fitting results obtained from the ICP algorithm. Fig. 4 (a) shows the wrong fitted result caused by uneven surface, while Fig. 4 (b) shows good matching result even with rough terrain because of compensation of roll and pitch angles.

The overall SLAM results are shown in Fig. 5. The red solid line is raw pose data from the odometry and the green square shows graph SLAM results with ICP without considering roll and pitch angles. The blue dashed line indicates the results of our approach. The error exists in the green square results because of the invalid constraint generated by ICP without consideration of gyroscopic data. In order to confirm performance of our method, Fig. 5 is enlarged. Fig. 6 shows the enlarged Section A of Fig. 5 and Fig. 7 shows the enlarged Section B of Fig. 5. As can be seen in Fig. 6 and Fig. 7, the proposed method seems to be more accurate. In Sections A and B, a loop closing occurred where the front laser scanner views the previously-seen feature data. The raw data accumulates the error in a long run time. However, proposed method generates a constraint by loop closing and the accumulated pose error is minimized by the optimization of the graph structure.

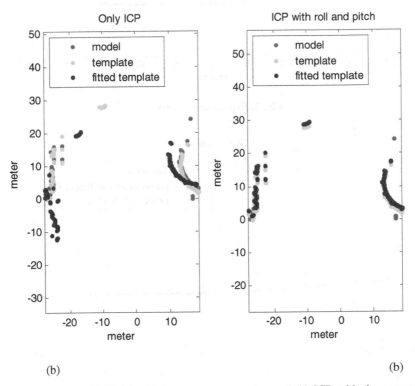

(b)   (b)

Fig. 4. The results of ICP (a) without any preprocessing and (b) ICP with the preprocessing considering roll and pitch angles from a gyroscope

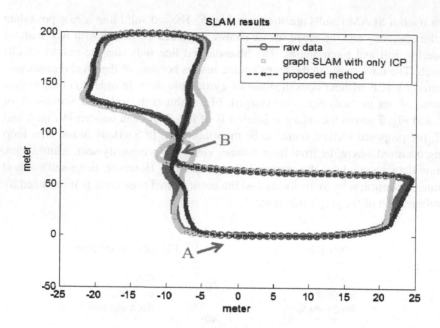

Fig. 5. Experiment results

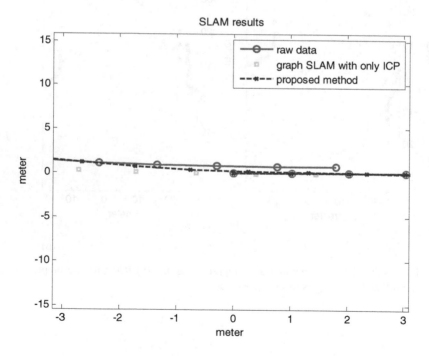

Fig. 6. Section A of Fig. 5

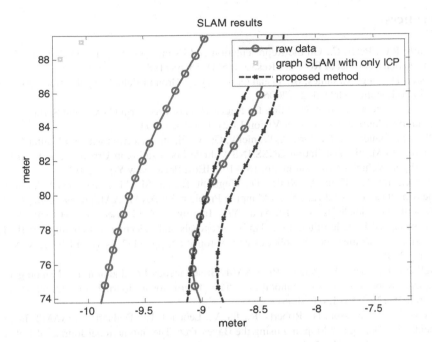

Fig. 7. Section B of Fig. 5

4   Conclusion and Future Work

We proposed a novel method for robot localization by incorporating gyroscopic data in the ICP scan-matching. We confirmed that the localization error is reduced even in a rough terrain. This method is efficient for the areas where GPS signal is denied. However, since this method is based on a 2D graph SLAM, the method can cause problems when the robot exhibits dynamic 3D motion. Furthermore, we did not consider a probabilistic model for building constraints.

For future works, we will expand our method to a 3D graph SLAM taking the probabilistic model into account when generating the constraints.

Acknowledgement. This work was supported by the R&D program of the Korea Ministry of Trade, Industry and Energy (MOTIE) and the Korea Evaluation Institute of Industrial Technology (KEIT). (The Development of Low-cost Autonomous Navigation Systems for a Robot Vehicle in Urban Environment, 10035354).

This work is financially supported by Korea Minister of Ministry of Land, Infrastructure and Transport (MOLIT) as 「 U-City Master and Doctor Course Grant Program」.

References

1. Smith, R.C., Peter, C.: On the Representation and Estimation of Spatial Uncertainty. The International Journal of Robotics Research 5(4), 56–68 (1986)
2. Nistér, D., Oleg, N., James, B.: Visual Odometry for Ground Vehicle Applications. Journal of Field Robotics 23(1), 3–20 (2006)
3. Zhang, Z.: Iterative Point Matching for Registration of Free-form Curves and Surfaces. International Journal of Computer Vision 13(2), 119–152 (1994)
4. Obst, M., Bauer, S., Reisdorf, P., Wanielik, G.: Multipath Detection with 3D Digital Maps for Robust Multi-constellation GNSS/INS Vehicle Localization in Urban Areas. In: IEEE Intelligent Vehicles Symposium, pp. 184–190. IEEE Press, New York (2012)
5. Montemerlo, M., Thrun, S., Koller, D., Wegbreit, B.: FastSLAM: A Factored Solution to the Simultaneous Localization and Mapping Problem. In: AAAI/IAAI, pp. 593–598 (2002)
6. Grisetti, G., Stachniss, C., Burgard, W.: Improving Grid-based SLAM with Rao-blackwellized Particle Filters by Adaptive Proposals and Selective Resampling. In: IEEE International Conference on Robotics and Automation, pp. 2432–2437. IEEE Press, New York (2005)
7. Dellaert, F., Kaess, M.: Square Root SAM: Simultaneous Localization and Mapping via Square Root Information Smoothing. The International Journal of Robotics Research 25(12), 1181–1203 (2006)
8. Kaess, M., Johannsson, H., Roberts, R., Ila, V., Leonard, J.J., Dellaert, F.: iSAM2: Incremental Smoothing and Mapping using the Bayes Tree. The International Journal of Robotics Research 31(2), 217–236 (2011)
9. Kim, H., Oh, T., Lee, D., Choe, Y., Chung, M.J., Myung, H.: Mobile Robot Localization by Matching 2D Image Features to 3D Point Cloud. In: Ubiquitous Robots and Ambient Intelligence (2013)
10. Kim, H., Lee, D., Oh, T., Lee, S.W., Choe, Y., Myung, H.: Feature-Based 6-DoF Camera Localization Using Prior Point Cloud and Images. In: Kim, J.-H., Matson, E., Myung, H., Xu, P. (eds.) Robot Intelligence Technology and Applications 2. AISC, vol. 274, pp. 3–12. Springer, Heidelberg (2014)
11. Lee, D., Jung, J., Myung, H.: Pose Graph-Based RGB-D SLAM in Low Dynamic Environments. In: IEEE International Conference on Robotics and Automation, IEEE Press, New York (2014)
12. Lee, D., Myung, H.: Solution to the SLAM Problem in Low Dynamic Environments Using a Pose Graph and an RGB-D Sensor. Sensors 14(7), 12467–12496 (2014)
13. Dellaert, F.: Factor graphs and GTSAM: A hands-on introduction (2012)

A Usability Study for Signal Strength Based Localisation

Tommi Sullivan, Jun Jo, and Michael Lennon

School of Information Communication and Technology,
Griffith University, Queensland, Australia
{tommi.sullivan,michael.lennon}@griffithuni.edu.au, j.jo@griffith.edu.au

Abstract. The Global Positioning System (GPS) may be one of the largest technological advances of the human beings. This technology became an essential component of most vehicles including airplanes, cars and ships. This device can be used anywhere around the world. While GPS was made for outdoors, there are major problems about indoor localisation. There are some useful indoor localisation tools that have been researched, such as WiFi and ZigBee technology. But due to the inaccuracy as well as interference that can occur in indoor locations, such as walls, other Wireless Sensor Networks (WSNs) and human interference, the recent systems that have been created, are still deemed to be unstable. Thus the increase of precision is required for these indoor systems. This research investigated the relationships between signal strengths and distance in the real situation. The data set obtained was analysed by using a regression method. Sensitivity was studied with Confidence Interval. The outcome of this research is expected to provide the users with a practical guideline for signal strength based localisation.

1 Introduction

Positioning or identifying the position of an object, in the space without a GPS guidance, is a challenging problem. Various localisation methods for indoor locations have been developed and applied to many areas including robotics, geography and construction over the past 20 years. Localisation tools generally use various sensors such as Laser range finders, Radar, Sonar, Ultrasonic, Infrared and Cameras. Recently wireless technologies are becoming popular tools based on the relationship between signal strength and distance. However, their inaccuracy and inconsistency are the major issues. Much research has been done to identify the relationship between the two parameters; distance and signal strength [2]. However these research outcomes were generated through simulation or in a set environment and are still far from the practical use. This research will review the existing techniques, conduct experiments in the field and investigate the practical usability for localisation. Two popular wireless techniques were used for this research: Wireless Fidelity (WiFi) and ZigBee.

This research will also make a state of the art comparison between these two devices. There have been major breakthroughs with multiple nodes of Wifi or

ZigBee to localise indoor positioning of robots or mobile devices. For example, assuming that the mobile phone is held in front of the body, an indoor positioning system based off Wifi can be accurate enough to provide location estimates.

ZigBee needs a ZigBee device to send the signal strength between each module, which uses the same signal frequency as Wifi. For example, ZigBee has been used to monitor the position of cattle in grazing fields, this can be achieved due to the longer distance of the signal the ZigBee can reach. These methods provide localisation precisely, and with less error than GPS that is used around the globe. Yet, there are some errors that need to be improved for accurate indoor or outdoor localisation. In this research, we compare the difference between WiFi and ZigBee for localisation and the accuracy of distance compared to signal strength.

Since the methods that have already been provided necessitate a number of assumptions such as needing the person to hold the mobile phone at the front of the body, the methods provide advantages but also disadvantages. While this research will first test both sensor nodes in an open area, we will be able to discover these problems and solve them in future research projects.

Kelly describes, due to the ubiquity, WiFi signals are a commonly employed indicator of location, and from this knowledge, the identity and intensity of these signals can provide an estimation of the devices location in the vicinity [6].

Regardless of the popularity, these indoor localisation methods have not attained precise positioning. Unlike these systems mentioned above, there are many conditions and methods that can be tested before using multiple WiFi or ZigBee nodes.

2 Related Work

2.1 WiFi Techniques

WiFi Localisation is one of the major indoor techniques that has been widely used due to the popularity in the device especially in densely populated urban areas where WiFi routers form tightly interconnected networks. Due to handheld mobile phones having WiFi modules installed, there is a large economy waiting to discover the most efficient and precise method for indoor localisation [8].

Different types of methods for WiFi can be used for localisation. One method is the Access Point Pre-localising Method.

This method pre-localises the Access Points (AP's) and makes a visual map of where these AP's are then tested with Test Points (TP's) that can either be sending ping data or RSSI data or even both to double check the accuracy. Because this method needs to know the location and distance between each node before receiving data, it can be very accurate. However it can only be used in areas that the user can place his own AP's [12]. Figure 1 shows how the AP's and Training Points(TP's) can be placed to be used for testing.

The AP's provide the signal strength and with that data the signal strength is put through a calculation to discover the relative distance. These methods have different mathematical models that can be used and some are the Received Signal

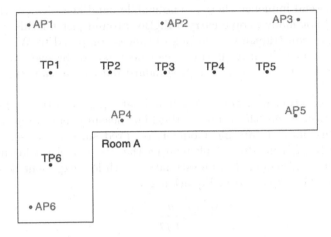

Fig. 1. Pre-localising Map Example

Strength Indicator (RSSI) map-based algorithms. This uses k-nearest neighbor (kNN) training algorithms to map the RSSI data.

The kNN algorithm uses a data set of beacon node RSSI values at different TP's to create the RSSI signature map. This method works well with the pre-localising method because beacon nodes have predefined locations, and creates the data set for the kNN algorithm to be used. Given a new device provides the system with a RSSI value, the system searches through the training data set to find the k-nearest data records [13].

The proximity can be define in Euclidian distance that can be calculated with equation 1.

$$D_t \sqrt{\sum_{j=0}^{N} (RSSI_{tj} - RSSI_{testpoint-j})} \tag{1}$$

Where N is the Nth-dimension, $RSSI_{tj}$ is the RSSI value read from beacon J at location t, $RSSI_{testpoint-j}$ is the RSSI value read from j at the target point and the unit of D_t is dbm, which is the decibels(dB) of the measured power.

There are several existing indoor localisation systems using the kNN algorithm such as, RADAR based on 802.11 RF based wireless technology [1], LAND-MARC: Indoor location Sensing Using Active RFID [15] and also Ekhau real time location system (RTLS) tested in indoor construction areas [3]. The best accuracy was Ekhau RTLS, which was 1.5-2m [7].

2.2 ZigBee Techniques

ZigBee can be very useful for indoor localisation due to its low power consumption. While the ZigBee device is used in other research areas[5][4], due to the lack of commercial influence there are no mobile hand held devices that have it installed, and not many people carry a ZigBee around just for this purpose.

Since ZigBee can transmit data long distances compared to WiFi, the Long Range Link Quality Indication (LQI) can be used to discover distance with ZigBee networks, where the LQI is a standard data type that ZigBee devices transmit.

The LQI measurement is the strength and data quality of the received packet from the ZigBee, where the range of the LQI measurements ranges from 0 to 255. With this data, a suitable model is required since it is used differently compared to RSSI. But after implementing the two models; a log-normal [14] and the exponential model [11], an estimation path loss exponent was derivable from the later [11], explained in Equation 2.

$$r_{ij} = \frac{a}{d_{ij}a} + n_i \tag{2}$$

Where r_{ij} is the sensed RSS value in the ith sensor from the jth transmitting node, d_{ij} is the distance between the nodes i and j, and a is path loss component. Finally n is white Gaussian noise.

2.3   Signal Strength Localisation Methods

While LQI is a data type that ZigBee only has, there are many other methods that both the WiFi and ZigBee modules can use for indoor localisation with signal strength. Such methods can be Trilateration, where Trilateration is sometimes confused with triangulation, and is very similar. In this case, it calculates the distance between multiple nodes but obtaining the intersection of three spheres. The distance however can be derived from either the Time of Arrival that is also known as Ping, or the RSSI. The two dimensional method can be obtained by using 3 transmitter nodes, while a three dimensional location estimation can be obtained by using four or more nodes. There is no significant difficulty with this algorithm, although it consumes a large amount of time due to the resources needed to implement [10].

Another method that is similar to the Trilateration method is the MinMax method. The MinMax algorithm uses the RSSI values from the target node and calculates the distance using the RSSI values based on the path-loss model. With the distance, a square is drawn around the beacon node, and from this derives the target node which lies within the overlapping are of all the squares drawn around the beacon nodes [13]. This algorithm is very simple to implement. However due to the squares giving extra distance at the corners, there is a larger error rate [9].

3   Usability Methods

In order to increase the accuracy of indoor localisation, it is necessary to discover the reliability of the WiFi and ZigBee devices individually preferably with close to no noise or interference from other devices. This can be achieved by obtaining test data outdoors in an open area where there is lower interference from other devices and obstacles such as walls and human bodies.

The data is obtained every 50cm by moving back with a proper measurement method. The transmitter node is left at 0cm and the receiver moves backwards. Everyone 50cm each node scanned for the signal strength 30 times to provide a confident result.

3.1   Experiment Settings

To obtain clear data, the experiments have been taken with two of the same wireless sensor nodes for both the WiFi and ZigBee. The WiFi Module that were used in the experiments is the WiFi Sheild V2.2 for Arduino that is based on the WizFi210 this model can use all 802.11b/g/n wireless modes. Table 1 shows this in further detail.

Table 1. Specifications of WiFi Modules Used

	Specifications
WiFi Chip	WizFi210
Radio Protocol	IEEE 802.11b/g/n
RF Frequency	2.4 2.497 GHz
Power Consumption	Standby = 34.0 vA
	Receive = 125.0 mA
	Transmit 135.0 mA
RF Output Power	8dBm ± 1dBm
Power Source	3.3V
Dimensions (excluding antenna)	59 x 54 mm
Range indoors	20 meters
Range outdoors	Greater than 20 meters
Supported Data Rates	11, 5.5, 2, 1 Mbps (802.11b)

The ZigBee device that we used for experiment conduction is the XBee ZB from Digi.com. Table 2 shows this in further details.

Both devices were set with different power inputs to set the maximum range of seven meters to imitate indoor areas and because ZigBee can reach up to 120 meters it was wise to fit the WiFi components for a comparison.

Table 2. Specifications of ZigBee Modules Used

	specifications
Radio Protocol	802.15.4 ZigBee
RF Frequency	2.4GHz
Power Consumption	power down = 1 vA
	Receive = 38 mA
	Transmit = 40mA
RF Output Power	N/A
Power Source	2.1 - 3.6 VDC
Dimensions (excluding antenna)	24.38mm x 27.61mm
Range Indoors	40m
Range Outdoors	120m
Supported Data Rate	

4   Data Analysis

We conducted a series of experiments to evaluate the reliability and difference of the WiFi and ZigBee nodes.

4.1   Regression Results

The Regression Method is used to give an estimate equation to find the relationship between different variables. In this research the two variables used are RSSI and distance in centimeters.

There are multiple types of regression methods; simple regression methods, such as linear regression types, basic average and moving average regression methods. The complex methods include logarithmic, polynomial where polynomial can be in the order of 2 and above, power regression and exponential regression. This research will focus on Cubic regression also called 3rd order polynomial regression, to obtain an estimation equation to insert test signal strength to discover the accuracy of this method. The basic cubic function can be represented as equation 3. The reason for using a cubic function and not a higher degree of regression, is to keep the real time calculation at a speed that does not slow down but still keep accuracy to a high expectation.

$$f(\chi) = a\chi^3 + b\chi^2 + c\chi + d \tag{3}$$

To obtain the regression estimation models, two thirds of the sample data from both WiFi and ZigBee were used and the other third were used to test the estimation equation to derive the estimated distance. Figure 2 shows both the WiFi and ZigBee data with the regression equation. R^2 is the coefficient of determination indicating how well the data points fit the regression curve. The value of the R^2 for WiFi is 0.91832, and ZigBee is 0.98958, where 1.00 is a perfect fit.

The regression equation was derived after obtaining the average of the sample data, and from this the test data was then input into the equation to provide the

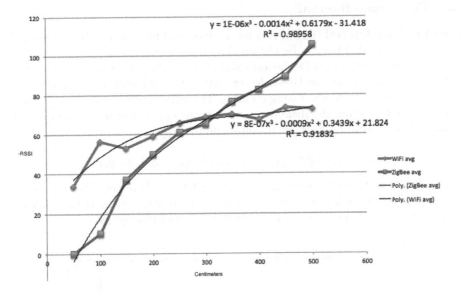

Fig. 2. WiFi and ZigBee data with regression equation

estimated distance at the locations tested. Table 3 shows each position tested with the real distance but also the estimated distance from the regression equation. Where WiFi has up to 83cm error distance at position 400cm and the ZigBee has a maximum of 42cm error at position 450cm. This proves that Zig-Bee has less Error compared to WiFi and the average error shows the same problem.

Table 3. Error Distance at Each Position for both WiFi and ZigBee

Real Distance	WiFi Error	ZigBee Error
50cm	+0.73663335cm	+14.895cm
100cm	+49.94882974cm	-30.46186527cm
150cm	-24.26846171cm	-4.802750355cm
200cm	-22.1273708cm	+4.325cm
250cm	+29.7431086cm	+10.27061595cm
300cm	+48.87606311cm	-16.93177146cm
350cm	+28.54710191cm	-5.838003906cm
400cm	-82.9434375	-24.08606665cm
450cm	+30.87137314cm	+41.85792096cm
500cm	-44.91091205cm	-19.1178708cm
Average	+36.29732919cm	+17.25868654cm

4.2   Confidence Interval

The Confidence Interval (CI) in this research is used to indicate the reliability of the data produced. This can be obtained by discovering the mean of each value, then finding the Standard Deviation (SD) from all the data from each distance.

SD is represented as the Greek letter sigma (σ), and shows how much variation of dispersion from the average exists. The average of the data is simple to calculate, but to show the dispersion with σ was calculated using the following formula 4 in Matlab to show each point side by side.

$$x = exp(-0.5 * ((y - M)/\sigma).^2)/(\sigma * sqrt(2 * pi)) \tag{4}$$

The following figure 3 shows this in further detail, where the sigma is showing the dispersion at each point by graphing the Gaussian curve. If the Gaussian curve is smaller, there is less noise at that point and vise versa.

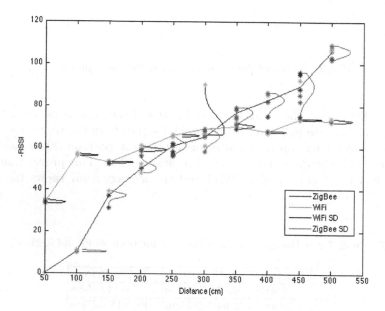

Fig. 3. Dispersion Graph of WiFi and ZigBee

Looking closer at this data, after 150cm on the ZigBee device,the noise variation increases the closer it is to 500cm where 450cm for ZigBee shows a large dispersion rate. At distance 300cm for WiFi has one point that is abnormally higher compared to the other 30 signal strengths. This can be one outlier on that position. Overall WiFi hs smaller noise than ZigBee.

With all the data included from each point with the SD implemented, it was possible to start the CI calculation. Equation 5 shows how this can be calculated.

$$x \pm z\left(\frac{\sigma}{\sqrt{s}}\right) \tag{5}$$

Where x is the mean of the data at a certain location, z is the confidence coefficient, a fixed value for the confidence percentage that we will use. For example, as the test will be finding the confidence for 95%, z becomes 1.96. σ is the SD number and s is the sample size. Everything to the right of x is also called the Margin of Error (MoE).

From this equation we get the maximum confidence for WiFi, $\pm$ 0.38 and for ZigBee, $\pm$ 2.71 RSSI. These results show that ZigBee has more noise coming from the device itself compared to WiFi when scanning for RSSI data, that is needed for distance retrieval. However ZigBee shows more obvious relationships between signal strength and distance compared to WiFi.

5 Conclusion

This research used cubic regression model and confidence interval to discover a relationship between signal strength and distance. To test the reliability, all experiments were tested outdoors to provide un-noisy data before testing indoors like other facilities. From the experiments and results from this research, we have proven that WiFi and ZigBee can be reliable in a way that WiFi has less noise at each distance and ZigBee has a cleaner linear trend line to be used for estimation models. One problem that can be discovered from this research is that the maximum distance of the signal strength was shortened to imitate indoor environments. Changing the power input can change the signal strength at different distances.

Testing indoors with multiple devices may further decrease these distance errors. Further research is needed to investigate what interference can change signal strength to provide this data.

References

[1] Bahl, P., Padmanabhan, V.N.: Radar: An in-building rf-based user location and tracking system. In: Proceedings of the IEEE INFOCOM 2000 Nineteenth Annual Joint Conference of the IEEE Computer and Communications Societies, vol. 2, pp. 775–784 (2000)

[2] Blumenthal, J., Reichenbach, F., Timmermann, D.: Minimal transmission power vs. signal strength as distance estimation for localization in wireless sensor networks. In: 2006 3rd Annual IEEE Communications Society on Sensor and Ad Hoc Communications and Networks, SECON 2006, vol. 3, pp. 761–766. IEEE (2006)

[3] Ekahau. Ekahau real-time location system (2014), http://www.ekahau.com/real-time-location-system/technology

[4] Huang, S.C., Hsiao, C.S.: Applications of zigbee on building a home automation system. In: Juang, J., Huang, Y.-C. (eds.) Intelligent Technologies and Engineering Systems. LNEE, vol. 234, pp. 535–541. Springer, Heidelberg (2013)

[5] Kadar, P.: Zigbee controls the household appliances. In: International Conference on Intelligent Engineering Systems, INES 2009, pp. 189–192. IEEE (2009)

[6] Kelly, D., Behan, R., Villing, R., McLoone, S.: Computationally tractable location estimation on wifi enabled mobile phones. In: IET Irish Signals and Systems Conference (ISSC 2009), pp. 1–6. IET (2009)

[7] Khoury, H.M., Kamat, V.R.: Evaluation of position tracking technologies for user localization in indoor construction environments. Automation in Construction 18(4), 444–457 (2009)

[8] Kothari, N., Kannan, B., Glasgwow, E.D., Dias, M.B.: Robust indoor localization on a commercial smart phone. Procedia Computer Science 10, 1114–1120 (2012)

[9] Langendoen, K., Reijers, N.: Distributed localization in wireless sensor networks: a quantitative comparison. Computer Networks 43(4), 499–518 (2003)

[10] Langendoen, K., Reijers, N.: Distributed localization algorithm. In: Embedded Systems Handbook. CRC Press, Boca Raton (2004)

[11] Lee, J., Cho, K., Lee, S., Kwon, T., Choi, Y.: Distributed and energy-efficient target localization and tracking in wireless sensor networks. Computer Communications 29(13), 2494–2505 (2006)

[12] Lim, C.B., Kang, S.H., Cho, H.H., Park, S.W., Park, J.G.: An enhanced indoor localization algorithm based on ieee 802.11 wlan using rssi and multiple parameters. In: 2010 Fifth International Conference on Systems and Networks Communications (ICSNC), pp. 238–242. IEEE (2010)

[13] Luo, X., O'Brien, W.J., Julien, C.L.: Comparative evaluation of received signal-strength index (rssi) based indoor localization techniques for construction jobsites. Advanced Engineering Informatics 25(2), 355–363 (2011)

[14] Mao, G., Anderson, B., Fidan, B.: Path loss exponent estimation for wireless sensor network localization. Computer Networks 51(10), 2467–2483 (2007)

[15] Ni, L.M., Liu, Y., Lau, Y.C., Patil, A.P.: Landmarc: indoor location sensing using active rfid. Wireless Networks 10(6), 701–710 (2004)

Path Mapping and Planning with Partially Known Paths Using Hierarchical State Machine for Service Robot

A.A. Nippun Kumaar and T.S.B. Sudarshan

Department of Computer Science and Engineering, Amrita Vishwa Vidyapeetham,
School of Engineering, Bangalore Campus, Karnataka - 560035, India
{nippun05,sudarshan.tsb}@gmail.com

Abstract. Path mapping is a very essential part of a mobile robot navigation system. In this work, a novel technique to map and plan path for a mobile service robot without any vision aids in indoor environment using hierarchical state machine with partially known paths is proposed. The known paths are taught to a robot using Learning by Demonstration technique (LfD). The first phase of the algorithm is to map the paths as a hierarchical state machine using the partially known paths. Second phase is to plan the path given the source and destination. The algorithm is implemented and tested using a 2D simulation environment platform, Player/Stage.

Keywords: Path Planning, Path Mapping, Service Robots, Mobile Robots, Navigation System, Hierarchical state machine.

1 Introduction

Navigation system is the most important component of mobile robot. Navigation system is responsible for the robot movement to reach from a start state to a goal state in its environment. Path planning plays a vital role in achieving this task. This is an active research area for the past few decades. There has been lot of techniques developed for path planning and efficient navigation in a mobile robot. These techniques can be broadly classified into three categories based on the awareness the robot has about the environment. They are path planning with known, unknown and partially known environment.

Path planning with known environment is also called as Deliberative planning [1]. In this category of algorithms the aim is to search for an optimal path and to generate a plan to reach the goal. It's a top down approach. It requires the environment map for path planning, in advance. Classical heuristic search algorithms like Dijkstra's [2] and A* [3] are the foremost techniques in this category [4]. Both the techniques search an optimum path between the start and end point. The notable difference being that A* algorithm directs the search towards more optimal states to reduce complexity. This kind of approach cannot be used in dynamic unknown environments, which is essential for service robots.

© Springer International Publishing Switzerland 2015
J.-H. Kim et al. (eds.), *Robot Intelligence Technology and Applications 3,*
Advances in Intelligent Systems and Computing 345, DOI: 10.1007/978-3-319-16841-8_5

Path planning with totally unknown environment is known as Reactive planning [1]. This technique overcomes the drawback of deliberative planning in unknown dynamic environment. The robot uses onboard sensors to sense the environment and decide upon the action to be taken. In this category of algorithms, path is planned on the fly depending on the environment. Various approaches like Motor Schema based approach [5, 6], Potential Field method, Improved potential field method [7], artificial potential field method [8], Modified flexible vector field method [9] etc., are proposed. The major drawback in these techniques is that the path may not be an optimum path as in the case of deliberative planning techniques.

Path planning with partially known environment is a combination of both deliberative and reactive planning techniques. This is a hybrid technique that first plans the path using deliberative planning and uses reactive planning if it encounters any obstacle in the previously planned path. Some heuristic approaches like D* [10], Field D* [11] belong to this category. D* algorithm is derived from A* algorithm which has a capability to dynamically re-plan the path if necessary. In the literature [12-14] global path planning algorithms that carries out path planning using A* deliberative planning technique and then using a reactive planning technique like conjugate gradient descent, improved potential field or D* Lite to locally improve the quality of the path has been reported.

Rashid et.al., proposed a path planning technique using visibility binary tree in which the algorithm constructs set of all possible paths and map a visibility binary tree based on obstacles in the environment [15]. This technique can either be used in known or unknown environment. Coverage Based Path Planning [16] method proposed by Galceran and Carreras helps in determining a path that passes over all points of an area or volume of interest while avoiding obstacles. Rapidly exploring Random Tree (RRT) [17] is a probabilistic technique to generate a path with local constraints. Many recent techniques have been proposed which uses AI and soft computing techniques for path planning [18-20].

In deliberative planning the environment space map has to be pre-loaded and known to the robot, which is not practically possible, as most of the environment in which indoor-based service robot work, are dynamic in nature. In reactive planning the route that the robot is taking may not be the shortest or most efficient one. In hybrid technique, the robot has to be pre-loaded with the environment map for planning the initial global route and then allowed to adapt to the changes in the environment.

Considering the above issues, this work proposes a novel path planning technique in which the robot will map the paths to various destinations with partially known or learnt paths. Here an environment map is not required and the path planned will be the shortest path considering all constrains in the environment. This technique can be used in a mobile service robot, were the robot is working in an indoor environment.

The prerequisite for the robot to map the path is, knowing the partial paths to all the possible destinations from one source. This is carried out by a technique called Learning from Demonstration (LfD) [21, 22], which is introduced in section 2 of this paper. Section 3 explains path-mapping technique using Hierarchical State Machine (HSM) and then path planning. This is followed in Section 4, explanation of experiments carried out and simulation results of the algorithm in an indoor scenario. Section 5 & 6 concludes the work and discuss the future direction.

2   Learning from Demonstration (LfD)

A basic prerequisite for the proposed algorithm is to acquire partial paths. Partial paths can be learnt using Learning from Demonstration (LfD). LfD is a technique in which the user teaches the robot to perform a task in teaching phase and later the robot performs the task autonomously, in execution phase. So, in this case the user will teach the robot to go to various points in an indoor environment known as destinations from a single point known as home position. This will be the seed path for the robot to map other paths. The learnt path will be stored in the form of a two dimensional path matrix as shown in equation 1.

$$\alpha = \begin{pmatrix} dij & cij & ... & dij & ci(j+n) \\ ... & ... & ... & ... & ... \\ d(i+m)j & c(i+m)(j+1) & ... & d(i+m)j & c(i+m)(j+n) \end{pmatrix}_{mxn} \tag{1}$$

where, α is the path matrix representation of the learnt path. In the matrix each row represents a path from home position to a destination. The number of rows in the matrix 'm' represents the total number of destinations. The path variables are 'd' and 'c' which is direction and distance/angle respectively. The direction variable is encoded as 1, 2, 3 and 4 for front, back, left and right respectively. If the direction 'd' is linear (front or back) then 'c' represents distance or if 'd' is angular (left turn or right turn) then 'c' represents angle. The matrix is framed in such a way that all the odd elements are direction values and the immediate even elements are their corresponding distance/angle values. Every 'd' and 'c' is called as pose pair. This will determine the robot pose in every state of the robot. Here, 'm' is total number of states, the robot will transit through, before getting to the corresponding destination.

3   Proposed Algorithm

The proposed algorithm has two phases. First phase is path mapping and the second phase is path planning. In the path mapping phase building the Hierarchical State Machine with the learnt state as a base is carried out. In the path planning phase, robot will use the acquired knowledge about the environment and will be capable of navigating from one point to another by traversing through intermediate states and plan the optimum path.

3.1   Hierarchical State Based Path Mapping

This phase of the algorithm maps the path from partially known path. Here the partially known paths are nothing but the paths that are learnt from home to various other destinations. This path-mapping algorithm will map the path between the destinations. Fig.1 illustrates the path mapping behavior.

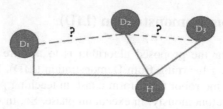

Fig. 1. Illustration of path mapping (solid lines are learnt path and dotted lines are unknown paths)

The algorithm for path mapping identifies Virtual LandMark (VLM) along the path being learnt. These VLMs are represented as the states in the finite state machine model. The VLMs are chosen such that it merges two are more paths in the path matrix so as to act, as via points between two or more destinations. This phase has two stages; they are preprocessing stage and state formation stage.

Preprocessing Stage: This stage is carried out in two phases first the path matrix α is sorted as per the direction variable "dn1". This is to sort the matrix based on the direction of the path from home position.

```
Algorithm
Preprocessing Phase - Sorting Stage

for(i=0; i<m; i++) {
   for (j=i+1; j<m; j++) {
     if (α[i][0] > α[j][0])
        swap(i,j);
     else
        continue;
   }
}
```

After sorting α, the second phase is the sub-path matrix phase. In this phase the path matrix α is split into three sub path matrix af, al and αr forward, left and right sub matrix.

```
Algorithm
Preprocessing Phase - Sub path matrix Stage
Int fc,rc,lc;
for(i=0; i<m; i++) {
  if(α[i][0] == 1) {
    for(j=0; j<n; j++)
      αf[fc][j] = α[r][j]; fc++;
  }
```

```
  else if(α[i][0] == 3) {
    for(j=0; j<n; j++)
      αf[lc][j] = α[r][j]; lc++;
  }
  else if(α[i][0] == 3) {
    for(j=0; j<n; j++)
      αf[lc][j] = α[r][j]; lc++;
  }
}
```

State Formation Phase. This stage decides the optimum VLM as per the learnt paths. The VLMs, Destinations and Home position are states in the state machine. Decision to go from one state to other, depends on the direction variable. From one state, traversal can be to next or previous state. The next state can be forward, left or right state while previous state can be a backward state.

In some cases sub destinations have to be specified. For example, if a robot has learnt to go from a home position to a living room of the house and in living room robot has to reach a television set to perform a task and then go to a table in the living room to pick up some object; these additional destinations are sub-destinations. In these cases, the algorithm will be represented as a Hierarchical State Machine (HSM) that can accommodate multiple sub destinations for each major destination. So the hierarchy is, home is in level 1 with all the destinations and living room is level 2 with television set and table as destinations in this level. In this case, living room acts as a major destination and also a VLM to go to either television set or table.

```
Algorithm
State Formation Phase
fc; // from sub path matrix phase
if(fc != 0) {
  for(i=0; i<m; i++) {
    for(j=0; j<n & αf[i][j] != 0; j++) {
      if(αf[i][j] == 1)
        create_forward_VLM;
      else if (αf[i][j] == 3)
        create_left_VLM;
      else if (αf[i][j] == 4)
        create_right_VLM;
      j += 2;
    }          // Similar algorithm is used to build state
machine from αr, αl.
```

By the end of the first phase a HSM will have all VLMs, Destinations and home as it states. The state machine will be fully connected such that from one state to other state a link can be found directly or through other states.

3.2   Path Planning – Merging Technique

In this work, path planning is carried out using a merging technique. Here, given destination1, which becomes a source for further navigation and destination2, which the robot has to reach, an optimum path can be found with the help of HSM. These two routes have to be merged via common VLMs that act as via points between destination1 and destination2. If there is no common VLM then the routes cannot be merged. In this case home has to be a via point.

```
Algorithm
Path Planning - Merging Technique

hmsrc[m] = Find_path(home to source);
hmdst[m] = Find_path(home to destination); // function
returns the names of the VLMs in the path in order
for(i=0; i<m; i++) {
  for(j=0; j<m; j++) {
    if(hmsrc[i] = hmdst[j])
      break;
    else
      continue;
  }
Merge_routes(i,j); \\This function will merge the route
with i and j as stop point for hmsrc and hmdst respec-
tively
```

4   Experiments and Results

Player/Stage, a 2D robotic simulation environment is used to implement and test the proposed algorithm. Fig. 2 shows a stage which is designed as a typical house layout with many rooms.

First phase will be a teaching phase in which the user will teach the robot from home position to all individual destinations. Fig. 3 depicts the paths robot learnt during the LfD phase. Here the robot learns to navigate to eight different destinations and two sub destinations from Room2, from its home position H, this way the path is partially learnt.

After the preprocessing and state formation HSM for the given environment is generated. Fig. 4 illustrates the Hierarchical state machine that is formed using the LfD paths. The first state is the home and leaf node states are the destinations. The intermediate states represent VLMs. There can be three states left, center and right for any node. This is a symbolic representation of the direction to be taken to navigate from one state to another.

Fig. 2. Player/Stage Environment as a house layout with a hexagonal shaped robot in bottom center (home position)

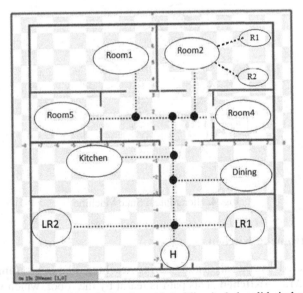

Fig. 3. Paths from home to various destinations (dotted lines), dark solid circles are the probable VLMs

As depicted in Fig. 4, Room 2 has two more destinations R1 and R2. So this will be the level 2 in the state machine. Room 2 will act as a destination as well as VLM for R1 and R2.

Now given any source and destination the robot should be able to plan the path using merging technique and navigate in an optimum way. For experimental purpose, two cases have been used and the algorithm has been tested for its function and accuracy.

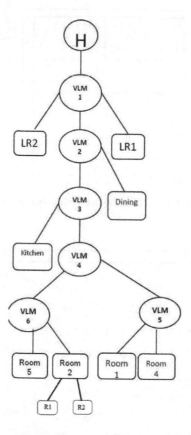

Fig. 4. Hierarchical State Machine

4.1   Case 1: Home to Kitchen

The path generated by the algorithm during teaching phase is **Home → VLM1 →VLM2 →VLM3 →Kitchen**. This path is learnt by the robot, which will start from home position and reach the goal kitchen by following the path generated. Fig. 5 depicts the same.

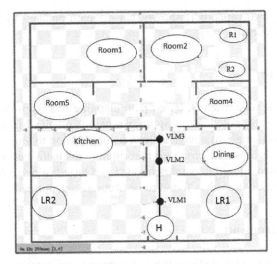

Fig. 5. Path followed by the robot to reach kitchen from home through VLMs

4.2   Case 2: Room 4 to LR1

In this case, during the teaching phase the robot learns two paths. One path from Home to Room4; **Home → VLM1 → VLM2 → VLM3 → VLM4 → VLM5 → Room4** and another path which is from Home to LR1; **Home → VLM1 → LR1.** As it is evident from both the paths the common via point is VLM1, so the algorithm merges the two paths at this point and the path generated for navigation will be **Room4 → VLM5 → VLM4 → VLM3 → VLM2 → VLM1 → LR1.** Fig. 6 shows the robot navigating the path.

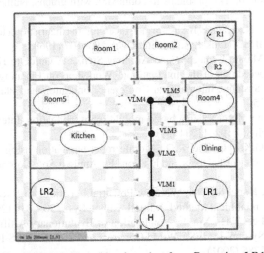

Fig. 6. Path followed by the robot from Room4 to LR1

From the above two cases, it is shown that the path planned by the algorithm is both accurate and optimal.

5   Conclusion

A novel hierarchical state machine based path mapping that maps the path from partially known paths is proposed. Partially known path is obtained from LfD technique in which the robot is taught to navigate from home position to various other destinations. With this as the base the proposed algorithm maps paths between all the destinations. The proposed algorithm is implemented and tested in player/stage simulation environment. The resultant path is found to be accurate and optimal.

6   Future Direction

As the robot navigates and explores the environment, orientation of the robot may change. This issue can be addressed by identifying several reference junctions, where the robots will self-evaluate to reorient themselves using proximity sensors. Also, a dynamic obstacle technique has to be developed to avoid obstacles on the fly and reorient the robot to its path.

References

1. Nakhaeinia, D., Tang, S.H., Mohd Noor, S.B., Motlagh, O.: A review of control architectures for autonomous navigation of mobile robots. International Journal of the Physical Sciences 6(2), 169–174 (2011)
2. Dijkstra, E.: A note on two problems in connexion with graphs. Numerische Mathematik, 269–271 (1959)
3. Hart, P., Nilsson, N., Rafael, B.: A formal basis for the heuristic determination of minimum cost paths. IEEE Trans. Sys. Sci. and Cyb., 100–107 (1968)
4. Dave Ferguson, Maxim Likhachev and Anthony Stenz.: A Guide to Heuristic-based Path Planning. In: Proceedings of the International Workshop on Planning under Uncertainty for Autonomous Systems, International Conference on Automated Planning and Scheduling (ICAPS) (2005)
5. Ronald, C.: Arkin.: Motor schema based mobile robot navigation. International Journal of Robotics Research 8(4), 92–112 (1989)
6. Huq, R., Mann, G.K.I., Gosine, R.G.: Mobile robot navigation using mo-tor schema and fuzzy context dependent behavior modulation. In: Applied Soft Computing, pp. 422–436. Elsevier (2008)
7. Lee, J., Nam, Y., Hong, S., Cho, W.: New Potential Functions with Random Force Algorithms Using Potential Field Method. J. Intell. Robot. Syst., 303–319 (2012)
8. Vadakkepat, P., Lee, T.H., Xin, L.: Application of Evolutionary Artificial Potential Field in Robot Soccer System. In: IFSA World Congress and 20th NAFIPS International Conference, pp. 2781–2785 (2001)

9. Hong, J., Choi, Y., Park, K.: Mobile robot navigation using modified flexible vector field approach with laser range finder and IR sensor. In: International Conference on Control, Automation and Systems, Seoul, Korea, pp. 721–726 (2007)
10. Stentz, A.: Optimal and Efficient path planning for partially known environments. In: Proceedings of the IEEE International Conference on Robotics and Automation (1994)
11. Ferguson, D., Stentz, A.: The Field D* Algorithm for Improved Path Planning and Replanning in Uniform and Non-Uniform Cost Environments. Tech. Report CMU-RI-TR-05-19, Robotics Institute, Carnegie Mellon University (2005)
12. Dolgov, D., Thrun, S., Montemerlo, M., Diebel, J.: Practical Search Techniques in Path Planning for Autonomous Driving. In: Proceedings of the First International Symposium on Search Techniques in Artificial Intelligence and Robotics, STAIR 2008 (2008)
13. Du, Z., School, G., Qu, D., Xu, F.: A Hybrid Approach for Mobile Robot Path Planning in Dynamic Environments. In: Proceedings of the 2007 IEEE International Conference on Robotics and Biomimetics, Sanya, pp. 1058–1063 (2007)
14. Xu, L., Zhang, L.-G., Chen, D.-G., Chen, Y.-Z.: The Mobile Robot Navigation In Dynamic Environment. In: Proceedings of the Sixth International Conference on Machine Learning and Cybernetics, Hong Kong, pp. 566–571 (2007)
15. Rashid, A.T., Ali, A.A., Farsca, M., Fortuna, L.: Path Planning with obstacle avoidance based on visibility binary tree algorithm. International Journal on Robotics and Autonomous Systems, 1440–1449 (2013)
16. Galceran, E., Carreras, M.: A Survey on Coverage Path Planning for Robotics. International Journal on Robotics and Autonomous Systems, 1258–1276 (2013)
17. LaValle, S.: Rapidly-exploring random trees: a new tool for path planning. Tech. Report. Computer Science Dept., Iowa State University (1998)
18. Yan, X., Wu, Q., Yan, J., Kang, L.: A Fast Evolutionary Algorithm for Robot Path Planning. In: IEEE International Conference on Control and Automation, China, pp. 84–87 (2007)
19. Lei, L., Wang, H., Wu, Q.: Improved Genetic Algorithms Based Path planning of Mobile Robot Under Dynamic Unknown Environment. In: Proceedings of the IEEE In-ternational Conference on Mechatronics and Automation, pp. 1728–1732 (2006)
20. Zhang, Y., Zhang, L., Zhang, X.: Mobile Robot Path Planning base on the Hybrid Genetic Algorithm in Unknown Environment. In: Eighth International Conference on Intelligent Systems Design and Applications, pp. 661–665 (2008)
21. Nippun Kumaar, A.A., Sudarshan, T.: Mobile Robot Programming By Demonstration. In: Fourth International Conference on Emerging Trends in Engineering & Technology (ICETET 2011), pp. 206–209. IEEE Computer Society, Mauritius (2011)
22. Nippun Kumaar, A.A., Sudarshan, T.S.B.: Learning from Demonstration with State Based Obstacle Avoidance for Mobile Service Robots. International Journal Applied Mechan-ics and Materials, 448–455 (2013)

Dual Multiobjective Quantum-Inspired Evolutionary Algorithm for a Sensor Arrangement in a 2D Environment

Si-Jung Ryu, Rituparna Datta, and Jong-Hwan Kim

Department of Electrical Engineering, KAIST,
291 Daehak-ro, Yuseong-gu, Daejeon 305-701, Republic of Korea
{sjryu,rdatta,johkim}@rit.kaist.ac.kr

Abstract. This paper proposes dual multiobjective quantum-inspired evolutionary algorithm (DMQEA) for a sensor arrangement problem in a 2D environment. DMQEA has a dual stage of dominance check by introducing secondary objectives in addition to primary objectives. In an archive generation process, the secondary objectives are to maximize global evaluation values and crowding distances of the non-dominated solutions in the external global population and the previous archive. The proposed DMQEA is applied to the sensor arrangement problem to allocate the sensors considering three objectives: coverage rate, interference rate of each sensor, and the number of the sensors. The result of the sensor arrangement was successful enough to satisfy user's preference for the objectives such that the sensors are placed on the proper positions.

1 Introduction

Multiobjective evolutionary algorithms (MOEAs) aim to solve optimization problems having multiple objectives and to achieve a wide spread non-dominatd solutions [1–4]. Another critical issue of MOEA is the selection of a preferable solution among the Pareto-optimal solutions. The preference based selection is a major concern of the decision makers in case of most real world optimization problems. To alleviate this issue, preference-based solution selection algorithm (PSSA) was proposed [5]. The PSSA algorithm selects a solution considering user's preference for each objective, which is represented by the fuzzy measures. Based on PSSA, multiobjective quantum-inspired evolutionary algorithm with preference-based selection (MQEA-PS) was proposed [5–7].

In this paper, dual multiobjective quantum-inspired evolutionary algorithm (DMQEA) is proposed by employing secondary objectives in addition to primary objectives. The primary objectives are original objective functions of the problem. The secondary objectives are the maximization of global evaluation values and crowding distances of the solutions during archive generation process.. The idea of incorporating secondary objectives is to induce better balanced exploration of the non-dominated solutions in terms of users preference and diversity. The proposed DMQEA has the dual-stage of dominance check respectively for the primary and secondary objectives. In the first stage, the dominated solutions with respect to primary objectives are obtained by primary objectives-based nondominated sorting (PONS). In the second stage,

© Springer International Publishing Switzerland 2015
J.-H. Kim et al. (eds.), *Robot Intelligence Technology and Applications 3*,
Advances in Intelligent Systems and Computing 345, DOI: 10.1007/978-3-319-16841-8_6

nondominated sorting is applied for the secondary objectives to generate an archive, which is called secondary objectives-based nondominated sorting (SONS). The secondary objectives are to maximize global evaluation values and crowding distances of the solutions in the previous archive and the external global population.

The effectiveness of the proposed DMQEA is applied to a sensor arrangement problem which is to allocate the sensors considering three objectives: coverage rate, interference rate, and the number of the sensors. The experimental results confirm that the proposed DMQEA generates the positions of the sensors, which satisfies user's preference for the objectives.

The rest of this paper is organized as follows: quantum-inspired evolutionary algorithm (QEA), preference-based solution selection algorithm (PSSA), and crowding distance are briefly described in Section II. Section III proposes dual multiobjective evolutionary algorithm (DMQEA). The results of sensor arrangement using DMQEA are presented in Section IV and concluding remarks follow in Section V.

2   Preliminaries

2.1   QEA

Quantum-inspired evolutionary algorithm (QEA) is an evolutionary algorithm, which employs the probabilistic mechanism inspired by the concept and principles of quantum computing, such as a quantum bit and superposition of states [8, 9]. Building block of classical digital computer is represented by two binary states, '0' or '1', which is a finite set of discrete and stable state. In contrast, QEA utilizes a novel representation, called a Q-bit representation, for the probabilistic representation that is based on the concept of qubits in quantum computing [10]. Since Q-bit individual represents the linear superposition of all possible states probabilistically, various individuals are generated during the evolutionary process. To solve multiobjective optimization problems, multiobjective quantum-inspired evolutionary algorithm (MQEA) is also developed [11].

2.2   PSSA

Preference-based solution selection algorithm (PSSA) selects a solution among the obtained nondominated solutions considering user's preference [5]. The nondominated solutions cannot be directly compared against each other, and therefore a multicriteria decision making (MCDM) algorithm is required to evaluate them. In PSSA, the global evaluation value of a candidate solution is calculated by the fuzzy integral, as an MCDM algorithm, of the partial evaluation values with respect to the fuzzy measures. The fuzzy measures represent the degrees of consideration for objectives, and the partial evaluation value indicates a normalized objective function value.

2.3   Crowding Distance

The crowding distance estimates the density of solutions surrounding a particular solution in the population [3]. The crowding distance is aimed to uniformly select the solutions in the front, making the solutions in the most dense areas less likely to be selected. The crowding distance is defined by the average distance of the closest points on either side of the point for each objective. Therefore, the crowding distance is inversely proportional to the density of solutions. Boundary points for each objective have the maximum crowding distance, and they are always selected.

3 DMQEA

Dual multiobjective quantum-inspired evolutionary algorithm (DMQEA) has the dual-stage of dominance check for the primary and secondary objectives. Primary objectives are the given objectives of the problem. The secondary objectives are to maximize both the global evaluation values and crowding distances of the solutions in the external global population obtained for the primary objectives and the previous archive. In each archive generation process, the secondary objectives are employed for sorting the solutions, which is called secondary objectives-based nondominated sorting (SONS). By the proposed SONS, the archive stores first-tier solutions.

3.1 SONS

SONS is to sort the solutions with the secondary objectives for maximizing the global evaluation value (GEval) and crowding distance (CD). The SONS is performed for the solutions in the external global population obtained for the primary objectives and the previous archive. The proposed SONS is depicted in Fig. 1. First, GEval and CD of every solution in the external global population and the previous archive are calculated. And then, the solutions that are not dominated by any other solutions are obtained as first-tier solutions that are stored in the archive. The solutions with higher values of GEval and CD are better in terms of user's preference and diversity. For example, in the figure, blue points are classified as first-tier solutions to be stored in the archive.

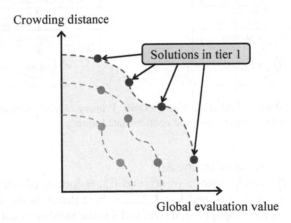

Fig. 1. Secondary objectives-based nondominated sorting

3.2   Procedure of DMQEA

In an archive generation process, MQEA employs dominance-based sorting for primary objectives of the solutions in the external global population and the previous archive. Most of them are nondominated by the other solutions because primary objectives-based nondominated sorting (PONS) or fast nondominated sorting is already performed in each subpopulation. It means that the dominance-based sorting for the primary objectives might be an ineffective operation in selecting solutions to be stored in the archive. To solve this problem, DMQEA employs SONS in the archive generation process. By SONS, each solution is classified into the corresponding tier and the solutions in the first tier are stored in the archive. These are used for reference solutions through the global random migration process. The overall procedure of DMQEA is summarized in Algorithm 1, and depicted in Fig. 2. Each step is described in detail in the following.

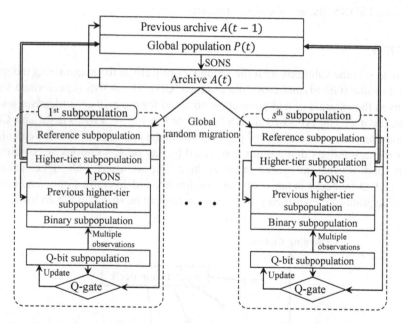

Fig. 2. Overall procedure of DMQEA, where PONS: Primary objectives-based nondominated sorting, SONS: Secondary objectives-based nondominated sorting

1. Initialize $Q_k(t)$ and generate archive $A(t)$

$Q_k(0)$ including $\mathbf{q}_j^0$, which consists of α_{ji}^0 and β_{ji}^0, is initialized with $1/\sqrt{2}$, where $i = 0, 1, ..., m-1, j = 1, 2, ..., n$, and $k = 1, 2, ..., s$. Note that m is the string length of Q-bit individual, n is the subpopulation size, and s is the number of subpopulations. It means that one Q-bit individual, $\mathbf{q}_j^0$, represents the linear superposition of all possible states with same probability. Binary solutions in $P_k(0)$ are produced by multiple observing the states of $Q_k(0)$, where $P_k(0) = \{\mathbf{x}_1^0, \mathbf{x}_2^0, ..., \mathbf{x}_n^0\}$ and $\mathbf{x}_j^0 = \{x_{j,m-1}^0, x_{j,m-2}^0, ..., x_{j0}^0\}$,

---

Algorithm 1. Procedure of DMQEA

- $P_k(t) = \{\mathbf{x}_1^t, \mathbf{x}_2^t, ..., \mathbf{x}_n^t\}$
- $\mathbf{x}_j^t = \{x_{j,m-1}^t, x_{j,m-2}^t, ..., x_{j0}^t\}$
- $Q_k(t) = \{\mathbf{q}_1^t, \mathbf{q}_2^t, \cdots, \mathbf{q}_n^t\}$
- $\mathbf{q}_j^t = \begin{bmatrix} \alpha_{j,m-1}^t & \alpha_{j,m-2}^t & \cdots & \alpha_{j,0}^t \\ \beta_{j,m-1}^t & \beta_{j,m-2}^t & \cdots & \beta_{j,0}^t \end{bmatrix}$
- $R_k(t) = \{\mathbf{r}_1^t, \mathbf{r}_2^t, ..., \mathbf{r}_n^t\}$
- s = No. of subpopulations
- n = Size of subpopulation
- m = Q-bit string length

---

1. Initialize $Q_k(t)$ and generate archive $A(t)$
   - 1: $t = 0$
   - 2: **for** $k = 1$ to s **do**
   - 3:   **for** $j = 1$ to n **do**
   - 4:   **for** $i = 0$ to $m-1$ **do**
   - 5: $\alpha_{ji}^t = \beta_{ji}^t = 1/\sqrt{2}$
   - 6:   **end for**
   - 7:   Make $P_k(t)$ by multiple observing the states of $Q_k(t)$
   - 8:   **for** each objective **do**
   - 9:   Evaluate the objective value from $\mathbf{x}_j^t$
   - 10:   **end for**
   - 11:   Copy all solutions in $P_k(t)$ into $P(t)$
   - 12:   Store first-tier solutions of $P(t)$ by SONS in the archive $A(t)$
   - 13:   **end for**
   - 14: **end for**

2. Generate global population $P(t)$
   - 1: $t = t + 1$
   - 2: **for** $k = 1$ to s **do**
   - 3:   **for** $j = 1$ to n **do**
   - 4:   Make $P_k(t)$ by multiple observing the states of $Q_k(t)$
   - 5:   **for** each objective **do**
   - 6:   Evaluate the objective value from $\mathbf{x}_j^t$
   - 7:   **end for**
   - 8:   **end for**
   - 9:   Run PONS for $P_k(t) \cup B_k(t-1)$
   - 10:   Store n higher-tier solutions of $P_k(t) \cup B_k(t-1)$ into $B_k(t)$
   - 11: **end for**
   - 12: Store all solutions in every $B_k(t)$ into $P(t)$

3. Update archive $A(t)$
   - 1: **for** each solution in $A(t-1) \cup P(t)$ **do**
   - 2:   Evaluate GEval and CD
   - 3: **end for**
   - 4: Run SONS
   - 5: Store the first-tier solutions into the archive $A(t)$

---

4. Migrate and update $Q_k(t)$

   1: **for** $k = 1$ to s **do**
   2:    **for** $j = 1$ to n **do**
   3:       Select a solution in $A(t)$ randomly
   4:       Store it into $\mathbf{r}_j^t$
   5:       Update $\mathbf{q}_j^t$ using Q-gates referring to the solutions in $\mathbf{r}_j^t$
   6:    **end for**
   7: **end for**

5. Go back to Step 2 and repeat

$j = 1, 2, ..., n$. A bit of one binary solution, x_{ji}^0, has a value either '0' or '1' according to the probability either $|\alpha_{ji}^0|^2$ or $|\beta_{ji}^0|^2$, where $i = 0, 1, ..., m-1$, $j = 1, 2, ..., n$, as follows:

$$x_{ji}^0 = \begin{cases} 0 \text{ if } \text{rand}[0,1] \geq |\beta_{ji}^0|^2 \\ 1 \text{ if } \text{rand}[0,1] < |\beta_{ji}^0|^2. \end{cases} \tag{1}$$

Multiple observation is performed on each and every Q-bit individual in subpopulations, $\mathbf{q}_j^0$ in $Q_k(0)$, $k = 1, 2, ..., s$. Each binary solution in $P_k(0)$ is decoded to a real number if necessary, and its objective value is calculated. All solutions in each binary subpopulation $P_k(0)$ are copied to the external global population $P(0)$ and store first tier solutions of $P(0)$ by SONS in the archive $A(t)$.

2. Generate global population $P(t)$

Binary solutions are generated by multiple observations of Q-bit individuals in Q-bit subpopulation $Q_k(t)$. Each bit of binary solution x_{jl}^t, $l = 1, 2, ..., o$, where o is the observation index is obtained. $\mathbf{x}_j^t$ is assigned by the best among the observed binary solutions x_{jl}^t, $l = 1, 2, ..., o$, from the multiple observations. And then, evaluation is performed to $P_k(t)$, where $k = 1, 2, ..., s$. Therefore, objective values of all solutions in each subpopulation are obtained. The solutions in the previous higher-tier subpopulation and the current binary subpopulation $P_k(t) \cup B_k(t-1)$ are sorted by PONS to select n solutions in order from the first tier to the lower tiers. The n higher-tier solutions form $B_k(t)$, where $B_k(t) = \{\mathbf{b}_1^t, \mathbf{b}_2^t, ..., \mathbf{b}_n^t\}$ that is to become the previous higher-tier subpopulation in the next generation. To update Q-bit individuals corresponding to higher-tier subpopulation later, Q-bit subpopulation $Q_k(t)$ is rearranged by replacing each $\mathbf{q}_j^t$ in the subpopulation by the Q-bit individual that has generated $\mathbf{b}_j^t$. All higher-tier solutions in each subpopulation $B_k(t)$ are copied to the external global population $P(t)$.

3. Update archive $A(t)$

Global evaluation values are calculated by the fuzzy integral and crowding distance is also calculated. The fuzzy integral reflects how much a user prefers the solution, and crowding distance denotes the density of the solutions. SONS with GEval and CD for the solutions in the external global population and the previous archive is performed. The nondominated solutions in the first tier are stored into the archive $A(t)$. The size of the archive might be different each generation.

4. Migrate and update $Q_k(t)$

The solutions in the archive $A(t)$ are randomly selected n times and they are globally migrated to each reference subpopulation $R_k(t)$, where $R_k(t) = \{\mathbf{r}_1^t, \mathbf{r}_2^t, ..., \mathbf{r}_n^t\}$. Note that the solutions in $R_k(t)$ are employed as references to update Q-bit individuals, each of which is corresponding to the solution in the higher-tier subpopulation. Global random migration procedure occurs at every generation. In the update process of Q-bit individuals, the rotation gate is employed. $\mathbf{r}_j^t$ and $\mathbf{b}_j^t$ in each subpopulation are compared bit-by-bit to decide the update directions of Q-bit individuals in the rotation gate $U(\Delta\theta)$, which is defined as follows:

$$\mathbf{q}_j^t = U(\Delta\theta) \cdot \mathbf{q}_j^{t-1} \tag{2}$$

with

$$U(\Delta\theta) = \begin{bmatrix} \cos(\Delta\theta) & -\sin(\Delta\theta) \\ \sin(\Delta\theta) & \cos(\Delta\theta) \end{bmatrix}$$

where $\Delta\theta$ is the rotation angle of each Q-bit.

5. Go back to Step 2 and repeat

Go back to Step 2 and repeat until a termination condition is satisfied.

4 Application to Sensor Arrangement Problem

4.1 Configuration for the Experiments

The proposed DMQEA was applied to sensor arrangement in 2D environment. Each sensor was allocated to a suitable positions considering user's preference for the objectives, which was generated by DMQEA. In the experiment, three objectives: the ratio of the sensing region to the feasible region (f_1), a ratio of the overlapped sensor range to overall sensing range (f_2), and the number of sensors (f_3) were employed. Note that f_1, f_2, and f_3 are related with coverage rate, interference rate of each sensor, and the effective number of the sensors. Therefore, the goal of this experiment is to maximize f_1, and minimize f_2 and f_3. We assume that all sensors are homogeneous with the same performance so that they have the same sensing range. Each objective is defined as follows:

$$f_1 = \frac{S_s}{S_f}, \qquad f_2 = \frac{R_o}{R_c}, \qquad f_3 = |N_s| \tag{3}$$

where S_s represents the area covered by the sensors in the feasible region and S_f represents the area of the feasible region. R_o and R_c represent the overlapped area of multiple sensor ranges and the total areas of every sensor range, respectively. $|N_s|$ represents the number of the sensors and was set to have a range of 6 to 10. Parameters for DMQEA are given in Table 1 and it was assumed that three objectives had the same interaction degree which was set to 0.25.

Table 1. Parameter setting of DMQEA

Parameters	Values
The population size ($N = n \cdot s$)	100
No. of generations	3000
Subpopulation size(n)	25
No. of subpopulations (s)	4
No. of multiple observations	10
The rotation angle ($\Delta\theta$)	0.23π
Decoding resolution	20

4.2 Experimental Results

Fig. 3 shows the results of the optimized sensor arrangement considering each objective using DMQEA. The preference degree for three objectives was set as $f_1 : f_2 : f_3 = 10 : 1 : 1$ for Fig. 3(a), $f_1 : f_2 : f_3 = 5 : 10 : 1$ for Fig. 3(b), and $f_1 : f_2 : f_3 = 5 : 1 : 10$ for Fig. 3(c). A small triangle represents a position of the sensor optimized by DMQEA and a gray region covered by a circle represents a detecting range of the sensor, which was set to 1.5. Three objective values that represent coverage rate, interference rate, and the number of the sensors are also presented in the figure. As shown in Fig. 3(a), when f_1 was mostly considered, the coverage rate is mostly maximized while other two objectives were less optimized. Also, the overlapped region and the number of the sensors had the most minimized values when considering mostly f_2 and f_3, respectively.

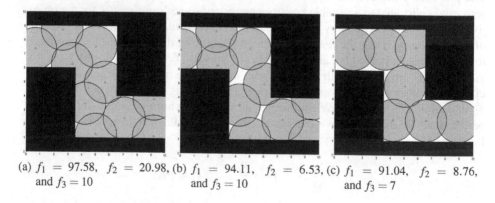

(a) $f_1 = 97.58$, $f_2 = 20.98$, and $f_3 = 10$ (b) $f_1 = 94.11$, $f_2 = 6.53$, and $f_3 = 10$ (c) $f_1 = 91.04$, $f_2 = 8.76$, and $f_3 = 7$

Fig. 3. The optimized sensor arrangements considering mostly (a) sensor coverage (b) interference rate (c) the number of sensors

5   Conclusion

In this paper, a sensor arrangement in a 2D environment using dual multiobjective quantum-inspired evolutionary algorithm (DMQEA) was proposed. DMQEA employs a dualstage of dominance check for both primary and secondary objectives. The secondary objectives are to maximize global evaluation values and crowding distances of the solutions in the archive generation process. The sensor arrangement problem which is to allocate the sensors considering three objectives was adopted to demonstrate the effectiveness of the proposed DMQEA. The objectives which are related to coverage rate, interference rate, and the number of the sensors were employed. The results show that DMQEA generated the positions of the sensors, which satisfies user's preference for the objectives.

Acknowledgement. This work was supported by the Technology Innovation Program, 10045252, Development of robot task intelligence technology, funded by the Ministry of Trade, Industry & Energy (MOTIE, Korea).

References

1. Ni, L.M., Liu, Y., Lau, Y.C., Patil, A.P.: Landmarc: indoor location sensing using active rfid. Wireless Networks 10(6), 701–710 (2004)
2. Kim, J.-H., Kim, Y.-H., Choi, S.-H., Park, I.-W.: Evolutionary Multi-objective Optimization in Robot Soccer System for Education. IEEE Computational Intelligence Magazine 4(1), 31–41 (2009)
3. Deb, K., Pratap, A., Agarwal, S., Meyarivan, T.: A fast and elitist multiobjective genetic algorithm: NSGA-II. IEEE Transactions on Evolutionary Computation 6(2), 182–197 (2002)
4. Zitzler, E., Laumanns, M., Thiele, L.: SPEA2: Improving the strength Pareto evolutionary algorithm. In: Proc. of EUROGEN, pp. 95–100 (2001)
5. Kim, J.-H., Han, J.-H., Kim, Y.-H., Choi, S.-H., Kim, E.-S.: Preference-based solution selection algorithm for evolutionary multiobjective optimization. IEEE Transactions on Evolutionary Computation 16(1), 20–34 (2012)
6. Lee, K.-B., Kim, J.-H.: Multiobjective Particle Swarm Optimization With Preference-Based Sort and Its Application to Path Following Footstep Optimization for Humanoid Robots. IEEE Transactions on Evolutionary Computation 17(6), 755–766 (2013)
7. Ryu, S.-J., Lee, K.-B., Kim, J.-H.: Improved Version of a Multiobjective Quantum-inspired Evolutionary Algorithm with Preference-based Selection. In: Proc. of IEEE World Congress on Computational Intelligence, pp. 1672–1678 (2012)
8. Han, K.-H., Kim, J.-H.: Quantum-inspired evolutionary algorithm for a class of combinatorial optimization. IEEE Transactions on Evolutionary Computation 6(6), 580–593 (2002)
9. Han, K.-H., Kim, J.-H.: Quantum-inspired evolutionary algorithms with a new termination criterion, $H\varepsilon$ gate, and two phase scheme. IEEE Transactions on Evolutionary Computation 8(2), 156–169 (2004)
10. Hey, T.: Quantum computing: an introduction. Computing and Control Engineering Journal 10(3), 105–112 (1999)

11. Kim, Y.-H., Kim, J.-H., Han, K.-H.: Quantum-inspired multiobjective evolutionary algorithm for multiobjective 0/1 knapsack problems. In: Proc. of IEEE Congress on Evolutionary Computation, pp. 2601–2606 (2006)
12. Marichal, J.: An axiomatic approach of the discrete Choquet integral as a tool to aggregate interacting criteria. IEEE Transactions on Fuzzy Systems 8(6), 800–807 (2000)
13. Zitzler, E.: Evolutionary algorithms for multiobjective optimization: methods and applications. In: Berichte aus der Informatik. Shaker Verlag, Aachen-Maastricht (1999)

RRT*-Quick: A Motion Planning Algorithm with Faster Convergence Rate

In-Bae Jeong, Seung-Jae Lee, and Jong-Hwan Kim

Department of Electrical Engineering, KAIST
335 Gwahangno, Yuseong-gu, Daejeon 305-701, Republic of Korea
{ibjeong,sjlee,johkim}@rit.kaist.ac.kr

Abstract. This paper proposes RRT*-Quick as an improved version of Rapidly-exploring Random Tree Star (RRT*). The proposed RRT*-Quick utilizes one of the characteristics of RRT* that nodes in local area tend to share common parents. It uses the ancestor nodes to efficiently enlarge the pool of parent candidates for a faster convergence rate. Branch-and-bound, one of the key extensions of RRT*, prunes the unuseful nodes from the tree to help the search algorithm focus on improving solutions. Since the proposed algorithm generates the initial solution with a lower cost, it can prune unuseful nodes earlier than the conventional RRT*.

1 Introduction

Motion planning algorithm that finds a series of motions to complete the given tasks, became essential part of robotics. As robotics technology develops and the degree of freedom of a robot or a manipulator increases, motion planning problem gets harder to solve because the dimension of problems also increases.

Conventional motion planning algorithms were mainly based on graph theories that needs to discretizing the solution space. However, the resolution of discretized solution space is limited by the memory requirement, since the requirement increases exponentially to the problem dimension [1–3].

One of the commonly used techniques to solve problems in a high-dimensional continuous solution space uses a random sampling-based algorithm. It selects a randomly chosen state and tries to iteratively solve the given problem by repeating the procedure. Two of the commonly used sampling-based algorithms are Probabilistic RoadMap (PRM) and Rapidly-exploring Random Tree (RRT) [4, 5]. These algorithms have been researched extensively; therefore various variants exist to fulfill the requirement of each domain [6, 7]. Among them, RRT* is a well known RRT variant. RRT* not only finds a feasible solution but also tries to improve the solution while a robot is following the generated path [8, 9].

In this paper, an algorithm that has a faster convergence rate is proposed. To have a faster convergence rate without causing much additional computation time, it utilizes the characteristics of RRT* that nodes in local area tend to share common parents. As the cost of the best solution gets lowered, more nodes are

© Springer International Publishing Switzerland 2015
J.-H. Kim et al. (eds.), *Robot Intelligence Technology and Applications 3*,
Advances in Intelligent Systems and Computing 345, DOI: 10.1007/978-3-319-16841-8_7

pruned by branch-and-bound, one of the key extensions of RRT*, therefore the solutions gets improved more.

This paper is organized as follows. Section 2 introduces preliminaries on RRT and RRT*. The proposed algorithm is explained in Section 3, and the simulation environment and the results are shown in Section 4. Concluding remarks follow in Section 5.

2   Preliminaries

RRT and RRT* are briefly described with their characteristics in this section.

2.1   RRT

RRT is an algorithm that searches for a feasible trajectory in a non-convex high-dimensional space by repeatedly connecting a randomly sampled state to the existing tree if there is no collision between them, as shown in Algorithm 1. It has been proven that RRT is probabilistically complete, which means that it fully covers the whole space as the number of samples increases infinitely. Since RRT focuses on finding a feasible solution but not improving the solution, it rarely generates an optimal solution.

Input: $T = (V, E)$
Output: T
while $i < N$ **do**
    $x_{rand} \leftarrow Sample(i)$;
    $x_{nearest} \leftarrow Nearest(T, x_{rand})$;
    $x_{new} \leftarrow Extend(x_{nearest}, x_{rand})$;
    $\sigma_{new} \leftarrow Steer(x_{nearest}, x_{new})$;
    if $CollisionFree(\sigma_{new})$ **then**
        $V \leftarrow V \cup x_{new}$;
        $E \leftarrow E \cup (x_{nearest}, x_{new})$;
    end
end
$T \leftarrow (V, E)$;
return T;

Algorithm 1. RRT

2.2   RRT*

RRT* is an algorithm based on RRT, which focuses on improving the quality of solutions. While RRT simply connects a random sample to the nearest node in the tree, RRT* searches the nodes in a volume sphere with a specific radius to find the node which makes the lowest cost to get to the random sample, as described in Algorithm 2. It also tries to rewire the existing node to the random sample if it lowers the cost. As the number of samples increases, the solution generated by RRT* asymptotically converges to the optimal solution.

There are two key extensions of RRT*, comitted trajectory and branch-and-bound.

Committed Trajectory. Comitted trajectory takes the initial portion of currently best solution then sets a new root node. While a robot follows the committed trajectory, RRT* tries to improve the new (uncommitted) tree.

Branch-and-Bound. Branch-and-bound technique is employed to reduce the number of nodes in the tree. Once a feasible solution is found, the cost of the solution is used as the upper bound. Every node in the tree which has higher cost than the upper bound is removed from the tree along with its descendents. It makes RRT* focus on improving the solution. More nodes are pruned as the upper bound gets lowered, so the computation time for processing an iteration decreases.

3   RRT*-Quick

In RRT*, the radius of the volume sphere decides the converge rate. The solution improves faster with a larger radius, but the number of nodes in the volume sphere also increases exponentially; therefore much more computation time is needed. RRT*-Quick that is proposed in this paper, utilizes the characteristics of the shape of the tree to efficiently increase the converge rate without causing much computation time.

The main idea of RRT*-Quick is based on the observation that nodes in local area tend to share a commmon parent because of the procedure 'ChooseParent' and the rewire operation of RRT*. Since commonly used metrics to measure the cost satisfies the triangular inequality, the parents are good candidates for the parent with the lowest cost.

As described in Algorithm 3, in addition to the nodes in the volume sphere, RRT*-Quick also takes account of the ancestors of them. The ancestors always provides a path with a lower (or equal) cost if the metric satisfies the triangular inequalty, even if it doesn't, the increased computation time will be negligible compared to that caused from a larger radius. In Algorithm 3, $Ancestors(x, degree)$ is a procedure that returns the set of the ancestors of x up to the given $degree$. Although three or four is enough for $degree$, setting $degree$ to infinite is fine because the number of siblings decreases exponentially as following up along the ancestry, which means it hardly affects the performance.

The shapes of trees of RRT, RRT*, and the proposed RRT*-Quick are depicted in Fig. 1.

Input: $T = (V, E)$
Output: T
while $i < N$ **do**
$\quad x_{rand} \leftarrow Sample(i);$
$\quad x_{nearest} \leftarrow Nearest(T, x_{rand});$
$\quad x_{new} \leftarrow Extend(x_{nearest}, x_{rand});$
$\quad \sigma_{new} \leftarrow Steer(x_{nearest}, x_{rand});$
$\quad$ **if** $CollisionFree(\sigma_{new})$ **then**
$\quad\quad X_{near} \leftarrow Near(T, x_{new});$
$\quad\quad x_{min} \leftarrow x_{nearest};$
$\quad\quad c_{min} \leftarrow Cost(x_{nearest}) + Cost(\sigma_{new});$
$\quad\quad$ **foreach** $x_{near} \in X_{near}$ **do**
$\quad\quad\quad \sigma_{near} \leftarrow Steer(x_{near}, x_{new});$
$\quad\quad\quad$ **if** $Cost(x_{near}) + Cost(\sigma_{near}) < c_{min} \wedge CollisionFree(\sigma_{near})$ **then**
$\quad\quad\quad\quad x_{min} \leftarrow x_{near};$
$\quad\quad\quad\quad c_{min} \leftarrow Cost(x_{near}) + Cost(\sigma_{near});$
$\quad\quad\quad$ **end**
$\quad\quad$ **end**
$\quad\quad V \leftarrow V \cup x_{new};$
$\quad\quad E \leftarrow E \cup \{(x_{min}, x_{new})\};$
$\quad\quad$ **foreach** $x_{near} \in X_{near}$ **do**
$\quad\quad\quad \sigma_{near} \leftarrow Steer(x_{new}, x_{near});$
$\quad\quad\quad$ **if** $Cost(x_{new}) + Cost(\sigma_{near}) < Cost(x_{near}) \wedge CollisionFree(\sigma_{near})$
$\quad\quad\quad$ **then**
$\quad\quad\quad\quad E \leftarrow E \setminus \{(Parent(x_{near}), x_{near})\};$
$\quad\quad\quad\quad E \leftarrow E \cup \{(x_{new}, x_{near})\};$
$\quad\quad\quad$ **end**
$\quad\quad$ **end**
$\quad$ **end**
end
$T \leftarrow (V, E);$
return $T;$

Algorithm 2. RRT*

Input: $T = (V, E)$
Output: T
while $i < N$ **do**
    $x_{rand} \leftarrow Sample(i)$;
    $x_{nearest} \leftarrow Nearest(T, x_{rand})$;
    $x_{new} \leftarrow Extend(x_{nearest}, x_{rand})$;
    $\sigma_{new} \leftarrow Steer(x_{nearest}, x_{rand})$;
    if $CollisionFree(\sigma_{new})$ **then**
        $X_{near} \leftarrow Near(T, x_{new})$;
        $x_{min} \leftarrow x_{nearest}$;
        $c_{min} \leftarrow Cost(x_{nearest}) + Cost(\sigma_{new})$;
        foreach $x_{near} \in X_{near} \cup Ancestors(X_{near}, degree)$ **do**
            $\sigma_{near} \leftarrow Steer(x_{near}, x_{new})$;
            if $Cost(x_{near}) + Cost(\sigma_{near}) < c_{min} \wedge CollisionFree(\sigma_{near})$ **then**
                $x_{min} \leftarrow x_{near}$;
                $c_{min} \leftarrow Cost(x_{near}) + Cost(\sigma_{near})$;
            end
        end
        $V \leftarrow V \cup x_{new}$;
        $E \leftarrow E \cup \{(x_{min}, x_{new})\}$;
        foreach $x_{near} \in X_{near}$ **do**
            $\sigma_{near} \leftarrow Steer(x_{new}, x_{near})$;
            $X_{candidates} \leftarrow Ancestors(x_{new}, degree) \setminus Ancestors(x_{near}, degree)$;
            $x_{min} \leftarrow x_{new}$;
            $c_{min} \leftarrow Cost(x_{new}) + Cost(\sigma_{new})$;
            foreach $x_c \in X_{candidates}$ **do**
                $\sigma_c \leftarrow Steer(x_c, x_{near})$;
                if $Cost(x_c) + Cost(\sigma_c) < c_{min}) \wedge CollisionFree(\sigma_c)$ **then**
                    $x_{min} \leftarrow x_c$;
                    $c_{min} \leftarrow Cost(x_c) + Cost(\sigma_c)$;
                end
            end
            $E \leftarrow E \setminus \{(Parent(x_{near}), x_{near})\}$;
            $E \leftarrow E \cup \{(x_{min}, x_{near})\}$;
        end
    end
end
$T \leftarrow (V, E)$;
return T;

Algorithm 3. RRT*-Quick

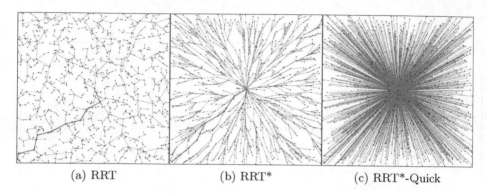

(a) RRT   (b) RRT*   (c) RRT*-Quick

Fig. 1. Extended trees after 1000 iterations

4   Simulation

The effectiveness of the proposed algorithm was verified through computer simulations. RRT, RRT* and RRT*-Quick were run 50 times for each with the same set of random samples.

4.1   Simulation Environment

The simulation environment is shown in Fig. 2 with the start point and the goal area colored with red and blue, respectively. The size of the environment is 600×600 and an obstacle is blocking the direct path from the start point to the goal point. The optimal path lies on the left side of the map, but a tree tends to grow towards wider area, which is the right side of the environment; therefore local minima exist on the right side. The environment also has a narrow passage that makes harder to find a feasible path. The cost of the optimal solution is 1349.69.

4.2   Simulation Results

Major performance measures of a sampling-based motion planning algorithm are the quality of the initial solution and the convergence rate. Since a sampling-based motion planning algorithm works in a random way, the generated solution might be good or not at all. So total 50 runs of simulations were carried out and the results were analyzed statistically.

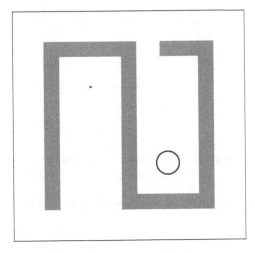

Fig. 2. Simulation environment

Initial Solution. The number of iterations and the time taken to find the initial solution were measured on each run with the cost of the initial solution.

Table 1 shows the summary of the number of iterations to get the initial solution. It is also plotted as shown in Fig. 3.

Table 1. Statistical summary of the number of iterations needed to find the initial solution

Iterations to get the initial solution	
Mininum	236.0
1st Quartile	546.2
Median	695.0
Mean	715.4
3rd Quartile	852.8
Maximum	1680.0

RRT found the initial solution most quickly, but the cost of the solution is the highest among the algorithms. RRT* and RRT*-Quick took more time to find the initial solution than RRT, but generated better solutions.

RRT*-Quick took slightly less time than RRT* and found much better solution than RRT*.

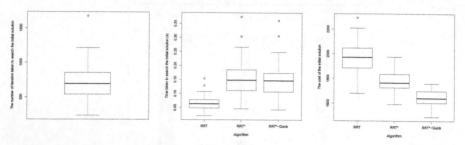

(a) Iteration taken to search the initial solution   (b) Time taken to search the initial solution   (c) Cost of the intial solution

Fig. 3. Statistical results of the initial solution

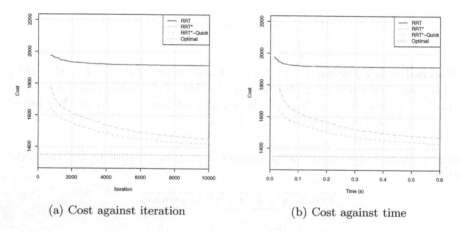

(a) Cost against iteration   (b) Cost against time

Fig. 4. Costs are plotted against iteration/time

Convergence Rate. The cost of the best solution was measured on each iteration step and time step. Fig. 4 shows the cost against iterations and time, respectively, with the optimal cost. In each iteration, RRT*-Quick did more collision check, which consumes the most computation time, than RRT*. However, it turned out that it took only a half of time that RRT* took to get solutions with cost 1500.

The tree of each algorithm was drawn at every 1200 iterations as shown in Fig. 5. It is noted that the proposed algorithm converges more rapidly to the optimal and escaped from the local minima earlier than RRT*.

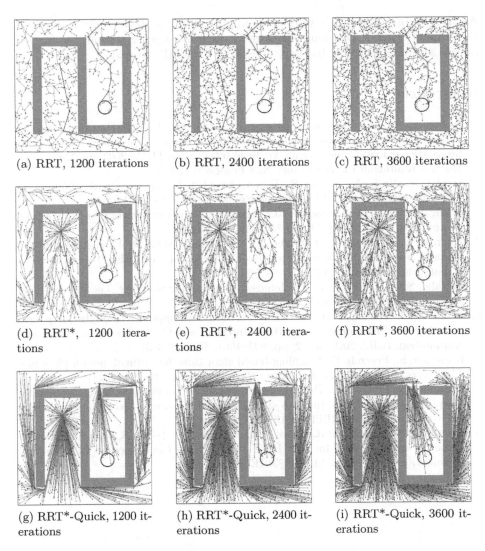

(a) RRT, 1200 iterations   (b) RRT, 2400 iterations   (c) RRT, 3600 iterations

(d) RRT*, 1200 iterations   (e) RRT*, 2400 iterations   (f) RRT*, 3600 iterations

(g) RRT*-Quick, 1200 iterations   (h) RRT*-Quick, 2400 iterations   (i) RRT*-Quick, 3600 iterations

Fig. 5. Generated tree

5   Conclusion

In this paper, RRT*-Quick algorithm that has a faster convergence rate than the conventional RRT* is proposed. The proposed algorithm enlarges the pool of effective parent candidates without increasing much computation time. Since it generates a motion plan with a lower cost, more samples are pruned and the computation time is reduced as a result. Simulations were performed to verify the performance of the proposed algorithm. It was demonstrated that the algorithm generated a shorter path compared to RRT and RRT*. Statistical analysis of the proposed algorithm and experiments with a robot in the real environment remain issues for further research.

Acknowledgement. This work was supported by the Technology Innovation Program, 10045252, Development of robot task intelligence technology, funded by the Ministry of Trade, Industry & Energy (MOTIE, Korea).

References

1. Bjornsson, Y., Enzenberger, M., Holte, R., Schaejfer, J., Yap, P.: Comparison of different grid abstractions for pathfinding on maps. In: Proceedings of the 18th International Joint Conference on Artificial Intelligence, IJCAI 2003, pp. 1511–1512. Morgan Kaufmann Publishers Inc., San Francisco (2003)
2. Daniel, K., Nash, A., Koenig, S., Felner, A.: Theta*: Any-angle path planning on grids. Journal of Artificial Intelligence Research 39(1), 533–579 (2010)
3. Nash, A., Koenig, S., Likhachev, M.: Incremental phi*: Incremental any-angle path planning on grids. In: IJCAI, pp. 1824–1830 (2009)
4. Kavraki, L.E., Svestka, P., Latombe, J.C., Overmars, M.H.: Probabilistic roadmaps for path planning in high-dimensional configuration spaces. IEEE Transactions on Robotics and Automation 12(4), 566–580 (1996)
5. LaValle, S.M., Kuffner, J.J.: Randomized kinodynamic planning. The International Journal of Robotics Research 20(5), 378–400 (2001)
6. Kuffner, J.J., LaValle, S.M.: Rrt-connect: An efficient approach to single-query path planning. In: Proceedings of the IEEE International Conference on Robotics and Automation, ICRA 2000, vol. 2, pp. 995–1001. IEEE (2000)
7. Karaman, S., Frazzoli, E.: Sampling-based algorithms for optimal motion planning. The International Journal of Robotics Research 30(7), 846–894 (2011)
8. Karaman, S., Frazzoli, E.: Optimal kinodynamic motion planning using incremental sampling-based methods. In: 2010 49th IEEE Conference on Decision and Control (CDC), pp. 7681–7687. IEEE (2010)
9. Karaman, S., Walter, M.R., Perez, A., Frazzoli, E., Teller, S.: Anytime motion planning using the rrt*. In: 2011 IEEE International Conference on Robotics and Automation (ICRA), pp. 1478–1483. IEEE (2011)

Wave-Based Tracking Control of a Flexible Arm Using Lumped Model

Tangwen Yang[1,*], Jianda Han[2], and Yong Qin[3]

[1] Institute of Information Science, Beijing Jiaotong University, Beijing, China
twyang@bjtu.edu.cn
[2] The State Key Laboratory of Robotics, Shenyang Institute of Automation,
Chinese Academy of Sciences, Shenyang, China
jdh@sia.cn
[3] The State Key Laboratory of Rail Traffic Control and Safety,
Beijing Jiaotong University, Beijing, China
qinyong2146@126.com

Abstract. A wave control is presented to suppress vibration at the distal end of a flexible robotic arm while achieving accurate tracking control. It assumes the arm with an actuator at one end and a load at the other, and interprets the arm's dynamics with a lumped model in terms of mechanical waves entering and leaving the arm at the actuator-arm interface. Control input to the actuator is then resolved into two superposed waves, which the actuator launches and absorbs simultaneously. From the motion start-up, the launch wave is assigned to the actuator, to which the absorbing wave is added. The absorbing wave is computed with the delayed tip motion. It absorbs the vibratory energy within the arm, and prevents the wave returning from the arm tip from entering the control system again. The control system's properties are investigated, and it works pretty well for trajectory tracking control, as demonstrated in the numerical results.

Keywords: Flexible robotic arm, lumped model, tracking control, wave absorption.

1 Introduction

Large, flexible robotic systems now feature in many applications. Not only are they playing a more important role in space station construction and maintenance, and EVA (extravehicular activity) support, but they can also be used to extend and fold the solar arrays on Mars rovers. In addition, large robotic systems find applications in the fields of aircraft and oil tanker clean, and nuclear waste clear-up, where long-reach manipulation is required. Usually, a large robotic system has long and slender arms. For example, the Canadian Mobile Servicing System in the International Space Station is approximately 17 meters in length when all the arms are fully extended.

---

*Corresponding author.

© Springer International Publishing Switzerland 2015
J.-H. Kim et al. (eds.), *Robot Intelligence Technology and Applications 3*,
Advances in Intelligent Systems and Computing 345, DOI: 10.1007/978-3-319-16841-8_8

Such a robotic system is far from being stiff, and compared with conventional heavy and bulky industrial robots, it is lightweight, and compliant, and has such advantages as larger work volume, lower energy consumption, etc.. But, flexible robotic systems suffer from large deformations and low-frequency vibrations, typically caused by structural flexibility. As a result, issues such as motion planning and dynamic modeling become notoriously difficult, and distal position and force control are more challenging.

Over the past decades, these issues have already received intensive study, with no generic solution to date [1]. To model a flexible robotic system, the energy principle with Lagrangian, Hamiltonian or Newton-Euler formulation, is frequently used. The derived dynamics equations with infinite vibration modes are usually truncated and remain the first several dominant modes with the assumed-mode method or the finite-element method, as introduced in [2]. In [3-5], lumped systems were used to model flexible-link robots as well. Banerjee and Singhose model a flexible-link robot with a series of rigid beam connected by rotational springs [6]. On the control issue, various approaches such as adaptive control [4], singular perturbation method [7], Lyapunov based controller [8], have been widely investigated. Unfortunately, for these model based controllers, accurate system models are required *a priori*, which have proven difficult to obtain. Dynamics coupling, nonlinearities, parameter variations and uncertainties, etc., contribute to this difficulty. Neurofuzzy appears to be promising to control a flexible manipulator [9], but its learning is time-consuming, and fails to provide fast and accurate response.

Due to the typically light structural damping of a flexible robotic arm, it takes considerable time for the vibrations at the tip end of the arm to die out after a maneuver. Vibration suppression is a fundamental problem, and viable solutions to the problem are required. To the authors' knowledge, the techniques currently used may be generally classified as follows: (a) structure and system design to modify the arm dynamics and make the fundamental natural frequency independent of position control [10,11]; (b) the addition of extra actuators, such as piezoelectric patch actuator to counteract vibration [8]; (c) optimal planning to design suitable actuator motion trajectory [12]; and (d) advanced control algorithms. Of these solutions, control technique is of wide engineering interest and is the focus of this paper as a generic problem. With or without feedback, control attempts to increase the system damping or cancel certain system poles and zeros, which dominate the easily exited vibration modes. As a substantially studied technique, input shaping has been advocated as a feasible solution to reducing vibration of a flexible system [6,13]. It works by shaping the reference input to the actuator with correctly chosen impulses, but requires knowledge of the flexible system *a priori,* such as the natural frequencies and damping ratios of the first several vibration modes, which however may be uncertain and may vary, and becomes problematic in practice.

From the point of wave motion, the vibration of a flexible structure, such as beam and string, can be interpreted in terms of wave propagating and decaying in waveguide [3,14-18], and wave based strategies have been introduced to suppress vibration over decades. Matsuda et al. [16] proposed a wave control, in which a compensator formulated as an H_∞ method, is used to minimise the reflective wave to the actuator, for the purpose of vibration control. In [17], a control scheme is developed based on waveform solution of flexible structures, with collocated rate and non-collocated position feedbacks, considering the time delay due to wave motion, and an observer based predictor is introduced to estimate the time-delay of the system state. Besides, some works [18-20] focus on active vibration control with the idea of an imaginary structure, which is used as an absorber to dissipate the wave energy inside the real flexible structure. In [18], an imaginary beam with finite length and distributed damping is assumed to be connected at the free end of a real beam to absorb the vibratory energy of traveling waves. A control scheme for a pendulum system is proposed based on wave absorption [19], where the lateral motion at the pendulum support end satisfies a wave-absorbing condition, and an imaginary counterpart is used to absorb the vibratory energy. In [20], a locally controlled absorber, which comprises of a passive absorber and an internal dynamic feedback, is used for a multi-mass system to tune the vibration characteristics. Wave absorption control needs limited sensing and knowledge of flexible systems. It is a promising solution to suppress the vibrations at the tip of a flexible arm.

In this paper, a new, practical, wave absorption based control is proposed not only to suppress the vibrations of a flexible arm, but simultaneously tracking the arm tip trajectory precisely. A lumped system is first used to model the dynamics of the flexible arm. Then, based on mechanical wave, the motion at the actuator-system interface is interpreted into superposed outward and returning waves. Control input to the actuator is thus resolved into two components, i.e., the launching and absorbing waves. The absorbing wave is applied to restrain the formation of the returning waves, while to absorb the vibratory energy inside the flexible system. It is obtained from the measured position at the tip with a specific delay, providing active vibration damping of the overall control system. Finally, numerical results are given to illustrate the effectiveness of the wave control strategy.

2   Lumped Dynamic Model of Flexible Arm

For decades, intensive work has been done to model the dynamics of a flexible robot. Ideally, the model derived is hoped to be accurate and simple enough for the purpose of a real-time control. However, it is actually a quite difficult task. Approximation is frequently made, as done in the assumed-mode method or the finite element method. Alternatively, lumped mass-spring method has been used to model a flexible robotic system, too. In [5], a flexible beam is modeled by a lumped mass and a weightless linear spring. The model is simple enough, and computationally efficient, but only one vibratory mode is considered. Feliu et al. [4] assumed the mass of a consecutive beam is concentrated in some fixed points, i.e., lumped masses, so that more vibration modes can be taken into account. In [21], a mass-spring model is used to represent a continuous non-linear flexible system, which is lumped parameter approximation. Since the lumped mass method is a practical way to derive the dynamic model of flexible structures, it is used below to observe the dynamics of a flexible robotic arm.

The flexible arm, shown in Fig. 1, is restrained to move in a horizontal plane, and a lumped system is used to approximate its dynamics behavior with n mass-spring-damper units. The first mass m_0 represents an actuator equivalently, and its position is directly controlled by the actuator sub-controller, for the sake of positioning the tip mass m_n. Here, the motions of the first and final masses correlate to those of the actuator and of the tip load in the real flexible arm. $x_i(t)$ represents the displacement of the ith mass. k_i and c_i are the spring and damping constants of mass m_i ($i=1,2,\cdots,n$), respectively. When an external force $f(t)$ is applied to the first mass, it pushes all masses to move rightwards in sequence.

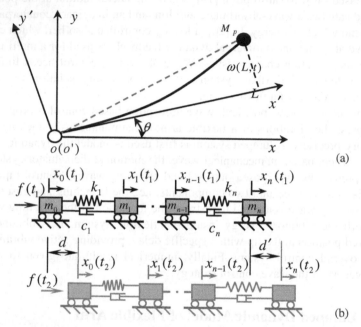

Fig. 1. A mass-spring model for a flexible robotic arm: (a) a flexible robotic arm, (b) a lumped mass-spring model

It is not assumed that the lumped system is uniform. The masses, spring constants and damping coefficients may have different values throughout the system. But, it is required that the vibration modes of the lumped system have a good match with a real flexible arm, and thus the tip position of the final mass is expected to be

$$x_n(t) = \theta + \omega(L,t)/L \tag{1}$$

where θ is the rotation angle of the actuator, and presumes that the tip deflection of the flexible arm $\omega(L,t)$ is far less than the arm length L.

If there is no deflection and vibration at the tip of the arm, the final mass displacement will be same as the actuator in the lumped system, so will the remained masses. Note that any mass in the middle of the string produces exactly the motion of a point along the arm, and the number of masses indicates the truncated oscillatory frequencies of the arm.

Now, the equations of motion of the lumped system can be derived by virtue of the Newton's law. For the first mass, it yields

$$m_0\ddot{x}_0(t) = f(t) - k_1(x_0(t) - x_1(t)) - c_1(\dot{x}_0(t) - \dot{x}_1(t)) \tag{2}$$

For any intermediate mass i, the equation of its motion is of the form

$$
\begin{aligned}
m_i\ddot{x}_i(t) &= k_i(x_{i-1}(t) - x_i(t)) - k_{i+1}(x_i(t) - x_{i+1}(t)) \\
&+ c_i(\dot{x}_{i-1}(t) - \dot{x}_i(t)) - c_{i+1}(\dot{x}_i(t) - \dot{x}_{i+1}(t))
\end{aligned}
\qquad i=1, 2, \cdots n\text{-}1 \tag{3}
$$

and the equation of motion for the final mass can be given by

$$m_n\ddot{x}_n(t) = k_n(x_{n-1}(t) - x_n(t)) + c_n(\dot{x}_{n-1}(t) - \dot{x}_n(t)) \tag{4}$$

If all the initial state values are zero, the Laplace transform of the equation of motion for the first mass is

$$(m_0 s^2 + c_1 s + k_1)x_0(s) - (c_1 s + k_1)x_1(s) = f(s) \tag{5}$$

Similarly, (3) and (4) can be transformed respectively to be

$$
\begin{aligned}
&(m_i s^2 + c_i s + k_i + c_{i+1}s + k_{i+1})x_i(s) \\
&- (c_i s + k_i)x_{i-1}(s) - (c_{i+1}s + k_{i+1})x_{i+1}(s) = 0
\end{aligned}
\qquad i=1,\cdots, n\text{-}1 \tag{6}
$$

and

$$(m_n s^2 + c_n s + k_n)x_n(s) - (c_n s + k_n)x_{n-1}(s) = 0 \tag{7}$$

Equation (7) is rewritten to the form of a transfer function as

$$G_n(s) = \frac{x_n(s)}{x_{n-1}(s)} = \frac{c_n s + k_n}{m_n s^2 + c_n s + k_n} \tag{8}$$

then, the transfer functions of all the adjacent mass pairs can be derived, from the second last mass back towards the actuator, with (6), to yield

$$G_i(s) = \frac{x_i(s)}{x_{i-1}(s)} = \frac{c_i s + k_i}{m_i s^2 + c_i s + k_i + (c_{i+1}s + k_{i+1})(1 - G_{i+1})} \qquad i=n\text{-}1,\cdots, 1 \tag{9}$$

and the transfer function $G_0(s)$, between the displacement of the first mass and the applied external force, can be given by

$$G_0(s) = \frac{x_0(s)}{f(s)} = \frac{1}{m_0 s^2 + (c_1 s + k_1)(1 - G_1)} \tag{10}$$

The dynamic model of the lumped system is used to interpret the dynamic characteristics of the flexible robotic arm. It characterizes the dynamics of a real system, and its simplicity makes the controller design easy, and enables to exam the wave behavior of flexible structures.

3   Wave Based Control Strategy

3.1   Launching and Absorbing Motion Definition

Based on wave motion analysis [22], the motion of any mass within a lumped system is interpreted as mechanical waves propagating in two directions. These waves within the system are induced by actuator forces or motion, which must also eventually bring the system to rest. Therefore, one of the boundary conditions for these waves can be defined by the actuator motion, the other by the tip or the load end. Enabling the actuator to absorb the waves returning from the final mass, an active vibration control can be achieved essentially.

To move the final mass a desired amount, a controller needs to simultaneously *"push"* the actuator to move half this amount while allowing itself to be *"dragged"* the other half displacement by the returning motion. It means the controller does two jobs, in other words, it gets the actuator to launch and absorb motion at the same time. Launching motion causes the final mass to move half the desired displacement. Absorbing motion counterbalances the waves returning from the tip, and absorbs the vibratory energy out of the system at the actuator interface, which leads to another half displacement. A control strategy with this idea elegantly reconciles the potentially conflict of position control and vibration suppression.

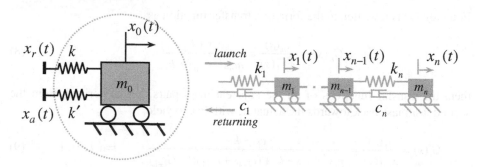

Fig. 2. Notional force components

To depict the control strategy, the actuator is deliberately isolated from the lumped system, as shown in Fig. 2. The force on its right side is presumed to be made up of two components: launching and returning force waves. This force is caused by the interaction between the actuator and the other masses in the system. At the same time, the actuator is driven by an external or "input" force, $f(t)$, determined by the controller. The drive force is shown on the left side of the actuator, as if produced by imaginary displacements $x_r(t)$ and $x_a(t)$ acting on two springs with stiffness k and k'. The first displacement, $x_r(t)$, is defined as reference input, committed to launch motion and push the system half the target displacement. The second, $x_a(t)$, is absorbing input, designed to counteract the returning wave energy by supplying the right amount of "give" to the actuator when it is pulled by the returning force wave. At steady state, the displacement associated with the absorption process, $x_a(t)$ will then equal the net or total displacement associated with the launch motion, $x_r(t)$. The notional spring k or k' is simply a device to convert $x_r(t)$ or $x_a(t)$ into appropriate input force component to the actuator.

The reference motion $x_r(t)$ could be a step, a ramp, or any trajectory input which is given *a priori*. It launches a wave into the system, and travels down to the final mass where the wave is reflected back towards the actuator. To prevent the reflected wave from returning into the system again, an absorbing input is immediately added to absorb its energy, which is computed from the tip displacement, $x_n(t)$ delayed by a particular time, so that

$$x_a(t) = x_n(t - T_d) \qquad (11)$$

where T_d is the time required for the wave to travel from the final mass to the actuator, and is a critical parameter to be determined.

Then, the Laplace transform of (11) is

$$x_a(s) = e^{-T_d s} x_n(s) \qquad (12)$$

where $x_n(s)$ can be obtained from the system dynamic model, given by

$$x_n(s) = G_1(s)G_2(s)\cdots G_n(s)x_0(s) = G'(s)x_0(s) \qquad (13)$$

where $G'(s)$ is the motion transfer function from the actuator to the final mass.

Now, the force produced by two notional spring inputs could be merged into an equivalent spring of stiffness k_e with a single input $x_e(t)$. This implies that

$$k_e(x_e - x_0) = k(x_r - x_0) + k'(x_a - x_0) \qquad (14)$$

Rearranging (14), we have

$$x_e(t) = \frac{k}{k_e} x_r(t) + \frac{k'}{k_e} x_a(t) + x_0(t)(1 - \frac{k}{k_e} - \frac{k'}{k_e}) \qquad (15)$$

Explicitly, the equivalent motion is defined by the reference input, the tip and the actuator motion. Set $k_e = k + k'$, the third term on the right-hand side of (15) will disappear. That is to say, position feedback is not required in the outer loop of the control system (introduced below), and if $k = k'$, we further have

$$x_e(t) = 0.5x_r(t) + 0.5x_a(t) \qquad (16)$$

or, in the Laplace domain,

$$x_e(s) = 0.5x_r(s) + 0.5x_a(s) \qquad (17)$$

Theoretically, the launch and absorb inputs, $x_r(s)$ and $x_a(s)$, are of the same importance. As to the equivalent motion $x_e(t)$, it is actually the required motion to the actuator, and is the combination of a pre-determined reference motion and a real-time position feedback from the tip.

3.2   Control Strategy Property

Equations (12) and (17) establish the framework of the wave absorption based control, which is illustrated by the block diagram representation of Fig.3. Overall, the control scheme comprises of inner and outer control loops. The inner is a negative feedback loop for the actuator control, in which the actuator position is fed back and obtained by collocated measurement, and the actuator servo controller could be PID law or other algorithm. In the outer closed loop, the absorbing wave is computed on the basis of the lumped dynamics model, which is used as an observer to estimate the dynamics of the real flexible arm, partly because it is difficult to accurately model the arm dynamics and measure the deflection at the arm tip. Integrated optic sensors (camera, PSD), accelerator, etc., are often used to obtain the deflection and vibration information at the tip of a flexible arm, but, these sensor data are unsuitable to compute the tip position because of their slow response time and/or noises.

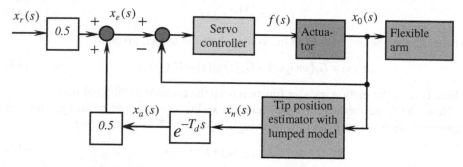

Fig. 3. Block diagram of the proposed wave-based control

This control strategy is stable and robust, and works significantly better than simple loop control. Its properties are analyzed herein, particularly the steady-state error. Let us presume that a classical PID law $G_c(s)$ is used to control the actuator, written as

$$G_c(s) = k_p + k_d s + k_i \frac{1}{s} \tag{18}$$

where k_p, k_d and k_i represent the proportional, derivative and integral constants, respectively.

Now, the transfer function between its input and output motions can be derived, and has the form of

$$G(s) = \frac{x_n(s)}{x_r(s)} = \frac{G_c(s)G_0(s)G'(s)}{2(1 + G_c(s)G_0(s)) - e^{-T_d s}G_c(s)G_0(s)G'(s)} \tag{19}$$

Without loss of generality, the lumped system with one actuator and three masses is taken as an example to investigate these controller properties. Although not necessary, for convenience, the system damping is deliberately neglected, and the last three mass-spring pairs are assumed to be identical.

In terms of (8) to (10), the transfer functions in (19) can be given by

$$G_0(s) = \frac{m^3 s^6 + 5km^2 s^4 + 6k^2 ms^2 + k^3}{P(s)}$$

$$G'(s) = \frac{k^3}{m^3 s^6 + 5km^2 s^4 + 6k^2 ms^2 + k^3}$$

thereby obtaining

$$G(s) = \frac{k^3 (k_p + k_d s + k_i \frac{1}{s})}{2[P(s) + (k_p + k_d s + k_i \frac{1}{s})(m^3 s^6 + 5km^2 s^4 + 6k^2 ms^2 + k^3)]} \tag{20}$$
$$- e^{-T_d s}(k_p + k_d s + k_i \frac{1}{s})k^3$$

where $P(s) = Mm^3 s^8 + (5km^2 M + km^3)s^6 + (6k^2 mM + 4k^2 m^2)s^4 + (k^3 M + 3k^3 m)s^2$ and M is the equivalent mass of the real actuator, and k and m are the equivalent spring stiffness and mass values to model the system dynamics of interest. The analytic mass-spring model is lumped-parameter approximation, which is used to virtually represent the continuous flexible robotic arm. The parameters of the lumped model can be experimentally estimated through vibration response analysis of a real flexible arm.

Then, by virtue of the final value theorem, the steady-state error of the control system is given by

$$e_{ss} = \lim_{s \to 0} sE(s) = \lim_{s \to 0} s(x_r(s) - x_n(s)) \tag{21}$$

Using the closed-loop transfer functions derived above, the steady-state error e_{ss} is found to be zero, regardless of the gain values chosen for the PID control law.

4   Numerical Simulation

The wave based control is used here for tip trajectory tracking control as well, and compared to a PD control law. The same four-mass lumped model is used in the simulations, and damping in this model is neglected. In fact, any damping presence will help reduce vibrations further, so that the absence of damping in a flexible system becomes a worst case. The parameters of the lumped system are given in Table 1.

Table 1. Model parameters in the four-mass lumped system

Parameter Symbol	Quantity	Value
M	Equivalent actuator mass	1 kg
m	equivalent mass	0.2 kg
k	equivalent spring stiffness	50 N/m
T_d	delay time	1.0 s

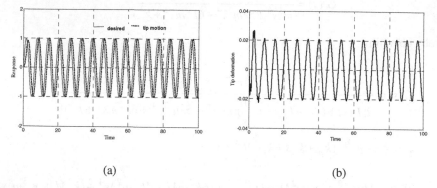

(a)   (b)

Fig. 4. The wave-based control results for tip tracking: (a) sine tip trajectories, (b) tip deformation

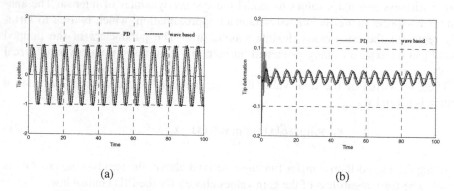

(a)   (b)

Fig. 5. Responses of the wave-based and PD controllers for tip tracking

Figure 4 shows the responses to a sine reference input. The control gains of the PD law in the inner loop are chosen to be k_p=4.0, and k_d=0.5. Figure 6(a) shows the sine trajectory tracking response at the tip, and (b) the tip vibration. The results of comparative study are given in Figures 5, as can be seen the wave absorption strategy nearly does not give rise to vibration at the tip end of this flexible system, whereas the

PD law causes large vibrations at the beginning of motion. The wave absorption approach is superior to the PD law, from the angle of vibration control, which is the major objective for many flexible systems control.

5   Conclusions

In this paper, a wave based control is proposed to suppress the vibration at the tip of a flexible robotic arm. The dynamics of the arm is first estimated with a lumped mass-spring system, which could be very flexible, of arbitrary length and with arbitrary component values. The control design is on the basis of interpreting the dynamic behaviour of the flexible system in terms of mechanical waves entering and leaving the system at the actuator-system interface. Therefore, from the motion start-up, a launching motion is assigned to the actuator, to which an absorbing motion is added, to realize active vibration damping and precise motion control, in a single motion. Numerical simulation results demonstrate that the wave absorption control works well for trajectory tracking control.

Acknowledgement. This work is supported in part by the National Natural Science Foundation of China under 61375109 and 61273356, and the China Railway Corporation Science and Technology Research and Development Program Grant 2013T002-A-3, and the State Key Laboratory of Robotics under 2013-012.

References

1. Robinett III, R.D., Feddema, J., Eisler, G.R., Dohrmann, C., Parker, G.G., Wilson, D.G., Stokes, D.: Flexible Robot Dynamics and Controls. Kluwer Academic/Plenum Publishers, Dordrecht (2002)
2. Junkins, J.L., Kim, Y.: Introduction to Dynamics and Control of Flexible Structures. AIAA Education Series, Washington DC (1993)
3. O'Connor, W.J.: Wave-based Analysis and Control of Lumped-modeled Flexible Robots. IEEE Trans. Robot. 23(2), 342–352 (2007)
4. Feliu, J.F., Feliu, V., Cerrada, C.: Load Adaptive Control of a Single-Link Flexible Arm Based on a New Modelling Technique. IEEE Trans. Robot. Autom. 15(5), 793–811 (1999)
5. Zhu, G., Ge, S.S., Lee, T.H.: Simulation Studies of Tip Tracking Control of a Single-Link Flexible Robot Based on a Lumped Model. Robotica 17(1), 71–78 (1999)
6. Banerjee, A.K., Singhose, W.E.: Command Shaping in Tracking Control of a Two-Link Flexible Robot. J. Guid. Control, Dyn. 21(6), 1012–1015 (1998)
7. Siciliano, B., Book, W.J.: A Singular Perturbation Approach to Control of Lightweight Flexible Manipulators. Int. J. Robot. Res. 7(4), 79–90 (1988)
8. Dadfarnia, M., Jalili, N., Xian, B., Dawson, D.M.: A Lyapunov-based Piezoelectric Controller for Flexible Cartesian Robot Manipulators. ASME J. Dyn. Syst., Meas., Control 126(2), 347–358 (2004)
9. Caswara, F.M., Unbehauen, H.: A Neurofuzzy Approach to the Control of a Flexible-Link Manipulator. IEEE Trans. Robot. Autom. 18(6), 932–944 (2002)
10. Wang, F.Y., Russell, J.L.: Optimum Shape Construction of Flexible Manipulator with Total Weight Constraint. IEEE Trans. Syst., Man Cybern. 25(4), 605–614 (1995)

11. Yang, T.W., Xu, W.L., Han, J.D.: Dynamic Compensation Control of Flexible Macro-Micro Manipulator Systems. IEEE Trans. Control Syst. Technol. 18(1), 143–151 (2010)
12. Lambeck, S., Sawodny, O.: Trajectory Generation and Oscillation Damping Control for a Flexible Link Robot. In: Proc. IECON, Taipei, pp. 2748–2753 (2007)
13. Singhose, W., Vaughan, J.: Reducing Vibration by Digital Filtering and Input Shaping. IEEE Trans. Control Syst. Technol. 19(6), 1410–1420 (2011)
14. Von Flotow, A.: Traveling Wave Control for Large Spacecraft Structures. J. Guid. Control Dyn. 9(4), 462–468 (1986)
15. Mei, C., Mace, B.R.: Wave Reflection and Transmission in Timoshenko Beams and Wave Analysis of Timoshenko Beam Structures. ASME J. Vibrat. Acoust. 127(4), 282–394 (2005)
16. Matsuda, K., Kanemitsu, Y., Kijimoto, S.: A Wave-Based Controller Design for General Flexible Structures. J. Sound Vibrat. 216(2), 269–279 (1998)
17. Halevi, Y.: Control of Flexible Structures Governed by the Wave Equation Using Infinite Dimensional Transfer Functions. ASME J. Dyn. Syst., Meas. Control 127(4), 579–588 (2005)
18. Sawada, Y., Ohsumi, A., Ono, A.: Wave Control of a Class of Flexible Beams by an Idea of Imaginary Beam. In: Proc. CDC, Kobe, Japan, pp. 4228–4233 (1996)
19. Saigo, M., Tani, K., Usui, K.: Vibration Control of a Travelling Suspended System Using Absorbing Wave Control. ASME J. Vibrat. Acoust. 125(3), 343–350 (2003)
20. Filipovic, D., Schroeder, D.: Control of Vibrations in Multi-mass Systems with Locally Controlled Absorber. Automatica 37(2), 213–220 (2001)
21. O'Connor, W.J., McKeown, D.J.: Time-Optimal Control of Flexible Robots Made Robust through Wave-based Feedback. ASME J. Dyn. Syst., Meas., Control 133(1) (2011)
22. Yang, T.W., O'Connor, W.J.: Wave Theory Applied to Vibration Control of Elastic Robot Arms. In: Proc. IASTED-MIC, Innsbruck, Austria, pp. 260–265 (2005)

Stable Modifiable Walking Pattern Generator with a Vertical Foot Motion by Evolutionary Optimized Central Pattern Generator

Chang-Soo Park and Jong-Hwan Kim

Department of Electrical Engineering, KAIST, 291 Daehak-ro,
Yuseong-gu, Daejeon, 305-701, Republic of Korea
{cspark,johkim}@rit.kaist.ac.kr

Abstract. This paper proposes a stable modifiable walking pattern generator with a vertical foot motion generated by a central pattern generator that is obtained from the evolutionary constrained optimization. A modifiable walking pattern generator is employed which generates sagittal and lateral position trajectories of center of mass of humanoid robot and a CPG generates the vertical foot trajectory of swing leg. The oscillation of the ground reaction forces causes the external disturbance while walking. To decrease the oscillation of the ground reaction forces, sensory feedback in the CPG is designed, which uses force sensing resistor signals. For the optimization of parameters in the CPG, two-phase evolutionary programming is employed. The effectiveness of the proposed scheme is demonstrated by simulations using a Webots dynamic simulator for a small sized humanoid robot, HSR-X, developed in the Robot Intelligence Technology Lab, KAIST.

1 Introduction

Despite the complexity of high-DOF systems of humanoid robots, these days various humanoid robots have been developed [1]–[3]. Research on generation of stable walking patterns of humanoid robots plays one of the important roles in this field. There are two typical approaches to bipedal walking of humanoid robot, such as dynamic model based approach and biologically inspired approach. In the former, a 3-D linear inverted pendulum model (3-D LIPM) is one of popular schemes, which decouples the sagittal and lateral motion equations of central of mass (COM) with constant COM height in single support phase [4], [5]. A modiable walking pattern generator (MWPG) extends the 3-D LIPM for a zero moment point (ZMP) variation by the closed form functions. Thus, it can change the walking pattern of a humanoid robot in real-time while walking [6]–[8]. On the other hand, a central pattern generator (CPG) is widely used as a biologically inspired approach [9]–[14]. It can generate rhythmic output signals without rhythmic input signals and modify generated signals to deal with environmental disturbance by sensory feedbacks.

This paper proposes stable MWPG with a vertical foot motion using the CPG that is obtained from the evolutionary constrained optimization. The MWPG is employed and it generates position trajectory of COM of humanoid robot and the CPG generates

© Springer International Publishing Switzerland 2015

J.-H. Kim et al. (eds.), *Robot Intelligence Technology and Applications 3,*
Advances in Intelligent Systems and Computing 345, DOI: 10.1007/978-3-319-16841-8_9

vertical swing foot trajectory. To decrease ground reaction forces, a sensory feedback in the CPG is designed, which uses force sensing resistor (FSR) signals. For the optimization of parameters in the CPG, two-phase evolutionary programming (TPEP) is employed [15], [16]. The effectiveness of the proposed scheme is demonstrated by simulations using a Webots dynamic simulator for a small sized humanoid robot, HSR-X, developed in the Robot Intelligence Technology (RIT) Lab, KAIST.

2   Stable MWPG with CPG

This section presents the proposed stable MWPG with the vertical foot motion by the evolutionary optimized CPG. In this paper, the sagittal and lateral COM motions are generated by the MWPG. Meanwhile, the vertical foot motion is generated by the proposed CPG for stable bipedal walking.

2.1   Vertical Foot Motion by CPG

The vertical foot trajectory of the swing leg, z_{foot} is generated by a cycloid function in the previous MWPG as follows:

$$z_{foot}(t) = R(1 - (cos^2(\frac{\pi t}{T_{ss}}) - sin^2(\frac{\pi t}{T_{ss}}))) \tag{1}$$

where R denotes the radius of the cycloid circle and T_{ss} denotes the single support time. However, this method is hard to modify the generated vertical foot trajectory to deal with environmental disturbance. To solve this problem, this paper proposes the vertical foot motion generation by using the CPG obtained from the evolutionary constrained optimization.

The CPG generates rhythmic signals without a rhythmic sensory or central input. In this paper, to generate the rhythmic signals for the CPG, neural oscillator is employed and the neuron is represented as follows [17]:

$$\tau \dot{u}_i = -u_i - \sum_{j=1}^{n(j \neq i)} w_{ij} o_j - \beta v_i + u_0 + feed_i \tag{2}$$

$$\tau' \dot{v}_i = -v_i + o_i \tag{3}$$

$$o_i = max(u_i, 0) \tag{4}$$

where n is the number of neurons. u_i, v_i and y_i, $i = 1, \cdots, n$, are the inner state, the self-inhibition state and the output signal of the ith neuron, respectively. u_0 is the input signal that affects the output amplitude and w_{ij} is the connecting weight which determines the phase difference between the ith and jth neurons. τ and τ' are time constants, which influence on the shape and frequency of output signal. β is the weight of self-inhibition. $feed_i$ is the sensory feedback signal. A biological rhythmic locomotion is performed by the sequence of extension and flexion of muscles. For the modeling of this biological system, the CPG structure was devised [9]. In this structure, the rhythmic locomotion is assumed to be generated by the neural oscillators, each of which is composed of two neurons: an extensor neuron (EN) and a exor neuron (FN). The phase difference between EN and FN is π.

Using the CPG, the vertical foot trajectory of swing leg is generated as follows:

$$z_{foot} = R(1 - ((o_1 - o_2)^2 - (o_3 - o_4)^2))\tag{5}$$

where o_1, o_2, o_3 and o_4 are the output signals of the neurons in the CPG. o_1 and o_3 perform the EN role, and o_2 and o_4 perform the FN role. The generated vertical foot trajectory by the CPG should be similar to the generated vertical foot trajectory by the cycloid circle. Thus, waveforms of $o_1 - o_2$ and $o_3 - o_4$ should be similar to the waveforms of $cos(\pi t/T_{ss})$ and $sin(\pi t/T_{ss})$, respectively. To satisfy these conditions, the time constants and connecting weights are optimized by TPEP.

While walking, the sum of the ground reaction forces (GRFs) on the feet oscillates around the weight of the robot, which causes the external disturbance. Thus, to improve the stability of the humanoid robot while walking, the sensory feedback in the CPG is designed to minimize the oscillation of the GRFs as follows:

$$feed_1 = k_f(F_L + F_R - mg)\tag{6}$$
$$feed_2 = -feed_1\tag{7}$$
$$feed_3 = -feed_1\tag{8}$$
$$feed_4 = -feed_3\tag{9}$$

where k_f is the scaling factor and it is to be optimized by TPEP. F_L and F_R denote the GRFs on the left and right feet, respectively. They are obtained by FSRs attached to the sole of foot.

2.2 Evolutionary Optimization for CPG

The objective of this evolutionary optimization is to obtain the desired output signal of the neural oscillators and to improve the stability of the humanoid robot while walking. When the magnitude of $o_1 - o_2$ reaches the maximum (minimum) value, the magnitude of $o_3 - o_4$ should reach the zero. Also, when $o_3 - o_4$ is positive, the corresponding time T^+ should be a desired single support time T_{ss}. Meanwhile, to improve the stability of the humanoid robot while walking, the oscillation of GRFs should be minimized.

To satisfy these constraints and the objective, the following objective function considering equality constraints is defined to obtain the time constants and connecting weights in the neural oscillators, and scaling factor in the sensory feedback by the TPEP [15], [16]:

$$\text{Minimize } f = \sum |F_L + F_R - mg| + P\tag{10}$$

subject to equality constraints

$$c1 : o_3 - o_4 = 0 \text{ (when } o_1 - o_2 \text{ reach the maximum value)}$$
$$c2 : o_3 - o_4 = 0 \text{ (when } o_1 - o_2 \text{ reach the minimum value)}$$
$$c3 : T^+ - T_{ss} = 0$$

where P is the penalty which is given if humanoid robot loses its balance and collapses.

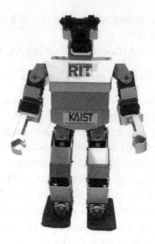

Fig. 1. HSR-X

Table 1. The obtained parameter values of the designed CPG by the TPEP

τ	0.32	τ'	0.21
$w_{\pi/2}$	0.35	$w_{-\pi/2}$	0.35
k_f	6.6×10^{-3}		

3   Simulations

The proposed algorithm was implemented into a small-sized humanoid robot, HanSaRam-X (HSR-X) (Fig. 1) [18]. HSR-X is the latest one of HSR-series. HSR-series has been in continual redesign and development in the RIT Lab, KAIST. Its height and weight are 45.2 cm and 2.7 kg, respectively. It has 20 DOFs that consists of 8 DOFs in the upper body and 12 DOFs in the lower body. Computer simulation was carried out using the simulation model of HSR-X by Webots [19].

3.1   CPG Parameters Optimization by TPEP

In the simulation, Z_c set as 23.35cm. β and u_0 in the neural oscillators were set as 2.5 and 2.9, respectively. $w_{1,2}$, $w_{2,1}$, $w_{3,4}$ and $w_{4,3}$ were set as 2.5 to make the phase difference between EN and FN to π. Single and double support times were set as 0.8s and 0.4s, respectively. Also, the phase differences between o_1 and o_3, o_2 and o_4, o_4 and o_1 and o_3 and o_2 are all equal to $\pi/2$. Thus, $w_{1,3}$, $w_{2,4}$, $w_{4,1}$ and $w_{3,2}$ are set as $w_{\pi/2}$. In the same way, $w_{1,4}$, $w_{2,3}$, $w_{3,1}$ and $w_{4,2}$ are set as $w_{-\pi/2}$. The time constants, τ and τ', the connecting weights, $w_{\pi/2}$ and $w_{-\pi/2}$, and the scaling factor, k_f, were obtained by TPEP as shown in Table 1.

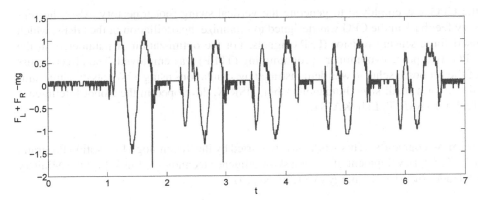

Fig. 2. Measured oscillation of the GRFs by the cycloid function

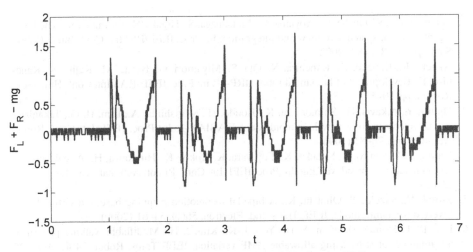

Fig. 3. Measured oscillation of the GRFs by the CPG

3.2 Simulation Result

In this simulation, the simulation model of HSR-IX by Webots was used. Figs. 2 and 3 show the measured oscillation of the GRFs while bipedal walking by the cycloid function and the proposed CPG, respectively, to generate vertical foot motion of swing leg. In this result, the sum of oscillation of the GRFs was reduced by 25.6% by the proposed CPG. It means the HSR-X could walk stably with the proposed algorithm.

4   Conclusion

This paper proposed stable MWPG with a vertical foot motion generated by the evolutionary optimized CPG. The MWPG was employed to generate the position trajectory of COM of humanoid robot. To improve the stability of humanoid robot while walking, the CPG was employed to generate the vertical swing foot trajectory. Also, the sensory feedback in the CPG was designed to minimize the oscillation of the GRFs, which used force sensing resistor (FSR) signals. For the optimization of parameters in the CPG, two-phase evolutionary programming (TPEP) was employed. The effectiveness of the proposed scheme was demonstrated by simulations using a Webots dynamic simulator for a small sized humanoid robot, HSR-X, developed in the Robot Intelligence Technology (RIT) Lab, KAIST.

Acknowledgments. This work was supported by the Technology Innovation Program, 10045252, Development of robot task intelligence technology, funded By the Ministry of Trade, Industry & Energy (MOTIE, Korea).

References

1. Sakagami, Y., Watanabe, R., Aoyama, C., Matsunaga, S., Higaki, N., Fujimura, K.: The intelligent ASIMO: system overview and integration. In: Proc. IEEE/RSJ Int. Conf. Intell. Robot. Syst., pp. 2478–2483 (2002)
2. Akachi, K., Kaneko, K., Kanehira, N., Ota, S., Miyamori, G., Hirata, M., Kajita, S., Kanehiro, F.: Development of humanoid robot HRP-3P. In: Proc. IEEE-RAS Int. Conf. Humanoid Robots, pp. 50–55 (2005)
3. Ogura, Y., Aikawa, H., Shimomura, K., Kondo, H., Morishima, A., Lim, H.-O., Takanishi, A.: Development of a new humanoid robot WABIAN-2. In: Proc. IEEE Int. Conf. Robot. Automat., pp. 76–81 (2006)
4. Kajita, S., Kanehiro, F., Kaneko, K., Fujiwara, K., Yokoi, K., Hirukawa, H.: A realtime pattern generator for biped walking. In: Proc. IEEE Int. Conf. Robot. Automat., vol. 1, pp. 31–37 (2002)
5. Motoi, N., Suzuki, T., Ohnishi, K.: A bipedal locomotion planning based on virtual linear inverted pendulum mode. IEEE Trans. Ind. Electron. 56(1), 54–61 (2009)
6. Lee, B.-J., Stonier, D., Kim, Y.-D., Yoo, J.-K., Kim, J.-H.: Modifiable walking pattern of a humanoid robot by using allowable ZMP variation. IEEE Trans. Robot. 24(4), 917–925 (2008)
7. Hong, Y.-D., Lee, B.-J., Kim, J.-H.: Command state-based modifiable walking pattern generation on an inclined plane in pitch and roll directions for humanoid robots. IEEE/ASME Trans. Mechatron. 16(4), 783–789 (2011)
8. Hong, Y.-D., Kim, J.-H.: 3-D command state-based modifiable bipedal walking on uneven terrain. IEEE/ASME Trans. Mechatron. (2011) (accepted)
9. Taga, G.: A model of the neuro-musculo-skeletal system for human locomotion. Biol. Cybern. 73, 97–111 (1995)
10. Nakamura, Y., Mori, T., Sato, M., Ishii, S.: Reinforcement learning for a biped robot based on a CPG-actor-critic method. Neural Networks 20(6), 723–735 (2007)
11. Righetti, L., Ijspeert, A.J.: Programmable central pattern generators: an application to biped locomotion control. In: Proc. IEEE Int. Conf. Robot. Automat., pp. 1585–1590 (2006)
12. Héliot, R., Espiau, B.: Multisensor input for CPG-based sensory—motor coordination. IEEE Trans. Robotics 24(1), 191–195 (2008)
13. Endo, G., Morimoto, J., Matsubara, T., Nakanishi, J., Cheng, G.: Learning CPG-based biped locomotion with a policy gradient method: application to a humanoid robot. Int. J. Robotics Research 27(2), 213–228 (2008)

14. Park, C.-S., Hong, Y.-D., Kim, J.-H.: Evolutionary Optimized Central Pattern Generator for Stable Modifiable Bipedal Walking. IEEE/ASME Trans. Mechatron. 19(6), 1374–1383 (2014)
15. Kim, J.-H., Myung, H.: Evolutionary programming techniques for constrained optimization problems. IEEE Trans. Evol. Comput. 1(2), 129–140 (1997)
16. Myung, H., Kim, J.-H.: Hybrid evolutionary programming for heavily constrained problems. BioSystems 38, 29–43 (1996)
17. Matsuoka, K.: Sustained oscillations generated by mutually inhibiting neurons with adaptation. Biol. Cybern. 52(6), 367–376 (1985)
18. Yoo, J.-K., Lee, B.-J., Kim, J.-H.: Recent Progress and Development of Humanoid Robot HanSaRam. Robotics and Autonomous Systems 57(10), 973–981 (2009)
19. Michel, O.: Cyberbotics Ltd. WebotsTM: Professional mobile robot simulation. Int. J. Advanced Robot. Syst. 1(1), 39–42 (2004)

Falling Prevention System from External Disturbances for Humanoid Robots

Gyeong-Moon Park, Seung-Hwan Baek, and Jong-Hwan Kim

Department of Electrical Engineering, KAIST
335 Gwahangno, Yuseong-gu, Daejeon, 305-701, Republic of Korea
{gmpark,shbaek,johkim}@rit.kaist.ac.kr

Abstract. Humanoid robot requires a robust prevention system against external disturbances to protect itself from falling to the ground and to perform its tasks completely. In this paper, a Falling Prevention System for humanoid robot is proposed to avoid falling from the disturbances, and helps humanoid robot recover its balance from external force by taking a step. The algorithm for the Falling Prevention System consists of two processes. First, humanoid robot can perceive whether it is falling or not by using an IMU sensor, and if falling, the center of mass (CoM) and swinging leg trajectories are calculated for the robot to take a step. The CoM and swinging leg trajectories are also used to acquire all joint angles of lower body by inverse kinematics. Furthermore, designed foot trajectory helps humanoid robot minimize its yawing moment. Next, mass-spring-damper system for the robot's legs is modeled to reduce large impact force from the ground. The effectiveness of the proposed method is demonstrated through computer simulations for a humanoid robot.

Keywords: Falling prevention system, mass-spring-damper modeling, humanoid robot.

1 Introduction

Humanoid robots have been developed to carry out a complex task by imitating human dynamic motions. Nowadays, these tasks are getting more complex and it is required for humanoid robot to have high stability. It is very important for robots to keep balance and safety in situations which can damage robots. In a real environment, it is common for humanoid robots to confront lots of obstacles and limitations such as collision with people, uneven terrain, and so on. These situations happen quite frequent, as a result of which humanoid robot should have ability to cope with these external disturbances. It is essential for them to be robust to all the emergent conditions. Since humanoid robots are high dimensional and nonlinear systems, it is critical for the robots to control themselves with stability under such harsh conditions.

To maintain the stability of humanoid robot, existing research has been focused on compensating zero moment point (ZMP) error for biped balance [1], [2].

© Springer International Publishing Switzerland 2015   97
J.-H. Kim et al. (eds.), *Robot Intelligence Technology and Applications 3*,
Advances in Intelligent Systems and Computing 345, DOI: 10.1007/978-3-319-16841-8_10

Also, humanoid robot locomotion studies have been conducted using linear inverted pendulum model (LIPM) [3–7]. These researches have heavily influenced lots of roboticists. Another approach is that humanoid robot took a step for keeping balance [8], [9]. Three steps for push recovery was introduced in [10]. First, humanoid robot used ankle torques against external force. Then, the robot moved internal joints and took a step. The capture point was proposed to calculate a capture region for the inverted pendulum with flywheel model [11]. Reactive stepping method was developed to prevent humanoids from falling by simultaneously planning the trajectories of the center of gravity and ZMP based on analytic solution of the LIPM [12].

In this paper, we propose an integrated system to prevent humanoid robots from falling to the ground. Humanoid robot can recognize whether it is falling or not using sensors, and if falling, it calculates the CoM and swinging leg trajectories in real time to take a proper step to keep balance. When a robot foot hits the ground, the yawing moment is produced by the ground reaction force. It might make humanoid robot unstable, so a modified swinging trajectory is designed for minimizing the yawing moment. Furthermore, to reduce the impact from the GRF, the mass-spring-damper system is designed for the robot legs. This is effective to maintain balance against the large GRF. The effectiveness of the proposed method is demonstrated through computer simulations for a humanoid robot.

This paper is organized as follows. Section 2 presents overall description of the proposed algorithm. First, it describes how to know whether it is falling to the ground or not, and how to calculate the CoM and swinging leg trajectories. The mass-spring-damper modeling is also discussed. In Section 3, simulation results are presented. Conclusion and further work follow in Section 4.

2   Falling Prevention System for Humanoid

In order to prevent humanoid robot from falling to the ground, robot should perceive whether it is falling or not. In this research, humanoid robot receives information from the sensors such as the inertial measurement unit (IMU) and gyro sensors, and takes a proper step to prevent falling down. Humanoid robot can avoid falling down by planning the steps and calculating the CoM and foot trajectories.

The algorithm for this Falling Prevention System consists of the two processes. First, the CoM trajectory of the humanoid robot in falling is calculated and simulated using the equation of motion of inverted pendulum. Since the robot's initial velocity, when robot starts falling down, directly affects angular speed and acceleration in the equation of motion of inverted pendulum, the initial velocity is required to calculate the CoM trajectory. Humanoid robots use the IMU and gyro sensors to acquire roll, pitch, and yaw angle values and angular velocities to calculate the CoM trajectory. Next, from the calculated CoM trajectory, the swinging leg position at which the robot stops, while keeping stability without falling down, is automatically determined. Using the swinging leg position, the swinging leg trajectory is calculated and every joint angles for the swinging leg is acquired by the inverse kinematics. For the humanoid robot with 12 degrees of freedom (DoF) in the lower body, the inverse kinematics calculates the 12 joint angles to take a step from the origin to the calculated swinging leg position. Finally, to reduce the yawing moment of humanoid robot as much as possible, the swinging leg is located nearby the CoM position on the ground.

2.1 Judgement of Humanoid Falling

When the external force is exerted on the back of humanoid robot in the sagittal plane, it produces the GRF which points to the CoM of robot. If a direction vector is newly defined from the toe to the CoM of humanoid, it is in the same direction as the GRF and initially tilted from normal vector to ground plane. This initially tilted angle is given as

$$\theta_i = \frac{\pi}{2} - atan\left(\frac{L_C}{L_A}\right) \tag{1}$$

where the θ_i, L_C, L_A are the initially tilted angle, the length from the CoM to sole, and the length from ankle to toe, respectively. L_C can be approximated by leg length. It means that the GRF has a force element in opposite direction to the external force, so it induces a repulsive torque. The external force and repulsive torque are shown in Fig. 1(a).

When the external force is large enough to tilt the humanoid robot over θ_i, the repulsive torque goes to zero and the robot starts falling down, as shown in Fig. 1(b). Since the humanoid robot has IMU and gyro sensors located nearby the CoM, the robot can perceive whether it is falling or not from the sensors' information.

2.2 The CoM and Swinging Leg Trajectories

Once the robot perceives it is falling, the CoM trajectory is calculated to predict the elapsed time until it hits the ground. Simplified to the linear inverted pendulum, the robot follows the CoM trajectory calculated by the following equation of motion of an inverted pendulum:

$$\left(ml^2 + I_{CoM}\right)\ddot{\theta} = mgl\sin\theta \tag{2}$$

where the θ, m, l, I_{CoM}, and g are the tilted angle of robot, the total mass of robot, the length from the CoM to toe, the robot's inertia moment of the CoM, and the gravity acceleration, respectively. Since the equation (2) is a differential equation, it needs initial values of angles and velocities to solve the equation. As previously mentioned, the initial angle is the same as θ_i and the initial velocity is obtained from a gyro sensor of robot. Then, the equation is solved by numerical solution. In this study, suitable impact angle, when robot impacts the ground, is calculated from this equation.

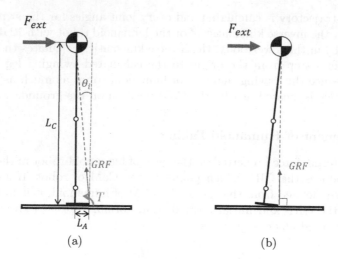

Fig. 1. External force (F_{ext}) is exerted on humanoid robot. (a) Repulsive torque occurs from the GRF. (b) If external force is large enough to overcome repulsive torque, then robot starts falling down.

Next, the swinging leg trajectory is calculated with respect to the CoM trajectory. In order to be stable after hitting the ground, humanoid robot takes a step to the projected position of the CoM onto the ground and the swinging leg trajectory is defined as

$$z(t) = \begin{cases} h \times sin\left(\frac{3\pi}{2T} \times t\right) + H(t) & (t \leq T/3) \\ \frac{h}{2} \times \left(1 + cos\left(\frac{3\pi}{2T} \times (t - T/3)\right)\right) + H(t) & (T/3 < t \leq T) \end{cases} \quad (3)$$

with

$$H(t) = sin(\theta_{impact}) \times l \times \frac{sin(\theta_{impact} - \theta_i)}{T} \times t \quad (0 \leq t \leq T) \quad (4)$$

where h, T, $H(t)$ are the maximum height of swinging leg trajectory, the duration time of stepping and the compensation height, respectively. Since the CoM height is gradually decreased, the swinging leg height should be increased to avoid a crash with the ground. While the time t is shorter than $T/3$, the swinging leg takes off the ground following the sinusoidal trajectory. At the time of $T/3$, a foot reaches its highest position. After $T/3$, the foot height is getting decreased smoothly to lessen the GRF as much as possible because the slope of $(1 + cos(t))/2$ is more moderate than that of the sinusoid. After time T, height of the swinging leg should be zero. The CoM and swinging leg trajectories are calculated by (2) and (3) (Fig. 2). Lastly, when humanoid robot impacts the ground, the inertia of the swinging leg causes increment in the yaw moment, which makes the humanoid unstable. If an external force is large, humanoid robot may fall down after hitting the ground. In this case, the robot's yaw moment can be significantly decreased by shifting the swinging leg trajectory in

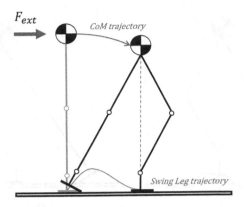

Fig. 2. The CoM and swinging leg trajectories

y-direction by a half of pelvis length within time T. It should be noted that the robot could experience self-collision between its own legs, while the swinging leg moves to the goal position that is in the same sagittal plane as the CoM. This issue is addressed in Section 3.

2.3   Mass-Spring-Damper Modeling

When humanoid robot hits the ground, the GRF is produced and affects humanoid robot and makes the robot unstable. In order to minimize the impact, a virtual mass-spring-damper model is adopted to humanoid robot's legs. The equation of motion of mass-spring-damper system is given as

$$m\ddot{x} + c\dot{x} + kx = F \tag{5}$$

where x, m, c and k denote the compressed distance of spring, the total mass of robot, the damping coefficient and the spring constant, respectively. Divided by m on both sides, it is represented as

$$\ddot{x} + 2\zeta\omega_0\dot{x} + \omega_0{}^2x = \frac{F}{m} \tag{6}$$

with

$$\zeta = \frac{c}{2\sqrt{km}}, \; \omega_0 = \sqrt{\frac{k}{m}} \tag{7}$$

where ζ, ω_0, and F denote the damping ratio, the natural frequency and the GRF which is measured by FSR sensors attached on humanoid's soles, respectively.

As shown in Fig. 3(a), humanoid robot bounces up because of the large GRF. To minimize the effect from the bouncing, a virtual mass-spring-damper is designed for humanoid robot's leg, as shown in Fig. 3(b). The spring-damper system makes hitting time longer, resulting in the maximum impact force decreased. Recalling the impulse is represented by $F\Delta t$, where F and Δt denote the impact

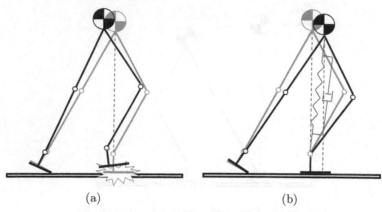

(a)   (b)

Fig. 3. Mass-spring-damper modeling. (a) No spring-damper system applied. Humanoid is bounced because of GRF. (b) Spring-damper system applied. Humanoid is not bounced and has more stability.

force and the duration of impact, respectively, the impact force is distributed over Δt. In this way, humanoid robot can reduce disturbances effectively not to be bounced after hitting the ground, which makes humanoid robot more stable. However, it should be seriously considered that the CoM of humanoid robot should be located in the allowable ZMP region, once the robot's leg is compressed to stably stop. When the humanoid leg is compressed, having limit on a compressed distance, the compressed distance becomes saturated with a large GRF. Therefore, the CoM of humanoid robot is always located inside the foot boundary, even though the CoM of humanoid robot moves in the forward direction. Another important point is that the damping ratio should have a value that is larger than one. It means the spring damper system is overdamped, which is suitable for this situation, because the underdamping causes an overshoot. Since the overshoot makes disturbances in the z-axis direction, underdamping is not suitable. Critical damping has a negative effect on the system because the response to it is too slow to effectively absorb impact energy. Therefore, this system is designed as an overdamped one.

3   Results

In this section, the simulation setup and results are discussed. The proposed method was tested for the humanoid robot, Mybot-KSR, which is a human-sized bipedal robot and it has 12 DoFs in the lower body. The Mybot-KSR was modeled by the robot simulator Webots [13], which provided physically and dynamically imitated environments.

As shown in Fig. 4, humanoid robot was pushed by an external force and it could not avoid falling, causing the humanoid robots fall to the ground directly. However, after adopting the Falling Prevention System, the humanoid robot could prevent falling down from the external force (Fig. 5). Figure 5(a) is the

Fig. 4. No Falling Prevention System applied. Arrow is the direction of external force. Humanoid fell down because of external force.

Fig. 5. Falling Prevention System applied. Arrow is the direction of external force. Humanoid perceived inclination of it and took a step to avoid falling. Also, robot could stand up after taking a step.

initial pose before being disturbed by the external force. The external force was applied to the humanoid robot from the back in the forward direction at a certain time. In Fig. 5(b), the humanoid robot was not tilted yet by the initial angle θ_i, leaving the Falling Prevention System inactivated. If the humanoid robot was tilted over θ_i, then it would perceive that it was falling down using the IMU sensor. In Fig. 5(c), the humanoid robot started to take a step just after it recognized that it is falling, and the swinging leg was stretched to the goal position which was already calculated by the CoM trajectory in (2). After the swinging leg arrived at the wanted position, the humanoid robot finally hit the ground, causing lots of disturbances because of the GRF in Fig. 5(d), and the robot bent its swinging leg to diminish the disturbances by the specific compressed distance which was calculated by design parameters, ζ and ω_0. Then, the robot could avoid falling down and stably stop, by virtue of the spring-damper system in Fig. 5(e). After that, the humanoid robot stood up to come back to initial pose by the Falling Prevention System algorithm. In Fig. 5(h), the humanoid robot stood up with the same pose as in Fig. 5(a). From this result, it was shown that the humanoid robot could be prevented from falling down by taking a step and maintain stability against a large collision force.

4   Conclusion

In this paper, the Falling Prevention System was proposed for humanoid robots to avoid falling against external disturbances. First, the CoM and swinging leg trajectories were calculated for the stepping. All the joint angles were calculated by inverse kinematics with the CoM and swinging leg trajectories. Also, the foot trajectory was shifted in the y-axis direction to reduce the yawing moment which caused the humanoid robot unstable. Once the humanoid robot hit the ground, the robot was affected by the GRF produced from collision. For this reason, mass-spring-damper system was introduced to lessen this force effectively. These procedures were simulated using Webots simulator and human-like robot, the Mybot-KSR, was modeled in the Webots to test this algorithm. The simulation result demonstrated the effectiveness of the proposed method and the humanoid robot could maintain stability after being pushed by external force and disturbed by the GRF as well.

The Falling Prevention System can be adopted for biped robots and it is helpful for the robot to maintain balance against external forces. Moreover, the proposed method can be adopted to any type of motions such as biped walking, hopping, running, etc. because it takes advantages of maintaining the balance of humanoid robot, in case it falls down. However, to prevent humanoid from falling with more perfection, it must be able to cope with omni-directional external forces. Humanoid robots may need to take more steps to maintain balance for a larger external force. These issues also will be covered in future work.

Acknowledgement. This work was supported by the Technology Innovation Program, 10045252, Development of robot task intelligence technology, funded by the Ministry of Trade, Industry & Energy (MOTIE, Korea).

References

[1] Vukobratovic, M., Frank, A., Juricic, D.: On the stability of biped locomotion. IEEE Transactions on Biomedical Engineering BME-17(1), 25–36 (1970)
[2] Vukobratovic, M., Borovac, B.: Zero-moment point - thirty five years of its life. I. J. Humanoid Robotics, 157–173 (2004)
[3] Kajita, S., Matsumoto, O., Saigo, M.: Real-time 3d walking pattern generation for a biped robot with telescopic legs. In: IEEE International Conference on Robotics and Automation, vol. 3, pp. 2299–2306 (2001)

[4] Kajita, S., Kanehiro, F., Kaneko, K., Fujiwara, K., Harada, K., Yokoi, K., Hirukawa, H.: Biped walking pattern generation by using preview control of zero-moment point. In: IEEE International Conference on Robotics and Automation, vol. 2, pp. 1620–1626 (September 2003)

[5] Lee, B.-J., Stonier, D., Kim, Y.-D., Yoo, J.-K., Kim, J.-H.: Modifiable walking pattern of a humanoid robot by using allowable zmp variation. IEEE Transactions on Robotics 24(4), 917–925 (2008)

[6] Komura, T., Leung, H., Kudoh, S., Kuffner, J.: A feedback controller for biped humanoids that can counteract large perturbations during gait. In: IEEE International Conference on Robotics and Automation, pp. 1989–1995 (April 2005)

[7] Tajima, R., Honda, D., Suga, K.: Fast running experiments involving a humanoid robot. In: IEEE International Conference on Robotics and Automation, pp. 1571–1576 (May 2009)

[8] Yamamoto, K., Nakamura, Y.: Switching control and quick stepping motion generation based on the maximal cpi sets for falling avoidance of humanoid robots. In: IEEE International Conference on Robotics and Automation, pp. 3292–3297 (May 2010)

[9] Jalgha, B., Asmar, D., Shammas, E., Elhajj, I.: Hierarchical fall avoidance strategy for small-scale humanoid robots. In: IEEE International Conference on Robotics and Biomimetics (ROBIO), pp. 1–7 (December 2012)

[10] Stephens, B.: Humanoid push recovery. In: IEEE-RAS International Conference on Humanoid Robots (2007)

[11] Pratt, J., Carff, J., Drakunov, S., Goswami, A.: Capture point: A step toward humanoid push recovery. In: IEEE-RAS International Conference on Humanoid Robots, pp. 200–207 (December 2006)

[12] Morisawa, M., Harada, K., Kajita, S., Kaneko, K., Sola, J., Yoshida, E., Mansard, N., Yokoi, K., Laumond, J.-P.: Reactive stepping to prevent falling for humanoids. In: IEEE-RAS International Conference on Humanoid Robots, pp. 528–534 (December 2009)

[13] Michel, O.: Cyberbotics Ltd. WebotsTM: Professional mobile robot simulation. Int. J. Advanced Robot. Syst. 1(1), 39–42 (2004)

Modifiable Walking Pattern Generator on Unknown Uneven Terrain

Young-Min Kim and Jong-Hwan Kim

Department of Robotics Program, KAIST
335 Gwahangno, Yuseong-gu, Daejeon, Republic of Korea
{ymkim,johkim}@rit.kaist.ac.kr

Abstract. This paper proposes a novel algorithm for humanoid robots to walk on unknown uneven terrain by using the Modifiable Walking Pattern Generator (MWPG). The proposed algorithm runs with a finite state machine including ascending and descending states. If the landing time of a swinging leg on unknown uneven terrain is shorter than the assigned single support time, the state is switched to the ascending state. If longer, the state is switched to the descending state. When a swinging leg lands on a surface of unknown uneven terrain, robot receives an impulsive contact force. The average impulsive contact force is reduced by expanding the duration of the contact time with respect to the impulse-momentum equation. According to the change of the step length due to unknown uneven terrain, the newly calculated center of mass (CoM) trajectory in the double support phase after landing the swinging leg is used. The proposed algorithm is used with the modified foot trajectory for adapting to the height of the uneven terrain. The effectiveness of the proposed algorithm is demonstrated through computer simulations using the simulation model of the small-sized humanoid robot, HanSaRam-IX (HSR-IX).

Keywords: Humanoid robot, uneven terrains, modifiable walking pattern generator (MWPG), 3-D linear inverted pendulum model (LIPM).

1 Introduction

Recently, there has been various research to develop humanoid robots. To enable the robot to walk, walking pattern generators of humanoid robots have been studied [1]–[3]. However, these algorithms effect not on uneven terrain but on flat terrain though our real world is composed of various land shapes. Therefore, research for walking on uneven terrain is also one of the key issues for humanoid robot [4]–[7].

This paper proposes the walking pattern generator on unknown uneven terrain based on MWPG, which allows the zero moment point (ZMP) variation in real-time by closed form functions [3]. The conventional MWPG can only be applied on the even terrain. If the exact terrain information is given, MWPG could be applied to both even and uneven terrain. However, it is hard to get the exact

© Springer International Publishing Switzerland 2015
J.-H. Kim et al. (eds.), *Robot Intelligence Technology and Applications 3*,
Advances in Intelligent Systems and Computing 345, DOI: 10.1007/978-3-319-16841-8_11

terrain information in a real world, which makes the robot unstable in walking. The proposed algorithm runs with a finite state machine even for the walking of humanoid robots on the terrain without the terrain information. It is composed of 4 states: Ascending, Descending, the conventional MWPG, and Reducing average impulsive contact force states. If the landing time of a swinging leg on unknown uneven terrain is shorter than the assigned single support time, the state is switched to the ascending state. If longer, the state is switched to the descending state.

When swinging leg lands on a surface of unknown uneven terrain, the robot receives an impulsive contact force, which may make the robot unstable. To overcome this situation, average impulsive contact force has to be reduced by expanding the duration of the contact time using the impulse-momentum equation. After landing, the height of the uneven terrain can be calculated by kinematics using the joint angles of the robot. Due to the unexpected sudden landing on uneven terrain, the step length could be changed. As the step length is changed, the center of mass (CoM) trajectory during the double support phase has to be re-calculated. After the double support phase is completed, the hind leg should be lifted up on the uneven terrain in the next single support phase. However, the collision between the robot and the terrain could happen when the conventional foot trajectory is used. Before applying the conventional foot trajectory, the foot perpendicularly ascends in a short time. This novel algorithm is implemented on a Webots simulation model of the small-sized humanoid robot, HanSaRam-IX (HSR-IX) that developed at the Robot Intelligence Technology laboratory, KAIST. The effectiveness of the proposed algorithm is demonstrated through computer simulations.

This paper is organized as follows. Section 2 presents the MPWG on unknown uneven terrain. In this section, reducing average impulsive contact force method is presented along with the modifying CoM trajectory in the double support phase and the modified foot trajectory for the hind foot. Section 3 presents the simulation results followed by the conclusions in Section 4.

2   The MWPG on Unknown Uneven Terrain

The algorithm of generating walking pattern for unknown uneven terrain is based on the MWPG. To apply the original extended MWPG based on 3D command state, the information about the uneven terrain as shown in Fig. 1 has to be given. In this paper, we propose a reducing average impulsive contact force method that could be applied without the terrain information. Assume that all part of the foot's sole contact with the upper side of the obstacle. Using this method, the swinging leg in the single support phase could land stably. It guarantees the stability of the robot based on impulse-momentum equation by extending the landing time. The next CoM trajectory in the double support time is newly calculated. After moving the CoM in the double support time, the hind foot trajectory is also modified to avoid collision with uneven terrain.

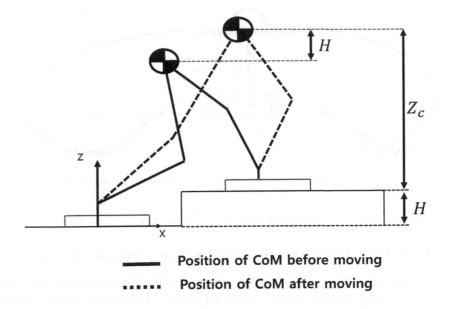

—— **Position of CoM before moving**

······ **Position of CoM after moving**

Fig. 1. Trajectory of CoM in the double support phase using the extended MWPG

2.1 Reducing Average Impulsive Contact Force during Landing Swinging Leg

In terms of the stability of the robot, reducing average impulsive contact force during the swinging leg is important on unknown uneven terrain. To solve this problem, a finite state machine in Fig. 2 is used. There are three important states. The following three states can be distinguished by the values of FSR sensor attached on the sole of the swinging leg.

1. Reducing average impulsive contact force state: extending contacting time of the swinging leg during around 0.2 s.
2. Ascending state: If the landing time on unknown uneven terrain is shorter than the assigned single support time (0.8 s), then the state is changed to ascending state.
3. Descending state: If the landing time on unknown uneven terrain is longer than the assigned single support time (0.8 s), then the state is changed to descending state.

The reducing average impulsive contact force method is based on the following impulse-momentum equation:

$$\int_{t_1}^{t_2} \vec{F} \, dt = m\Delta v. \tag{1}$$

If the terrain comes back to even terrain again, the walking pattern returns to the conventional MWPG.

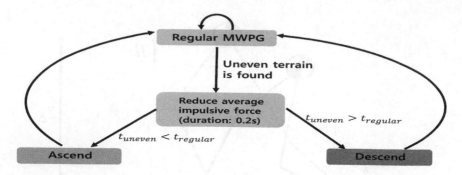

t_{uneven}: time during swing leg lands on uneven terrain surface
$t_{regular}$: regular single support time

Fig. 2. A finite state machine for MWPG on unknown uneven terrain

2.2   CoM Trajectory Generation in the Double Support Phase

The conventional method of generating trajectory of CoM in the double support phase cannot be applied in unknown uneven terrain. The conventional method is assumed that the travel distance of the swinging leg is the same as the normal step length. As the swinging leg lands on unknown uneven terrain unexpectedly, the travel distance of the swinging leg is shorter or longer than the normal step length, as shown in Fig. 3. Thus, the trajectory of CoM in the double support time should be modified. The total travel distance of CoM in the single and double support phases must be the same as step length. This method makes Zero Moment Point (ZMP) being included in the convex hull of the foot supporting

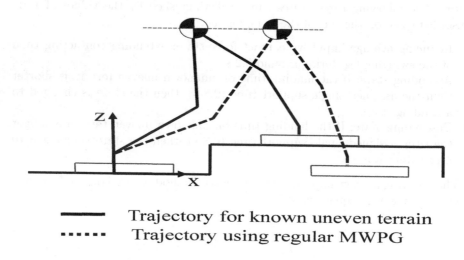

—— Trajectory for known uneven terrain
······ Trajectory using regular MWPG

Fig. 3. Trajectory comparing between even and uneven terrains

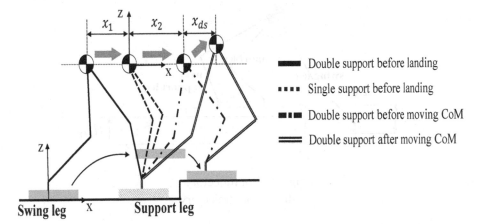

Fig. 4. The trajectory of CoM in the single and double support phases before and after landing

area. Note that step length is also modified due to the uneven terrain. For z-axis, the trajectory of CoM is generated from Z_c to $Z_c + H_{l/r}$ by the cubic spline interpolation.

In Fig. 4, x_1 is the travel distance of CoM in the single support phase before passing z-axis, x_2 is the travel distance of CoM in the single support phase after passing z-axis and x_{ds} is the travel distance of CoM in the double support time.

2.3   The Hind Leg Trajectory after Landing

When the conventional foot trajectory from MWPG is used, the hind foot has a high probability of collision on uneven terrain. In this situation, the heading angle of the robot can be altered or the stability of the robot can't be guaranteed. According to this, the horizontal movement of foot stops until z_{foot} is larger than $H_{l/r}$ first. Afterward, the trajectory is followed by the cubic spline interpolation function as shown in Fig. 5. $H_{l/r}$ can be obtained using the kinematics of the support leg. The vertical foot trajectory of the swinging leg, z_{foot} is generated by a cycloid function.

The foot trajectory of the swinging leg is as follows:

$$x_{foot}(t) = \begin{cases} -2\dfrac{S_{l/r}+S_{l/r}^{pre}}{T_{l/r}^{ss\,3}}t^3 + 3\dfrac{S_{l/r}+S_{l/r}^{pre}}{T_{l/r}^{ss\,3}}t^2 - S_{l/r}^{pre} & \text{if } z_{foot}(t) > H_{l/r} \\ 0 & \text{otherwise} \end{cases} \tag{2}$$

where $S_{l/r}^{pre}$ represents sagittal step length at $t = 0$, $T_{l/r}^{ss}$ represents single support time during left/right support phase and $S_{l/r}$ is sagittal step length at $t = T_{l/r}^{ss}$.

When the robot on the descending state because of the absence of FSR value, the foot trajectory is modified to be lifted down vertically. Vertical movement let the robot avoid the unsolved inverse kinematics problem.

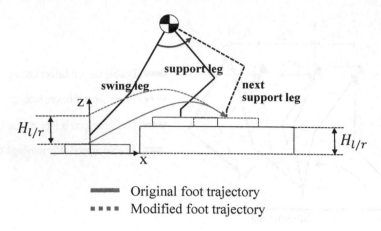

Fig. 5. Original and modified foot trajectories on ascending state

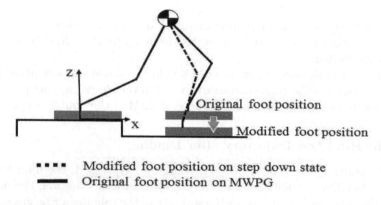

Fig. 6. Original and modified foot trajectories on descending state

3   Simulation Results

3.1   HanSaRam-IX

The proposed algorithm was implemented on the simulation model of the small-sized humanoid robot, HSR-IX. The simulation model of HSR-IX was modeled by Webots that is the 3-D robotics simulation software and robot behavior can be tested in physically realistic worlds[9]. HSR has been in continual development and research by the Robot Intelligence Technology laboratory, KAIST. Its height and weight are 52.8 cm and 5.5 kg, respectively. It has 26 DOFs which consist of 12 DC motors with harmonic drivers in the lower body and 16 RC servo motors in the upper body (two servo motors in each hand control). To measure ground reaction forces on the feet and the real ZMP trajectories while walking, four force sensing resisters are equipped on each foot.

3.2 Simulation Results on Unknown Uneven Terrain

Fig. 7 shows ascending and descending state simulation for unknown uneven terrain using the Webots simulator. The height of the terrain is about 1.0 cm. However, this information is not provided to the robot. The normal step length is 12.0 cm, the step length for uneven terrain is 7.0 cm. the duration of the double support time is 0.4 s and the duration of the single support time is 0.8 s. But, the modified single support time is about 0.485 s after landing on uneven

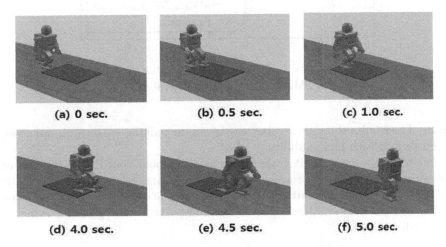

(a) 0 sec.   (b) 0.5 sec.   (c) 1.0 sec.

(d) 4.0 sec.   (e) 4.5 sec.   (f) 5.0 sec.

Fig. 7. Simulation results on unknown uneven terrain

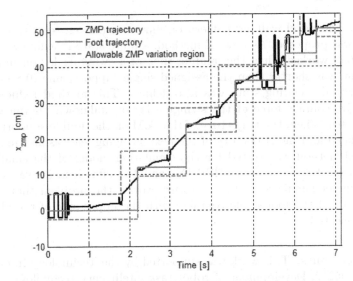

Fig. 8. ZMP trajectory of the sagittal motion

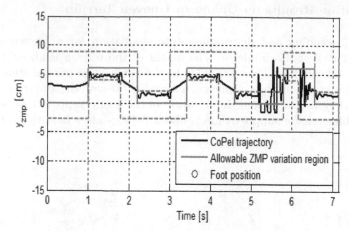

Fig. 9. ZMP trajectory of the lateral motion

terrain. The results show that robot can generate walking pattern on unknown uneven terrain with high stability.

Figs. 8 and 9 show ZMP trajectories to the global coordinate systems of sagittal motion and lateral motion on ascending state, respectively. The ZMP trajectory is within an allowable ZMP variation region so that the stability of the walking pattern is guaranteed. Due to the unexpected landing of the swinging leg, overshoot occurs.

4   Conclusion

In this paper, the novel walking pattern on unknown uneven terrain based on the MWPG was proposed. As a novel method, a finite state machine was used. Ascending state worked when the landing time of a swinging leg on unknown uneven terrain was shorter than the assigned single support time (0.8 s). As a similar manner, descending state was also defined. This method reducing average impulsive contact force of the swinging leg on unknown uneven terrain was implemented. Then, the trajectory of the CoM on the double support phase was re-calculated and the trajectory of the next swinging leg was also modified. The proposed method was tested for the simulation model of the small-sized humanoid robot, HanSaRam-IX (HSR-IX) and the effectiveness was verified through computer simulations. In our future works, the problem that not all but partial part of the foot's sole contacts with the upper side of the obstacle will be implemented.

Acknowledgements. This work was supported by the Technology Innovation Program, 10045252, Development of robot task intelligence technology, funded by the Ministry of Trade, Industry & Energy (MOTIE, Korea).

References

1. Kajita, S., Kanehiro, F., Kaneko, K., Fujiwara, K., Yokoi, K., Hirukawa, H.: A realtime pattern generator for biped walking. In: Proc. IEEE Int. Conf. Robot. Autom., vol. 1, pp. 31–37 (2002)
2. Kajita, S., Kanehiro, F., Kaneko, K., Fujiwara, K., Harada, K., Yokoi, K., Hirukawa, H.: Biped walking pattern generation by using preview control of zero-moment point. In: Proc. IEEE Int. Conf. Robot. Autom., vol. 2, pp. 1620–1626 (2003)
3. Lee, B.-J., Stonier, D., Kim, Y.-D., Yoo, J.-K., Kim, J.-H.: Modifiable Walking Pattern of a Humanoid Robot by Using Allowable ZMP Variation. IEEE Transactions on Robotics 24(4), 917–923 (2008)
4. Hong, Y.-D., Kim, J.-H.: Walking Pattern Generation on Inclined and Uneven Terrains for Humanoid Robots. In: Proc. International Conference on Robot Intelligence Technology and Applications, RiTA (December 2012)
5. Hong, Y.-D., Kim, J.-H.: 3-D Command State-Based Modifiable Bipedal Walking on Uneven Terrain. IEEE/ASME Transactions on Mechatronics 18(2), 657–663 (2013)
6. Takubo, T., Imada, Y., Ohara, K., Mae, Y., Arai, T.: Rough terrain walking for bipedal robot by using ZMP criteria map. In: Proc. IEEE Int. Conf. Robot. Autom., pp. 788–793 (2009)
7. Sato, T., Ohnishi, K.: Walking Trajectory Modification with Gyroscope for Biped Robot on Uneven Terrain. IEEE Transactions on Industrial Electronics, 969–974 (June 2011)
8. Wei, H., Shuai, M., Wang, Z.: Dynamically Adapt to Uneven Terrain Walking Control for Humanoid Robot. Chinese Journal of Mechanical Engineering 25(2) (2012)
9. Kang, S.-C., Komoriya, K., Yokoi, K., Koutoku, T., Kim, B.-C., Park, S.-S.: Control of Impulsive Contact Force between Mobile Manipulator and Environment Using Effective Mass and Damping Controls. International Journal of Precision Engineering and Manufacturing 11(5), 697–704 (2010)
10. Michel: Cyberbotics Ltd. WebotsTM: Professional mobile robot simulation. Int. J. Advanced Robot. Syst. 1(1), 39–42 (2004)

A Four-Legged Social Robot Based on a Smartphone

David A. Diano and David Claveau

CSU Channel Islands, Camarillo, CA USA
david.diano751@myci.csuci.edu, david.claveau@csuci.edu

Abstract. This paper presents a simple four-legged robot platform that can transform a smartphone or tablet into an autonomous walking social robot. Smartphones are already designed to be easy to interact with and to assist humans in everyday activities. They are also equipped with impressive computational resources and powerful sensors. As such they can serve as a powerful controller for a robot. By giving them four legs they have the added abilities to walk about our environments and express themselves using posture and body language. An example of such a four-legged robot is described in this paper along with a preliminary exploration of its capabilities as a social robot. Simple walking and posturing are demonstrated in ways that show how such a robot can better play the role of companion or assistant in the office or home. The platform is designed and built using inexpensive off-the-shelf components and can serve as an affordable development system for students and practitioners who wish to study social robotics.

Keywords: social intelligence, human-robot interaction, education.

1 Introduction

Powerful mobile devices such as smartphones and tablets have become commonplace items. These devices have been designed to allow humans to interact with them in rich ways using visual display, touch and voice. However, they remain devices that must be carried around, locked in a body without mobility. This despite the fact that they contain the computational resources and the sensors needed to control an articulated robot body. Even the least expensive of these devices contain cameras, accelerometers and GPS. An internet search for "smartphone robot" is likely to find some interesting examples such as Romo [1], a programmable robot companion that uses an iPhone as a controller and that can respond to humans and roam around a desktop using tracked wheel locomotion. Another example is the SmartBot [2] which is a small robot that can use Android or iOS-based smartphones to control its behavior. Both are shown in Figure 1. While these robots add mobility to a smartphone they are limited in their ability to move; they cannot move in ways that dogs, cats and other four-legged creatures do. We have become very familiar and comfortable with four-legged creatures and we easily understand their body language. If we could plug our smartphone into a four-legged body it would allow the device to more closely embody an assistant or perhaps a robot 'familiar' such as a magician's black cat.

© Springer International Publishing Switzerland 2015
J.-H. Kim et al. (eds.), *Robot Intelligence Technology and Applications 3,*
Advances in Intelligent Systems and Computing 345, DOI: 10.1007/978-3-319-16841-8_12

Your smartphone could now climb onto your shoulder and alert you by whispering in your ear, perhaps helping you to cross a street if you have a visual impairment. It could also use a form of body language to express itself more clearly.

Fig. 1. Commercially available smartphone robots include (a) Romo [1] and (b) SmartBot [2]

Here we present a simple four-legged robot platform that can transform a smartphone or tablet into an autonomous walking robot. A smartphone can be simply 'plugged into' the body and then control its own motion. An important aspect of our design is that only inexpensive off-the-shelf components are needed along with open source software. This makes it a convenient platform for students who want to explore social robotics. Just about every university student has a smartphone or tablet and soon even young children will have such devices. This opens some exciting possibilities in education since students can simply plug their device into the robot platform and program new behaviors. In this paper we give some simple examples of walking and posturing behaviors. An informal survey was performed to explore their effectiveness and their reception by an average smartphone user. In the rest of this paper, section 2 discusses some related work and following that the smartphone robot platform is described in section 3 with implementation and preliminary results in sections 4 and 5.

2   Background and Related Work

The personal nature of our relationship with smartphones and tablets is a topic of current exploration [3]. These devices contain everything from our memories of happy and sad experiences to our preferences in music and film. Our physical relationship with these devices is rather limited however. Since visual interaction is confined to the display we can think of these devices as 'virtually embodied'. Studies have shown that a 'physically embodied' robot that can move its own body is more engaging [4][5][6]. These studies have also shown that physical embodiment leads to more feelings of enjoyment, trust and respect toward the robot. An example of a socially expressive robot based on a smartphone is MeBot [7]. This is a telepresence robot that allows an operator to express nonverbal behavior such as hand and head gestures. It combines video and audio of the remote operator's face with mechanical arms and wheeled desktop mobility.

There are several examples of mobile robots based on smartphones that bear some similarity to our work but they have generally been wheeled robots and have not addressed the social interaction issues. A popular option is to use an Android-based smartphone directly connected to an IOIO (or the newer IOIO-OTG) circuit board that controls servos and other sensors [8]. The IOIO board can be connected to the smartphone via Bluetooth or USB and can read values from digital/analog sensors using the Java API provided. For example, an Android smartphone can be programmed to send commands through an IOIO board that is connected to an inexpensive remote controlled car [8][9]. The smartphone can read sensors, such as the built-in GPS or an infrared sensor connected to the IOIO, and send commands to the servos and electronic speed controller to navigate the car. Other options for building smartphone robots include using the LEGO Mindstorms system [10] which is used in schools and universities because of its affordable yet flexible building options.

While wheels are fine for the desktop, legs are needed for unstructured surfaces such as cluttered desktops and lumpy beds. They are also needed to allow the robot to assume expressive postures and poses. Legged robots are a very active research topic. A visible example is the BigDog robot [11] which can achieve animal-like mobility on uneven terrain with a focus on search-and-rescue applications. The robot is very impressive but is very sophisticated and expensive. The robot platform in this paper is an attempt to combine the mobility and expressiveness of legs in an open and affordable platform that can be used by students and researchers to explore physically embodied social robotics.

3   The Robot Platform

This section describes the design and implementation of the robot platform. The design had to meet the following basic requirements:

1. The platform should have four legs similar to a dog or cat.
2. The platform should be comparable in size and weight to a smartphone.
3. The platform should have an interface that allows a smartphone to be easily plugged into it.
4. The platform should allow for easy programming of behaviors.

3.1   Design

The design of the platform is a tradeoff between simplicity and naturalistic embodiment. It should be simple and inexpensive to build but still able to perform natural cues and exhibit personality and character. The main body of the platform, as shown in Fig. 2, is designed as a simple rectangular base made of a stiff material such as lightweight wood or plastic. Four legs are attached to the bottom of this base and a smartphone can be placed face-up on top of it. The legs were designed to each have two revolute joints of 1-DOF. This is sufficient for simple walking gaits and simple postures and makes the legs very easy to design. In fact, each leg only used one 'femur' and used the servo-motor's body as the end-effector that touches the ground.

Fig. 2. The base and legs of the robot platform

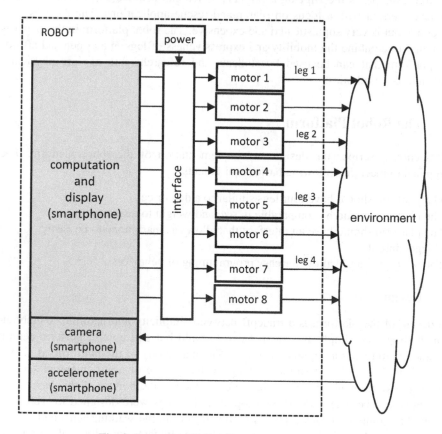

Fig. 3. System block diagram for the robot platform

Standard off-the-shelf servo-motors and brackets can easily be used for these legs. A hole is made in the base for the camera on the back of the phone; a small periscope can be mounted over the hole so that the camera can be used for navigation.

The smartphone connects to the robot platform via a simple interface. The complete system diagram is shown in Fig. 3. The smartphone is on the left and contributes computation, display and sensing. There are eight motors (two for each leg) connected to the smartphone through the interface. A second power supply is needed for the motors. The interface needs to drive the eight motors independently and receive commands from the phone. Smartphones typically have a micro-usb or similar port that can be used for this. The implementation section describes how we were able to use an off-the-shelf interface to meet all of these needs.

3.2   Implementation

The smartphone used in this project is the Android-based LG Google Nexus 4. It has a 1.512 GHz quad-core Krait CPU, 2 GB of RAM, a 3-axis accelerometer, an 8-megapixel camera and several other sensors. These are typical features for many smartphones and they are well-suited for robotics applications. The smartphone is connected to the other components via USB as shown in the hardware block diagram of Fig. 4. Most smartphones at least support the USB 2.0 standard that can read/write on an average speed greater than 30MBps. Some Android-based devices do not provide the correct voltage required to power USB devices such as a keyboard or mouse. This seems to be a rare case and unfortunately a 'hack' to the Android kernel and a three-way Y cable connecting a USB power pack is usually required to enable USB support.

The Phidgets Advanced Servo Controller 1061[12] is a motor controller that uses a mini-USB cable connection to control the position, acceleration, and velocity of up to 8 servo motors in a compact module that requires 6-15 volts to power. The circuit board can be purchased individually or in a kit that comes with 4 Hitec HS-422 Deluxe servo motors and a power supply. Specifications state that each servo is powered at up to 3.4 amps using a switching regulator that protects the motors from overvoltage and maintains a high resolution of 125 steps per degree The position of each servo is controlled by the smartphone using an Android application and a library of functions provided by Phidgets. Also included with the installation of the drivers and libraries is the Phidget Control Panel which can help test your Phidget board to make sure it is working properly, calibrate servos, and update the firmware.

The software stack for the robot is shown in Fig. 5. At the bottom is the Android operating system which is open source, making it possible for developers to modify the kernel as mentioned above to enable USB support on the Nexus 4. It provides all of the tools needed to create applications that take full advantage of the hardware capabilities for each device. Using Android Studio we developed a small Java application that can run on any device with Android Honeycomb 3.1 or higher. The application has a user interface on the phone that displays simple controls to move the servos and show accelerometer readings for testing purposes. In addition to this we used the OpenCV computer vision library to test the camera and use some simple machine vision algorithms for navigation. For the final implementation the total cost of materials turned out to be around $250 (USD) without the cost of the smartphone (See Table 1).

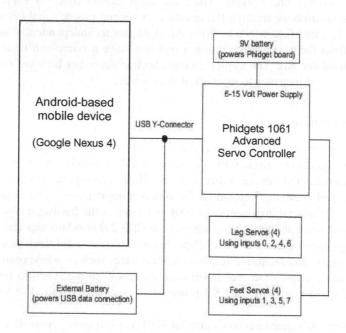

Fig. 4. Hardware block diagram for the robot platform

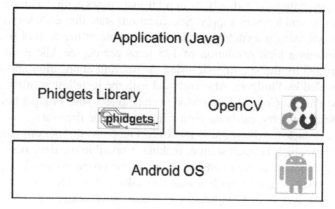

Fig. 5. Software stack for the robot platform

Table 1. Prices paid (USD) for robot platform components

Google Nexus 4	$200
PhidgetsAdvanceServo Kit (4-motors)	$140
Hitec HS-422 Deluxe Servo (4)	$50
Lynxmotion Femur for legs (4)	$25
Photojojo Smartphone Spy Lens	$25
Stiff board for base	$10
TOTAL COST(USD)	$450

4   Experiments and Results

So far we have created two demonstrations of the robot's social capabilities as described in the following two subsections. The first is a simple demonstration of the robot's ability to move about our environment and physically interact with us. The second is an exposition of a sequence of body postures designed to enhance conversations with humans.

4.1   Good Morning Robot!

The first demonstration involves the smartphone robot playing the role of a personal assistant. The robot is 'sleeping' on the bed and awakens when the smartphone's alarm rings. It proceeds to walk over the bed to tap the arm of its boss and announce any schedule commitments. This gives us a hint at the possibilities of having a robot assistant that has all of the information of a smartphone coupled with a familiar and capable four-legged body. The video for this demonstration is available at:

https://www.youtube.com/watch?v=ErMmKLlDkp4&feature=youtu.be

We have presented this video at seminars and talks and it has been well received without any mention of the creepy feelings that accompany some artificial moving creatures as in the so-called uncanny valley effect. As an example of how easy it is to program, the walking was accomplished through observation and trial-and-error. Our first attempt at walking was having the robot push off its hind legs like a frog but the movement would jostle the smartphone and circuit board, and then eventually offset the balance of the robot causing it to falling over. Walking needed to either have one leg move at a time so that there are three contact points to keep the robot stable or alternate a pair of legs similar to the movement of a dog.

4.2   Robot Body Postures for Conversation

The second demonstration consisted of having the robot assume some set postures that would be intended to communicate something conversational to a human.

This type of body language would enhance typical verbal or visual messages, making the robot seem more lifelike. Fig. 6 shows an example of two such postures with the accompanying facial expressions. A video sequence of four such postures was created in order to survey a typical user's interpretation of each posture without the accompanying face. The video is available here:

https://www.youtube.com/watch?v=PtCIbGjJV4c&feature=youtu.be

The survey question was as follows:

You are interacting with the robot from the video. What do you think it is trying to convey after each posture?
Please put the posture number with the interaction that you think matches:

a)   It is confused
b)   It is ready to listen
c)   It understands what you said
d)   It is about to leave

We did a small, informal study of ten participants gathered from faculty and staff close to our offices. The results of the survey are shown in Fig. 7. The first two postures were accurately interpreted by most people in the survey. The last two had some problems but this was attributable to their ambiguous nature when not combined with verbal and facial cues. A full video with facial expressions is here:

https://www.youtube.com/watch?v=B43FYX4GtXs&feature=youtu.be

The survey participants also viewed the robot postures with the facial expressions and all agreed that the combination was clear and added to the feeling of having a real conversation.

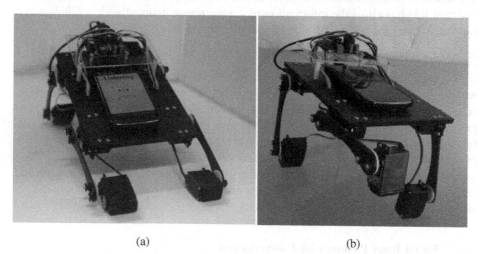

(a)   (b)

Fig. 6. Example postures (a) "ready to listen" and (b) "waving goodbye, leaving"

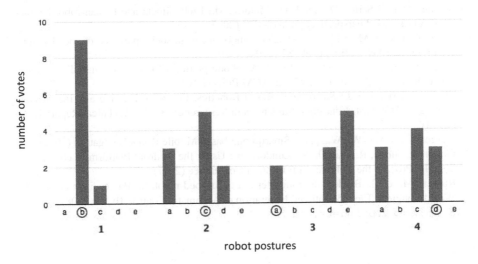

Fig. 7. Survey results for robot postures (intended response is circled for each posture)

5   Conclusion and Future Work

We have presented a simple four-legged robot platform that can transform a smart-phone or tablet into an autonomous, walking social robot. Simple walking and postur-ing have been demonstrated to show how such a robot can better play the role of companion or assistant in the office or home. The platform offers students and re-searchers an inexpensive development system to explore social robotics. Our prelimi-nary tests have shown that the system is easy to program and is well received by those interacting with it. The open nature of the software makes it possible to develop low-cost, downloadable apps by anyone who wishes to add to the system and customize it. In the future, we hope to develop a better vision system which can help to achieve richer interactions in which the robot can interpret human gestures. This type of soft-ware is also openly available and should only have to be ported and adjusted to fit the platform. There are many similar possibilities and we look forward to experimenting with the robot and working with others who are also interested.

References

1. http://www.romotive.com/
2. http://www.overdriverobotics.com/
3. Vincent, J.: Is the mobile phone a personalized social robot? In: Intervalla, vol. 1 (2013), http://www.fc.edu/intervalla/images/pdf/6_vincent.pdf
4. Wainer, J., Feil-Seifer, D.J., Shell, D.A., Mataric, M.J.: The role of physical embodiment in human-robot interaction. In: IEEE Proceedings of the International Workshop on Robot and Human Interactive Communication, Hatfield, UK, pp. 117–122 (2006)

5. Wainer, J., Feil-Seifer, D., Shell, D., Mataric, M.: Embodiment and human-robot interaction: A task-based perspective, pp. 872–877 (2007)
6. Jung, Y., Lee, K.M.: Effects of physical embodiment on social presence of social robots. In: Presence 2004, Spain, pp. 80–87 (2004)
7. Adalgeirsson, S., Breazeal, C.: MeBot: A robotic platform for socially embodied telepresence. In: Proc. HRI 2010, pp. 15–22. ACM Press (2010)
8. Oros, N., Krichmar, J.: Smartphone Based Robotics: Powerful, Flexible and Inexpensive Robots for Hobbyist, Educators, Students and Researchers. CECS Technical Report 13-16 (2013)
9. Herget, N., Keyes, W., Wang, C.: Smartphone-based Mobile Robot Navigation (2012)
10. Gobel, S., Jubeh, R., Raesch, S., Zundorf, A.: Using the Android Platform to control Robots. In: Proc. of the Robotics in Education Conference (2011)
11. Raibert, M., et al.: Bigdog, the rough-terrain quadruped robot. In: Proceedings of the 17th World Congress The International Federation of Automatic Control, IFAC (2008)
12. http://www.phidgets.com

Behavior Selection Method of Humanoid Robots to Perform Complex Tasks

Woo-Ri Ko and Jong-Hwan Kim

Department of Electrical Engineering, KAIST,
355 Gwahangno, Yuseong-gu, Daejeon, Republic of Korea
{wrko,johkim}@rit.kaist.ac.kr
http://rit.kaist.ac.kr

Abstract. This paper proposes a behavior selection method of humanoid robots to perform complex tasks using the degree of consideration-based mechanism of thought (DoC-MoT). The four input (context) symbols and seven target (atom behavior) symbols are defined to perform five complex tasks. The degree of consideration (DoC) for each input symbol is represented by the λ-fuzzy measure and the knowledge link strengths between input and target symbols are represented by the partial evaluation values. Each target symbol is globally evaluated by the fuzzy integral of the partial evaluation values with respect to the fuzzy measure values. Then, one target symbol with the highest evaluation value is selected and activated. To make corrections to the robot's wrong behaviors, a learning process from a human's behaviors is employed to update the DoCs and the knowledge link strengths. To show the effectiveness of the proposed method, simulations are performed in a text-based simulator developed in Visual Studio 2012. The results show that the proposed method can generate human-like behaviors.

Keywords: Behavior selection algorithm, learning algorithm, complex tasks, humanoid robots, human-like mechanism of thought.

1 Introduction

A complex task of a humanoid robot can be executed by a series of atom behaviors. For example, "toasting bread" task can be executed by approaching, grasping and moving bread behaviors. To do so effectively, the atom behaviors should be described as descriptive symbols not low-level motor actions [1]. Also, a humanoid robot should select a behavior considering the context information and learn which behavior is more appropriate for a certain situation based on the user's feedback [2], [3].

In this regards, this paper proposes a behavior selection method using the degree of consideration-based mechanism of thought (DoC-MoT) [4]. The DoC for each input symbol is represented by the λ-fuzzy measure and the knowledge link strengths between input and target symbols are represented by the partial evaluation values. Each target symbol is globally evaluated by the fuzzy integral

© Springer International Publishing Switzerland 2015
J.-H. Kim et al. (eds.), *Robot Intelligence Technology and Applications 3*,
Advances in Intelligent Systems and Computing 345, DOI: 10.1007/978-3-319-16841-8_13

of the partial evaluation values with respect to the fuzzy measure values, and after that, one target symbol with the highest evaluation value is selected and activated. To make corrections to the robot's wrong behaviors, a learning process from human behaviors is employed to update the DoCs and the knowledge link strengths. The effectiveness of the proposed method is demonstrated by simulations carried out with a text-based simulator developed in Visual Studio 2012.

This paper is organized as follows. Section II presents the DoC-MoT, which is a well-modeled mechanism of human thought. Section III proposes the behavior selection method to perform complex tasks using the DoC-MoT. Section IV presents the simulation results to demonstrate the effectiveness of the proposed method. The concluding remarks and future works follow in Section V.

2   Degree of Consideration-Based Mechanism of Thought

The DoC-MoT was proposed to explain how humans make a conclusion from several perceived information [4]. A human brain is formed of a number of well-connected neurons and each set of adjacent neurons represents an input symbol, i.e. perceived information, or a target symbol, i.e. conclusion. The link between input and target symbols represent the knowledge about perceived entities, and therefore, it is called a knowledge link. Since the human thought process is largely affected by personal biases [5], the degree of consideration (DoC) or importance for each input symbol was represented by the λ-fuzzy measure [6]. The knowledge link strength between input and target symbols represents a partial evaluation value of the target symbol over the input symbol. Then, each target symbol is globally evaluated using the Choquet fuzzy integral aggregating the DoCs and partial evaluation values and one target symbol with the highest evaluation value is selected as a conclusion [7].

The λ-fuzzy measure of a subset of input symbols is calculated as follows:

$$g(A \cup B) = g(A) + g(B) + \lambda g(A)g(B), \tag{1}$$

where $g(A)$ and $g(B)$, $A, B \subset X$ represent the DoCs of the subsets A and B, respectively, and $\lambda \in [-1, +\infty]$ denotes an interacting degree index. If the two subsets A and B have negative (positive) correlation, (1) becomes a plausible (belief) measure and λ is a positive (negative) value so that $g(A \cup B) < g(A) + g(B)$ $(g(A \cup B) > g(A) + g(B))$. If the two subsets are independent, (1) becomes a probability measure and the value of λ is zero so that $g(A \cup B) = g(A) + g(B)$.

The evaluation value of each target symbols is calculated by the Choquet fuzzy integral as follows:

$$I(z) = \sum_{i=1}^{n} g(A_i)\{h(x_i) - h(x_{i-1})\}, \tag{2}$$

where the input symbol set X is sorted so that $h(x_i) \geq h(x_{i+1}), i = \{1, \ldots, n-1\}$ and $h(x_0) = 0$, $g(A_i)$ is the fuzzy measure value of A_i, $A_i = \{x_i, x_{i+1} \ldots, x_n\}$ is the subset of X, and $h(x_i)$ is the partial evaluation value of the target symbol z over the ith input symbol x_i. Note that the value of $I(z)$ also represents the evaluation value of z to be a conclusion of the information-processing.

3   Behavior Selection Method

In this section, the DoC-MoT is applied to the behavior selection to perform complex tasks. Fig. 1 shows the overall architecture of the proposed method, which is composed of eight modules: sensor, perception, context, behavior selection, actuator, user command, learning and memory modules. The context module identifies the current environmental context based on the perceptions from the perception module. The memory modules stores all necessary memory contents including the DoCs for contexts and the knowledge link strengths between the contexts and atom behaviors. The behavior selection module evaluates each atom behavior using the DOC-MoT and one atom behavior with the highest evaluation value is generated through the actuator module. The learning algorithm based on a user's feedback is executed in the learning module. The key modules for behavior selection, namely context, behavior selection and learning modules, are described in the following.

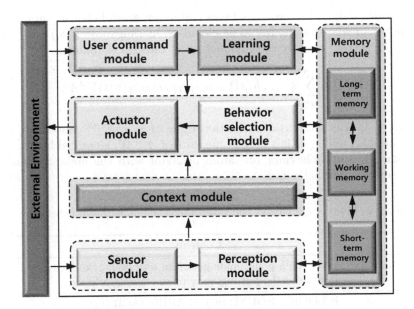

Fig. 1. Overall architecture of the proposed method

3.1 Context Module

As shown in Table 1, four contexts are defined in the context module; the distance between the robot's upper body and the target object, the distance between the robot's hand and the target object, the distance between the current and target positions of the grasping object and the difference between the current and target rotations of the grasping object. Note that the context information is used as the criteria for behavior selection in the behavior selection module.

Table 1. Four input symbols (contexts)

	Context	Unit	Range	DoC
1	The distance between the robot's upper body and the target object	m	[0,5]	4
2	The distance between the robot's hand and the target object	m	[0,1]	3
3	The distance between the current and target positions of the grasping object	m	[0,1]	2
4	The difference between the current and target rotations of the grasping object	°	[0, 360]	1

3.2 Behavior Selection Module

As shown in Table 2, five complex tasks and seven atom behaviors are defined in the behavior selection module. The global evaluation value $E(b_i)$ of ith atom behavior b_i, is computed by the Choquet fuzzy integral as follows:

$$E(b_i) = \sum_{j=1}^{4} \{w_{ij} \cdot h_{ij}(t) - w_{i(j-1)} \cdot h_{i(j-1)})t(t)\}g(A), \qquad (3)$$

where $h_{ij} \in [0,1]$ is the partial evaluation value of b_i over j^{th} context c_j, w_{ij} is the weight of h_{ij} and $g(A)$ is a λ-fuzzy measure of $A \subset X$, identified by (1). After evaluating all atom behaviors, one atom behavior with the highest evaluation value is selected and activated.

Table 2. Five complex tasks and seven target symbols (atom behaviors)

	Contents
Complex tasks	Watering a plant, finding stuff in a drawer, taking contents from a bottle, putting away toys, toasting a bread
Atom behaviors	B_LOOK_AROUND (b_1), B_APPROACH (b_2), B_BEND_BODY (b_3), B_GRASP (b_4), B_MOVE (b_5), B_TILT (b_6), B_TAKE_CONTENT (b_7)

3.3   Learning Module

In the learning module, learning from the user's feedback signals is performed for humanoid robots. The learning process is executed in real time and the user's feedback signals cause the change of the DoCs for contexts and the weights of the knowledge link strengths. The DoC $d_j(t)$ for c_j at time t, is calculated as follows:

$$d_j(t+1) = d_j(t) + k_1 \cdot h_{ij}, \qquad (4)$$

where $k_1 \in [0,1]$ is the learning rate and h_{ij} is the partial evaluation value of the recommended behavior b_i^r over c_j.

The weight w_{ij} of knowledge link strength h_{ij} is changed at the same time as the DoCs for contexts are updated as follows:

$$w_{ij}(t+1) = w_{ij}(t) + k_2 \cdot h_{ij}(t), \qquad (5)$$

where $k_2 \in [0,1]$ is the learning rate.

4   Simulations

To show the effectiveness of the proposed behavior selection method, we developed a text-based simulator on Visual Studio 2012 (C++). To use as an feedback signals in the learning module and compare with the robot's behaviors, we stored

Table 3. The context information and selected behaviors for "toasting a bread" task

	Target object	Context				Selected behavior		
						human	Humanoid robot	
		c_1	c_2	c_3	c_4		Before learning	After learning
1	Table	-	-	-	-	B_LOOK_AROUND	B_LOOK_AROUND	B_LOOK_AROUND
2	Table	4.0	4.1	-	-	B_APPROACH	B_APPROACH	B_APPROACH
3	Table	2.6	2.6	-	-	B_APPROACH	B_APPROACH	B_APPROACH
4	Table	1.1	1.2	-	-	B_APPROACH	B_APPROACH	B_APPROACH
5	Bread	0.5	0.6	-	-	B_BEND_BODY	B_BEND_BODY	B_BEND_BODY
6	Bread	0.3	0.6	-	-	B_GRASP	B_GRASP	B_GRASP
7	Bread	0.3	0.1	-	-	B_GRASP	B_MOVE	B_MOVE
8	Bread	0.3	0.0	0.3	0	B_MOVE	B_MOVE	B_MOVE
9	Toaster	0.6	0.5	-	-	B_APPROACH	B_BEND_BODY	B_APPROACH
10	Bread	0.4	0.0	0.0	90	B_TILT	B_TILT	B_TILT
11	Bread	0.4	0.0	0.1	0	B_MOVE	B_TAKE_CONTENT	B_MOVE
12	Lever	0.3	0.1	-	-	B_GRASP	B_GRASP	B_GRASP
13	Lever	0.3	0.0	0.1	0	B_MOVE	B_TAKE_CONTENT	B_MOVE
14	None	-	-	-	-	B_LOOK_AROUND	B_LOOK_AROUND	B_LOOK_AROUND
15	Toast	-	-	-	-	B_LOOK_AROUND	B_LOOK_AROUND	B_LOOK_AROUND
16	Toast	0.3	0.1	-	-	B_GRASP	B_GRASP	B_GRASP
17	Toast	0.3	0.0	0.1	0	B_MOVE	B_TAKE_CONTENT	B_MOVE
18	Plate	0.6	0.6	-	-	B_APPROACH	B_BEND_BODY	B_APPROACH
19	Toast	0.4	0.0	0.0	90	B_TILT	B_TILT	B_TILT
20	Toast	0.4	0.0	0.3	0	B_MOVE	B_MOVE	B_MOVE

Table 4. The behavior similarities with a human

	Complex task					Total
	1	2	3	4	5	
P-MoT	70%	45%	55%	30%	55%	51%
DoC-MoT w/o learning	75%	95%	100%	70%	70%	82%
DoC-MoT with learning	75%	100%	100%	75%	95%	89%

100 events for five complex tasks and each event has context information and the selected behavior by a human. Table 3 shows the stored events related to "toasting a bread" task. If the selected behavior by the robot was different from the human behavior for the same event, the learning process was performed and the human behavior was used as the user's feedback signal.

After learning from the stored events, the robot selected behaviors again using the updated DoCs and knowledge link strengths. The behavior similarities with a human in three behavior selection methods were compared, as shown in Table 4. In the behavior selection using the probability-based mechanism of thought (P-MoT) [3], the behavior similarity was 51%. In the behavior selection methods using the DoC-MoT, the behavior similarities were 82% without the learning process and 89% with the learning process. The results show that the generated behaviors by the proposed method were the most similar to human behaviors.

5   Conclusions and Future Works

The behavior selection method of humanoid robots to perform complex tasks was proposed using the DoC-MoT. The four input (context) symbols and seven target (atom behavior) symbols were defined to perform five complex tasks. The degree of consideration (DoC) for each input symbol was represented by the λ-fuzzy measure and the knowledge link strengths between input and target symbols were represented by the partial evaluation values. Each target symbol was globally evaluated by the fuzzy integral of partial evaluation values with respect to the fuzzy measure values. Then, one target symbol with the highest evaluation value was selected to be activated. To make corrections to the robot's wrong behaviors, a learning process from human behaviors was performed to update the DoCs and the knowledge link strengths. To show the effectiveness of the proposed method, the behavior similarities with a human were compared in the three behavior selection methods. The results showed that the generated behaviors by the proposed method were most similar to human behaviors. The future work focuses on applying the proposed method to a real humanoid robot to perform complex tasks, e.g. finding an object in a drawer, toasting a bread, etc. The robot will select behaviors based on the past experiences, including the success/fail of the generated behaviors.

Acknowledgment. This work was supported by the Technology Innovation Program, 10045252, Development of robot task intelligence technology, funded by the Ministry of Trade, Industry & Energy (MOTIE, Korea).

References

1. Stulp, F., Beetz, M.: Combining Declarative, Procedural, and Predictive Knowledge to Generate, Execute, and Optimize Robot Plans. Robotics and Autonomous Systems 56, 967–979 (2008)
2. Ko, W.-R., Kim, J.-H.: Organization and Selection Methods of Composite Behaviors for Artificial Creatures Using the Degree of Consideration-based Mechanism of Thought. In: Proc. International Conference on Robot Intelligence Technology and Applications (RiTA), Denver, USA (2013)
3. Kim, J.-H., Cho, S.-H.: Two-layered Confabulation Architecture for an Artificial Creatures' behavior selection. IEEE SMC-C 38, 834–840 (2008)
4. Kim, J.-H., Ko, W.-R., Han, J.-H., Zaheer, S.A.: The Degree of Consideration-based Mechanism of Thought and Its Application to Artificial Creatures for Behavior Selection. IEEE Computational Intelligence Magazine 7, 49–63 (2012)
5. Gilovich, T., et al.: Heuristics and Biases: The Psychology of Intuitive Judgment. Cambridge Univ. Press (2002)
6. Sugeno, M.: Theory of Fuzzy Integrals and Its Applications. Ph.D. dissertation, Tokyo Institute of Technololy (1974)
7. Sugeno, M.: Fuzzy Measures and Fuzzy Integrals - A survey. In: Fuzzy Automata and Decision Processes (1977)

A Research Platform for Flapping Wing Micro Air Vehicle Control Study

Hermanus V. Botha[1], Sanjay K. Boddhu[1], Helena B. McCurdy[1],
John C. Gallagher[1], Eric T. Matson[2], and Yongho Kim[2]

[1] Department of Computer Science & Engineering, Wright State University, Dayton, USA
{botha.2,boddhu.2,mccurdy.18,john.gallagher}@wright.edu
[2] Computer and Information Technology, Purdue University, West Lafayette, USA
{ematson,kim1681}@purdue.edu

Abstract. The split-cycle constant-period frequency modulation for flapping wing micro air vehicle control in two degrees of freedom has been proposed and its theoretical viability has been demonstrated in previous work. Further consecutive work on developing the split-cycle based physical control system has been targeted towards providing on-the-fly configurability of all the theoretically possible split-cycle wing control parameters with high fidelity on a physical Flapping Wing Micro Air Vehicle (FWMAV). Extending the physical vehicle and wing-level control modules developed previously, this paper provides the details of the FWMAV platform, that has been designed and assembled to aid other researchers interested in the design, development and analysis of high level flapping flight controllers. Additionally, besides the physical vehicle and the configurable control module, the platform provides numerous external communication access capabilities to conduct and validate various sensor fusion study for flapping flight control.

Keywords: Flapping Wing Control Testbed, Split-Cycle Control, PID Control, Gumstix Control.

1   Introduction

There exists growing interest and necessity to the study of flapping wing flight [1] [6] [5] and the development of flapping wing robots, in vein of their potential applications in various domains such as defense, surveillance and rescue and agriculture. Currently, the engineering challenges associated with realizing the flapping wing based vehicles at different sizes is still under active investigation [1] [7]. One of the revolutionary research activities in the field [1] [7] that is aimed at implementing a self-sustainable flapping wing vehicle, had provided the research community with a prototype flapping wing vehicle [7]. Via the manipulation of a single actuator in each wing, it is possible to achieve two degrees of freedom control over a flapping wing vehicle. The prototype of a flapping wing aircraft, shown in figure 1, was built using a four-bar linkage assembly.

© Springer International Publishing Switzerland 2015
J.-H. Kim et al. (eds.), *Robot Intelligence Technology and Applications 3*,
Advances in Intelligent Systems and Computing 345, DOI: 10.1007/978-3-319-16841-8_14

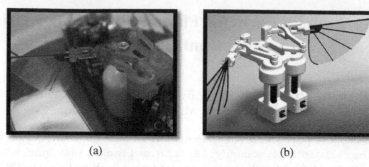

(a)   (b)

Fig. 1. Vehicle's four-bar linkage assembly. This mechanism can allow two degree control with using only one actuator per wing as described in [7]. Figure 1(a) show a mounted vehicle built using the mechanism and figure 1(b) shows the CAD simulation model of the mechanism.

A single rotation of an actuator is transformed to flapping motion of the respective wing. A control system was designed to manipulate the motion of flapping wings, such that differential upstroke and downstroke velocities can be generated over the course of a single wing beat cycle. The drag generated through the designed control system was tested to demonstrate the ability to control two degrees of freedom, namely, roll and horizontal translation of the vehicle that corroborated with previous studies mentioned in [2] [3] [6]. The designed control system provided a mechanism to configure a variety of the parameters of the wing motion during the vehicle's flight such as flapping rate and stroke active periods. However, the parameters provided limited configurability of the shape parameter of each stroke phase. The low level controllers adhere to the shape parameter via a pre-determined table containing the shape stored in the control system. This limited ability to configure the shape of the wing strokes was justifiable for the project's vision [7] in the context that it had to effectively reduce the communication overhead of the system. This resulted as an ideal deployable system for specific vehicle structure and mission requirements. However, it was demonstrated in varied research efforts [6] [5], that the configurability of the shape parameter of the wing's trajectory would provide interesting insights into the flapping wing flight mechanism. Consecutive work by authors presented in [9], improved the control system to provide configurability of all the three wing's split-cycle parameters such as flapping rate, stroke active periods, and shape. Further, in the work presented by authors in [10], an improved mechanism was designed for the four bar linkage, and that increased the ruggedness in the vehicle and its wing design to sustain longer flight times. It is crucial to assemble a platform that enables extensive research towards the understanding of fundamental flapping wing flight control both at lower and higher levels. This is now possible with the two vital components, vehicle and the control system, readily available. The study of wing aerodynamics at lower level, the control at higher level, as well as vehicle dynamics is possible using the right platform. This is important for the flapping wing flight control research community and also for robotic control researchers in general.

Similar to other research platforms, the proposed platform should provide modularity, durability, safety and intuitive accessibility, besides providing the domain specific functionality. In this vein, the second section of the paper provides a brief overview of the split-cycle frequency modulation theory, followed by which the proposed platform modules are described along with implementation and validation details. Lastly, the paper ends with a discussion section that would outline current implications of the proposed research platform for the interested research community and also provides insights into the possible future improvements.

2   Split-Cycle Frequency Modulation Theory

A split-cycle cosine model is used to generate force from the wings. The model is intended to produce a frequency wave that maintains a constant period. The complete cycle is a combination of two different waves that maintain a similar frequency. Split-cycle modulation aims to produce more force in one direction than the other. A modulation which has symmetrical cycles would result in the force generated from the wings during an upstroke negating the force generated from a downstroke.

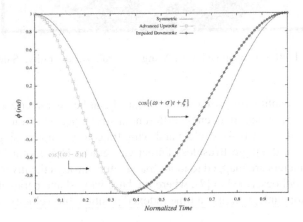

Fig. 2. A Sample Split Cycle Cosine

The wingbeat motion is dictated by the shape of a single split-cycle motion for both upstrokes and downstrokes, a sample cosine shape is shown in figure 2. The two different cosine waves that define the full split-cycle are respectively responsible for the upstroke and downstroke. For example, as shown in figure 2, to produce aerodynamic forces, a wing must advance one stroke phase velocity and impede the other stroke phase to complete the wingbeat cycle. The split-cycle cosine wave that describes an advancing wingbeat motion contains a half wave cosine (1 radian to -1 radian). The first half of the cycle finishes before the first half of the period is complete. The second section of the split-cycle cosine contains another half wave cosine that completes the cycle period.

Fig. 3. FWMAV Platform integrated with Wing Control, Accessibility, Safety and Power modules

Thus, in brief, a split-cycle wing motion control can be characterized by three parameters, namely wing beat frequency, differential stroke period or ratio (the split between upstroke and downstroke cycles) and wing trajectory shape. Using the four-bar linkage assembly with single Brushless Direct Current(BLDC) motor based actuator setup, these parameters are mapped to motor parameters through the inverse kinematics mathematical model described in [7], where the base wing shape parameter is mapped into a commutation table, which is a time indexed position and velocity of a wing.

3   FWMAV Research Platform Modules

The proposed FWMAV platform conceptually consists primarily of a vehicle, wing control module, accessibility module, safety and power module, and sensor module as shown in figure 3 and figure 8. The vehicle employed in the platform has been described earlier in the paper with reference to figure 1, interested readers can refer to [10] for more details regarding its implementation and design. The next four sub sections provide the design decisions and implementations details of the remaining four modules of the platform.

3.1   Wing Control Module

The wing-level control module is the lowest level control module, the objective of which is to produce user desired wing stroke motion. The motion must satisfy the split-cycle parameters and produce apt traversal of upstroke and downstroke period cycles with varying velocities. Further, this module should also support configurability of split-cycle parameters during the active flight motion of the vehicle.

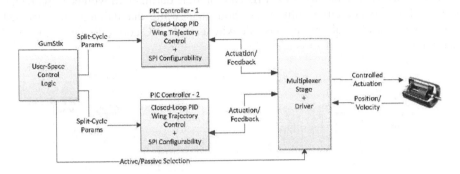

Fig. 4. Wing-level Control Multiplexing Architecture

In brief, the first challenge is solved by employing a proportional, derivative, and integral (PID) control, which has shown to effectively perform actual wing trajectory tracking. Using a relative position encoder, fitted to a BLDC motor to detect the crank position and velocity [7], the PID control tracks the actual wing trajectory against and reference trajectory which is stored in a time indexed position/velocity table (commutation table). These commutation tables define the shape of the wing motion and are designed to be configurable between the wing beats. Figure 4 provides an overview of the wing-level control actuation module for fixed split-cycle parameters, which are wingbeat rate, downstroke and upstroke periods and the commutation tables. As seen in figure 4, the encoder fitted to the motor, provides pulses that indicate relative position of the motor shaft. These encoder pulses are used as position measurements via a counter and the rate of encoder pulse transitions is used as the velocity. The measured position and velocity are used to generate deviations (cumulative errors) against the expected position and velocity values in lookup-tables as per the given timer index. Here, the timer period determines the wing beat rate, which can be reconfigured at end of a wing beat. The incremental index values of the commutation tables are determined based on the user configured stroke periods. These increments might differ from upstroke to downstroke appropriately to traverse the complete shape of the wing trajectory. Further, due to the periodic nature of the control system, wrapping would be required on the computed position and velocity errors. The generated errors are multiplied with the pre-computed gains for the given vehicle assembly and the subsequent PID signal is used to tune the duty-cycle of the pulse width modulation (PWM) generator that is used to energize the motor's coil. Based on sign of the error from PID control, the motor is commutated by choosing the appropriate motor coil, based on the current hall sensors reading. The above described closed-loop trajectory has been successfully implemented on a PIC-based microcontroller, which is equipped with PWM generation modules. The PIC software control logic uses timer interrupts and level-change interrupts to read hall sensors and encoder signals aptly within the wingbeat period. Besides the PID and interrupt-driven sensor feedback logic, the motors are driven by driver circuitry, built with H-bridges, as described in [7]. The current closed-loop trajectory implementation on the PIC controller is tightly coupled due to the time critical nature of the PID control logic during high wingbeat frequency control. Thus, to configure split-cycle parameters while maintaining uninterrupted closed loop control would not be possible with a single PIC controlling each motor of each wing.

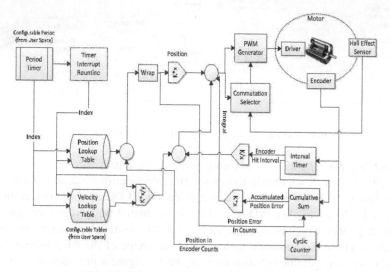

Fig. 5. Motor-level Feedback Control Flow

Therefore, a control multiplexing architecture has been proposed, as shown in figure 5, where each motor is equipped with two PIC controllers with similar control logic as described earlier along with a controllable multiplexing stage connecting to the motor. Besides the control logic, the PIC controllers use Serial Peripheral Interface (SPI) logic to receive split-cycle parameters commanded from user space. Given the time-critical nature of the closed-loop control, only one PIC is actively controlling a motor, while the other PIC is in a passive mode and receptive for user space configurability. The above described Wing-level and User-level modules have been implemented on respective platforms, towards an aim of creating a stand-alone control module on a compact single Printed Circuit Board (PCB), as shown in figure 6.

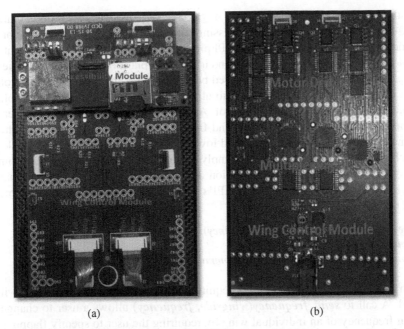

Fig. 6. Wing-Level Control Module of the FMAV Platform. (a) is the top view, that also holds the Gumstix for accessibility module and (b) is the bottom view, that shows the multiplexing PICs and motor drivers.

3.2   Control Experimentation Accessibility Module

A user has access to a variety of tools to allow for research and experimentation. The higher level abstraction of the vehicle platform includes the ability to interact with the lower level control system without changing the fundamental functionality. This is made available by allowing the higher level systems to communicate with the lower level control systems via SPI. The messages that the higher level transmits to the lower level are parameterized to force the lower control level to adjust either the frequency of flapping and/or the flapping shape parameters. Effectively, the higher level is responsible for the adjustment of commutation tables which reside in the lower levels as well as the speed at which flaps occur. The higher level control preferred in this design resides within the capabilities provided by a computer on board module namely Gumstix Overo® COM[8]. The module provides the ability to run a linux distribution containing any needed packages that a user might require such as python, gcc, and any package that can cross-compile in the ARM architecture. The module controls two SPI busses which allow for communication to both left and right wing lower level controllers individually. A user is therefore provided with the capability to compute and command the lower level controllers with new wing shape parameters and flapping frequencies. The Gumstix Overo®[8] module conveniently provides wireless access, allowing a user to communicate with the vehicle control platform wirelessly. Combining the system with batteries produces a system which is completely wireless and untethered.

Another optional approach that a user might take is to disregard the built in Gumstix Overo® support and use another processing platform. The control platform for this implementation provides the user with SPI pin ribbon connectors. Any system that supports SPI can interface with the vehicle and it's control platform since a small variety of serial text based parameters are transferred between higher and lower level controllers. As part of designing a system for users to utilize in machine learning and control, this implementation provides a small set of API's necessary to apply newly learned commutation tables, delta parameters, and frequencies to vehicle controllers. The parameter transmission between higher and lower controllers is conveniently handled by an interfacing API library. A user can simply call a required transmit function from the library which will handle the transmission as well as any state maintenance that the system requires. The following are the API's that are provided that handle fundamental control transmissions:

send_frequency(int wingside, double frequency);
send_delta(int wingside, double delta);
send_commutation(int wingside, double* commutation);

Each of the above mentioned calls require a user to specify a wing which will be targeted. A call to **send_frequency(wingside, frequency)** allows a user to change the flapping frequency of an individual wing by requiring the user to specify flapping frequency parameter. This call is useful to change the rate of flapping of one wing or both wings without effecting the flapping shape of either wing. A call to **send_delta(wing-side, delta)** allows a user to change the delta parameter of the split cycle flapping motion and requires a user to specify the desired delta parameter. This call will make the wing shape either more or less aggressive in the downward or upward flapping motion. This call can also equalize the flapping shape, meaning that the downward and upward flapping speed is equal. A call to **send_commutation (wingside, commutation)** allows a user to load a new wing shape parameter to the lower control system. The transition from an older to newer wing motion is handled by the transmission library, meaning that the transition is smooth and not fragmented. In this implementation the library maintains state information which is important for smooth transition during transmission as well as avoiding fragmented commutation tables for the lower level control systems. The primary focus of this module is to provide discrete (manual) and continuous (automated) control of vehicle's motion for different experiments. Thus, this module should at the very least provide a user with the ability to program the vehicle's high-level controller as well as record experimental data. Further, for the purpose of experimentation, the vehicle must remain free of unnecessary obstructions such as electrical cabling. A viable option is a wireless communication interface. The chosen Gumstix Overo® COM[8] module provides built in wifi that allows a user to remotely log into the Linux system and transfer code that would likely have been written on another Linux system. The intent of use of this module in the design is to provide accessibility and modularity working towards the aim of providing an intuitive interface for integration into a higher research framework. The implemented module on Gumstix Overo® COM [8], provides a user with the option to program the device on-the-fly which allows for ease of use and simplicity when performing and recording experiments.

3.3   Safety and Power Module

The above described control module in this current implementation requires 5V and 1A of electricity. This is enough to power the Gumstix Overo®[8] module as well as the underlying lower level electronics. The Faulhaber Series 1028 Brushless DC-Servomotors function nominally at 6V and can require between 1.5A -3A depending on torque requirements. To satisfy the power requirements 2 x 3.7V Lithium Polymer batteries are combined while producing 2100mAh. Each motor is supplied with 5V and up to 3A with a drop down converters and the control module is supplied with 5V and up to 1A with a drop down converter.

As the vehicle body and motor housings are 3D printed from plastic, there is some concern for temperature constraints [10]. Additionally, some battery packs have temperature constraints which is a combustible risk. To assist in safety of users as well as the electronics of the control platform, some precautions are implemented to the power rail of the vehicle. A safety switch is implemented which monitors temperatures across batteries as well as vehicle motor housings, using a negative temperature coefficient (NTC) thermistor. The switch will only allow the control platform to be powered when the temperature levels are safe. This means that when a battery becomes close to dangerously hot or if the motor housing becomes close to hot enough to melt, the system.

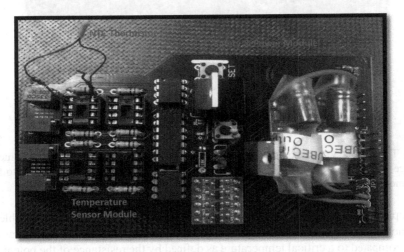

Fig. 7. A Safety and power Regulator Module of the FMAV Platform. Also shown is the NTC thermistor employed.

Fig. 8. The FWMAV is accommodated in constrained casing that can provide a structural reference frame for learning and also to integrate any external sensors like "camera" to build sensor module of the platform.

will stop functioning to avoid a fire or a melted vehicle body. By using adjustable potentiometers, a voltage divider, and an operational amplifier, the current temperature can be compared to a critical temperature as defined by the resistance of the potentiometer. Once the temperature measured surpasses the critical temperature, a logical low signal is produced, which will cause the safety to cut power to the control module. A completed implemented safety and battery pack power regulator module is shown in figure 7.

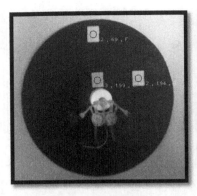

Fig. 9. Sampled Images of the Vehicle Platform with "L" shaped RGB dot structure forVehicle Position Tracking and Heading.

3.4   Position and Heading Tracking (Sensor) Module

Since the Gumstix Overo®[8] is a small portable unit with noticeable resource constraints, having multiple sensors controlled directly from the Gumstix Overo®[8] could be detrimental to the performance of the device specifically when paired with a machine learning algorithm. Therefore, using the device's wireless network capabilities, a large framework of sensors can be added to the environment. Working towards functioning closed-loop control learning capability, the framework should provide vehicle's position and heading tracking module, thus a tracking module has been designed using image features tracking mechanism described in the tutorial [12]. In this implementation blob tracking of colored dot objects using Open Source Computer Vision (OpenCV) and a simple webcam was selected. OpenCV is an open source library containing predefined API's which allow for efficient real-time computer vision. OpenCV provides a function to filter out a specific Hue Saturation Value (HSV) from the captured frames of the video feed. This allows for the tracking of colored dots to be possible. A simple webcam that can provide 24 frames per second would provide reasonable tracking, when mounted over the moving vehicle in a constrained environment as shown in figure 8.

To track the vehicle, three dots each a different color, were used for this implementation, as shown in figure 9. One for the front of the vehicle, one for the center, and the last for the right hand side, forming an "L" shape. The front dot and center dot are used to determine the orientation of the vehicle, while the addition of the dot on the right is used to determine direction of travel. Using the position of the center dot, the speed over multiple image captures can also be determined. It can be determined by using the camera's capture rate in frames per second and the change in position of the dot over a specified number of frames. The color of the dots are important as each color has its own specific HSV range that are set before running the tracking system. This value is required by built in OpenCV functions to determine which objects in a captured frame is a tracking dot. It also allows for easy identification of which dot is front, center, and right as these are also set before running the tracking system. In this implementation, each frame captured is converted from a Red-Green-Blue (RGB) color space to HSV.

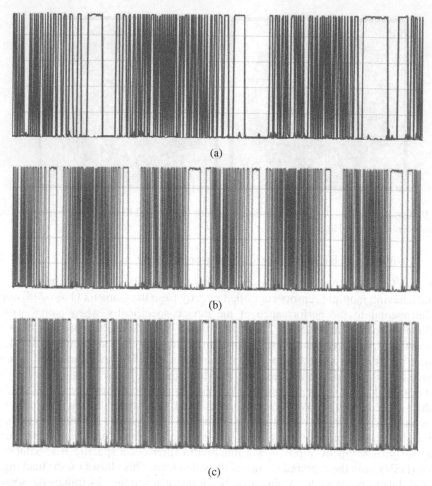

Fig. 10. Normalized sampled encoder data for flapping wing frequencies of the FMAV at (a) 2Hz, (b) 4Hz and (c) 8 Hz

The frame is then scanned to see if the desired HSV range is located in the capture and the rest of the image is filtered out. For this process each dot should be a minimum of 15 by 15 pixels in size for OpenCV to properly identify. If the dot size is smaller, OpenCV will identify it as noise or as some form of interference. OpenCV's morphological operations to dilate and erode the found area are then performed on the filtered image to clean up the lines of the object (dot) located in the image and to filter out any noise. A centralized moment, or weighted average of the filtered image pixels' intensities is then created for each object via its own identifiable HSV value. The x and y positions for that centralized moment are then stored in the dot object. All three of the dot objects found in the captured image are stored in a vector each containing an x and y value. Each dot object is also representative of front, center, and right respectively, thus providing a means to track the vehicle position and heading information.

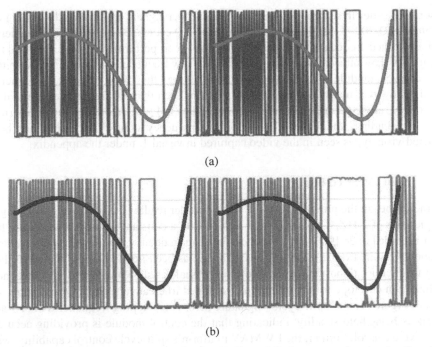

(a)

(b)

Fig. 11. Re-scaled encoder data and qualitative "split-cycle" signature estimation for flapping wing frequencies (a) 8Hz, and (b) 12Hz

4   FWMAV Research Platform Testing

Individual modules of the platform have been unit and integrated tested rigorously for durability and functionality. But, the quality of the integration test would be better determined by the accuracy of the platform to achieve the theoretical behavior of the split-cycle control and also steady and commanded flapping frequency. Thus, as described in section 3.1, the control module of the vehicle has been developed to accurately achieve the split-cycle control over various wing flapping frequency ranges. Though, the employed control technique is scalable in theory over different flapping frequency ranges, a maximum flapping frequency would be limited by the practical constraints of the underlying PIC microcontroller's processing power. Two tests were conducted below to verify if the vehicle can sustain a desired flapping frequency in the wings consistently over long periods of time, I.E. Flapping Frequency Testing and also if the desired split-cycle is accurately maintained within a given flap cycle across multiple frequencies.

4.1   Flapping Frequency Testing

This test has been performed by collecting motor's encoder readings at each wing, by dictating the flapping frequencies of 2 Hz, 4 Hz, 8 Hz and 12 Hz via the control APIs described earlier in section 3.2. Further, the encoder signals are collected by using the Hantek DSO-2090 USB oscilloscope and over 100 cycles for each flapping frequency and automated frequency period estimation has been performed as described in [11]. Figure 10, shows the encoder data collected for few frequencies and it can be seen that the control module is consistently maintaining the wing flapping frequency consistently over multiple cycles. Also, it can be seen that the frequencies are scaling accurately , as the flap periods in 8 Hz encoder data is 4 times more that the flap periods in 2 Hz. Further, the FWMAV platform's frequency control capability has been verified visually, as seen in the video captured in visual 1, under the appendix.

4.2   Split Cycle Testing

As mentioned in the previous test, the same encoder readings at each wing, flapping at frequencies of 2 HZ, 4 HZ, 8HZ and 12 HZ with a constant split-cycle signature has been recorded over 100 cycles for each flapping frequency. But, unlike performing a frequency estimation, a re-scaling mechanism has been applied across the three flapping frequencies to normalize them to a single scale as shown in figure 11. Once, normalized, an envelope trajectory path is estimated to create a split-cycle signature. As seen in Figure 11, the "split-cycle" signature across the flapping frequencies of 8Hz and 12 Hz is being held steadily, indicating that the control module is providing accurate split-cycle control. Further, the FWMAV platform's split-cycle control capability, with different wing motion speeds in downstroke and upstroke, has been verified visually, as seen in the slow-motion video captured in visual 2, under the appendix.

Additionally, completely FWMAV platform has been tested for remote control accessibility, as shown in the video captured in visual 3, under the appendix.

5   Conclusion

As part of an ongoing effort towards the research of flapping wing micro air vehicles, in this paper we presented a research platform which is modular in design and provides the capability to test, develop and validate flight controllers based on split-cycle flapping wing theory. The platform aims to provide researchers with the ability to perform on the fly machine learning experiments targeted towards FWMAV flight controllers while maintaining easy access to sensor data and untethered communication.

As stated earlier, the presented research platform is the result of an effort to demonstrate the theoretical application of split-cycle constant-period frequency modulation in FWMAVs. The intended modular design allows researchers to make use of a variety of high level tools to develop and test higher level control and learning algorithms. The platform includes the hardware and software needed for researchers to continue experimentation and development of high level controllers for FWMAVs while also serving as a platform to test and learn more towards lower level wing control and aerodynamics. Further, the platform provides structural framework and computational resources to integrate a multitude of external sensors with real time data capture and processing to perform online closed loop machine learning experiments. Currently, the designed platform is being employed to validate theoretical controllers and perform high level control learning experiments. Further, the current platform focuses primarily on FWMAV's that have wings with 2 degrees of freedom. In future implementations and alternate adaptations of this platform, researchers might require accessibility to control 3 degrees of freedom in vehicle wings. This adaptation can be possible on the proposed platform, as the primary control board and the microprocessors that exist in this implementation still have some open pin access and features that can be employed to add the third degree of freedom control.

Acknowledgements. This material is based upon work supported by the National Science Foundation under Grant Numbers CNS-1239196, CNS-1239171, and CNS-1239229.

References

1. Doman, D.B., Oppenheimer, M.W., Sigthorsson, D.O.: Wingbeat Shape Modulation for Flapping-Wing Micro-Air Vehicle Control during Hover. Journal of Guidance, Control and Dynamics 33(3), 724–739 (2010)
2. Anderson, M.L.: Design and Control of Flapping Wing Micro Air Vehicles, Ph.D. thesis, Air Force Institute of Technology (2011)
3. Finio, B.M., Shang, J.K., Wood, R.J.: Body Torque Modulation for a Microrobotic Fly. In: 2010 IEEE International Conference on Robotics and Automation (May 2010)
4. Oppenheimer, M.W., Doman, D.B., Sigthorsson, D.O.: Dynamics and Control of a Biomimetic Vehicle Using Biased Wingbeat Forcing Functions. Journal of Guidance, Control and Dynamics 34(1), 204–217 (2010)
5. Gallagher, J.C., Oppenheimer, M.: An improved evolvable oscillator for all flight mode control of an insect-scale flapping-wing micro air vehicle. In: Proceedings of the 2010 Congress on Evolutionary Computation. IEEE Press (2011)
6. Boddhu, S.K., Gallagher, J.C.: Evolving neuromorphic flight control for a flapping-wing mechanical insect. International Journal of Intelligent Computing and Cybernetics 3(1), 94–116 (2010)
7. Oppenheimer, M.W., Sigthorsson, D.O., Weintraub, I.E., Doman, D.B., Perseghetti, B.: Wing Velocity Control System for Testing Body Motion Control Methods for Flapping Wing MAVs. In: 51st AIAA Aerospace Sciences Meeting including the New Horizons Forum and Aerospace Exposition (2013)
8. Gumstix development tools (2013), https://www.gumstix.com/ (retrieved July 20, 2014)
9. Boddhu, S.K., Botha, H.V., Perseghetti, B.M., Gallagher, J.C.: Improved Control System for Analyzing and Validating Motion Controllers for Flapping Wing Vehicles. In: Kim, J.-H., Matson, E., Myung, H., Xu, P. (eds.) Robot Intelligence Technology and Applications 2. AISC, vol. 274, pp. 557–568. Springer, Heidelberg (2014)
10. Perseghetti, B.M., Roll, J.A., Gallagher, J.C.: Design Constraints of a Minimally Actuated Four Bar Linkage Flapping-Wing Micro Air Vehicle. In: Kim, J.-H., Matson, E., Myung, H., Xu, P. (eds.) Robot Intelligence Technology and Applications 2. AISC, vol. 274, pp. 545–556. Springer, Heidelberg (2014)
11. Boddhu, S.K., Gallagher, J.C., Vigraham, S.A.: A commercial off-the-shelf implementation of an analog neural computer. International Journal on Artificial Intelligence Tools 17(02), 241–258 (2008)
12. OpenCV Tutorial: Multiple Object Tracking in Real Time, by Kyle Hounslow (2013), https://www.youtube.com/watch?v=4KYlHgQQAts (retrieved July 20, 2014)

Appendix: Visuals of FWMAV Platform Testing

Below are the links to video visuals captured during the FWMAV platform testing:

Visual 1: http://www.youtube.com/watch?v=gwr7zXrkbWs_

Visual 2: http://www.youtube.com/watch?v=j8Y_xJdSmy8

Visual 3: http://www.youtube.com/watch?v=-ftJOrw_-gs

Multi-Behaviour Robot Control using Genetic Network Programming with Fuzzy Reinforcement Learning

W. Wang[1], N.H. Reyes[1], A.L.C. Barczak[1], T. Susnjak[1], and Peter Sincak[2]

[1] Massey University, Albany, New Zealand
N.H.Reyes@massey.ac.nz
[2] Technical University of Košice, Slovakia
peter.sincak@tuke.sk

Abstract. This research explores a new hybrid evolutionary learning methodology for multi-behaviour robot control. The new approach is an extension of the Fuzzy Genetic Network Programming algorithm with Reinforcement learning presented in [1]. The new learning system allows for the utilisation of any pre-trained intelligent systems as processing nodes comprising the phenotypes. We envisage that compounding the GNP with more powerful processing nodes would extend its computing prowess. As proof of concept, we demonstrate that the extended evolutionary system can learn multi-behaviours for robots by testing it on the simulated Mirosot robot soccer domain to learn both target pursuit and wall avoidance behaviours simultaneously. A discussion of the development of the new evolutionary system is presented following an incremental order of complexity. The experiments show that the proposed algorithm converges to the desired multi-behaviour, and that the obtained system accuracy is better than a system that does not utilise pre-trained intelligent processing nodes.

1 Introduction

Fuzzy logic, reinforcement learning and evolutionary algorithms constitute a handful of complementary family of algorithms that found enormous successes in building cores of intelligence for robots. Firstly, fuzzy logic is a computational paradigm that provides a mathematical facility for transforming vague linguistic rules into precise real-time control systems. To mention a few, it was used effectively for smoothly navigating robots and vehicles [2,3], autofocusing cameras, colour correction for object recognition tasks [4] and even automated space docking of satellites. As the control requirements grow, however, more fuzzy systems needs to be constructed and calibrated. As demonstrated in [2,5], steering a robot for target pursuit with obstacle avoidance behaviours and automatic speed adjustment can be achieved using a cascade of four fuzzy systems with the guidance of an informed optimal search algorithm, called A*. Building fuzzy systems necessitate expert knowledge in terms of control rules, and efficient calibration techniques tailored particularly for the control problem at hand. For calibrating fuzzy systems for robot navigation, an example can be found in [6].

© Springer International Publishing Switzerland 2015   151
J.-H. Kim et al. (eds.), *Robot Intelligence Technology and Applications 3*,
Advances in Intelligent Systems and Computing 345, DOI: 10.1007/978-3-319-16841-8_15

Secondly, reinforcement learning is an inherently on-line learning paradigm that allows robots to learn how to behave well through repeated interaction with the environment, as conditioned by a reward function. In the course of its training, a robot is able to learn dynamically a policy for maximising its rewards in the long run. By combining fuzzy logic and reinforcement learning together into one hybrid system, the fuzzy rules can be determined automatically on-line using the mechanics of the RL, without any prior knowledge of the environments dynamics, while paving the way for attaining optimal behaviour.

Lastly, Genetic Network Programming is a recent advancement over Genetic Programming [7], allowing for a population of individuals (genotypes) to be represented using directed graph structures. A phenotype is comprised of three node types, namely, a start node, processing nodes and judgement nodes. GNP was introduced in [7] and was previously fused with Fuzzy Logic and RL algorithms for learning a wall-following behaviour of a Khepera robot [1]. In [1], the fuzzy logic component was used merely for implementing the fuzzy judgement nodes of the Fuzzy GNP-RL algorithm. The degree of firing of each membership function in the judgement node is used as a measure of probability for selecting the next node during node transitions. Furthermore, the processing nodes tested in [1] did not make use of more sophisticated trained intelligent systems. Studies in the literature also indicate that a GNP-RL integration was investigated, but with findings showing less efficient exploration ability.

This work is an extension of the work presented in [1], and the main challenge addressed is that of building a hybrid evolutionary learning methodology for generating intelligence with multi-objective behaviours that allows for the utilisation of any pre-trained intelligent sub-systems as processing nodes inside a phenotype. We envisage that compounding the GNP with more powerful processing nodes would extend its computing prowess. As proof of concept, we demonstrate that the extended evolutionary system can learn multi-behaviours by testing it on the robot soccer domain to learn both target pursuit and wall avoidance simultaneously during training.

2   Methods

A schematic diagram of the new hybrid evolutionary learning algorithm is depicted in Fig.1.

2.1   Modified GNP Individual (Phenotype)

The first major change made to the original Fuzzy GNP-RL[1] is that the new proposed architecture can now accommodate any pre-trained complete intelligent systems as processing nodes within a GNP individual. A sample GNP individual is given in Fig.2, detailing its composition: start node, simple action-generator processing nodes, intelligent system processing nodes and judgement nodes. In the experiments, as a proof of concept, we have set some of the processing nodes to be a Fuzzy-RL system that was pre-trained to learn the ball pursuit behaviour. This intelligent processing node feeds on an input coming from the environment (i.e. angle of the robot from the ball) and outputs a precise steering angle for controlling the robot. The rest of the processing nodes were set to be simple action-generator nodes. During the GNP evolutionary learning phase, the Fuzzy-RL processing node is set to run only the Greedy policy for action selection. The rationale behind this approach is that if you have previously developed a well-trained intelligent system that can efficiently implement a particular behaviour, it can be readily integrated into the hybrid architecture as a processing node within a GNP individual without requiring any further training for that particular intelligent processing node. It is the role of the evolutionary mechanisms of the GNP to alter the connections between the nodes, and allow for mutations. On the other hand, during the testing phase, the Fuzzy-RL processing node employs only the e-greedy policy for action selection; thus, allowing for immediate adaptation to a dynamically changing environment.

The design of the composition of the processing nodes was suited to the problem domain. As can be viewed from an example in Fig.3, six different processing node types were defined corresponding to six different behaviours. The first four processing nodes are all related to wall avoidance actions, one per wall. On the other hand, the fifth processing node is for adjusting the speed accordingly, having the ball distance as input. Finally, the last processing node type corresponds to the the Fuzzy-RL algorithm trained for ball pursuit.

There are two types of judgement nodes used within the GNP individual (see the hexagonal nodes in Fig.3). The first one is used for judging the angle of the robot in relation to the ball, while the second type is used for finding which of the four walls is closest to the robot. An execution time of zero is set for the judgement nodes as they do not execute any robot actions - based on the decision made, the next node can be executed soon after the judgement node finishes. The problem specific settings can be viewed from Fig.3.

2.2   Node Execution Time Component

The second major change done to the system is the new implementation of the execution time (referred to as "'time delay"' in the original GNP [8]) of each node comprising a GNP individual. To illustrate how the execution time of nodes works, referring to the example in Fig.2, according to the time allotment, the Fuzzy-RL node will execute 5 complete times of ball pursuit behaviour before transitioning to the next node (which is the judgement node, in the example). It is important to note that the actual nodes executed within an individual GNP relies solely on what the robot is experiencing while interacting with the environment. The decisions selected by the judgement nodes dictate the flow of control. As for evaluating the performance of a GNP individual, there is a pre-defined maximum training time used.

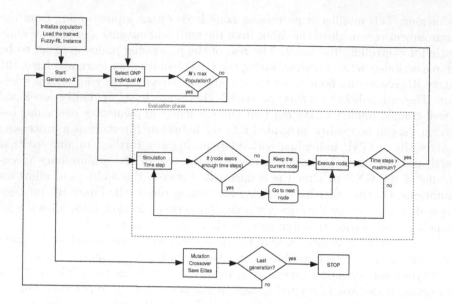

Fig. 1. Schematic diagram of the GNP with Fuzzy-RL node

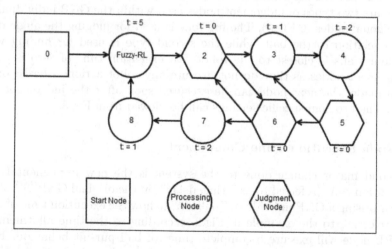

Fig. 2. A sample of the modified GNP individual used in this research

2.3   Fitness of a GNP Individual

The fitness function (Algorithm 1) was tailored specifically to the desired multi-behaviour, as it is the basis for selecting the best GNP individuals. It consists of three parts: the ball pursuit behaviour fitness, the speed control fitness and the wall avoidance fitness. Note that the parameters, (e.g. thresholds for Distance-FromBall and speed) can all be adjusted in the algorithm.

Require: Thresholds for: Speed S_1, S_2, S_3, S_4; Distance from ball D_{b1}, D_{b2}, D_{b3};
       Fitness adjustment $F_{a1}, F_{a2}, F_{a3}, F_{a4}, F_{a5}, F_{a6}$; Distance from Wall D_{w1}, D_{w2};
1: individual starts
2: $Fitness = 0$
3: **repeat** {each time step}
4:   **if** $DistanceFromWall < D_{w1}$ **then**
5:     **if** $AngleFromWall <= -90$ or $AngleFromWall > 90$ **then**
6:       $Fitness+ = (F_{a1} + BallPursuitReward)$
7:     **else**
8:       $Fitness+ = BallPursuitReward$
9:     **end if**
10:   **end if**
11:   **if** $(DistanceFromBall < D_{b1}$ and $speed == S_1)$ or
      $(D_{b1} <= DistanceFromBall < D_{b2}$ and $speed == S_2)$ or
      $(D_{b2} <= DistanceFromBall < D_{b3}$ and $speed == S_3)$ or
      $(DistanceFromBall >= D_{b3}$ and $speed == S_4)$ **then**
12:     $Fitness+ = F_{a2}$
13:   **end if**
14:   **if** $DistanceFromWall < D_{w2}$ **then**
15:     $Fitness- = F_{a3}$
16:   **end if**
17:   **if** Robot moves away from the ball **then**
18:     $Fitness- = F_{a4}$
19:   **end if**
20: **until** time steps exceeded the maximum training time for one individual
21: $Fitness+ = F_{a5} - F_{a6} * DistanceFromBall$
22: individual training ends

Algorithm 1. Fitness Function. The following thresholds were found empirically: $S_1{=}0.5, S_2{=}1.0, S_3{=}1.5, S_4{=}2.0, D_{b1}{=}10, D_{b2}{=}20, D_{b3}{=}50, F_{a1}{=}20,$ $F_{a2}{=}6, F_{a3}{=}50, F_{a4}{=}15, F_{a5}{=}500, F_{a6}{=}10, D_{w1}{=}20, D_{w2}{=}15.$

Lastly, a hill climbing algorithm is employed in tandem with the evolutionary learning of GNP (see line 15 of Algorithm 2). Its job is to explore the different ways of connecting the nodes using a greedy approach, to further improve an individual's fitness. As this greedy search is very time-consuming, it is employed only to a few top GNP individuals, after every 10 generations, or so. This can also be employed when the top fitness of the population is no longer improving after a number of generations.

2.4 Experiment Results and Discussion

The experiments were carried out using our own simulation of the robot soccer platform, using simplified physics, which includes both collision and friction. Utilising the proposed fitness function (Algorithm 1), and the following GNP parameters: population size: 200, number of mutations: 77, number of crossovers: 120, tournament size: 5, probability of mutation: 0.1 and probability of crossover: 0.5.

```
 1: Load trained Fuzzy-RL instance into the GNP individual
 2: Initialise the population
 3: repeat {for each generation}
 4:    repeat {for each individual}
 5:       repeat {for each time step}
 6:          Execute current node
 7:          Update environment
 8:          Update fitness of individual
 9:          if there is enough time steps for this individual then
10:             go to next node
11:          end if
12:       until time steps exceeded the max value of individual
13:       Calculate final fitness of individual
14:    until all individuals have been evaluated
15:    Apply hill-climbing algorithm for elites {e.g., top 3 individuals}
16:    Keep elites, select more individuals using tournament selection
17:    Apply Mutation Operation
18:    Apply Crossover Operation
19: until maximum generation is reached
```

Algorithm 2. GNP with Trained Fuzzy-RL

The hybrid evolutionary algorithm was run for at least 50 generations onwards, per experiment.

Our first initial training results did not readily give us a good solution. Even after 200 generations, the best individual ran in circles. This compelled us to increase the number of instances for each type of nodes within a GNP individual, to allow the GNP to generate more variations in the phenotypes.

Consequently, after many experiments, it was observed that identifying the final best GNP individual necessitates picking at least the top 5 elites and evaluating them visually. The best performing individual that completely succeeded in meeting the desired behaviour; that is, ball pursuit with wall avoidance behaviour and automatic speed adjustment, garnered a fitness of at least 4,000, using Algorithm 1. Interestingly, as depicted in Fig. 4, the best trained robot never makes a mistake of colliding against any of the four walls, while following the ball. The robot is the red square (with yellow lines) and the ball is the little red circle.The dotted red line indicates the path of the ball, while the dotted white lines indicate the path of the robot. Although it is difficult to see in the static pictures, the robot smoothly follows the ball and adjusts speed very rapidly. The space between the dots allows for tracing the variations in speed. In general, the robot behaves in such a way that when the ball gets too close to one of the walls, the robot avoids it; nevertheless, it kept trying to hit the ball, in a somewhat Brownian movement in the region where the ball is. A recorded video of the experiment testing how well the trained robot behaves can be viewed from the following link: http://youtu.be/woqMnbO-CKg

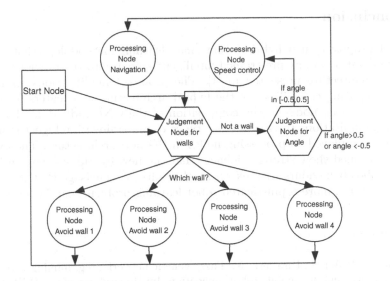

Fig. 3. Sample GNP individual with minimum number of nodes

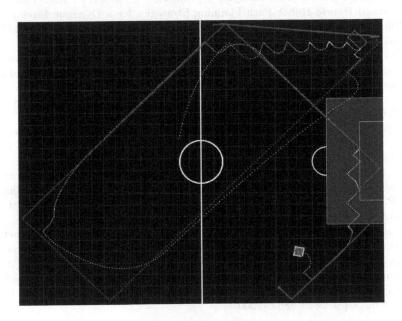

Fig. 4. An example of the general performance of a good individual

3   Conclusion

This work proposes a new hybrid evolutionary learning methodology that allows for the integration of any pre-trained intelligent system into the previous Fuzzy GNP-RL architecture presented in [1]. The proposed modifications strengthens the computing capabilities of the hybrid architecture as it caters for more complex processing nodes in the composition of the GNP individual. There is virtually no limitations to the different types of pre-trained intelligent systems that can be employed as processing nodes in the new architecture. The experiments performed show strong evidences that the new hybrid system can learn multiple robot behaviours with exceptional results. We envisage that using the new architecture, more sophisticated robot learning problems can be solved.

References

1. Sendari, S., Mabu, S., Hirasawa, K.: Fuzzy genetic network programming with reinforcement learning for mobile robot navigation. In: Proceedings of the IEEE International Conference on Systems, Man and Cybernetics, Anchorage, Alaska, USA, pp. 2243–2248. IEEE (2011)
2. Reyes, N.H., Barczak, A.L., Susnjak, T., Sincák, P., Vaščák, J.: Real-Time Fuzzy Logic-based Hybrid Robot Path-Planning Strategies for a Dynamic Environment. In: Efficiency and Scalability Methods for Computational Intellect, pp. 115–141. IGI Global (2013)
3. Vascák, J., Reyes, N.H.: Use and perspectives of fuzzy cognitive maps in robotics. In: Papageorgiou, E.I. (ed.) Fuzzy Cognitive Maps for Applied Sciences and Engineering. ISRL, vol. 54, pp. 253–266. Springer, Heidelberg (2014)
4. Reyes, N.H., Dadios, E.P.: Dynamic color object recognition using fuzzy logic. JACIII 8(1), 29–38 (2004)
5. Gerdelan, A., Reyes, N.: Synthesizing adaptive navigational robot behaviours using a hybrid fuzzy a* approach. In: Reusch, B. (ed.) Computational Intelligence, Theory and Applications, vol. 38, pp. 699–710. Springer, Heidelberg (2006)
6. Reyes, N.H., Barczak, A.L.C., Susnjak, T.: Tuning fuzzy-based hybrid navigation systems using calibration maps. In: Kim, J.-H., Matson, E., Myung, H., Xu, P. (eds.) Robot Intelligence Technology and Applications. AISC, vol. 208, pp. 713–722. Springer, Heidelberg (2013)
7. Hirasawa, K., Okubo, M., Katagiri, H., Hu, J., Murata, J.: Comparison between genetic network programming (gnp) and genetic programming (gp). In: Proceedings of the 2001 Congress on Evolutionary Computation, vol. 2, pp. 1276–1282 (2001)
8. Katagiri, H., Hirasama, K., Hu, J.: Genetic network programming - application to intelligent agents. In: IEEE International Conference on Systems, Man and Cybernetics, vol. 5, pp. 3829–3834 (2000)

Robust and Reliable Feature Extractor Training by Using Unsupervised Pre-training with Self-Organization Map

You-Min Lee and Jong-Hwan Kim

Department of Electrical Engineering, KAIST 291 Daehak-ro,
Yuseong-gu, Daejeon 305-701, Republic of Korea
{ymlee,johkim}@rit.kaist.ac.kr

Abstract. Recent research has shown that deep neural network is very powerful for object recognition task. However, training the deep neural network with more than two hidden layers is not easy even now because of regularization problem. To overcome such a regularization problem, some techniques like dropout and de-noising were developed. The philosophy behind de-noising is to extract more robust features from the training data. For that purpose, randomly corrupted input data are used for training an auto-encoder or Restricted Boltzmann machine (RBM). In this paper, we propose unsupervised pre-training with a Self-Organization Map (SOM) to increase robustness and reliability of feature extraction. The basic idea is that instead of random corruption, our proposed algorithm works as a feature extractor so that corrupted input maintains the main skeleton or structure of original data. As a result, our proposed algorithm can extract more robust features related to input data.

1 Introduction

In the image recognition task, deep neural network has shown good performance than shallow one, since the deep network can represent hierarchical feature of objects. However, until early 2000, there has not been an efficient training algorithm for deep architecture. The conventional neural network algorithm, the supervised error back-propagation (EBP) [4] does not suffice for training deep architecture because stochastic gradient decent (SGD) effect would be very weak at frontal hidden layer and the overall complexity of deep neural network is so high. However, after greedy layer wise pre-training was introduced, training the deep neural network with more robust and efficient could be possible [8][9]. Among many greedy layer wise methods, Auto-encoder (AE) has been a popular one because it can be implemented easily and training time is shorter than another famous algorithm, Restricted Boltzmann machine (RBM) [7]. Also, there are variants of the AE that were proved better than the conventional one. De-nosing Auto-encoder (dAE) was proposed to extract more robust features than the conventional AE [6]. The dAE uses corrupted data for auto-encoder input instead of original data; therefore, the dAE can detect more various and robust features than the conventional AE. In the dAE, the corruption method relies on randomness such that, for example, it randomly discards

© Springer International Publishing Switzerland 2015
J.-H. Kim et al. (eds.), *Robot Intelligence Technology and Applications 3*,
Advances in Intelligent Systems and Computing 345, DOI: 10.1007/978-3-319-16841-8_16

30% pixels on the image. Therefore, it is expect that the obtained corrupted input by using the dAE may lose the main skeleton or structure of the original data. Therefore, the dAE may cause neural network to learn features less related to given input data.

To deal with such a weak point, this paper proposes to use the Self-Organization Map (SOM) [3] for corrupt input data to pre-train the neural network. The SOM has been used for classifying or clustering unlabeled data in an unsupervised manner. However, in this paper, the training algorithm does not use the SOM for classifying. Instead, the SOM is used for detecting the main skeleton or structure of input data. If the greed size of the SOM is smaller than the number of data pixels, the SOM can corrupt input data without much loss of its main structure. For the dAE, if input destruction rate goes up more than 80%, input data lose most of the main information and as a result, it cannot extract good features. However, for the AE with SOM, even if the greed size is 3x3, which is only 5% of average number of input data pixels, it can detect meaningful features.

The paper is organized as follows. Section 2 describes the basic theory and related works. In Section 3, the performance of the proposed AE with the SOM is demonstrated through the comparison with the original AE and the dAE. Section 4 presents the experimental results and concluding remarks follow in Section 5.

2   Related Work

2.1   The Auto-encoder and Stacked Auto-encoder

To train the deep architecture efficiently, a generative model has been developed for many years. Two popular algorithms for the generative model are Restricted Boltzmann machine (RBM) and Auto-encoder (AE). The structure of AE is one hidden layer neural network and can be trained by error back-propagation (EBP) algorithm. As the purpose of using AE is to reconstruct input data at the output layer, the desired target output of the AE should be input data itself. Since there is one hidden layer between input and output layers, we can regard weights located between input and hidden layers as an encoder and weights located between hidden and output layers as a decoder. The EBP learning rule is applied to optimize weights in auto-encoder.

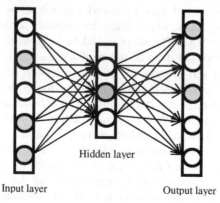

Hidden layer

Input layer   Output layer

Fig. 1. The architecture of the auto-encoder

Through the forward propagation, the network output is obtained as

$$y = f(W_2 \cdot f(W_1 \cdot x + b_1) + b_2). \tag{1}$$

where W_1 and W_2 are weight vectors located between input and hidden layers, and hidden and output layers, respectively. Also, b_1 and b_2 are bias terms. The function f is a neural activation function. Neural activation functions mainly used are sigmoid and linear rectified unit (ReLU). Since the neural network output should be input data itself, the desired target output d is the same as x. Therefore, the training rule is to minimize reconstruction error $E = \frac{1}{2}\|x - y\|^2$ and find parameters to minimize a reconstruction error and it can be represented as.

$$W^*, b^* = \arg min_{W,b} E(x, y). \tag{2}$$

It should be noted that training the AE neural network can be regarded as increasing the conditional probability of getting x at the output layer when the input data is x, which is $P(x|x)$.

Since the number of neurons of hidden layer is smaller than that of input and output neurons, the hidden layer acts as a kernel, which can extract informative features of data. However, previous research showed that even if the number of hidden nodes is larger than that of input and output nodes, network can still extract good features [1].

To include auto-encoder training rule in the deep neural network, the greedy layer wise training method can be used. The training step is as follow [6][10]:

1) Train the auto-encoder that uses original input data as its input. And then, use weights located between auto-encoder input layer and hidden layer as the first layer of the deep neural network.
2) Using the output of the first hidden node of the deep neural network as input, train the second auto-encoder. And then, use weight located between auto-encoder input layer and hidden layer as the second layer of the deep neural network.
3) Using the output of the kth hidden node of the deep neural network as input, train the $(k + 1)^{th}$ auto-encoder. And then, use weights located between auto-encoder input layer and hidden layer as the $(k + 1)^{th}$ layer of the deep neural network.
4) Repeat 3) until all the deep neural network layers are trained.
5) For classification, insert single regression or the softmax layer on the output node and do the fine-tuning.

2.2   The De-noising Auto-encoder

To extract robust features, de-noising auto-encoder (dAE) was developed [6]. The difference between the conventional AE and dAE is that the AE uses original input itself as auto-encoder input and the dAE uses corrupted version of original input as auto-encoder input. Because of corruption, the dAE tries to recover original data from distorted data and such a constraint forces the dAE to extract more robust features. There are various corruption methods. The most popular one is to use randomness;

set input pixel value to 0 at some probability. Depending on corruption rate, the features of dAE can be different. The dAE learning rule is to minimize the reconstruction error $E = \frac{1}{2}\|x - y\|^2$. The dAE output y is represented as:

$$y = f(W_2 \cdot f(W_1 \cdot \tilde{x} + b_1) + b_2) \tag{3}$$

where $\tilde{x}$, W_* and b_* are corrupted input, weight vectors and bias terms, respectively. The dAE architecture is shown in Fig. 2.

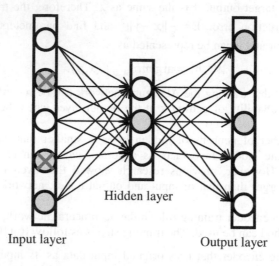

Input layer   Output layer

Fig. 2. The architecture of the De-noising auto-encoder

2.3   Self-Organization Map

The Self-organization map (SOM) is unsupervised learning based neural network, which has been used for data clustering and classification [3]. The SOM learning rule is to update weight that is closer to input x than other weights. Therefore, winner takes all. Weights are updated as follow:

$$W_{k+1} = W_k + \eta \cdot h_{ij} \cdot (x - W_k) \tag{4}$$

where η is a learning-rate and h_{ik} is a neighborhood equation, respectively. The parameter h_{ik} forces weight that is located near the winner weight to update strongly and forces weight, which is located far from the winner weight to update weakly. The parameter h_{ik} is represented as

$$h_{ij} = \exp\left(-\frac{d_{ij}^2}{2\sigma^2}\right) \tag{5}$$

where parameter d_{ij} represents the distance between two weights and the parameter σ is a standard deviation. After training, the SOM network learns the distribution of input data; such a property can be used to corrupt input data without much loss of the important information.

3   Models

3.1   Architecture

The Self-organization map (SOM) should be included between original input and the auto-encoder (AE) input layers to corrupt input data. The AE with SOM architecture is shown in Fig. 3. The sigma and eta value used in the SOM was 200 and 0.1, respectively.

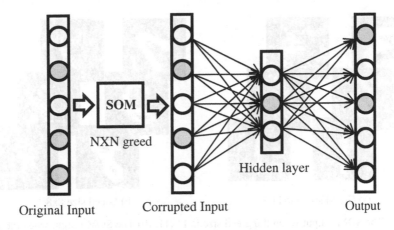

Fig. 3. The AE with SOM architecture

By adjusting the SOM greed size, we can change the corruption rate. If the greed size is small, the corruption rate is big. On the other hand, if the greed size is big, the corruption rate is small. Before corrupt the input data, the SOM constructs a self-organization map. Here are some examples. The input grey image is a 28x28 size digit from the Mixed National Institute of Standards and Technology (MNIST).

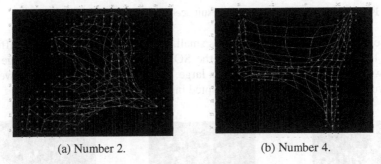

(a) Number 2.   (b) Number 4.

Fig. 4. (a) shows the SOM result when input data is number 2 and (b) shows the SOM result when input data is 4

In Fig. 4, red points represent pixel location of the original image and white map shows self-organized map after training the SOM. As shown in Fig. 4, the SOM greed is well distributed on the original input data pixels. However, some points in the greed are out of the input data region. Therefore, it is required to remove those pixels. The removing is implemented as follow: after training the SOM, the white pixels located out of the red pixel region are moved to the nearest red pixel point to reconstruct the valid input format for the AE.

3.2   Input Corruption Using the SOM

Before the experiments, the comparison between the dAE corrupted input and the SOM corrupted input is needed. SOM corrupted input examples are provided in this section.

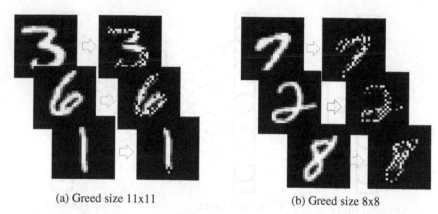

(a) Greed size 11x11   (b) Greed size 8x8

Fig. 5. (a) The SOM output when the greed size is 11x11. (b) The SOM output when the greed size is 8x8.

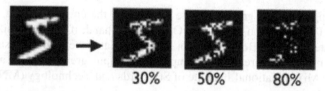

Fig. 6. The dAE corrupted input result according to different corruption rate

In case that the dAE corrupted rate is small, it is hard to distinguish the difference between the dAE corrupted input and the SOM corrupted input with a large greed size. However, if the corruption rate is a large, there is a clear difference between the dAE corrupted input and the SOM corrupted input with a small greed size.

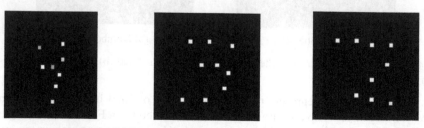

(a) The SOM corrupted input when the greed size is 3x3.

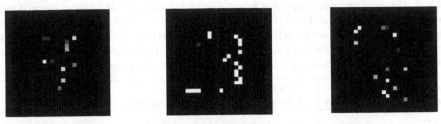

(b) The dAE corrupted input when the corruption rate is 90%.

Fig. 7. The comparison between the dAE corrupted input and the SOM corrupted input. Original input numbers are 4, 3 and 2, respectively. We can see that even though the SOM corrupted input has a few pixel points compared to the dAE corrupted input, it maintains the main skeleton of input data. For the dAE corrupted input, it is almost impossible to answer what the original number is.

4   Experiments

Experiments were conducted using a deep neural network with three hidden layer, each layer has 1000 nodes and the number of input nodes is 28x28. At the output layer, 10 nodes softmax classifier was used for digits classification. To compare the accuracy of the AE, dAE and AE with SOM, we trained weights located between input and first hidden layers with different methods. For the conventional AE, no input distortion was used to train auto-encoder. For the dAE, various distortion rates were tested for the first hidden layer and fixed distortion rate for second hidden layer and third one to 20% and 30%, respectively. For the AE with SOM, various greed sizes were tested for the first hidden layer and fixed distortion rate for second hidden layer and the third one to 20% and 30%, respectively. Because of limited time, we use only 10,000 training data. The error rate was obtained from 10,000 test data. The number of pre-training epoch was 12 for each layer and 20 epoch was used for the fine tuning. Learning rate were 0.1 and 0.3 for the pre-training and fine-tuning, respectively. At the end of experiments using 60,000 training data, we compared the accuracy between the dAE and the AE with SOM. In these experiments, parameters like the number of nodes and learning rate are chosen based on trial and error.

The programing environment was MATLAB 2013a and to increase learning speed, GPU was used. By using GPU operator, the program learning time was three times faster than that without GPU operator. For each task, elapsed time was recorded. However, different computers were used to test the dAE and the AE with SOM. For the dAE, 3.2GH CPU, GeForce 560, and 8GH RAM computer was used. For the SOM, 2.8GH CPU, GeForce 440 and 4GH RAM computer was used. Therefore, the SOM elapsed time is much longer than the dAE. The reason using two different computers is because of very long learning time.

Table 1. The error rate was checked at before fine tuning and after fine tuning. Elapsed time was checked for the AE and the dAE using 3.2GH CPU, GeForce 560 and 8G RAM computer. Elapsed time was checked for the AE with SOM using 2.8GH CPU, GeForce 440 and 4G RAM computer. At glance, the learning time of the AE with SOM looks like taking very longer than that of the dAE. However actually there is very little difference between them.

Method (10,000 training data)	Before Fine tuning	After Fine tuning	Elapsed time(sec)
AE	7.66%	3.97%	5266
dAE(10%)	4.44%	2.96%	4909
dAE(30%)	4.44%	2.73%	4878
dAE(50%)	4.44%	3.09%	4984
dAE(80%)	5.27%	3.9%	4998
AE with SOM(13X13) Distortion rate 10%	4.72%	3.15%	7800
AE with SOM(10X10) Distortion rate 35%	4.5%	3.02%	7743
AE with SOM (7X7) Distortion rate 70%	4.35%	2.93%	7723
AE with SOM (3X3) Distortion rate 95%	5.09%	4%	7800

As shown table 1, the best error rate for the dAE and the AE with SOM was better than the conventional AE. Which justifies our hypothesis; the dAE and the AE with SOM can extract more robust features than the conventional AE. For the dAE and the SOM, if the input data corruption rate was small, there was not a clear visual difference between the dAE and the SOM features obtained. However, if the input data corruption rate was high, the SOM showed much better performance than the dAE. These results imply that when corrupted data is used for an auto-encoder input, the main skeleton of input data should be maintained to extract meaningful features. When distortion rate was small, the dAE as well as the SOM corrupted input data can maintain their main structure; therefore, there was not a big difference. Even though their best error rate was similar, there were differences between features learned. When we saw the dAE features, most of them look like local features. For the SOM, however, their features look like the mixture of local filter and the conventional auto-encoder features. At this moment, we cannot interpret what do that means, but it may be worth thinking about it. In terms of learning time, the AE with SOM takes longer than that of the dAE because there is a SOM network in the input to first hidden layer. However, the elapsed time is almost same. For 60,000 training data, the SOM network tasks only about 5 minute. This is very small compared to the total elapsed time which is for 60,000 training cases, about 6 hours.

Fig. 8 shows each feature obtained at the first hidden layer with different methods.

(a) The features learned from the conventional auto-encoder.

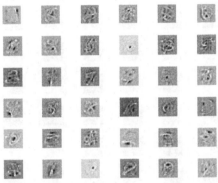

(b) The features learned from the auto-encoder with 30% distortion.

(c) The features learned from the auto-encoder with 80% distortion.

Fig. 8. Obtained features from the different methods

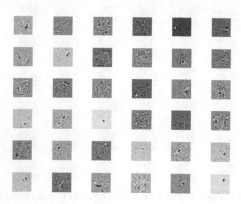

(d) The features learned from the auto-encoder with 10% distortion.

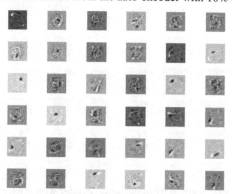

(e) The features learned from the auto-encoder with SOM greed size of 7x7.

Fig. 8. *(Continued)*

Table 2 shows the best performance achieved by training 60,000 data. Before fine tuning, the AE with SOM's error rate was 2.99% that is almost 0.5% lower than that of the dAE with any distortion rate. After fine tuning, the dAE error rate was 1.43% that is the best of all experiments in this paper.

Table 2. The error rate test when the number of training data is 60,000. When test the dAE, distortion rate for each layer was well tuned according to other researcher's paper. The greed size and the distortion rate for each layer were randomly chosen for the AE with SOM.

Method (60,000 training data)	Before Fine tuning	After Fine tuning	Elapsed time(sec)
dAE(**10%**)	3.54%	**1.43%**	22,600
AE with SOM(7X7) Distortion rate **70%**	**2.99%**	1.61%	23,000

5   Conclusion

To train the deep neural network, we need good representations for input data. If we can insert good representations about the world in the deep neural network, we may not have to worry about pre-training the network and fine-tuning procedure. To obtain robust features that represent the world more efficiently, a generative model was proposed. Also, to reduce the regularization problem, some techniques such as the de-noising auto-encoder and the dropout were proposed. The de-noising auto-encoder shows how we can get good representations related to the environment data. However, the random method may not guarantee reliable performance. The AE with SOM concept was started with the thinking "If the data's main skeleton is maintained, human can recover original data even if the original data was largely corrupted." With this in mind, this paper tried to verify that 1) The AE with SOM shows reliable performance than the dAE in terms of error rate standard deviation. 2) The AE with SOM's performance does not much depend on its distortion rate. For 2), the AE with SOM shows that even for high distortion, it can extract meaningful features. Therefore, we can justify that distorted auto-encoder input data should maintain the main skeleton or structure. For 1), we need more time for experiments. This remains a topic for future research.

Acknowledgment. This work was supported by the Technology Innovation Program, 10045252, Development of robot task intelligence technology that can perform task more than 80% in inexperience situation through autonomous knowledge acquisition and adaptational knowledge application, funded by the Ministry of Trade, industry & Energy (MOTIE, Korea).

This work was also supported by the Software Computing Technology Development Program, 14-824-09-012, Technology Development of Virtual Creatures with Digital Emotional DNA of Users, funded by the Ministry of Science, ICT and Future Planning.

References

1. Bengio, Y.: Learning deep architectures for AI. Foundations and Trends® in Machine Learning 2(1), 1–127 (2009)
2. Aarts, E.H., Jan, H.M.: Boltzmann machines and their applications. In: Treleaven, P.C., Nijman, A.J., de Bakker, J.W. (eds.) PARLE 1987. LNCS, vol. 258, pp. 34–50. Springer, Heidelberg (1987)
3. Kohonen, T.: The self-organizing map. Proceedings of the IEEE 78(9), 1464–1480 (1990)
4. Rumelhart, D.E., Hinton, G.E., Williams, R.J.: Learning representations by back-propagating errors. Cognitive Modeling (1988)
5. Rojas, R.: Unsupervised Learning and Clustering Algorithms. In: Neural Networks, pp. 99–121. Springer, Heidelberg (1996)

6. Vincent, P., et al.: Extracting and composing robust features with denoising autoencoders. In: Proceedings of the 25th International Conference on Machine Learning. ACM (2008)
7. Hinton, G.: A practical guide to training restricted Boltzmann machines. Momentum 9(1), 926 (2010)
8. Le, Q.V.: Building high-level features using large scale unsupervised learning. In: 2013 IEEE International Conference on Acoustics, Speech and Signal Processing (ICASSP). IEEE (2013)
9. Lee, H., et al.: Convolutional deep belief networks for scalable unsupervised learning of hierarchical representations. In: Proceedings of the 26th Annual International Conference on Machine Learning. ACM (2009)
10. Hinton, G.E.: Deep belief networks. Scholarpedia 4(5), 5947 (2009)

Self-organization in Groups of Intelligent Robots[*]

Anatoliy Gaiduk[1], Sergey Kapustyan[1], and Igor Shapovalov[2]

[1] Scientific Research Institute of Multiprocessor Computer Systems
of Southern Federal University, Taganrog, Russia
gaiduk_2003@mail.ru, kap56@mail.ru
[2] Department of Automatic Control Systems
of Southern Federal University, Taganrog, Russia
shapovalovio@gmail.com

Abstract. The designation principles and the creation methods of conditions, which provide an opportunity of the self-organization processes in groups of the intelligent robots, are considered. The local rules of the self-organization in these groups are offered. Efficiency of proposed approach is shown on the example of moving a body on a horizontal surface. Results of computer simulation are given.

Keywords: intelligent robots, groups of the intelligent robots, self-organization, local rules of self-organization.

1   Introduction

Now the groups of the non-crew robots are applied very often because the possibilities of the one robot, even the intelligent robot, are not sufficient for decision of many practical tasks.

In general groups of intelligent robots contain robots of different types. It can be transport or communication robots, robots for chemical, radiation or topographical reconnaissance, and so on [1-5]. Naturally, such group has complex purposes. Achievement of these purposes demands participation of several types of robots or polytypic robots. Complexity of a situation requires new ways, methods and means, which will provide achievement of the purposes independently, i.e. without participation of the person [2]. Therefore the group of intelligent robots must be able to estimate purposes to find conditions of their achievement to analyze the situation and to create the plan of actions. Hence, the group of robots should be capable to self-organization [5, 6]. So the aim of this work is development of principles and methods which provide conditions for self-organization in a group of intelligent robots. The result of this self-organization is the group structure and the algorithm of actions for the group or its part (cluster), focused on achievement of the group purpose.

A group of intelligent robots is a set of robots which operate independently (without person or with his minimum participation) in any environment. These robots are capable to collect information about environment, they can react on changes of the

---

* This work is supported by RFBR, projects number 13-08-00249 and № 13-08-00794.

environment and collaborate with each other for common target problem solution. The collaboration of the robots is interaction of the robots with each other (by means at physical fields, electric signals, informational channels, etc.) [2].

2   Self-organization in Technical Systems

Let us briefly consider basic features of self-organization processes. Originally, the self-organization phenomenon was observed in the natural and experimental physico-chemical systems: Turing structures, Belousov-Zhabotinskiy reactions, Benar cells, etc [6, 7]. The certain structure of these systems arises spontaneously under influence of external disturbances. Changes of external conditions cause respective alterations of the system structure and parameters. The result of the certain combination of external conditions and system parameters is spontaneous occurrence of a new organization (structure) of this system. The arisen structure is kept in the given conditions and disappears when they are changed. This phenomenon is called self-organization.

In control systems the self-organization problem was considered by U.R. Ashby in 1959 for the first time [8]. Further it was considered in many papers, for example in [6, 9 – 12] and others. It was U.R. Ashby who defined the term "self-organization" as "process, duration which complex system organization creates, reproduces or improves itself" [5]. Let us note this definition does not contain the reasons which cause self-organization, and its purposes. Only the fact of creation or change of the system organization is fixed.

At present the definition of self-organization, made by M. Bushev, is used very often. According to this definition "self-organization is the process, in which global external influences stimulate the start of internal mechanisms system, which in turn produce the specific structures in the system" [11] (fig. 1).

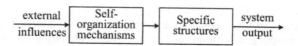

Fig. 1. Self-organization by Bushev

Apparently, in the Bushev definition, unlike the Ashby definition, causes of the self-organization process are emphasized. However both of these definitions are partial, because they don't contain the main component of the self-organization process. It's the self-organization aim, which is inherent to any process both in natural and in technical systems.

In our opinion, the P.K. Anokhin definition of self-organization, presented schematically in fig. 2, is more correct. Analyzing biological systems, P.K. Anokhin supposed that the "desired result" is "the compulsory and decisive component of the system; it is the tool to create the necessary interaction between all components of the system". "The self-organization mechanism consist in exception of the redundant degrees of freedom of the system components which are not necessary for a certain result and, vise versa, in preservation of those degrees of freedom, which promote creation of the result" [6, p. 30]. As we can see, hear we have same point which is formation of the "specific" or "organized" structure as a result of the ordered interaction of the system components.

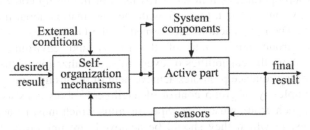

Fig. 2. Self-organization by Anokhin

But, how the "necessary" degrees of freedom are chosen, how the "specific structures" are formed; what is "the self-organization mechanisms"? Answers to these questions are absent in the considered definitions. Description of "the self-organization mechanisms" is absent and in the other definitions of the self-organization [7, 9 – 11].

Conceptions of the "natural" and "artificial" self-organization were introduced in the monograph [6], perhaps, in the first time. In opinion of the authors of this book the process of natural self-organization which takes place in natural systems is formed by laws of nature. The artificial self-organization proceeds in the technical systems according to the local self-organization rules, which are developed by the system designer.

To be more specific let us accept the following definition: self-organization in the technical system a process of independent formation of the optimal structure and optimal algorithm of interaction of the system elements according to the system purpose, some criterion and the external conditions. The self-organization process corresponding to the given definition is similar to the transient processes in the control systems (CS). The distinctive feature is new system structure which is the result of the self-organization process.

In relation to same group of robots the essence of this definition is the following. The results of the self-organization process should be: a new structure of the group, a structure of clusters, which focused on achievement of same specific purposes and the algorithms of robot actions to achieve of these purposes.

So, according to all abovementioned, for initiation of self-organization in the group of intelligent robots oriented on achievement of same purpose requires local self-organization rules.

3   Problem Statement

Let us assume, the group includes intelligent robots $R_1 \div R_N$ and a group control system (GCS) with processes of self-organization and control of robot actions [1, 2]. In general case the GCS can arise (as a result of self-organization also), in the control subsystem of the group leader [1] or by integration of several robots' control subsystems. In the first case it will be a centralized system, in the second case – a distributed control system [1].

The robot group operates as follows. The robots of the group collect the information about the environment and, together with the information about their own state, send in the GCS. The robots actions are formed by the GCS on basis of the information about the group aims, state of the robots and environment, and also the information about the capabilities of every robot. Realization of these actions by the robots in certain sequence promotes achievement of the group aims.

The main complexity of group control of intelligent robots consists in variety of problems which such group must solve. Specific aims which must be achieved by the group, conditions, in which they should be achieved, are unknown. Therefore any certain algorithms cannot be defined beforehand. These algorithms must be created by the group itself or by its leader according to the present situation.

Obviously, each group of robots is oriented on the achievement of some certain range of aims. Hence, the range of algorithms which define necessary actions, generally speaking, is limited. It gives the basis to suppose that the considered problem can be solved. Therefore self-organization in the robots' group should be focused, firstly, on formation of the clusters which are oriented on achievement of the certain aims [1], and secondly, on the formation of the algorithms of the robots' actions of the cluster which will provide achievement of the corresponding aims. Certainly, one cluster can include all robots of the group in some cases.

To make it more clear, we suppose in this paper, that the robot group has one purpose. Self-organization of a cluster in the robot group can be described as follows. Each robot R_i, $i = \overline{1,N}$ can perform a certain set of actions $\mathbf{A}_i = \{A_{1i}, A_{2i}, \ldots, A_{m_i i}\}$ [1, 2]. In the other side, some aim $\mathbf{T}_\mu$ can be achieved in the conditions $\mathbf{f}_\mu^\circ = \{f_{1\mu}^\circ, f_{2\mu}^\circ, \ldots\}$ as a result of performing some certain set of actions $\mathbf{T}_\mu = \{T_{1\mu}, T_{2\mu}, \ldots, T_{p_\mu \mu}\}$. Every action $T_{\nu\mu}$ is described by two features: its type and intensity.

A set of requirements $\mathbf{Q}_\mu = \{J(\mu), t_{ach}(\mu), n(\mu), \ldots\}$ always accompanies the process of achievement of the purpose $\mathbf{T}_\mu$ [2, 3, 5]. Here $J(\mu)$ is the criterion of efficiency; $t_{ach}(\mu)$ is the time of achievement of this aim, $n(\mu)$ is the quantity of cluster robots, etc. If all elements $T_{\nu\mu}$, $\nu = \overline{1, p_\mu}$ are available among the elements of the set A_{ji^*} of the robot R_{i^*}, i.e. $T_{\nu\mu} \in \mathbf{A}_{i^*}$, $\nu = \overline{1, p_\mu}$ achievement of the aim $\mathbf{T}_\mu$ in "supposed" conditions $\mathbf{f}_\mu^\circ$, can obviously, provided by this only robot. Hence, the cluster $\mathbf{K}_\mu$ oriented on achievement of the aim $\mathbf{T}_\mu$ will consist from the single robot R_{i^*}.

But usually a single robot is not capable enough, and only several robots can provide achievement of the aim $\mathbf{T}_\mu$. These robots satisfy the condition:

$$T_{v\mu} \in \{A_{i_1(\mu)}, A_{i_2(\mu)} \cdots A_{i_{n(\mu)}(\mu)}\}, \quad v = \overline{1, p_\mu}, \tag{1}$$

where $n(\mu)$ is the quantity of the cluster robots. And all the p_μ actions of the set $\mathbf{T}_\mu$ are among the possible actions of these robots. It's clear that if the condition (1) isn't satisfied relative to the certain robot group $R_1 \div R_N$ then this group can't provide aim $\mathbf{T}_\mu$ achievement in the conditions $\mathbf{f}_\mu^\circ$. If every action $T_{v\mu}$ can be performed by one robot R_i, then the robots quantity in the cluster $\mathbf{K}_\mu$ oriented on the aim $\mathbf{T}_\mu$ will be equal to the number of the actions p_μ, i.e. $n(\mu) = p_\mu$. In opposite case $n(\mu) < p_\mu$.

As a rule, the same action $T_{v\mu}$ is performed by different robots R_i with different efficiency [1 – 5]. Let $q_{iv}(\mu)$ be the efficiency of the action $T_{v\mu}$ execution by a robot R_i. Obviously, in the cluster $\mathbf{K}_\mu$ oriented on the aim $\mathbf{T}_\mu$ it's reasonable to include those robots from the group $R_1 \div R_N$, which provide aim achievement with the maximum efficiency, i.e. those robots, which satisfy the condition:

$$J = \sum_{v=1}^{p_v} q_{iv}(\mu) \rightarrow \max_{i=1,N} \tag{2}$$

On the other side, the number of robots in the cluster $\mathbf{K}_\mu$, oriented on the aim $\mathbf{T}_\mu$, should be minimal by rated conditions, i.e.

$$n(\mu) \rightarrow \min. \tag{3}$$

In fact, the conditions (1) – (3) are local self-organization rules of the cluster $\mathbf{K}_\mu$ oriented to the aim $\mathbf{T}_\mu$ achievement. Fulfilment of these conditions by robots is the algorithm of the self-organization of the cluster. Creation of the algorithmic support and software of intelligent robots for realization of this algorithm does not meet any problems [1].

However realization of the algorithm at conditions (1) – (3) is possible, only if the sets $\mathbf{A}_i$ and $\mathbf{T}_\mu$ are created according to the sets $\mathbf{f}_\mu^\circ = \{f_{1\mu}^\circ, f_{2\mu}^\circ, \ldots\}$ and $\mathbf{Q}_\mu = \{J(\mu), t_{ach}(\mu), n(\mu), \ldots\}$. The sets $\mathbf{A}_i$ can be generated easily since the components A_{ji} of these sets are represented in the technical certificates of the robots. The basic difficulties of the self-organization algorithm are generation of the sets $\mathbf{T}_\mu$, $\mathbf{f}_\mu^\circ$ and $\mathbf{Q}_\mu$ caused by a priori unknown purpose for the components of these sets.

The robot group is always created to achieve a certain set Ξ of aims, therefore $\mathbf{T}_\mu \in \Xi$. The actions $T_{v\mu}$, $v = \overline{1, p_\mu}$, execution of which provides achievement of the aim $\mathbf{T}_\mu$, and the order of their execution at the conditions $\mathbf{f}_\mu^\circ$, $\mathbf{Q}_\mu$ are defined by known laws of nature. There are two possible approaches to create the sets $\mathbf{T}_\mu$, $\mathbf{f}_\mu^\circ$, $\mathbf{Q}_\mu$. The first approach, when the sets $\mathbf{T}_\mu$, $\mathbf{f}_\mu^\circ$, $\mathbf{Q}_\mu$ are created by means of the methods of the intelligent system theory by the robots themselves in real time [2]. At the second approach, when sets are created by experts as the ontological models which will be stored in the knowledge bases of the robots [1, 13]. The experts approach is used further.

Let the aim $\mathbf{T}_\mu$ be defined for the robots group. First of all, the robots of this group create the cluster $\mathbf{K}_\mu$ according to the conditions (1) - (3) for achievement of this aim. Then its robots form the action algorithm $\mathbf{L}_\mu = \mathbf{L}_\mu(t, \mathbf{T}_\mu, \mathbf{f}_\mu^\circ, \mathbf{R}_\mu)$ for achievement of this aim at the conditions which are reflected by the sets $\mathbf{f}_\mu^\circ$, $\mathbf{Q}_\mu$ created by the experts. This algorithm is a sequence of the actions $T_{\nu\mu}$ in time, i.e. $\mathbf{L}_\mu(t, \mathbf{T}_\mu, \mathbf{f}_\mu^\circ, \mathbf{R}_\mu) = \{t, T_{\nu_1(\mu)}, T_{\nu_2(\mu)}, ..., T_{\nu_\mu(\mu)}, \mathbf{f}_\mu^\circ, \mathbf{R}_\mu\}$. The real conditions $\mathbf{f}_\mu$ will most likely differ from the conditions assumed by experts, i.e. $\mathbf{f}_\mu \neq \mathbf{f}_\mu^\circ$, therefore the algorithm $\mathbf{L}_\mu(t, \mathbf{T}_\mu, \mathbf{f}_\mu^\circ, \mathbf{R}_\mu)$ should be adaptive. It should provide achievement of the aim $\mathbf{T}_\mu$ in changed conditions [5, 6, 12 and 14]. The local self-organization rules are needed to create the algorithm $\mathbf{L}_\mu(t, \mathbf{T}_\mu, \mathbf{f}_\mu^\circ, \mathbf{R}_\mu)$ by the cluster robots. But these rules can be formed just for a certain aim.

Let us assume the local self-organization rules for achievement of the single aim $\mathbf{T}_\mu$ by the robots group $R_1 \div R_N$ are created. Then the process of self-organization proceeds in this group as follows:

- on the basis of expressions (1) – (3) the robots $R_1 \div R_N$ create the cluster $\mathbf{K}_\mu = \{R_{i_1(\mu)}, R_{i_2(\mu)}, ..., R_{i_{n(\mu)}(\mu)}\}$. The quantity of the cluster robots $n(\mu)$ is defined by the quantity of the actions $T_{\nu\mu}$, which are necessary for achievement of the aim $\mathbf{T}_\mu = \{T_{1\mu}, T_{2\mu}, ..., T_{p_\mu\mu}\}$ at the conditions $\mathbf{f}_\mu^\circ$, $\mathbf{Q}_\mu$. The cluster can be created by the algorithms of the collective interaction [1];
- the robots $R_i \in \mathbf{K}_\mu$ extract the algorithm $\mathbf{L}_\mu(t, \mathbf{T}_\mu, \mathbf{f}_\mu^\circ, \mathbf{R}_\mu)$ from their knowledge bases and adapt it to the current conditions $\mathbf{f}_\mu$;
- the robots $R_i \in \mathbf{K}_\mu$ carry out the actions $A_{i_1(\mu)} = T_{\nu_1(\mu)}$, $A_{i_2(\mu)} = T_{\nu_2(\mu)}$, ..., $A_{i_q(\mu)} = T_{\nu_q(\mu)}$ according to the algorithm $\mathbf{L}_\mu(t, \mathbf{T}_\mu, \mathbf{f}_\mu^\circ, \mathbf{R}_\mu)$.

Apparently, in the suggested approach to self-organization the robots carry out all actions depending on the group aim and external conditions. The processes which take place in the group of the intelligent robots correspond to the definition of artificial self-organization.

4   Methodological Example

Let us consider the proposed approach on an example of the round body moving on a horizontal surface from an initial point to the given point when obstacles are absent [1]. This problem is a mathematical model of the load moving, when one or several driving means unite with the freight and move the freight to the given point. For example, this can be a big ship or a sea oil-extracting platform, which is moved by one or several tugboats without crew along unrestricted sea area.

Let us suppose, that the transport robot group $R_1 \div R_N$ join the body M as shown in fig. 3, where φ_i is a corner between the axis x and the draft direction of the robot R_i. The robots $R_1 \div R_N$ should move the body M from the initial point x_0, y_0 to the aim point x_a, y_a along the given trajectory $y = f(x)$; $f(x)$ is a differentiate function; $y_0 = f(x_0)$ and $y_0 = f(x_0)$. The point C with the coordinates x_c, y_c is the center of gravity of the body M.

When the robot R_i is active, its draft is $P_i = P_0 \neq 0$, and when it is passive, its draft is $P_i = 0$, $i \in [1, N]$. The resistance force of the body movement depends on the speed of the body. Complexity of the considered task is that robots join a body any way without the coordination with each other and with a forthcoming direction of the body moving. However the robots have an opportunity to determine their own position, the position of the body in some system of coordinates and they have information channels for connection with each other.

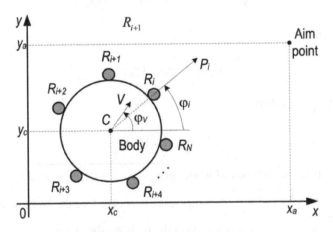

Fig. 3. Body and robots group

In the beginning the body is motionless, i.e. $\dot{x}_{c0} = 0$, $\dot{y}_{c0} = 0$, $\dot{\psi}_{c0} = 0$ and $P_0 > \sqrt{\varphi_x^2(\dot{x}_c) + \varphi_y^2(\dot{y}_c)}$, where $\varphi_x(\dot{x}_c)$, $\varphi_y(\dot{y}x_c)$ are the projections of the resistance force on the axes of coordinates.

The projection of each robot R_i draft on a tangent to the given trajectory is an efficiency measure of the robots R_i actions. This tangent to a trajectory undertakes in the point nearest to the robot R_i. From here follows, that the necessary cluster includes two active robots R_{i^*} and R_{i^*+1} between which the given trajectory passes at each moment of time. Therefore the local rule of the formation of the necessary cluster, following from (1) – (3) and look like:

1.
$$tg(\varphi_{i^*}) \leq \frac{df(x)}{dx} = f'(x) \leq tg(\varphi_{i^*+1}) \tag{4}$$

The cluster R_{i^*}, R_{i^*+1} is replaced by the new cluster, when the condition (1) is not fulfilled.

2. The robot R_{i^*} begins to move the body. The active robot of the cluster R_{i^*}, R_{i^*+1} is replaced by an other robot of this cluster when $d_c \geq d_{all}$. Here d_{all} is an allowable distance; d_c is the real distance of the body from the given trajectory. The constant d_c is defined by the following expressions:

If $\left| x_c - f^{-1}(y_c) \right| \leq l_m$ and $\left| y_c - f(x_c) \right| \leq l_m$, then

$$d_c = \frac{\left| [x_c - f^{-1}(y_c)][y_c - f(x_c)] \right|}{\sqrt{[x_c - f^{-1}(y_c)]^2 + [y_c - f(x_c)]^2}} ; \tag{5}$$

If $f^{-1}(y)$, then

$$d_c = x_c - f^{-1}(y_c) ; \tag{6}$$

If $\left| x_c - f^{-1}(y_c) \right| > l_m$, then

$$d_c = y_c - f(x_c) . \tag{7}$$

3. The robots move the body M while $\sqrt{(x_a - x_c)^2 + (y_a + y_c)^2} > l_{brake}$;

4. If $\sqrt{(x_a - x_c)^2 + (y_a + y_c)^2} \leq l_{brake}$, the robots reduce the body speed;

5. If $\sqrt{\dot{x}_c^2 + \dot{y}_c^2} \leq V_{stp}$, then robots stop the body by the brake.

In the expressions (4) – (7) $f^{-1}(y)$ is an inverse function to the functions $f(x)$; l_m is the given constant; l_{brake} is the braking distance of the body by the robots of the cluster R_{i^*}, R_{i^*+1}; V_{stp} is the speed of the body stop.

Thus, in this case the local rules of the cluster self-organization (1) – (3), are concretized by the inequality (4). The expressions (5) – (7) describe the local rules of formation of robots actions. The suggested algorithms (4) – (7) are formal and, obviously, can be realized by the group of the intelligent, non-crew robots.

Efficiency of the suggested approach to condition creation of the self-organization occurrence in the intelligent robot group was estimated by simulation in MATLAB. Movements of the body by the group of robots are described by the equations

$$m\ddot{x}_c = [P_i - F_{fr}(V_c)]\cos\varphi_i \, , \tag{8}$$

$$m\ddot{y}_c = [P_i - F_{fr}(V_c)]\sin\varphi_i \, , \tag{9}$$

$$J\ddot{\psi} = 0 \, , \tag{10}$$

where m and J are the mass and the moment of the body inertia with the attached robots; $i = i^*$ or $i = i^* + 1$; $V_c = \sqrt{(\dot{x}_c)^2 + (\dot{y}_c)^2}$ is the body speed [15].

The group includes 23 robots. The coordinates of the points: initial $x_0=1$, $y_0=7,5$; aim $x_a=10$, $y_a=3$. The given trajectory is described by the function $y = f(x) = 3 + 5x - 0,5x^2$. The result of simulation with $d_{all} = 0,1$ is shown in fig. 4 where 1 is the given trajectory, 2 is the received trajectory of the body moving.

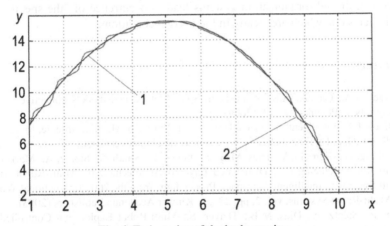

Fig. 4. Trajectories of the body moving

If the given trajectory is a direct line, the body is moved (Fig. 5) by one cluster, which was automatically formed according to the rule (4).

Certainly, the intelligent robots can create the algorithms (4) – (7) using AI-approach with the equations (8) – (10), but it is more reasonable to store these algorithms in the knowledge bases of robots, because these algorithms are very simple. The above mentioned expressions are analytical, therefore they can be realized by computer complexes of the intellectual robots with use of sensor systems.

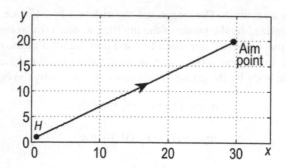

Fig. 5. Body moving by single robot

5   Conclusion

The carried out analysis gives possibility to conclude, that the processes of the artificial self-organization in technical systems proceed according to the local rules of self-organization which are created by developers of the systems. These rules are determined basically by the type of the system elements and physics processes in them. The local rules of artificial self-organization are similar to the logic control. Transients in the self-organization systems lead to occurrence of "the specific structures", i.e. clusters, which are steady in the current conditions.

References

1. Kaliaev, I.A., Gaiduk, A.R., Kapustyan, S.G.: Method and models collective control iv group robots. Phizmathgiz, M. (2009)
2. Kaliaev, I.A.: Intelligent robot / Under edit. E.I. Yurevich. Machinostroenie, M. (2007)
3. Redko, V.G. (ed.): From behavior model to artificial intellect (2006)
4. Goldberg, D., Cicirello, V., Dias, M.B., Simmons, R., Smith, S., Stentz, A.: Market-Based Multi-Robot Planning in a Distributed Layered Architecture. In: Multi-Robot Systems: From Swarms to Intelligent Automata: Proceedings from the 2003 International Workshop on Multi-Robot Systems, vol. 2, pp. 27–38. Kluwer Academic Publishers (2003)
5. Zlot, R., Stentz, A., Dias, M.B., Thayer, S.: Multi-Robot Exploration Controlled By A Market Economy. In: IEEE International Conference on Robotics and Automation (ICRA) (May 2002)
6. Prigogin, I., Kondepudi, D.: Modern thermodynamics. From heart engines to dissipative structure. John Wiley & Sons, Toronto (2000)
7. Ashby, W.R.: Introduction to cybernetics. Univ IEI. Lit., New York (1959)
8. Novikov, D.: Mathematic models of organization and operating command. Phizmathlit, M. (2008)
9. Bushev, M.: Synergetics: Chaos, Order, Self-Organization. World Scientific Publisher (1994)
10. Saridis, G.N.: Self-Organization Control Stochastic Systems. Marcel Dekker, New York (1977)
11. Fradkov, A.L.: Adaptive control in complexity systems. Nauka, M. (1990)

12. Gaiduk, A.R.: Algorithmic maintenance of self-organization controller with extrapolation. In: Proceedation RAS. Theory and Control Systems, vol. 3, pp. 56–63 (2002)
13. Kolchin, A.F., Eliseeva, N.V.: Representation of model of knowledge of the expert - designer on the basis of the ontological approach. Information Technology in Design and Production 3, 66–69 (2006)
14. Kohanen, T.: Self-Organization and Associative Memory. Springer, Berlin (1984)
15. Shapovalov, I.O., Kapustyan, S.G.: The organization and control of intellectual robots group with the purpose of moving a body. In: Robotics. A Sight in the Future: Works of the International Scientific and Technical Seminar, pp. 212–214. Publishing House "Polytechnic-Service", Saint Petersburg (2010)

Formation Task in a Group of Quadrotors[*]

Donat Ivanov, Igor Kalyaev, and Sergey Kapustyan

Scientific research institute of multiprocessor computer systems
of Southern Federal University, Taganrog, Russia
donat.ivanov@gmail.com

Abstract. The paper is dedicated to a formation task in a group of quadrotors. It contains a brief analysis of existing methods for the formation task. The authors also propose a new method for solving the formation task, which make it possible to ensure accurate compliance with distances between quadrotors in the formation, as well as featuring low computational complexity.

Keywords: formation task, group robotics, quadrotor, quadrocopter , unmanned aerial vehicle.

1   Introduction

There are a lot of types of unmanned aerial vehicles (UAVs). The progress in a microelectronics and a computing make it possible to produce a small-sized UAV, which can be cheap and easily accessible in case of mass production. However, practical possibilities of a single UAV are limited. As widely shown in the recent literature [1], robustness and flexibility constitute the main advantages of multiple-robot systems vs. single-robot ones. Also the use of a group of UAVs opens wide perspectives for an unmanned aircraft [2, 3].

Micro Aerial Vehicle (MAV) can be classified on: airplane, helicopter, bird-like, insect-like, autogiro and blimp. There is a comparison of various types of MAV in [4], where showed that quadrotors are the most universal type of MAV.

Groups of unmanned quadrotors can be used for a video monitoring, forming phased antenna arrays and other applications (see for example [5] and the references therein). This applications requires mutual position relative to each other quadrocopters in a group. The target location of quadrotors in space is named "formation", and the task of forming the formation is named "formation task".

2   Known Methods for Solving Formation Task

There are some well-known methods for solving a formation task in a group of mobile robots, including a behavior based [6,7], leader-follower approach [8-12], a

---

[*] This work was supported in part by the Russian Foundation for Basic Research under Grants №14-08-01176, №13-08-00794.

© Springer International Publishing Switzerland 2015
J.-H. Kim et al. (eds.), *Robot Intelligence Technology and Applications 3*,
Advances in Intelligent Systems and Computing 345, DOI: 10.1007/978-3-319-16841-8_18

virtual structure/virtual leader approach [13], based on the game theory [14] etc. However some of the methods make it possible to form a formation by only a certain set of shapes. Some others require considerable computing resources, but there are not such resources on-board of quadrotors.

Many of the methods are aimed at positioning quadrotors in absolute coordinates, but in practice compliance required distances between the quadrotor is more important than the quadrotor's positioning in absolute coordinates.

Thus the formation task in a group of quadrotors needs a computationally simple method, which provide to derive desired target formation of various sharps and precise distances between quadrotors.

3   Formal Statement of the Formation Task in a Group of Quadrotors

There is a set $\mathbf{R}$ of controlled quadrotors $r_i \in \mathbf{R}(i = \overline{1, N})$, where N – is the number of quadrotors in a group. The status of each quadrotor $r_i \in R$ is described by the vector of status $\mathbf{s}_i(t) = [s_{i,1}(t), s_{i,2}(t), \ldots, s_{i,h}(t)]^T$, where the state variable $s_{i,h}(t)$ mean quadrotor's coordinates $x_i(t), y_i(t), z_i(t)$, current speed, acceleration, roll $\phi_i(t)$, pitch $\theta_i(t)$ and yaw $\psi_i(t)$ angles, the remaining board energy reserves, etc.

Current mutual arrangement of quadrotors in a group is described by the matrix

$$\mathbf{D}(t) = \begin{bmatrix} 0 & d_{1,2}(t) & d_{1,3}(t) & \cdots & d_{1,N}(t) \\ - & 0 & d_{2,3}(t) & \cdots & d_{2,N}(t) \\ - & - & 0 & \ddots & \vdots \\ - & - & - & 0 & d_{N-1,N}(t) \\ - & - & - & - & 0 \end{bmatrix},$$

where each element $d_{i,j}(t)$ of matrix $\mathbf{D}(t)$ represents the distance between quadrotors r_i and r_j the current time.

Each quadrotor $r_i \in \mathbf{R}$ available information about their own condition $\mathbf{s}_i(t)$, and information about distances $d_{i,j}(t)(i, j = \overline{1, N}, i \neq j)$ between quadrotors $r_i \in \mathbf{R}$ and other quadrotors $r_j \in \mathbf{R}(j = \overline{1, N}, j \neq i)$.

Each quadrotor $r_i \in \mathbf{R}$ have a control system, which make it possible to change coordinates $\mathbf{x}_i(t) = x_i(t), y_i(t), z_i(t)$ according to control inputs $u_i(t)$ based on the mathematical model considered in [4].

In order to prevent collisions and a mutual interference of quadrotors, the limit of positions is introduced:

$$\left| \mathbf{x}_i(t) - \mathbf{x}_j(t) \right| \geq \Delta_r, (i \neq j; i, j = \overline{1, N}) \tag{1}$$

where Δ_r – is a minimal acceptable distance between quadrotors, excludes mutual interference of quadrotors.

The target formation is a set $\mathbf{V}$ of target positions $v_\mu \in \mathbf{V}(\mu = \overline{1, N})$ of single quadrotors. Each target position $v_\mu \in \mathbf{V}$ is described by a point $p_\mu(\mu = \overline{1, N})$ with coordinates (x_μ, y_μ, z_μ). But there is not information about point's $p_\mu(\mu = \overline{1, N})$ coordinates and about assignments between quadrotors $r_i \in \mathbf{R}$ and target positions $v_\mu \in \mathbf{V}$.

The only one available information about the target formation is a matrix

$$\mathbf{D}_f = \begin{bmatrix} 0 & d_{1,2} & d_{1,3} & \cdots & d_{1,N} \\ - & 0 & d_{2,3} & \cdots & d_{2,N} \\ - & - & 0 & \ddots & \vdots \\ - & - & - & 0 & d_{N-1,N} \\ - & - & - & - & 0 \end{bmatrix},$$

where each variable $d_{i,j}$ of $\mathbf{D}_f$ is a distance between points p_i and p_j of target positions v_i and v_j in a target formation.

Formation task in the group of quadrotors is to determine a sequence of controls $\mathbf{u}(t) = [u_1(t), u_2(t), \ldots u_N(t)]^T$ which lead a group from start formation with distances $\mathbf{D}(t_0)$ to desired target formation with distances $\mathbf{D}_f$ for minimum time and with a restrictions on the quadrotor's positions (1).

4   The Proposed Method

We propose the following method for solving formation task, which is named "methods of circles".

At the first step we need to choose the target position v_μ and the quadrotor r_i, which are used for the beginning of formation building. Point p_c is the center of the group. The point p_c is described by radius vector $\vec{l}_c$:

$$\vec{l}_c = \frac{1}{N} \sum_{i=1}^{N} \vec{l}_i$$

The nearest for the point p_c quadrotor r_i get an assignment with target position v_1. And the current coordinates of the quadrotor r_i is a coordinates p_1 of the target position v_1.

At the second step the assignment for target position v_2 and its location p_2 is determined. For this construct a circle $c_{1,2}$ with center in point p_1 and radius $d_{1,2}$ (from the matrix $\mathbf{D}_f$). Then construct $N-1$ straight lines, each one passing through the point p_1 and the current position of quadrotors $r_i(i = \overline{2,N})$.

To determine the coordinates of the intersection points of these lines and the circle $c_{1,2}$ for each quadrotor $r_i(i = \overline{2,N})$ use the system of equations:

$$\begin{cases} (y_1 - y_i)x_2 - (x_1 - x_i)y_2 + (x_1 y_i - x_i y_1) = 0; \\ (x_2 - x_1)^2 + (y_2 - y_1)^2 = d_{1,2}^2. \end{cases} \quad i \in \left[\overline{2,N}\right]$$

and get equation roots (x_2, y_2) for each $r_i(i = \overline{2,N})$.

Then calculate distances between quadrotors $r_i(i = \overline{2,N})$ and their points (x_2, y_2):

$$l_{2,i} = \sqrt{(x_i - x_2)^2 + (y_i - y_2)^2} \; i \in \left[\overline{2,N}\right]$$

The quadrotor with the minimal length of distance $\min(l_{2,i}), i \in \left[\overline{2,N}\right]$ get an assignment target position v_2 and point p_2 with coordinates (x_2, y_2), like it is showed at Fig. 1.

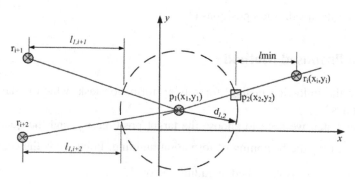

Fig. 1. Definition of the target position's point $p_2(x_2,y_2)$

At the third step the coordinates $p_3(x_3, y_3)$ of the target position v_3 is determined. For this construct two circles $c_{1,3}$ and $c_{2,3}$. The circle $c_{1,3}$ has the center in point p_1 and the radius $d_{1,3}$. The circle $c_{2,3}$ has the center in point p_2 and the radius $d_{2,3}$. Then find points of intersections of circles $c_{1,3}$ and $c_{2,3}$. There are two points of intersections in general, but there is only one in some cases.

To determine the coordinates of the intersection points of circles $c_{1,3}$ and $c_{2,3}$ for each quadrotor $r_i(i = \overline{3,N})$ use the system of equations:

$$\begin{cases} (x_3 - x_1)^2 + (y_3 - y_1)^2 = d_{1,3}^2; \\ (x_3 - x_2)^2 + (y_3 - y_2)^2 = d_{2,3}^2. \end{cases} i \in \overline{[3,N]}$$

There are two equation roots (x_3, y_3) in general. Then calculate distances $l_{3,i}[i = \overline{3,N}]$ between quadrotors $r_i(i = \overline{3,N})$ and each of roots (x_3, y_3)

$$l_{3,i} = \sqrt{(x_i - x_3)^2 + (y_i - y_3)^2} \; i \in \overline{[3,N]}$$

The quadrotor with the minimal length of distance $\min(l_{3,i}), i \in \overline{[3,N]}$ get an assignment target position v_3 and point p_3 with coordinates (x_3, y_3), like it is showed at Fig. 2.

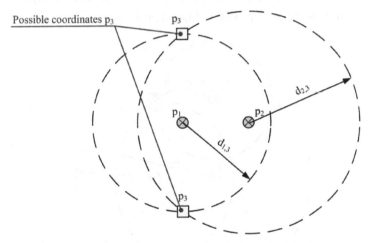

Fig. 2. Definition of the point p₃(x₃,y₃) of target position v₃

At the each next step the coordinates $p_\mu(\mu = \overline{4,N})$ of target position v_μ is determined. For this construct three circles $c_{\mu-3,\mu}$, $c_{\mu-2,\mu}$ and $c_{\mu-1,\mu}$. The first circle $c_{\mu-3,\mu}$ has the center in point $p_{\mu-3}$ and the radius $d_{\mu-3,\mu}$. The second circle $c_{\mu-2,\mu}$ has the center in the point $p_{\mu-2}$ and the radius $d_{\mu-2,\mu}$. And the third circle $c_{\mu-1,\mu}$ has the center in the point $p_{\mu-1}$ and the radius $d_{\mu-1,\mu}$.

Then find points of intersections of circles $c_{\mu-3,\mu}$, $c_{\mu-2,\mu}$ and $c_{\mu-1,\mu}$. There are two points of intersections in general, but there is only one in some cases.

To determine the coordinates of the intersection points of circles $c_{\mu-3,\mu}$, $c_{\mu-2,\mu}$ and $c_{\mu-1,\mu}$ for each quadrotor $r_i(i = \overline{k,N})$ use the system of equations:

$$\begin{cases} (x_k - x_{k-1})^2 + (y_k - y_{k-1})^2 = d_{k-1,k}^2; \\ (x_k - x_{k-2})^2 + (y_k - y_{k-2})^2 = d_{k-2,k}^2; \\ (x_k - x_{k-3})^2 + (y_k - y_{k-3})^2 = d_{k-3,k}^2. \end{cases}$$

Then calculate distances $l_{k,i}[i = \overline{k,N}]$ between quadrotors $r_i(i = \overline{k,N})$ and each of roots (x_k, y_k)

$$l_{k,i} = \sqrt{(x_i - x_k)^2 + (y_i - y_k)^2} \; i \in \overline{[k,N]}$$

The quadrotor with the minimal length of the distance $\min(l_{k,i}), i \in \overline{[k,N]}$ get an assignment target position v_k and the point p_k with coordinates (x_k, y_k). Fig. 3 shows a definition of the point $p_4(x_4,y_4)$ of target position v_4.

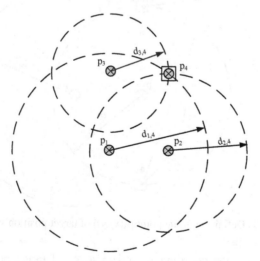

Fig. 3. Definition of the point $p_4(x_4,y_4)$ of target position v_4

Further calculations for target positions $(v_5, v_6, \ldots v_N)$ is similar. Fig. 4 illustrate derive desired target formation contains five quadrotors.

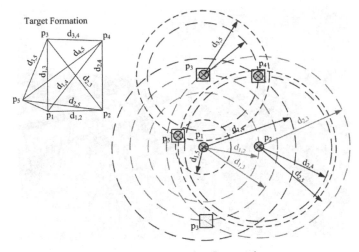

Fig. 4. Formation contains five quadrotors

When all assignments and coordinates of all target positions are obtained, each quadrotor moves to its own target position by straight line, except the case with threat of collisions between quadrotors. In latter case quadrotors circumnavigates each other along the arc.

5   Computer Modelling and Experiments

The proposed approach to derive desired target formation was tested by using the software model (Fig. 5) and the experimental stand, using quadrotors Ar.Drone (Fig. 6).

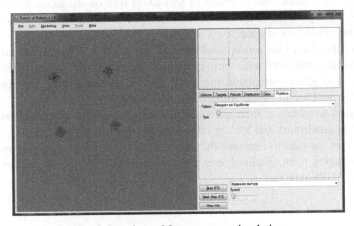

Fig. 5. Interface of for computer simulation

Fig. 6. Practical experiments

An information exchange between quadrotors into a group and with the operator's control panel implemented by Wi-Fi. Inertial navigation system is used. Formation task was solved on a horizontal plane, when all quadrotors into a group fly on at the same altitude (~1 meter).

In contrast to the well-known quadrotors formation projects [15], in that experiments external cameras or sensors were nor used. Quadrotors determined its positions only by on-board sensors.

6   Conclusions and Future Work

In this paper we consider the problem of formation task on the plane for those cases where the mutual position of quadrotors is more important rather than absolute coordinates of their positions in space.

The algorithm based on a proposed method has a low computational complexity, allows to create formations with various shapes, and opens up opportunities for practical application groups of quadrotor UAV for video monitoring, forming phased antenna arrays and mobile telecommunication systems.

If you need to obtain a three-dimensional formation, you should consider three coordinates of quadrotors and look at intersections of spheres (not circles).

In the future we plan to increase the quadrotor's positioning accuracy by aggregation of information from on-board sensor devices and analysis of video from on-board cameras and   GPS-navigation.

References

1. Franchi, A., Secchi, C., Ryll, M., Bulthoff, H.H., Giordano, P.R.: Shared control: Balancing autonomy and human assistance with a group of quadrotor UAVs. IEEE Robotics & Automation Magazine 19(3), 57–68 (2012)
2. Schwager, M., Julian, B., Angermann, M., Rus, D.: Eyes in the sky: Decentralized control for the deployment of robotic camera networks. Proceedings of the IEEE 99(9), 1541–1561 (2011)
3. Fink, J., Michael, N., Kim, S., Kumar, V.: Planning and control for cooperative manipulation and transportation with aerial robots. International Journal of Robotics Research 30(3), 324–334 (2010)
4. Bouabdallah, S.: Design and control of quadrotors with application to autonomous flying. Lausanne Polytechnic University (2007)
5. Tonetti, S., Hehn, M., Lupashin, S., D'Andrea, R.: Distributed control of antenna array with formation of uavs. In: World Congress, vol. 18(1), pp. 7848–7853 (2011)
6. Balch, T., Arkin, R.C.: Behavior-based formation control for multirobot teams. IEEE Transactions on Robotics and Automation (6), 926–939 (1998)
7. Lawton, J.R.T., Beard, R.W., Young, B.J.: A Decentralized Approach to Formation Maneuvers. IEEE Transactions on Robotics and Automation 19(6), 933–941 (2003)
8. Wang, P.K.C.: Navigation strategies for multiple autonomous mobile robots moving in formation. J. Robot. Syst. 8(2), 177–195 (1991)
9. Desai, J.P., Ostrowski, J., Kumar, V.: Controlling formations of multiple mobile robots. In: Proc. IEEE Int. Conf. Robotics and Automation, Leuven, Belgium, pp. 2864–2869 (May 1998)
10. Mesbahi, M., Hadaegh, F.Y.: Formation flying control of multiple spacecraft via graphs, matrix inequalities, and switching. AIAA J. Guidance, Control, Dynam. 24(2), 369–377 (2000)
11. Wang, P.K.C., Hadaegh, F.Y.: Coordination and control of multiple microspacecraft moving information. J. Astronaut. Sci. 44(3), 315–355 (1996)
12. Desai, J., Ostrowski, J., Kumar, V.: Control of Formations for Multiple Robots. In: Proceedings of the IEEE International Conference on Robotics and Automation, Leuven, Belgium (1998)
13. Lewis, M.A., Tan, K.-H.: High precision formation control of mobile robots using virtual structures. Auton. Robot. 4, 387–403 (1997)
14. Erdoğan, M.E., Innocenti, M., Pollini, L.: Obstacle Avoidance for a Game Theoretically Controlled Formation of Unmanned Vehicles. In: 18th IFAC 2011 (2011)
15. Flying Machine Arena, http://www.flyingmachinearena.org/

Cooperative Control of UAVs Using a Single Master Subsystem for Multi-task Multi-target Operations

Jae Chung

Department of Mechanical Engineering
Stevens Institute of Technology
Castle Point on Hudson
Hoboken, NJ 07030

Abstract. In this paper, a semi-autonomous control method is discussed for remote operation of multiple aerial vehicles through teleoperation to perform a task of pursuing multiple targets. The UAV team is formed by different automated aircrafts. A potential field algorithm is employed to implement a leader-follower formation control approach to guide the team of UAVs. A leader is selected and teleoperated by a human operator while all other UAVs follow the leader autonomously in a formation. An algorithm for paring the UAVs and targets is derived from an auction algorithm for a multi-task multi-target case, which optimizes effects-based vehicle-task-target pairing based on a heuristic algorithm. The pairing method produces a weighted attack guidance table (WAGT), which includes the benefits of assignments of intelligent combinations of tasks and targets. Finally, simulation studies were performed to illustrate the efficacy of the developed control method and highlight the improvement of the teleoperation of the UAVs in terms of task efficiency.

1 Introduction

Obvious advantages of collaborative multiple Unmanned Aerial Vehicle (UAVs) systems are to enhance working efficiency when tasks need to be accomplished by collaboration that cannot be performed by a single UAV. One of the important issues in collaborative control of multiple UAVs is formation control. A significant amount of research work has been proposed to resolve the issue of the formation control of a group of vehicles [2-4]. As found from the research papers [2-4], many control approaches, e.g. leader-follower strategy [2] virtual structure approach [3], and behavior-based method [4], have been developed to enable the UAV team to keep a formation while avoiding obstacles and tracing targets. In the virtual structure approach, all the UAVs have a rigid geometric relationship based on virtual points or virtual agents. Due to its rigid geometric relationship, the environment where the UAV team goes through should be large enough; otherwise, the UAV team would be jammed. However, in practical applications, e.g. life rescues, military surveillance, nuclear plant repair, and etc, the environment is so highly unstructured that the UAV team formation is greatly distorted to penetrate a very narrow passage to reach targets. In the leader-follower strategy, one UAV is considered as a leader and the others are

© Springer International Publishing Switzerland 2015
J.-H. Kim et al. (eds.), *Robot Intelligence Technology and Applications 3*,
Advances in Intelligent Systems and Computing 345, DOI: 10.1007/978-3-319-16841-8_19

193

followers which track the leader [6]. The UAV team formation can be changed when the followers are able to move regarding the leader's positions and velocities. Nonetheless, because the UAV team movement is totally guided by the single leader motion, the UAV team fails if the single leader breaks down. Hence, the robustness and reliability of the multi-UAV system are decreased. In the behavioral structure, tracing a target with obstacle or collision prevention is an important issue. In addition, each of the UAVs in the team is properly controlled to achieve its desired behavior [6]. Several control methods such as potential field, neural network, fuzzy control, and so on have been proposed to deal with this issue [6-8]. Potential field based control method about repulsion from obstacles and attraction to targets is well known in this area of research. In [6], the repulsion is adopted to make UAVs avoid obstacle or collision with other objects. Nevertheless, when UAV approaches the obstacle or objects in the potential field, since the resultant vector of the UAV motion is equal to the repulsive vector from the obstacle or the objects, it may not get around the obstacle or the objects and may reach a local minima position.

Besides, another issue about collaboration of the UAV team in the unknown environment is the lack of the capability of highly adaptation to different environments. This is very important for field applications such as life rescue, outer space operation, military reconnaissance, and etc because the applications require a team of two or more UAVs to navigate for a single mission about multi-target and multi-task scenarios in the unknown environments. Furthermore, in the applications, due to the limitations of the current sensors and computer decision-making technologies, navigation of the UGV team in the unknown environments prohibits the use of fully autonomous systems [7]. Therefore, it is required that human decision making be involved in the UAV systems. Teleoperators, in which a human operator is an integral part of the control, are suggested to integrate the human decisions to the control loop of the systems. The teleoperator consists of master and slave subsystems. The human operator remotely controls the UAV team, i.e. the slave subsystem that locally works on tasks in the unknown environment, via another subsystem, the master subsystem that reads human commands and reflects sensory data from the slave subsystem to the human operator. The teleoperator that involves cooperative multiple slave UAVs can be classified into two types of the teleopeartion systems based on different numbers of the master-slave subsystems, Single-master Multi-slave System (SMMS) [7] and Multi-master Multi-slave System (MMMS) [8]. However, only the SMMS teleoperation system [7] can be used to minimize the required human resources but amplify the human effort; therefore, it has been considered in this paper.

Besides, from a practical point of view, the SMMS teleoperation is so complicated that it is always performed to fulfill a single mission that has multi-tasks and multi-targets. In order to improve the work efficiency, the slave UAV team is suggested to be regrouped into subteams, which enables the slave UAVs to execute multi-tasks on multi-targets properly and simultaneously. Some methods and solution algorithms have been developed in research work [8] of task-target-vehicle allocations. Nonetheless, most of the research work was not able to produce optimal solutions of the task-target-vehicle allocations for existence of multiple tasks and targets within one mission in the unknown environments.

Therefore, in this paper, the primary objective is to propose a control method for a SMMS teleoperation system to control the UAV team to execute multiple tasks on multiple targets in the unknown environments. The proposed control method is able to integrate Component (1) modified potential field based leader-follower formation control method [9-10] and Component (2) vehicle-task-target pairing method based on weighted attack guidance table (WAGT) . The proposed control method with Component (1) and (2) enables the SMMS teleoperator to solve the multi-task multi-target problem of the UAV team when the work environment is uncertain. Component (1) computes UAV team leader-follower formation in which a human operator can concentrate on controlling a leader UAV when follower UAVs autonomously move regarding the positions and velocities of the leader. Moreover, Component (1) enables the UAV team to move in a leader-follower formation to avoid obstacles as positive ions and attract to its target positions as negative ions as if UAVs as positive ions were bonded together with bonding strengths varying according to their roles in an artificial potential field. For example, the bonding strength for leader-follower 1 is stronger than that for follower 1-follower 2 and so on. As soon as the leader UAV teleoperated by the human operator is approaching the region of interest, Component (2) dynamically calculates the weighted attack guidance table (WAGT) in which optimal solutions of vehicle-task-target assignments are found and generated. With the optimal solutions of vehicle-task-target assignments, the UAV team excluding the leader UAV is automatically split into many UAV subteams, each of which has the most suitable functionalities of UAVs to deal with its assigned tasks and the most vehicles closest to its assigned target positions. After all the subteams are formed with Component (2), all subteam UAVs approaches their assigned target positions while avoiding obstacles with Component (1) until all tasks are completed.

The rest of this paper is organized as follows. Section 2 develops the leader-follower formation control method for multiple UAVs based on an artificial potential field algorithm. Section 3 proposes the vehicle-task-target pairing method to assign UAVs to different tasks for different targets. In Section 4, the control method is proposed to integrate the proposed formation control and pairing algorithm for the SMMS teleoperation in multi-task multi-target scenarios. In Section 5, the work efficiency of the SMMS teleoperation system with the proposed control method was evaluated through simulation studies. Section 4 concludes this paper and shows future research directions.

2   Leader-Follower Formation Control Based on Artificial Potential Field

Formation is defined as a geometric disposition of vehicles in the environment. A common reference in order to specify the position of every vehicle that forms a team in the formation is prescribed. Three techniques, (1) leader-reference, (2) neighbor-reference, and (3) unit-center reference, have been proposed in the following ones explained in [3], as described in Figure 1.

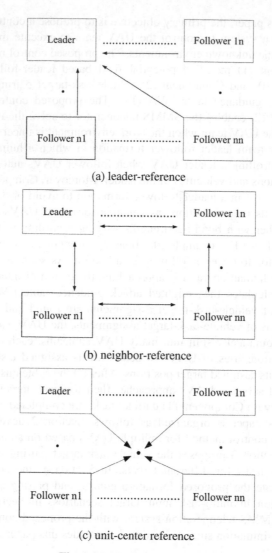

(a) leader-reference

(b) neighbor-reference

(c) unit-center reference

Fig. 1. Multi-vehicle formations

(1) leader-reference: the position of one of the vehicles is taken as the reference for the others, in order to calculate the desired position for each one. (2) neighbor-reference: the vehicles assume the position of the closet partner as the reference to calculate its own position. (3) unit-center reference: the unit-center location of the group is calculated by each one of the vehicles and is taken as the reference to calculate its desired position.

Global information about the positions of all vehicles is required in advance to form (3), but the global information does not always exist in a real system. On contrary, for teleoperation, only (1) and (2) are suitable for formation control of the team of vehicles because they have the benefit that local information is needed to be fed back from local sensors mounted on the vehicles to sense their positions when the local information is available. Moreover, (1) is better than (2) because more communication links between the vehicles are needed in (2) than (1), which causes larger transmission delays in the system. Therefore, in this paper, the leader-follower formation is proposed; nonetheless, as mentioned previously, the whole team fails if only one vehicle, the leader, in the team breaks down. Thus, this paper proposes that the leadership is able to be online switched over between vehicles in the team wirelessly by a human operator at a remote site.

Besides, most of the UAV team remotely controlled by the human operator who identifies the obstacle location on-line by receiving the delayed feedbacks can avoid the obstacles. However, the human operator feels fatigue while maintaining situational awareness for a long time. Moreover, the team formation is needed to be automatically adapted when the team penetrates a narrow area where only one vehicle is allowed to pass through each times. Hence, the robotic intelligence is needed and added to make the formation control intelligent enough to move the UAV team in changing environments even if the human operator is not aware of the UAV team when the UAV team is remotely controlled by him or her. Several robotic intelligences, such as artificial potential field, virtual structure, neural network, fuzzy control, and so on, have been proposed to deal with the problem as mentioned previously. Nevertheless, only the artificial potential field enables the UAV team to adapt the formation properly to avoid obstacles and get through a narrow passage by only having local feedbacks from sensors mounted on UAVs when UAVs do not necessarily spend a long time learning their desired behavior. Therefore, in the following, the artificial potential field based formation control is described in detail to explain how it is supposed to work in our approach.

3   Artificial Potential Field Based Formation Control

In order that the UAV team autonomously avoids the obstacles and keeps a distance from other neighboring vehicles simultaneously while tracking the target, the approach that the most commonly has been used is potential field based formation control. Nonetheless, the potential field based formation control has the local minima problem [5], which can hold the vehicles in a specified formation while in motion. Therefore, the potential field with a prioritized bonding between team vehicles is proposed in this paper. The strength of the bonding to keep a constant distance between vehicles varies depending on which two vehicles are connected. For example, the bonding between Leader and Follower 1 is the strongest when the one between Follower n and Leader is the weakest if there are the n number of vehicles. Thus, the formation becomes more adaptive due to the attraction from the target and repulsion from the obstacle, but no vehicle in the team is left behind due to the vehicle-vehicle bonding with different strengths.

At any time instant, the vehicles can be visualized as particles with negative charge moving in a potential field. Each obstacle is viewed as a particle with negative charge while the target is seen as a particle with positive charge. Hence, the vehicle is repulsed from the obstacle while attracted to the target in the potential field. In addition, with the prioritized vehicle-vehicle bonding, the follower keeps a constant distance from the neighboring vehicles when the leader tracks the human position commands closely. In the potential field, the vehicle path is generated by summing up the attraction, repulsion, and prioritized bondings if the vehicle is a follower or human position commands if the vehicle is a leader.

Repulsion from Obstacles

The repulsive force U_o derived from the obstacle potential functions is written in Eq. (1).

$$U_o = \varphi\left(-k_e \partial D_1 - b_e \partial V_1\right) \tag{1}$$

where $\partial V_1 = \dot{x}_{si}\, G/\partial x_o^2$, $\partial D_1 = H/\partial x_o^2$. ∂x_o is the distance between the obstacle and vehicle. x_{si} is the position of the vehicle. φ, G, H, k_e, and b_e are positive parameters.

Attraction to Targets

Besides the repulsive force, the attractive force U_T derived from the target potential functions for the ith vehicle is written in Eq.(2).

$$U_T = \gamma_i \partial x_T \tag{2}$$

where for $i = 1....n$, n is the number of UAVs, and γ_i is the positive tuning gains for the ith vehicle. ∂x_T is the distance between the vehicle and target(s). The ith vehicle can be a leader or one of the follower vehicles in the team.

Bonding between Vehicles

Except the attraction and repulsion functions, for the followers, the prioritized vehicle-vehicle bonding U_f is needed to maintain the formation, as shown in Eq. (3).

$$U_f = -k_{ij}\left(r_{s\min i} - \partial x_{si}\right) \tag{3}$$

where for $i = 1....n$ and $j = 1....n$, $i \neq j$, k_{ij} is the positive parameter. The bonding strength varies because of different vehicle combinations according to the two vehicle priority numbers. For instance, the Leader-Follower 1 bonding, k_{L1}, is larger than the Follower 1 - Follower 2 bonding k_{12} when the Follower n - Leader bonding, k_{nL} is the smallest. $r_{s\min i}$ is the desired distance that the vehicle needs to keep from the neighboring vehicles. ∂x_{si} is the measured distance between vehicle and vehicle. During the teleoperation, the leader is online selected, and the target is identified by the human operator based on his or her experience. The obstacle(s) are detected and avoided by UAV according to the repulsive function in the potential field. The target is tracked by the vehicle according to the attraction function in the potential field. Follower's path is generated by combining the vehicle-obstacle repulsion, vehicle-target attraction, and vehicle-vehicle bonding forces. In addition, Leader's path is produced by integrating the vehicle-obstacle repulsion, vehicle-target attraction, and human position commands. Due to the vehicle-vehicle bonding forces with their varying strength, the probability of occurrences of the local minima problem [1] is reduced.

Nonetheless, as said in the previous section, even if a team of UAVs is being teleoperated in the leader-follower formation based on the artificial potential field, the team cannot be enabled to handle a multi-task multi-target scenario efficiently by tackling all tasks on all targets successfully and simultaneously. Therefore, the vehicle-task-target pairing method is proposed to enable the teleoperation system to split the team into subteams that especially perform the assigned task on the assigned target. In the following section, how to adopt and formulate the proposed vehicle-task-target pairing method in the team is described in details.

4   Vehicle-Task-Target Pairing Method

In fact, the concept of the proposed vehicle-task-target pairing method is that all team vehicles act largely independently in terms of planning for themselves but are able to take into account team resources by working on the tasks with other team members. With the pairing method proposed in this paper, the leader vehicle in the teleoperation system not only takes any human command from the master vehicle but also works as an auctioneer to send and show all bid data e.g. target locations and their base prices that are also online shared by all other vehicles called team followers. Any team vehicle is online appointed as the leader vehicle by the human operator via the master subsystem if the original one fails. All follower vehicles act as bidders to form a subteam by themselves in order to maximize a sum of all follower bid values and bid on the targets when the corresponding task on the targets is performed by the cooperation of the subteam. In the subteam, the bidder with the maximum bid value is selected as a subteam leader that is responsible for monitoring and coordinating all subteam member actions. According to the largest bid proposed by the subteam, the auctioneer, the leader vehicle, decides which subteam wins the bid with the restriction that each task and target are only gained by one subteam per auction. If all subteam bid values are smaller than the base price, or any slave vehicle cannot compute the bid value due to lacking of the information surrounding the targets, the auctioneer obtains the bid. If any one of the subteams of the bidders already completes the task on the target, it will inform the auctioneer to cancel the bid.

After being paired to the task and target by using the proposed pairing method, the subteam vehicles will move to the targets in the artificial potential field based leader-follower formation mentioned in previous section and perform the paired task, e.g. the transportation or capture of the target.

Vehicle-Task-Target Pairing Formulation

Consider such a scenario, in a two-dimensional and limited rectangular environment X with n_c square cells, n_p slave vehicles pursue n_e targets, for $n_p > n_e$ The set of the vehicles is denoted by a matrix of $A = \left[a_1, a_2, \ldots a_j, a_{n_p}\right]$ where a_j is the jth vehicle matrix. The jth vehicle capability vector for the tth task is denoted by $\hat{C}_j^t$ $1 \le j \le n_p$, and the set of targets is expressed as a target matrix of $T = \left[T_1, T_2, \ldots T_{n_e}\right]$ where T_{n_e} is the n_e^{th} target matrix. The vector representing the capability required to accomplish the tth task on the Tth target is denoted by $\overline{C}_t^T$, $1 \le T \le n_e$. Agent $A \cup T$ denotes the teams of vehicles and targets. For simplification, we assume that both space and time can be quantized, therefore the environment can be regarded as a finite collection of cells, denoted by $X_c = \left[1, 2, \ldots n_c\right]$. There exist some static obstacles with fixed sizes and regular shapes, and their locations are determined by the mapping $M : X_c \rightarrow \{0,1\}$ for $\forall x \in X_c$, $M(x) \ge 1$ indicates that the cell x is occupied by obstacles. $\forall x \in X_c, M(x) \le 0$ indicates that the cell x is free. Each of the heterogeneous team vehicles needs different capabilities to complete different tasks on different targets, such as the target capture and transportation.

Vehicle Capability

The weighted capability vectors of the jth vehicle with u functionalities to complete the ith task can be defined as

$$\hat{C}_j^i = w_j^t diag\left\{b_{j1}^i, b_{j2}^i, b_{j3}^i, \cdots b_{ju}^i\right\}\left[c_{j1}^i \cdots c_{ju}^i\right]^T \tag{4}$$

where u is the maximum number of the individual functionality with which the jth vehicle can complete the ith task. c_{jk}^i is a capability value for the jth vehicle with k^{th} functionality to do the ith task. w_j^t is a positive integer such that for the given target t and vehicle j, the following is satisfied. If the vehicle is assigned to the target, $w_j^t = 0$, otherwise, $w_j^t = 1$. The $u \times u$ dimension diagonal matrix of b_{ju}^t is used to estimate the percentage of possibility of using the $u \times 1$ dimensional capability vector c_j^i to do the ith task by the jth vehicle successfully. However, if Vehicle j does not have the capability c_{jk}^i, then the b_{jk}^t is 0.

Capability Required to Execute Tasks on Targets

It is assumed that there are p tasks which need to be done independently and simultaneously. All tasks are represented by the matrix of t that contains a set of the separate task matrices of $\left[t_1, \dots t_p\right]$ in the system for $p \le n_e$, i.e. one task can be paired to two or more targets, but each target can only be paired to one task. The capability vector that is required to accomplish Task i on Target k is defined as

$$\overline{C}_k^i = diag\left\{\beta_{i1}^k, \beta_{i2}^k, \cdots \beta_{iu}^k\right\}C_{ku} \tag{5}$$

where the $u \times u$ dimension diagonal matrix of β_{iu}^k is used to describe the percentage of possibility of using the $u \times 1$ dimension capability vector C_{ku} with which the vehicle can finish the ith task on the kth target. $C_{ku} = \left[c_{kt1} \dots \dots c_{ktu}\right]^T$ when the total number of the functionalities is u. c_{ktu} is the capability vector that is required to complete the tth task with the ith functionality. However, if the tth task can not be done successfully by any vehicle with the capability c_{ktu} on the kth target, then the β_{tu}^k is 0. Otherwise, β_{tu}^k is 1.

Subteam Capability

For Subteam f, the subteam capability vector is a sum of capability vectors of all subteam vehicles, Vehicle $a - b$, from the team to do the ith task cooperatively. It is defined that Subteam f is formed by the subteam vehicles, i.e. Vehicle $a - b$ by assuming that all subteam vehicles have a capability to do the ith task on the kth target. For Subteam f made by the subteam vehicles from the team to do i^{th} task, if $\hat{C}_j^i \ge 0$, then the subteam capability vector $Q((a-b), i, f)$ is defined as

$$Q((a-b),i,f)=\sum_{j=a}^{b}\hat{C}_j^i \qquad (6)$$

where the $a-b \; \forall a \geq b$ is the total number of the vehicles in Subteam f. a is the first and b is the last indices of the elements in the matrix $Q((a-b),i,f)$ for Task i and Subteam f. Subteam f is able to perform Task i on Target k if the condition, $\overline{C}_k^i \leq Q((a-b),i,f)$, is satisfied. Vehicle j is selected as a subteam leader when its magnitude of the capability vector $\hat{C}_j^i$ is largest in Subteam f. The subteam leader knows all capability information about its subteam members.

Bidding Winner Determination

In order to determine that the capability of each subteam is the most suitable for a specific task and target, the following equation is written to compute the bid value of Vehicle j in Subteam f for Task i and Target k.

$$\hat{B}(k,i,j)=\left(Q((a-b),i,f)-\overline{C}_k^i\right)\left(1-X_{ij}^k\right) \qquad (7)$$

where X_{ij}^k is the positive integer weight for Vehicle j in Subteam f for Target k. If Task i is the most preferred by Vehicle j in Subteam f to be done on Target k when $\hat{B}(k,i,j)$ is the maximum bid value of Vehicle j in Subteam f for Task i and Target k by comparing the bid values for other tasks and targets, then $X_{ij}^k=0$. Otherwise, $X_{ij}^k=1$. Different subteams are formed by different combinations of the slave team vehicles with their bid values computed with Eq (7), all of which are placed into Table 1.

Table 1. Weighted attack guidance table (WAGT)

Subteam 1	$m_{k,1}$	...	Subteam f	$m_{k,f}$
$\hat{B}(1,1,1),\cdots\hat{B}(1,i,1)$	$m_{1,1}$	...	$\hat{B}(1,1,f),\cdots\hat{B}(1,i,f)$	$m_{1,f}$
....	...	...	....	...
$\hat{B}(k,1,1),\cdots\hat{B}(k,i,1)$	$m_{k,1}$	...	$\hat{B}(k,1,f),\cdots\hat{B}(k,i,f)$	$m_{k,f}$

In Table 1, $m_{k,f}$ is the positive integer weight for Subteam f to bid on all tasks on Target k. If $Q((y_a-y_b),i,f)$ is smaller than the base price which is a positive integer, or Target k has already been assigned to Subteam f, $m_{k,f}$ is 0.

Otherwise, $m_{k,f}$ is 1. By arranging $m_{k,f}$ and $\hat{B}(k,i,f)$ into Table 1, called Weighted Attack Guidance Table (WAGT), each row of WAGT corresponds to a target with Tasks (1 to i) and Vehicle Subteam (1 to f) when i is the total number of the tasks, and f is the total number of the subteams formed in the team. In addition, each column of WAGT corresponds to Subteam that accomplishes Tasks (1 to i) on Targets (1 to k) when k is the total number of the targets. Therefore, there are the k rows and $2f$ columns in WAGT. The scanning proceeds from the first to the last column. Hence, Subteams specified in column i takes precedence over Subteams specified in column $i+2$. The maximum value in each row of WAGT is the most preferred bid for the subteam corresponding to the column number to do the task on the target corresponding to the row number. Therefore, with WAGT, the optimization of the vehicle-target pairing can be formulated in the following to determine the bidding winner.

Given Subteam f, Target k, and Task i in WAGT, an assignment of a subteam is found in such a format that WAGT is satisfied, and its objective function in Eq. (8) is maximized within the given constraints in Eqs. (9) and (10) where the decision variables are the magnitudes of the subteam and task capability vectors, i.e. Q_f and $\overline{C}_k^i$, and X_{ij}^k and $m_{k,f}$. Eq. (9) is used to ensure that all subteams formed by the team vehicles can obtain a corresponding task and target when Eq. (10) is used to ensure that all bid values submitted by the subteams are valid values, based on which all targets and tasks are successfully paired to the subteams. Therefore, for Vehicle $p_1 - p_n$ in Subteam f to work on Task i for Target k as seen in Table 1, the objective function is to find the maximum of the magnitude of

$ObjFun(k,f)$, where $ObjFun(k,f) = \sum\limits_{j=p_1}^{j=p_n} \sum\limits_{i=0}^{t} \hat{B}(k,i,j)m_{k,f}$.

$$\text{Maximize } Magnitude(ObjFun(k,f)) \tag{8}$$

Subject to

$$\sum_{f=1}^{f=n} m_{k,f} > 0 \tag{9}$$

$$\sum_{f=p_1}^{f=p_n} \hat{B}(k,i,j) \geq 0 \tag{10}$$

where $m_{k,f}$ is the positive integer weight for Subteam f and Target k. Initially, $m_{k,f}$ is equal to one if no subteam is assigned to any target. However, if Subteam

f is assigned to Target k, $m_{k,f}$ is equal to zero. Hence, Subteam f that proposes the maximum affordable value of the magnitude of $\hat{B}(k,i,j)m_{k,f}$ can win Task i on Target k by solving Eqs (8) within the constraints Eqs. (9) and (10). By using the vehicle-task-target pairing method, the optimal pairs are computed based on given WAGT. In order to make the system be able to split its team into some subteams to execute different tasks on different targets simultaneously, our proposed control method is improved by integrating the vehicle-task-target pairing method into the teleoperation system. The vehicle-task-target pairing method is created to enable the system based on the found pairs to form subteams, appoint the vehicles as a subteam leader and followers, pair the tasks to the subteams, and generate the position reference inputs to the subteams to work on the given targets.

However, how to apply the artificial potential field based leader-follower formation control mentioned in the previous section and the vehicle-task-target pairing method developed in this section into a teleoperation system, has not been well found in some of the research papers [1-10] about the development of the teleoperators. Hence, this paper develops a control method that integrates Component (1) the artificial potential field based leader-follower formation control mentioned in the previous section and Component (2) the vehicle-task-target pairing method developed in this section is proposed to enable the teleoperation of multi-vehicle system to handle multi-tasks on multi-targets effectively. In this paper, the Single-Master Multi-Slave (SMMS) teleoperator is suggested to be used because only SMMS teleoperator can maximize human effort but minimize required human resources when it still keeps the human operator from entering into a hazardous work-site or an inaccessible area. Then, this proposed control method that combines Component (1) and (2) is formulated and described in details in the following section.

5   SMMS Teleoperator with Potential Field Based Leader-Follower Formation Control and Vehicle-Task-Target Pairing Method

The SMMS teleoperator with the proposed control method is presented in Figure 2 Figure 2 represents the overall architecture of the SMMS teleoperation system integrating the (1) modified potential field based leader-follower formation control and (2) vehicle-task-target pairing method. In the SMMS teleoperation configuration, the master and slave subsystems in Figures 2(a) and (b), respectively, are connected over the wireless internet. The difference from the ones that were proposed in [3],[4] is that the slave subsystem with the proposed control methods is operated fully autonomously for the following reason. In order to enhance its working efficiency, the slave vehicle team excluding a team leader is able to form different subteams to simultaneously perform multiple tasks on multiple targets while the team leader is teleoperated by the human operator.

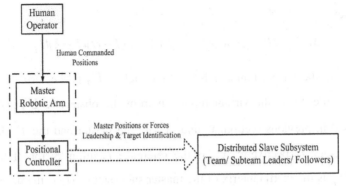

(a) Master subsystem

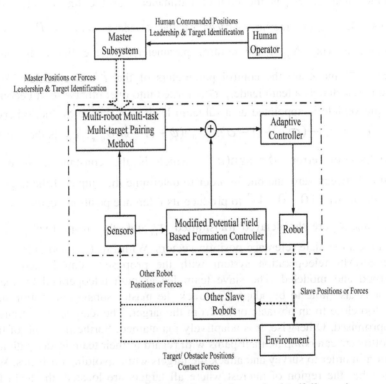

(b) Distributed slave subsystem (team/subteam leaders/followers)

Fig. 2. Proposed SMMS Teleoperation System

The SMMS teleoperation system with the proposed control method in Figure 2 can be formulated into the following equations of motion.

Master:

$$M_m \ddot{e}_m + B_m \dot{e}_m + K_m e_m = 0 \tag{11}$$

i^{th} Slave:

$$M_{si}\ddot{e}_{si} + B_{si}\dot{e}_{si} + K_{si}e_{si} = U_T + U_o + (I - \sigma)(I - \lambda)U_f \tag{12}$$

where U_f is the virtual bonding between vehicles. U_T is the virtual attraction to the target while U_o is the virtual repulsion from the obstacles. $U_f, U_T,$ and U_o are proposed in previous section. x_m and x_{si} are the master and the ith slave vehicle position vectors, respectively. x_{sdi} is the reference position vector of the ith slave vehicle. M_m is the inertia matrix of the master subsystem. B_m is the master adaptive impedance matrix. K_m is the control parameters for the linear diagonal master matrices. M_{si} is the inertia matrices of the ith slave vehicles. B_{si} is the slave impedance matrix. K_{si} is the control parameters for the linear diagonal slave matrices. σ and λ are the control parameters of the ith slave vehicle. When the vehicle is selected as a team leader, σ is turned into one; otherwise, it becomes zero. When the vehicle is appointed as a subteam leader, λ becomes one; otherwise, λ is zero. $e_{si} = x_{si} - (\sigma x'_m + (1 - \sigma)x'_{ideal})(\phi + (1 - \phi)\theta_{pos}$. e_{si} is the slave current vehicle position error. $\phi = \text{sgn}(e_{si}^2)$ which is the constant positive integers switching between zero and one in order to determine the output of the target matrix. θ_{pos} is the matrix, $[0 \quad 0 \quad 1]^T$ to produce its reference position vectors transformed from x'_m and x'_{si} are the delayed transmitted x_m and x_{si}, respectively. x'_{ideal} is the slave subteam vehicle reference position vectors. With Eqs.(11) and (12), the motion of the SMMS teleoperation system with the proposed control method can be understood and modeled. The slave team leader path teleoperated by the human operator is also able to be adapted to track the master subsystem motion unless its path is too close to an obstacle or far from the target. The team leader's transparency is compromised. Otherwise, it is adaptively maintained. Furthermore, all of the slave team followers can compute their paths with regard to their team leader path to form a formation in order to survey and attack the targets while avoiding obstacles. When the team reaches the region of interest where all targets are located, the team follower vehicles become independent of the team leader vehicle by changing their reference positions x'_{ideal} in Eqs. (11) and (12). Then it is split into subteams paired to tasks and targets with the vehicle-task-target pairing method by solving Eq (8) within the constraints in Eqs (9) and (10) With the optimization equations Eq (8) by satisfying Eqs (9) and (10) the target and task matrices from which the reference positions to the slave subteam vehicle are extracted are produced. Transforming those matrices into the reference positions can enable the slave subteam vehicle to work on the paired tasks for the paired targets automatically.

During navigation to the targets, only the vehicle-obstacle, vehicle-target, and vehicle-vehicle distances can be sensed. If too long vehicle-target distances and/or too short vehicle-obstacle distances are found, all slave team vehicles autonomously adjust their routes to adapt the formation to approach the target and avoid the obstacles. Furthermore, the subteam leader-follower formation can be maintained or adapted by integrating (1) the virtual vehicle-vehicle bondings with different strengths based on which two team vehicles are connected, (2) the attraction to the target with regard to vehicle-target distances, and (3) the repulsion from the obstacles with regard to vehicle-obstacle distances. In such a formation, all followers in the subteam move with regard to the subteam leader's motion. After the target is reached, the slave vehicles will perform the assigned tasks, such as target captures or attacks relying on the task and target matrices.

6   Simulation Results

Table 2. SMMS simulations for a multi-target operation

Simulations	Control Objectives
Sim (1)	Position Control Method
Sim (2)	Proposed Control Method

The scenario for simulation studies is that UAVs form a team to handle a multi-target mission in the presence of time-varying communication delays, 0.1 - 0 seconds [1] with the proposed control method, as described in Table 2. In simulations, the master subsystem was a joystick connected to a laptop that reads motion commands from a human operator, and each of the slave UAVs were modeled in Matlab as a holonomic mobile UAV. Six static obstacles and two targets were simulated as mass-spring-damper systems [2]

The simple tasks, target captures, were performed by the slave team vehicles simultaneously. TB and TA were static and captured by being surrounded by at least three mobile vehicles. The simulations were set up with the following parameters. The desired distance between robots was 3m. The minimum vehicle-obstacle distance was 5m. Six obstacles with the radii of 5m were used in each simulation. Two targets with the radii of 5m were represented by TA and TB. Only two directions on x-y plane parallel to the ground were considered in the simulation. The following parameters were used in the simulations:

$M_m = 3$ kg, $K_m = 6$ Ns/m, $M_{si} = 30$ kg, $B_{si} = 1.0$ Ns/m, $K_{si} = 60$ N/m, $M_T = 60$ kg, $B_T = 0.0$ Ns/m, $K_T = 800$ N/m, $M_o = 6000$ kg, $B_o = 0.0$ Ns/m, $K_o = 1000$ N/m, $/mu = 10$, $k_e = 100$, $b_e = 60$, $r_{imin} = 5$, $r_{smin} = 5$, $\phi = 100$, and $\gamma_i = 1$

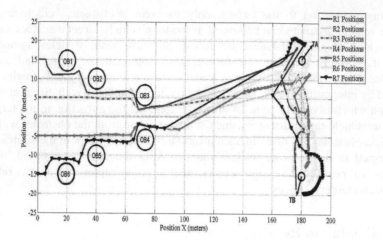

Fig. 3. <u>Sim (1)</u> - team positions

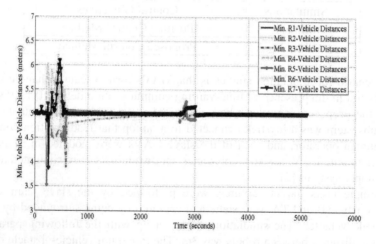

Fig. 4. <u>Sim (1)</u> - minimum vehicle-vehicle distances

Simulation - Sim(1)

In Sim (1), R4 was teleoperated by a human operator via the master subsystem to reach TA when R1-3 and R5-7 were coordinated with R4 to surround and capture it. After TA was captured, the human operator commanded R4 to move to TB while R1-3 and R5-7 were also moving with regard to R4 motion to approach and then capture TB. During the team navigations to catch TA and TB in Figure 3, all team vehicles were able to keep a constant distance, 5 meters, from each other as shown in Figure 4 and avoid the obstacles, Ob1-6, with maintaining at least 5 meters between a vehicle

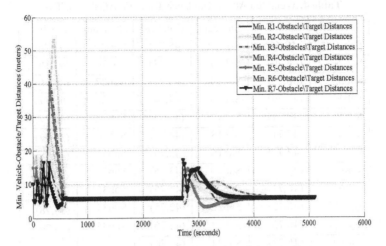

Fig. 5. Sim (1) -minimum vehicle-obstacle/target distances

and each of them in Figure 5. All tasks were completed in more than 1.38 hours (5000 seconds).

Simulation - Sim(2)

In Sim (2), the same task as in Sim (1), target captures, was performed on two targets TA and TB. However, in this simulation, the two targets should be captured at the same time by the team of R1-7 while the team was being teleoperated by the human operator. Since these two targets were required to be tackled simultaneously, R1-7 should form two subteams. Based on our developed vehicle-task-target pairing method, R1-7 is able to form 35 types of Subteams (Sub1-35) as shown in Table 3.

Table 3. Vehicle Subteams

Subte	Comb	Subte	Com	Subte	Com
Sub1	R1R2	Sub1	R1R	Sub2	R2R6
Sub2	R1R2	Sub1	R1R	Sub2	R3R4
Sub3	R1R2	Sub1	R1R	Sub2	R3R4
Sub4	R1R2	Sub1	R2R	Sub2	R3R4
Sub5	R1R2	Sub1	R2R	Sub2	R3R5
Sub6	R1R3	Sub1	R2R	Sub3	R3R5
Sub7	R1R3	Sub1	R2R	Sub3	R3R6
Sub8	R1R3	Sub2	R2R	Sub3	R4R5
Sub9	R1R3	Sub2	R2R	Sub3	R4R5
Sub1	R1R4	Sub2	R2R	Sub3	R4R6
Sub1	R1R4	Sub2	R2R	Sub3	R5R6
Sub1	R1R4	Sub2	R2R		

Table 4. Weighted Attack Guidance Table (WAGT) (Ta, Tb)

Subteam	Combos	Subteam	Combos	Subteam	Comboo
Sub1	(48,69)	Sub13	(45,73)	Sub25	(41,76)
Sub2	(48,70)	Sub14	(45,74)	Sub26	(41,76)
Sub3	(48,70)	Sub15	(45,74)	Sub27	(40,76)
Sub4	(47,70)	Sub16	(44,74)	Sub28	(40,77)
Sub5	(47,71)	Sub17	(44,74)	Sub29	(40,77)
Sub6	(47,71)	Sub18	(44,74)	Sub30	(39,77)
Sub7	(47,71)	Sub19	(43,75)	Sub31	(39,78)
Sub8	(46,71)	Sub20	(43,75)	Sub32	(39,78)
Sub9	(46,71)	Sub21	(43,75)	Sub33	(38,78)
Sub10	(46,72)	Sub22	(42,75)	Sub34	(38,78)
Sub11	(46,72)	Sub23	(42,76)	Sub35	(38,79)
Sub12	(46,72)	Sub24	(41,76)		

According to the proposed pairing method, bids needed to be calculated in Eq. (7) as an inverse of the sum of target-vehicle distances in a subteam minus the base price when the base price for capturing was 10. The reason was that in order to start with the tasks, the vehicles needed to maintain at least 10(m) from the target for capturing because the vehicles need some space to do capturing. In addition, all bid values for capturing on TA and TB were zero. Therefore, in order to simplify the WAGT table, those were not shown in Table 4. The bids (Ta, Tb) in Table 4 were written where Ta was the bid values calculated for capturing on TA when Tb was the bid values calculated for capturing on TB.

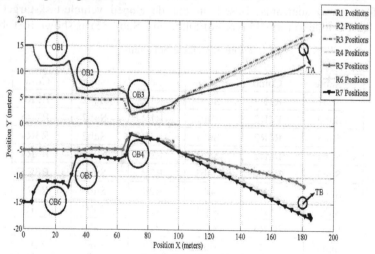

Fig. 6. <u>Sim (2)</u> - team positions

As shown in Figure 6, only R4, the team leader, was teleoperated by the human operator when R1-3 and R5-7 automatically formed two subteams to capture TA and TB simultaneously in 0.25 hours (900 seconds). All tasks were done by Sub1 and Sub35 fully autonomously. By comparing the results in Figures 8 and 5, the performance of the system in Sim (2) was better than that in Sim (1) because the tasks were successfully finished more quickly in Sim(2) than Sim (1). The reasons were that forming the subteams could save all seven vehicles from visiting all targets to complete two tasks. R1-7 were regrouped into two subteams to perform the task on each target simultaneously as shown in Figure 6. By taking advantage of the task planning independently done by each subteam, the work efficiency was enhanced when the operation time was decreased to 900 seconds in Sim(2) from 5000 seconds in Sim(1) in Figures 8 and 5 as the vehicle average speed was almost the same.

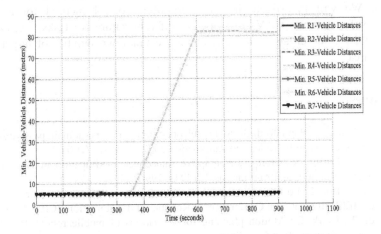

Fig. 7. Sim (2) - minimum vehicle-vehicle distances

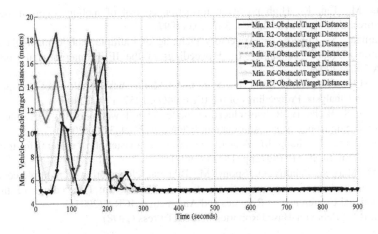

Fig. 8. Sim (2) -minimum vehicle-obstacle/target distances

7   Conclusions and Future Work

The control method with artificial potential field based the leader-follower formation control and vehicle-task-target pairing method is developed for the SMMS teleoperation system composed of a team of UAVs for multi-task multi-target scenario and improves the work efficiency as seen from the simulation results. Moreover, UAVs with the proposed control method can complete assigned tasks on specific targets while avoiding obstacles because the main components in the method enable all robots to generate their paths based on sensed vehicle-obstacle/target/vehicle distances. However, the proposed vehicle-task-target pairing method could generate a suboptimal solution in general since it is heuristic.

Therefore, our future work will be to further evaluate the performance of the SMMS system with the proposed control method to verify the quality of the pair solutions. We will investigate into the proposed control method for a SMMS teleoperator for multiple UAVs to work on different natures of tasks, e.g. captures, rescues, attacks, surveillance, and etc, about the unknown number of targets in an uncertain area.

References

1. Ryu, J., Kwon, D., Hannaford, B.: Stable teleoperation with time-domain passivity control. IEEE Transactions on Robotics and Automation 20(2), 365–373 (2004)
2. Rubio, A., Avello, A., Florez, J.: Adaptive impedance modification of a master-slave manipulator. Robotics and Automation 3, 1794–1799 (1999)
3. Gustavi, T., Hu, X.: Formation control for mobile robots with limited sensor information. In: Proc. IEEE International Conference on Robotics and Automation, ICRA 2005 (2005)
4. Barfoot, T., Clark, C.: Motion planning for formations of mobile robots. Robotics and Autonomous Systems 46, 65–78 (2004)
5. Lee, G., Chong, N.: Decentralized formation control for small-scale robot teams with anonymity. Mechatronics 19, 85–105 (2009)
6. Lewis, M., Tan, K.: High precision formation control of mobile robots using virtual structures. Utonomous Robotics 4, 387–403 (1997)
7. Yamamoto, Y., Fukuda, S.: Trajectory planning of multiple mobile manipulators with collision avoidance capability. In: Proceedings of the IEEE International Conference, ICRA 2002 (2002)
8. Dias, M.B., Zlot, R.M., Kalra, N., Stentz, A.: Market-based multirobot coordination: a survey and analysis. Proceedings of the IEEE 94(7), 1257–1270 (2006)
9. Monteiro, S., Bicho, E.: Robots allocation and leader-follower pairs. In: IEEE International Conference on Robotics and Automation, ICRA, pp. 3769–3775 (2008)
10. Lee, H., Shin, M., Chung, M.: Adaptive controller of master-slave systems for transparent teleoperation. In: ICAR 1997, pp. 14–23 (1997)
11. Lee, D., Martinez-Palafox, O., Spong, M.: Bilateral teleoperation of multiple cooperative robots over delayed communication networks: Theory. In: Proceedings of the IEEE International Conference on Robotics and Automation, ICRA (2005)
12. Arkin, R.C.: Behavior-Based Robotics. The MIT Press (1998)

Applying Self Organization Maps to Cluster Analysis of Regional Industrial Structures

Hou Aoyu[1], He Bin[2], Cui Yong[2], and Cai Zhonghua[1]

[1] School of Economics and Management, Beijing University of Chemical Technology, Beijing, China
houaoyu1201@163.com, caizh@mail.buct.edu.cn
[2] Sinopec International Petroleum Exploration and Production Corporation, Beijing, China
{Bhe.sipc,Ycui.sipc}@sinopec.com

Abstract. Using the SOM method, this paper mainly makes the clustering analysis on industrial structures of Chinese provinces and the United States. As the analysis results shown, regional industrial structures can be significantly clustered into several areas with different characteristic. Further analysis results indicate that industrialization process leads to the differences of Chinese regional industrial structures, while such differences in the United States depends on the tertiary industry level. These statistics results show the correlation between national industrial structure and economic development.

Keywords: Self Organization Maps (SOM), industrial structure, clustering analysis.

1   Introduction

Self Organization Maps (SOM), which is a self-organizing and self-learning network developed by Kohonen [1-2], has been widely used in various clustering analysis. Compared with other clustering analysis methods, SOM shows several advantages, including realization of real-time learning, auto stability of network, independence from external evaluation function, recognition of the most meaningful features within vector space, etc. [3-5], which make it widely applicable for unsupervised clustering analysis of high dimensional data [6-7].

In recent years, neural network has been gradually applied in economic structure researches [8-9], with good outcomes in terms of pattern recognition. Some researches attempts to evaluate and categorize regional industrial strengths applying BP neural network [10], yet there are several limitations. For instance, this attempt requires sufficient samples when training network as such arithmetic belongs to supervised learning. Meanwhile, convergence and holistic optimality of BP arithmetic during the training process are unable to be strictly guaranteed. In addition, the mechanism and selecting options of BP neural network's hidden neurons still remain puzzled. All these factors result in BP neural network's limited application in analyzing and categorizing economic structures. However, the self-organizing map neural network developed by Kohonen can help to not only classify input information automatically but

© Springer International Publishing Switzerland 2015
J.-H. Kim et al. (eds.), *Robot Intelligence Technology and Applications 3*,
Advances in Intelligent Systems and Computing 345, DOI: 10.1007/978-3-319-16841-8_20

also realize the spatial gather of neurons with identical functions. As an unsupervised clustering method, SOM method covers the limitations and enables the conduction of evaluating as well as categorizing economic structures.

In this paper SOM method is adopted to conduct clustering analysis on the industrial structure of China and the United State, in order to explore both identical and different features on this issue, as well as proceed further analysis basing on clustering results. After an introduction of SOM model in section 2, clustering analysis and discussions on industrial structures of China and America will be presented respectively in section 3 and 4, with a conclusion on section 5.

2   Self Organization Maps

Self Organization Maps is an unsupervised learning network which is developed by Kohonen from fundamental competitive network model. SOM network consists of one input layer and one output layer. To be specific, node number of input layer equal to dimensionality of input vector and each node connects one component of input vector. While output layer is commonly in form of array of one- or two-dimension, for two-dimension output layer, the topological structure varies in form of rectangle, hexagon, random connection, etc. Each node of input layer connects with that of output layer in form of weights.(Shown in figure 1)

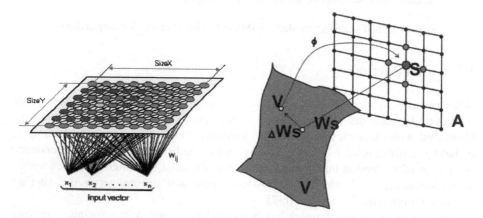

Fig. 1. SOM output layer structure and SOM weight vector structure

SOM actually is a kind of nonlinear mapping from discrete or continuous space V to one- or two-dimension discrete space A. For vector $\vec{v}$ from V, best match unit (BMU) from output space A is set according to characteristic mapping Φ, and synaptic weight vector W_S from S can be regarded as projection coordinate of S in input space. Adjusted weight matrix W enables output space A to simulate input space V through bellowing arithmetic.

There are 5 steps of the training process of SOM:

1. Build the network A based on the dimension and quantity of simples to be trained and make the first weight W.
2. Suppose that the input space is m dimensions from which the training samples are chosen randomly as $X = [x_1, x_2, ..., x_m]^T$.
3. Choose m dimension as the synaptic weight vector of the output layer node which matches the unit optimally.

   The weight of node j is $W_j = [w_{j1}, w_{j2}, ..., w_{jm}]^T$. Then the BMU of X is $i(X)$. Then define (1) as below:

$$i(X) = \arg \min \|X - W_j\|, \, j = 1, 2, ..., l \tag{1}$$

   Then the competing mechanism makes the vector of high dimension into low and scattered one.
4. Choose the node of which the weight should be adjusted. Based on the principle of Lateral interaction effect, the winner neuron would activate the close neuron and the range of activation is showed by the parameter σ which would fall over time gradually. The degree of activation is demonstrated by $h_{j,i}$, which is as below:

$$\begin{cases} h_{j,i} = \exp(-\dfrac{|d(j,i)|^2}{2\sigma^2}) & d(j,i) \le \sigma \\ h_{j,i} = 0 & d(j,i) > \sigma \end{cases} \tag{2}$$

   Whereas $d(i, j)$ is the Euclidean Distance between i and j in scattered space, $\sigma(t) < \sigma(t - \tau)$.
5. The adjustment of weight. This process is based on Hebb hypothesis, which is a positive feedback process. The adjustment is as below:

$$\frac{dW_j}{dt} = \eta(t) \cdot h_{j,i}(t) \cdot (X - W_j) \tag{3}$$

   Whereas $\eta(t)$ is parameter vector rate which falls over time.

Repeat the process 2 to 5 above until W is convergent.

3   Clustering Analysis Results on Chinese Regional Industrial Structures

Relevant data and information on industrial structures of Chinese provinces in 2012 are derived from category of gross regional domestic products in Chinese statistical yearbook 2013. After further processing, industrial structures are clustered as agriculture, manufacture, construction, transportation, catering and retail, finance and insurance, real estate, social service (combined by forestry, animal husbandry, sideline occupations and fishery, geological examination, government agencies and other industries). In this paper, Y_i represents GDP in region i, while Y_{ij} stands for output of industry j in region i, thus $\alpha_{ij} = Y_{ij} / Y_i$ can be regarded as the contribution ratio of different industries to certain region's GDP.

Each region's industrial structure is converted in to an input layer of 11-dimension vector, and processed by SOM kit in Matlab, during which process, learning efficiency varies from 0.8 to 0.02 with 800 training times. Each region's industrial structure vector is clustered through unsupervised learning method, with the clustering results listed in two-dimension 10*10 grid, as demonstrated in figure 2.

Clustering results of output layer show stability under different training times and learning efficiency. Majority of regional industrial structures are clustered into 5 except the region of Beijing, Jiangxi and Guangxi. Combined with data of regional industrial structures and Per Capita GDP, analysis results of each region are listed below:

Region a (Jilin, Anhui, Sichuan, Guizhou, Yunnan, Hunan, Gansu, Inner Mongolia and Xinjiang): the proportion of manufacture is between 30% and 35%, which is relatively low, and Per Capita GDP also remains at a low level of less than RMB 6500.

Region b (Jiangsu, Zhejiang, Shanxi, Heilongjiang, Guangdong, Shandong, Tianjin and Shanghai): the proportion of manufacture is above 45% and Per Capita GDP also remains at a high level of more than RMB 10000 (except Shanxi).

Region c (Henan, Hubei, Hebei, Fujian, Liaoning): the proportion of manufacture is at a relatively high level of 40% to 45%.

Region d (Chongqing, Ningxia, Shaanxi, Qinghai): both the proportions of manufacture and primary industry are relatively low.

Region e (Tibet, Hainan): the proportion of manufacture is small yet the proportion of agriculture is remain at high level.

Provinces fell in the same region shows similarities in terms of industrial structures, yet provinces of different regions have significant differences in industrial structures and economic development.

One thing that should be mentioned is, because of Beijing's particular industrial structure (its proportion of service industry is 61.6%, much higher than other provinces in China), its scores both in one-dimension clustering and two-dimension clustering are remarkably different from other provinces', which separates Beijing as a specific region.

Jilin; Anhui				Henan; Hubei	Hebei			Jiangsu	Shanxi; Heilongjiang
Sichuan; Guizhou; Yunnan					Fujian	Liaoning		Zhejiang	Guangdong; Shandong
Hunan	Gansu								Tianjin; Shanghai
Inner Mongolia; Xinjiang			Chongqing; Ningxia						
			Shaanxi						
Jiangxi		Qinghai			Beijing				
Hainan; Tibet									Guangxi

Fig. 2. Two-dimension clustering results of regional industrial structures (Source: Chinese statistical yearbook 2013)

Similarly, clustering results are also mapped in one-dimension 50*1 grid and converted into a Euclidean Distance rank of different region's industrial structures vector compared to one specific region.

Based on previous findings, certain region's distribution in one-dimension clustering results is compared with industrialization level (ratio of manufacture to GDP) in this region, resulting in significant linear correlation, as illustrated in figure 3.

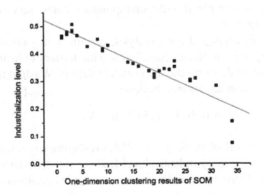

Fig. 3. Correlation analysis on one-dimension clustering results of regional industrial structures and industrialization process

Linear-regression analysis on the data results in regression equation as bellow:

$$Y = 0.502 - 8.6 * 10^{-3} X \qquad (4)$$

Under the significance level of 0.05, R^2 =0.92, presenting expected linear correlation, which indicates that industrialization process leads to the differences of Chinese regional industrial structures under current economic conditions.

4   Clustering Analysis Results on American Regional Industrial Structures

In order to examine the reliability of SOM's application in regional industrial structure research, spatial clustering analysis on the United States industrial structures is conducted, followed by a comparison with the clustering results of China.

Relevant data and information on industrial structures of American states in 2012 are derived from category of gross regional domestic products in Statistical Abstracts 2013, with a nine-cluster categorization of agriculture, manufacture, construction, transportation, retail, finance and insurance, social service and government expenditures.

Each region's industrial structure is then converted in to an input layer of 9-dimension vector, and processed by SOM kit in Matlab, during which process, learning efficiency varies from 0.8 to 0.02 with 800 training times. Each region's industrial structure vector is clustered through unsupervised learning method, and clustering results of output layer show stability under different training times and learning efficiency. Clustering results show that majority of regional industrial structures are clustered into 6 except several states.

Similarly, clustering results are mapped in one-dimension output layer. Resulted one-dimension scores are compared with corresponding states' service industry, with the results showed in figure 4.

Based on previous findings and data analysis, certain state's one-dimension clustering results have significant linear correlation with tertiary industry level of this state. Linear-regression analysis on the data (except data of Washington DC, Hawaii, Alaska) results in regression equation as bellow:

$$Y = 0.863 - 1.54 * 10^{-3} X \qquad (5)$$

Under the significance level of 0.05, R^2 =0.958, presenting expected linear correlation, which indicates that tertiary industry level leads to the differences of the United States regional industrial structures under current economic conditions.

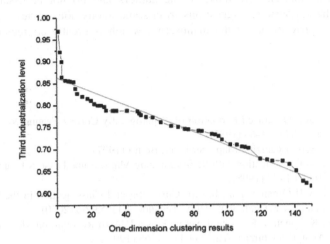

Fig. 4. Correlation analysis on clustering results of the United States regional industrial structures and tertiary industrialization level

5   Conclusion

In this paper, regional industrial structures of China and the United States are analyzed and compared through SOM clustering method, with key findings listed below:

1. According to the clustering results on regional industrial structures of China and the United States, SOM method are appropriate. Clustering results show expected stability under different parameters.
2. Clustering results of output layer on Chinese regional industrial structures are clustered into 5 regions, among which provinces in the same region shows similarities in terms of industrial structures, yet provinces of different regions have significant differences in industrial structures.
3. Clustering results indicate that under current economic condition, industrialization process leads to the differences of Chinese regional industrial structures, while such differences in America depend on the tertiary industry level.

By comparing results of clustering analysis, this paper suggests that dominant factor leading to differences of national industrial structures varies from nation to nation, which also reflects the correlation between national industrial structure and corresponding economic development.

Acknowledgments. This research has been supported by a grant from Ministry of Education, China Project of Humanities and Social Sciences (Project No.11YJCZH004) and the National Soft Science Research Program (Project No.2012GXS5DB3). Any opinions, findings, and conclusions or recommendations expressed in this material are those of the authors and do not necessarily reflect the views of the supporters. Suggestions from some anonymous reviewers have led to significant improvements of the manuscript, which is greatly appreciated by the authors.

References

1. Kohonen, T.: Self Organized Formation of Topologically Correct Feature Maps. Biological Cybernetics 43(1), 59–69 (1982)
2. Kohonen, T.: Self Organizing Maps. Springer, Berlin (1997)
3. Dieter, M.: Text Classification with Self-organizing Maps: Some Lessons Learned. Neurocomputing 21(1), 61–77 (1998)
4. Dimuthu, C.: Self Organization Map for Clustering and Classification in the Ecology of Agent Organizations. J. Cent. South Univ. Technol. 7(1), 53–56 (2000)
5. Song, X.-H., Kohonen, P.K.: Neural Network as a Pattern Recognition Method Based on the Weight. Analytica Chimica Acta 334(1), 57–66 (1996)
6. Yen, C.-L., Lu, M.-C., Chen, J.-L.: Applying the self-organization feature map (SOM) algorithm to AE-based tool wear monitoring in micro-cutting. Mechanical Systems and Signal Processing 34, 353–366 (2013)
7. Cheng, G., et al.: Gear fault identification based on Hilbert–Huang transform and SOM neural network. Measurement 46, 1137–1146 (2013)
8. Cai, Z., et al.: Self Organization Maps for Clustering Regional Industrial Structure in China. Complex Systems and Complexity Science 1(4), 62–66 (2004)
9. Hao, J., Wang, Y.: Self-organizing Feature Map for the Evaluation of Regional Economy–A Case Study of Yan'an City. Areal Research and Development 24(2), 58–61 (2005)
10. Feng, L.: Analysis of Artificial Neural Network of Comprehensive Strength of Area. Economic Geography 23(1), 9–11 (2003)

Research of a Self-adaptive Robot Impedance Control Method for Robot-Environment Interaction

Zhengyi Li[1,2,], Dandan Yang[1,2], Hui Zhou[1,2], and Huimin Cao[1,2]

[1] College of Biomedical Engineering, South-Central University for Nationalities,
Wuhan, People's Republic of China
[2] Key Laboratory of Cognitive Science (South-Central University for Nationalities),
State Ethnic Affairs Commission, Wuhan, People's Republic of China
{lizhengyixt,cao_huimin}@163.com, ryycq@126.com,
zhouhui@mail.scuec.edu.cn

Abstract. The robot impedance control performance decreases with unknown or changing environmental stiffness and damping parameters, in order to resolve this problem, this paper designs a self-adaptive robot impedance control method, which is characterized by integration of off-line learning and on-line adjustment to afford the stiffness and damping of the robot control system's impedance model competent for unknown or changing environment. For the off-line learning, defining the robot impedance control performance criterion, and establishing the geometric representation of the varying stiffness parameter, we derive the initial values of stiffness for the impedance model. Further, a neural network is designed to estimate the environmental effective stiffness, and combined with critical damping condition of the second-order robot-environment interaction system, we solves the initial value of damping for the impedance model. During the on-line adjustment, a rule self-tuned fuzzy controller is dedicated to adjust the stiffness and damping of the impedance model based on robot real-time contact force and position feedback. At last, experiments demonstrate the excellent stability and accuracy for the robot-environment contact force tracking control.

Keywords: Robot force control, self-adaptive impedance control, parameters estimation, rule self-tuned fuzzy control.

1   Introduction

Robot impedance control is to apply a desired profile of force in the constrained degrees of freedom while following the reference trajectory in the unconstrained degrees of freedom, it provides a unified approach by controlling both free motion and contact tasks using a single impedance algorithm, essentially, the impedance control strategy aims to maintain a desired dynamic relationship between the position error, velocity and the external force exerted by the robot's end-effector at the environment[1]. At present, the commercial robot impedance control system is mainly based on position error feedback in practice; we regard it as the position-based impedance control[2]. Robot impedance control is especially beneficial for the force and position synchronous

© Springer International Publishing Switzerland 2015   221
J.-H. Kim et al. (eds.), *Robot Intelligence Technology and Applications 3,*
Advances in Intelligent Systems and Computing 345, DOI: 10.1007/978-3-319-16841-8_21

control of counterpart systems, such as co-operating robots[3], the robot-human co-operation [4] and robotic assembly tasks[5].

As for the robot impedance control applied in the contact force and position control of the robot–environment interaction, if adjusting the impedance model parameters timely with changing environments, the system performance (force and position tracking stability and accuracy) will significantly superior to the system with constant impedance model parameters [6-7], this is non-trivial for the physically interaction between the robot and environments (elastic environment, as the spring-damper system), as different environments exhibit diverse damping and stiffness, even though the same environment, different forces exerted by the robot at the environment results in various environmental stiffness and damping, moreover, the multi-link articulated robot has different end-effector's effective stiffness and damping when its posture changes. Therefore, ways must be found for adjusting the stiffness and damping of the robot impedance model to accommodate environmental effective stiffness and damping varying in actual application. At present, one of the hot topics for robot impedance control is to improve the stability of contact force tracking control during robot-environment interaction, considered that environmental effective stiffness and damping are unknown or changing [8-9].

In recent years, researchers have tried to introduce the intelligent control methods, such as the fuzzy logic control (FLC), neural networks (NN) and the artificial intelligence, to adjust impedance model parameters in accordance to varying or unknown environmental stiffness and damping. For example, Nagata F. employed records of the contact force and environmental stiffness from the robot-environment interaction to train the NN off-line, and the trained NN was used to estimate the environmental stiffness on-line, it established impedance model based force control adaptive to various environmental stiffness [9]. Mallapragada V. adopted the recorded displacement and contact force during the robot-environment interaction to train the NN, and the trained NN afforded the adjustment for the coefficients of the proportional-differential (PD) control according to different environments, it established the impedance model based force control[10]. Some only adopted the fuzzy control, Xu G. applied the fuzzy control method to modulate the impedance model parameters of the rehabilitation robot control system based on estimation of the impaired limb's mechanical impedance characteristic parameters on-line[11]. Burn K designed the Sunderland fuzzy adaptive controller and fuzzy model reference adaptive control method to adjust the proportion between the contact force error and the control system output (robot joint angular displacement) on-line, it established the impedance model based force tracking control adaptive to varying environmental stiffness[12]. Also, some designed the comprehensive impedance control scheme combined with the neural network, genetic algorithm and fuzzy control, for instance, Cojbašić Ž. M. devised the fuzzy neural network integrated with the genetic algorithm to compensate the uncertainties of the environmental damping and stiffness[13]. Depending on the expert's experience of skilled operators, Xu Z. L. tried to establish the fuzzy relationship between the inputs and outputs of the desired robot impedance controller, and the fuzzy relationship was then used for training the neural network to afford impedance model parameters adjustment on-line, it established the robot adaptive impedance control [14].

The researches about the robot impedance based force position control adaptive to environmental stiffness and damping unknown or changing has made great achievements till now, but researches mainly concentrate in theoretical verification of the complex algorithms with perfect performance or establishing the simplified robot control system adaptive to engineering realization at the expense of degraded performance, there are short of achievements about counterbalancing well the permissible error between the force and position tracking control according to actual applications, and much efforts needs to simplify the complex algorithm of the robot impedance control adequate to varying environments in practical application. In this article, combining the off-line learning with on-line adjustment for the stiffness and damping parameters estimation, we implement a self-adaptive robot impedance model based force and position tracking control, the research work is characterized with the implementation of the robot-environment interaction experiment with excellent contact force tracking control stability.

2   Geometric Representation of the Varying Stiffness for Robot Impedance Control

In this paper, the environmental effective stiffness and damping represent the total stiffness and damping including the characteristics of the control system, the contacted object/surface, the robot mechanical body, the wrist force sensor, fixtures and tools. For the simplicity, the working environment in stable robot-environment contact could be modeled as the linear spring model with the stiffness k_e, $f = k_e x$, when the robot's end-effector interacts with the environment and the system is in steady state, the dynamic equation of the robot impedance control can be written as [15]:

$$f - f_d = -k_d(x - x_d) \tag{1}$$

where k_d denotes the stiffness of the desired impedance model.

As the base, defining the performance criterion of the robot impedance control in steady state:

$$\min_{k_d} J = (f - f_d)^2 + k_m^2 (x - x_d)^2 \tag{2}$$

where k_m represents the weighting distribution between the permitted position and force tracking error, its value is user-specified. According to Eq.(2), if $k_d = k_e^{-1} k_m^2$, the value of J is the smallest, the performance of the robot impedance control is optimal.

For the convenience of understanding, the geometric representation of k_d variation in Eq.(2) with $k_m = 1$ is shown in Fig. 1, the three solid lines denote the environmental dynamics model $f = k_e x$ and the impedance control model with different k_d respectively. The intersection point (x, f) of the environmental dynamics model and impedance control model represents the contact force and position when the robot's end-effector is in interaction with the environment, $f = k_e (f_d + k_d x_d)(k_e + k_d)^{-1}$,

$x = (f_d + k_d x_d)(k_d + k_e)^{-1}$, with k_d increasing from $k_d = 0$ to $k_d = \infty$, the intersection point moves along the environmental dynamics model from $(f_d k_e^{-1}, f_d)$ to $(x_d, k_e x_d)$. When the distance from the desired contact position/force point (x_d, f_d) to the environmental dynamics model ($f = k_e x$) is the shortest, the robot impedance control performance Eq. (2) is optimal, as shown in Fig. 1, when the line with the slope $k_d = k_e^{-1}$ is perpendicular to line $f = k_e x$ with an intersection (x_{opt}, f_{opt}), the distance between points (x_{opt}, f_{opt}) and (x_d, f_d) is the smallest, $k_d = k_e^{-1}$ is the optimal value of k_d when the system is in steady state, point (x_{opt}, f_{opt}) indicates the position and the exerted contact force of the robot's end-effector when $k_d = k_e^{-1}$. For Eq. (2) with $k_m \neq 1$, if we modify the coordinate frame in Fig. 1 as $(k_m x, f)$, the analysis is similar to $k_m = 1$.

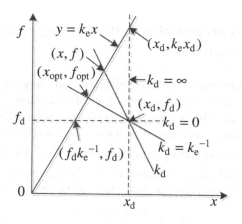

Fig. 1. Geometric representation for the variation of the impedance model stiffness k_d (system in steady state)

3   Estimation of the Environmental Effective Stiffness by Neural Network

Compared to all other impedance model parameters, the stiffness affects evidently the accuracy of impedance based robot force and position tracking control[16]. In this paper we design a neural network to estimate the environmental effective stiffness.

The experiment about the 6-DoF serial-link robot in interaction with the wooden board is depicted in Fig. 2, a force sensor is mounted at the wrist plate of the robot, it is used to monitor the contact force and close the loop in the force control, an aluminum small-ball probe, with a smooth tip, is attached to the force sensor. In order to have a better understanding of this robot-environment interaction, a decoupled, elastically compliant environment model is considered, it is described as $F_e = K_e(X - X_e)$, and the interaction is regarded as the point contact [15].

The experiment of the robot interacting with the wooden board starts with the probe slowly moving forward and to press the board surface, accompanied with records of the contact force and displacement every 0.03mm the probe moves ahead until the contact force grows up to 32N, next, the probe moves backwards and the pressure releases slowly, along with recording the contact force and displacement every 0.03mm until the contact force decreases to nearly zero. The measured contact force and displacement of the probe are plotted in Fig. 3, the circles present the press movement, and the stars refer to the release process. Fig. 4 depicts a plot of the contact force and environmental effective stiffness during the interaction, the calculation of the stiffness is $k_i = (f_{i+1} - f_i)(x_{i+1} - x_i)^{-1}$, $i=1,2, 3 \ldots$, where (f_i, x_i) and (f_{i+1}, x_{i+1}) are neighboring points in Fig. 3, the value of 0.1mm in Fig.3 occurred due to the backlash of the robot.

We adopt the multi-layer neural network depicted in Fig. 5 to fit the stiffness-contact force nonlinear relationship. The neural network learning is based on the back-propagation (BP) algorithm, the input of the proposed neural network is the contact force $f(t)$, and the output is environmental effective stiffness $k(t)$. The neural network is composed of one input layer, two hidden layers and one output layer, the input layer contains only one neuron, the hidden layers contain thirty neurons, the terms, y_1, y_i and y_j are the outputs of each corresponding layer, w_{1i}, w_{ij} and w_{j1} are the connection weights between neighboring layers, the output layer contains only one neuron without any excitation function, it means the value of the environmental effective stiffness is the sum of the output layer weights. The relationships between inputs and outputs of the output layer are described as (omit the variable t):

$$k = \sum_{j=1}^{15} w_{j1} y_j \qquad j = 1,2,3,\ldots \tag{3}$$

For the hidden layer 2, the relationships between inputs and outputs are:

$$y_j = f(net_j), \quad net_j = \sum_{i=1}^{15}\sum_{j=1}^{15} w_{ij} y_i \tag{4}$$

For the hidden layer 1, the relationships between inputs and outputs are:

$$y_i = f(net_i), \quad net_j = \sum_{i=1}^{15} w_{1i} y_i \quad i=1,2,3,\ldots \tag{5}$$

Fig. 2. Experiments for the robot interacting with the wooden board

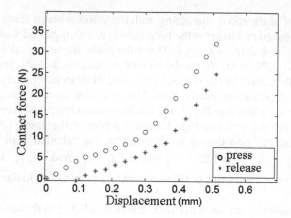

Fig. 3. Contact force-displacement of the interaction between the probe and the wooden board

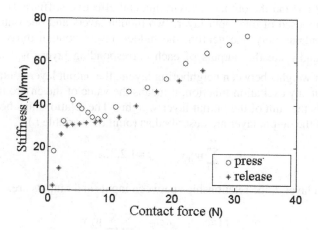

Fig. 4. Environmental effective stiffness-contact force of the robot-environment interaction

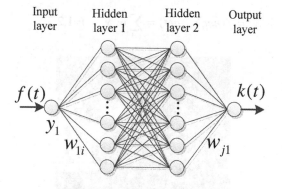

Fig. 5. The NN for fitting the contact force-stiffness relationship

The excitation function of the hidden layer is $f(x) = (1+e^{-x})^{-1}$, and the excitation functions of the input and output layers are linear. For the neural network learning by BP algorithm, the cost function is defined as $E(t) = 2^{-1}[k_d(t) - k(t)]^2$, this study minimizes the cost function using the steepest descent method and the chain rule to achieve the weighting corrections (Δw_{j1}, Δw_{ij}, Δw_{1i}), these corrections are :

$$\Delta w_{j1} = -\eta \, \partial E(t) \big/ \partial w_{j1} = -\eta \, \partial E(t) \big/ \partial k \, y_j = \eta[k(t)-k_d(t)]y_j$$

$$\Delta w_{ij} = -\eta \, \partial E(t) \big/ \partial net_j * \partial net_j \big/ \partial w_{ij} = \eta[k_d(t)-k(t)]w_{j1}y_i$$

$$\Delta w_{1i} = -\eta \, \partial E(t) \big/ \partial w_{1i} = \eta[k_d(t) - k(t)]w_{ij}y_1(1 - y_1)f(t)$$

where $k_d(t)$ is the environmental effective stiffness available in Fig. 4, $k(t)$ is the output of the neural network, t is the number of the sampling interval, the learning rate η is restricted to (0, 1).

Before real-time control, the neural network is trained off-line by the stiffness-contact force data sets shown in Fig. 4, and the training result of the neural network is shown in Fig. 6 and Fig. 7, with the neural network learning cycles increasing, the accuracy of the stiffness-contact force curve fitting is improving, the training process would not conclude until the outputs and inputs of the neural network could fit the given stiffness-contact force relationship with negligible difference. After that, with the contact force as the inputs, the neural network can predict the environmental effective stiffness k_e on-line corresponding to the environmental mechanical impedance characteristic.

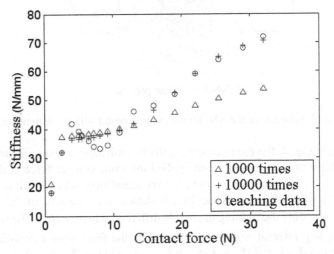

Fig. 6. NN learning result of the contact force-stiffness relationship of the press process

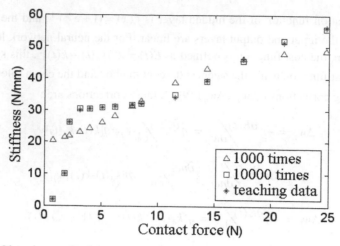

Fig. 7. NN learning result of the contact force-stiffness relationship of the release process

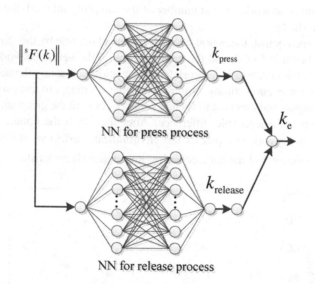

Fig. 8. Structure of the NNs for the environmental stiffness estimation

According to Fig. 4, the environmental effective stiffness reveals difference for the press and release process even though applied the same contact force at the environment, therefore, it is necessary to provide two neural networks to estimate the environmental effective stiffness. As the Fig. 8 shown, the structure of the two neural networks is same, only the training data set is different, where $^sF(k)$ is the contact force at k-step sampling interval, superscript s presents the force sensor coordinate frame. Defining $\Delta\|^sF(k)\| = \|^sF(k)\| - \|^sF(k-1)\|$, and $\Delta\|^sF(k)\| > 0$ indicates the press process, the neural network trained by the contact force and stiffness of the press motion is selected, $k_e = k_{press}$, if $\Delta \| {}^s F(k) \| < 0$, the neural network learning with the contact force and stiffness of the release process is adopted, $k_e = k_{release}$. After obtained the environmental effective stiffness k_e, substituting k_e into the impedance model stiffness expression $k_d = k_e^{-1} k_m^2$ derived in Sec. 2 yields the initial value of impedance model stiffness $k_d(0)$:

$$k_d(0) = k_e^{-1} k_m^2 \tag{6}$$

where k_m is defined in Eq. (2).

4   Calculating the Initial Values of Stiffness and Damping for the Robot Impedance Model

The damping of the impedance model parameters has explicitly influence on the stability of the impedance based force and position tracking control, and in order to improve the control system stability, the initial value of the impedance model damping is calculated based on the dynamic stability analysis of the second-order robot-environment interaction system.

The goal of robot impedance control is to endow the robot's end-effector with the desired impedance relationship, as specified by the robot impedance model equation in Cartesian space[1]:

$$M_d(\ddot{X} - \ddot{X}_d) + B_d(\dot{X} - \dot{X}_d) + K_d(X - X_d) = F - F_d \tag{7}$$

where M_d, B_d and K_d denote diagonal symmetric positive definite matrices of the inertia, damping, and stiffness gains of the desired impedance model, respectively. X, $\dot{X}$ and $\ddot{X}$ are the position, velocity and acceleration vectors of the robot's end-effector in the task space, the subscript d refers to the desired values. F and F_d are the actual and desired force vectors applied by the robot's end-effector to the environment, and F can be expressed by environmental dynamics model:

$$F = -B_e \dot{X} - K_e X \tag{8}$$

where matrices B_e and K_e are the environmental damping and stiffness coefficients, the term, X, is the environment deformation due to the external applied force F. Combining Eq. (8) and Eq. (7), and given that $\ddot{X}_d$, $\dot{X}_d$, X_d and F_d are constant, it yields the characteristic equation for the robot-environment interaction:

$$s^2 + M_d^{-1}(B_d + B_e)s + M_d^{-1}(K_d + K_e) = 0 \tag{9}$$

The dynamic stability (position tracking accuracy) of the second-order robot-environment interaction system Eq. (9) is optimal when it is in critical damping (some practical application maybe needs the overdamping) condition, it is possible to derive the optimal value of B_d for a single degree of freedom interaction between the robot and environment:

$$B_d = 2\sqrt{M_d(K_d + K_e)} - B_e \tag{10}$$

where, the value of M_d is user-specified, its value ensure Eq. (10) to be plus, environmental effective stiffness K_e is predicted by the neural network described in Sec. 3, K_d is the initial value of the impedance model stiffness calculated by Eq. (6), the environmental effective damping B_e can be estimated[15].The Eq. (10) determines the theoretical value of impedance model damping B_d, and its result is regarded as the initial value of the impedance model damping $B_d(0)$.

In summary, the flow for calculating the initial stiffness $K_d(0)$ and damping $B_d(0)$ of the impedance model is illustrated in Fig. 9, it is a preprocessing unit for adjusting impedance model parameters adaptive to varying environment, $K_d(0)$ and $B_d(0)$ must be modified on-line to afford the contact force and position tracking control during the practical robot-environment interaction.

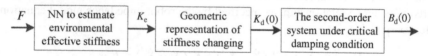

Fig. 9. The flow of the impedance model stiffness and damping initial value calculation

5   Design of the Self-adaptive Robot Impedance Control

The schematic diagram of the overall self-adaptive impedance control framework is illustrated in Fig. 10, gravity compensation is to subtract the end-effector's gravity from the measured resultant force ${}^B F_e(k)$ to obtain actual contact force, matrix ${}^B_T R$ is the transformation matrix from the robot's base coordinate frame B to robot's tool coordinate frame T, the self-tuned fuzzy control module shown in Fig.10 contains the fuzzy various stiffness controller (FVSC) and fuzzy various damping controller (FVDC).

As for the system impedance model of Eq. (7), it is assumed that there is no interference of the impedance characteristic between three axes, defining $M_d = m_d I^{l*l}$, $B_d = b_d I^{l*l}$, $K_d = k_d I^{l*l}$, l=3, and rewriting the impedance model in component form (along contact surface normal direction) yields:

$$e_f = m_d \ddot{e}_x + b_d \dot{e}_x + k_d e_x \tag{11}$$

where $e_x = x - x_d$ and $e_f = f - f_d$ are the position and contact force error of the robot's end-effector in contact surface normal direction, and the terms m_d, b_d and k_d are the inertia, damping and stiffness elements of the impedance model in contact surface normal direction. If m_d, b_d and k_d are regarded as coefficients of the $\ddot{e}_x$, $\dot{e}_x$ and e_x contribution to the contact force error e_f, then we can regulate m_d, b_d and k_d to optimize force tracking control performance. For the robot force control application, as the robot movement inclines to keep in a low translational speed, inertia $m_d\ddot{e}_x$ can be regarded as a constant, only b_d and k_d need to be adjusted in practice.

The designed self-adaptive impedance control establishing the force control in the contact surface normal direction and position control along the contact surface tangent direction is explained below. At first, it needs to estimate the ranges of the damping and stiffness (b_d, k_d) of the impedance model for the robot-environment interaction by trials and error in the whole workspace, $k_{min} \leq k_d \leq k_{max}$, $b_{min} \leq b_d \leq b_{max}$, defining λ_b and λ_k as correction factors to regulate the damping and stiffness of the impedance model, $\lambda_b, \lambda_k \in (-1, 1)$, and arranging the damping b_d adjustment:

$$\begin{cases} b_d(k+1) = b_d(k) + \lambda_b [b_{max} - b_d(k)], \lambda_b \geq 0 \\ b_d(k+1) = b_d(k) + \lambda_b [b_d(k) - b_{min}], \lambda_b < 0 \end{cases} \tag{12}$$

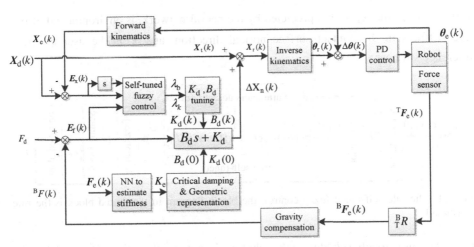

Fig. 10. Schematic diagram of the self-adaptive robot impedance control

Defining the stiffness k_d adjustment expression:

$$\begin{cases} k_d(k+1) = k_d(k) + \lambda_k [k_{max} - k_d(k)], \lambda_k \geq 0 \\ k_d(k+1) = k_d(k) + \lambda_k [k_d(k) - k_{min}], \lambda_k < 0 \end{cases} \tag{13}$$

where $b_d(k)$ and $k_d(k)$ are the damping and stiffness of the k-step sampling period, $k=0,1,2,\ldots$, the initial value $k_d(0)$ and $b_d(0)$ are calculated by Eq.(6) and Eq. (10).

Next, applying the FVSC with inputs of e_f, e_x to modulate λ_k, and the FVDC with inputs of $\dot{e}_x$, e_f to adjust λ_b. Owing to the relationship of weights dependence between e_x and e_f ($\dot{e}_x$ and e_f), the adjustment factor α is introduced to perform the fuzzy rule modification, by regulating the proportions of various contributory inputs (e_f, e_x, $\dot{e}_x$) corresponding to the practical application, the mechanism of the rule self-tuned FLC is depicted in Fig.11.

The formula for adjusting λ_k is

$$\begin{cases} U = \langle \alpha(t)E_x + [1-\alpha(t)]E_f \rangle \\ \alpha(t+1) = \alpha(t) - \gamma S[e_x \bullet e_f] \end{cases} \tag{14}$$

where U, E_x and E_f are the fuzzy quantities of λ_k, e_x and e_f separately, t is control cycles, the norm, $\langle x \rangle$, stands for the quantization operator, and the value of $\langle x \rangle$ is an integer such that its absolute value equals the minimum integer which is larger than or equal to the absolute value of x. The adjustment factor $\alpha \in (0\ 1)$, $S[x] = \begin{cases} 1, & \|x\| > 1 \\ x, & -1 \le x \le 1 \end{cases}$, $\gamma = S[k_0^{-1}k_e]$ reflects the current environmental effective stiffness level, the term, k_e, predicted by the neural network, is environmental effective stiffness in the contact surface normal direction, and k_0 is the user-specified reference stiffness.

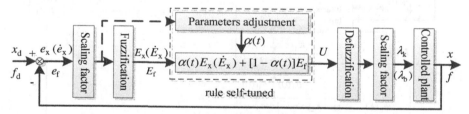

Fig. 11. The rule self-tuned fuzzy control, the block diagram in the dotted block is the rule self-tuned

Eq. (14) is used to adjust k_d during the robot-environment interaction, if $\|e_x \bullet e_f\| > 1$, the system's force or position tracking deviates from the desired value, hence, it needs to enlarge stiffness k_d according to Eq. (11), while $k_0^{-1}k_e$ is larger (the environmental stiffness is larger), α decreases rapidly, and the weight of E_f in U becomes large, as E_f is larger and E_x is smaller, the value of U increases relatively, the value of λ_k rises, it results in enlarging k_d to improve the system force/position tracking precision. While $k_0^{-1}k_e$ is smaller (the environmental stiffness is smaller), α decreases slowly, the weight of E_x in U is larger, as E_f is smaller and E_x is larger, then U increases relatively, the value of λ_k rises, that is enlarging k_d to improve system force/position tracking control precision. Whereas, if the system's force and position tracking errors are small ($\|e_x \cdot e_f\| < 1$), the analysis is similar to the above, but the terms α, U and λ_k have little changes, λ_k changes slightly, the system keeps in steady state.

Expression for adjusting λ_b is

$$\begin{cases} U = \langle \alpha(t)\dot{E}_x + [1-\alpha(t)]E_f \rangle \\ \alpha(t+1) = \alpha(t) - \gamma S[\dot{e}_x \cdot e_f] \end{cases} \tag{15}$$

where U and $\dot{E}_x$ are the fuzzy quantities of λ_b and $\dot{e}_x$, $\gamma = S[b_0^{-1}b_e]$, b_e is defined as the components of the environmental effective damping in the contact surface normal direction, b_0 is the user-specified reference damping, the remained parameters are similar to Eq. (14). For using Eq. (15) to adjust b_d during the robot-environment interaction, when the system is at serious oscillation state ($\|\dot{e}_x \cdot e_f\| > 1$),it needs to increase damping b_d according to Eq.(11), if $b_0^{-1}b_e$ is lager (the environmental damping is larger), α decreases rapidly, the weight of E_f in U increases, as the E_f is larger and $\dot{E}_x$ is smaller, then U or λ_b increases, it results in enlarging damping b_d to stabilizes the system. If $b_0^{-1}b_e$ is smaller (the environmental damping is smaller), α decreases slowly, the weight of $\dot{E}_x$ in U increases, as E_f is smaller and $\dot{E}_x$ is larger, then U, λ_b and b_d increase to suppress system's oscillation. On the other hand, when the system is almost stable with low force tracking error ($\|\dot{e}_x \cdot e_f\| < 1$), similar to the analysis above, it concludes that the changes of α, U or λ_b is small, then b_d has little change, the system is almost keep stable.

At last, substituting the newly λ_b and λ_k obtain from Eq. (14) and Eq. (15) into Eq. (12) and Eq. (13) to update k_d and b_d . The updated (k_d , b_d) is substituted into impedance model Eq. (11) to generate new reference trajectory for compensating environmental parameters changing.

6   Robot Force Control Experiments

To demonstrate the performance of the proposed robot impedance control strategy, experiments were conducted using the National Numerical Control System Engineering Research Center experimental test-bed consisting of a 6-DoF industrial robot fitted with an ATI Delta (SI-165–15) wrist force sensor, and a smooth aluminum probe is mounted at the sensor in sequence, the position control cycle is 4ms.The experimental set-up is shown in Fig.12.

Mounting the smooth elastic wooden flat board tightly at the vertical wall and neighboring with the industrial robot fixed in a horizontal plane. In the experiment, the probe is tasked to slide on the contacted flat surface with the desired contact force 2N perpendicular to the contacted surface. For simplify of the force calculation, the orientation of the probe keeps vertically to the surface (+y-direction of the robot base coordinate frame), the movement along the contacted surface is controlled by the direction buttons on the controller panel. In actual experiments, the probe is at rest in free space firstly, with the probe is almost vertical to the flat surface. Next, command the control system to start force control function, as the contact force is lower than 2N, the probe moves towards and presses the wooden board surface until the contact force rises up to 2N. At last, manual control the probe sliding on the board surface by pushing the buttons on the controller panel: moving along −x-direction at first, then along +x-,-z-directions simultaneity (in the robot base coordinate frame), it means the robot reference trajectory is user-specified in the experiment, the reference translational velocity is 20mm/s.

In the experiment, the desired neural network is trained to estimate the environmental effective stiffness ahead, the probe's position error e_x and its derivative $\dot{e}_x$ lie in [-1 1]mm and [-100 100]mm/s, respectively, the contact force error e_f belongs to [-0.2 0.2]N. Defining the domains of e_x, $\dot{e}_x$, e_f, λ_b and λ_k are the same:{-1,-0.8,-0.6,-0.4,-0.2, 0,0.2,0.4,0.6,0.8,1}, fuzzy partitioning of the controller input variables (e_x, $\dot{e}_x$, e_f) and output variables (λ_b, λ_k) has been performed by choosing five fuzzy sets for each variable, marked with linguistic labels (NB, NS,ZE, PB, PS) that appear in the fuzzy control rules, the Gaussian membership functions of the five fuzzy subsets are defined as follows, $\mu_{NB}(x) = [1 + e^{-15(x+0.7)}]^{-1}$, $\mu_{NS}(x) = e^{-25(x+0.5)^2}$, $\mu_{PB}(x) = [1 + e^{-15(x-0.7)}]^{-1}$, $\mu_{PS}(x) = e^{-25(x-0.5)^2}$ and $\mu_{ZE}(x) = e^{-25x^2}$. The applied rule base consists of all fifty rules, and each rule produces one controller output, and the fuzzy rules are adjusted timely by modifying factor α along the task, the utilization of different tables of rules accordingly the task to be performed, for example, the Table 1 is the fuzzy rule base while $\alpha = 0.5$.

During the experiment, as time evolves, the contact force in the surface normal direction measured by the force sensor is shown in Fig.13 (the sampling period is 1KHz), notice that the contact force is null during the motion in free space and rises rapidly after the contact happens, the contact force converges to the desired value 2N during the interaction with trivial oscillation as expected, the steady-state error of the force tracking is bounded in [-0.2 0.2]N (oscillations are ascribed to the effect of the irregularities on the surface and measurement noise). The probe's position is calculated by the measured robot joint angle displacement, and Fig.14 is the probe's displacement over time in x-, y-, and z-direction of the robot base coordinate frame, the displacements in all direction fluctuate around the desired trajectory slightly (excluding the deviation due to the irregularities on the surface), it means that the interaction of the probe sliding on the wooden board is stable. As the damping and stiffness of the fixed wooden board is unknown at first, the experiments demonstrate that the proposed robot impedance control method can afford stable contact force/position tracking control for the interaction between the robot's end-effector and the environment with unknown damping and stiffness.

Contrasted with the designed fuzzy controller described in [17], the fuzzy control directly modifies the reference trajectory by the contact force deviation and this easily induces the reference trajectory unsmooth, the fuzzy controller designed in this paper has advantage of smooth reference trajectory and it conduces the contact force, position tracking control more stable, the reason is that, the designed rule self-tuned fuzzy control uses the position and contact force error to adjust the impedance model stiffness, and applies the velocity, contact force error of the end-effector to modulate the impedance model damping, then it produces the correction of the reference trajectory through the updated impedance model. In [18], the designed neural network established the contact force and position tracking control for the robot-environment interaction, with the robot's end-effector position error and its derivative, the contact force error as the inputs and the robot joint driving torque as the outputs, large numbers of inputs result in difficulty for system programs development, in contrast, the proposed neural network in this paper is dedicated for fitting contact force-stiffness relationship consisted of only one input and the rule self-tuned fuzzy controller for adjusting impedance model parameters are convenient for engineering practice.

Fig. 12. Picture of experimental set-up

Table 1. Representation of the fuzzy rule base ($\alpha = 0.5$)

$\lambda_k(\lambda_b)$		e_f				
		NB	NS	ZE	PS	PB
$e_x(\dot{e}_x)$	NB	NB	NS	ZE	PS	PS
	NS	NS	NS	ZE	ZE	PS
	ZE	NS	ZE	ZE	ZE	PS
	PS	NS	ZE	ZE	PS	PS
	PB	NS	NS	PS	PS	PB

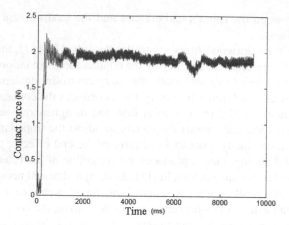

Fig. 13. Contact force in the contact surface normal direction

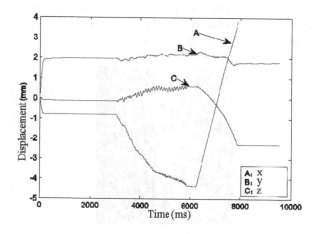

Fig. 14. Probe displacement in x-,y- and z-direction

7   Conclusions

Owing to the proposed robot impedance control method, we achieve a satisfactory performance for constant contact force control between the robot end-effector and environments with unknown or changing damping and stiffness, the effectiveness of the proposed control method is proved through experiments.

For the unknown or changing damping and stiffness of the environment, given the performance criterion of the robot impedance control, we obtain the optimal initial value of the stiffness and damping for the robot's impedance model. Based on robot's real-time contact force and position feedback errors, a sophisticated rule self-tuned fuzzy controller is design to adjust the stiffness and damping of the control system's impedance model on-line to achieve excellent robot force/position tracking stability

and accuracy. Further experimental tests on other nonlinear and complicated ro-
bot-environments interaction using the proposed control method will be performed in
the near future.

Acknowledgments. This work was supported by the National Natural Science Foundation of China (No.61178087) and the Special Fund for Basic Scientific Research of
Central Colleges, South-Central University for Nationalities.

References

1. Seraji, H., Colbaugh, R.: Force Tracking in Impedance Control. The International Journal of Robotics Research 16(1), 97–117 (1997)
2. Huang, L., Ge, S.S., Lee, T.H.: An Adaptive Impedance Control Scheme for Constrained Robots. International Journal of Computers, Systems, and Signals 5(2), 17–26 (2004)
3. Rastegari, R., Moosavian, S.A.: Multiple impedance control of space free-flying robots via virtual linkages. Acta Astronautica 66(5-6), 748–759 (2010)
4. Tsuji, T., Yoshiyuki, T.: Tracking Control Properties of Human–Robotic Systems Based on Impedance Control. IEEE Transactions on Systems, Man, and Cybernetics 35(4), 523–535 (2005)
5. Chen, H.P., Wang, J.J., Zhang, G., Fuhlbrigge, T., Kock, S.: High-precision assembly automation based on robot compliance. Int. J. Adv. Manuf. Technol. 45(9/10), 999–1006 (2009)
6. Valency, T., Zacksenhouse, M.: Accuracy/Robustness Dilemma in Impedance Control. Transactions of the ASME 125(3), 310–319 (2003)
7. Katić, D.: Advanced Connectionist Control Algorithm for Robotic Compliance Tasks based on Wavelet Network Classifier. Scientific Technical Review 2, 24–28 (2006)
8. Owen, W., Croft, E., Benhabib, B.: Stiffness optimization for two-armed robotic sculpting. Industrial Robot 35(1), 46–57 (2008)
9. Nagata, F., Mizobuchi, T., Hase, T., Haga, Z., Watanabe, K., Habib, M.K.: CAD/CAM-based force controller using a neural network-based effective stiffness estimator. Artificial Life and Robotics Archive 15, 101–105 (2010)
10. Mallapragada, V., Erol, D., Sarkar, N.: A new method of force control for unknown environments. International Journal of Advanced Robotic 24(3), 313–322 (2007)
11. Xu, G., Song, A.: Fuzzy variable impedance control for upper-limb rehabilitation robot. In: 5th International Conference on Fuzzy Systems and Knowledge Discovery, pp. 216–220. Inst. of Elec. and Elec. Eng. Computer Society (2008)
12. Burn, K., Short, M., Bicker, R.: Adaptive and nonlinear fuzzy force control techniques applied to robots operating in uncertain environments. Journal of Robotic Systems 20(7), 391–400 (2003)
13. Cojbašić, Ž.M., Nikolić, V.D.: Hybrid Industrial Robot Compliant Motion Control. Automatic Control and Robotics 7(1), 99–110 (2008)
14. Xu, Z.L., Fang, G.: Fuzzy-neural impedance control for robots. Robotic Welding, Intelligence and Automation 299, 263–275 (2004)

15. Fanaei, A., Farrokhi, M.: Robust adaptive neuro-fuzzy controller for hybrid position/force control of robot manipulators in contact with unknown environment. Journal of Intelligent and Fuzzy Systems 17(2), 125–144 (2006)
16. Nagata, F., Mizobuchi, T., Tani, S., Watanabe, K., Hase, T., Haga, Z.: Impedance model force control using a neural network-based effective stiffness estimator for a desktop NC machine tool. Journal of Manufacturing Systems 28(2/3), 78–87 (2009)
17. Zhongxu, H., Bicker, R., Marshall, C.: Position/force control of manipulator based on force measurement and its application to gear deburring. Journal of Intelligent & Fuzzy Systems 14(4), 215–223 (2003)
18. Abu-Mallouh, M., Surgenor, B.: Force/velocity control of a pneumatic gantry robot for contour tracking with neural network compensation. In: ASME International Manufacturing Science and Engineering Conference, pp. 11–18. IEEE Press, Illinois (2008)

Gaze Control Factors for Natural Human Robot Interaction from Scanpath Comparisons

Bum-Soo Yoo and Jong-Hwan Kim

Department of Electrical Engineering, KAIST, 291 DaeHak-ro,
Yuseong-gu, Daejeon, Republic of Korea
{bsyoo,johkim}@rit.kaist.ac.kr

Abstract. Human-like gaze control of robots is essential for natural human robot interaction (HRI), and thus it is necessary to identify the gaze control factors that affect human gaze. In this paper, the gaze control factors are derived inductively from the observation of human scanpaths. Human scanpaths are measured from movie clips and are transformed into vector-based representation that contains spatio-temporal information. The transformed scanpaths are compared to find the most common scanpath. The selected scanpath is analyzed to identify the gaze control factors. The derived factors are anticipation, knowledge, and attractions in a static image such as salient regions, humans, objects, and center of visual inputs.

Keywords: Eye tracking, gaze, scanpaths, similarity measure, vector-based similarity measure.

1 Introduction

Gaze plays an important role in human-to-human interaction, as it represents paying attention [1]. With adequate mutual gaze, people feel comfortable and tend to maintain interactions with others.

To be companions of humans, robots need to make human-like gaze control that will improve interactions between humans and robots [2]. To make human-like gaze control, considerable research has been conducted on identifying what affect human gaze. Researchers introduced several factors based on their hypotheses and demonstrated it with experiments [3], [4]. However, these experiments are susceptible to be biased in testing hypotheses. Even though gaze control factors were derived based on the observation of humans scanpaths in free-viewing of images, the factors derived from the static images need to be verified because unlike the real environment, there was no spatio-temporal information [5]. Thus, in this paper, human scanpaths from free-viewing of movie clips are observed and are analyzed. Since visual inputs are continuously changed in movie clips, there exists spatio-temporal information, which allows us to derive gaze control factors in an environment similar to the real environment. Scanpaths are recorded from 53 subjects with 31 movie clips, and they are compared to each other from five aspects: shape, length, direction, position and duration.

© Springer International Publishing Switzerland 2015
J.-H. Kim et al. (eds.), *Robot Intelligence Technology and Applications 3,*
Advances in Intelligent Systems and Computing 345, DOI: 10.1007/978-3-319-16841-8_22

Based on the comparisons, a scanpath with the largest similarity is selected and analyzed to identify the gaze control factors.

The paper is organized as follows. Section II describes a vector-based scanpath comparison method. Section III describes comparisons among human scanpaths from free-viewing of movie clips and explains the gaze control factors. Lastly, we conclude with important remarks in Section IV.

2   Similarity Measurement

It is necessary to quantitatively measure similarity among human scanpaths to find the most common scanpath in each movie clips. In the previous research, the earth mover's distance (EMD) was used to compare scanpaths from free-viewing of images [5]. The EMD has a strong point in representing spatial deviation. However, it has a weakness in losing sequential information that is important when a scanpath is measured from free-viewing of movie clips where visual inputs are changing continuously as time passes by.

A vector-based comparison method is introduced to compare scanpaths [6]. It represents a scanpath with a combination of fixations and saccadic movements, and compares a scanpath with others from the five viewpoints: shape, length, direction, position, and duration. A strong point of this method is containing spatio-temporal information. However, it cannot represent smooth pursuing, and it needs an assumption that gaze consists of fixations and saccadic movements.

2.1   Simplification

A scanpath, represented by a combination of fixations and saccadic movements, contains noise from measurements. It is necessary to filter noise while maintaining its properties. Noises are filtered with two steps simplification. First, gaze direction may vibrate in a small area, and the vibration makes several small saccadic movements in a small area with different directions. They can be considered as one direction. Thus, a group of consecutive vectors $\{u_1, u_2, \ldots, u_m\}$, $m > 1$, are merged to one vector if $\|\sum_{i=1}^{n} u_i\| < T_{amp}$ for all $n \leq m$. In this paper, T_{amp} is assigned as 50.0 pixels. Second, gaze direction may move continuously to the same direction, and it makes several consecutive saccadic movements having similar directions. They can be considered as one large saccadic movement. Thus, a group of consecutive vectors $\{u_1, u_2, \ldots, u_m\}$, $m > 1$, are merged to one vector if $\max |\angle u_n - \angle u_{avr}| < \angle T$ for all $n \leq m$, where $\angle u_{avr}$ is the average angle from u_1 to u_n. In this paper, $\angle T$ is assigned as 30.0 °. Fig. 1 shows examples of a scanpath simplification.

After the simplification, a simplified scanpath $U = \{u_1, u_2, \ldots, u_m\}$, should be temporally aligned with other simplified scanpath $V = \{v_1, v_2, \ldots, v_n\}$ based on their shapes. From the two scanpaths, a comparison matrix M is produced, which is defined as follows:

$$M(k, l) = \frac{u_k \cdot v_l}{|u_k||v_l|} \tag{1}$$

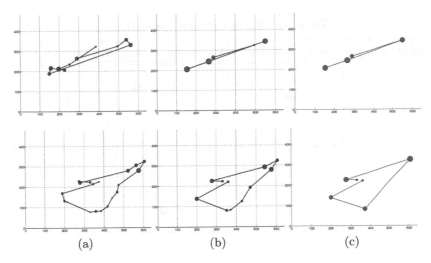

Fig. 1. Scanpaths simplification. (a) The scanpaths are measured from free-viewing of movie clips. (b) Scanpaths are simplified with magnitude. (c) Scanpaths are simplified with angles.

where $k < m$ and $l < n$. The comparison matrix is transformed into a graph, and the Dijkstra algorithm is used to find a shortest path from $(1,1)$ to (m,n). According to the shortest path, two vectors are temporally aligned. Fig. 2 shows examples of the Dijkstra algorithm with the comparison matrix.

With the aligned scanpaths, $A = \{a_1, a_2, \ldots, a_r\}$ and $B = \{b_1, b_2, \ldots, b_r\}$, each vector is compared to each other from the five view points as in the following.

- Shape of the vector: $\left(\sum_{i=1}^{r} \frac{|a_i - b_i|}{L_{diag}} \right)/r$

- Length of the vector: $\left(\sum_{i=1}^{r} \frac{|a_i| - |b_i|}{\max\{a_i, b_i\}} \right)/r$

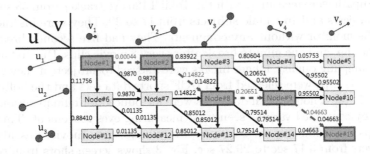

Fig. 2. The Dijkstra algorithm is applied to the comparison matrix. The shortest path is marked with dot-lines. u_1 is aligned with $\{v_1, v_2\}$, u_2 is aligned with $\{v_3, v_4\}$, and u_3 is aligned with v_5 temporally.

Table 1. Aligned vectors in scanpaths

Simplified Scanpath	Aligned vector no. 1	Aligned vector no. 2	Aligned vector no. 3
	u_1	u_2	u_3
	v_1 v_2	v_2 v_3 v_4	v_5

- Direction of the vector: $\left(\sum\limits_{i=1}^{r} \dfrac{cos^{-1}\frac{a_i \cdot b_i}{|a_i||b_i|}}{180} \right)/r$

- Position of fixation: $\left(\sum\limits_{i=1}^{r+1} \dfrac{|F_P(a_i) - F_P(b_i)|}{L_{diag}} \right)/(r+1)$

- Duration of fixation: $\left(\sum\limits_{i=1}^{r+1} \dfrac{|F_D(a_i) - F_D(b_i)|}{\max\{F_D(a_i), F_D(b_i)\}} \right)/(r+1)$

Note that L_{diag} represents the diagonal length of input space, $F_P(a_i)$ represents a position of fixation before the saccadic movement a_i, and $F_D(a_i)$ represents a duration of fixation before the saccadic movement a_i.

3   Experiments

3.1   Eye Tracking Experiments

The scanpath were measured with the Tobii T120 eye tracker from 53 subjects, 37 male subjects and 16 female subjects from 17 to 30. They were instructed to look at the monitor without any comments. They had five seconds break after each movie clip, and one minute break after every 8 movie clips. Due to fatigue of eyes and sampling quality, longer breaks were allowed to subjects if they wanted.

Test movie clips were selected from UCF sports data sets and the hollywood2 data sets [7], [8]. Though both data sets were introduced for human action recognition, they contained various scenes for measuring eyes movement. Thirty-one movie clips were randomly selected except black-and-white movie clips, and their lengths were from 3.17 sec to 29.27 sec. Fig. 3 shows screen shots from selected movie clips.

Table 2 shows the result of the scanpaths comparisons. The smaller value represents the larger similarity. The scanpath with the largest similarity showed

Fig. 3. Screen shots from the selected movie clips

Table 2. Results of the scanpaths comparisons

No. Movie	Min (Std. Dev.)	No. Movie	Min (Std. Dev.)	No. Movie	Min (Std. Dev.)
1	0.2333(0.0214)	12	0.2435(0.0160)	23	0.2149(0.0312)
2	0.2279(0.0230)	13	0.2472(0.0187)	24	0.2265(0.0267)
3	0.2317(0.0220)	14	0.2357(0.0229)	25	0.2078(0.0212)
4	0.2518(0.1027)	15	0.2302(0.0185)	26	0.2466(0.0233)
5	0.2218(0.0245)	16	0.2500(0.0175)	27	0.2043(0.0299)
6	0.2129(0.0221)	17	0.2292(0.0234)	28	0.2017(0.0229)
7	0.2246(0.0225)	18	0.2249(0.0222)	29	0.2313(0.0180)
8	0.1542(0.0361)	19	0.2139(0.0237)	30	0.2787(0.0222)
9	0.2075(0.0240)	20	0.2461(0.0203)	31	0.1890(0.0277)
10	0.2296(0.0309)	21	0.2143(0.0278)		
11	0.2039(0.0361)	22	0.2345(0.0232)		

0.2237 in average, the minimum was 0.1542, and the maximum was 0.2787. Note that a scanpath of the 17-th subject from the 4-th movie clip was excluded because it consisted of one fixation point without any saccadic movement.

3.2   Analysis

In each movie clip, the scanpath with the largest similarity was selected and observed to derive the gaze control factors.

Anticipation. From the visual inputs, subjects could recognize situations, and they could anticipate what would happen next. Then subjects will gaze at points according to their anticipation. Fig. 4 shows an example of gaze control with anticipation, where red dots represent fixation points with magnitude and a larger circle represents a longer duration. In Fig. 4(a), a man is shooting a basketball. Subjects gazed at the basketball stand even before the ball reached at the basketball stand, as shown in Fig. 4(b), because they could understand the situation and anticipate that the basketball was heading toward the basketball stand. In Fig. 4(c), subjects gazed at the basketball stand continuously to identify whether the shoot was successful or not.

(a)   (b)   (c)

Fig. 4. (a) Subjects gazed at a man throwing a basketball. (b) Subjects gazed at the basketball stand before the ball reached. (c) Subjects continued to gazed at the basketball stand to identify whether the shoot is successful or not.

(a)   (b)   (c)

Fig. 5. (a) Subjects gazed at the man playing golf. (b) Subjects tried to find a golf ball at the ground in front of the man. (c) Subject gazed at the man again.

Knowledge. Human behaviors are highly related with objects, and according to humans behaviors, a specific object is required. After subjects recognized human behaviors, they tried to find objects related to the human behaviors based on their knowledge. Fig. 5 shows an example of gaze control with knowledge. A man swung a golf club in Fig. 5(a). Subjects recognized his behavior and found the golf club easily, and then they tried to find a golf ball in Fig. 5(b). They gazed at the ground in front of him to find a golf ball, not in the sky based on their knowledge. In Fig. 5(c), subjects gazed at the person again.

Since subjects knew where a golf ball is usually placed, they gazed at the ground in front of him, not the sky or beside him. This is an example of gaze control based on scene schema knowledge. There are four kinds of scene knowledge in human gaze control system: short-term episodic scene knowledge, long-term episodic scene knowledge, scene schema knowledge, and task knowledge [9]. Table 3 shows the meaning of each scene knowledge with an example. In the experiment, not only scene schema knowledge, but also other kind of knowledge was observed. Fig. 6 shows a gaze control occurred by the short-term episodic scene knowledge. In Fig. 6(a), a man used a phone. He put the phone at the lower left corner of the visual input as shown in Fig. 6(b). When a woman moved to the left side as shown in Fig. 6(c), subjects gazed at the lower left corner because they knew where the phone was based on their short-term episodic scene knowledge. She took the phone in Fig. 6(d).

Table 3. Four kinds of scene knowledge in human visual system

Name	Explanation	Example
Short-term episodic scene knowledge	Specific knowledge about a particular scene at a particular time	I just put a key on the table
Long-term episodic scene knowledge	Specific knowledge about a particular scene that is stable over time	The TV is besides the drawer
Scene schema knowledge	Generic knowledge about a particular category of scene	Cars are on ground, generally on a road, not sky
Task knowledge	Generic knowledge about a particular scene at a particular time	Before changing lanes, check the side-view mirror

(a)   (b)   (c)   (d)

Fig. 6. Examples of the short-term episodic scene knowledge. (a) The man used a phone. (b) The man put the phone at the lower left corner. (c) and (d) Subjects knew where the phone was placed and gazed at the lower left corner before the woman used the phone based on the short-term episodic scene knowledge.

Attractions in Static Images. Targets that attract human gaze in images also attracted human gaze in movie clips such as salient regions, object, humans, center of image and so on [5]. Subjects gazed at humans continuously because humans are inherently attracted human gaze. In case of objects, objects that were related with human behaviors attracted human gaze. Sometimes meaningless objects became meaningful objects when they were related with human behaviors.

Salient regions were not apparent in movie clips compared to static images. Since movie clips had a limited time for expressions, they usually had focused targets, and subjects tended to gaze at focused targets. When visual inputs were fixed without any movements, subjects sometimes gazed at salient regions.

The centers of visual inputs in movie clips attracted human gaze stronger than in images. In images, subjects had a time to gaze at boundaries of visual inputs because they were not changed. However, in the movies, subjects had less time to gaze at boundaries of visual inputs because they were continuously changed. Also, focused targets were usually placed at the center of visual inputs to express them effectively.

4 Conclusion

This paper identified the gaze control factors by the observation of human scanpaths from free-viewing of the movie clips. The observed scanpaths were represented by the vector-based representation containing spatio-temporal information, and they were temporally aligned through the dijkstra algorithm.

The aligned scanpaths were compared from the five viewpoints of shape, length, direction, position, and duration to find the most common scanpath. Based on the comparisons, anticipation, knowledge, and attractions in static images were identified as the gaze control factors. These factors could be useful if they were included in a design of robots' gaze control algorithm.

Acknowledgement. This work was supported by the Technology Innovation Program, 10045252, Development of robot task intelligence technology, funded by the Ministry of Trade, Industry & Energy (MOTIE, Korea).
This work was supported by the Software Computing Technology Development Program, 14-824-09-012, Technology Development of Virtual Creatures with Digital Emotional DNA of Users, funded By the Ministry of Science, ICT and Future Planning.

References

1. Pease, B., Pease, A.: The Definitive Book of Body Language. Bantam (2008)
2. Kim, J.-.H., Choi, S.-H., Park, I.-W., Zaheer, S.A.: Intelligence Technology for Robots That Think. IEEE Comput. Intelli. Mag. 8(3), 70–84 (2013)
3. Itti, L., Koch, C.: A Saliency-based Search Mechanism for Overt and Covert Shifts of Visual Attention. Vision Res. 40(10), 489–1506 (2000)
4. Cerf, M., Harel, J., Einhäuser, W., Koch, C.: Predicting Human Gaze Using Low-Level Saliency Combined with Face Detection. Adv. Neur. 20, 241–248 (2008)
5. Yoo, B.-S., Kim, J.-H.: Scanpaths analysis with fixation maps to provide factors for natural gaze control. In: Kim, J.-H., Matson, E., Myung, H., Xu, P. (eds.) Robot Intelligence Technology and Applications 2. AISC, vol. 274, pp. 361–368. Springer, Heidelberg (2014)
6. Jarodzka, H., Holmqvit, K.: A Vector-based, Multidimensional Scanpath Similarity Measure. In: Proc. Symp. Eye-Tracking Research and Application, pp. 211–218 (2010)
7. Marszalek, M., Laptev, I., Schmid, C.: Actions in Context. In: Proc. IEEE Conf. Computer Vision and Pattern Recognition, pp. 2929–2936 (2009)
8. Rodrigues, M.D., Ahmed, J., Shah, M.: Action MACH a Spatio-Temporal Maximum Average Correlation Height Filter for Action Recognition. In: Proc. IEEE Conf. Computer Vision and Pattern Recognition, pp. 1–8 (2008)
9. Herderson, J., Ferreira, F.: The Interface of Language Vision and Action: Eye Movements and the Visual World, Psychology Press (2004)

The Affective Loop: A Tool for Autonomous and Adaptive Emotional Human-Robot Interaction

Maria Vircikova, Gergely Magyar, and Peter Sincak

Center for Intelligent Technologies, Dpt. of Cybernetics and Artificial Intelligence,
Technical University of Kosice, Slovakia
{Maria.Vircikova,Gergely.Magyar,Peter.Sincak}@tuke.sk

Abstract. The paper presents an affective model for social robotics, where the robot is capable of behavior adaptation, in accordance with the needs and preferences of a particular user. The proposed approach differs from other studies in human-robot interaction as these usually have been using the 'Wizard of Oz' technique, where a person remotely operates a robot. On the other side, simulated robots are not able of personalized behaviors and behave according to the preprogrammed set of rules. We provide a tool to personalize affective artificial behaviors in cooperative human—robot scenarios, where human emotion recognition, appropriate robotic behavior selection and expression of robotic emotions play a key role. The preliminary experiments show that the personalized affective robotic behavior can achieve better results in a scenario in which a robot motivates children in learning. We believe that human—robot interfaces which mimic how humans interact with one another in an empathic way could ultimately lead to robots being accepted in the wider domain.

Keywords: Affective Robotics, Personalization, Social Human—Robot Interaction, Social Robots, Subjective Computing.

1 Introduction

1.1 Social Human-Robot Interaction

Social robotics is an emerging part of service robotics, where social robots can engage people in natural exchanges with the focus on human interaction. In contrast to service robots, social robots are explicitly developed for the interaction of humans and robots to support human—like interaction. The current goal of the human-robot interaction(HRI) is to identify methods to promote longer—lasting interactive relationships. According to Ros et al.[1] endowing robots with capabilities for contingent and appropriate handling of user responses, plus some capacity for the prediction of responses to their subsequent actions, is one of the main challenges in HRI. General problems of HRI to be solved can be grouped into the following: to develop intelligent behavior in a complex, unpredictable environments; to sense and recognize emotion and affect in others; to express affect and internal states to others; and to respond to humans with social adeptness and appropriateness.

© Springer International Publishing Switzerland 2015
J.-H. Kim et al. (eds.), *Robot Intelligence Technology and Applications 3*,
Advances in Intelligent Systems and Computing 345, DOI: 10.1007/978-3-319-16841-8_23

Social machines should understand not only human speech, but also human non—verbal expressions, human emotions, needs and preferences, and should adjust their functions and features. They should be able to change their performance to the expectations of a particular human, in order to cooperate with this human more efficiently, to help him better, to motivate him in doing something, to be from the user's point of view empathic—like real social companions.

There is a large personality difference in people´s expectations and preferences, not only ranging from environments such as hospitals, schools, and homes, but from person to person. We are far from building universal—type robots. The solution to Bill Gates[2]´ idea of "a robot in every home" can be a personalization. Robots can learn from humans – from interacting with partners or even from social psychologists, even if they do not have programming or technical skills.

1.2   Building Emotional Intelligence

Human intelligence expressed in both verbal and non—verbal communication has not only attributes and attainments; it has affect. Empathy, as defined by Keen[3], means to recognize feelings of others, the causes of these feelings, and to be able to participate in the emotional experience of an individual without becoming part of it. This work studies the emotional interaction that considers empathy, and, as Kozima[4] showed, empathy plays an essential role in social communication. Empathy is considered one of the key elements of emotional intelligence, the ability to recognize one´s own emotions, understand what they're expressing, and realize how they affect others. It involves a perception of others and understanding how other people feel, which allows managing relationships.

Cramer et al.[5] studied how empathy affects people's attitudes towards robots. In this study, there was a significant negative effect on user's trust towards a robot that displayed inaccurate empathic behavior. Conversely, participants who observed the robot displaying accurate empathic behaviors perceived their relationship with the robot as closer. A relevant finding in empathy research is that empathy is correlated with social supportive behavior, as described by Hoffman[6].

Up to now, in empathic social robots usually an operator stands in for the socially cognitive capacities of the robot, or robots deliver human emotions to people as a mediator. The other approach in state—of—the—art emotional social robots are simulated robots, with own personality to recognize human emotions and express its own emotions. These empathic robots have a limited set of output behavior or expressions, and users in long—term interaction can start to find them boring. This paper proposes an approach in which the simulated, autonomous, emotional robot becomes personalized during the interaction. It gives robots ability to learn how their users want them or need them to behave. Users teach a robot to express affect, without any programming skills, according to their preferences.

2   Design of the Affective Loop

This work proposes a model of emotional intelligence which operates during the interaction with users of the robot called "affective loop". It is a process that runs in a cycle from start to end of the program. Following Goleman´s definition of human emotional intelligence[7], the proposed affective loop has the following parts that can be grouped into three abilities of the system: perception, management of emotions (control of emotions) and expression. The Affective Loop is described and developed in our previous work[17].

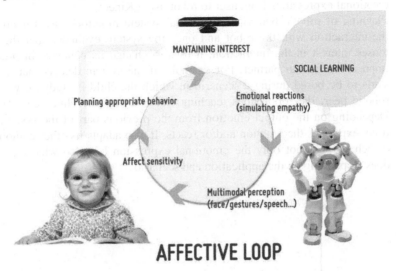

Fig. 1. The Affective Loop. The first parts of the model stand for "perception of human," where a system for body—based emotion recognition using Kinect is implemented. The next part represents the internal mode of the robot. This gives a smooth change in behavior, which may lead towards believability in the robotic reactions. It is also a significant tool for the visualization of the internal states of the robot, where emotional reactions can be monitored. For experimental purposes, this work uses Plutchik´s emotional model, but this can be replaced with another model. The last part of the system stands for the output, the affective behavior of the robot.

2.1 Subjectivity Incorporation

The proposed emotional model implements the subjectivity in different stages:

1. Perception of users ´emotional states. As people express emotions in different ways, there is a need to build a recognition system able to deal with it. Such system for emotion recognition was proposed in our previous work [8].
2. Management of artificial emotions. This part of the system manages internal states of the robot. An emotional model is constructed, where basic emotions are combined into a complex emotional spectrum using Plutchik's psychoevolutionary theory of emotions which combines eight basic emotions into the entire emotional spectrum. This part of affective loop is also a tool for visualization of the robotic emotional model. A tool that provides

personalization is the graph that represents the internal model and the parameters that represent each emotion; users can adjust the duration of the emotion and its intensity.

3. Learning emotional expressions from demonstration. A tool that provides subjective adjustement is a framework for mapping of body—based emotional expressions from user to robot using Kinect.

4. Planning of robotic behavior. The visual system monitors user´s interest in the interaction with the robot and once the system evaluates that the user looses interest in the interaction, it tries to change its behavior in order to impress the human partner. For example, if the system detects that a child starts to be bored during a scenario in which the child is studying with the robotic peer, the robot stops teaching and entertain the child (e.g. dance). Depending on the output emotion from the previous part of the system, the robot expresses the emotion and/or reacts. It also adapts its other component of behavior — not only the emotional expression but also what the robot does, depending on the application and scenario.

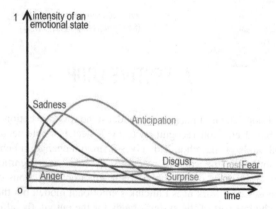

Fig. 2. The internal model in a form of a graph where the intensity of each emotion and its duration can be adjusted.

Fig. 3. User is able to create his/her body-based emotional expression without any programming knowledge using the Kinect sensor.

2.2   Behavioral Models

The aim of the personalization is not only to adjust the form of the emotional expressions but also their timing based on stimuli—when and which emotion has to be expressed in order to increase the believability of the robot and the effectiveness of the affective HRI. To achieve this, the system adjusts the behavioral model based on emotional mimicry and the behavioral model based on monitoring the goal of the human—robot cooperation.

During the emotional mimicry, the robot copies the emotional expressions of user. This idea is not new—even the first empathy studies in HRI focused on mimicking the user's affective state, a particular aspect of empathy also designated as emotional contagion. Hess[9] defines mimicry as "The tendency to imitate facially, vocally or posturally people with whom we are interacting." Many studies agree that mimicry improves the quality of interactions, e.g. Bernieri and Rosenthal[10], Chartrand and Bargh[11], LaFrance and Broadbent[12], Yabar and Hess[13]. Mimicry tends to be shown preferentially to in—group members or others with whom we cooperate, as, arguably, we are generally more likely to affiliate with those towards whom we are positively inclined as shown by Bourgeois and Hess[14], or with whom we expect to cooperate rather than to compete, shown by Lanzetta and Englis[15], Weyers et al.[16].

In case of the behavioral model based on monitoring the goal of cooperation, the robot acts depending on the evolvement of the scenario in the role of a motivator—this kind of behavior can be called "social supportive behavior." The robot monitors the development of user´s achievements to finish the given task. The toy task to demonstrate the usability of the system is to remember English words for non-English speakers. Then, the goal is to choose the actions of the robot which contribute to, for example, that the child remembers all the English words that are presented to him/her during the interaction. Then, after the child learns new skills successfully, the robot expresses its emotion of joy.

The behavioral model is being personalized by using a user specific strategy, taking into consideration what kind of behaviors are the most appropriate and motivating for a particular child. The system monitors the goal of the interaction in a cooperative task and adapts its affective behaviors. The robot reflects detected user's emotions and, at the same time, follows the goal of the task. During the experiment, a proper combination of the designed models should be found to adapt to particular users.

3   Experiments and Results

3.1   Design of Experiments

The system was tested during child—robot interactive scenarios in which the human is in the centre of the HRI - the robots adapt to their users and become a responsive entity. The evaluation of the effectiveness of HRI is proposed within the task collaboration environment. We selected as a test-bed scenario in order to fuse playing and rehabilitation/education techniques by using a robotic design to induce child—robot interaction, in which the criterion is to achieve an entertaining and effective communication with the children. We placed Nao robots to school environment and did the following steps: 1. The robot is introduced to the children; 2. Children can freely interact with it and 3. Educational part. Children aged 5—7 were chosen as subjects for these studies deliberately, as children of this age are playful, interested in the unknown and, thus, are not limited by fear of the unknown. On the other hand, the problem of children of this age is their concentration. The majority of children are not able to repeat the same activity during a longer time. However, the observations show that interacting with robots make the educational process attractive for children.

In this case the robot is in a role of peer who helps children to learn new words in a foreign language. Our motivation was principally to motivate children to learn, as there is a wide application potential. According to Oh and Kim[17], focusing too much on the efficiency of education, conventional 'Information, Communication and Technology (ICT)' education in elementary schools has its own limitations, so high expectations are being placed upon robotic products, which can be used to promote students' independent learning experiences in school. In this sense, the school environment provides a favorable climate and serves a good starting point for HRI experimentation and research.

Several educational games based on a similar basis were designed. The robot randomly picks an object from a known set and asks the user to show the appropriate object. The robot uses a simple visual and command recognition system to check whether the user provides correct answers. The design of the application was discussed with a professional pedagogic psychologist. The algorithm is as follows:

1. The robot asks the user to show a picture of an object/color/letter (depends on the game),
2. The user selects the picture from the set and shows it to the robot,
3. The robot recognizes whether the selected picture is correctly chosen or not,
4. The user can show the robot a picture and ask it to say the name of the object. The user is asked to repeat the word. In this case, the robot checks the pronunciation of the user,
5. The words to learn can be easily added to the database without any programming knowledge.

Fig. 4. Photos illustrating the experiment

3.2   Preliminary Results

During the HRI, the robot tries to adjust its simulation of affect (e.g. it recognizes that user is happy, it can adopt the emotion of joy and express it also or it can select the emotion of sadness if user is not remembering the English word). The system evaluates its effectiveness by measuring the deviation from the overall goal (whether children are remembering the English words) after every step and adjusts. The goal is a desired final state of a robot, achieved in a finite number of steps and the states can be types of ideal robotic affective behaviors or actions that lead to the maximal performance of the user. This measurement of quality of the interaction brings an engineering approach for the HRI community in scenarios in which the goal of the human—robot cooperation is known.

The preliminary results with 10 children during 3, 20-minutes long sessions show that the personalization is needed and contributes to the efficiency of the human-robot cognitive training. The adaptation of the robot to its user should lead to maintenance of engagement with users over the time, which is the topic of central importance in HRI: the multi—modal approach must support an engagement, naturalistic interaction. The questionnaire was discussed with psychologists and the results show that all subjects evaluated the interaction with the robot positively. We observed that Nao robot is suitable for the children-robot interaction: children like his friendly design and there are were no safety issues reported.

4   Conclusion

In order for robots to be truly integrated into humans' every—day environment to provide services such as company, care—giving, entertainment, patient monitoring, aids in therapy, etc., they cannot be simply designed and taken off the shelf to be directly embedded into a real—life setting. Adaptation to incompletely known and changing environments and personalization to their human users and partners, are necessary features to achieve successful long—term integration.

An important note regarding the implementation of the model is that the system was autonomous during interaction. That is a key quality of the proposed system, and can make social robots independent from any operator usually presented in human-robot interaction experiments. We think that a software AI module capable of personalization and affective interaction can move social robotics forward in terms of commercial issues. The novelty of the model consists in its ability of personalization in each part of the model. It means that the proposed methodology for machine personalization/empathy contains elements of learning. The learning tools can be divided into three categories: from data – mainly based on neural networks, from humans in the form of experience – using fuzzy rules and learning by demonstration, using Kinect and optimized by interactive evolutionary computation.

Acknowledgments. This publication/paper/poster/article is the result of the Project implementation: University Science Park TECHNICOM for Innovation Applications Supported by Knowledge Technology, ITMS: 26220220182, supported by the Research & Development Operational Program funded by the ERDF.

References

1. Ros, R., Nalin, M., Wood, R., Baxter, P., Looije, R., Demiris, Y., Pozzi, C.: Child—robot interaction in the wild: advice to the aspiring experimenter. In: Proceedings of the 13th International Conference on Multimodal Interfaces, pp. 335–342. ACM (2011)
2. Gates, B.: A robot in every home. Scientific American 296(1), 58–65 (2007)
3. Keen, S.: Empathy and the Novel. Oxford University Press (2007)
4. Kozima, H., Nakagawa, C., Yano, H.: Can a robot empathize with people? Artificial Life and Robotics 8, 83–88 (2004)
5. Cramer, H., Goddijn, J., Wielinga, B., Evers, V.: Effects of (in)accurate empathy and situational valence on attitudes towards robots. In: ACM/IEEE Int. Conf. on Human—Robot Interaction, pp. 141–142. ACM (2010)
6. Hoffman, M.: Empathy and moral development: Implications for caring and justice. Cambridge Univ. Press (2001)
7. Goleman, D.: The Brain and Emotional Intelligence: New Insights, More Than Sound (2011) ISBN 978-1-93444-115-2
8. Vircikova, M., Pala, M., Smolar, P., Sincak, P.: Neural Approach for Personalised Emotional Model in Human-Robot Interaction. In: WCCI 2012: IEEE World Congress on Computational Intelligence, Brisbane, Australia, June 10-15, pp. 970–977. IEEE (2012) ISBN 978-1-4673-1489-3
9. Hess, U., Blairy, S.: Facial mimicry and emotional contagion to dynamic emotional facial expressions and their influence on decoding accuracy. International Journal of Psychophysiology 40, 129–141 (2001)
10. Bernieri, F.J., Rosenthal, R.: Interpersonal coordination: behavior matching and interactional syn-chrony. In: Feldman, R.S., Rimé, B. (eds.) Fundamentals of Nonverbal Behavior, pp. 401–432. Cambridge University Press, Cambridge (1991)
11. Chartrand, T.L., Bargh, J.A.: The chameleon effect: the perception–behavior link and social interac-tion. Journal of Personality and Social Psychology 76, 893–910 (1999)
12. LaFrance, M., Broadbent, M.: Group report: posture sharing as a nonverbal indicator. Group and Organization Studies 1, 328–333 (2009)
13. Yabar, Y., Hess, C., Display, U.: of empathy and perception of out—group members. New Zealand Journal of Psychology 36, 42–50 (2006)
14. Bourgeois, P., Hess, U.: The impact of social context on mimicry. Biological Psychology 77, 343–352 (2008)
15. Lanzetta, J., Englis, T., Expectations, B.G.: of cooperation and competition and their effects on observ-ers' vicarious emotional responses. Journal of Personality and Social Psychology 56, 543–554 (1989)
16. Weyers, P., Muehlberger, A., Kund, A., Hess, U., Pauli, P.: Modulation of facial reactions to avatar emotional faces by nonconscious competition priming. Psychophysiology 46, 328–335 (2009)
17. Oh, K., Kim, M.: Social Attributes of Robotic Products: Observations of Child— Robot Interactions in a School Environment. International Journal of Design 4(1) (2010)
18. Vircikova, M.: Machine Empathy: Towards Artificial Emotional Intelligence with Active Personalization in Social Human Robot Interaction. PhD. Thesis, Technical University of Kosice, Slovakia (2014)

3D Visibility Check in Webots for Human Perspective Taking in Human-Robot Interaction

Ji-Hyeong Han and Jong-Hwan Kim

Department of Electrical Engineering, KAIST, 291 Daehak-ro, Yuseong-gu, Daejeon,
305-701, Republic of Korea
{jhhan,johkim}@rit.kaist.ac.kr

Abstract. The rapid development of intelligent robotics would facilitate humans and robots will live and work together at a human workspace in the near future. It means research on effective human-robot interaction is essential for future robotics. The most common situation of human-robot interaction is that humans and robots work cooperatively, and robots should give proper assistance to humans for achieving a goal. In the workspace there are several objects including tools and a robot should identify the human intended objects or tools. There might be situational differences between a robot's perspective and a human perspective because of several obstacles in environment. Thus, a robot needs to take the human perspective and simulates the situation from the human perspective to identify the human intended object. For human perspective taking, first of all a robot needs to check its own visibility for the environment. To address this challenge, this paper develops a 3D visibility check method by using a depth image in Webots. By using the developed method, a robot can determine whether each point in the environment is visible or invisible at its posture and detect objects if they are visible.

Keywords: Human-robot interaction, 3D visibility check, human perspective taking, Webots.

1 Introduction

Since robot technology and intelligence technology have been matured, robots will come into our daily lives in the near future. Thus, the effective human-robot interaction (HRI) is needed especially from human-robot cooperation point of view. The research dealing with HRI problems has been intensively studied in various applications [1]-[7].

Among the various HRI problems, the human-robot cooperation problem should be solved first since getting a robot into a human workspace is the most possible and helpful way to get robots involved in human life. When a robot and a human work cooperatively, they should share the context information to achieve the goal successfully. The usual human-robot cooperation situation is that a robot assists a human partner and they work together using some tools or objects. Therefore, a robot needs to identify the human intended object to provide the proper assistance to a human.

© Springer International Publishing Switzerland 2015
J.-H. Kim et al. (eds.), *Robot Intelligence Technology and Applications 3,*
Advances in Intelligent Systems and Computing 345, DOI: 10.1007/978-3-319-16841-8_24

There might be situational differences between a robot's perspective and a human perspective because of several obstacles in environment; therefore, a robot should take the human perspective and consider the situation from the human perspective to identify the human intended object [8]. There can be four kinds of object states based on the robot's perspective and human perspective: i) an object that is visible from both perspectives, ii) an object that is visible from a robot's perspective and invisible from a human perspective, iii) an object that is invisible from a robot's perspective and visible from a human perspective, or iv) an object that is invisible from both perspectives. Among the objects of all the cases, the objects of the first and third cases might be the human intended objects since the object that a human can not see might not be the human intended one. Therefore, to identify the human intended object, first of all a robot needs to check its own visibility for the objects and also needs to check the human visibility by taking the human perspective.

To deal with the above mentioned issue, this paper develops a 3D visibility check method. The developed method considers 3D visibility rather than 2D visibility since a robot needs to consider the heights of obstacles. We develop the 3D visibility check method by using an RGB-D sensor in Webots. Webots is a robot simulator that has been widely used in robotics since 1998 [9]. In recent years, an RGB-D sensor has been used to wide robotics field for 3D modeling of environments, object and gesture recognition, simultaneous localization and mapping (SLAM), etc [10]-[12]. By using the proposed method, a robot can check which points in the environment are visible or invisible such that it can detect the objects if they are visible.

This paper is organized as follows. Section 2 presents the developed 3D visibility check method. In Section 3, experimental results are discussed and concluding remarks follow in Section 4.

2   The Developed 3D Visibility Check Method

In this section, the developed 3D visibility check method in Webots is explained along with step by step calculation. The developed method uses a depth image from an RGB-D camera sensor which locates at the top of a simulated mobile robot in Webots.

2.1   The Developed 3D Visibility Check Method

Fig. 1 shows coordinates of a depth image obtained from an RGB-D camera sensor in Webots and relationship between a depth image and real world. Since an RGB-D camera in Webots is a standard pinhole camera and it is implemented using OpenGL, the relationship between a depth image and real world can be modeled. The pixels in a depth image mean they are visible from the robot's perspective; thus, a robot can determine whether each point in real world is visible or invisible from its posture by using the relationship between the depth image and real world. The detailed calculations are explained in the following.

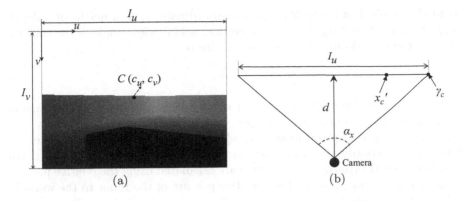

Fig. 1. (a) A depth image obtained from an RGB-D sensor in Webots. I_u and I_v are width and height of the image, u and v are pixel coordinates of the image, and $C(c_u, c_v)$ is a center of the image. (b) The relationship between a depth image in u-coordinate and real world in camera's x-coordinate. α_x is a horizontal field of view angle of the camera, d is a depth value obtained from a depth image, and x_c' and γ_c are points of the real world in camera's x-coordinate.

The relationship between a pixel of a depth image, $p(u', v')$ and a corresponding point of real world based on a camera posture, $q_c(x_c', y_c', z_c')$ can be represented as a proportional expression. The relationship between u-coordinated of a depth image and x-coordinate in real world camera posture is represented as follows:

$$\gamma_c : \frac{I_u}{2} = x_c' : (u' - c_u) \tag{1}$$

where u' is a pixel of a depth image in u-coordinate, x_c' is a corresponding point of real world in camera's x-coordinate, I_u is a width of a depth image, and $c_u = I_w/2$. γ_c is a point of real world in camera's x-coordinate corresponding to a right-hand end pixel of a depth image in u-coordinate and it is represented as follows:

$$\gamma_c = d \tan \frac{\alpha_x}{2} \tag{2}$$

where d is a depth value of p obtained from a depth image and α_x is a horizontal field of view angle of the camera. Therefore, a point of real world in camera's x-coordinate is represented as follows:

$$x_c' = \gamma(u' - c_u)\frac{2}{I_u} = \frac{2d(u' - c_u)\tan(\alpha_x/2)}{I_u}. \tag{3}$$

In the same manner as the calculation of x_c', y_c', which is a point of real world in camera's y-coordinate corresponding to a bottom end pixel of a depth image in v-coordinate, is represented as follows:

$$y_c' = \frac{2d(v' - c_v)\tan(\alpha_y/2)}{I_v} \tag{4}$$

where v' is a pixel of a depth image in v-coordinate, I_v is a height of a depth image, $c_v = I_v/2$, and $\alpha_y = \alpha_x \frac{I_v}{I_u}$ is a vertical field of view angle of the camera. A point of real world in camera's z-coordinate is

$$z'_c = d. \tag{5}$$

By using the above calculation, a robot can match a pixel of a depth image in (u, v)-coordinate to a point of real world in camera's (x, y, z)-coordinate.

The next step is transforming $q_c(x'_c, y'_c, z'_c)$ to $q_g(x'_g, y'_g, z'_g)$ that is based on global (x, y, z)-coordinate. To transform the camera's coordinate to the global coordinate, the transformation matrices are calculated using the relative posture of the camera to the robot and the relative posture of the robot to the world in Webots. The transformation is done as follows:

$$\mathbf{q}_g = {}^g_r\mathbf{T} \, {}^r_c\mathbf{T} \, \mathbf{q}_c \tag{6}$$

where ${}^r_c\mathbf{T}$ is a transformation matrix from camera's coordinate to robot's coordinate and ${}^g_r\mathbf{T}$ is a transformation matrix from robot's coordinate to global coordinate.

Objects in the environment can be detected if a robot identify they are visible at robot's posture by using the developed 3D visibility check method. Each point in the environment is under three cases. The first case is that the point has a depth value exceeding the maximum depth range of an RGB-D camera. In this case, the point is considered as empty like the sky and ignored. The second case is that y-value of the point is close to zero, i.e. $y < \sigma$, where σ is a small value like 1.0 cm. It means that the point belongs to the ground and the reason using σ is that an RGB-D camera has noise. The last case is that the point is a part of an object, when satisfying $||\mathbf{C}_{obj} - \mathbf{P}||_2 < ObjectSize/2$, where $\mathbf{C}_{obj}$ is a center point of an object and $\mathbf{P}$ is a considered point.

3   Experiment

This section presents the experimental environment and results that were conducted to show the effectiveness of the developed method.

3.1   Experimental Environment

A differential wheel robot with an RGB-D camera sensor at its top was used to show the effectiveness of the developed method. Fig. 2 shows the experimental environment. There were two obstacles with different heights, one object (a ball), and a simulated human. The height and width of ground were 4.0 m. The parameters of the RGB-D camera mounted on the robot were a horizontal field of view angle as 0.994837 radian, width as 320, height as 240, minimum and maximum depth range as 0.3 m and 4.5 m, respectively, in the same manner as Kinect.

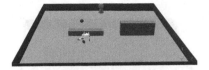

Fig. 2. The experimental environment in Webots

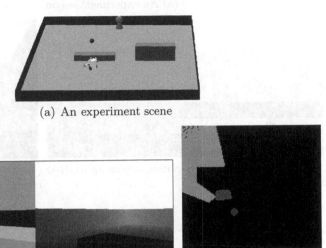

(a) An experiment scene

(b) RGB and depth images obtained from an RGB-D (c) 3D visibility check
camera.   result

Fig. 3. Results of experiment 1

3.2   Experimental Result

To show the effectiveness of the developed method, four different experiments
with different robot's postures and object positions were conducted. Figs. 3-
6 show experiment scenes, RGB and depth images obtained from an RGB-D
camera mounted on the robot, and 3D visibility check results of experiments 1,
2, 3, and 4, respectively. The 3D visibility check result figures are projections of
3D points of real world to 2D plane to show the visibility check result effectively.
In the figures of 3D visibility check result, the circle with gray is a robot, black
means invisible area, green means visible area, brown means the visible upper
part of obstacles, red means a visible object, and blue means a visible human.

As shown in the figures, the robot could check its 3D visibility well and detect
the object and human if they are visible. Also, the robot could identify the empty
space like the sky and ground as well. The robot recognized that the area behind
the obstacles or objects is invisible. By the developed 3D visibility check rather
than 2D visibility check, the robot could recognize the situation difference caused

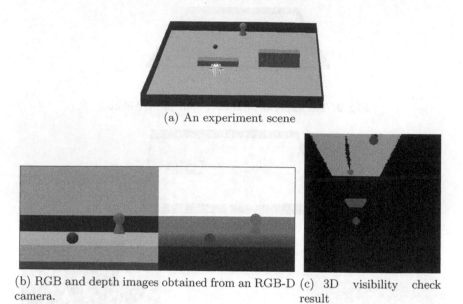

(a) An experiment scene

(b) RGB and depth images obtained from an RGB-D camera.

(c) 3D visibility check result

Fig. 4. Results of experiment 2

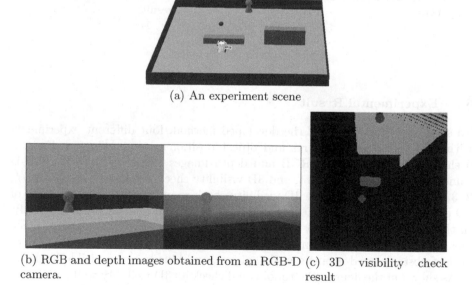

(a) An experiment scene

(b) RGB and depth images obtained from an RGB-D camera.

(c) 3D visibility check result

Fig. 5. Results of experiment 3

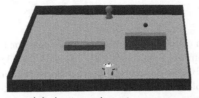

(a) An experiment scene

(b) RGB and depth images obtained from an RGB-D camera. (c) 3D visibility check result

Fig. 6. Results of experiment 4

by heights of obstacles as shown in Fig. 6(c). In Fig. 6(c), a robot identified that the upper part of lower obstacle is visible, but the upper part of higher obstacle is invisible.

4 Conclusion

This paper presents the development of a 3D visibility check method using a depth image in Webots. By using the relationship between a depth image and real world, a robot could identify whether each point in real world is visible or invisible at its posture. Further more, a robot could detect objects based on the result of visibility check. The developed method dealt with 3D visibility rather than 2D visibility to consider the heights of obstacles and objects. This method can be used to make a robot take a human perspective for effective human-robot interaction. Because a robot needs to determine its own visibility first and then take a human perspective by simulating the situation assuming that a robot is located at human posture and calculating the human visibility in the same manner as its own visibility calculation. As further research, we will make a robot find out the human intended object by human perspective taking even when there are situational difference between robot's perspective and human perspective.

Acknowledgment. This work was supported by the Technology Innovation Program, 10045252, Development of robot task intelligence technology, funded by the Ministry of Trade, Industry & Energy (MOTIE, Korea).

References

1. Arkin, R.C., et al.: An ethological and emotional basis for human-robot interaction. Robotics and Autonomous Systems 42(3), 191–201 (2003)
2. Ros, R., et al.: Adaptive human-robot interaction in sensorimotor task instruction: From human to robot dance tutors. Robotics and Autonomous Systems 62(6), 707–720 (2014)
3. Severinson-Eklundh, K., et al.: Social and collaborative aspects of interaction with a service robot. Robotics and Autonomous Systems 42(3), 223–234 (2003)
4. Sorbello, R., et al.: Telenoid android robot as an embodied perceptual social regulation medium engaging natural human-humanoid interaction. Robotics and Autonomous Systems (2014), http://dx.doi.org/10.1016/j.robot.2014.03.017
5. Schmidt, P.A., et al.: A sensor for dynamic tactile information with applications in human-robot interaction and object exploration. Robotics and Autonomous Systems 54(12), 1005–1014 (2006)
6. Fritsch, J., et al.: Multi-modal anchoring for human-robot interaction. Robotics and Autonomous Systems 43(2), 133–147 (2003)
7. Cifuentes, C.A., et al.: Human-robot interaction based on wearable IMU sensor and laser range finder. Robotics and Autonomous Systems (2014), http://dx.doi.org/10.1016/j.robot.2014.06.001
8. Trafton, J.G., et al.: Enabling effective human-robot interaction using perspective-taking in robots. IEEE Trans. Systems, Man, and Cybernetics-Part A: Systems and Humans 35(4), 460–470 (2005)
9. Webots: robot simulator, http://www.cyberbotics.com
10. Henry, P., Krainin, M., Herbst, E., Ren, X., Fox, D.: RGB-D mapping: Using depth cameras for dense 3D modeling of indoor environments. In: Khatib, O., Kumar, V., Sukhatme, G. (eds.) Experimental Robotics. STAR, vol. 79, pp. 477–491. Springer, Heidelberg (2012)
11. Ramey, A., et al.: Integration of a low-cost RGB-D sensor in a social robot for gesture recognition. In: Proc. 6th International Conference on HRI, pp. 229–230 (2011)
12. Hu, G., et al.: A robust RGB-D slam algorithm. In: Proc. IEEE/RSJ International Conference on Intelligent Robots and Systems, pp. 1714–1719 (2012)

Part II
Computational Intelligence and Intelligent Design for Advanced Robotics

Obstacle Avoidance and Dribbling Strategy for Humanoid Soccer Robots

Ping-Huan Kuo, Ya-Fang Ho, Chih-Chieh Sun, and Tzuu-Hseng S. Li

aiRobots Laboratory, Department of Electrical Engineering
National Cheng Kung University, Tainan City, Taiwan, R.O.C.
{col122000,makinosakuya}@hotmail.com,
zpsyhapcst@gmail.com, thsli@mail.ncku.edu.tw

Abstract. This paper mainly investigates the obstacle avoidance and dribbling strategy for humanoid soccer robot by using only monocular vision, where the local object is used for positioning and recognizing the obstacles. Then by the relation between the object and the robot, the position of the robot at the world coordinate could be figured out with the ameliorative particle filter. It would be more robust and perform well for the obstacle avoidance and dribbling strategy to apply the above information. An enhancement scheme of the particle filter is also proposed in the paper. Furthermore, a real-time control strategy is developed for dribbling ball and avoiding obstacles. Finally, experiments verify the accuracy and robustness of the local positioning system, and show the feasibility and validity of the proposed obstacle avoidance and dribbling strategy for humanoid soccer robots.

Keywords: obstacle avoidance, dribbling strategy, humanoid soccer robot, localization.

1 Introduction

In recent years, there is a wide variety of researches and developments of robots. It seems that the researches of robots have been the trend of the scholars and industry in the world. The integrated system of robots includes the vision system, the motion control system, power supply system, and the communication system. Because the computational ability for the small devices has become well-developed, establishing a system of the robot is not an impossible thing anymore.

It is challenging that in the complex and dynamic environment of the robot soccer game there are multi-agent researches, involving many topics such as motion control, image processing algorithm [1]-[11], localization [12]-[20], and collaboration strategy. RoboCup and FIRA are two major international robot-soccer associations [21]-[22]. They concentrate in promoting AI (Artificial Intelligence), robotics, and related fields. And the final goal of the RoboCup is to arrange a team of fully autonomous humanoid robots which can get the better of the human world champion soccer team. The competitions are attended to learn from others and enhance our technique of the humanoid robot.

© Springer International Publishing Switzerland 2015
J.-H. Kim et al. (eds.), *Robot Intelligence Technology and Applications 3*,
Advances in Intelligent Systems and Computing 345, DOI: 10.1007/978-3-319-16841-8_25

In the RoboCup, the robot should only be equipped with the vision system to detect the ball and other robots, so the image processing algorithm is very important. Some of the humanoid robots integrate the vision system and the motion control system into one system. This architecture makes the computing inefficiency. A better architecture is parallel processing to increase the efficiency of the system. Therefore, the computation of the vision of robots is based on the pico820 and the system of the motion control is based on Arduino MEGA. The image information is captured by the CMOS sensor. And the communication system via simple instruction between the pico820 and Arduino MEGA is the RS232.

To demonstrate the autonomy of the robot in the competition, the strategies need to be developed by considering many cases. Besides, the walking motion is obtained from the previous works [23]-[25]. The aim of this paper is to develop implementation of obstacle avoidance and dribbling strategy for humanoid soccer robots.

2   The Algorithm for Localization

To navigate reliably in the competition field, the robot must know where it is. Therefore, reliable position estimation is an important issue. There have been many studies on localization methods for mobile robots in recent years [26]-[27]. All these methods are generically known as particle filters. Recent work on Bayesian filtering with particle-based density representations has introduced a new approach for robot localization, based on these principles. In this paper, the particle filter method is introduced, where we represent the probability density involved by maintaining a set of samples that are randomly drawn from it. By using a sampling-based representation, we obtain a localization method that can represent arbitrary distributions.

2.1   Grid Map

The robot only gets the information from the environment by the camera which is equipped on its head, so the distance calculated may lack accuracy due to the head vibrating when the robot performs some motions. Furthermore, the feedback on the estimated turning angle and walking distance of the motion control might not be precise. However, the robots track the ball most of the time, so it is hard to get information on the local points.

Hence, a trustworthy method is needed, so the position of the robot is represented by a grid map. The RoboCup field is divided into a grid of 600×400 cells, whose area is 1×1 cm^2 as shown in Fig. 1 (a). The origin of the coordinate is in the bottom left of the field. By the particle filter, each particle has pose information $p_i = \left\langle \tilde{x}_i, \tilde{y}_i, \tilde{\theta}_i \right\rangle$,

where $\tilde{x}_i$ and $\tilde{y}_i$ are the center coordinates of the cells. $\tilde{\theta}_i$ represents the orientation, and i identifies the particle. However, there is a fitness function to decide which the best particle is, and the particle will correspond to the map regarding the pose of the robot.

2.2   Line Contribution

The data on the lines is very useful for localization because the lines are observed by the robot the most frequently. Therefore, the lines are defined by using a grid map. These lines are established in the grid map as shown in Fig. 1 (b). When the robot sees the line, the contribution of the line would be set at a value. If the color of the grid is all white, then the contribution is set at 4. But if the color of the grid is not white, the contribution is set at 0. If the color of the grid is between empty and full, the contribution is proportional to the area of white, as shown in Fig. 2.   By defining the contribution of the lines, if the robot sees the lines, it could calculate the distance and angle of the lines. A greater contribution means that the estimated position is more accurate.

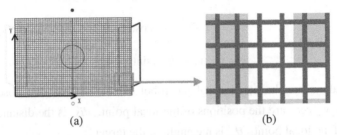

(a)   (b)

Fig. 1. (a) The resolution of the grid map corresponding to the 4m×6m field size (b) The lines in the grid map

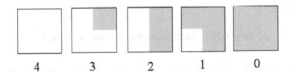

4   3   2   1   0

Fig. 2. The different contribution of the grid

2.3   Fitness Function

How to select the best particle to decide the position of the robot is a very important issue. In the research, an easy method is used to solve this problem. Through the image process, some data on the localization can be obtained, such as the distance and angle between the robot and the local point. Therefore, the observed position calculated by this data can be derived, and it is the key information needed to ensure a correct position for the robot.

Suppose that the data of the feature is observed by the robot at time t. The feature of the observed distance and orientation is represented by particle p. Then, the degree of the belief of p could be represented by the fitness function (1). The larger the fitness, the greater the degree of the belief.

$$fitness(p_i) = \exp(-error)$$ (1)

where $p_i = (\tilde{x}, \tilde{y}, \tilde{\theta})$; $\tilde{x}$ and $\tilde{y}$ are the positions of the particle and $\tilde{\theta}$ is the angle of the particle.

However, in equation (1), the error could be replaced with the following values, the relative distance between the local objects and robots (e_rdo), the relative position between the robot and lines (e_line), or the angle of the robot facing (e_rao).

The e_rdo could be calculated with the information of the target position. When the image data is captured, the observed position of the robot could be derived as shown in equation (2)-(4). And the relation of e_rdo between the position of the particle and the observed position is shown in Fig 3 (a).

$$e\_rdo = (x - \tilde{x})^2 + (y - \tilde{y})^2 \tag{2}$$

$$x = x\_ref - dis \times \cos(\tilde{\theta} + \theta_o) \tag{3}$$

$$y = y\_ref - dis \times \sin(\tilde{\theta} + \theta_o) \tag{4}$$

where x and y are the positions of the robot calculated from the observed data. x_ref and y_ref are the positions of the local point. dis is the distance between the robot and the local point. θ_o is the angle of the target.

And the e_line is easily derived as shown in equation (5). When the robot sees the line, the position of the line can be obtained. By the grid map, the contribution of the line can be obtained.

$$e\_line = 4 - contribution(x\_line, y\_line) \tag{5}$$

where the x_line and y_line are the positions of the line calculated by the observed data. The schematic diagram of e_line is shown in Fig. 3 (b). Finally, the e_rao is the most important because the direction the robot is facing is up to the e_rao. The e_rao is shown as follows:

$$e\_rao = \left| \arccos(v_1 \cdot v_2 / |v_1||v_2|) \right| \tag{6}$$

$$v_1 = (\ x\_ref - \tilde{x}, \ y\_ref - \tilde{y}\ ) \tag{7}$$

$$v_2 = (\ \cos(\theta + \theta_o), \ \sin(\theta + \theta_o)\ ) \tag{8}$$

where v_1 and v_2 mean the vector of the robot and local point. And v_1 is obtained by the grid map; v_2 is calculated by the angle of the particle and the angle of the local point. The schematic diagram of e_rao is shown in Fig. 3 (c).

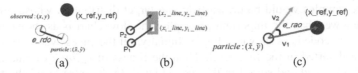

Fig. 3. The schematic diagram of (a) e_rdo, (b) e_line, and (c) e_rao

2.4 Threshold for Resample

In this work, the threshold for the resample was computed according to the formulation of Doucet et al. [28] as follows.

$$\begin{cases} N_{\text{eff}} = \dfrac{1}{\displaystyle\sum_{i=1}^{N}(\tilde{w}^{(i)})^{2}} \\ \tilde{w}^{(i)} = \dfrac{fitness(p_{i})}{\displaystyle\sum_{i=1}^{N} fitness(p_{i})} \end{cases} \tag{9}$$

where $\tilde{w}^{(i)}$ refers to the normalized weight of particle i, and p_{i} refers to particle i; N_{eff} is the threshold for the resample.

By the approach proposed in [29], the proposed algorithm could determine whether the resample should be carried out or not. When N_{eff} drops below the threshold of $N/2$ where N is the number of particles, the resample will execute. In some experiments, this method drastically reduces the risk of resample because the resample operation is performed when needed.

2.5 Hierarchy Particle Filter

In a real situation, if the distance between the robot and local points is great, the distance is not accurate or may even be completely incorrect. Therefore, the hierarchy particle filter is proposed. This means using hierarchy architecture to eliminate the incorrect particles. First, the e_rao of the particles is calculated and the fitness of the particles is obtained in (1). And only a quarter of the credible particles are chosen. Then, the e_line is calculated to follow the same step as before. The final result is up to the fitness of the e_rdo of the particles. Through the experiment, a good performance has been shown to obtain the right position.

For example, at first, the data of the particles should be initialed. Particle i has the pose information $p_{i} = \langle x_{i}, y_{i}, \theta_{i} \rangle$, and the initial data is $p_{i} = \langle 300, 200, 0 \rangle$. With the motions of the robot, the p_{i} would be updated by the localization data of the motion controller board. With the hierarchy particle filter, the best particle would be selected from the information of the robot. However, if the threshold of the resample is reached, the samples would be drawn from the target distribution whose importance weights are lower than the important principle. The flow chart is as shown in Fig. 4. And the pseudo code is as shown in Fig. 5. Initially, all the fitness values of *e_rao* are calculated and sorted in the first level. Then the fitness of *e_line* of a quarter of the particles are obtained and sorted in the second level. In the third level, the fitness of *e_rdo* of rest of the particles are evaluated and sorted. If the threshold for resample is achieved, all the particles will be regenerated; else the particle with the best fitness will be outputted in the final step.

Fig. 4. Flow chart of the particle filter

```
1st  level
Calculate fitness of e_rao of all the particles
Using Quick_Sort to order the particles


2nd level
Calculate fitness of e_line of a quarter of the particles
Using Quick_Sort to order the particles


3rd level
Calculate fitness of e_rdo of rest of the particles
Using Quick_Sort to order the particles


If n_eff_caculate() < NUMBER_PARTICLE/2
{
        Generate particle;
        Calculate fitness of the particles;
        Using Quick_Sort to order the particles
}
else
{
                Using Quick_Sort to order the particles
}

Ouput the best fitness of the particle
```

Fig. 5. Pseudo code of the particle filter

3   The Control Strategy for Obstacle Avoidance and Dribbling

According to the rule of the technical challenge of RoboCup, it is not allowed to use robots which are specialized for a specific technical challenge. Only the robots used for the soccer games are allowed to participate in the technical challenges. No hardware modifications on the robots are allowed for the Technical Challenges (i.e. a robot cannot be modified from the configuration during the soccer games). Therefore, a good strategy would be important for the technical challenge.

3.1   The Direction of the Forward

Just stable motions are not enough to solve the challenge, the dribbling direction should be known. The direction relative the robot of the interval and the goal can be detected with the vision of the robot. But which direction is more important? There never was a doubt that the direction of the goal because the final result is to dribble the ball to the goal. There is an assumption that if the robot can see the goal and the interval is enough to pass. In fact, the assumption would be right due to the image processing. When the width of the goal is not wide enough, the goal would not be detected in the vision system so the direction of the interval would replace the direction of the goal. Then what direction the robot faces has been easily to solve.

3.2   Obstacle Avoidance and Dribbling

The dribbling direction has been solved. But when the robot dribbling, there would be a situation that it is too close to the obstacles to touch the obstacles. It is not only just dribbling toward the direction, but also avoiding the obstacles. The problem should be solved by some special strategies. When the robot dribbles the ball, it should also notice the obstacles. If the obstacles are too close the ball, the robot should take a turn motion and dribble as shown in Fig. 6. Like Fig. 7, the robot should turn to the left side of the ball to avoid the obstacle and dribble the ball for a proper distance through the interval.

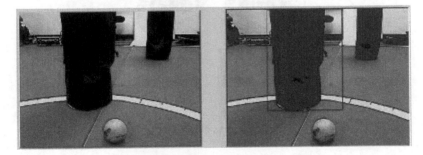

Fig. 6. The ball is close to the obstacle

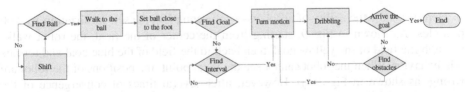

Fig. 7. Flow chart of the strategy for obstacle avoidance and dribbling

3.3   The Strategy for Obstacle Avoidance and Dribbling

By integrating above three parts, the strategy could be got and the flow chart is shown in Fig. 7. The steps are addressed as follows:

1. If the information of the ball is known, go to step 2. Else the robot shifts to find the ball.
2. Make the ball in the proper position. If the ball is at the proper position, go to step 3.
3. Find the right dribbling direction and turn the robot to face the direction. If the direction of the robot is right, go to step 4.
4. Dribble the ball and avoid the obstacles.   If the robot arrives the goal, go to end. Else go to step 3.

4   Experimental Results

The experiment results evaluating the foot-eye coordination control are presented in this section. We show the performance and effectiveness of the proposed control scheme through the humanoid robots, "aiRobots-V," as shown in Fig. 8. The experimental results for localization are illustrated first. Then, the results of the strategies for the throw-in challenge and for obstacle avoidance and dribbling are presented. Finally, the results for the strategies for the double passing challenge are demonstrated.

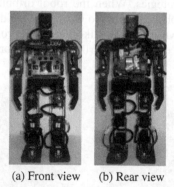

(a) Front view   (b) Rear view

Fig. 8. Pictures of aiRobot-V

4.1   Experimental Results for Localization

Here we show the ability of aiRobots-V to accomplish the self-localization by 500 particles. As shown in Fig. 9, starting from the center of the field, the robot walks through the field of the yellow goal, then back to the field of the blue goal and repeats this behavior. When the robot cannot see the local point, the positions of the robot are wrong, as shown in Fig. 9 (j). However, after several times of convergence of the localization algorithm, the position of the robot comes back to the right position.

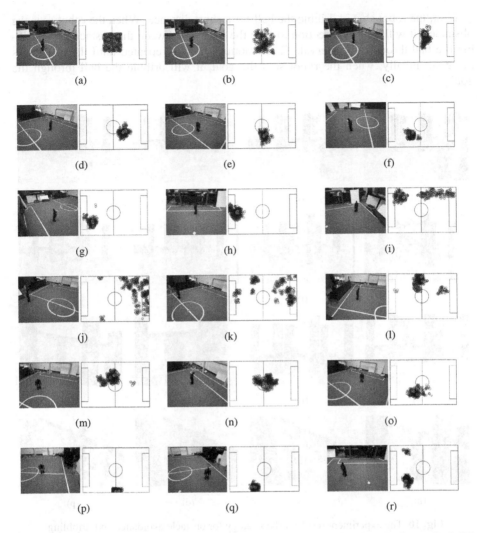

Fig. 9. The experiments of the localization. The red circles are in the top 5 of a total of 500 particles, the green circle is in the top 1, and the blue circle is the mean of the red and green ones. The line means the direction faced by the robot.

4.2   Experimental Results for Obstacle Avoidance and Dribbling Challenge

Here, there are six obstacles set in the field as shown in Fig. 10, and the obstacles are put in arbitrary positions just before the start of each trial. The distance between the obstacles is at least 50 cm, and no obstacles are placed in the center circle. The ball is placed at the center mark and the robot directly behind it facing the yellow goal. However, the robot is not allowed to touch the obstacles. The result of the strategy for obstacle avoidance and dribbling is shown in Fig. 10, and the experiment time for obstacle avoidance and dribbling challenge is about 5 minutes.

First, the robot should dribble the ball for a short distance. When the robot sees the obstacles, it will turn its direction to face the interval between the obstacles and dribble the ball through the interval. These motions will be repeated until the robot sees the goal. Finally, when the robot sees the goal, it will dribble the ball through the goal.

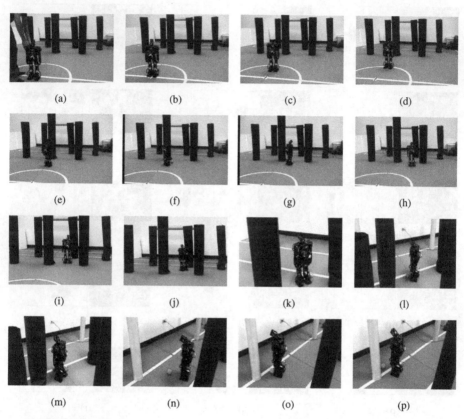

Fig. 10. The experiment result of the strategy for obstacle avoidance and dribbling

5   Conclusions

In this paper, the localization method and strategy systems have been developed and implemented for obstacle avoidance and dribbling for humanoid soccer robot. Also, the robot does not need to see the local point and it still can localize its position by the algorithm of the particle filter. Furthermore, the control strategy for obstacle avoidance and dribbling is introduced. Finally, the experiment results demonstrate the capability of vision and strategy system, and the efficiency and validity in the humanoid soccer challenge of RoboCup.

Acknowledgment. This work was supported by Ministry of Science and Technology, R.O.C, under Grants NSC100-2221-E006-055-MY3 and NSC 101-2221-E-006 - 193 -MY3.

References

1. Messom, C.H., Sen Gupta, G., Demidenko, S.: Hough Transform Run Length Encoding for Real-Time Image Processing. In: Proceedings of the IEEE Instrumentation and Measurement Technology Conference, IMTC 2005, pp. 2198–2202 (2005)
2. Davison, A.J., Reid, I.D., Molton, N.D., Stasse, O.: MonoSLAM: Real-Time Single Camera SLAM. IEEE Transactions on Pattern Analysis and Machine Intelligence 29, 1052–1067 (2007)
3. Schmudderich, J., Willert, V., Eggert, J., Rebhan, S., Goerick, C., Sagerer, G., Korner, E.: Estimating Object Proper Motion Using Optical Flow, Kinematics, and Depth Information. IEEE Transactions on Systems, Man, and Cybernetics, Part B: Cybernetics 38, 1139–1151 (2008)
4. Gevers, T., Stokman, H.: Classifying Color Edges in Video into Shadow-Geometry, Highlight, or Material Transitions. IEEE Transactions on Multimedia 5, 237–243 (2003)
5. Sen Gupta, G., Bailey, D.: Discrete YUV Look-Up Tables for Fast Colour Segmentation for Robotic Applications. In: Canadian Conference on Electrical and Computer Engineering, CCECE, pp. 963–968 (2008)
6. Chang, C.-M.: Design and Implementation of Vision and Strategy System for Humanoid Robot Soccer Competition. Master Thesis, Department of Electrical Engineering (2009)
7. Lee, J.-S., Jeong, Y.-H.: CCD Camera Calibrations and Projection Error Analysis. In: Proceedings of the 4th Korea-Russia International Symposium on Science and Technology, KORUS, vol. 2, pp. 50–55 (2000)
8. Ernst, S., Stiller, C., Goldbeck, J., Roessig, C.: Camera Calibration for Lane And Obstacle Detection. In: Proceedings of IEEE/IEEJ/JSAI International Conference on Intelligent Transportation Systems, pp. 356–361 (1999)
9. Yong-Sheng, C., Shen-Wen, S., Yi-Ping, H., Chiou-Shann, F.: Camera Calibration with A Motorized Zoom Lens. In: Proceedings of 15th International Conference on Pattern Recognition, vol. 4, pp. 495–498 (2000)
10. Daniel: Camera calibration tools, http://ubimon.doc.ic.ac.uk/dvs/m581.html/
11. Zhang, Z.: A Flexible New Technique for Camera Calibration. IEEE Transactions on Pattern Analysis and Machine Intelligence 22, 1330–1334 (2000)
12. Röfer, T., Laue, T., Thomas, D.: Particle-Filter-Based Self-Localization Using Landmarks and Directed Lines. In: Bredenfeld, A., Jacoff, A., Noda, I., Takahashi, Y. (eds.) RoboCup 2005. LNCS (LNAI), vol. 4020, pp. 608–615. Springer, Heidelberg (2006)
13. Lenser, S., Veloso, M.: Sensor Resetting Localization for Poorly Modelled Mobile Robots. In: Proceedings of the IEEE International Conference on Robotics and Automation, ICRA 2000, pp. 1225–1232 (2000)
14. Harmati, I., Skrzypczyk, K.: Robot Team Coordination for Target Tracking Using Fuzzy Logic Controller in Game Theoretic Framework. Robotics and Autonomous Systems 57, 75–86 (2009)

15. Ueda, R., Fukase, T., Kobayashi, Y., Arai, T., Yuasa, H., Ota, J.: Uniform Monte Carlo Localization - Fast and Robust Self-Localization Method for Mobile Robots. In: Proceedings of the IEEE International Conference on Robotics and Automation, ICRA 2002, pp. 1353–1358 (2002)
16. Buschka, P., Saffiotti, A., Wasik, Z.: Fuzzy Landmark-Based Localization for A Legged Robot. In: Proceedings of IEEE/RSJ International Conference on Intelligent Robots and Systems (IROS), vol. 2, pp. 1205–1210 (2000)
17. Herrero-Perez, D., Martinez-Barbera, H.: Fast and Robust Recognition of Field Line Intersections. In: IEEE 3rd Latin American Proceedings of Robotics Symposium, LARS 2006, pp. 115–119 (2006)
18. Dellaert, F., Fox, D., Burgard, W., Thrun, S.: Monte Carlo Localization for Mobile Robots. In: Proceedings of IEEE International Conference on Robotics and Automation, vol. 2, pp. 1322–1328 (1999)
19. Doucet, A., De Freitas, N., Gordon, N.: Sequential Monte Carlo Methods in Practice. Springer (2001)
20. Grisetti, G., Stachniss, C., Burgard, W.: Improved Techniques for Grid Mapping with Rao-Blackwellized Particle Filters. IEEE Transactions on Robotics 23, 34–46 (2007)
21. RoboCup, http://www.robocup.org/
22. FIRA, http://www.fira.net/
23. Li, T.-H.S., Su, Y.-T., Lai, S.-W., Hu, J.-J.: Walking Motion Generation, Synthesis, and Control for Biped Robot by using PGRL, LPI and Fuzzy Logic. IEEE Trans. on System, Men, and Cybernetics, Part B 41, 736–748 (2011)
24. Su, Y.-T., Chong, K.-Y., Li, T.-H.S.: Design and Implementation Fuzzy Policy Gradient Gait Learning Method for Walking Pattern Generation of Humanoid Robots. International Journal of Fuzzy Systems 13, 369–382 (2011)
25. Li, T.-H.S., Su, Y.-T., Liu, S.-H., Hu, J.-J., Chen, C.-C.: Dynamic Balance Control for Biped Robot Walking Using Sensor Fusion, Kalman Filter, and Fuzzy Logic. IEEE Transactions on Industrial Electronics 59, 4394–4408 (2012)
26. Röfer, T., Laue, T., Thomas, D.: Particle-Filter-Based Self-localization Using Landmarks and Directed Lines. In: Bredenfeld, A., Jacoff, A., Noda, I., Takahashi, Y. (eds.) RoboCup 2005. LNCS (LNAI), vol. 4020, pp. 608–615. Springer, Heidelberg (2006)
27. Harmati, I., Skrzypczyk, K.: Robot Team Coordination for Target Tracking Using Fuzzy Logic Controller in Game Theoretic Framework. Robotics and Autonomous Systems 57, 75–86 (2009)
28. Doucet, A., De Freitas, N., Gordon, N.: Sequential Monte Carlo Methods in Practice. Springer (2001)
29. Grisetti, G., Stachniss, C., Burgard, W.: Improved Techniques for Grid Mapping with Rao-Blackwellized Particle Filters. IEEE Transactions on Robotics 23, 34–46 (2007)

Path Planning of Nonholonomic Mobile Robot for Maximum Information Collection in Dynamic Environment

Rui Zhang, Weirong Liu, Xiaoyong Zhang[*], Fu Jiang, and Bin Chen

School of Information Science and Engineering, Central South University,
Changsha, HUNAN 410075, China
zhangxy@csu.edu.cn

Abstract. In this paper, an optimal path planning method is proposed for mobile robots to maximize the collected information from object regions in a 2D space with static and movable obstacles. Therein, the path planning problem is solved by improved genetic algorithm, which introduces the attractive operator and repulsive operator besides the conventional crossover and mutation operators. The initial population path is obtained by solving the traveling salesman problem with pattern search method. The proposed approach can generate the optimal trajectory planning while meeting nonholomic and dynamical constraints. And the robot could real-time change its moving path in the local areas based on the artificial obstacle repulsive force and object region attractive force. Simulation results verify the effectiveness of the proposed approach.

Keywords: path planning, nonholonomic mobile robot, movable obstacles, evolutionary operators, dynamic environment.

1 Introduction

The path planning problem is a key issue for mobile robots in the field of autonomous navigation. The problem is aimed at designing a path for robot to follow in such a way that given mission and the goal are achieved, which should consider kinematic, collision avoidance criterion, optimization on energy or path length at the same time. In this paper, the topic that the nonholonomic robot is to maximize the collected information while considering the algorithm feasibility and path optimality in the fixed time and complex dynamic environment.

Many methods such as the roadmap methodology, the artificial potential field, and cell decomposition have been proposed to solve the basic trajectory planning problem. However, few of them focus on the information collecting efficiency, a key issue of path planning. Currently, the authors of the fields attempt to minimize the total mission time instead of maximizing the collected information [1]. But in many

---

[*] Corresponding author.

© Springer International Publishing Switzerland 2015
J.-H. Kim et al. (eds.), *Robot Intelligence Technology and Applications 3*,
Advances in Intelligent Systems and Computing 345, DOI: 10.1007/978-3-319-16841-8_26

practical applications, the mission time always is given, and the main objective is to maximize the collected information in the fixed time. In this article, we concern the information maximization in the given the task time as in [1].

The literature [2] introduces how to deal with dynamic obstacles. However, these methods in [2] cannot handle moving obstacles while guaranteeing the maximum collected information. It is shown that parametric method also can better make real-time path planning and handle dynamic obstacles [3], but the parameter optimization is complex.

Also, the potential field is a widely used method of avoiding obstacles for robot. And the robot can achieve the target under the force of the potential field in theory. However, the robot may be trapped beyond the target point in the complex environment and cannot reach the desired point. Such as the typical example, when the robot goes into the obstacles of U shape, the robot will be in the local minimum point of the potential field and cannot go away from the region, then the robot cannot reach the goal, as Fig.1. But, this article proposes an algorithm that combines the genetic algorithm and the artificial potential field. It utilizes the idea of global planning and local optimization to plan the robot's path, which is superior to the single potential field.

The genetic algorithm is a heuristic search that mimics the process of natural evolution [4]. It imitates natural selection and survival of the fittest, and could solve the path optimization problem effectively. Genetic algorithm is a multi-point search algorithm, and it can be better able to solve the global optimal solution [5].

In this paper, the problem of real-time path planning in dynamic environments for mobile robot is addressed with three steps. First, the sequence of several object regions which the robot will access is determined by solving the traveling salesman problem [6] with pattern search method. Second, the initial path population need to determine and the initial point of the path need to be given. Finally, two new mutation operators are introduced, namely, obstacle-exclusion and object-region-attractive, which combine with standard operators to achieve the real-time path planning and avoid obstacles.

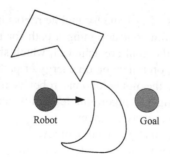

Fig. 1. A case of local minimum point

2   Problem Statement

The goal of the path planning in this paper is to find an optimal or suboptimal trajectory that is from the initial position to final position for the robot in the dynamic environment with moving obstacles [7]. This path should be satisfied with the following conditions: (1) The path can avoid obstacles, and meet the safety distance constraint; (2) The path passes through all of the object regions as far as possible to collect the maximum information [8]; (3) The length of path should be as short as possible.

The operating environment for the nonholonomic robot is two-dimensional flat space, which concludes static and mobile obstacles. And the mobile obstacles move at a constant speed along a broken line [9]. The position of the mobile obstacles can be real-time acquired [10], so the information of environment can be measured.

2.1   The Kinematical Model of A Robot

The motion model of the robot is important to its path planning on avoiding the dynamic obstacles [11]. See Fig.2.

The vector $q=[x_p,y_p,\theta]$ is the generalized coordination of the robot. The coordinate of the system's reference point P is (x_p, y_p) in the 2D image, and θ is the heading angle (also heading direction). In the Fig.1, the distance between the left wheel and the right wheel is recorded as d, the radiuses of the two wheels both are r, $M(x_M, y_M)$ is the center of the two wheels, l is the distance between P and M. β is the angle between PM and the axis of the mobile robot. From Fig.2, formula (1) can be obtained as follows.

$$\begin{cases} x_p = x_M + l \cdot \cos(\theta + \beta) \\ y_p = y_M + l \cdot \sin(\theta + \beta) \end{cases} \tag{1}$$

Make derivation of t for the left and right in (1).

$$\begin{cases} \dot{x}_p = \dot{x}_M + l \cdot \dot{\theta} \cdot \cos(\theta + \beta) \\ \dot{y}_p = \dot{y}_M + l \cdot \dot{\theta} \cdot \sin(\theta + \beta) \end{cases} \tag{2}$$

Two wheels can be simplified as one single wheel located M for Fig.1, so nonholonomic constraints of the single wheel system as formula (3).

$$\dot{x}_M \cdot \sin\theta - \dot{y}_M \cdot \cos\theta = 0 \tag{3}$$

So the constraint equation for nonholonomic motion of the system is as (4).

$$\dot{x}_p \cdot \sin\theta - \dot{y}_p \cdot \cos\theta + l \cdot \dot{\theta} \cdot \cos\beta = 0 \tag{4}$$

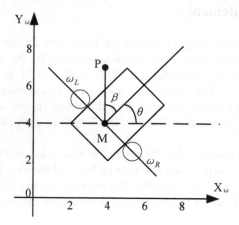

Fig. 2. Schematic of robot motion

$$\begin{cases} \dot{x}_M = \dfrac{1}{2} \cdot (r \cdot \omega_L + r \cdot \omega_R) \cdot \cos\theta \\[2mm] \dot{y}_M = \dfrac{1}{2} \cdot (r \cdot \omega_L + r \cdot \omega_R) \cdot \sin\theta \\[2mm] \dot{\theta} = \dfrac{r \cdot \omega_R - r \cdot \omega_L}{d} \end{cases} \tag{5}$$

Where ω_L is angular velocity of the left wheel, also ω_R is the right wheel.

$$\begin{cases} \dot{x}_p = \left[\dfrac{1}{2} \cdot r \cdot \cos\theta + \dfrac{r \cdot l}{d} \cdot \sin(\theta + \beta) \right] \cdot \omega_L + \left[\dfrac{1}{2} \cdot r \cdot \cos\theta - \dfrac{r \cdot l}{d} \cdot \sin(\theta + \beta) \right] \cdot \omega_R \\[2mm] \dot{y}_p = \left[\dfrac{1}{2} \cdot r \cdot \sin\theta - \dfrac{r \cdot l}{d} \cdot \sin(\theta + \beta) \right] \cdot \omega_L + \left[\dfrac{1}{2} \cdot r \cdot \sin\theta + \dfrac{r \cdot l}{d} \cdot \cos(\theta + \beta) \right] \cdot \omega_R \\[2mm] \dot{\theta} = \dfrac{r \cdot \omega_R - r \cdot \omega_L}{d} \end{cases} \tag{6}$$

Formula (6) is the kinematical model of the robot. And linear velocity and angular velocity of the point M can be expressed as (7).

$$v_M = \sqrt{x_M + y_M} = \frac{r}{2} \cdot (\omega_R + \omega_L)$$

$$\omega_M = \theta = \frac{r}{d} \cdot (\omega_R - \omega_L) \tag{7}$$

There are three components of the state vector from the formula (6), namely, x, y and θ, and the system is controlled by ω_R and ω_L. The moving direction is changed by the difference speed between the left wheel and the right wheel.

2.2   Collect Information

A camera is installed on the top of the nonholonomic robot, which can collect image information from the regions of interest. And its height and depression angles are fixed, as see Fig.3.

It is assumed that the regions of interest of the camera in the object region is denoted as a rectangle, and the camera can obtain four vertices of the rectangle, then difference of two vertices can get the side length of the rectangle, namely, l_1 and l_2. So the collected area information(AI) is AI=$l_1 \times l_2$, and the collected information(CI) is the area that AI is projected onto the camera, such as formula (8).

$$CI=a \times (l_1 \times l_2 \times sin\theta) \qquad (8)$$

$$a = \begin{cases} 1 & \textit{robot is in the object region} \\ 0 & \textit{otherwise} \end{cases} \qquad (9)$$

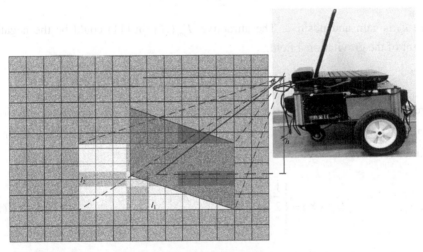

Fig. 3. Information collected by camera

where the variable a represents whether the robot is in the object regions. And the maximum information of the article is the sum of each object region's collected information, that, $\sum_{i=1}^{n} CI_i$, where n is the number of object regions.

When the robot enters the same region again, it can estimate the current information by using previous image information [12]. Besides, if the robot can get higher resolution image than previous image, then the latter instead of the former is to structure objective information and the attractive force of the object region could decrease when the robot closes to it again.

2.3   Attractive Force and Repulsive Force

If the robot is moving in the blank regions (beyond the obstacles and object regions), there should be attractive force and repulsive force like in the artificial potential field [13]. And the attractive force made the robot along the planned path to enter the object regions, the repulsive power make it away from the obstacles. Thus, these behaviors make the path better, which can collect maximum information in the shorter time.

According to the definitions of the artificial potential field method, it is assumed that the location of the robot is $X = [x \quad y]^T$ in the workspace. The attractive function could be defined as $U_{att}(X)$ at location X as formula (10), which relates to the location of objects $X_g = [x_g \quad y_g]^T$, that, the center of the object region in this paper. As reference [4],

$$U_{att}(X) = \frac{1}{2}k_p(X - X_g)^2 \tag{10}$$

where k_p is gain and positive. The attractive $F_{att}(X)$ in (11) could be the negative gradient of the gravitational field.

$$F_{att}(X) = -\Delta U_{att}(X) = -k_p(X - X_g) = k_p \rho_1 \overrightarrow{n_{RG}} \tag{11}$$

where $\overrightarrow{n_{RG}}$ is the unit vector which from the robot to the object regions, and $\rho_1 = \|X - X_g\|$ is the distance between the robot and the object regions. Defining the function of repulsion field as $U_{rep}(X)$. As formula (12).

$$U_{rep}(X) = \begin{cases} \frac{1}{2}(\frac{1}{\rho} - \frac{1}{\rho_0})^2(X - X_g)^n & \rho \le \rho_0 \\ 0 & \rho > \rho_0 \end{cases} \tag{12}$$

where ρ is a positive gain that is the shortest distance between the robot and the edge of the obstacles. The formula (12) means that the robot suffers the repulsive inside the distance of ρ_0 between the robot and the edge of the obstacles, otherwise, the robot suffers zero. And $\forall n > 0$. The relative distance between the robot and the object could be $(X - X_g)^n = |(x - x_g)^n| + |(y - y_g)^n|$.

Then the repulsive can be described as formula (13).

$$F_{rep}(X) = -\Delta U_{rep}(X) \tag{13}$$

And the resultant force can be divided into two parts.

$$F_{rep} = \begin{cases} F_{rep1} + F_{rep2} & \rho \le \rho_0 \\ 0 & \rho > \rho_0 \end{cases} \qquad (14)$$

$$= \begin{cases} \|F_{rep1}\| \overrightarrow{n_{OR}} + \|F_{rep2}\| \overrightarrow{n_{RG}} & \rho \le \rho_0 \\ 0 & \rho > \rho_0 \end{cases}$$

therein, $F_{rep1} = (\frac{1}{\rho} - \frac{1}{\rho_0}) \frac{1}{\rho^2} (X - X_g)^n$, $\|F_{rep1}\| = (\frac{1}{\rho} - \frac{1}{\rho_0}) \frac{1}{\rho^2} \rho_1^n$,

$$F_{rep2} = -\frac{n}{2}(\frac{1}{\rho} - \frac{1}{\rho_0})^2 (X - X_g)^{n-1}, \quad \|F_{rep2}\| = -\frac{n}{2}(\frac{1}{\rho} - \frac{1}{\rho_0})^2 \rho_1^{n-1}.$$

where F_{rep1} and F_{rep2} are components of F_{rep}, and their directions are $\overrightarrow{n_{OR}}$ and $\overrightarrow{n_{RG}}$. $\overrightarrow{n_{OR}}$ represents the unit vector which from the point nearest to the robot of the obstacle to the robot and $\overrightarrow{n_{RG}}$ represents the unit vector which from the robot to the object region. $\|F_{rep1}\|$ and $\|F_{rep2}\|$ are the module of the vectors.

The repulsive force will change with the distance between the robot and the obstacles, but there is always a stable safety distance. If the robot is too close to the obstacles beyond the safety distance, it could suffer certain penalties that the collected information may decrease.

So from the section, the evaluation function of the topic, that, maximize, can be described as:

$$b_1 \times CI - b_2 \times Penalty\_ Safety\_Distance \qquad (15)$$

b_1 will decrease if the robot along best path at the end of a certain number of generations does not enter all the object regions. b_2 will increase if the robot along best path at the end of a certain number of generations beyond the safety distance.

3   Path Planning Using Evolutionary Computations

3.1   The Discretization of Given Time

In order to simulate the robot's moving path clearly and really, the given time should be discreted firstly. The continuous time $[t_0, t_f]$ is divided into N equal time subintervals of 1s. All variables of the robot are assumed constant in each subinterval, such as the robot's marching direction [14].

3.2   Visiting Sequence of Object Regions for the Robot

There are several object regions that the robot should reach in the motion region of the robot. However, the optimal and shortest path should be considered while the robot enters each object region and collects maximum information. This paper determines the visiting sequence of object regions by solving the travelling salesman problem with PatternSearch.

Calculation of Distance between Two Object Regions

The length of straight line for two object regions' center is treated as the distance between any two object regions. But, if there are obstacles between the two object regions, PatternSearch should be used to find a vertex or tangent of the obstacles, which is used as turning point on the initial straight line. Then a new line is constructed, so the length of the new line is the distance between two object regions. The process of calculation is illustrated in Figs.4-6. OR means object region, and red regions means obstacles.

Determining The Visiting Sequence of Object Regions

Computing the distance of any two object regions as described in the last section and Figs.4-6. Then comparing the distances and selecting the shortest paths to connect a closed broken ring. As the travelling salesman problem, the robot departures at the initial point, and moves at the closed broken ring to the final point at last.

3.3   Searching The Initial Path

The initial population (initial path) is generated by changing the heading direction of the robot, and based on solving the travelling salesman problem with PatternSearch, then the chromosome framework and operators as below sections.

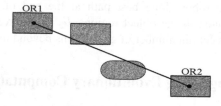

Fig. 4. Initial state

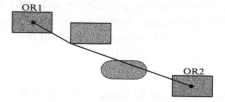

Fig. 5. First segmentation

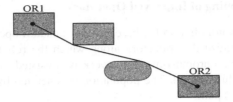

Fig. 6. Second segmentation

Chromosome Framework

In this section, population's chromosome framework can be divided into three parts. The first part is the robot's marching time which is according to the second part. The second part is the robot's heading direction. And the last part is the the robot's swerve time which robot changes direction in order to move to the next object region. See Fig.7. α is 30 degree north by east, β is 40 degree west by south. According to the figure, the robot moves with the direction of α for 80×0.9=72s, and takes 0.2s from α to β.

Mutation

Mutation adopted is single substitution mutation in this article, which select a random location points in chromosome to generate an effective random number to replace the position of the point [15] and its rate is 0.003. For example, the last part of chromosome can randomly chooses between 0~0.1 to be changed.

Crossover

The crossover operation is a process which randomly selects two parent chromosomes to generate children chromosome by crossover operator. In this article, the crossover is achieved by exchanging the second part of two parent chromosomes. Thus the new children chromosome is completed . The rate of the crossover is 0.5 in this article. As in Fig.8.

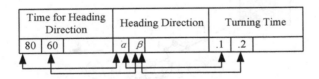

Time for Heading Direction		Heading Direction		Turning Time	
80	60	a	β	.1	.2

Fig. 7. Chromosome framework

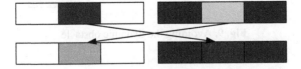

Fig. 8. Crossover of chromosomes

3.4   The Path Planning of Improved Operators

Improved genetic operators like in [1] based on the standard operators are introduced in this section to optimize the trajectory along which the robot can hide static and dynamic obstacles. The chromosome framework is consisted of the robot's heading direction and its marching speed. The population is generated by randomly changing the position of previous paths.

Operators of Obstacle Repulsive

The operation of obstacle repulsive is making a safety distance between the robot and the obstacles by the repulsive that the robot suffers from the obstacles. With the moving direction of the obstacle, the heading direction of robot can be changed a little to avoid the obstacle at any time. Such as Fig.9. Assuming that the speed of round obstacle is v, and the obstacle's moving direction is from left to right, that is, number 1 to number 8. The speed of robot is variable and its heading direction is from OR1 to OR2. Its moving number is consistent with the obstacle's number. From Fig.9, the robot moves along the previous dashed line until it encounters the round obstacle. When the distance between the two individuals is a certain value, the robot breaks away from the dashed line to move at the broken line and avoid the obstacle by suffering the repulsive. After avoiding the obstacle, the robot reenters the previous dashed line.

Operators of Object Region Attractive

The operation of object region attractive is to make the robot return into the previous planned trajectory after having avoided the dynamic obstacles such as the Fig.9. But the case in Fig.9 just is an ideal situation. The path of the robot's returning could be an arc close to the planned straight line.

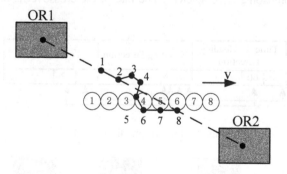

Fig. 9. Avoiding the dynamic obstacle

4   Simulation

The simulation results are obtained in the MATLAB environment. In Figs.10-12, OBS is the obstacle and OR is the object region and the robot achieved the path in the given time 100s. The coordinate of the robot's initial position and final position is (-1750mm, -1450mm). Then, the coordinates of the object regions which are OR1, OR2, OR3, OR4 and OR5, are (-2500, -750), (-1250, 1000), (250, 1750), (1700, 750) and (1500, -1750). And the coordinates of the obstacle regions which are OBS3, OBS4 and OBS5, (-2000, 750), (0, -200) and (900, -1900). Dynamic OBS1 and Dynamic OBS2 are the moving robots. In Fig.10 the worst results are shown while having no any algorithms and the robot can enter each object region at any time, but the trajectory is so long and goes through the obstacles, which is forbidden.

In Fig.11, after having solved the traveling salesman problem with pattern search method and using the standard genetic operators, the path can be better generated. And due to this step, the robot can avoid being trapped because of the current planned initial path in the obstacles of circle or U shape in the next step as Fig.12. But there is still a serious problem about the path, that, the path also passes the obstacles including the static and the dynamic ones.

In order to solve the above problem, it needs to introduce the new operators to optimize the path of the robot. When the robot suffers from the attractive force of the object regions and the repulsive force of the obstacles, the robot can avoid the obstacles as Fig.12. In Fig.12 the result of the optimal trajectory is shown. About dealing with the moving obstacles, when the Dynamic obstacle OBS1 and Dynamic obstacle OBS2 arrive at the light blue points, the robot will move along the path of the new generation using improved operators and the robot can pass through the previous paths with the moving obstacles and enter the next object region at the next time. Comparing the three simulation results, the improved operators have a better effect.

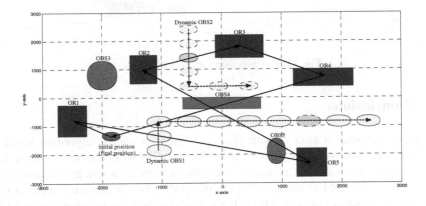

Fig. 10. The initial paths

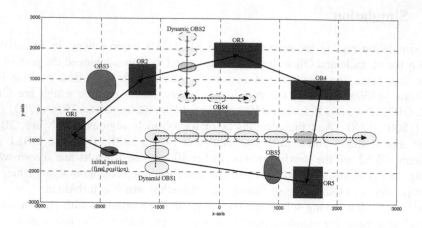

Fig. 11. The path using TSP and conventional operators

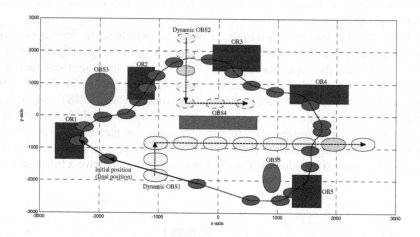

Fig. 12. The path using improved operators

5   Conclusions

This paper proposes a kind of path planning based on genetic algorithm which is introduced new operators to optimize the trajectory of the mobile robots for collecting the maximum information. The method uses operators of attractive and repulsive like artificial potential field to change the previous path which is constructed by solving the traveling salesman problem with PatternSearch method, so the robot can avoid the obstacles in real-time. The simulation results verify the effectiveness of the method and the improved operators.

Acknowledgement. The authors would like to acknowledge that this work was partially supported by the National Natural Science Foundation of China (Grant No. 61379111, 61402538, 61202342, and 61403424) and Research Fund for the Doctoral Program of Higher Education of China (Grant No. 20110162110042).

References

1. Ergezer, H., Leblebicioğlu, K.: Path planning for UAVs for maximum information collection using evolutionary computation. IEEE Trans. Aerosp. Electron. Syst. 49(1), 502–520 (2013)
2. Bullo, F., Frazzoli, E., Pavone, M., Savla, K., Smith, S.L.: Dynamic vehicle routing for robotic systems. Proceedings of the IEEE 99(9), 1482–1504 (2011)
3. Luo, W., Peng, J., Wang, J.: Optimal Real-time Trajectory Planning for a Fixed Wing Vehicle in 3D Dynamic Environment. In: Proceeding of the IEEE International Conference on Robotics and Biomimetics (ROBIO), Shenzhen, China, pp. 710–715 (2013)
4. Zheng, C., et al.: Evolutionary route planner for unmanned air vehicles. IEEE Transactions on Robotics 21(4), 609–620 (2005)
5. Zhang, R., Zheng, C., Yan, P.: Route planning for unmanned air vehicles with multiple missions using an evolutionary algorithm. In: Proceedings of the IEEE 3rd International Conference on Natural Computation, Haikou, China, vol. 3, pp. 1499–1506 (2007)
6. Bekta,s, T.: The multiple traveling salesman problem: an overview of formulations and solution procedures. Omega 34, 209–219 (2006)
7. Yuan, H., Shim, T.: Model based real-time collision-free motion planning for nonholonomic mobile robots in unknown dynamic environments. Int. J. Precis. Eng. Man. 14(3), 359–365 (2013)
8. Liu, S., Sun, D., Zhu, C.: Coordinated motion planning for multiple mobile robots along designed paths with formation requirement. IEEE/ASME Transactions on Mechatronics 16(6), 1021–1031 (2011)
9. Yang, J., Qu, Z., Wang, J., Hull, R.A.: Real-time optimized trajectory planning for a fixed-wing vehicle flying in a dynamic environment. J. Aerospace. Eng. 22(4), 331–341 (2009)
10. Gökto˘gan, A.H., Sukkarieh, S., et al.: Airborne vision sensor detection performance simulation. In: SimTecT 2005, Simulation Conference and Exhibition, Sydney, Australia (2005)
11. Zhong, L., Luo, Q., Wen, D., Qiao, S.D., Shi, J.M., Zhang, W.M.: A task assignment algorithm for multiple aerial vehicles to attack targets with dynamic values. IEEE Transactions on Intelligent Transportation Systems 14(1), 236–248 (2013)
12. Klesh, A.T., Kabamba, P.T., Girard, A.R.: Path planning for cooperative time-optimal information collection. In: 2008 American Control Conference, Seattle, pp. 1991–1996 (2008)
13. LaValle, S.M.: Planning algorithms. Cambridge Univ. Press (2006)
14. Latombe, J.C.: Robot Motion Planning. Kluwer Academic Press, Norwell (1990)
15. Pehlivanoglu, Y.V., Baysal, O., Hacioglu, A.: Vibrational genetic algorithm based path planner for autonomous UAV in spatial data based environments. In: Proc. 3rd Int. Conf. Recent Adv. Space Technol., vol. 7, pp. 573–578 (2007)

D-NSGA-II: Dual-Stage Nondominated Sorting Genetic Algorithm-II

Ki-Baek Lee

Department of Electrical Engineering, Kwang Woon University,
20 Kwangwoon-ro, Nowon-gu, Seoul, Republic of Korea
kblee@kw.ac.kr

Abstract. This paper proposes a novel multi-objective optimization algorithm, dual-stage nondominated sorting genetic algorithm-II (D-NSGA-II) for many-objective problems. Since the percentage of the nondominated solutions increases exponentially with the increasing number of objectives, just finding the nondominated solutions is not enough for solving many-objective problems. In other words, it is necessary to discriminate more meaningful ones from the other nondominated solutions by additionally incorporating user preference into the algorithms. The proposed D-NSGA-II can obtain not only user preference oriented, but also diverse nondominated solutions by introducing an additional stage of multi-objective optimization. The second stage employs the corresponding secondary objectives, global evaluation and crowding distance which were proposed in the previous research for representing the user's preference to a solution and the crowdedness around a solution, respectively. To demonstrate the effectiveness of the proposed algorithm, some benchmark functions are tested and the outcomes of the proposed D-NSGA-II and the NSGA-II are empirically compared. Experimental results show that D-NSGA-II properly reflects the user's preference in the optimization process as well as the performance in terms of the diversity and solution quality is competitive with the NSGA-II.

Keywords: Multi-Objective Evolutionary Algorithm, Dual-Stage, Nondominated Sorting Genetic Algorithm-II, User Preference, Crowding Distance.

1 Introduction

Most of the engineering problems require the optimization of some parameters with multiple objectives conflicting with each other. Thus, there have been a lot of studies for finding candidate solutions as closely as possible to the Pareto-optimal front [1–8]. Among them, nondominated sorting genetic algorithm-II (NSGA-II) [1] has been widely used for its highly competitive and stable ability to find diverse nondominated solutions.

However, with the increasing number of objectives, the percentage of the nondominated solutions increases. Especially if the number of the objectives is over 10, over 90% of the possible solutions are the nondominated solutions [9]. Thus, just finding the nondominated solutions is not effective enough for many-objective problems; it is also necessary to determine which candidate solutions are more preferable than the others.

© Springer International Publishing Switzerland 2015   291
J.-H. Kim et al. (eds.), *Robot Intelligence Technology and Applications 3*,
Advances in Intelligent Systems and Computing 345, DOI: 10.1007/978-3-319-16841-8_27

Moreover, when we consider the user preference, the balance between exploration and exploitation should be carefully considered.

Therefore, in this paper, I propose dual-stage NSGA-II (D-NSGA-II) by introducing secondary objectives in addition to the primary objectives. The primary objectives are the given objectives of the problem and as the secondary objectives, the crowding distance (CD) and the global evaluation value (GEval) are employed. The CD is for the diversity of the solutions and the GEval is for considering the user preference. In the first stage, the nondominated solutions with respect to the primary objectives are obtained. Upon the obtained solutions, an additional stage of the nondominated sorting is conducted with the secondary objectives. The parent population is determined based on the second stage outcome. Through the dual-stage, not only more preferable, but also less crowded solutions are selected to generate the offspring. As a result, well balanced exploration is induced in terms of user preference and diversity throughout the optimization process. To demonstrate the effectiveness of the proposed DMO-NSGA-II, it is empirically compared with NSGA-II for DTLZ functions which are test problems for multi-objective optimization [10].

The remainder of this paper is organized as follows. Section II explains the proposed DNSGA. In Section III, experimental results are discussed. Finally, conclusion follows in Section IV.

2 Dual-Stage Nondominate Sorting Genetic Algorithm-II

In this section, the proposed two-tier optimization approach is explained in detail. The key idea of this approach is that the decision making process is supplemented with one more tier of the optimization. In addition to the first tier of the optimization where the candidate solutions are obtained with respect to the given objectives of the problem, i.e. primary objectives, the second tier of the optimization is performed with respect to the secondary objectives, a global evaluation and a crowding distance. They stand for the users preference to a solution and the crowdedness around a solution, respectively.

The global evaluation (GEval) of each solution is carried out based on the fuzzy integral in this paper. It represents the quality of the solution according to the user's preference. The user's preference which indicates the degree of consideration for each objective is represented using the fuzzy measure. The GEval of a solution is calculated by the fuzzy integral that integrates its partial evaluation values (PEvals) with respect to the degree of consideration. The PEvals are carried out through the objective functions of the problem. Algorithms 1 shows the overall procedures for calculating GEvals. Note that the detailed information of the fuzzy measure and fuzzy integral is given in [7, 8].

The crowding distance of each solution is calculated based on the crowdedness in the solution space [11]. As shown in Fig. 1, the crowding distance (CD) of the point i is an estimate of the size of the largest cuboid enclosing i without including any other point. Algorithms 2 shows the pseudo code for computing the crowding distances. Note that more information is described in [11].

Finally, Fig. 2 shows the overall flow diagram of the proposed approach. At first, the Pareto optimal solutions for the primary objectives are obtained through the first tier optimization. And then, among them, the Pareto optimal solutions for the secondary

---

Algorithm 1. Global Evaluation

- M: The number of objectives
- N: The number of solutions
- $O = \{o_1, o_2, \cdots, o_M\}$: A set of objectives
- $P(O)$: A power set of O
- $h_i(\mathbf{x}_k)$: Partial evaluation value of the k-th solution over the i-th objective
- $e(\mathbf{x}_k)$: Global evaluation value of the k-th solution

1. Fuzzy measure identification
   1: **for** each $A \in P(O)$ **do**
   2:     Calculate fuzzy measure $g(A)$
   3: **end for**
2. Global evaluation of the solutions
   4: **for** $k = 1, 2, \ldots, N$ **do**
   5:     **for** $i = 1, 2, \ldots, M$ **do**
   6:         Calculate $h_i(\mathbf{x}_k)$
   7:     **end for**
   8: **end for**
   9: **for** $k = 1, 2, \ldots, N$ **do**
   10:     $e(\mathbf{x}_k) = \int g \circ h$ (Choquet fuzzy integral)
   11: **end for**

Algorithm 2. Crowding Distance Computation

- M: The number of objectives
- N: The number of solutions
- $O = \{o_1, o_2, \cdots, o_M\}$: A set of objectives
- $f_i(\mathbf{x}_k)$: The objective function value of the k-th solution over the i-th objective
- CD_k: The crowding distance of the k-th solution

1: **for** $k = 1, 2, \ldots, N$ **do**
2:     $CD_k = 0$
3:     **for** $i = 1, 2, \ldots, M$ **do**
4:         Sort the solutions with respect to $f_i(\mathbf{x}_k)$
5:         $CD_1 = CD_1 + 100$
6:         $CD_N = CD_N + 100$
7:         **for** $j = 2, 2, \ldots, N - 1$ **do**
8:             $CD_j = CD_j + f_i(\mathbf{x}_{j+1}) - f_i(\mathbf{x}_{j-1})$
9:         **end for**
10:     **end for**
11: **end for**

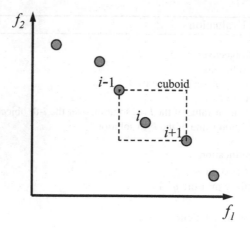

Fig. 1. Crowding distance computation

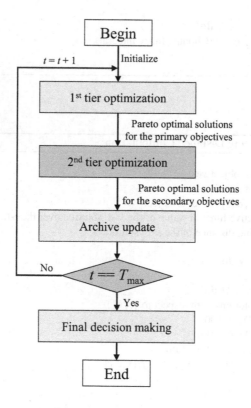

Fig. 2. The overall flow diagram of the proposed approach

Table 1. The parameter settings for the D-NSGA-II and NSGA-II

Parameters	Values
Population size (N)	100
No. of generations	3000
Crossover probability (p_c)	0.9
Mutation probability (p_m)	0.1

objectives are discriminated by the second tier optimization. These steps are repeated until the number of iterations reaches to a certain value T_{max}. After the two-tier optimization process is finished, a decision making is conducted on the finally obtained candidate solutions according to their global evaluation values.

3   Experimental Results

To demonstrate the effectiveness of the proposed algorithm, DTLZ functions [10] were tested with the proposed D-NSGA-II and the NSGA-II. And then the outcomes of the algorithms were empirically compared and analyzed. For the GA, binary-coded GA was employed with the simulated binary crossover (SBX) operator and polynomial mutation [12]. The parameters used in the experiment are given in Table 1. Every DTLZ function is a minimization problem and the number of objectives was set to 5. The number of variables of each DTLZ function was set to 9 for DTLZ1, 16 for DTLZ2 - DTLZ6, and 26 for the DTLZ7 function.

As the performance metrics, the size of dominated space and diversity were employed. The size of dominated space, S is defined by the hypervolume of the finally obtained candidate solutions [13]. The reference point to calculate S was set to (10.0, 10.0, 10.0, 10.0, 10.0). The quality of the obtained solution set is proportional to the hypervolume. Diversity, D is for evaluating the spread of the finally obtained candidate solutions [14], which is defined as follows:

$$D = \frac{\sum_{k=1}^{n}(f_k^{(max)} - f_k^{(min)})}{\sqrt{\frac{1}{|N_0|}\sum_{i=1}^{|N_0|}(d_i - \bar{d})^2}} \tag{1}$$

where N_0 is the set of nondominated solutions, d_i is the minimal distance between the i-th solution and the nearest neighbor, and $\bar{d}$ is the mean value of all d_i. $f_k^{(max)}$ and $f_k^{(min)}$ represent the maximum and minimum objective function values of the k-th objective, respectively.

To verify that the comparison result is statistically significant, statistical hypothesis testing was used [15]. For every pair of comparison, X_1 and X_2, a null hypotheses, $\mathcal{H}_0$ was defined as $\bar{X}_1 - \bar{X}_2 = 0$ which means that there is no significant difference between X_1 and X_2, where $\bar{X}$ is the mean value of X. Through the 50 pairs of sample data, Welch's t-value and the corresponding p-value were calculated [16]. The null hypothesis was rejected and the corresponding alternative hypothesis, $\mathcal{H}_1$ is claimed if

Table 2. The hypothesis testing on S and D with and without the proposed approach

	$\mathcal{H}_0 : \bar{S}_{\text{D-NSGA-II}} - \bar{S}_{\text{NSGA-II}} = 0$		
	t-value (p-value)	Reject	$\mathcal{H}_1$
DTLZ1	2.211 (0.031)	YES	$\bar{S}_{\text{D-NSGA-II}} - \bar{S}_{\text{NSGA-II}} > 0$
DTLZ2	2.085 (0.042)	YES	$\bar{S}_{\text{D-NSGA-II}} - \bar{S}_{\text{NSGA-II}} > 0$
DTLZ3	-0.332 (0.741)	NO	N/A
DTLZ4	3.787 (0.000)	YES	$\bar{S}_{\text{D-NSGA-II}} - \bar{S}_{\text{NSGA-II}} > 0$
DTLZ5	4.913 (0.000)	YES	$\bar{S}_{\text{D-NSGA-II}} - \bar{S}_{\text{NSGA-II}} > 0$
DTLZ6	3.629 (0.001)	YES	$\bar{S}_{\text{D-NSGA-II}} - \bar{S}_{\text{NSGA-II}} > 0$
DTLZ7	3.223 (0.002)	YES	$\bar{S}_{\text{D-NSGA-II}} - \bar{S}_{\text{NSGA-II}} > 0$

	$\mathcal{H}_0 : \bar{D}_{\text{D-NSGA-II}} - \bar{D}_{\text{NSGA-II}} = 0$		
	t-value (p-value)	Reject	$\mathcal{H}_1$
DTLZ1	1.904 (0.063)	NO	N/A
DTLZ2	2.313 (0.025)	YES	$\bar{D}_{\text{D-NSGA-II}} - \bar{D}_{\text{NSGA-II}} > 0$
DTLZ3	1.981 (0.053)	NO	N/A
DTLZ4	3.635 (0.001)	YES	$\bar{D}_{\text{D-NSGA-II}} - \bar{D}_{\text{NSGA-II}} > 0$
DTLZ5	3.148 (0.003)	YES	$\bar{D}_{\text{D-NSGA-II}} - \bar{D}_{\text{NSGA-II}} > 0$
DTLZ6	2.932 (0.005)	YES	$\bar{D}_{\text{D-NSGA-II}} - \bar{D}_{\text{NSGA-II}} > 0$
DTLZ7	5.176 (0.000)	YES	$\bar{D}_{\text{D-NSGA-II}} - \bar{D}_{\text{NSGA-II}} > 0$

the calculated p-value is below the significance level of 0.05. Note that in this paper, the normality of every sample data set was verified through Jarque-Bera test [17].

Table 2 shows the pairwise hypothesis testing results on the both S and D, respectively. As shown in the table, in terms of S, the obtained solutions with the proposed D-NSGA-II outperformed those with the NSGA-II. This was due to the additional stage of optimization where the user preference was incorporated. Moreover, also in case of D, the obtained solutions with the D-NSGA-II were better than those with the NSGA-II. This was because the balance between diversity and user preference was efficiently managed by the secondary objectives.

4 Conclusions

In this paper, a dual multi-objective optimization algorithm, dual-stage nondominated sorting genetic algorithm-II (D-NSGA-II) for many-objective problems was proposed. In the D-NSGA-II, an additional stage of nondominated sorting with the secondary objectives was introduced. The most important advantage of the proposed algorithm was that both the user preference and solution crowdedness could be incorporated into the optimization process. The effectiveness of the proposed algorithm was demonstrated by the empirical comparison between the outcomes of the proposed D-NSGA-II and the NSGA-II. The comparison results indicated that the D-NSGA-II was considerably effective to reflect user preference in the many-objective problems while maintaining the diversity and solution quality.

References

1. Deb, K., Pratap, A., Agarwal, S., Meyarivan, T.: A fast and elitist multiobjective genetic algorithm: NSGA-II. IEEE Transactions on Evolutionary Computation 6(2), 182–197 (2002)
2. Coello, C., Lechuga, M.: MOPSO: A proposal for multiple objective particle swarm optimization. In: Proceedings of IEEE Congress on Evolutionary Computation, pp. 1051–1056 (2002)
3. Hu, X., Eberhart, R.: Multiobjective optimization using dynamic neighborhood particle swarm optimization. In: Proceedings of IEEE Congress on Evolutionary Computation, pp. 1677–1681 (2002)
4. Coello, C., Pulido, G., Lechuga, M.: Handling multiple objectives with particle swarm optimization. IEEE Transactions on Evolutionary Computation 8(3), 256–279 (2004)
5. Köppen, M., Yoshida, K.: Substitute distance assignments in nsga-ii for handling many-objective optimization problems. In: Obayashi, S., Deb, K., Poloni, C., Hiroyasu, T., Murata, T. (eds.) EMO 2007. LNCS, vol. 4403, pp. 727–741. Springer, Heidelberg (2007)
6. Bader, J., Zitzler, E.: Hype: An algorithm for fast hypervolume-based many-objective optimization. Evolutionary Computation 19(1), 45–76 (2011)
7. Kim, J.H., Han, J.H., Kim, Y.H., Choi, S.H., Kim, E.S.: Preference-Based Solution Selection Algorithm for Evolutionary Multiobjective Optimization. IEEE Transactions on Evolutionary Computation 16(1), 20–34 (2012)
8. Lee, K.B., Kim, J.H.: Multiobjective particle swarm optimization with preference-based sort and its application to path following footstep optimization for humanoid robots. IEEE Transactions on Evolutionary Computation 17(6), 755–766 (2013)
9. Ishibuchi, H., Tsukamoto, N., Nojima, Y.: Evolutionary many-objective optimization: A short review. In: Proceedings of IEEE Congress on Evolutionary Computation, pp. 2419–2426 (2008)
10. Deb, K., Thiele, L., Laumanns, M., Zitzler, E.: Scalable multi-objective optimization test problems. In: Proceedings of IEEE Congress on Evolutionary Computation, vol. 1, pp. 825–830 (2002)
11. Raquel, C.R., Naval Jr., P.C.: An effective use of crowding distance in multiobjective particle swarm optimization. In: Proceedings of the 2005 Conference on Genetic and Evolutionary Computation, pp. 257–264. ACM (2005)
12. Agrawal, R.B., Deb, K., Agrawal, R.B.: Simulated binary crossover for continuous search space (1994)
13. Zitzler, E.: Evolutionary algorithms for multiobjective optimization: Methods and applications. Doctoral dissertation ETH 13398, Swiss Federal Institute of Technology (ETH), Zurich, Switzerland (1999)
14. Li, H., Zhang, Q., Tsang, E., Ford, J.: Hybrid estimation of distribution algorithm for multiobjective knapsack problem. In: Gottlieb, J., Raidl, G.R. (eds.) EvoCOP 2004. LNCS, vol. 3004, pp. 145–154. Springer, Heidelberg (2004)
15. Lehmann, E., Romano, J.: Testing Statistical Hypotheses. Springer (2006)
16. Welch, B.L.: The generalization of 'student's' problem when several different population variances are involved. Biometrika 34(1/2), 28–35 (1947)
17. Judge, G.G., Hill, R.C., Griffiths, W., Lutkepohl, H., Lee, T.C., Tuladhar, J., Banerjee, B., Kelly, V., Stevens, R., Stilwell, T., et al.: Introduction to the theory and practice of econometrics. Economic Development and Cultural Change 32(4), 767–780 (1984)

Practical Real-Time System for Object Counting Based on Optical Flow

Jacky Baltes[1], Amirhossein Hosseinmemar[1], Joshua Jung[1],
Soroush Sadeghnejad[2], and John Anderson[1]

[1] University of Manitoba, Winnipeg, MB, R3T 2N2, Canada,
jacky@cs.umanitoba.ca
http://www.cs.umanitoba.ca/~jacky
[2] Amirkabir University of Technology, No. 424, Hafez Ave., Tehran, Iran,
s.sadeghnejad@aut.ac.ir
http://autman.aut.ac.ir

Abstract. This paper describes a simple and effective system for counting the number of objects that move through a region of interest. In this work, I focus on the problem of counting the number of people that are entering or leaving an event. I design a pedestrian counting system that uses a dense optical flow field to calculate the integral of the optical flow in a video sequence. The only parameter used in the system is the the estimated integral flow for a single person. This parameter can be easily calculated from a short training sequence. Empirical evaluations show that the system is able to provide accurate estimates even for complex sequences in real-time. The described system won 2nd place in the pedestrian counting computer vision competition at the IEA-AIE 2014 conference.

1 Introduction

This paper describes a practical real-time system for counting objects that pass through a region of interest ROI. There are many applications for these type of systems in security, urban planning, and crowd control. This paper focuses on the specific problem of counting the number of people that enter or exit via a doorway (i.e., a sub-problem of pedestrian counting). Estimating the number of people entering or leaving a major event or venue (e.g., Disneyland) is important to ensure the safety of visitors. Overcrowding may lead to injuries or even deaths when trying to evacuate (e.g., in case of a bomb threat or fire) or when visitors panic into a stampede. Unfortunately, there are many cases in recent years where people have been injured or died due to overcrowding at an event. Famous examples of human stampedes due to overcrowding are the Who Concert in Cincinnati, U.S.A. on 3rd Dec. 1979, the Hillsborough Stadium, Sheffield, England, on 15th April 1989, and the Love Parade in Duisburg, Germany, on 24th July 2010 [1].

Figure 1 shows some images of the type of pedestrian counting that I address in this research. The first row shows a simple scenario where people enter from

© Springer International Publishing Switzerland 2015
J.-H. Kim et al. (eds.), *Robot Intelligence Technology and Applications 3*,
Advances in Intelligent Systems and Computing 345, DOI: 10.1007/978-3-319-16841-8_28

the top of the image through a well defined gate area and exit at the bottom. The people move in different sized groups (shown are a single person (left), two people (middle), and three people (right).

The second row shows a more complex scenario. The region of interest is larger and the perspective distortion of the camera is more pronounced. Some people enter or leave on the edges of the field of the view of the camera, which results in only legs or arms being seen in the video. Furthermore, the sun casts long and hard shadows across the ground. There are several non-people objects that cross the scene. For example, baby carriages are shown in the middle and left image and an umbrella in the right image. There are also people moving in the opposite direction as shown in the right image.

The images shown in Fig. 1 were taken from sample videos provided by the organizers of the computer vision competition of the 27th International Conference on Industrial, Engineering & Other Applications of Applied Intelligent Systems (IEA-AIE) [2,3], held on June 6th 2014 in Kaohsiung, Taiwan.

The images in the first row were taken at the Sun Yat Tse Memorial Hall in Taipei, Taiwan. The images in the second row were taken at the Flora Expo in Taipei, Taiwan.

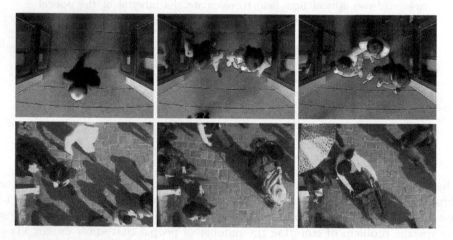

Fig. 1. Sample scenes for pedestrian counting as used in this research. Images taken from the competition website. Videos were sample videos for the IEA-AIE-2014 Computer Vision Competition.

2   Design

As can be seen from the sample images, the difficulty of the problem can vary greatly between image frames and a 100% accurate solution would require an extensive collection of background knowledge to deal with the various objects that may be present in the scene such as baby strollers, umbrellas, and packages. The large number of possible models leads to increased processing times for such systems and they often are not able to maintain real-time performance, i.e., in and out counts can be updated at 15 frames per second or higher frame rate. Furthermore, these models would require extensive calibration and tuning of the algorithm, which makes them unsuitable to be used in situations where setup needs to be done quickly and where only untrained personal is available.

However, the main goal of our system was to help in crowd control during large events, so I believe that: (a) real-time monitoring, (b) fast setup and (b) setup by untrained workers are essential in those circumstances.

I therefore decided to focus on a system that provides accurate estimates in a majority of likely scenarios while maintaining real-time performance and to ignore errors introduced by outliers.

The processing pipeline of our system consists of four steps:

1. **Pre-processing**: the image is cropped to the region of interest and the images are converted into grey-scale.
2. **Optical Flow**: a dense optical flow field of the images is calculated.
3. **Noise Removal**: flow vectors below a threshold are discarded.
4. **Optical Flow Field Discretization**: The flow field direction is discretized into the two directions perpendicular to the gate axis.
5. **Optical Flow Field Integration**: The flow field is integrated according to the two directions perpendicular to the gate axis.
6. **Pedestrian Counting**: The average integral of the flow field is used to estimate the number of people that entered or left the area.

Each processing step is described in more detail in the following subsections.

2.1   Cropping and Conversion to Grey-Scale

In the first part of my system, the images are cropped to a region of interest representing the gate area. Furthermore, the axis of the gate is input as a parameter thus defining the In and Out directions of my pedestrian counter.

Next the image is converted into grey-scale to be used as input for the optical flow computation.

In this step, I have also considered the possibility of adding a pre-processing step to enhance the contrast of images. In practice, that did not improve the results in our evaluation.

2.2   Farnebäck Optical Flow Computation

At the heart of my system is the computation of dense optical flow field. There are many algorithms described in the research literature to compute a dense optical flow field starting from seminal work by Horn and Schunk [4,5].

In this system, we used a dense optical flow algorithm developed by Farnebäck [6]. The main reason for choosing this algorithm is that a robust and efficient implementation of this algorithm is available in the OpenCV library [7]. OpenCV is a open-source library which includes implementations of many different image capture, low level processing, and higher level processing algorithms. OpenCV also provides bindings to many different programming languages such as C++ and Python.

In my pedestrian counting system, I used the OpenCV library with the Python bindings.

The images in Fig. 2 show the output of the Farnebäck optical flow algorithm on the sample video sequences. The direction of the flow is indicated by the direction of the small lines and the magnitude of the flow is given by the length of the line segment where a 10 pixel line segment corresponds to a 1 pixel inter frame flow.

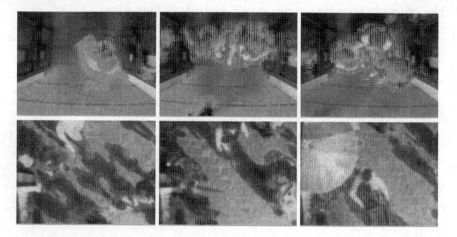

Fig. 2. Dense optical flow field calculated from the sample sequences. Direction and magnitude of the optical flow is shown by the direction and length of the line segments.

2.3   Noise Removal

As can be seen in the images, the direction and magnitude of the optical flow is more prominent in the simple images in the first row. The images in the second row show that the shadows of the people move as well and therefore generate an optical flow.

In some of the images, noise introduced by the camera resulted in small movement of features. I added a noise removal step where optical flow fields that were below a threshold were removed.

2.4   Optical Flow Field Discretization

In this step, the direction of the optical flow vectors are being discretized into one of eight directional buckets with major axis (0 deg., 45.0 deg., 90 deg., 135 deg., 180.0 deg., 225 deg., 270 deg., and 315 deg.). The width of each bucket is 45 degrees which leads to the following boundaries for the buckets: (-22.5 deg.

to 23.0 deg., 22.5 deg. to 68.0 deg., 67.5 deg. to 113.0 deg., 112.5 deg. to 158.0 deg., 157.5 deg. to 203 deg., 202.5 deg. to 248.0 deg. 247.5 deg. to 293.0 deg., 292.5 deg. to 338.0 deg.). The bucket boundaries are overlapping by 0.5 deg. to avoid sudden cutoffs during the integration along the boundary of a bucket.

2.5   Optical Flow Field Integration

The integral of the optical flow field of bucket is denoted as R_i. R_i is calculated given the following formula where i corresponds to a bucket in the optical flow field discretization, and t is time. B_i is the major axis of bucket i (e.g., 0 deg., 45 deg., ...). B is the set of all buckets. T is the length of the entire video sequence. The dot product calculates the projection of the flow onto the major axis of its corresponding bucket(s).

$$\forall i \in B : R_i = \int_0^T F_i(t)\, dt$$

where

$$F_i(t) = F(t) \bullet B_i \text{ if } F(t) \text{ assigned to bucket } i, 0 \text{ else.}$$

The current value of the integral flow field for all buckets for the video sequences shown in Fig. 1 are shown in Fig. 3.

2.6   Pedestrian Counting

As can be seen, individuals or groups of people result in a sharp increase in the integral of the optical flow. The top figure of Fig. 3 shows timeline for a video showing three people entering (90 deg.) and 23 people leaving the gate. The timeline in the top figure shows that the three people that in two groups: a single person and a group of two.

Furthermore, I estimate the average integral flow per person by diving the total integral of the flow field by the number of people that crossed the gate. In this case, the average flow of the people moving through the gate.

The estimate of the integral flow per person for video 1 (E_{V1} is given as the average flow for the up and down directions (I_{Up}

$$E_{V1} = \frac{I_{Up} + I_{Down}}{N_{Up} + N_{Down}} \approx \frac{0.5 * 10^7 + 3.5 * 10^7}{3 + 23} = 1.5 * 10^6$$

The second video shows approximately 20 people entering and approximately 64 people leaving. The exact number of people is impossible to determine since some are hidden behind an umbrella and some are only indicated by their shadow. I use a similar method to determine an average integral flow for the second video: $E_{V2} = (0.1 * 10^8 + 0.9 * 10^8)/(20 + 64) \approx 1.1 * 10^6$.

So for each video, a short training sequence is used to determine the average integral flow field E for a person.

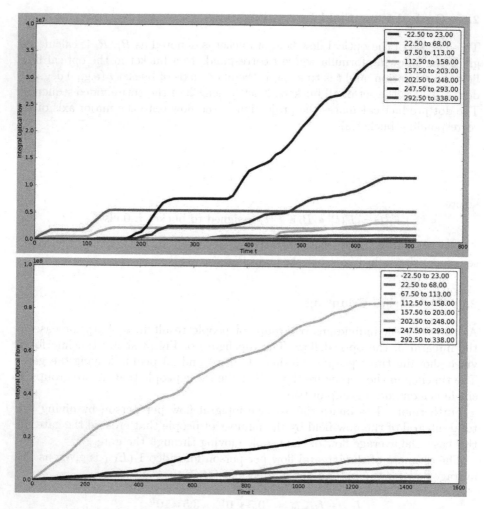

Fig. 3. Integral Optical Flow Field for sample video sequences. Top figure shows 3 people entering and 23 people leaving. Bottom figure shows approximately 20 people entering and approximately 64 people leaving.

The total number of pedestrians that entered or left the gate are determined by dividing the total integral flow field by the estimated flow field for each person. This simple measure turned out to be a reasonable good estimator for our pedestrian counter as is described in the next section.

3   Evaluation

I evaluated the system on the benchmark videos provided by the organizers of the IEA-AIE 2014 computer vision competition.

Video	Error in percent
SYS Memorial Hall normal	3.80
SYS Memorial Hall advanced	8.00

The system was also used in the IEA-AIE-2014 Computer Vision Competition on June 6th 2014 in Kaohsiung, Taiwan. My system won 2nd place in the competition against twelve other teams.

4   Conclusion

This paper describes the implementation of a practical system for counting objects (e.g., pedestrians or cars) as they traverse a ROI. The system is based on calculating the integral of the optical flow perpendicular to traversal axis. Some limitations of the system may be addresses in the future.

One assumption of the system is that the objects have approximately the same size, so that the average of the integral of the optical flow field can be used as an estimator. This assumption was warranted when the system was used to count pedestrians since there was not much difference in the integral of the flow field for small children or large adults. However, a more sophisticated classification algorithm may be necessary when the system is used to count vehicles that pass through an intersection. In this case, the integral of the optical flow field would be very different for a car or a truck. Note that the ability to distinguish between types of vehicles is not always important. The standard way of estimating traffic uses induction loops (wires) that are placed on top of the lane. Similar to our system, these systems are also unable to distinguish between one truck with a trailer or two cars driving closer together. In many cases, manual counting is used when the type of traffic is important.

Another weakness of the system is the fact that the integral of the optical flow for an object remains constant. This assumption is problematic in some circumstances. As can be seen in the images in the second row of Fig. 2, a person that casts a shadow on the ground will have a larger optical flow integral as one that follows closely behind a second person. The additional component of the integral of the optical flow due to the shadow is not taken into consideration in our system. Improvements to our system could be made by removing optical flows that are most likely associated with shadows. This may require providing some information about the lighting condition in the scene (e.g., direction of the sun) to the system.

However, as can be seen in our evaluation, the pedestrian counting system worked reasonable well over a large set of benchmark problems without dealing with shadows separately.

References

1. Weidinger, P.: 10 tragic human panics and stampedes. WWW (November 2010) (visited August 1, 2014)
2. 2014 international computer vision competition with iea-aie 2014. WWW (June 2014) (visited July 1, 2014)
3. The 27th international conference on industrial, engineering & other applications of applied intelligent systems (iea-aie). WWW (June 2014) (visited July 1, 2014)
4. Horn, B.K., Schunck, B.G.: Determining optical flow. In: 1981 Technical Symposium East, pp. 319–331. International Society for Optics and Photonics (1981)
5. Barron, J.L., Fleet, D.J., Beauchemin, S.S.: Performance of optical flow techniques. International Journal of Computer Vision 12(1), 43–77 (1994)
6. Farnebäck, G.: Two-frame motion estimation based on polynomial expansion. In: Bigun, J., Gustavsson, T. (eds.) SCIA 2003. LNCS, vol. 2749, pp. 363–370. Springer, Heidelberg (2003)
7. The opencv library. WWW (2014) (visited August 1, 2014)

A Study on Motion Energy of Humanoid Robot in Different Walking Gaits or Postures*

Kuo-Yang Tu[1], Pohsu Lin[2], and Chun-Chi Peng[3]

[1] Institute of Electrical Engineering,
National Kaohsiung First University of Science & Technology,
2 Jouyue Road, Nantsu, 811 Kaohsiung City, Taiwan, R. O. C.
tuky@nkfust.edu.tw
[2] MitraStar Technology Corporation, Taiwan, R. O. C.
[3] Industrial Technology Research Institute, Taiwan, R. O. C.

Abstract. It is worthy to study the motion energy of humanoid robots, but it is not easy to calculate for their complicated dynamics. In this study, the trajectories of voltage and current used in walking are stored to calculate actual motion energy for comparing the humanoid robot in different walking gait. Four walking gaits, covering the situation of humanoid robot walking are selected to compare their motion energy. After sorting the walking gait of wasting least motion energy based on experiments, we focus on the study of motion energy influenced by walking gait parameters including pace length, pace height and body location. In addition, a proper experiment is designed to study the energy wasted by humanoid robot join friction. The energy is compared with the walking energy to understand whether it is reasonable that the humanoid robot dynamics often neglect friction force or not.

Keywords: Humanoid Robot, Motion Energy, Walking Gait, 4-3-4 Polynomial Trajectory.

1 Introduction

In recent decades, humanoid becomes the most attention in popular robot research. The international teams collect more research for the top humanoid robot. For example, HONDA developed famous ASIMO [1].

In recent years, robot competitions are active for shorter learning curve to solve the bottleneck of humanoid research such as HuroCup [2] and RoboCup [3]. RoboCup focuses on soccer to develop better humanoid robots. HuroCup focuses on the development of full capability humanoid robot. Thus HuroCup is organized by eight events including spring, basketball, penalty kick, weight lifting, lift & carry, obstacle run, wall climbing and marathon. The humanoid successfully finishing the eight events is the full capability to have strong enough overall parts.

* This research was supported by National Science Council, Taiwan, Rep. of China under grant NSC 101-2221-E-327-023-.

In humanoid robot manipulation or motion, energy is core and unique resource. How to make use of the energy efficiently becomes an important issue on humanoid robot development. In spite of stability, the energy usage efficiency is a key point to evaluate the performance of humanoid robot walking gait. In this paper, the exactly energy used for humanoid robot motion is thus studied. In the study, the different walking gaits and postures are counted by the voltage and current during the humanoid robot motion.   The power is calculated by voltage and current to compare for the most efficient walking gait and posture.

2   Humanoid Robot Kinematics

The kinematic model of a humanoid robot is shown in Fig. 1.   In the kinematic model, two degrees of freedom are on ankle, rotating to X and Y axes, respectively. One rotating to Y axis is on knee.   Three rotating to X, Y, and Z, respectively, are on hip. In general, humanoid robot makes use of legs for motion, and hands for balance.

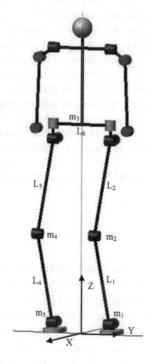

Fig. 1. The kinematic model of a humanoid robot

To derive the kinematic equations easily, the kinematic model is usually separated into two 2-dimensian models as shown in Figs. 2 and 3.   These two models are called sagittle and front planes, respectively.

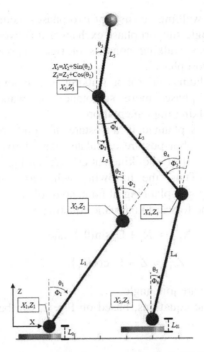

$X_3=X_2+\text{Sin}(\theta_2)$
$Z_3=Z_2+\text{Cos}(\theta_2)$

X_3,Z_3

X_2,Z_2

X_4,Z_4

X_1,Z_1

X_5,Z_5

Fig. 2. The humanoid robot in sagittle plane (X-Z plane)

Fig. 3. The humanoid robot on front plane (Y-Z plane)

The humanoid robot walking consists of two phases including single support and double phases.   The single support phase exchanges it between right and left.   Thus, the humanoid continuous walking consists of right support, double support, left support and double support phases.

Every phase of the humanoid robot walking is combined with many postures. More posture for one phase, more smoothness in walking motion, but more computation load for walking implementation.

When the humanoid is planned in a posture, its joints need to be controlled at angles.   The humanoid robot posture calculated for relative joint angles is inverse kinematics relationship.   But the kinematics relationship is derived by forward kinematics equations.   Thus, the following will derive the forward kinematics equations for the walking gait planning of humanoid robot.

As shown in Fig. 2, the forward kinematic equations are:

$$X_{i+1} = X_i + L_i \sin(\theta_i), \text{ and} \tag{1}$$

$$Z_{i+1} = Z_i + L_i \cos(\theta_i), \tag{2}$$

where θ_i (for i = 1, ..., 5) are join angle

The forward kinematic equations, based on Fig. 2, can be used to derive inverse kinematic equations.

3   Trajectories of Walking Gaits

The walking gaits of a humanoid robot are consisted of many postures in a time trajectory at Cartesian coordinate system. Via inverse kinematics, the time trajectory posture can be transferred into the joints of time trajectory for walking gait implementation. The time trajectories include the walking gaits in front and sagittle planes, respectively. In this section, many different kinds of time trajectories are defined for energy analysis.

In this study, the time trajectory of walking gait in the front plane as usual makes use of trapezoid trajectory as shown in Fig. 4.

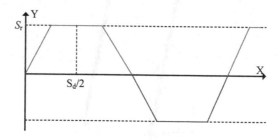

Fig. 4. The time trajectory of trapezoid shape for the walking gait in the front plane

In the sagittle plane, a walking gait can be consisted of waist and heel trajectories. Let the waist and heel trajectories be C_w and C_s.   Fig. 5 is the definition of walking gait parameters at the sagittle plane.   The parameters are explained by the following definition.

Definition 1: The key parameters of C_w and C_s are:

H_d: The move distance of C_w for a walking gait, that is the move distance of hip.

H_h: The maximum height of C_w in a walking gait.

S_d: The move distance of C_s for a walking gait, that is the step of a walking gait.

S_h: The maximum height of C_s in a walking gait.

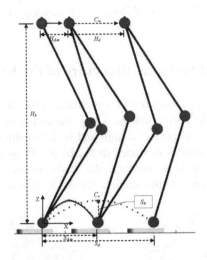

Fig. 5. The definition of walking gait parameters at the sagittle plane

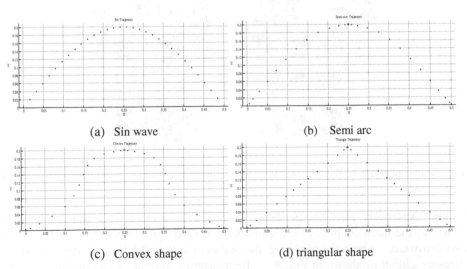

(a) Sin wave   (b) Semi arc

(c) Convex shape   (d) triangular shape

Fig. 6. Four kinds of time trajectories for the walking gait at the sagittle plane

In general, the trajectory C_w is a straight line for simple and easy move humanoid robot body. In this study, the energy waste of walking gait only compares different trajectories about C_s. There are three kinds of trajectories C_s for the energy waste comparison, including sin wave, triangular shape and 4-3-4 polynomial. In special, double 4-3-4 polynomials are consisted of the trajectory. For studying different lifting off speed, two kinds of double 4-3-4 polynomials are used for the time trajectories. In summary, there are four kinds of trajectories C_s, as shown in Fig. 6, for the study of walking gait energy waste. As shown in Fig. 6, (a) – (d) are sin wave, semi arc, convex shape and triangular shape, respectively. Notice that Figs. (b) and (c) are the 4-3-4 polynomial trajectories at high and low lifting speed, respectively. In this study, the walking gait of wasting energy will compare with these four C_s trajectories.

4   Hardware and Software Structure of the Experiments

The study of wasting energy makes use of a commercial humanoid robot, Bioloid, made of Robotis company. The reason that the study makes use of a commercial product is that such product usually has perfect design and is demonstrated by more users. The comparison of wasting energy is grounded in fair base. However, ever if the humanoid robot, Bioloid, is a commercial product, the study needs construct hardware and software for the comparison. In this section, the description focuses on this part.

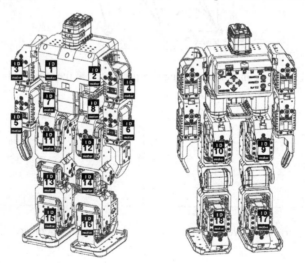

Fig. 7. The platform of experiments to compare the waste energy of humanoid in different walking gaits

The experiment platform as shown in Fig. 7 is Bioloid. For wasting energy measure, the main controller of Bioloid, CM-5, needs to be revised by its hardware and firmware. About the hardware, the main controller CM-5 is added by a current measure circuit as shown in Fig. 8. The measuring current can be produced to the battery voltage during humanoid walking for the power consumption of walking gait. Unfortunately, the battery voltage always drops down during walking because the humanoid walking gaits consumes too much power. Thus, in addition to current, the battery voltage is also measured for accurate power consumption.

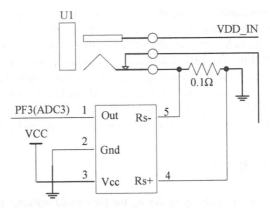

Fig. 8. The circuit added on CM-5 main controller for measure current wasted by Boiloid

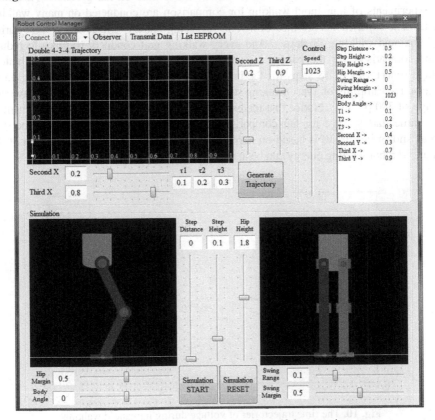

Fig. 9. The operation and setting interface of the experiments

In addition to revised hardware, the main controller firmware is changed to have four functions as follows. 1. Communicate with personal computer; 2. Calculate the inverse kinematics for joint angle trajectories; 3. Control joint motors for walking gait implementation; 4. Read ADC to storage battery voltage and current during humanoid walking.

However, the experiments still need more works. Thus, an operation interface as shown in Fig. 9 is implemented for the experiment operation and setting. The operation interface includes the simulation of walking gaits on sagittle and front planes for demonstration. The time trajectory of simulation walking gait is shown on the left-top side in Fig. 9. After examining, the walking gait is transferred to CM-5 for implementation. The operation interface can set walking gait parameters, and read the EEPROM data in CM-5. In addition the menu item observer can show the time trajectories of battery voltage and current consumed for humanoid walking.

5   Experiment Results

Because the battery voltage is changing during the humanoid walking, the comparison of four different walking gaits based on power source is hard on fair ground. Therefore, the experiments of humanoid walking for comparison are conducted on many walking steps. The experiments are measuring and recording battery voltage and current during the humanoid walking 48 steps. And every walking gait conducts 10 times to reduce unfair comparison for the battery on unfair initial level. Fig. 10 shows the battery voltage trajectories of 48 steps for the humanoid robot on four different walking gaits. Because the sampling of every walking step has 30 points, the every point is the average voltage of 300 points. Notice that the experiments for different walking gait are hard to have same start voltage for fair comparison. The battery at same voltage on the start step as shown in Fig. 10 is based on normalizing by shift. Then, the comparison of the humanoid robot wasting energy can be at the fair base.

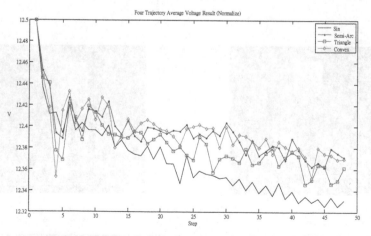

Fig. 10. The time trajectories of voltage during humanoid walking

Fig. 11 is the current trajectories of the humanoid robot on four different walking gaits.   Let the definition of wasting energy during the humanoid on four different walking gaits be

$$P = \sum_{t=0}^{m} V(t) \cdot I(t) \cdot T_s \ . \tag{3}$$

Then the wasting energy can be calculated by both Figs. 10 and 11 to get Fig. 12. Thus the total energy of the humanoid robot walking 48 steps can be calculated from Fig. 12.   As shown in Fig. 12, the total energy of walking gaits on sin wave, semi arc, triangular shape and convex shape is 268.0675 J, 216.4583 J, 224.8621 J and 240.9342 J, respectively.   Thus, the walking gait of semi arc wastes smallest energy, and that of sin wave wastes biggest energy.

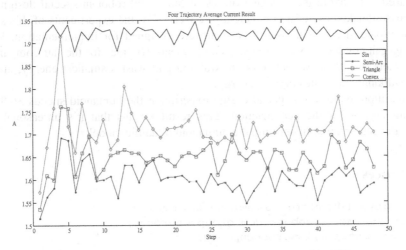

Fig. 11. The time trajectories of current during humanoid walking

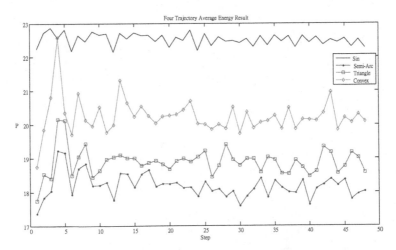

Fig. 12. The time trajectories of energy during humanoid walking

In order to measure the energy wasted by joint rotation at no load, a simple experiment is planned. The experiment lets the humanoid robot lie down on the ground, and rotates 48 times between -5° and 5°. The battery voltage and current are measured for the calculation of wasting energy. The total energy is 3.7764 J. When the humanoid is walking, ten servo motors need to run. Compared with the humanoid robot wasting 102.6217J for walking, the no load rotation spends about 36.8 %. It's not small ratio.

6   Conclusions

In this study, the wasting energy of humanoid robot in different walking gait is compared. To avoid the special features of humanoid robot in special design, this comparison makes use of a commercial product. For the experiments of this study, the electronic circuit and control software are designed for implementation. Every walking gait conducts 48 walking steps and repeats 10 times for the fair comparison. Experiment results show that semi arc walking gait wastes smallest energy, and sin wave walking gait wastes biggest energy.

In addition, based on a special design experiment the humanoid spends 36.8% of walking energy for the gear friction. This result reveals that it's unreasonable to ignore the gear friction in deriving the humanoid dynamic model.

References

1. HONDA-ASIMO, http://asimo.honda.com/
2. https://www.facebook.com/groups/hurocup/
3. RoboCup, http://www.robocup.org/

Dense 3D Mapping for Indoor Environment Based on Kinect-Style Depth Cameras

Yalong Wang, Qizhi Zhang, and Yali Zhou

School of Automation, Beijing Information Science &Technology University, Beijing, China
Wangyalong1989@163.com

Abstract. Kinect style depth cameras provide RGB images along with pre-pixel depth information, the richness of their data and recent development of low-cost sensors have made them more popular in mobile robotics research. In this paper, we present a framework of dense 3D mapping. Sparse visual features are used to determine an initial rough transformation, then it is refined by color-GICP (General iterative closest point). We employ a window sparse bundle adjustment to optimize the local map after it is constructed and a new keyframe is created at the same time. Visual features and dense information are also used in loop closure detection, following by a globally consistent optimization based on graph. Moreover, we introduce a user interaction to improve the map building progress. This proposed approach is evaluated by the RGB-D benchmark and two real indoor environments, and experiment results show the feasibility and effectiveness of this approach.

Keywords: RGB-D, SLAM, 3D Mapping, graph optimization, point cloud.

1 Introduction

It is an important task for mobile robots to build rich 3D model of environments with applications in navigation, path planning, semantic mapping, autonomous exploration and manipulation. To be truly useful, such systems require the fast and accurate estimation of the robot pose and scene geometry of the environment all around. This problem is well known as SLAM (Simultaneous Localization and Mapping), whose goal is to estimate both the camera poses and map at the same time.

Mobile service robots, for example, require a detailed model of their workspace for collision-free motion planning, and also need detailed maps for localization and navigation. Previously, many 3D mapping approaches have relied on expensive and heavy laser scanners, the commercial launch of RGB-D cameras based on structured light such as Microsoft Kinect sensor and Asus Xtion Pro Live sensor has provided an attractive, powerful alternative.

Most 3D Mapping systems are consist of three main components: first, the spatial alignment of consecutive data frames; second, the detection of loop closures for scenes are already viewed; third, after detecting all the loop closures, a globally consistent optimization is applied to the complete data sequence. 3D point clouds provided by

© Springer International Publishing Switzerland 2015
J.-H. Kim et al. (eds.), *Robot Intelligence Technology and Applications 3,*
Advances in Intelligent Systems and Computing 345, DOI: 10.1007/978-3-319-16841-8_30

laser scanners are well used for frame-to-frame alignment and 3D reconstruction but they ignore valuable information contained in RGB images. RGB images, on the other hand, contain rich visual information which is useful for loop closure detection [1]. However, the lack of depth information makes it hard to extract dense depth data.

RGB-D cameras can capture RGB images and per-pixel depth information at the same time which provide much convenience for robots to build 3D model of environments. This makes it widely used in the SLAM technology of mobile robots.

In this paper, we present a framework of building 3D model for indoor environment using a RGB-D camera with the advantage of dense color and depth information. Visual features are used to find corresponding 3D points in different frames, and the alignment between frames is computed by integrating the shape and appearance information provided by the RGB-D sensors. Visual appearance is incorporated by extracting sparse feature points from the RGB images combining with depth information and matching them via a 3D RANSAC procedure. The resulting feature matches are then combined with a least squares solver to determine an initial alignment between the frames by minimizing reprojection error of the correspondences. Finally, the best alignment is computed by a color supported GICP. Our approach detects loop closures by matching data frames against a subset of previously collected frames. The local map and global map are optimized via sparse bundle adjustment and efficient pose graph optimization respectively.

We also introduce a user interaction approach to provide more information about the process during map building. This interaction makes it easy to complete the creation even for novice users and also improves the accuracy of the final map.

After discussing related work, we introduce our approach in Section 3, and experiments are presented in Section 4.

2   Related Work

Many techniques and algorithms have been developed for 3D mapping with the develop of robots and computer vision and there are many sensors used by SLAM systems, for example: rang-scans(e.g. May et al. [2] ,Newman et al.[3]), stereo cameras (e.g. Konolige et al.[4]), monocular cameras (e.g. Strasdat et al.[5], Clemente et al.[6]) and also photos collected for recover 3D structures (e.g. Snavely et al.[7]). Most SLAM systems follow a common architecture, which contains the spatial alignment of consecutive frames, the detection of loop closures and the global consistent alignment of all data frames. And a large number of variations and specific approaches for different part exist, each with its own attributes.

For 2D/3D lasers which commonly provide quite accurate depth information of the environment at a high frequencies, the iterative-closest-point (ICP) algorithm and variants are widely used techniques to compute the rigid transformation by minimizing distance between corresponding points. Some novel variants such as point-to-plan associations[8] or point-to-line metric[9] are robust and suit for 3D man-made environment. However, ICP also has a disadvantage that depends on a good initial guess to avoid getting stuck in a local minimum.

Vision-based SLAM techniques have developed for many years and been gaining popularity recently. Vision SLAM approaches, estimate the motion of robots and build map of the environment at the same time using vision information provided by cameras, which also referred to as "structure and motion estimation". Different with laser-based SLAM, most vision SLAM systems take the advantage of color data, extract sparse keypoints from RGB image and match them in different frames to find the location of the same landmarks in pixels. Some typically keypoint detectors and descriptors such as SIFT, SURF and ORB are widely used in lots of existing SLAM algorithms. There are also many analyses based on different keypoint detectors combined with different descriptors[10]. In our experiment, we have compared the performance of different detectors in our SLAM system.

With the consumer RGB-D cameras being more and more popular, they provide a valuable alternative to lasers. Most RGB-D cameras depend on active stereo (e.g. Microsoft Kinect, ASUS Xtion Pro Live) or time-of-flight (e.g. Cobra) to generate depth information for every pixel together with RGB image. In this work, we use Kinect Xbox as our depth sensor which provides 640x480 pixels color image and depth image at the same time at 30fps. However, Kinect-style cameras have some disadvantages with respect to 3D mapping: their valid value of depth is limited to less than 5m, their depth estimates are noisy and a constrained field of view (~60。) also make it a challenge to build 3D map for large environment.

The RGB-D Mapping system proposed by Henry et al. [11] who combine sparse keypoints with GICP creates a pose graph and uses SBA (sparse bundle adjustment) and TORO to optimize it is one of the first RGB-D SLAM systems. Particularly, the KinectFusion algorithm, introduced by Newcombe et al., is an impressive approach to produce a volumetric reconstruction of a scene in real-time with an unprecedented level of accuracy, however it does not take account of loop detection which may lead to accumulating drift and also requires high performance graphics hardware. Hao Du et al. developed an interactive 3D mapping system which can be controlled by users during mapping and the accuracy is improved by user interaction, they also described the capability of detailed 3D modeling in many promising applications such as accurate 3D localization, measuring dimensions, and interactive visualization[12]. Just like KinectFusion and Kinfu provided by PCL (Point Cloud Library), Bylow et al. present a novel approach to directly estimate the camera movement using a signed distance function to the surface[13], but it is drift-free only on small workspace.

3   Approach

The objective of our whole work is to design a framework of a RGB-D SLAM system. We follow a common architecture as many RGB-D SLAM systems were designed: Frontend, backend, and final map representation. For the frontend, frame alignment is done by GICP between current frame and keyframe in a local map, if the current frame is accepted, it is added into the dense 3D local map otherwise a new local map is created. For the backend, loop closure detection step uses sparse feature points to match the new keyframe against previous candidate keyframes after a new local map is created.

If a loop closure is detected, a constraint is added to a graph, whose vertexes are key-frames and edges are the constraints between them, and a global optimization of the graph is triggered. Finally, 3D point cloud map is used to represent the model of the environment once camera poses are determined. A schematic representation of our system is shown in Figure 1.

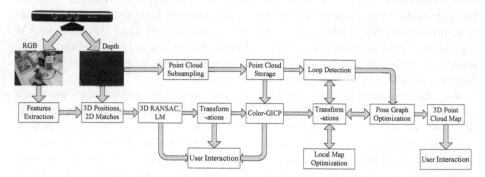

Fig. 1. The framework of our proposed approach

3.1   Visual Odometry

Our odometry algorithm is based on a standard feature-based methods with frame-to-frame matching, The goal of visual odometry is to compute geometric relation z_{ij} which is used to estimate the motion of the camera between state X_i and X_j given the position of landmarks $P=p_1, p_2,..., p_n$ in different frames. 3D coordinates of the landmarks are generated by extracting sparse visual features in the color frame as implemented in OpenCV and associating them with their corresponding depth value captured by RGB-D camera. Recently, there are many visual features such as SIFT, SURF, ORB or FAST that can be used to implement these steps.

Descriptors are needed to describe the visual features after extracting, which is necessary for the keypoints matching. A pair of keypoints descriptors (d_i, d_j) are often matched by computing the distance between them in the descriptor space, where d_i and d_j are always high dimension vectors. For SIFT and SURF, the proposed distance is Euclidean and Hamming for ORB and FAST. However, it is generally feasible to choose a rejection threshold due to the high dimensionality. We use the feature ratio test: first, compute two nearest neighbors of each descriptor. Under the assumption that a keypoint only matches to exactly one other keypoint in another image, the second nearest neighbor should be much further away. Thus, we set a threshold base on the ratio between the distance of nearest and second nearest neighbor combining with a cross check filter, which is used to make sure that the nearest neighbor of two descriptors are cross to decide whether a match is a inlier or not.

RANSAC algorithm has been proved to be an effective method to prune outliers. In this work, we employ 3D points RANSAC when estimate full 6D transformations between frame pairs with initial transformations computed by SVD from three feature correspondences or LM (a nonlinear least-squares solver) from four. We also compare it with the classical 7-point solution (see Section 4), 3 (4)-point solution is more robust and efficient due to the depth information.

Consider that we have N pairs of initial feature matches between frame F_1 and F_2 with respective 3D homogeneous coordinates (X_1^i, X_2^i) in their reference systems according to the match steps, then RANSAC algorithm samples 3 pairs of the feature correspondences and computes the rigid transformation T consist of rotation R and translation t using SVD or LM. The fitness of each solution is computed by counting the number of inliers n and reprojection error e.

$$n(F_1, F_2, T) = \sum_i^N f(X_1^i, X_2^i, T) \tag{1}$$

where

$$f = \begin{cases} 1, & e = \left\| TX_1^i - X_2^i \right\| < \theta \\ 0, & otherwise \end{cases} \tag{2}$$

A feature match (X_1^i, X_2^i) is considered as a inlier if their corresponding error e is less than a given threshold θ, transformation with the largest number of inlier matches is chosen as final one which is used in next steps. After RANSAC steps, transformation T is then refined by minimizing feature reprojection error of inliers using LM with the cost function:

$$T^* = \arg\min_T \left(\frac{1}{n} \sum_{i=0}^n \left(\left\| TX_1^i - X_2^i \right\| + \left\| T^{-1}X_2^i - X_1^i \right\| \right) \right) \tag{3}$$

where n is the number of inliers, T is the initial value of LM algorithm which is provided by RANSAC and T^* is the optimized transformation. In contrast to undirectional motion refinement, which minimizes the reprojection error of newly detected features onto the reference frame, we take advantage of bidirectional refinement, which additionally minimizes the reprojection error of reference features projected onto the new frame. The efficiency of this procedure has been proved by Huang in his recent work [14].

3.2   Color-GICP

ICP algorithm has been widely used in the field of 3D reconstruction due to its dominance for dealing with the rang image alignment. All ICP and its variants need a good initial guess of the transformation between two frames since they might get stuck in a local minimum, this guess can be provided by sparse features matching and transformation estimate described in 3.1. Generally, ICP algorithm is often used to refine a result which is already convergeed to the global minimum but with lower precision. However, sometimes the guess is needless when the difference between measurements is sufficiently small benefit from a high rate (up to 30Hz) provided by sensors.

Consider two pointclouds A={a_1,...,a_M} and B={b_1,...,b_N}, and also a initial transformation T_0, the standard ICP algorithm can be described as follows:

Algorithm 1 Standard ICP

Input : Pointclouds: $A = \{a_1,...,a_M\}, B = \{b_1,...,b_N\}$
      An intinal transformation T_0
Output : The transformation T which aligns A and B

1 : $T \leftarrow T_0$
2 : **while** not converged **do**
3 :   **for** $i \leftarrow 1$ **to** N **do**
4 : $m_i \leftarrow$ Find closest point in $A(Tb_i)$
5 :   **if** $\|m_i - Tb_i\|_2 \leq d_{max}$ **then**
6 : $w_i \leftarrow 1$
7 :   **else**
8 : $w_i \leftarrow 0$
9 :   **end if**
10 :   **end for**
11 : $T \leftarrow \arg\min_T \sum_i w_i \|Tb_i - m_i\|_2^2$
12 : **end while**

However, the standard ICP algorithm considers only depth information when finding correspondence between A and B. Korn et al. propose a novel variants of ICP by integrating color into the points matching step[15] (line 4) which is proved more effective and robust. In this work, our final transformation between different frames is refined by color-GICP with an initial value from the visual odometry.

3.3   Loop Closure Detection and Global Optimization

The alignment between successive frames provides useful visual odometry information. However, the individual estimations are noisy and the quantization in depth values also causes the estimation of camera to drift over time, particularly in man-made environments with few features or features are out of range. This cumulative error may lead to a map with two or more representations of the same region in different locations when the camera comes back to a visited location after a long path. This is known as loop closure problem, which is quite important in SLAM. To dealing with this problem, the solution should contain two parts: loop closure detection and error correction.

Generally, the simplest approach is to compare a new location with all previously visited locations, if no loop is detected, then the new location is added to the map. However, the computation time required for a new frame increases quickly with the number of locations in the map which makes it not feasible to compute in real time when builds large maps.

In this work, we exploit a keyframe technology, and keyframes are defined as a subset of the aligned frames which are used in the loop detection steps. We first initialize a local map consist of a keyframe (initial one is the first frame) and a active frame, a new frame is compared with the keyframe and active frame respectively using the matching algorithm and 3D RANSAC provided by the visual odometry. If the new frame is accepted, it is added into the local map and the active frame is replaced by it, otherwise, a new local map is created with a new keyframe which is the active frame of the previous local map and current frame becomes the active frame of this new local map.

Our approach makes it very easy to select the acceptance criteria of deciding when to determine a keyframe. The simplest is to choose every n-th frame which is a rude way taking no account of visual odometry information. Another option is to establish a new keyframe whenever the accumulated rotation or translation is above a threshold[16]. We have found a more robust way based on visual overlap: after we align a frame F, the number of 3D RANSAC inliers is N_k and N_a compared with keyframe and active frame respectively. As the camera continues to move, its view contains progressively fewer 3D feature point matches with the previous keyframe. A local map creation along with the new keyframe determination is triggered when the average fitness represent by the Mahalanobis distance between the correspondences exceeds a given threshold or the average ratio α is below another threshold.

$$\alpha = \frac{N_k}{N} + \frac{N_k}{N_a} \tag{4}$$

where, N is the number of features in current frame.

Each time we create a new local map (determine a keyframe), and we first employ sparse bundle adjustment[17, 18] to optimize the finished one. This step minimizes the re-projection error of feature points across all frames by adjusting the estimated 3D locations of feature points along with the camera poses. We just take into account features matched with the keyframe since all frames need to compare with it. This procedure has also been independently proposed by Henry et al. [16] in his most recent work and referred to as two-frame sparse bundle adjustment, and we call it window sparse bundle adjustment used in our approach.

After the optimization, we attempt to detect loop closures with previous keyframes. We implement this by applying a metrical nearest neighbor search, thereinto, all key-frames are used to build a k-d-tree which is a very efficient data structure for our searching step. Loop closure candidates are searched in a sphere with pre-defined radius around the keyframe position then we compute the rotation between each candidate and the keyframe. Candidates with rotation below a given threshold are used to compare with the keyframe to compute the relative transformation and the associated covariance matrix use the 3D RANSAC. To validate a candidate we employ the same acceptance criteria for keyframe selection. However, instead of considering both N_k and N_a, we now only use N_k, so α is redefined as $\alpha = \frac{N_k}{N}$ where N is the number of features in the candidate keyframe.

We represent the global map as a graph of camera poses where every vertex is a pose (rotation and translation) of a keyframe. The edges represent constraints (relative transformations) between the vertexes with associated covariance matrices. Each time we add a keyframe to the map, the graph also creates a chain of the keyframe linked by relative transformations w.r.t other keyframes. Valid loop closures also add new edges to the graph then the accumulated error can be reduced by optimizing the graph use a non-linear least squares solver. We use the graph framework of MRPT[19] as the implementation for the map representation and optimization.

3.4   Map Representation

The system described so far provides a globally consistent trajectory which can be used to general a global map by projecting the original point clouds into a common coordinate frame, thereby we can create a 3D model of the world represented by a point cloud.

However, one frame from the RGB-D camera contains about 300,000 points, which will need large memory resources to store such a map with increasing frames. In our implementation, we use octree, an efficient tree structure, to compress the map data. For a 3D point cloud map we have constructed, an unfiltered map requires about 500 Megabytes, in contrast, the octree makes it reduce to only 50 Megabytes with a resolution of 5mm.

3.5   User Interaction

User interaction has been utilized in some SLAM systems to improve the accuracy and avoid the break of automatic systems in situations with some failure occurred[20]. Most frame-to-frame approaches have no ability to deal with serious cases cause it always has to handle a new frame. There are also existing work who regards user interaction as a post-processing step, in those approaches, maps are generated offline with the user's control.

In this work, we provide an interface for users followed [20] to view the current reconstruction in real time, view the up-to-date 3D model. From this interface, users can know which parts of the scene have been captured and choose a better path to reconstruct the world.

In many times, failure may be occurred in match steps such as when users move too fast, motion blur or there are too few features. Our application is designed to be robust with those failures. Once the new frame fails to register with the previous frames, the system shows user an alert and ignores invalid frames until a new frame that can successfully register. Moreover, the last valid frame is shown to the user which can be a guidance for moving the camera back to a new valid frame. This interaction can improve the robust and accuracy of the map building, and also provide an easy way for novice users to use our application.

4   Experiments

In this section, a series of experiments have been performed to show the effect of our system. We analyze the present system using a public benchmark dataset by means of the quantitative results. Note that the result was obtained offline and depended on the used hardware. However, we also provide successful online mapping. All experiments are tested on a 2.10GHz laptop computer with windows7 system, Intel Core i3 CPU and Nvidia GeForce GT520M graphics card at 5 to 6 frames per second.

The RGB-D benchmark provides an RGB-D dataset of several sequence captured with Microsoft Kinect and Asus Xtion pro Live sensor, Synchronized ground truth data for the sensor trajectory is available for all sequences obtained from a high-accuracy motion-capture system with eight high-speed tracking cameras (100 Hz)[21]. We use the absolute trajectory error (ATE) which directly measures the difference between points of the true and the estimated trajectory described in the RGB-D benchmark to evaluate our system.

Feature detectors and descriptors are one of the most influential choices for accuracy and runtime performance, in this work we evaluated SURF, ORB and FAST features use the OpenCV implementations and employ a GPU-based SURF which has sped up 6 times compared with the CPU-SURF. 600-700 keypoints have been extracted for SURF per frame will perform a good result, however, we deal with frames only when the number of matched features is below 200. Figure2 shows the matched SURF features between two frames and the depth image of the keyframe. Not all RGB points have the corresponding depth information according to the depth image, so points without depth should be rejected after detecting features and before matching. In our experiments, about 150~200 pairs of points were retained as inliers with the 3D RANSAC threshold setting to 0.03m.

Fig. 2. SURF features and matches between two frames, the right shows pixels that are matched between frames, where features are drawn with red circles and matches are drawn with green lines. The left is the depth image of keyframe, pixels without depth information are black.

After applying 3D RANSAC and LM refinement, we have computed an accurate transformation which is used as an initial guess for color-GICP. In our experiment, the color-GICP always run into a convergence in less than ten iterations according to the initialization.

Keyframes were selected together with the local map creation when the relative keypoints ratio is small enough or the fitness is not good to register two frames. Whenever a new local map is created, we optimize the old local map which contains two keyframes with a window sparse bundle adjustment. The covariance matrix of each keypoints in the image which is needed in the sparse bundle adjustment optimization is computed by the approach proposed in [22]. Figure 3 shows the localization uncertainty of every keypoints represented by ellipses in the gray image. The employ of covariance matrixes has significantly improved the accuracy of our optimization.

We compared the performance of different features on several datasets, Table 1 shows the evaluation of accuracy with SURF feature, our system can achieve a quite good trajectory reconstruction, and the comparison results with different feature types are illustrated in Table 2. As shown, ORB and GPU-based SURF may be good choices for 3D mapping with limited computational resources or for applications that require real-time performance.

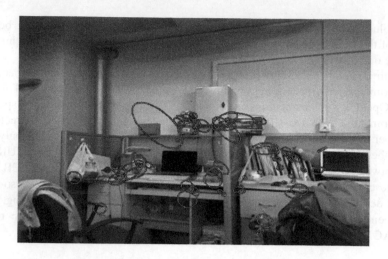

Fig. 3. Covariances estimated for SURF features

Figure 4 shows us a pose graph created by our approach, this graph is optimized using a graph optimization after loop closure is detected. Transformations between different poses are represent as edges in the graph which are used as constraints during optimizing. Figure 5 describes the comparison of trajectory between the ground truth and our estimation.

Table 1. Evaluation of different datasets with SURF features

Sequence	Fr1 desk	Fr1 room	Fr1 floor	Fr1 xyz	Fr1 desk2
ATE RMSE	0.0490m	0.0532m	0.0572m	0.0588m	0.0468m
ATE Median	0.0237m	0.0406m	0.0229m	0.01432m	0.0256m
ATE Mean	0.0339m	0.0434m	0.0344m	0.0282m	0.0310m
ATE Min	0.0100m	0.0079m	0.0022m	0.0034m	0.0020m
ATE Max	0.1754m	0.1586m	0.2646m	0.2504m	0.2128m

Table 2. Compaison of accuracy and runtime with respect to feature type. Times do not consider graph optimization.

Feature Type	SURF	ORB	FAST+ORB	FAST+BRIEF
ATE RMSE	0.0237m	0.0333m	0.0367m	0.0481m
ATE Median	0.0186m	0.0229m	0.0316m	0.0323m
Time/frame	0.23s	0.15s	0.17s	0.13s
Test dataset	Fr3 structure, time: 31.55s, length: 5.884m, frames: 907			

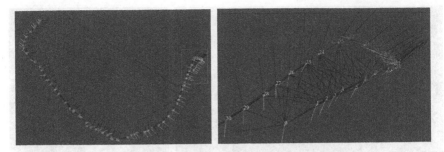

Fig. 4. Pose graph created by our approach (right) with poses represented by orange points and edges represented by blue lines, detail of the graph (left) shows the loop detection information. Each time a new loop is detected, an edge is add into the graph.

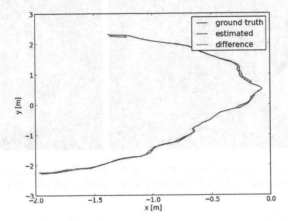

Fig. 5. Comparison of trajectory between the ground truth and our estimation with SURF feature on dataset fr3 structure

Since the dataset of the RGB-D benchmark is offline, so we cannot perform our implementations of user interaction. Therefore, we built maps online with a Kinect Xbox360. We choose two indoor environments due to the limit of the depth range of Kinect. During mapping, the Kinect was hold by a person and generally pointed in the direction of travel. If there occurred a failure, our system shows a warning and pauses the mapping process. Then the user should go back to the early position where may view a valid frame. We have found that failures occurred often when many textureless areas or motion blur exists and also the users failed early in the process using the offline system. On the other hand, our online interactive system could do better cause users can successfully recover from failures and complete the map building easily with the help of the interaction. 3D maps generated with our system is shown in Figure 6. Our system can reconstruct the 3D model of indoor environment accurately.

Fig. 6. 3D map of two indoor environments created online by our approach. The lab (upper) map is built in daylight and the bedroom (lower) is built in night with low lighting.

5   Conclusions

In this paper, we introduce a framework of 3D mapping for indoor environment using RGB-D cameras that can be used for robot localization, navigation, and also path planning. Visual features are extracted to perform 2D matching between frames and combined with depth image to localize them in 3D, then we use 3D RANSAC and a least squares optimization (LM) to estimate the transformations between associated keypoints which are then refined by a color supported GICP. Keyframes are selected along with the local map creation which also referred to as vertexes in a global graph. Finally, the global graph is optimized by a nonlinear least squares solver to generate the 3D point cloud map.

We also introduce an implementation of user interaction which makes our approach better use and more robust. 3D maps of indoor environment created by this implementation bring us a view evaluation. Our approach has been evaluated using the RGB-D benchmark on different offline datasets. Results show the ability of robustly dealing with challenging situations. Moreover, different features have been implemented and compared in online and offline mapping.

Acknowledgment. This research is supported by the National Natural Science Foundation of China (NSFC) (11172047) and Funding Project for Academic Human Resources Development in Institutions of Higher Learning under the Jurisdiction of Beijing Municipality (PHR201106131).

References

1. Labbe, M., Michaud, F.: Appearance-Based Loop Closure Detection for Online Large-Scale and Long-Term Operation. IEEE Transactions on Robotics 29(3), 734–745 (2013)
2. May, S., Droeschel, D., Holz, D., et al.: Three-dimensional mapping with time-of-flight cameras. Journal of Field Robotics 26(11-12), 934–965 (2009)
3. Newman, P., Sibley, G., Smith, M., et al.: Navigating, recognizing and describing urban spaces with vision and lasers. The International Journal of Robotics Research 28(11-12), 1406–1433 (2009)
4. Konolige, K., Agrawal, M.: FrameSLAM: From bundle adjustment to real-time visual mapping. IEEE Transactions on Robotics 24(5), 1066–1077 (2008)
5. Strasdat, H., Montiel, J.M.M., Davison, A.J.: Scale Drift-Aware Large Scale Monocular SLAM. Robotics: Science and Systems 2(3), 5 (2010)
6. Clemente, L.A., Davison, A.J., Reid, I.D., et al.: Mapping Large Loops with a Single Hand-Held Camera. In: Robotics: Science and Systems, vol. 2 (2007)
7. Snavely, N., Seitz, S.M., Szeliski, R.: Photo tourism: exploring photo collections in 3D. ACM transactions on graphics (TOG) 25(3), 835–846 (2006)
8. Segal, A., Haehnel, D., Thrun, S.: Generalized-ICP. Robotics: Science and Systems 2(4) (2009)
9. Censi, A., An, I.C.P.: variant using a point-to-line metric. In: IEEE International Conference on Robotics and Automation, ICRA 2008, pp. 19–25. IEEE (2008)
10. Siegwart, R., Nourbakhsh, I., Scaramuzza, D.: Introduction to Autonomous Mobile Robots, 2nd edn. MIT Press, Cambridge (2011)
11. Henry, P., Krainin, M., Herbst, E., Ren, X., Fox, D.: RGB-D mapping: Using depth cameras for dense 3D modeling of indoor environments. In: Khatib, O., Kumar, V., Sukhatme, G. (eds.) Experimental Robotics. STAR, vol. 79, pp. 477–491. Springer, Heidelberg (2012)
12. Du, H., Henry, P., Ren, X., et al.: Interactive 3D modeling of indoor environments with a consumer depth camera. In: Proceedings of the 13th International Conference on Ubiquitous Computing, pp. 75–84. ACM (2011)
13. Bylow, E., Sturm, J., Kerl, C., et al.: Direct Camera Pose Tracking and Mapping With Signed Distance Functions. In: RGB-D Workshop on Advanced Reasoning with Depth Cameras, RGB-D 2013 (2013)
14. Huang, A.S., Bachrach, A., Henry, P., et al.: Visual odometry and mapping for autonomous flight using an RGB-D camera. In: International Symposium on Robotics Research (ISRR), pp. 1–16 (2011)
15. Korn, M., Holzkothen, M., Pauli, J.: Color Supported Generalized-ICP 0. In: International Conference on Computer Vision Theory and Applications (2014)
16. Henry, P., Krainin, M., Herbst, E., et al.: RGB-D mapping: Using Kinect-style depth cameras for dense 3D modeling of indoor environments. The International Journal of Robotics Research 31(5), 647–663 (2012)
17. Zhang, L., Koch, R.: Structure and motion from line correspondences: representation, projection, initialization and sparse bundle adjustment. Journal of Visual Communication and Image Representation 25(5), 904–915 (2014)

18. Lu, Y., Song, D., Yi, J.: High Level Landmark-Based Visual Navigation Using Unsupervised Geometric Constraints in Local Bundle Adjustment. In: IEEE International Conference on Robotics and Automation (ICRA), Hong Kong, China (2014)
19. http://www.mrpt.org/
20. Du, H., Henry, P., Ren, X., et al.: Interactive 3D modeling of indoor environments with a consumer depth camera. In: Proceedings of the 13th International Conference on Ubiquitous Computing, pp. 75–84. ACM (2011)
21. http://vision.in.tum.de/data/datasets/rgbd-dataset/

Exponential Backoff-Sampling RRT for Smart Carpet

Hao Sun and Xiaoping Chen

Dept. of Computer Science, University of Science and Technology of China
Hefei, Anhui, 230027, PRC
hhsun@mail.ustc.edu.cn, xpchen@ustc.edu.cn

Abstract. With the developing of sensor technology, smart materials and devices get increasing attention. Unlike desktop computing, smart devices typically have limited computing power and often need to react in real-time. So it's difficult to directly apply complex algorithms on it due to the restriction in computational capability. Planning algorithm is one of these complex algorithms. In this paper, we proposed a new planning algorithm based on Rapid-exploring Random Trees (RRTs), which are currently widely used in robotics. In this paper, firstly we introduced the RRT algorithm and its analysis. Secondly, based on the analysis, we proposed an improved algorithm - Exponential Backoff sampling Rapid-exploring Random Tree (EB_RRT). The main idea in EB_RRT is to estimate the most efficacious sampling area to avoid useless sampling. In our experiments, we constructed a map with randomly placed obstacles. In the map, EB_RRT and several currently best algorithms were compared through solving a path planning problem. The results showed that our method makes an order of magnitude improvement than the existing ones and the low time consuming characteristic makes it a promising technology for smart devices.

Keywords: Rapid-exploring Random Tree, Planning Algorithm, Smart Carpet.

1   Introduction

In recent years, more and more materials/devices around us become "smart". The smart things have sensors, brains and can interact with people. It seems like in the future, computing will appear everywhere and anywhere, that is a concept called "Ubiquitous Computing". To convert the concept into a reality, people are now equipping the smart devices with more and more intelligent algorithms. Planning algorithms are one of these useful intelligent algorithms, which have a big opportunity of applying in smart material/device. For example, we have an idea of using smart cloth to make a smart carpet. The carpet, which is able to detect obstacles on it, may use planning algorithm to show us a path to the destination we want. It will be quite useful in a big plaza or a library (Fig.1). But it's hard to apply these complex planning algorithms on the smart devices directly, because that these smart things often have limited computing power. We surely can use WiFi or other wireless communication means to send all the informations of sensors to a central computer and process them,

© Springer International Publishing Switzerland 2015
J.-H. Kim et al. (eds.), *Robot Intelligence Technology and Applications 3,*
Advances in Intelligent Systems and Computing 345, DOI: 10.1007/978-3-319-16841-8_31

but due to the time lag and unstableness of wireless network, the best way is relying on its own computing power. An important work is to simplify and improve these complex algorithms so they can perform good on embedded computing devices.

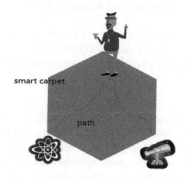

Fig. 1. Smart carpet in library (concept)

This paper is mainly about an improvement on a planning algorithm. We wish to develop a low-time-consuming path planning algorithm, and in the future, apply it on a smart carpet. The rest of the paper are organized as follows. In Chapter 2 we introduce RRT algorithm and its in-depth analysis. In Chapter 3, we analyze the relationship between collision event (a collision may happens in a step of RRT algorithm) and search completeness. Based on the analysis, we propose an improved RRT-like algorithm - Exponential Backoff RRT (EB_RRT). In Chapter 4, we compare EB_RRT with previous RRT algorithms. We construct a "carpet" and randomly put obstacles on it, then use our algorithm to calculate a path from starting point to destination point. The results showed us that EB_RRT is much faster than the existing ones. In Chapter 5, we summarize our work and present the idea about making a smart carpet in the future.

2   Background

2.1   Smart Carpet

Nowadays, with the developing of sensor technology, sensors are becoming much cheaper, more precise and more flexible than before and even can be easily incorporated into textiles. Typical smart textiles use capacitive sensors, resistance sensors and optical fiber sensors[1-2] to perceive pressure or movement on it. These techniques give us a new opportunity to attack low-cost localization and tracking problem[3]. Guided by these previous work, we have an idea making a carpet not only can track people's footstep but can show us the road to the destination.

2.2   Rapid Random Tree

Rapid Random-exploring Tree (RRT) algorithm[4] is a kind of path generating method which has found wide application in path planning. It can quickly find a collision-avoid path in searching space, so it's quite suitable for smart carpet. Naive RRT is a single tree which rooted from the starting configuration. As the tree randomly filling the space and finally reaches the end configuration, a path from the start point to end point can be easily constructed using a backtracking approach(Fig.2). RRT use a uniform sampling tactic to expand itself to ensure that the probability of expanding an existing state is proportional to the size of its Voronoi region.

For a general configuration space C, the algorithm in pseudocode is as follows:

```
Algorithm 1: Naive RRT
Begin
    V_rrt • x_init;E_rrt • Ø;i • 0
    While i<N do
        G • (V_rrt,E_rrt)
        X_rand• Sampling(i);i • i+1
        (V,E) • Extend(G,x_rand)
    Return G
```

In the pseudocode:

X_{init} is the starting point; V_{rrt} is a set of RRT nodes; Errt is a set of all edges of RRT; G is a graph of all the nodes and edges of RRT; "←" is a shorthand for "changes to"; x_{rand} is a random point in the configuration space C generated by function Sample; Extend is a function that expand the tree to the direction of x_{rand}.

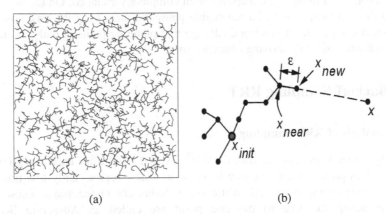

(a)   (b)

Fig. 2. Rapid-exploring Random Tree (pictures from[5])

Although RRT algorithm has a probability completeness of coverage of the search space which can guarantee that it will finally find a path if the path exists, but it has two shortcomings[5]. First, it searches the space uniformly and as the tree grows huge the search efficiency decreased (Long Tail Effect)[6]. Especially in the maze-like map, the search efficiency of RRT is particularly low. Second,it can't find the shortest path[7]. A recent widely accepted improved RRT algorithm is RRT*[8], which has been proved that it can find the shortest path at a slow rate of convergence. But RRT* is based on the nodes of RRT and add complex path optimization adjustments after each extending step thus increasing the time cost, so it can find a better path but much slower than RRT.To the blindness in the growth of RRT, some researchers proposed biased-RRT. In each sampling step the target point is taken as a sampling point with a certain probability (typically 5-10%) to let the tree grow tends to RRT goal. Some researchers focused on the maze-like map which has narrow passages. They use some tactics to detect narrow passages. Once a narrow passage is detected, we can just let the tree grow through along the passage[9-10]. It saves much time in the maze-like maps but has no effect on a map without too many passages.

2.3   Challenges On Smart Carpet Path Planning

RRT is the most efficacious and popular planning algorithm in path planning now, but in some applications it still can not satisfy the users. For example, a tunnel-like map often cause RRT consuming several hundreds seconds to approach the destination[10]. Even a small 2-D map (the step size is 0.5 and the map size is 10*10.) cost a RRT* 1500 steps to approach the destination area and 20000 steps to optimize the path (that means about 0.5s -6s on a desktop computer)[8].

In some application area such as robotics, several seconds processing time is bearable. For example we often see a robot "thinking" several seconds before moving and operating and that does not matter. But in our smart carpet scenario, people are changing their location swiftly all the time. When we show them the paths after several seconds "thinking", the map has been completely changed. On the other hand, if we don't want people to feel a noticeable pause we must keep the time cost below 0.2s in each computing. It is a big challenge that planning paths in real-time in a big map with many randomly moving obstacles on it.

3   Backoff-Sampling RRT

3.1   Analysis of RRT Sampling

From Algorithm 1 we can see that the RRT expand itself after each sampling step. Every random point attracts the tree to grow to a new region. Only few points can attract the tree to the right path while other points are unnecessary. These points which attracting the tree to the end point are called an Attraction Sequence $A=\{A_0,A_1,...A_n\}$. The question is after how many times sampling we will find a attraction sequence.

Let C_1 C_2,....C_i be random variables whose common distribution is Bernoulli distribution, [7] has proved that:

Lemma 1. If a feasible solution exists, the probability that the RRT fails to find a path is less than $e^{-1/2(ip_k-2k)}$:

$$\Pr(V_i^{rrt} \cap X_{goal} \neq \varnothing) \leq e^{-1/2(ip_k-2k)}$$

In this inequality, k is the length of the attraction sequence; i is the number of sampling times; p_k is a variable determined by the map which implies the difficulty to follow the attraction sequence, let $\mu(x)$ be the Lebesgue measure of x, p_k is defined as

$$p_k = \min_{i \in \{1,2,...k\}} (\mu(A_i) / \mu(X_{samp}))$$

The right part of Lemma 1 implies that as we repeat the sampling step, we will finally find the path if it exists. Let i tend toward infinity and we can easily get

Lemma 2

$$Lim_{i \to \infty} \Pr(V_i^{rrt} \cap X_{goal} \neq \varnothing) = 1$$

This lemma can guarantee the probabilistic completeness of the algorithm.

According to the above analysis, we know that RRT expand in all direction but only step into next A_i zone is "on the right way". So it's important for us to analyze how the RRT "jump" from A_{i-1} into A_i. For every A_i zone, there is a basin noted as B_i. Any points in B_i can be attracted to A_i.

$B_i \in X$ is a basin of A_i when

a. any points $x \in A_{i-1}, y \in A_i, z \in X-B_i, \|x-y\| < \|x-z\|$.
b. Any points in B_i can be freely connected

Now we can see the "jump" process more clearly. When RRT grow to A_{i-1}, with several sampling process, the tree fill the space of B_i and finally arrive at A_i. Only points in B_i can help the jump process because points outside B_i is farther (see Fig.3). So if we want to accelerate the growth, we should let more points fall into B_i.

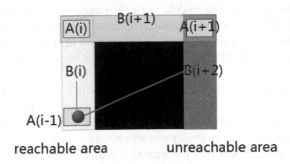

reachable area   unreachable area

Fig. 3. Jumping process of RRT

Theorem 1

Extend(G,x_{rand}) will not cause collision when

$$B_i \cap G^{rrt} \neq \varnothing, x_{rand} \in (B_1 \cup B_2 ... \cup B_i)$$

Proof: let $x_{rand} \in B_j$, $j \in \{1,2....i\}$. Because of that G has arrived B_i, the attraction sequence $A_1, A_2 ... A_{i-1}$ has established. According to the definition of B_j, it will be connected with points in B_j freely.

Theorem 2

Extend(x_{rand}) may cause a collision when

$$B_i \cap G^{rrt} \neq \varnothing, x_{rand} \in (B_{i+1} \cup B_{i+2} ... \cup B_k)$$

Proof: if $x_{rand} \in B_j$, $j \in \{i+1,i+2,...k\}$ and no collision event happens, then the tree can be extend to B_j, The tree will jump to B_j and the attraction sequence will change.

3.2   Backoff Sampling RRT

From the analysis in section 3.1, we know that the process of the growth of RRT is a process that tries to connect with other regions, and the sampling area is the whole map, so the RRT grow with blindness or "random". From Lemma1 we know to let the RRT grow to next B_{i+1}, the best sampling area is near the collision area. If we sampling biased to the area near the collision area, the tree will grow much faster. Base on this idea, we designed a new sampling strategy. During the extend operation, if no collision event occurs we decrease the sampling area to the end point to accelerate the growth; if a collision occurs, we increase the sampling area so that more area can be contained.

To make the abstract concept pictorial, we can re-express it. The big Voronoi regions are the undeveloped regions and the small Voronoi ones are developed regions. Undeveloped regions have more probability so we should try to connect them first. That is RRT. But when we can not connect the undeveloped regions, it implies that between the developed region and undeveloped region, there are some undeveloped regions called the middle zone. We may give a developed the middle zone. That is EB_RRT. When we increase the sampling area, the tree will grow back to those developed zone, that called backoff sampling operation. Fig 4 illustrate the process that how an EB_RRT go round an obstacle between the tree and the end point.

Fig.4.a shows that when EB_RRT try to connect with a point near the end point but failed, a backoff sampling operation processed. As we increase the sampling area, the tree has more chance to grow back showed in Fig.4.b. After this operation, we decrease the sampling area in attract the tree to grow to the end point because no collision occurs. We can see that the tree has go around the obstacle in Fig.4.c and Fig.4.d . The tree will not easily grow back only if collision event occurs again.

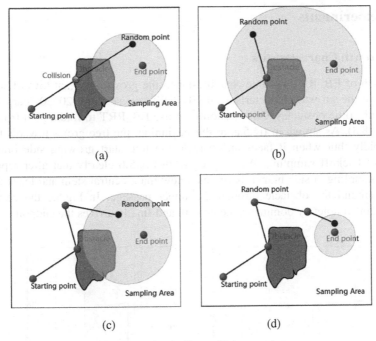

Fig. 4. The backoff sampling operation

The EB_rrt algorithm in pseudocode is as follows:

Algorithm 2:   Exponential Backoff-sampling RRT
Begin
  X_{min}, X_{max}
  X_{est} • X_{max};
  While i<N do
    Sampling(Xest)
    GROWTH()
    If CollisionDetected() **Then**
      Extend(Xest)
      If X_{est} > X_{max} **Then**
        X_{est} • X_{max}
    Else
      Reduce(Xest)
      If X_{est} < X_{min} **Then**
        X_{est} • X_{min}
  Return G

In the pseudocode, X_{min} and X_{max} are predefined sampling area bounds; Function "Extend" execute the backoff sampling operation; Function "Reduce" decrease the sampling area. Operator > and < compares the Lebesgue measure of their left and right sides.

4   Experiments

4.1   Growth Characteristics

The growth of EB_RRT is somewhat similar to the growth of plant toward sunlight. To observe the growth characteristic of EB_RRT, we set up a 20*20 "carpet map" placed with some obstacles. Our mission is to use EB_RRT to find a path from point (0,0) to (18,8). As shown in fig.5.a, at the beginning, the tree grows towards the end point rapidly, but when it faces an obstacle the tree start growing side branchlets because of backoff sampling. We can see from Fig.5.b clearly that after repeatedly back off sampling, a side branch becomes a new main central stem and leads the tree to grow around the obstacle through the upper passage. In Fig.5.c the new main central stem completely dominates the growth and finally arrives the endpoint.

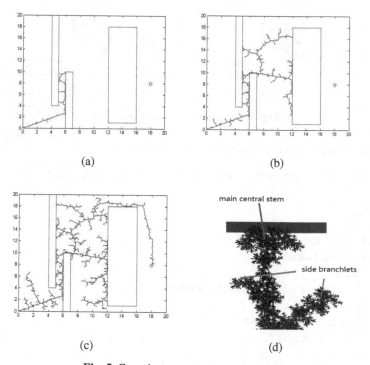

Fig. 5. Growth characteristics of a EB_RRT

This process is very similar to the growth of plants towards sunlight and its apical dominance mechanism(Fig.5.d). Shoot apex inhibits the growth of lateral buds so the plant grows to the sunlight rapidly, but when the plant's shoot apex faces obstacle, its lateral buds will get more resources and grow faster until one of them becomes a new main stem. This mechanism is a key different from RRT, which's all nodes extend with equal opportunity.

4.2 Random Obstacle Benchmark Test

In order to verify the robustness of our algorithm and compare it with existing RRTs, we propose a benchmark test - ROT(random obstacle test) to test the planner. Firstly we generate several rectangular blocks of random length and width, then we place the blocks into a random location of the map as obstacles. Lastly we use this planner to search a path from starting point to the end point which are predefined. This scenario is similar to that several people are randomly moving on the smart carpet, and the planner must find a path through the crowd. People have different sizes so do the random rectangular blocks have different width and length. Or we can envisage there is a spaceship traveling through space debris randomly distributed around it. In these scenarios, many random blocks are moving all the time so we must find the path in very limited time (several milliseconds) because of the quick changes of the environment.

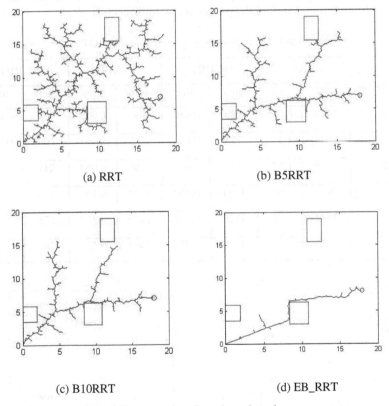

(a) RRT

(b) B5RRT

(c) B10RRT

(d) EB_RRT

Fig. 6. Four trees in a 3 random obstacles map

We use a program to automatically generate random obstacles in a 20*20 map. The number of obstacles is set from 5 to 30 with length and width from 1 to 5. This method can't guarantee maps generated are always have a feasible solution so we only count those can be solved in 10 seconds. Step size is set to 0.2 so each algorithm extends less than 0.2 in an iteration. In biased-RRT, biased factor is often set to 5% or 10% so we test biased-RRT twice with the two different bias factors. We call them B5RRT and B10RRT respectively.

Let's see a 3 random obstacles test (Fig.6). There are three obstacles randomly distributed in the map. RRT grow to the end point after 1258 iterations and 0.388 seconds, B5RRT use 519 iterations and 0.091 seconds, B10RRT use 400 iterations and 0.057 seconds. Our algorithm, EB_RRT, finds the path in only 262 iterations and 0.029 seconds.

If we add more obstacles to the map, the search difficulty increases. Table.1 and Fig.7 show us the performances of the 4 algorithms in more obstacles map. We can see that EB_RRT always has a huge speed advantage. Even if the map has many obstacles, the algorithm can find the path in a few milliseconds.

Table 1. Four algorithms' performances in ROT

The number of obstacles	RRT		B5RRT		B10RRT		EB_RRT	
	step	time	step	time	step	time	step	time
5	1424	0.385	529	0.089	406	0.059	174	0.014
10	2614	1.335	553	0.094	411	0.061	204	0.017
15	2685	1.573	632	0.096	524	0.064	211	0.018
20	3948	3.565	1561	0.587	1125	0.154	254	0.019
25	4584	4.674	1828	0.796	1479	0.333	276	0.023
30	5023	5.9	3037	1.771	2465	0.524	281	0.032

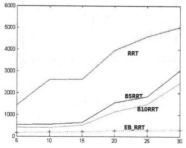
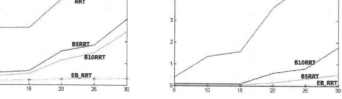

(a) Average number of iterations   (b) Average time consuming

Fig. 7. Increase the number of obstacles

The point to emphasize here is that the shape of the map may greatly influence the performance of an algorithm. Our benchmark test can only reflect the excellent performance of the algorithm in random obstacle map. For example in a maze-like map, the performance of EB_RRT may not be better than RRT.

5   Conclusion

In this paper, we proposed a new planning algorithm called EB_RRT which is designed for our "smart carpet" concept. EB_RRT is based on RRT but using a different sampling tactic. This tactic is based on the analysis on the sampling process of RRT, which implies us that collision event has a significant relationship with the searching completeness of "developed region". To accelerate the sampling process, we increase the sampling area when collision event occurs and whereas decrease sampling area. The simple change in sampling process results in interesting changes of the growth of the tree. EB_RRT grows like a plant and it has main central stem and side branchlets. When the main central stem encountered serious obstruction, side branchlets get more resources to grow and finally becomes a new central stem. Compared with RRT, we can see that RRT grows evenly and fill the space without adaptive.

We also made a preliminary analysis of the EB_RRT and our experiment has shown that, in a 20m*20m carpet map with 5-30 random obstacles. The EB_RRT can easily plan a path in 10^{-2} seconds. Our method makes an order of magnitude speed improvement than the existing ones, and this characteristic suggests that it would be a promising technology to be applied on an embedded device.

In the future work, we will implement a smart carpet and apply our planner on it. The smart carpet coped with projection technique can be used in library and show a path helping you to find the right book. Meanwhile, further acceleration of the planning algorithms is also needed.

Acknowledgments. This research is supported by the National Natural Science Foundation of China under grant 61175057 and the Research Fund for the Doctoral Program of Higher Education of China under grant 20133402110026, as well as the USTC Key-Direction Research Fund under grant WK0110000028.

References

1. Cantoral Ceballos, J., et al.: Smart carpet for imaging of objects' footprint by photonic guided-path tomography. In: AFRICON 2011. IEEE (2011)
2. Jayaraman, S.: A Note on Smart Textiles. IEEE Pervasive Computing 13(2), 5–6 (2014)
3. Savio, D., Ludwig, T.: Smart carpet: A footstep tracking interface. In: 21st International Conference on Advanced Information Networking and Applications Workshops, AINAW 2007, vol. 2. IEEE (2007)
4. LaValle, S.M.: Rapidly-Exploring Random Trees A New Tool for Path Planning (1998)

5. Kuffner, J.J., LaValle, S.M.: RRT-connect: An efficient approach to single-query path planning. In: Proceedings of the IEEE International Conference on Robotics and Automation, ICRA 2000, vol. 2. IEEE (2000)
6. Guitton, J., Farges, J.-L., Chatila, R.: Cell-RRT: Decomposing the environment for better plan. In: IEEE/RSJ International Conference on Intelligent Robots and Systems, IROS 2009. IEEE (2009)
7. Karaman, S., Frazzoli, E.: Sampling-based algorithms for optimal motion planning. The International Journal of Robotics Research 30(7), 846–894 (2011)
8. Karaman, S., Frazzoli, E.: Incremental sampling-based algorithms for optimal motion planning. arXiv preprint arXiv:1005.0416 (2010)
9. Pan, J., Zhang, L., Manocha, D.: Retraction-based RRT planner for articulated models. In: 2010 IEEE International Conference on Robotics and Automation (ICRA). IEEE (2010)
10. Lee, J., et al.: Sr-rrt: Selective retraction-based rrt planner. In: 2012 IEEE International Conference on Robotics and Automation (ICRA). IEEE (2012)

The Mobile Robot SLAM Based on Depth and Visual Sensing in Structured Environment

Yezhang Tu, Zhiwu Huang, Xiaoyong Zhang[*], Wentao Yu, Yang Xu and Bin Chen

School of Information Science and Engineering, Central South University,
Changsha, HUNAN 410075, China
hzw@csu.edu.cn

Abstract. Simultaneous localization and mapping (SLAM) is becoming one of the most attractive research focuses of robot control and visual processing. In this paper, robot performs a SLAM mission in an unknown and structured indoor environment with the Microsoft XBOX Kinect obtaining visual and depth information. The line features of the object as mapping elements are extracted from visual images, and according to the extraction result, the distance between line features of the object and the robot can be obtained to portray the object edge onto the map in 2D, and the distance is provided by depth data from the Kinect. Meanwhile the robot motion model is created for the trajectory plan. A method using the extended Kalman filter (EKF) is applied to provide a pose estimate for the robot motion trajectory. The SLAM strategy is demonstrated in simulation and experimental environment.

Keywords: Kinect, depth, line feature, structured 2D environment, the extended Kalman filter.

1   Introduction

The researches around mobile robot navigation are always of attraction. Navigation needs directions, which is given from a map containing elements of the surroundings. Map can be prepared or, under many circumstances, created. It's extremely applicable for what needs to perform in unknown environment without pre-knowledge if the mobile machine is capable of localizing itself and mapping simultaneously. As the robot proceeds, with every step it collects the elements from the environment and takes measurements between them and itself. Then a map expands with certain methods based on the elements and measurements taken from previous steps, and the position of the robot is also updated. The robot is performing SLAM [1].

The robot SLAM problems are variable, in which there must be a method that's multi-effective:

- The robot moves in the environment and acquires position where it gets data from the sensor system [2].
- Features are extracted from fuzzy detection under certain circumstances.

---

[*] Corresponding author.

© Springer International Publishing Switzerland 2015   343
J.-H. Kim et al. (eds.), *Robot Intelligence Technology and Applications 3*,
Advances in Intelligent Systems and Computing 345, DOI: 10.1007/978-3-319-16841-8_32

- Features extracted with previous steps are reserved as map landmarks to be associated with those with the next step.
- Map is updated correctly every step, revising coordinate value of the robot and the landmarks via the particle filter or extended Kalman filter (EKF).
- Invalid data is left out, and the valid becomes the new landmarks added in the map.

SLAM in a structured environment relieves the work of complex feature detection. The using of laser range finder makes contribute in the measurement. With the every detail of the object detected by a range finder or depth extractor, an edge of inaccessible area is portrayed according to the object profile.

Visual capture depends on camera, which is mainly used for detecting interest points and some other features which can't be extracted by depth sensor alone like color, gray level, and histogram of the object.

A method presented in [3] is introduced to solve the SLAM problem, taking the advantages of a particle filter for path estimation and a Kalman filter for landmark location estimation. It features that the SLAM problem is decomposed into a robot localization problem and a landmark extraction problem. Though efficient, it causes massive false features and data link. A wrong trajectory or coordinate can be identified valid. Cindy Leung et al [4] present an approach to SLAM problem where a line feature extraction algorithm is introduced. In structured environment, or more simplified, indoor, most features are continuous and well-regulated. Line features like corner and fork can be quickly acquired. With the depth sensing follow line feature, the edge detection is of high application and efficiency.

2   Prerequisites and Problem Formulation

This section presents the platform and equipment supported in this work, including the mobile robot platform, the XBOX360 Kinect and within the inflexible limitation, the SLAM problems that need to be solved.

2.1   Kinect and Robot Platform

Kinect is developed by Microsoft in 2009, a motion sensing peripheral based on Xbox 360 and Xbox One video game consoles. It's originally designed for motion gamming, and with its powerful function it has been used in many applications and researches like surface fitting, tracking, dense reconstruction [5] etc.. It can also do a great job on SLAM problem due to its visual-depth combination.

Kinect has three lenses, one in the middle for an RGB color camera and two on each side for a depth sensor consisting of an infrared transmitter and an Infrared CMOS camera. The RGB camera supports resolution up to 1280×960, while the depth sensors support 640×480 imaging. 57 degree of horizontal angle range from left to right, and 43 degree from above to below. However, there is a vital limitation for extracting distance due to depth sensing range for 0 to 4095 mm. As the Microsoft official recommendation, it's accurate and reliable to keep extracting distance within 1220 to 3810 mm.

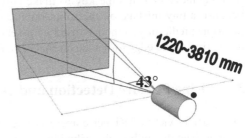

Fig. 1. XBOX Kinect Peripheral   **Fig. 2.** Depth Extracting Ranges

As the robot moves on, the distance between line features and itself changes respectively. It's of importance that the robot keeps the feature under radar and tracks it.

Fig. 3. Experimental Robot Platform

The robot platform is two-wheel-driven, each wheel with one shaft driven by different motor. The differential driving output makes it change its heading expediently. Encoder of each wheel is equipped to record the movement.

2.2   Problem Formulation

SLAM problems in structured indoor environment focus on the feature detection and trajectory plan. Since the Kinect camera is the only data source to achieve localization and mapping, and is fixed to the robot platform without freely moving, the problems below should be paid more attention to:

- The platform keeps the camera capable of tracking the feature with an appropriate view of observation as well as an ideal distance for depth.
- The position of line features are predicted via filter techniques like the EKF if camera loses the feature on the imaging plane.
- The motion control model is created, and use of the extended Kalman filter is adopted to correct the robot's pose.
- The record of two encoders is used for avoiding duplicate feature detection.

To achieve that, it's the key to make sure the robot movement controllably stable because a tiny mistake on the robot pose causes massive loss of feature. The feature position predicting method is used as a secondary solution in this case. Meanwhile, derivative of the edge of detected object is calculated for mapping and prediction.

3   Line Feature Detection and Depth Capture

The vertical view of Kinect camera is unchangeable. According to the experimental environment that will be described in the following, it's assumed that the line feature is horizontal and continuous. Line features on each object are formed into edges of close loops. And the depth data is captured from these features.

3.1   Line Feature Detection

The line feature extraction is executed via the Canny edge detecting method. Before that, a two dimensional Gaussian filter is applied to eliminate the noise:

$$G(x, y) = \frac{1}{2\pi\sigma^2} \cdot e^{-\frac{x^2+y^2}{2\sigma^2}} \tag{1}$$

And, the discrete form of convolution kernel H that's $(2k+1) \times (2k+1)$:

$$H_{i,j} = \frac{1}{2\pi\sigma^2} \cdot e^{-\frac{(i-k-1)^2+(j-k-1)^2}{2\sigma^2}} \tag{2}$$

If $k = 1$, we can get:

$$H = \frac{1}{2\pi\sigma^2} \cdot e^{\frac{1}{\sigma^2}} \cdot \begin{bmatrix} e^{-1} & e^{-0.5} & e^{-1} \\ e^{-0.5} & 1 & e^{-0.5} \\ e^{-1} & e^{-0.5} & e^{-1} \end{bmatrix} = \frac{1}{2\pi\sigma^2} \cdot e^{\frac{1}{\sigma^2}} \cdot \begin{bmatrix} 0.3679 & 0.6065 & 0.3679 \\ 0.6065 & 1 & 0.6065 \\ 0.3679 & 0.6065 & 0.3679 \end{bmatrix} \tag{3}$$

As noted, Gaussian filter is a weighted filter. All the weighted values add up to 1, thus H needs to be normalized:

$$H_N = H / \left(\sum_{i=1}^{3} \sum_{j=1}^{3} H_{i,j} \right) = \begin{bmatrix} 0.0751 & 0.1238 & 0.0751 \\ 0.1238 & 0.2043 & 0.1238 \\ 0.0751 & 0.1238 & 0.0751 \end{bmatrix} \tag{4}$$

The image matrix of the gray value $(M \times N)$ captured originally P , is divided into many smaller 3×3 pieces like $T_{i,j}$:

$$P = \begin{bmatrix} T_{1,1} & \cdots & T_{1,n} \\ \vdots & \ddots & \vdots \\ T_{m,1} & \cdots & T_{m,n} \end{bmatrix}, \ T_{i,j} = \begin{bmatrix} P_{i-1,j-1} & P_{i,j-1} & P_{i+1,j-1} \\ P_{i-1,j} & P_{i,j} & P_{i+1,j} \\ P_{i-1,j+1} & P_{i,j+1} & P_{i+1,j+1} \end{bmatrix} \tag{5}$$

In (5), $m = M/3$; $n = N/3$ (Assume that M and N can be divided by 3.) and $0 \le P_{i,j} \le 255$.

The noise filtering begins:

$$T_{i,j}^* = H_N T_{i,j} H_N^T, \quad P^* = \begin{bmatrix} T_{1,1}^* & \cdots & T_{1,n}^* \\ \vdots & \ddots & \vdots \\ T_{m,1}^* & \cdots & T_{m,n}^* \end{bmatrix} \tag{6}$$

After the elimination of noise, the image gradients of x, y axis (f_x and f_y), and their norms and direction (M_f and θ_f) are calculated by means below:

$$\begin{cases} f_x(i,j) = (P_{i,j+1}^* - P_{i,j}^* + P_{i+1,j+1}^* - P_{i-1,j}^*)/2 \\ f_y(i,j) = (P_{i,j}^* - P_{i+1,j}^* + P_{i,j+1}^* - P_{i+1,j+1}^*)/2 \\ M_f(i,j) = \sqrt{\left[f_x(i,j)\right]^2 + \left[f_y(i,j)\right]^2} \\ \theta_f(i,j) = \arctan\left[f_y(i,j)/f_x(i,j)\right] \end{cases} \tag{7}$$

Though the larger $M_f(i,j)$ means the larger gradient, it's not convincible that pixel (i,j) is the edge point. The points in P^* need to be non-maxima suppressed.

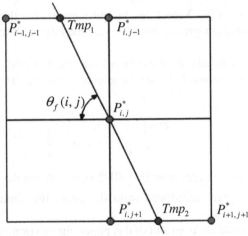

Fig. 4. Non-maxima suppression

In Fig.4, we need to compare the gray value of (i,j), $P_{i,j}^*$ with Tmp_1 and Tmp_2, which are in the direction of gradient (i,j) and on different side. Tmp_1 and Tmp_2 are evaluated by using weighted interpolation

$$\begin{cases} Tmp_1 = P_{i,j+1}^* \cdot \cot\theta_f(i,j) + P_{i-1,j+1}^* \cdot [1 - \cot\theta_f(i,j)] \\ Tmp_2 = P_{i,j-1}^* \cdot \cot\theta_f(i,j) + P_{i+1,j-1}^* \cdot [1 - \cot\theta_f(i,j)] \end{cases} \tag{8}$$

If $P_{i,j}^*$ is larger or less than both of Tmp_1 and Tmp_2, $P_{i,j}^*$ is considered not to be an edge point.

For now the edge points are roughly extracted, and need to be denoised and connected into contour. Two thresholds are introduced here, Th_1 and Th_2, $Th_1 > Th_2$:

a) Firstly, the edge points whose gradients are larger than Th_1 are extracted, while the rest are eliminated. Those points are marked convinced edge point.

b) Starting from a convinced edge point, the 8 neighbor points are detected that if their gradients are larger than Th_1, they join with the convinced into a contour line. The detection goes over again and again until there is an endpoint whose neighbor points are eligible. Those points are marked convinced edge point.

c) When getting to the endpoint using Th_1, the 8 neighbor points are detected that if their gradients are larger than Th_2, they're marked convinced edge point.

Like step b, until there is no eligible neighbor or it's formed a loop, it stops.

The convinced edge is the extraction of the line feature. With appropriate adjustment of Th_1 and Th_2, the extraction of line feature can be more precise.

3.2   Depth Capture and Measurement

The depth capture using Kinect can be quite easier, for the calibration of the RGB camera and infrared camera can be done with the internal library function, no need for extra work.

The line feature in this paper can be seen as the same height as the Kinect camera within the margin of error. Let D be the matrix to reserve the depth data of the image.

$$D = \begin{bmatrix} d_{1,1} & \cdots & d_{1,N} \\ \vdots & \ddots & \vdots \\ d_{M,1} & \cdots & d_{M,N} \end{bmatrix} \tag{9}$$

The element $d_{i,j}$ ranges from 0 to 4095 (mm). When the line feature is extracted, according to the pixel coordinate from the edge, the depth of the feature is also extracted.

In the structured environment of this paper, the environment and map are the form of vertical projection in 2D. One of the important tasks of the feature processing is the measurement between the features, and the measurement of the feature itself [6]. The measurement methods are various in many different cases.

Facing Straight. It's the most common situation in this paper. The situation can be described as Fig.5 below:

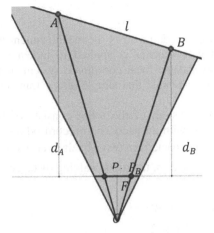
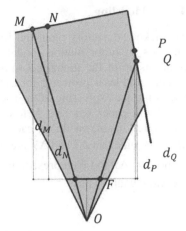

Fig. 5. Facing Straight to the Line Feature   **Fig. 6.** Meeting Corner in the Environment

In Fig.5, $\overline{OF}$ represent the focal length of the RGB camera which can be acquired from the camera parameters [7], d_A and d_B are the depth captured by infrared camera. According to their geometric relationship, the measurement between A and B is:

$$l = \sqrt{\left[\left|x_{P_A} - x_F\right| \cdot (1 + d_A / \overline{OF}) + \left|x_{P_B} - x_F\right| \cdot (1 + d_B / \overline{OF})\right]^2 + \left|d_A - d_B\right|^2} \tag{10}$$

Corner. The corner is detected via the change of differential polarity of the object. Described as Fig.6, the criterion is formed into the inequation (11):

$$\left(d_M - d_N\right) \cdot \left(d_P - d_Q\right) < 0 \tag{11}$$

Assuming corner edge as a set of edge points containing depth data:

$$D_{crn} = \left\{d_1\left(x_1, y_1\right), d_2\left(x_2, y_2\right), \ldots\right\} \tag{12}$$

The inquation (11) is typically applied where Canny edge of corner is extracted over the neighbor of $\left(x_i, y_i\right)$.

The robot also needs to avoid extracting inaccurate data like the depth that is too deep or too small. Distance control is added in the control strategy to make sure the object within the right range.

4   Indoor SLAM Strategy Based on the Extended Kalman Filter

The objects in structured indoor environment are often possessed of smooth and regular outline which can be added in the SLAM map quickly, and the outline can be modeled for prediction in the process of SLAM. In this section, strategies for trajectory and map construction are provided for the simulation and experiment.

4.1   Modeling

To make a trajectory plan, the first thing that needs to do is the establishment and analysis of robot motion model. In [8], tensor transfer is applied to indicate the relationship of the global coordinate system and the local coordinate system of the camera and laser projector for 2D SLAM. It's feasible that simple tensor transfer is used onto the expressing of the robot pose.

As the paper has said before, the driving system of the robot consists of two uncoupled engine modules, each one equipped with an encoder to record odometric data. Assume that ΔS_l and ΔS_r are the encoder feedback from two wheels over one step, while $\Delta \theta$, ΔS and W_R represent the robot body rotation, the robot walking distance and the track width. Thus:

$$\begin{cases} \Delta \theta = (\Delta S_l - \Delta S_r) / W_R \\ \Delta S = (\Delta S_l + \Delta S_r) / 2 \end{cases} \tag{13}$$

Let r be the turning radius, reaching that:

$$r = \Delta S / \Delta \theta \tag{14}$$

The relationship between the global coordinate and the robot coordinate is shown in Fig.7 below:

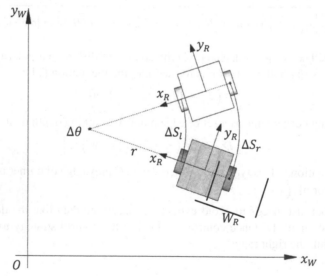

Fig. 7. The Global Coordinate and the Robot Coordinate.

In the robot coordinate system, the origin is on the current position of the robot; axis y_R is in the direction of robot motion; axis x_R is perpendicular to that. According to the coordinate setting, we can get:

$$\begin{cases} x_R = r \cdot \sin \Delta \theta \\ y_R = r \cdot [1 - \cos \Delta \theta] \end{cases} \tag{15}$$

Assuming the global pose of the robot is $X_k = [x_k, y_k, \theta_k]^T$ at step k, by the next step k+1 the position relative to the robot coordinate at step k is $(x_R, y_R) = (\Delta x, \Delta y)$. The global pose at step k+1 $X_{k+1} = [x_{k+1}, y_{k+1}, \theta_{k+1}]^T$ is found out:

$$
\left\{
\begin{aligned}
\begin{bmatrix} x_{k+1} \\ y_{k+1} \\ 1 \end{bmatrix} &= \begin{bmatrix} \cos\theta_k & -\sin\theta_k & x_k \\ \sin\theta_k & \cos\theta_k & y_k \\ 0 & 0 & 1 \end{bmatrix} \begin{bmatrix} \Delta x \\ \Delta y \\ 1 \end{bmatrix} \\
\theta_{k+1} &= \theta_k + \Delta\theta
\end{aligned}
\right.
\tag{16}
$$

Given the control input $u_k = [v_k, \omega_k]^T$, the noise $\boldsymbol{\varphi}_k = [\varphi_x, \varphi_y, \varphi_\theta]^T$ which is temporally independent, zero-mean and Gaussian-distributed and its covariance matrix Q_k, the state of the robot motion is:

$$
X_k = f(X_{k-1}, u_k) + \boldsymbol{\varphi}_k \tag{17}
$$

$f(X_{k-1}, u_k)$ is the nonlinear state transition function of the system.

$$
f(X_{k-1}, u_k) = \begin{bmatrix} f_1(X_{k-1}, u_k) \\ f_2(X_{k-1}, u_k) \\ f_3(X_{k-1}, u_k) \end{bmatrix} = \begin{bmatrix} x_{k-1} + \dfrac{v_k}{\omega_k} \cdot [\sin(\theta_{k-1} + \omega_k \Delta t) - \sin\theta_{k-1}] \\ y_{k-1} - \dfrac{v_k}{\omega_k} \cdot [\cos(\theta_{k-1} + \omega_k \Delta t) - \cos\theta_{k-1}] \\ \theta_{k-1} + \omega_k \cdot \Delta t \end{bmatrix} \tag{18}
$$

In (18), Δt is the time period of every step.

Assume that $Z_{k,i}$ is the function to describe the depth observation, $Z_{k,i} = [\rho_{k,i}, \theta_{k,i}]^T$, in which i is the label of feature, $\rho_{k,i}$ is the distance between robot and the feature, and $\theta_{k,i}$ is the direction angle. Taking the polar coordinate of the feature as observation, the position of feature i in global coordinate is (X_i, Y_i). According to the geometric relationship, the observation model definition of the robot pose is:

$$
Z_{k,i} = \begin{bmatrix} \rho_{k,i} \\ \theta_{k,i} \end{bmatrix} = \begin{bmatrix} \sqrt{(X_i - x_k)^2 + (Y_i - y_k)^2} \\ \arctan\left(\dfrac{Y_i - y_k}{X_i - x_k} - \theta_k\right) \end{bmatrix} + \boldsymbol{\gamma}_{k,i} \tag{19}
$$

The observation model (19) describes the relationship between the observation $Z_{k,i}$ and the robot pose. The observation equation is:

$$
Z_{k,i} = h(X_k) + \boldsymbol{\gamma}_{k,i} \tag{20}
$$

$\boldsymbol{\gamma}_{k,i}$ in (19) and (20) is temporally independent noise like $\boldsymbol{\varphi}_k$, $\boldsymbol{\gamma}_{k,i} \sim N(0, R_{k,i})$, in which $R_{k,i}$ is the covariance matrix of $\boldsymbol{\gamma}_{k,i}$.

4.2   SLAM Strategy

The extend Kalman filter (EKF) is one of the most significant theoretic method for SLAM problems [9], which is a recursive process consisting of four steps. Define matrixes A_{k-1} and H_k as the Jacobian matrixes for partial derivative $f(X_{k-1}, u_k)$ and $h(X_k)$ on global pose X respectively:

$$A_{k-1} = \begin{bmatrix} \dfrac{\partial f_1}{\partial x_{k-1}} & \dfrac{\partial f_1}{\partial y_{k-1}} & \dfrac{\partial f_1}{\partial \theta_{k-1}} \\[2mm] \dfrac{\partial f_2}{\partial x_{k-1}} & \dfrac{\partial f_2}{\partial y_{k-1}} & \dfrac{\partial f_2}{\partial \theta_{k-1}} \\[2mm] \dfrac{\partial f_3}{\partial x_{k-1}} & \dfrac{\partial f_3}{\partial y_{k-1}} & \dfrac{\partial f_3}{\partial \theta_{k-1}} \end{bmatrix}, H_k = \begin{bmatrix} \dfrac{\partial h_1}{\partial x_k} & \dfrac{\partial h_1}{\partial y_k} & \dfrac{\partial h_1}{\partial \theta_k} \\[2mm] \dfrac{\partial h_2}{\partial x_k} & \dfrac{\partial h_2}{\partial y_k} & \dfrac{\partial h_2}{\partial \theta_k} \end{bmatrix} \tag{20}$$

And the four specific steps are shown below:

Prediction. According to the estimation of the system state $X_{k|k}$ and control input u_k, the state and covariance matrix, $X_{k+1|k}$ and $E_{k+1|k}$, are predicted:

$$X_{k+1|k} = f(X_{k|k}, u_k) \tag{21}$$

$$E_{k+1|k} = A E_{k|k} A^T + Q_k \tag{22}$$

Observation. When robot acquires the real observation value of the ith feature $Z_{k,i}$, the prediction valued is obtained:

$$\gamma_{k+1,i} = Z_{k,i} - h(X_{k+1|k}) \tag{23}$$

And its error covariance matrix $S_{k+1,i}$:

$$S_{k+1,i} = H E_{k+1|k} H^T + R_{k+1,i} \tag{24}$$

Data Association. Before updating, data association is necessary, for inaccurate association causes data divergence. Detect the association between step $k+1$ and the previous ones $(0, 1, 2, ..., k)$ using inequation (25) where G is constant. If it's false, abandon the current observation value.

$$\gamma_{k+1,i} S_{k+1,i} \gamma_{k+1,i}^T \leq G \tag{25}$$

Update. After data association, use the formulas below to update the state estimation and its covariance matrix in step $k+1$.

$$X_{k+1|k+1} = X_{k+1|k} + W_{k+1,i} \cdot [Z_{k,i} - h(X_{k+1|k})] \tag{26}$$

$$E_{k+1|k+1} = (I - W_{k+1,i} H) \cdot E_{k+1|k} \tag{27}$$

And $W_{k+1,i}$ is the Kalman gain:

$$W_{k+1,i} = E_{k+1|k} H^T S_{k+1,i}^{-1} \tag{28}$$

The four procedures run in circles to recurse the pose estimation of the robot and the line features.

5   Simulation Result

The task of the mobile robot is to explore the simulated indoor environment. In this section, the strategies presented above are applied and performed with time step for 0.1s, execution time for total 18s. The simulation environment is constructed structured and multi-forked to imitate the real experimental environment as far as possible.

The initial pose of the robot is $X_0 = [0,0,0]^T$ with a small primary velocity at the center of the environment relative to the global coordinate system.

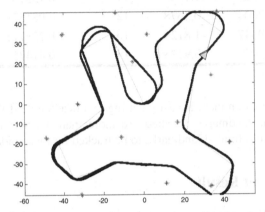

Fig. 8. The SLAM Map and the Path and Line Features of Simulation Environment

The robot starts from the map center and heads for the top left corner alongside then turns around to the top right side. In the following steps, the robot search for every corner and staple the line feature onto the map. And the black line is the real trajectory. The coordinates of the robot position every 1s in are shown below

Table 1. The Coordinates of the Robot

Time (s)	X(cm)	Y(cm)	Time(s)	X(cm)	Y(cm)
0	0	0	9	44.58525	-31.4327
1	-16.2442	34.94152	10	21.31336	-14.1813
2	-30.7604	27.04678	11	25.92166	18.85965
3	-20.1613	-2.48538	12	39.51613	35.23392
4	-42.7419	-20.0292	13	24.76959	46.05263
5	-35.1382	-37.2807	14	7.258065	30.84795
6	-14.1705	-29.6784	15	9.331797	11.54971
7	13.70968	-31.4327	16	3.571429	1.900585
8	33.29493	-45.4678	17	2.683106	2.856319

And the average error in each second of the coordinates of the robot in 14 trails:

Table 2. The Error of the Coordinates in 14 Trails

No	X(cm)	Y(cm)	No	X(cm)	Y(cm)
1	-0.6682	0.64912	8	0.52535	-0.3743
2	-0.7097	5.76023	9	42.97235	-0.4444
3	-0.8341	-0.1462	10	0.70507	-2.50292
4	-0.7327	-1.1053	11	0.75576	0.18129
5	-3.7604	-0.4678	12	0.98157	0.05263
6	-1.9447	-1.8129	13	0.728111	1.64912
7	1.1512	-0.9123	14	0.41475	0.90643

During the simulation there is easily getting to a dead circle if the trajectory is too bending where the landmarks detected are more than the normal. The simulation environment provides limited landmarks to be tracked for the avoidance.

6   Experiment Result

The SLAM using Kinect to extract the depth information needs appropriate observation view. And the SLAM efficiency and accuracy are in a degree affected due to the light condition and other irrelevant object distraction in the environment.

The acquirement of visual information in this environment is quite regular to extract feature. And the extraction images are shown in Fig.7.

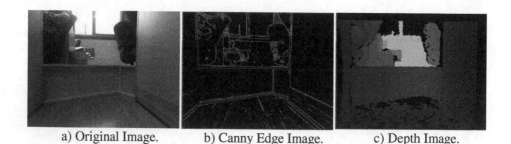

a) Original Image.   b) Canny Edge Image.   c) Depth Image.

Fig. 9. The Extraction Images.

The indoor environment is constructed below with the available object of the office to form the line feature and the boundary.

Before the experiment, it's necessary to debug the best match of *Th*1 and *Th*2 for Canny edge detection in the experimental environment, and the values of *Th*1 and *Th*2 are 50 and 100 respectively. And the feature

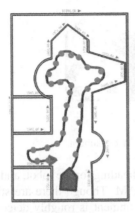

a) The Experiment Environment
Ichnography and Trajectory Plan

b) The Real Experiment Environment
for SLAM

Fig. 10. Experimental Environment.

The movement of the robot in the process of SLAM is shown in Fig.9 below.

Fig. 11. The Robot Movement in the Structured Environment

When the robot proceeds in the environment, the real trajectory and the estimated trajectory are recorded with the map constructing.

After the robot returns to the start point and detect the edge surrounding, the SLAM is completed. The SLAM map is constructed below:

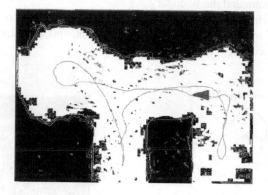

Fig. 12. Map Obtained Using the Kinect Camera.

The red triangle stand for the original point and heading for the robot, and the blue line represents the trajectory of the robot when SLAM. Though there are still many noise points in this map, the boundary of the environment is roughly detected. The robot in SLAM process avoids the barriers and edges and collects them into the map. SLAM in this paper takes about five minutes.

7   Conclusion

Research for SLAM for mobile robots is considered that it needs multi-academic work to be done, including the techniques of sensors and automatic detection, digital image processing as well as artificial intelligence. Throughout the history of SLAM for mobile robots, among tons of research solution, SLAM strategy based on visual and range sensing has become undeniably attractive, which is almost indispensable for the automatic movement and positioning of robots. The accuracy and efficiency of visual capture are the first to be considered for the SLAM process.

In this paper, a visual camera gets the object's feature from the image, and depth extractor take the depth data within the extracted pixel edge of the object on image. Line features are portrayed onto the map after the associated depth extraction is done. The EKF filter is used correcting the trajectory and position with every step.

The Kinect depth sensor has narrow view of angle and short detection range, and it can be used in SLAM problem typically in the small or narrow space. The Canny edge detection is limited somehow by the light condition, the filtering of the noise needs amelioration. In the simulation, the dead circle of the robot movement is because of the trajectory too bending for the robot to follow and the misjudgment to the new features.

The method of mobile robot SLAM is various. There are many alternative solutions available to make effort on, and they are also the inspiration for us to find out a better SLAM method.

Acknowledgment. The authors would like to acknowledge that this work was partially supported by the National Natural Science Foundation of China (Grant No. 61379111, 61071096, 61003233 and 61073103) and Research Fund for the Doctoral Program of Higher Education of China (Grant No. 20110162110042).

References

1. BoninFont, F., Ortiz, A., Oliver, G.: Visual Navigation for Mobile Robots: A Survey. Journal of Intelligent and Robotic Systems 53(3), 263–288 (2008)
2. Zureiki, A., Devy, M.: Appearance-based Data Association for 3D and Multisensory SLAM in Structured Environment. In: Information and Communication Technologies, Damascus, April 7-11 (2008)
3. Montemerlo, M., Thrun, S., Koller, D., et al.: FastSLAM: A Factored Solution to the Simultaneous Localization and Mapping Problem. In: The AAAI National Conference on Artificial Intelligence, Edmonton, Canada (2002)
4. Leung, C., Huang, S., Dissanayake, G.: Active SLAM in Structured Environments. In: IEEE International Conference on Robotics and Automation, Pasadena, CA, USA, May 19-23 (2008)
5. Newcombe, R.A., Izadi, S., Hilliges, O., et al.: KinectFusion: Real-Time Dense Surface Mapping and Tracking. In: IEEE International Symposium on Mixed and Augmented Reality, Basel, Switzerland, October 26-29 (2011)
6. Davison, A.J., Reid, I.D., Molton, N.D., et al.: MonoSLAM: Real-Time Single Camera SLAM. IEEE Transactions on Pattern Analysis and Machine Intelligence 29(6), 1052–1067 (2007)
7. Mariottini, G.L., Oriolo, G., Prattichizzo, D.: Image-Based Visual Servoing for Nonholonomic Mobile Robots Using Epipolar Geometry. IEEE Transactions on Robotics 23(1), 87–100 (2007)
8. Jung, M., Myung, H., Hong, S., et al.: Structured Light 2D Range Finder for Simultaneous Localization and Map-building (SLAM) in Home Environments. In: The International Symposium on Micro-Nanomechatronics and Human Science, Japan, October 31-November 3 (2004)
9. Leonard, J., Durrant-Whyte, H.F.: Dynamic map building for an autonomous mobile robot. International Journal of Robotics Research 11(4), 286–298 (1992)

Application Exploring of Ubiquitous Pressure Sensitive Matrix as Input Resource for Home-Service Robots

Jingyuan Cheng, Mathias Sundholm, Marco Hirsch, Bo Zhou,
Sebastian Palacio, and Paul Lukowicz

German Research Center for Artificial Intelligence (DFKI),
Trippstadter Straße 122, D-67663, Kaiserslautern, Germany

Abstract. We present how ubiquitous pressure sensor matrix can be used as information source for service-robots in two different applications. The textile pressure sensor, that utilizes the ubiquitousness of gravity, can be put on most surfaces in our environment to trace forces. As safety and human robot interaction are key factors for daily life service robots, we evaluated the pressure matrix in two scenarios: on the ground with toy furnitures demonstrating its capability for indoor localization and obstacle mapping, and on a sofa as an ubiquitous input device for giving commands to the robot in a natural way.

1 Introduction

Population ageing is taking place in nearly all countries in the world. According to the United Nations, "The global share of older people (aged 60 years or over) increased from 9.2 per cent in 1990 to 11.7 per cent in 2013 and will continue to grow as a proportion of the world population, reaching 21.1 per cent by 2050" [1]. A consequence of this phenomenon is the lack of workforce. Robots have already been widely used in industry for repetitive work and are now moving closer to humans in form of service robots, mainly for regular tasks such as floor cleaning, lawn mowing, and pool maintenance. According to the World Robotics study [2], 159,346 units of industry robots were sold in 2013, 16,067 professional service robots and about 3,000,000 personal and domestic use robots were sold in 2012. However, to fulfil the service sector demands, service robots still have a long way to go. The reason lies in the challenges of operating safely in both dynamically changing environments and environments with humans that can move unpredictably. The close contact and communication between human beings and robots are still limited, but these however, are key factors for service robots (e.g. support elderly at home, in healthcare, or in restaurant and hotel service). For example, a servant robot at a birthday party shall understand the requests from different guests, bring drinks and snacks to the correct guest as quickly as possible and meanwhile avoiding collision with "randomly" re-arranged tables, sofas, chairs, the undeterministically moving human beings, and possibly also

© Springer International Publishing Switzerland 2015
J.-H. Kim et al. (eds.), *Robot Intelligence Technology and Applications 3*,
Advances in Intelligent Systems and Computing 345, DOI: 10.1007/978-3-319-16841-8_33

running pets. The scientific aspects lying behind are collision avoidance (tracking, indoor/outdoor localization, distance measuring, loop control and etc.) and human robot interaction.

These are also key scientific aspects in ubiquitous computing, a fast developing field focusing on the shift from one central powerful computer, towards multiple smaller computing units, embedded into the environment and onto human beings.

We have designed an ubiquitous pressure sensing platform [3] (as shown in Fig. 1), which is unobtrusive, flexible, thin, cheap, and large scale. The platform is composed of a piece of smart textile and a smart sensing board. The pressure distribution on the smart textile is given as a heat map that refreshes at a rate of ~40Hz. Application scenarios include a smart gym-mat for recognition and counting of sport exercises [4], a smart floor for recognition of subtle user activities and person identity [5], and a smart table cloth for nutrition monitoring.

Merging ubiquitous computing and service robotics could bring benefit to both research areas, as the former emphasizes on information retrieval from humans and their environment, to which the latter gives active feedback. Both areas also find important applications in service and support. As an initial try, we apply this ubiquitous textile pressure sensing matrix to two scenarios, both designed to provide live information to service robots.

The contribution lies mainly in that we demonstrate that two different questions in robotics can be solved with the same ubiquitous system:

1. We demonstrate with a 1:10 down scaled floor prototype that an in-door obstacle mapping can be constructed using a pressure sensitive matrix. The system can build up a live obstacle map that can be used by robots to avoid collisions, not only with the environment, but also with randomly moving human beings. Chairs and tables can be detected and a walking human can be accurately tracked (two legs simulated by two fingers).
2. We demonstrated that the same structure can be used as a pervasive natural gesture input interface for sending commands to a robot. A subject independent leave-one-subject-out cross-validation showed that 15 gestures can be distinguished with an accuracy of 0.90.

1.1 Related Work

Pressure Sensing Matrix. A variety of pressure sensor matrices have already been proposed and demonstrated. An overview of existing approaches to such a textile device is given in Table 1. Compared with existing systems, our textile pressure matrix utilizes a general hardware architecture topology which is large-scale (maximum channel number can be more than 10^6), with high sensitivity and precision (24-bit ADCs, structural separation between the digital and analogue modules, and a specially designed power managing block minimizes noise level) and is suitable for a broad range of applications including on body sensing, smart table cloths, and carpet like structures.

Table 1. Overview on existing digital pressure matrices

application	function	parameters
chair surface	user posture [6], seat comfortness [7], anti-theft car seat [8]	84-by-48 nodes, 8-bit, 6Hz
bed	breath, heartbeat, sleeping posture [9] [10]	16-by-16 nodes, 12Hz
shoes	gait analysis [11]	64 nodes, 14-bit, 1.8kHz
clothes	body posture [12]	16-by-16 nodes, 10Hz

Localization and Tracking. Global Positioning System (GPS) is a common means for outdoor localization. Indoor localization using time of arrival of ultrawideband (UWB) signals [13] leads to a very high precision but is limited to simple room setups because reflections from furnitures and people are disturbing the original signal. WiFi based indoor solutions locate the user by matching the local signal strength to a pre-built signal strength map [14]. Inertial measurement units (accelerometer and gyroscope) combined with WiFi signals and GPS (for a concrete coordinate when entering and leaving the building) can be used to further improve the in-door map and localization precision [15] [16]. Magnetic coupling sensors can replace the map from WiFi signals with a field generated by the coils at a certain frequency which is hardly influenced by the normal environment, thus is being very robust [17]. RFID tags embedded into the environment can serve as beacons for an indoor map, too [18]. However, these can be used for tracking single objects only. For each object, an additional set of beacons must be added and the system complexity grows linearly.

Visual information can be used for building up a map and tracking objects in parallel. Cameras have already been used on robots in the last century for building up environment maps [19]. 3-D motion input devices based on multiple infra-red projectors and cameras used for gesture control with limbs and fingers [20] can be used for building a depth map. However, cameras raise privacy concerns and can not work under all light condition. Visible or infra-red light can be blocked, and the processing is normally complex and time consuming. Bränzel, et.al. recognize people and objects by sensing the pressure distribution in a room based on optical interference. For this, a transparent glass floor with a large space underneath the testing room is required by the camera [21]. Clearly such a solution involves much more installation effort than using a carpet-like textile.

Ubiquitous Input Interface. Currently, the most used ubiquitous input interface may well be the touch screen of mobile phones and tablets [22]. Other methods include voice recognition, body gesture detection with 2-D or 3-D cameras or wearable inertial measurement units, normal or wearable keyboards, electromyographic or even electroencephalographic signals, and more [23]. They can be grouped into wearable or environmental interfaces. The latter do not have to be carried or suffer from the battery run, but being limited to a certain area. For realizing a real pervasive input interface and at the same time reducing the user's effort to a minimum, the system has to be accessible everywhere. Carpets and furniture covers are made from textile materials. Hence it is likely the material which covers the largest area in rooms. Moreover, sitting or sleeping are activities which occupy at least half of a person's life time and are performed on textile covered chairs and beds. To give another concrete example, if an elderly who lives alone falls onto the ground and gets his leg injured, he might not be able to call out loud enough for the robot in the other room and can rely only on the smart textile floor as an emergent input interface.

2 Textile Pressure Sensitive Matrix

The sensor utilizes the property of elastic polymer materials that are electrically conductive. The core of the sensor fabric (produced by Sefar [24]) is a sheet made of such fabrics that have around 2 $k\Omega$ volume resistance (from top to bottom) in an 1 cm^2 area. As the material is pressed, this resistance decreases. The resistance-pressure relationship is predictable and repeatable, and resistance is insensitive to other physical variables, to the best of our knowledge, such as magnetic field or temperature. Therefore, the resistance distribution of the fabric sheet represents the pressure distribution across the surface. We apply an array of metallic conductive parallel stripes on both sides of the fabric sheet. The two arrays are perpendicular to each other. By addressing each row and column of the conductive stripes, we are able to measure the resistance at every crossing node. The result of a complete scan of the surface forms a frame of a 2-dimensional matrix. The electronics structure along with more details about the sensing principle is described in [3]. With dedicated 24-bit analog-digital converters, high speed analog signal routing circuitry and FPGAs as the controllers, the current hardware is capable of scanning such resistive matrix of up to 128-by-128 spatial resolution at 40 frames per second. The data is sent to a computer through a USB cable, in which up to 4 parallel COM ports at 8M baud are emulated. The electronics consist of high performance dedicated components, because we would like to investigate how much information is needed respectively in a variety of applications. If the specific application does not require either high speeds, nor the high resolution data acquisition, the hardware can be cut down significantly. Photos of the hardware prototypes and the data acquisition board are shown in Fig. 1.

3 Scenario 1: Indoor Localization and Obstacle Mapping

3.1 Experiment Setup

To evaluate the feasibility of our pressure sensor matrix for indoor localization and obstacle mapping, we build up a test environment where the floor is covered with the sensor, simulating a restaurant scenario where tables and chairs are randomly placed, and humans are moving following non-predefined paths. Since building up a full scale room is both complicated and expensive, we simulate the environment with a 1:10 down-scaled prototype. We use 80 × 80 cm^2 sensor

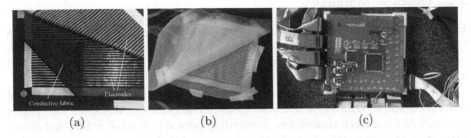

(a)   (b)   (c)

Fig. 1. (a) Structure of prototype sensor, (b) a smaller sensor prototype where the electrodes are integrated in the textile, and (c) data acquisition board.

matrix to cover the floor area. We use toy tables and chairs that are 10×10 cm^2 and 4×4 cm^2 in size respectively. The contact area of the legs for both chairs and tables are 0.8×0.8 cm^2. A picture of the test environment is shown in Fig. 2a. To simulate a person walking across the room we use walking fingers. Five rounds of data were recorded, simulating a person walking from one edge of the room to another between the tables and chairs placed across the room. For the experiment we use 2 tables and 8 chairs that are positioned in different locations at the beginning of each round. We also move certain chairs and tables within each round. Since the toy tables and chairs are too light to be detected by the sensor, we add additional weights of $0.5 - 1$ kg to each of them. Each round takes 1-2 minutes. We aim at detecting the obstacles that the robot should avoid, based on the measured pressure distribution. This includes detecting of the table and chairs, and more interestingly, tracking the position of the moving person.

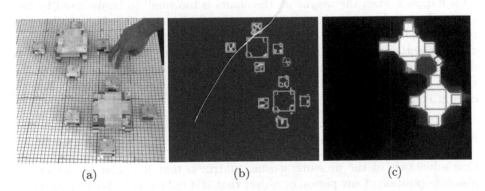

(a)   (b)   (c)

Fig. 2. (a) Photo of test environment, (b) preprocessed heat map recorded by pressure sensor matrix, and (c) derived obstacle map for the environment. Green squares show the positions of the tables, and chairs and the red circle shows a standing person.

3.2   Obstacle Mapping

The current pressure sensor prototype can accurately detect objects as small as 1 cm, (which would correspond to 10 cm for a full scale room). This is enough to detect the absolute positions tables and chairs legs in the room. Standing and walking persons can also be detected.

Fig. 2b shows a pressure image as captured by the sensor matrix after initial processing steps (removal of DC and noise). The positions of tables and chairs are marked with green rectangles. A standing person is marked with a red circle. The original sensor data has also been upsampled 4 times to create a higher resolution image that can detect objects more precisely. Intermediate pixels are calculated using bilinear interpolation. By using the pressure information we can construct an obstacle map for the room. The robots can use the obstacle map to safely navigate through the room avoiding collisions with moving humans and objects.

In order to derive the obstacle map we use morphological closing [25], a efficent and well-known algorithm used image processing. By applying morphological closing on the preprocessed pressure image (in this case a disk with diameter 32), the gaps between legs of the chairs and the tables are closed, which prevents a robot from moving there. The obstacle map using morphological closing is fast to compute and can be updated in real-time to take moving objects into account. An example of the derived obstacle map from Fig. 2b is shown in Fig. 2c.

Tables and chairs are detected and covered completely by the obstacle map. As it can be seen also areas between chairs and tables are covered by the obstacle map. This is in fact desired because the distance between tables and chairs around is very narrow. In a real scenario a robot should try to avoid such narrow areas. In case a person stands close to a table he will also be connected with the table, preventing the robot from navigating between him/her and the table. In a few cases we discovered that the obstacle map did not cover objects completely. This happens when the weight on the chairs is too small to be detected by the sensor. In a real scenario however the weight of an empty chair is several orders of magnitude larger which can easily be detected by the sensor. Our algorithm can efficiently create an obstacle map from the sensor data without having to detect exact positions every object in the scene. Such detection of objects would require more sophisticated object detection algorithms such as those used in computer vision [26,27].

3.3   Indoor Localization

One advantage of the pressure sensing matrix is that it is able to detect the absolute position of any person or object that is standing on it. We can therefore use it for accurately tracking the position of a walking person. For our tracking algorithm we seek inspiration from object tracking algorithms that are already well established in computer vision applications [27,28].

In order to separate moving persons from static obstacles we create a background model using consecutive frames. The background model is constructed using the minimum value of each pixel in frames from the last 5 seconds. Objects that moved during the last 5 seconds, like a walking person, will not be included into the background model. The background model is updated two times per second. The foreground is extracted from the image frame using background subtraction where the background model is subtracted from the current frame.

To remove noise that can arise around furnitures we construct an obstacle map using the background model. The map is created in the same manner as described in the previous section. The obstacle map is then used to filter out the area around obstacles in the foreground image. A 'Laplacian-of-Gaussian'-filter is convoluted with the foreground image in order to detect blobs that have the size of a persons feet. The position of a moving human can then be accurately detected by thresholding the convoluted foreground image. If multiple points are detected, the position with the highest peak is selected as the measurement. To prevent "jumping" of the position estimate, a Kalman-filter [29] in x- and y-directions is applied to the measurements. When the sensor gets new measurements from a position, the Kalman estimate will move in the direction of the measurement. When the Kalman estimate ranges within a threshold ϵ of the measurement tracking mode is enabled. The threshold ϵ is configured to be larger than the maximum size of a human step. When tracking is enabled enhanced filtering is applied and measurement points that are further than ϵ from the Kalman estimate are ignored. This allows us to remove measurements that originate from the other side of the room that could not possibly come from the same person. When we don't get any new measurements within the radius ϵ for a few seconds (for example if the signal was too weak to be detected) the tracking mode is turned off which prevents the tracking to get stuck if the trace is lost. Fig. 3 shows the results of five recordings of a person walking across the room. Green markers show the predicted position while red markers indicate the actual position.

This method enables accurate and robust tracking of a person across the room even if there are distractions such as moving of furniture. In our experiments we calculate an average positioning error of 1.23 cm, which is around the same size as the spatial resolution of the sensor. Our implementation is currently limited to tracking only one person, but since the sensor can track multiple points simultaneously, it is also possible to implement tracking of multiple persons in the future.

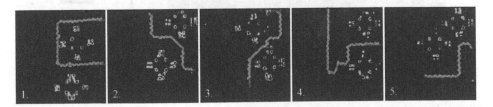

Fig. 3. Results of conducted tracking experiments. Green markers shows the predicted position, red markers is the actual position.

4   Scenario 2: Ubiquitous Input Device for Home-Service Robots

In the second scenario we evaluate the potential of the pressure sensor matrix as an ubiquitous touch input device for controlling home-service robots. The sensor can be woven into fabric or mounted on virtually any surface. While being unobtrusive, the interface can potentially be available everywhere in the users home, for example on bed sheets, on tables cloths, or in the shower. This is important since home-service robots are often targeted to support elderly or disabled people with moving limitations.

Moreover, our system has high fidelity reflections in Buxton's three metaphoric mirrors for interface design [30]. Touching physical objects with fingers or the hand is easy and natural to do (physical). Symbols of objects linked to the task as control gestures are language independent and can easily be remembered (cognitive). The pressure sensors are invisible and don't raise privacy concerns like camera or voice based systems (social).

4.1   Experiment Setup

A $30 \times 30\,\mathrm{cm}$ version of the sensor array was placed on one side of a two seater sofa covered by fabric. Subjects were asked to sit on the other side of the sofa and perform different touch gestures on the fabric called out by a software via voice command supported by on screen text and illustrations. To ensure natural behaviour, we gave examples of the gestures to the subjects but no further restrictions. Each successfully performed gesture was confirmed to the software by a button press afterwards. A picture of the scenario is shown in Figure 4a.

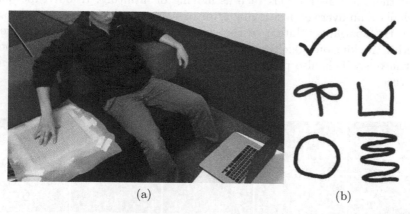

(a)   (b)

Fig. 4. Experiment setup for the touch input scenario (a) and example drawings of the symbols shown to the subjects (starting from top left: tick, cross, plant, glass, plate, water) (b).

For this feasibility study we choose six symbols as gestures which are easy to remember and can be linked to useful tasks in home robotics. The symbols are a tick, a cross, a glass, a plate, a plant, and water. Example drawings of those symbols can be seen in Figure 4b. Additionally, we choose eight control gestures, which are commonly used with touch control interfaces, for example on smart phones. They are tap, double tap, swipe left/right/up/down, and two finger pinch in/out. For the last gesture we defined, the whole hand should be laid down on the fabric for a moment.

Each of the 15 gestures was performed ten times in random order by six different participants (five male, one female, aged from 22 to 35 years). After conducting each experiment, we asked the subjects to give us feedback, ideas, and their opinion about the interface.

4.2 Data Evaluation

For recognizing the gestures, we used a feature based classification approach on isolated segments. The gestures were segmented automatically based on labels generated by the experiment software.

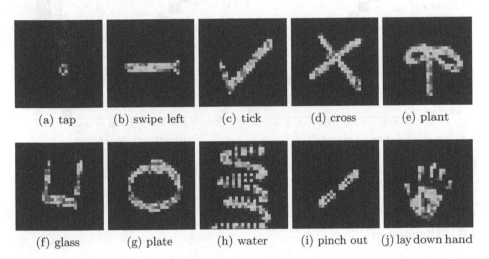

(a) tap   (b) swipe left   (c) tick   (d) cross   (e) plant

(f) glass   (g) plate   (h) water   (i) pinch out   (j) lay down hand

Fig. 5. Aggregated heat maps of the normalised 30×30 sensor value matrix from ten different example gestures.

After running all data through a noise filter, we extracted a total of 17 features from each gesture segment. Seven Hu image moments which are invariant under translation, scale, and rotation [31] were calculated on the aggregated sensor value matrix. Figure 5 shows heat maps of those matrices for some example gestures. For the other features, a threshold was set to get a binary image for every sample during the gesture segment. Based on these images, the following features were calculated

- No. of threshold crossings
- total No. of frames including a value over the threshold
- average blob count
- derivative of blob location deviation
- center of mass start-end location difference in two dimensions
- size of aggregated image bounding box in two dimensions
- zero crossing rate of center of mass location derivative in two dimensions

We used those features to train a bagging ensemble of 100 classification and regression trees (CARTs) and decided on accuracy as a performance measure since all gesture classes are evenly distributed in the data set. A subject independent leave-one-subject-out cross-validation showed that the 15 gestures can be distinguished with an accuracy of 0.90. A subject dependent 5-fold cross-validation lead to an average accuracy of 0.96. The respective confusion matrices are shown in Figure 6.

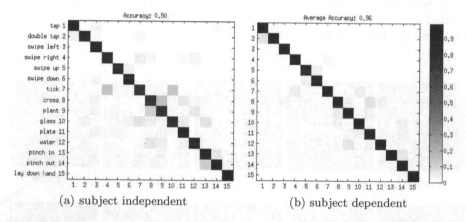

(a) subject independent   (b) subject dependent

Fig. 6. Confusion matrix for (a) leave-one-subject-out cross-validation and (b) subject dependent 5-fold cross-validation average.

4.3   Discussion

A subject independent gesture recognition rate of 0.90 with the underlying basic feature set shows the potential of the system for the proposed task. It can be safely assumed that recognition rate will improve by using finger tracking and more sophisticated algorithms even when allowing users to extend the set with their own custom gestures [32]. Moreover, in home settings normally only a few different people would use the system and can potentially be distinguished by larger pressure sensitive matrix through their body weight and shape. A person dependent approach further improves accuracy significantly, which is shown by the achieved average accuracy of 0.96.

Based on the feedback given during the interviews, all participants liked using the interface in general, but most would find it tiring to perform the relatively large gestures continuously over a long period of time. This would not be a problem when controlling, for example a home-service robot or a TV. Therefore only sporadic gestures have to be done. Some subjects liked the position of the sensor and the feeling of touching the silk but some proposed a different fabric and locating the sensor on arm rest rather than the seating area. Having only a defined area for touch input is a restriction specific to the experiment setup. Considering the low cost and flexibility of the sensor array, covering the whole sofa would be possible and the fabric, in which the sensor is embedded can be chosen freely.

Regarding to the participants, the used gesture symbols were easy to remember and intuitive to perform. However, they gave many ideas and different preferences for additional symbols suggesting user defined gestures.

5   Conclusion

In this paper we explored and evaluated two application of using ubiquitous pressure sensing matrix based on smart textile, as information resource for service robots. In the first scenario we place the sensor on the floor for generating a realtime updated obstacle map and for tracking of a person. In the second scenario we place the sensor on sofa and use it as an ubiquitous input interface. Both applications are based on a small prototype systems, but evaluated with multiple rounds or test subjects. Robust person tracking with an average precision of 1.23 cm and over 90% recognition accuracy for 15 gestures are both promising results that serves as a base for larger systems in the future and further merging of ubiquitous computing and robotics.

Acknowledgement. This work was partially supported by the collaborative project SimpleSkin under contract with the European Commission (#323849) in the FP7 FET Open framework. The support is gratefully acknowledged.

The authors would like to thank our assistant scientist Mingjie Lu for labeling in the 1st experiment and all the students who took part in the 2nd one.

References

1. World Population Ageing 2013 report. Technical report, United Nations, Department of Economic and Social Affairs, Population Division (2013), http://www.un.org/en/development/desa/population/publications/pdfageing/WorldPopulationAgeingReport2013.pdf
2. World Robotics Industrial Robots 2013 - Summary; Service Robots 2013 - Summary. Technical report, IFR Statistical Department, VDMA Robotics and Automation association (2013), http://www.worldrobotics.org/uploads/media/Executive_Summary_WR_2013.pdf
3. Zhou, B., Cheng, J., Sundholm, M., Lukowicz, P.: From smart clothing to smart table cloth: Design and implementation of a large scale, textile pressure matrix sensor. In: Maehle, E., Römer, K., Karl, W., Tovar, E. (eds.) ARCS 2014. LNCS, vol. 8350, pp. 159–170. Springer, Heidelberg (2014)

4. Sundholm, M., Cheng, J., Zhou, B., Sethi, A., Lukowicz, P.: Smart-mat: Recognizing and counting gym exercises with low-cost resistive pressure sensing matrix. In: The ACM International Joint Conference on Pervasive and Ubiquitous Computing (UbiComp 2014) (to appear)
5. Cheng, J., Sundholm, M., Zhou, B., Kreil, M., Lukowicz, P.: Recognizing subtle user activities and person identity with cheap resistive pressure sensing carpet. In: International Conference on Intelligent Environments (IE 2014) (to appear)
6. Tan, H., Slivovsky, L., Pentland, A.: A sensing chair using pressure distribution sensors. IEEE/ASME Transactions on Mechatronics 6(3), 261–268 (2001)
7. Verver, M., Van Hoof, J., Oomens, C., Wismans, J., Baaijens, F.: A finite element model of the human buttocks for prediction of seat pressure distributions. Computer Methods in Biomechanics and Biomedical Engineering 7(4), 193–203 (2004)
8. Xie, X., Zheng, B., Xue, W.: Object identification on car seat based on rough sets. In: 2011 IEEE 3rd International Conference on Communication Software and Networks (ICCSN), pp. 157–159 (2011)
9. Kortelainen, J., van Gils, M., Parkka, J.: Multichannel bed pressure sensor for sleep monitoring. In: Computing in Cardiology (CinC), pp. 313–316 (2012)
10. Lokavee, S., Puntheeranurak, T., Kerdcharoen, T., Watthanwisuth, N., Tuantranont, A.: Sensor pillow and bed sheet system: Unconstrained monitoring of respiration rate and posture movements during sleep. In: 2012 IEEE International Conference on Systems, Man, and Cybernetics (SMC), pp. 1564–1568 (2012)
11. De Rossi, S., Lenzi, T., Vitiello, N., Donati, M., Persichetti, A., Giovacchini, F., Vecchi, F., Carrozza, M.: Development of an in-shoe pressure-sensitive device for gait analysis. In: 33rd Annual International Conference of the IEEE EMBS (2011)
12. Xu, W., Huang, M.C., Amini, N., He, L., Sarrafzadeh, M.: ecushion: A textile pressure sensor array design and calibration for sitting posture analysis. IEEE Sensors Journal 13(10), 3926–3934 (2013)
13. Alavi, B., Pahlavan, K.: Modeling of the toa-based distance measurement error using uwb indoor radio measurements. IEEE Communications Letters 10(4), 275–277 (2006)
14. Chintalapudi, K., Padmanabha Iyer, A., Padmanabhan, V.N.: Indoor localization without the pain. In: Proceedings of the Sixteenth Annual International Conference on Mobile Computing and Networking, pp. 173–184. ACM (2010)
15. Evennou, F., Marx, F.: Advanced integration of wifi and inertial navigation systems for indoor mobile positioning. Eurasip Journal on Applied Signal Processing 2006, 164–164 (2006)
16. Leppäkoski, H., Collin, J., Takala, J.: Pedestrian navigation based on inertial sensors, indoor map, and wlan signals. Journal of Signal Processing Systems 71(3), 287–296 (2013)
17. Pirkl, G., Lukowicz, P.: Robust, low cost indoor positioning using magnetic resonant coupling. In: Proceedings of the 2012 ACM Conference on Ubiquitous Computing, pp. 431–440. ACM (2012)
18. Sanpechuda, T., Kovavisaruch, L.: A review of rfid localization: Applications and techniques. In: 5th International Conference on Electrical Engineering/Electronics, Computer, Telecommunications and Information Technology, ECTI-CON 2008, vol. 2, pp. 769–772. IEEE (2008)
19. DeSouza, G.N., Kak, A.C.: Vision for mobile robot navigation: A survey. IEEE Transactions on Pattern Analysis and Machine Intelligence 24(2), 237–267 (2002)

20. Song, P., Yu, H., Winkler, S.: Vision-based 3d finger interactions for mixed reality games with physics simulation. In: Proceedings of The 7th ACM SIGGRAPH International Conference on Virtual-Reality Continuum and Its Applications in Industry, p. 7. ACM (2008)
21. Bränzel, A., Holz, C., Hoffmann, D., Schmidt, D., Knaust, M., Lühne, P., Meusel, R., Richter, S., Baudisch, P.: Gravityspace: Tracking users and their poses in a smart room using a pressure-sensing floor. In: CHI 2013, Paris, France, April 27-May 2 (2013)
22. Ballagas, R., Borchers, J., Rohs, M., Sheridan, J.G.: The smart phone: a ubiquitous input device. IEEE Pervasive Computing 5(1), 70–77 (2006)
23. Jaimes, A., Sebe, N.: Multimodal human–computer interaction: A survey. Computer Vision and Image Understanding 108(1), 116–134 (2007)
24. Sefar (July 2014), http://www.sefar.com/
25. Serra, J.: Image Analysis and Mathematical Morphology. Academic Press, Inc., Orlando (1983)
26. Papageorgiou, C.P., Oren, M., Poggio, T.: A general framework for object detection. In: Proceedings of the Sixth International Conference on Computer Vision, ICCV 1998, p. 555. IEEE Computer Society, Washington, DC (1998)
27. Yilmaz, A., Javed, O., Shah, M.: Object tracking: A survey. ACM Comput. Surv. 38(4) (December 2006)
28. Wren, C., Azarbayejani, A., Darrell, T., Pentland, A.: Pfinder: Real-time tracking of the human body. IEEE Transactions on Pattern Analysis and Machine Intelligence 19, 780–785 (1997)
29. Kalman, R.E.: A new approach to linear filtering and prediction problems. Transactions of the ASME–Journal of Basic Engineering 82(Series D), 35–45 (1960)
30. Buxton, W.A.S.: The three mirrors of interaction: a holistic approach to user interfaces. In: MacDonald, L.W., Vince, J. (eds.) Interacting with Virtual Environments, pp. 1–12. Wiley (1994)
31. Hu, M.K.: Visual pattern recognition by moment invariants. IRE Transactions on Information Theory 8(2), 179–187 (1962)
32. Trier, O.D., Jain, A.K., Taxt, T.: Feature extraction methods for character recognition-a survey. Pattern Recognition 29(4), 641–662 (1996)

Arm Trajectory Generation Based on RRT* for Humanoid Robot

Seung-Jae Lee, Seung-Hwan Baek, and Jong-Hwan Kim

Department of Electrical Engineering, KAIST
335 Gwahangno, Yuseong-gu, Daejeon 305-701, Republic of Korea
{sjlee,shbeak,johkim}@rit.kaist.ac.kr

Abstract. In this paper, an arm trajectory generation method based on the Rapidly-Exploring Random Tree Star (RRT*) is proposed for humanoid robot. The RRT* is one of anytime motion planning algorithms. The RRT* adopts the three fundamental components from the RRT, the preceding version of RRT*: the state variables, local planner, and cost function. The end effector of humanoid robot is positioned on the desired point by manipulating the joint angles of the arm, which are the state variables. The Minimum-jerk method is applied as a local planner for more realistic trajectory and the local planner fulfills collision detection test. The cost taken to transit between two points is defined as the sum of angle differences of motor corresponding to the two points. Also, there has been the need for real time control and it is taken care of by introducing a multi-thread system. The arm under control initiates motioning, once the first trajectory that meets the target zone is constructed. While the arm is on the move, the RRT* continuously updates the trajectory. The effectiveness of the proposed method is demonstrated by simulating the 7 DOF robot arm, which has been performed under two environments: the obstacle-free and obstacle-constrained cases. Simulation is carried out for the humanoid robot, Mybot-KSR, developed in the RIT Lab., KAIST.

1 Introduction

As the science and technology become highly advanced, people's perspective to the robot has changed. They now consider robots as companions for the human beings, rather than the things merely fulfilling tasks [1]. The robots, as the companion, need to communicate and interact with human beings and this request has led to the development of human-like robots. This, in turn, calls for the development of algorithms to operate robots.

Among the indefinitely many researches, especially the researches regarding the development of arm and hand of the robot have drawn great attention and diverse types of platform have been developed [2]- [4]. The research on the arm and hand can be mainly classified into two categories: the grippers focusing on grasping objects and human-hand shaped hands. Humanoid robots generally adopt the latter type, the human-hand shaped hand. For this type of robot hand

© Springer International Publishing Switzerland 2015
J.-H. Kim et al. (eds.), *Robot Intelligence Technology and Applications 3*,
Advances in Intelligent Systems and Computing 345, DOI: 10.1007/978-3-319-16841-8_34

to perform various tasks, it is essential to be capable of generating trajectory for the arm to reach the target object. Conventional method to solve this problem is to introduce the inverse kinematic for the joint motor angles and the B-spline to generate the trajectory; the Moore Penrose Pseudo Inverse, Damped Least Square (DLS) Pseudo Inverse are the variants of such method. This sort of method has the advantage of being straightforward, but, in return, it requires all the parameters and conditions be given, or it fails working. Most of all, it is difficult to solve the problem with the Inverse Kinematics by itself, when it has obstacles on its trajectory.

Other algorithms were proposed to cover this disadvantage; the Evolutionary Computation can solve problems without given all the conditions, the Neural Network and the Genetic Algorithm is utilized to address the problem. But still, these kinds of approaches cause instability and the arm moves off the optimized course.

Popularly used algorithm in recent research is the sampling based algorithm for optimized dynamic planning [5]. One of this approach is the RRT (Rapidly Random Tree), which led into the variants of the RRT: RRT-connect [6], IK-RRT [7], Grasp-RRT [8], Multi-Goal RRT-Connect [9], etc. However, despite of these efforts, disadvantage of the RRT has not been covered; the trajectory does not converge to the optimized path even with indefinitely many points. The approach to address this problem is the RRT* [10]. The RRT* was developed to converge to the optimized path using the "selecting parent" and "rewiring". Additionally, it could be combined with the "Committed Trajectories" to provide a great feature, the online.

In this research, we aim to develop a robot arm with a human-hand-shaped robot hand and the algorithm that enables the anytime motion planning so that it provides a better performance than the conventional arm trajectory generation algorithms using the IK or RRT. In consideration of designing robot hand, it is focused to possess the capability to grasp objects of various shapes and sizes with less DOFs and to correspond to the requirements for the task intelligence project of the RIT Laboratory at KAIST. The RRT* is adopted to generate arm trajectory, and the motions of grasping and avoiding obstacles are verified by multiple sets of simulation. The effectiveness of the proposed method is demonstrated by simulating the 7 DOF robot arm, which has been performed under two environments: the obstacle-free and obstacle-constrained cases. Simulation is carried out for the humanoid robot, Mybot-KSR, developed in the RIT Lab., KAIST. It turns out that the controlled arm has successfully followed the real-time-designed trajectory to reach the target object. Even though the success rate depends on the shape of the objects in terms of grasping objects, the result demonstrates the trajectories are successfully updated, while the arm is on the move and completes its task.

This paper is organized as follows. Section II describes the preliminaries to comprehend the following sections. Section III presents theoretical background. In Section IV, experimental result is presented. In the final section, the meaning of this research is discussed and this paper is concluded.

2   Preliminaries

2.1   Rapidly-Exploring Random Tree

The RRT (Rapidly-exploring Random Tree) is an algorithm that is designed for efficiently searching non-convex high-dimensional spaces, having state variables, local planner, cost function as essential components [11]. The state variables represent the point created at every moment, and the local planner is utilized to define the path connecting randomly selected two points and to detect collision. The cost function returns the cost to transit to the randomly selected point. As indicated in the Algorithm 1, at every cycle, a random state variable is created and local planner selects the closest point to current one. Once local planner performs collision test and no any collision is detected, a path between the two points is created. Has been put a limitation on the maximum length of this path, it is named the "extended range". In general, the state variables are defined as the location of points, the local planner as the path between two points, and the cost function as the Euclidian distance between those points.

Algorithm 1. GENERATE RRT(x_{init}, K, Δt)

1: T.init($\mathbf{x}_{init}$);
2: **for** $k = 1$ to K **do**
3:     $x_{rand} \leftarrow$ RANDOM STATE();
4:     $x_{near} \leftarrow$ NEAREST NEIGHBOR(x_{rand}, T);
5:     $u \leftarrow$ SELECT INPUT(x_{rand}, x_{near});
6:     $x_{new} \leftarrow$ NEW STATE($x_{near}, u, \Delta t$);
7:     T.addvertex(x_{new});
8:     T.addedge(x_{near}, x_{new}, u);
9: **end for**
10: Return T

2.2   Anytime Motion Planning Using the RRT*

The RRT* is one of the sampling-based algorithms that fortifies the RRT's disadvantages. It is represented by the optimization capability with fast searching algorithm as in the RRT. While, once the path has been created, the RRT has a disadvantage of not being able to upgrade to optimized path even with the increased number of points, the RRT* searches to track down the optimized path with indefinite number of points. The RRT* can be represented by the two processes: finding the parent points and rewiring the paths among the points. As indicated in Algorithms 2 and 3, once a new point is created, the nearest point and the cost to reach the nearest point are set as z_{min} and c_{min}, respectively. Then a set of comparison is made for the points around the new point for the cost taken to create a path to the new point. With completion of the comparison, the point with the lowest cost is selected as a parent point and connected to the new point (Fig. 1 (a)). Once this process is completed, it is followed by another

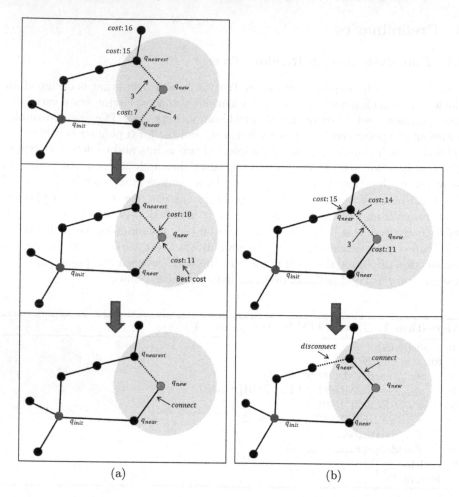

Fig. 1. RRT* methods (a) finding the parent point (b) rewiring

process, the rewiring. In the rewiring process, the cost for the pre-existing path and the cost for the path through the new point are compared and the one with lower cost becomes selected, while the previous path is removed (Fig. 1 (b)). Unlikely the RRT, the RRT* sets target as a range, instead of a point, leaving uncertainty on reaching target. Accordingly, it creates a path in short time but floppy targeting with wider range, while it takes longer time but tighter aiming with smaller range.

In addition to the conventional RRT*, one unique feature is implemented in this paper. The Committed Trajectory method improves the path by updating it at every point to point transition. It generates the path before triggering the robot's motion and updates the path using the RRT* on the way to the next point on the path. By this method, it becomes possible to reduce the total cost taken to reach the final destination. After all, the robot performs online updating.

Algorithm 2. $z_{min} \leftarrow$ ChooseParent($Z_{near}, z_{nearest}, z_{new}, x_{new}$)

1: $z_{min} \leftarrow z_{nearest}$;
2: $c_{min} \leftarrow$ Cost($z_{nearest}$) + c(x_{new});
3: **for** $z_{near} \in Z_{near}$ **do**
4:     $(x', u', T') \leftarrow$ Steer(z_{near}, z_{new});
5:     **if** ObstacleFree(x') and $x'(T') = z_{new}$ **then**
6:         $c' =$ Cost(z_{near}) + c(x');
7:         **if** $c' <$ Cost(z_{new}) *and* $c' < c_{min}$ **then**
8:             $z_{min} \leftarrow z_{near}$;
9:             $c_{min} \leftarrow c'$;
10:        **end if**
11:    **end if**
12: **end for**
13: Return T

Algorithm 3. $T \leftarrow$ ReWire($T, Z_{near}, z_{min}, z_{new}$)

1: **for** $z_{near} \in Z_{near} \setminus \{z_{min}\}$ **do**
2:     **if** ObstacleFree(x') and $x'(T') = z_{near}$ and Cost(z_{new}) + c(x') < Cost(z_{near}) **then**
3:         $T \leftarrow$ ReConnect(z_{new}, z_{near}, T);
4:     **end if**
5: **end for**
6: Return T

3 Trajectory Generation Algorithm

3.1 Necessary Components

The state vector q_i is a vector comprised of the state variables, which are the angles of motors at each joint, to represent the location of elbow and hand. The cost for the entire path is represented by the sum of change in the joint angles. By notating the start point and end point by q_{start} and q_{end}, the total cost is calculate by

$$cost_{total} = \sum_{j=start+1}^{end} \sum_{i=0}^{6} (q_{ji} - q_{(j-1)i}). \tag{1}$$

The local planner completes the path between the current and next points by interpolating the corresponding angles. In the Mybot-KSR system, the Minimum-Jerk method is used to interpolate the two points as a local planner. This method helps generating the trajectory by mimicking motion of human arm and minimizing the error at the target point by minimizing the acceleration rate [12].

Additionally, the committed trajectory is an essential feature to realize the online update for the path. Unlike the conventional research, it utilizes the RRT* with the infinite extended range to generate path offline, while the motor sets limit on its maximum rotation angle. In this way, a new point is created on the limited path and new update is performed to get to the next point (Fig. 3) .

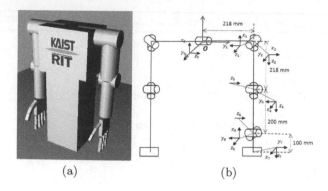

Fig. 2. (a) Mybot-KSR model for the simulation (b) D-H parameter

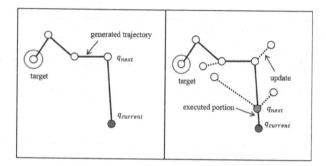

Fig. 3. Committed Trajectories for the experiment. Each motor's rotation is limited by the Committed Trajectory, not able to reach the optimized value; for example, even though a motor is supposed to rotate 12 degree at current cycle, if the executed portion is set to only 4 degree by the Committed Trajectory, the motor can rotates only 4 degree.

3.2   Arm Trajectory Algorithm

Fig. 4 shows the arm trajectory generating algorithm flow. It mainly consists of two threads: Motion planning thread and Control thread. The motion planning thread runs in the main computer, while the control thread runs in the control board attached on the back of the Mybot-KSR. At every control cycle, the motor angle values and control time for the next position are transmitted into the control board from the main computer and the minimum jerk method is used to calculate the next joint motor angles based on the values.

The overall procedure of the algorithm is follows:

1. Recognize the object.
2. Obtain the 3D coordination of the location from which the gripper approaches to the target object.
3. Sort out the cases into two categories: with limited approaching angle, without limited approaching angle.

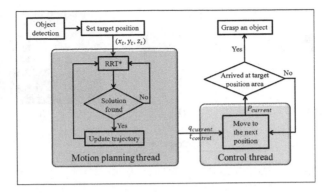

Fig. 4. The flow of arm trajectory generating algorithm. It mainly consists of two threads: Motion planning thread and Control thread. The motion planning thread runs in the main computer, while the control thread is run in the control board attached on the back of the Mybot-KSR.

4. Reset the target point to which the end-effector approaches.
5. Update path by the RRT*.
6. Once the feasible path is found, initiate motion of the arm.
7. Keep updating the path by running the RRT*; the control thread receives the required motor angles to reach the next point and runs the motors, while the motion planning thread keeps updating the path.

After repeating this procedure, if the last point settles in the target zone, the robot performs the process to check whether it is the case with the limited approaching angles or case with no limit on the approaching angles. In the former case, the robot directly transits to the next process of grasping the object, while, in the latter case, it re-directs the end-effector towards the point at which it can attain the feasible grasping posture and proceeds to grasp the object.

4 Experiments and Results

The simulation using the Webots v.6.4.1 has been performed on the Windows system, before implementing the algorithm into the Mybot-KSR for the experiment under the real environment. The robot arm is designed to have 7 DoFs in the simulation and the D-H parameters are represented as follows:

- L(x,y,z): target point of the left arm
- R(x,y,z): target point of the right arm
- Control cycle: 32ms
- Reachable (x,y,z): direction vector pointing to the target from the feasible direction.

Multiple sets of simulation were performed under the circumstances with obstacles on the 2 cases: approaching from the feasible directions and approaching from the infeasible directions.

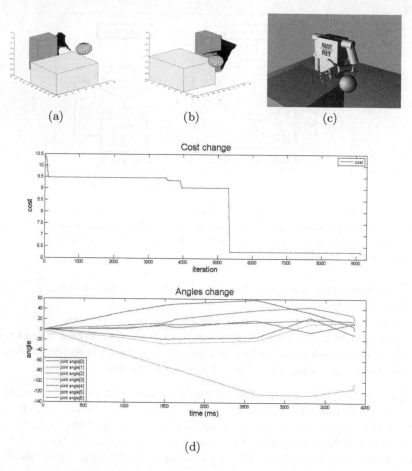

Fig. 5. Simulation for feasible direction (a), (b) Arm trajectory in the 3D space (c) Captured final position (d) The joint angle and cost change with time.

4.1   Arm Trajectory Generation with Obstacle

A table and a ball were setup as obstacles. Table's center coordination and dimension were set as (-0.5, 0.5, 0) and (0.4, 0.6, 0.6), respectively. The ball's center coordination and radius were set as (-0.25, 0.3, 0.4) and R=0.08, respectively.

Basically, this simulation consists of two cases. The first is the case all approaching from feasible direction (L=(-0.2, 0.3, 0.2)). Arm trajectory in the 3D space is shown in Figs. 5 (a) and (b). The joint angle and cost change in time is shown in Fig. 5 (d). Also, the final position is captured in Fig. 5 (c). In this case, about 10000 times of repetition was run and the result showed good avoidance maneuver to reach the target zone. One of the advantage of the RRT* augmented with the Committed Trajectory method is clearly shown in the graph representing the cost change (Fig. 5 (d)). The cost gets reduced during its maneuver, which means the trajectory is optimized online.

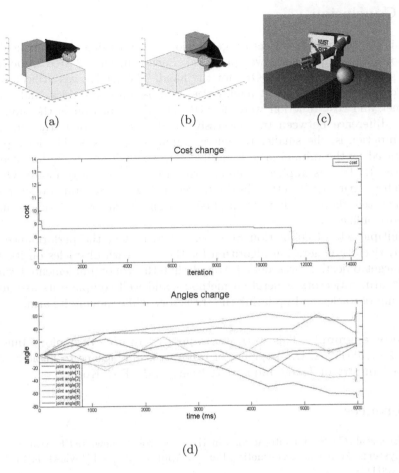

Fig. 6. Simulation for infeasible direction (a), (b) Arm trajectory in the 3D space (c) Captured final position (d) The joint angle and cost change with time.

The second simulation is for the case of approaching from the infeasible direction (L=(-0.1, 0.5, 0.218), Reachable=(0, -0.05, 0.1)). Arm trajectory in the 3D space is shown in Figs. 6 (a) and (b). The joint angle and cost change in time are shown in Fig. 6 (d). Also, the final position is captured in Fig. 6 (c). In this case, about 8000 times of repetition was run and the robot well avoided the obstacles to reach the first target point and successfully approached to the final target object. In this case as well, it is shown in the graph that the cost was optimized online (Fig.6 (d)). A sudden soaring at the end of the cost graph is caused by the change in the targets, from the approach-feasible position to the target object.

5   Conclusion

In this paper, an arm trajectory generation method using the humanoid robot Mybot-KSR's upper body was proposed. Each arm has 12 DOFs, including 7 DOFs for the arm and 5 DOFs for the hand. The RRT* was utilized for arm trajectory generation. Joint angles was used as state variables, and the local planner and cost function were defined. The cost function is the sum of the angle difference between two successive angles of each joint; the smaller the cost function is, the smaller the energy input to motors is. The local planner, generated by the minimum jerk method, ran the collision detecting algorithm, and the RRT* was applied to update the trajectory of every motor. Once the trajectory that finally settles in the target zone with acceptable tolerance level is generated, the robot initiates its first motion and keeps updating to improve the cost online.

Multiple sets of simulation were performed to test the performance of the arm trajectory generation algorithm for the case with obstacles on its way to the target object. It was demonstrated that the robot implemented with the RRT* arm trajectory generation method could well complete its assignments: avoiding obstacles and updating the trajectory while moving its arm.

Acknowledgment. This work was supported by the Technology Innovation Program, 10045252, Development of robot task intelligence technology, funded by the Ministry of Trade, Industry & Energy (MOTIE, Korea).

References

[1] Breazeal, C.: Social interactions in HRI: the robot view. IEEE Transactions on Systems, Man, and Cybernetics, Part C: Applications and Reviews 34(2), 181–186 (2004)

[2] Park, I.W., Kim, J.Y., Lee, J., Oh, J.H.: Mechanical design of humanoid robot platform KHR-3 (KAIST humanoid robot 3: HUBO). In: The 5th IEEE-RAS International Conference on Humanoid Robots, pp. 321–326. IEEE (2005)

[3] Sakagami, Y., Watanabe, R., Aoyama, C., Matsunaga, S., Higaki, N., Fujimura, K.: The intelligent ASIMO: System overview and integration. In: The IEEE/RSJ International Conference on Intelligent Robots and Systems, vol. 3, pp. 2478–2483. IEEE (2002)

[4] Yoo, J.K., Lee, B.J., Kim, J.H.: Recent progress and development of the humanoid robot HanSaRam. in Robotics and Autonomous Systems 57(10), 973–981 (2009)

[5] Karaman, S., Frazzoli, E.: Sampling-based algorithms for optimal motion planning. The International Journal of Robotics Research 30(7), 846–894 (2011)

[6] Kuffner Jr., J., LaValle, S.: RRT-connect: An efficient approach to single-query path planning. In: IEEE International Conference on Robotics and Automation, vol. 2, pp. 995–1001. IEEE (2000)

[7] Vahrenkamp, N., Berenson, D., Asfour, T., Kuffner, J., Dillmann, R.: Humanoid motion planning for dual-arm manipulation and re-grasping tasks. In: IEEE/RSJ International Conference on Intelligent Robots and Systems, pp. 2464–2470. IEEE (2009)

[8] Vahrenkamp, N., Asfour, T., Dillmann, R.: Simultaneous Grasp and Motion Planning. IEEE Robotics and Automation Magazine 19(2), 43–57 (2012)

[9] Hirano, Y., Kitahama, K.I., Yoshizawa, S.: Image-based object recognition and dexterous hand/arm motion planning using rrts for grasping in cluttered scene. In: IEEE/RSJ International Conference on Intelligent Robots and Systems, pp. 2041–2046. IEEE (2005)

[10] Karaman, S., Walter, M.R., Perez, A., Frazzoli, E., Teller, S.: Anytime motion planning using the RRT*. In: IEEE International Conference on Robotics and Automation, pp. 1478–1483. IEEE (2011)

[11] LaValle, S.M.: Rapidly-Exploring Random Trees A New Tool for Path Planning. TR 98-11, Computer Science Department, Iowa State University (1998), http://janouiec.cs.iastate.edu/papers/rrt.ps

[12] Flash, T., Hogan, N.: The coordination of arm movements: an experimentally confirmed mathematical model. The Journal of Neuroscience 5(7), 1688–1703 (1985)

Generation of Natural Dancing Motions for Small-sized Humanoid Robot

Young-Jae Ryoo

Dept. of Control Engineering and Robotics, Mokpo National University,
1666, Youngsan-ro, Cheonggye-myeon, Muan-gun, Jeonnam, 534-729, South Korea
yjryoo@mokpo.ac.kr

Abstract. In this paper, a generation of natural dancing motions for a small-sized humanoid robot using a motion editing tool is described. In order to build performances or dancing of a humanoid robot, natural motions and smooth movements should be generated. So a motion editing tool to generate specific motions is necessary. We used the proposed motion editing tool to generate robot's motions composed of several steps which are captured from every joint while the robot plays. The motion editing tool generates the continuous motion interpolated between each steps. We developed a small-sized humanoid robot of 44 cm tall to test the generated motions. The robot using the generated motions was demonstrated the natural dancing performance.

1   Introduction

Recently, there has been a growing interest in humanoid robotics. Some researchers designed robots for entertainment and used for promotion of their companies. Also, robotics applications grow daily and the creation of realistic motion for humanoid robot increasingly play's a key role. Since the motion can be regarded as a form of interaction and expression that allow to enrich communication and interaction, expressiveness and realism of improving humanoid robot motion is a form to accomplish better and richer human-robot interaction[1]. Considering dance as a rich and expressive type of motion, constituting a form of non-verbal communication in social interactions, and transmitting emotion, it imposes a good research of human-like motion[2, 3].

The dancing performance of humanoid robot is related to generate human-like motion referred to as human motion imitation[4]. Researches on the field of human motion imitation have focused mainly on generating human-like motions as close to human as possible. Because of its complexity, a human motion is hardly planned using a mathematical formula. To resolve this difficulty, researchers obtain human motions using a motion capture system and convert the captured human motions for a humanoid robot.

Ikeuchi proposed a framework of learning from observation[5]. A motion model which represented essential elements of the motion could be defined by a user within

© Springer International Publishing Switzerland 2015   385
J.-H. Kim et al. (eds.), *Robot Intelligence Technology and Applications 3*,
Advances in Intelligent Systems and Computing 345, DOI: 10.1007/978-3-319-16841-8_35

that framework. Observation and reproduction of the motion was based on the motion model. As another work on learning, a model of an assembly motion for robotic manipulators was proposed by Takamatsu[6] Using this model, a robot observed a human motion and extracted it. Then the robot rearranged motions using as a sequence of motion primitives. In the process, the essential problem was how to extract the elements of the framework and how to recreate natural motion with the framework. Imamura have proposed a 'mimesis loop' which dealt with general motions and used common motion primitives for all the types of motions[7][8]. In their method, motions could be properly classified according to the properties of motions by a hidden-markov of primitive sequences. Since the method was limited to simple motion, generating a precise and complex motion operation was difficult.

Nakaoka introduced the dance imitation of the robot using a motion model for a dance[9]. The main element of this model was to use a primitive motion, which is a minimal unit of choreography. A whole dance motion can be composed of a sequence of primitive motions. The same kind of primitive motion assumed to appear several times in the entire dance. However, since the dance motion was creative and was unique, their motion model for a dance may miss the peculiar characteristics of the motion.

Ryoo proposed a motion editing tool and algorithm to create the natural motions with intuitive method[10-12]. The motion editing tool can generate the continuous motion interpolated between each steps.

Therefore, we used the motion editing tool proposed before to generate robot's motions composed of several steps which are captured from every joint while the robot plays. We captured the existing motions from the humanoid robot directly and combined or rearranged them as a new motion. The method is good not to develop just a motion capture and player, but also to build the motion editor for the robot to perform dancing motions naturally satisfying user's requests. In order to test the generated motions, a small-sized humanoid robot of 44 cm tall was developed. The robot using the generated motions was demonstrated the natural dancing performance.

2   Structure of Humanoid Robot TOTO

TOTO stands for Training Oriented Test Object robot. It is a small-sized humanoid robot developed in the intelligent Space Laboratory(iSL), Mokpo National University, South Korea. It has 44cm tall and weighs 2.4kg. Fig. 1 shows the mechanism of the humanoid robot TOTO. The lower body of the TOTO is composed of 12 degrees of freedom totally and the upper body is composed of 8 degrees of freedom.

The Table 1 shows the brief specification of the humanoid robot. We tested the TOTO with the proposed motion editing tool. The main controller for the TOTO is a mini PC and the sub controller has used the MCU that plays the role of communication for the servo motors.

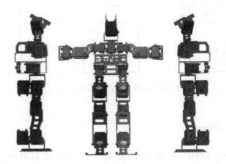

Fig. 1. Mechanism of humanoid robot TOTO

Table 1. Specification of humanoid robot TOTO

Item	Value
Height	44cm
Weight	2.4kg
Degrees of Freedom	18
OS	Linux
Electrical and Electronic Hardware	Camera, A secondary controller, status LED, Such as gyro sensor etc

3   Motion Editing Tool

3.1   Motion Capturing

Fig. 2 show the procedure to capture still postures of humanoid robot and store as a motion. In this procedure, each *still posture* of humanoid robot captured by the motion editing tool directly. The tool gathers the angle data of each joint of the robot at the moment of the still posture. Numbers of still postures can be captured and saved as a *Motion* for the robot.

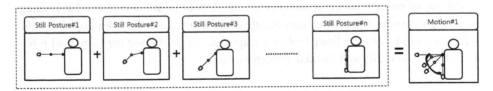

Fig. 2. Mechanism of humanoid robot TOTO

3.2   Action Generation

Fig. 3 explains the connecting numbers of *Motions* to generate an *Action*. The saved *Motion* can be connected to the other *Motion* respectively like a chain and stored as an *Action*. The motion editing tool should have the menu to connect each *Motions*.

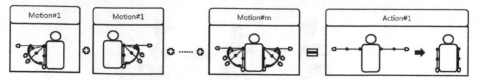

Fig. 3. Connection numbers of Motions to generate Action.

3.3   Action Generation

Fig. 4 shows the timing chart of motion play by motion editing tool. Each still Posture has pause time and play time. When all of still posture are connected, a motion can play in *Motion Time*.

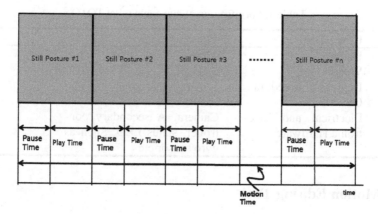

Fig. 4. Timing chart of playing motion

4   Conclusion

In this paper, we proposed the structure of the motion editing tools of the humanoid robot for the performance. Among various parameters of motion editing tools, we experimented on matching real motion after the flow of joint's waveform that is using various parameters for natural movement, and output applied to developed robots. When proposed motion editing tools were applied, we confirmed the humanoid robot moved more naturally and showed smooth move.

Acknowledgments. This research was financially supported by the Ministry of Education(MOE) and National Research Foundation of Korea(NRF) through the Human Resource Training Project for Regional Innovation (No. 2012H1B8A2026068)

This paper was supported in part by Research Funds of Mokpo National University in 2014.

References

1. Kim, C.H., Kim, D.I., Oh, Y.H.: Adaptation of human motion capture data to humanoid robots for motion imitation using optimization. In: ICINCO, vol. 4, pp. 85–92 (2006)
2. Sousa, P., Oliveira, J.L., Reis, L.P., Gouyon, F.: Humanized robot dancing: Humanoid motion retargeting based in a metrical representation of human dance styles. In: Antunes, L., Pinto, H.S. (eds.) EPIA 2011. LNCS, vol. 7026, pp. 392–406. Springer, Heidelberg (2011)
3. Jimmy, O.: Towards the development of emotional dancing humanoid robots. International Journal of Social Robotics 1(4), 367–382 (2009)
4. Miura, M., Sugiura, M., Takahashi, M., Sassa, Y., Miyamoto, A., Sato, S., Horie, K., Nakamura, K., Kawashima, R.: Effect of motion smoothness on brain activity while observing a dance: An fMRI study using a humanoid robot. Social Neuroscience 5(1), 40–58 (2010)
5. Ikeuchi, K., Suehiro, T.: Toward an assembly plan from observation, Part I: Motion recognition with polyhedral object. IEEE Trans. Robotics and Automation 10(3), 368–385 (1994)
6. Takamatsu, J.: Symbolic representation of trajectories for skill generation. In: ICRA, vol. 4, pp. 4076–4081 (2000)
7. Inamura, T., Nakamura, Y., Toshima, I.: Imitation and primitive symbol acquisition of humanoids by the integrated mimesis loop. In: ICRA, vol. 4, pp. 4208–4213 (2010)
8. Inamura, T., Nakamura, Y., Toshima, I.: Acquisition and embodiment of motion elements in closed mimesis loop. In: IGARSS, vol. 1, pp. 24–28 (2002)
9. Nakaoka, S., Nakazawa, A., Yokoi, K., Hirukawa, H., Ikeuchi, K.: Generating whole body motions for a biped humanoid robot from captured human dances. In: ICRA, vol. 3, pp. 3905–3910 (2003)
10. Ryoo, Y.-J.: Development of child-sized humanoid robot for dance performance. In: Kim, J.-H., Matson, E., Myung, H., Xu, P. (eds.) Robot Intelligence Technology and Applications 2. AISC, vol. 274, pp. 725–732. Springer, Heidelberg (2014)
11. Lim, D.-Y., Kwak, H.-J., Ryoo, Y.-J.: Motion editing tool to create dancing motions of humanoid robot. Inter. Jour. of Humanoid Robotics 11(4), 1442002-1–1442002-10 (2014)
12. Choi, J., Lee, Y., Ryoo, Y.-J., Choi, J., Choi, J.: Action petri net for specifying robot motions. Inter. Jour. of Humanoid Robotics 11(4), 1442004-1–1442004-12 (2014)

A Colour Detection and Connected-Objects Separation Methodology for VEX Robotics

Changjuan Jing, Johan Potgieter, and Frazer K. Noble

School of Engineering and Advanced Technology, Massey University,
Auckland 0632, New Zealand
{C.J.Jing,J.Potgieter,F.K.Noble}@massey.ac.nz

Abstract. One of the issues associated with programming a VEX Robotics Competition (VRC) robot for the autonomous period is providing it with enough information regarding its environment so that it can move about the field intelligently. As such, an objective of our research was to develop a series of machine vision tools so that a VRC robot could identify VRC field elements, other robots, and field perimeters; responding appropriately. We have carried out a review of relevant literature and identified a number of algorithms, image processing tools, and control paradigms, as well as, developed our own approaches, which we have implemented in C++ using the OpenCV library. Here we present the results of our initial efforts, namely our implemented colour identification, connected-object separation, and multiple connected-objects separation methodologies.

Keywords: VEX, robotics, machine vision, algorithms, image processing.

1 Introduction

The Robotics Education & Competition (REC) Foundation's VEX Robotics Competition (VRC) is, arguably, one of the largest, fastest growing education robotics programmes in the world, and in New Zealand alone there's over 100 teams actively participating in the programme. The VRC programme is designed to "increase student interest and involvement in science, technology, engineering, and mathematics (STEM)" [1] and with a particular focus in New Zealand to "inspire a passion for Science and Technology" [2].

Each VRC competition consists of a number of matches played out on a 12' x 12' field; where, each match consists of two teams – the RED and BLUE alliances – competing against each other for two minutes. The first 15 seconds of each match is dedicated to an autonomous period, where each teams' robots are programmed to play the game on their own. The remaining 75 seconds is dedicated to a driver period, where the teams' robots are programmed to responds to a driver's controls.

Most robots rely on non-visual sensors such as tactile, sonar, and laser sensors [3]. As visual input carries a rich source of information in the form of colour images from a camera, a lot of algorithms have been developed in the field of computer vision and

© Springer International Publishing Switzerland 2015

J.-H. Kim et al. (eds.), *Robot Intelligence Technology and Applications 3,*
Advances in Intelligent Systems and Computing 345, DOI: 10.1007/978-3-319-16841-8_36

machine vision to extract information from images. Vision Guided Robotic (VGR) is a technology guiding control mechanisms by implementing image processing skills. Feature detection is a method that aims at computer abstraction of image information for decision making on object verification. As colour is main difference between two teams and is highly robust to geometric variations of the shape, we use colour and shape feature for object detection in VEX game.

Segmentation is an important step in the automated analysis of image data. We implement segmentation for object location. To get the contour of the object of interest, active contours is parametric method to highlight a continuous curve through use of an energy function [4].However, this method is not suitable for shape detection, as it gets one smooth regular shaped contour losing the feature of peak points, thus unusable in analysing a polygon's shape. Watershed transformation is another segmentation method, which based on topological gradient [5]. However, this approach does not work well when two objects overlap. We will illustrate this result on paper. Threshholding is a simple, but effective tool to separate objects from the background. This is widely used in document image analysis, where lines, figure legends, and characters are searched for [6]. M. Sezgin and B. Sankur made a categorized survey of various threshold-based morphological approaches [7]. To obtain better thresh-holding results, we should consider application-specific information.

Most state-of-the-art approaches to solving computer vision problems, such as feature analysis [8], segmentation [9], blob clustering [10], object recognition [11]–[13], and illumination invariance [14]–[16] require a substantial amount of computational and/or memory resources. However, mobile robotic systems typically have strict constraints on the computational and memory resources available, but still demand real-time processing.

With respect to robotics, Mohan Sridharan and Peter Stone proposed a colour learning method, where a set of images were labelled and a robot learned the range of pixel values that map to each colour [17]. Sridhahan and Stone's results showed the robot can plan its motion autonomously; however, the pixel-colour mapping was dependent on the original training image set and should have the prior knowledge made available of the scene. Brett Browning and Dinesh Govindaraju presented two algorithms, Histogram Threshold Vector Projection and Histogram Threshold Ellipsoid Distance, for fast colour-based detection of objects under variable illumination [18]. Their algorithms are for a soccer ball detection, which requires lower computation.

In the 2013 VRC competition: "Toss Up", the game elements were large, semi-transparent and small, solid coloured balls. The normal classifier, i.e. colour thresh-holding, in computer vision easily mistakes the light reflections on the transparent balls as separate targets. One reason is that profile of a ball is too simple or not unique, to distinguish balls from other similar round objects. Another cause is the nature of semi-transparent object, which is made of thin material and reveals the background through reflection. In this paper, we propose a K-Means clustering method, and accumulate the various colour tints so we can identify the semi-transparent ball. Also the image of the balls can overlap, which is a common; we propose an objects separation methodology, which outperforms the general Watershed method. The approach used utilises the C++ OpenCV library.

The organisation of this paper is as follows: Section 1 introduces the paper, Section 2 presents our object colour identification methodology, Section 3 presents our connected-objects separation methodology, Section 4 presents the results of testing our methodologies, and Section 5 concludes this paper; discussing our approach and results.

2   Object Colour Identification

In this section we present our colour identification methodology.

A general approach for object colour identification is to capture a Red-Green-Blue (RGB) image, convert it into the Hue-Saturation-Value (HSV) colour space, and threshold for a specific hue value, resulting in a binary image, where only the desired hue is evident.

The objects we have used to test our object colour identification are the field elements from the VRC Toss Up season: a small solid coloured ball and large transparent coloured ball (see Fig. 1.). The general approach would work well for the small solid ball; but not for the large transparent ball due to the hue value being a combination of the foreground and background colours; widening the corresponding hue range necessary to threshold. Furthermore, the nature of the transparent material, is reflective therefore results in low saturation spots.

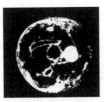

Fig. 1. VRC Toss Up field elements. (Left) the small solid coloured ball, and (Right) the large transparent coloured ball

Our approach is to accumulate the various colour hues, and classify the large transparent balls as separate objects. Our approach utilises the K-Means clustering method, which finds partitions such that objects within each cluster are as close to each other as possible and as far from objects in other clusters as possible. For example, given a set of observation $(x_1, x_2, x_3 ... x_n)$, where each observation is d-dimensional real vector, K-Means clustering aims to partition the n observations into k sets (k ≤ n), $= \{S_1, S_2, ..., S_k\}$, so as to minimize the within-cluster sum of squares:

$$\arg \min \sum_{i=1}^{k} \sum_{x_j \in S_i} \left\| x_j - \mu_i \right\|^2 \qquad (1)$$

Here μ_i is the mean of points in S_i.

As aforementioned, we have used the OpenCV library to implement our object colour identification; OpenCV uses the MAT data type to store images. When dealing with a RGB image, we have three channels, i.e. one per colour. In order to use

OpenCV's K-Means function, we first convert the MAT data from three channels into one. And then we perform K-Means on these data and get several cluster centres, as well as label of each pixel. The label indicates which cluster the pixel belongs to. Thus, we can get the interested cluster by indexing the label.

The following flowchart (see Fig. 2) provides an overview of our implementation in C++ using OpenCV's library:

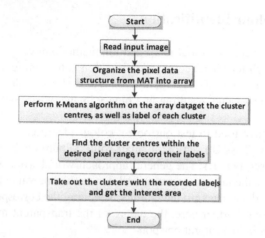

Fig. 2. Flowchart of K-Means procedure

3   Connected-Object Separation

This section describes the connected-object separation methodology.

A general approach for connected-object separation is to use a Watershed method, where an image is threshold for a specific hue and then a morphological filter is applied to erode/dilate the image, resulting in two close objects being separated.

As before, the objects we have used to test our connected-object separation approach are the field elements from the VRC Toss Up season: a small solid coloured ball and large transparent coloured ball. The general Watershed would work well when the view angle is nearly vertical; however, this situation is quite rare, because a VRC robot keeps moving during a match. The Watershed would not work very well for most cases, as shown in Fig. 3, where the balls overlap due to proximity and angle of the view.

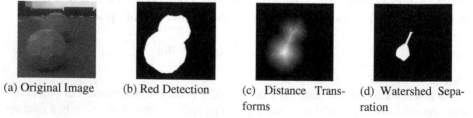

(a) Original Image   (b) Red Detection   (c) Distance Transforms   (d) Watershed Separation

Fig. 3. Watershed Separation Transform

Our approach is to use a defect as a clue for object separation. A defect is the point on concave hull, and the defects can be derived by thresh-holding the distance from concave hull to convex, which is smallest region enclosing the given polygon. And separation can be done by connecting the defect pair. The following flowchart (see Fig. 4) provides an overview of our methodology. We loop every contour to find the defects. We set a flag for multi-object separation, and it is used to decide whether it needs another separation. If the flag equals 0, it is the last instance of separation; otherwise we will loop through the process again.

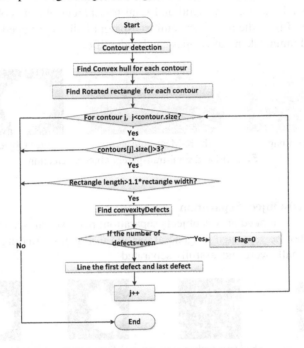

Fig. 4. Flowchart of object separation procedure

Also, we define rules for defect pair selection, and for locating missing defects, which created by occluding. The first is that the defect pairs should be the first point and the last point of the array, or the second point and the second to last point of the array. As defect points are stored in a sequence of array, starting from the lower left travelling anti clockwise to end at the upper left. The rule makes sure the defects are from the same objects. The other rule is that the direction of the line of the defect should be perpendicular to the ground. For the direction choice, we draw a rectangle enclosing the target region, and the longer side of the rectangle will contain the orientation to the ground. This is followed up with dot production to check if either direction is vertically aligned.

4   Results

In this section, we present the results of our implemented colour identification and connected-object separation methodologies. As discussed, the objects we have used to test our methodologies are the field elements from the VRC Toss Up season: a solid coloured ball and a transparent coloured ball.

a.   Solid and Transparent Object Colour Identification

The first case we looked at was solid and semi-transparent objects colour detection. As we can see in Fig.5, the pixels on semi-transparent ball are grouped together, after the pixels in red range taken out, shapes can be identified.

(a) Original image   (b) K-Means clustering   (c) Object detection

Fig. 5. Solid and semi-transparent objects detection

b.   Connected-Object Separation

The second case we looked at was objects separation; here we can see two overlapped balls which cannot be separated by general method in Fig.3, by finding the defects in Fig.6, those two balls were successfully separated.

(a) Original image   (b) Defects detection   (c) Separation   (d) Shape detection

Fig. 6. Connected-objects separation

c.   Multiple Connected-Object Separation

The third case we looked at was multiple connected-object separation. Fig.7. (a) is two red balls and one blue ball in contact, which is a plausible possibility in reality. After detecting the area of interest, the contact area with blue ball left an addition defect, and our method correctly selected the defect pair. In Fig.7. (b), there are three balls, as only three defects were found in the first separation loop; we need to do a second separation procedure. And every ball is identified after the second operation.

(a) Find defect pair in occluding condition (b) Twice separation for three connected objects

Fig. 7. Multiple Connected-objects separation

Fig.8 shows the combination of our colour detection and separation method, experimenting on both solid balls and semi-transparent balls, the results were successful and shows the robustness of this method.

Fig. 8. Five balls and semi-transparent balls separation results

d.   Objects in Noisy Environment

The last case we looked at was objects in a noisy environment. In Fig. 9, there are balls of different sizes and shape in the court. The feature used in detection is colour and shape. In Fig. 9 (c), Three balls with no background, shape detection was performed next which recognized the circular shape and ignored the blue mat.

(a) Original image   (b) K-Means feature selection   (c) Object shape detection

Fig. 9. Object detection in noisy environment

5   Conclusions and Discussion

This paper proposed a colour clustering method for semi-transparent ball detection, which is a new novel method to solve problems of this aspect. We have improved upon the traditional colour detection methods, the colour thresh-holding usually miss a large portion of objects body as discussed, and the classifier method in computer vision normally deals with complicated features with area caused by light refraction to be wrongly classified as objects in some conditions. While the K-Means clusters similar red pixels of a ball into one group, the results in Fig. 5 shows how semi-transparent balls can be detected successfully.

A methodology for objects separation based on concave defects was also developed, that outperforms the Watershed method, when the balls overlap in the image due to proximity and angle of the view. Fig. 6 proves the effectiveness compared to the result in Fig. 3, where the Watershed separation failed. In addition, this methodology can work on multi-object separation, even when the object is occluded. Furthermore, we combine the colour clustering and object separation methods together for a complete image processing technique. The methodology will work in noisy environments, with robustness. Thus, this work provides visual information for robotics, equip robots with autonomous detection to meet the challenges in a VEX competition.

Acknowledge. This work is supported in part by the scholarship from China Scholarship Council (CSC) under the Grant CSC201206890061.

References

1. VEX Robotics. A STEM education revolution (2014),
   http://www.vexrobotics.com/vexiq (last accessed October 2, 2014)
2. Kiwibots. New Zealand VEX Robotics (2014), http://www.kiwibots.co.nz/ (last accessed October 2, 2014)
3. Thrun, S., Fox, D., Burgard, W., Dellaert, F.: Robust monte carlo localization for mobile robots. Journal of Artificial Intelligence (2001)
4. Leymarie, F., Levine, M.D.: Tracking deformable objects in the plane using an active contour model. IEEE Transactions on Pattern Analysis and Machine Intelligence 15, 617–634 (2002)
5. Vincent, L., Soille, P.: Watersheds in digital spaces: an efficient algorithm based on immersion simulations. IEEE Transactions on Pattern Analysis and Machine Intelligence 13(6), 583–598 (1991)
6. Trier, O.D., Jain, A.K.: Goal-directed evaluation of binarization methods. IEEE Trans. Pattern Anal. Mach. Intell. PAMI-17, 1191–1201 (1995)
7. Sezgin, M., Sankur, B.: Survey over image thresholding techniques and quantitative performance evaluation. Journal of Electronic Imaging 13, 146–165 (2004)
8. Comaniciu, D., Meer, P.: Mean shift: A robust approach toward feature space analysis. IEEE Transactions on Pattern Analysis and Machine Intelligence 24(5), 603–619 (2002)
9. Sumengen, B., Manjunath, B.S., Kenney, C.: Image segmentation using multi-region stability and edge strength. In: The IEEE International Conference on Image Processing (ICIP) (September 2003)
10. Fred, A.L.N., Jain, A.K.: Robust data clustering. In: The International Conference of Computer Vision and Pattern Recognition, pp. 128–136 (2003)
11. Lowe, D.: Distinctive image features from scale-invariant keypoints. International Journal of Computer Vision (IJCV) 60(2), 91–110 (2004)
12. Belongie, S., Malik, J., Puzicha, J.: Shape matching and object recognition using shape contexts. Pattern Analysis and Machine Intelligence (April 2002)
13. Torralba, A., Murphy, K.P., Freeman, W.T.: Sharing visual features for multiclass and multiview object detection. In: The IEEE Conf. on Computer Vision and Pattern Recognition (CVPR), Washington D.C. (2004)

14. Finlayson, G., Hordley, S., Hubel, P.: Color by correlation: A simple,unifying framework for color constancy. IEEE Transactions on Pattern Analysis and Machine Intelligence 23(11) (November 2001)
15. Forsyth, D.: A novel algorithm for color constancy. International Journal of Computer Vision 5(1), 5–36 (1990)
16. Rosenberg, C., Hebert, M., Thrun, S.: Color constancy using kldivergence. In: IEEE International Conference on Computer Vision (2001)
17. Sridharan, M., Stone, P.: Structure-Based Color Learning on a Mobile Robot under Changing Illumination. Autonomous Robots Journal (2007)
18. Browning, B., Govindaraju, D.: Fast, Robust Techniques for Colored Object Detection in Variable Lighting Conditions

14. Thillainayagi, C., Handley, S., Ho, et al.: Colour by correlation: a simpler unifying framework for colour constancy. IEEE Transactions on Pattern Analysis and Machine Intelligence 23(11) (November 2001)

15. Forsyth, D.: A novel algorithm for colour constancy. International Journal of Computer Vision 5(1), 5–36 (1990)

16. Rosenberg, C., Hebert, M., Thrun, S.: Colour constancy using KL-divergence. In: IEEE International Conference on Computer Vision 2001

17. Sapiro, M., Storga: Bayesian colour constancy for outdoor object recognition. In: Illumination from Multiple Views. Image Series (2001)

18. Freeman, R., Grimson, F.: Blur Robust Techniques for Colour Object Detection in various images (2001)

Drawing Pressure Estimation Using Torque Feedback Control Model of A 4-DOF Robotic Arm

Meng Cheng Lau[1], Chi-Tai Cheng[2], Jacky Baltes[3], and John Anderson[3]

[1] Center for Artificial Intelligence Technology, Universiti Kebangsaan Malaysia,
43600 Bangi, Selangor, Malaysia
[2] Department of Electrical Engineering, Tamkang University, New Taipei City 25137,
Taiwan (R.O.C.)
[3] Autonomous Agent Lab, University of Manitoba, Winnipeg, Manitoba R3T 2N2,
Canada
mengcheng.lau@gmail.com,124646@mail.tku.edu.tw,{jacky,andersj}@cs.
umanitoba.ca

Abstract. This paper we introduce a torque feedback control (TFC) model to estimate pressure of the hand on a 4-DOF robotic arm of Betty, a humanoid robot. Based on several preliminary experiments of different stroke patterns, we measured and analysed the torque replies of Betty's servos in order to model the torque feedback. We developed a robust humanoid system to create sketch like drawing with limited hardware which has no force sensor but basic torque feedback from servos to estimate pressure apply on drawing pad. We investigated the efficiency of using the TFC in the drawing task based on several different stroke patterns. The experimental results indicate that the TFC model successfully corrected the errors during the drawing task.

Keywords: Torque Feedback Control, Robotic Arm, Drawing Pressure Estimation.

1 Introduction

In many applications [1,2,3,4,5], a robot must explicitly control the force it applies to the object it is manipulating, i.e., the actuators must be controlled to achieve the desired forces. Force control using feedback of joint torques is limited by the accuracy over a resistor to estimate the input current of the motors [6,7]. This indirect measurement has several errors, including the variations in the losses in the motor itself and the gearbox. To obtain accurate control of the force vector at the end effector, a wrist force sensor is placed between the tool plate and the end effector to measure end-effector force [8,9,10]. The force transform from the sensor to the end effector is simple but these load cell sensors are usually expensive. Torque in joint space is controlled by controlling the torque applied by each actuator (servo). Torque can be measured using a sensor (accurate) or estimated from armature current (simple). In this paper, we proposed

© Springer International Publishing Switzerland 2015   401
J.-H. Kim et al. (eds.), *Robot Intelligence Technology and Applications 3*,
Advances in Intelligent Systems and Computing 345, DOI: 10.1007/978-3-319-16841-8_37

a torque feedback control (TFC) to provide significant feedback for pressure estimation throughout the drawing process. The evaluation will be discussed in Section 4.

2   Hardware Configuration

The upper body of Betty consists of a twelve-revolute joint system with 12 degrees of freedom (DOF). Its head has 4 DOFs which are pan, tilt, swing and one DOF for the mouth. Each of its arms provides 4 DOFs, shoulders allow lateral and frontal motions, elbow gives lateral motion and a wrist motion. These joints are constructed by four Dynamixel RX-64 servos in the head and four Dynamixel RX-64 servos for each of its arms. Fig.1 shows the overview of Betty's upper body.

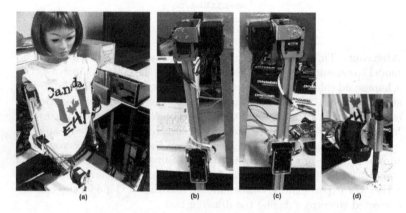

Fig. 1. Overview of Betty's 12-DOF upper body

The main reason we chose RX-64 to construct Betty's arms is it has higher final maximum holding torque, 64.4~77.2 kgf.cm compared to only 12~16.5 kgf.cm for the AX-12 [11,12] which were used in Betty's previous design [13,14]. This improvement allows the RX-64 servos to generate sufficient torque to support the weight of Betty's arms and head. In order to control the servos, we use Dynamixel's dedicated controller, CM-2+ from Robotis as the central control unit with its AVR ATmega128 microcontroller. The real-time OS establishes communication with the servos and its implementation can be seen in our previous publications [13,14].

3   Torque Feedback Control (TFC) Model

For Dynamixel RX-64 servos, torque can be measured on a 0-1023 (0x3FF) scale, based on its maximum holding (stall) torque 64 kgf-cm (6.92 Nm) at 18V. As seen in Tira-Thompson's [15] study, a Dynamixel servo can read back the current position, temperature, load (torque), etc.. For instance, position read back can measures to within 0.5 degrees of accuracy at full communication speed in every 130ms. However, reading the present load would not provide precise torque measurement. Robotis posted a comment on their official website that the "Present Load" or torque measurement is not a real torque or electrical current measurement. It is actually just based on the difference between current position and goal position. It has a control loop to make sure the motor actually got to the goal position. Hence, the "Present Load" value is the torque that the servo is *applying,* not the torque that it is *experiencing.* Therefore, it is not possible to perform "Torque Control" of a servo [16].

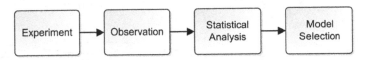

Fig. 2. Designing of torque feedback control (TFC) model

However, in the brief tutorial book of Anderson [17], he suggested the reality can be estimated based on the empirical evidence hypotheses and statistical models. In our approach, we designed a simple torque feedback control (TFC) model based on two simplest strokes, horizontal and vertical. Some preliminary experiments are performed to obtain suitable sets of data from servos' torque feedback (right arm). By using this data, a simple TFC model is created and the suitable set of candidate models is selected based on statistical hypothesis testing to estimate stroke pressure. "Given candidate models of similar predictive or explanatory power, the simplest model is most likely to be correct" [17]. Fig.3 shows the design of Betty's TFC model.

This TFC approach is implemented to observe pen pressure during the drawing process to prevent overpressure or no-touch errors. Several experiments were

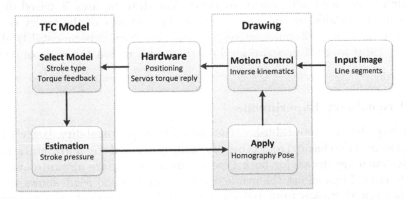

Fig. 3. Processing flow of torque feedback control (TFC) for drawing task

conducted to obtain the statistical model. It provides optimum thresholds to estimate pen pressure. In these experiments, the pressure applied on the Wacom Bamboo tablet (underneath the paper) are recorded.

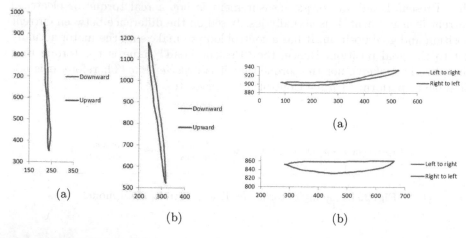

Fig. 4. (a):Medium pressure vertical strokes and (b): High pressure vertical strokes

Fig. 5. (a):Medium pressure horizontal strokes and (b): High pressure horizontal strokes

3.1 Experimental Setup

Fig.6 shows the experimental setup of Betty and the drawing pad. During the experiments, Betty used its right arm to draw based on sinusoidal interpolation motion trajectory. We placed a Wacom Bamboo tablet (CTH670M) under the paper to measure position and pressure on the drawing pad. The measures of tablet were only used to established the reliability of the experiments but did not provide any feedback control to Betty. The drawing area is based on the specification provided by the manufacturer, the accuracy of the tablet is ±0.5 mm on a 217 mm x 137 mm active area [18]. For each experimental trial, we recorded the 0 to 1024 pressure level on the tablet within the scale of 0.00 to 1.00.

3.2 Preliminary Experiments

In this experiment, we established the tests for torque reliability. Betty's right arm performed two basic stroke types are, namely vertical and horizontal strokes. Two direction are drawn for both type of strokes which are upward-downward and left-right. Fig.4 shows the average vertical strokes and Fig.5 shows the average horizontal strokes from 100 recorded strokes respectively in different directions. Medium pressure is the normal pressure read from the tablet in the

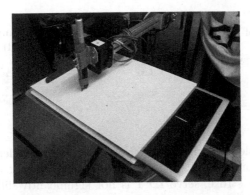

Fig. 6. Experimental setup using a CTH640M tablet, Betty, and pen

range of 0.7 ± 0.1 and high pressure represents 1.0 (highest) pressure measurement from the tablet. From these preliminary experiments we observed that high pressure caused significant errors in both horizontal and vertical strokes compares to normal pressure. So, it makes the estimation of stroke pressure essential. We recorded the torque measurements of each servos on Betty's right arm. By using this torque data as seen in the following tables, a simple TFC model is created. Hence, the suitable set of candidate models can be selected based on statistical hypothesis testing to estimate stroke pressure.

Table 1 shows the torque measurements from four servos on the right arm namely ID 1, 2, 3 and 4 for shoulder-frontal, shoulder-lateral, elbow and wrist respectively. Table 2 shows the model of vertical downward stroke which suggests how significant torque measurements are affected by different stroke pressure. As seen in Table 2, servo ID 2 is not significant to differentiate normal pressure stroke from no-touch conditions. However, servos ID 1 and 3 torques are highly significant to be used as the evaluators to estimate the pen's tip pressure for this stroke pattern.

Table 1. Results of vertical stroke (downward) tests

Vertical Strokes (Downward)											
Average Pressure	No			0.67				1.00			
Servo ID	1	2	3	4	1	2	3	4	1	2	3	4
Average	16.88	0.02	52.42	0.12	32.32	0.84	20.7	23.6	200.44	11.82	93.1	1.42
Std. Dev.	12.21	0.14	7.65	0.33	16.24	5.37	8.95	2.83	12.13	7.01	8.92	4.36

Table 2. Comparison of vertical (downward) stroke model

	Servo ID			
Pressure Comparison	1	2	3	4
No vs. Normal	2.96e-07	0.1430[1]	1.24e-34	3.59e-48
No vs High	6.73e-89	2.26e-16	2.65e-43	0.0202
Normal vs. High	3.27e-74	4.05e-14	2.57e-63	3.41e-47

[1] Not significant to identify different pressure on pen tip.

Table 3 and Table 4 show the model of vertical upward stroke. As seen in Table 4, servo ID 2 has no significant difference between normal pressure stroke and no-touch conditions. But servos ID 1, 3 and 4 torques are highly significant to be considered as the evaluators to estimate the pen's tip pressure for vertical upward stroke.

Table 3. Results of vertical stroke (upward) tests

Vertical Strokes (Upward)											
Average Pressure	No			0.67				1.00			
Servo ID	1	2	3	4	1	2	3	4	1	2	3	4
Average	76.04	0.12	33.36	0.2	32.42	1.08	11.96	23.52	40.96	58.58	46.42	18.78
Std. Dev.	11.65	0.33	6.08	0.4	18.22	3.99	10.12	3.39	3.08	18.42	8.34	5.21

Table 4. Comparison of vertical stroke (upward) model

Pressure Comparison	Servo ID			
	1	2	3	4
No vs. Normal	2.25e-24	0.048072[1]	2.08e-21	3.81e-44
No vs High	4.90e-28	9.74E-28	2.32e-14	3.67e-30
Normal vs. High	0.000964	2.19e-28	1.31e-33	3.20e-07

[1]Not significant to identify different pressure on pen tip.

Table 5 and Table 6 show the model of horizontal right-left stroke. As seen in Table 6, it is not identical to the previous stroke types we discussed. For horizontal right-left stroke, servo ID 1 is not significantly different to identify high pressure stroke and no-touch conditions. However, it is highly significant to differentiate these conditions from normal pressure. Servos ID 3 and 4 torques are also highly significant to be used as the evaluators to estimate the pen's tip pressure.

Table 5. Results of horizontal stroke (left to right) tests

Horizontal Strokes (Left to right)											
Average Pressure	No			0.67				1.00			
Servo ID	1	2	3	4	1	2	3	4	1	2	3	4
Average	47.26	30.04	66.24	26.52	23.64	36.34	21.86	23.06	47.44	54.74	41.02	11.66
Std. Dev.	7.77	12.87	7.76	4.68	1.79	14.99	4.04	5.41	2.55	12.12	3.18	5.82

Table 6. Comparison of horizontal stroke (left to right) model

Pressure Comparison	Servo ID			
	1	2	3	4
No vs. Normal	5.90E-28	0.013217	9.69E-49	0.00046
No vs High	0.498413[1]	1.15E-16	2.76E-31	3.77E-25
Normal vs. High	1.56E-69	6.08E-10	3.48E-45	3.05E-17

[1]Not significant to identify different pressure on pen tip.

Table 7 and Table 8 show the model of horizontal right-left stroke. As seen in Table 8, servo ID 2 is not significant to differentiate normal pressure stroke from no-touch conditions like the other vertical strokes we have seen. Similar to vertical upward stroke, servos ID 1, 3 and 4 torques are highly significant to be used as the evaluators to estimate the pen's tip pressure for horizontal right-left stroke.

Table 7. Results of horizontal stroke (right to left) test

Horizontal Strokes (Right to left)											
Average Pressure	No			0.67				1.00			
Servo ID	1	2	3	4	1	2	3	4	1	2	3	4
Average	45.14	35.34	66.24	26.36	18.52	35.56	20.84	17.06	25.72	11.88	88.86	1.06
Std. Dev.	6.55	15.14	7.76	6.62	4.21	13.84	4.1	5.45	14.91	7.07	13.83	3.82

Table 8. Comparison of horizontal stroke (right to left) model

Pressure Comparison	Servo ID			
	1	2	3	4
No vs. Normal	8.33e-40	0.469855[1]	1.33e-49	7.56e-12
No vs High	1.94e-12	2.95e-15	4.88e-16	2.03e-37
Normal vs. High	0.000871	4.93e-17	1.25e-39	7.85e-30

[1]Not significant to identify different pressure on pen tip.

Based on these preliminary experiments, Table 9 shows the general TFC model to estimate the pen's tip pressure from servos' torque feedback. For simplicity of the implementation Servo ID 1 and 3 (shoulder and elbow) are mostly used. Fig.7 illustrates the control loop of the TFC model. It showed that by way of different stroke patterns, a significant difference in torque feedback can be measured. It allowed for a significant estimate of the pressure of the pen's tip.

Table 9. Torque feedback control model

Stroke Type	Torque (τ_i) threshold conditions for normal pressure	
	Low Pressure (No-touch)	High Pressure
Vertical downward	ID1: $\tau_1 < 25$	ID1: $\tau_1 > 50$
Vertical upward	ID1: $\tau_1 > 50$	ID1: $\tau_1 < 50$ and ID3: $\tau_3 > 40$
Horizontal left-right	ID1: $\tau_1 > 30$ and ID3: $20 < \tau_3 < 35$	ID3: $\tau_3 > 20$
Horizontal right-left	ID1: $\tau_1 > 40$ and ID3: $20 < \tau_3 < 35$	ID2: $\tau_2 > 35$ and ID3: $\tau_3 > 20$

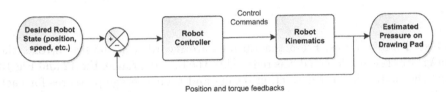

Position and torque feedbacks

Fig. 7. Closed-loop control diagram for TFC

4   Torque Feedback Control Experiment

In the experiments, we evaluated the feasibility and efficiency of our TFC model implementation based on the generic drawing tasks. We used two initial conditions, no-touch (0.0) and high pressure (1.0) on the tablet to justify the feasibility of TFC to correct these errors. Betty repeated 30 times for each vertical (upward and downward), horizontal (left-right and right-left) and diagonal lines (left-right with downward) based on these two initial conditions with total of 150 trails. For these trials, the average pressure on the tablet and measured torque reply for each stroke were recorded. We measured the error of pressure for each stroke based on the TFC model using initial errors of "high pressure", "low pressure" (no-touch) and desired "normal pressure" as seen in Section 3.2. The ideal average "normal pressure" experience on the tablet should be at the level of 0.7±0.1. A correction of 10 mm was applied when the measured torque was too low or too high. The total number of required correcting strokes to finish a drawing task and the changes in error over time, were sampled at some defined interval.

Fig.8 shows the different pressure applied to the tablet. In Fig.8(a), it started with initial lower pressure strokes until it was corrected by the TFC model in two strokes. Similarly, the system corrected the initial high pressure error in two strokes as seen in Fig.8(b). An real world example of the pressure correction is illustrated in Fig.9.

Fig.10 and 11 show how different stroke types corrected the high pressure error with their average pressures for each stroke. These figures clearly show that the high pressure error could be corrected at average of three strokes in

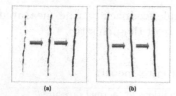

Fig. 8. Stroke pressure correction based on torque feedback control. **(a)**: low to normal **(b)**: high to normal. Green, blue and red were used to indicate low, normal and high pressure measurements respectively.

Fig. 9. Real world example of stroke pressure correction. A line segment is completed in few trial after depth correction based on torque feedback from servos.

all tested stroke types. From the 1.0 pressure recorded in the first stroke, the average of pressure detected was reduced to the 0.7±0.1 range. Fig.12 and Fig.13 show the no-touch error correction results and their average pressures for each stroke. As we have seen in the high pressure error results, the no-touch error could also be corrected at average of three strokes in those tested stroke types. From the initial no-touch stroke (no pressure is read from the tablet) the average of pressure detected increased to the 0.7±0.1 range.

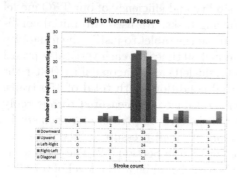

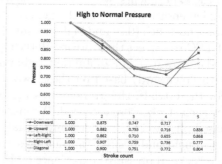

Fig. 10. Number of stroke occurrence: High to normal pressure

Fig. 11. Average pressure for each strokes: High to normal pressure

In this experiment, we labelled a trail as failure if no successful correction after the first stroke and the last stroke (fifth stroke). Hence, the TFC model provides 92.7% and 94.0% of successful detection of high pressure and no-touch respectively. Fig.14 shows the averages of trails needed to make the correction of each stroke type in both the high pressure and the no-touch conditions.

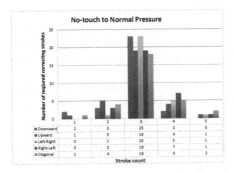

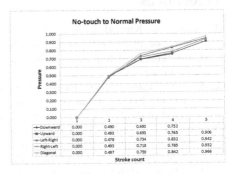

Fig. 12. Number of stroke occurrence: No-touch to normal pressure

Fig. 13. Average pressure for each strokes: No-touch to normal pressure

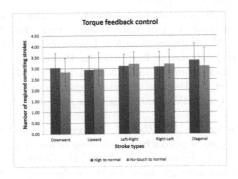

Fig. 14. Number of strokes for different stroke type based on TFC model. Error bars indicate the standard deviation.

5   Conclusion

The capability to evaluate the efficiency of different drawing implementations is very important. Unfortunately, none of the current research done in the field has provided a conclusive platform for drawing output evaluation. In our work, we used a Bamboo tablet and a series of experiments to provide the capability of drawing evaluation. This may lead to a better understanding of the various components, such as the correlation between pressure and deviation in different implementations. In this paper, we have described our methods and hardware that we used to evaluate the implementations of TFC model for a 4-DOF robotic arm. Based on the experimental results, they show that the TFC model successfully corrected the errors during the drawing task when no accurate force sensor feedback at the end effector is used. We established that TFC in the drawing task are reliable to pressure error deviation of strokes more than 90% of the time.

References

1. Kim, I.M., Kim, H.S., Song, J.B.: Design of joint torque sensor and joint structure of a robot arm to minimize crosstalk and torque ripple. In: 2012 9th International Conference on Ubiquitous Robots and Ambient Intelligence (URAI), pp. 404–407 (November 2012)
2. Ott, C., Eiberger, O., Englsberger, J., Roa, M.A., Albu-Schäffer, A.: Hardware and control concept for an experimental bipedal robot with joint torque sensors. Journal of the Robotics Society of Japan 30(4), 378–382 (2012)
3. Kim, T.K., Kim, D.Y., Cha, D.H., van Choi, S., Kim, B.S., Hwang, J.H., Park, C.W.: Development of joint torque sensor and calibration method for robot finger. In: 2013 10th International Conference on Ubiquitous Robots and Ambient Intelligence (URAI), pp. 161–162 (October 2013)
4. Tian, L., Goldenberg, A.: Robust adaptive control of flexible joint robots with joint torque feedback. In: Proceedings of the 1995 IEEE International Conference on Robotics and Automation, vol. 1, pp. 1229–1234 (May 1995)
5. Zhang, G., Furusho, J.: Control of robot arms using joint torque sensors. IEEE Control Systems 18(1), 48–55 (1998)
6. Luh, J., Fisher, W., Paul, R.: Joint torque control by a direct feedback for industrial robots. IEEE Transactions on Automatic Control 28(2), 153–161 (1983)
7. Aghili, F., Buehler, M., Hollerbach, J.M.: Dynamics and control of direct-drive robots with positive joint torque feedback. In: Proceeding of IEEE Internatinal Conference Robotics and Automation, pp. 2865–2870 (1997)
8. Leite, A.C., Lizarralde, F., Hsu, L.: Hybrid adaptive vision-force control for robot manipulators interacting with unknown surfaces. Int. J. Rob. Res. 28(7), 911–926 (2009)
9. Baeten, J., de Schutter, J., Schutter, J.: Integrated Visual Servoing and Force Control: The Task Frame Approach. Engineering online library. Springer (2004)
10. Olsson, T., Bengtsson, J., Johansson, R., Malm, H.: Force control and visual servoing using planar surface identification. In: Proceedings of the 2002 IEEE International Conference on Robotics & Automation, pp. 4211–4216. IEEE, Washington, DC (2002)
11. Robotis: Robotis: User's manual dynamixel rx-64 (2007)
12. Robotis: Robotis: User's manual dynamixel ax-12 (2006)
13. Lau, M.C., Baltes, J.: The real-time embedded system for a humanoid: Betty. In: Vadakkepat, P., et al. (eds.) FIRA 2010. CCIS, vol. 103, pp. 122–129. Springer, Heidelberg (2010)
14. Baltes, J., Cheng, C.T., Lau, M.C., Anderson, J.E.: Cost oriented automation approach to upper body humanoid robot. In: Proceedings of the 18th IFAC World Congress, Milano, Italy (September 2011)
15. Tira-Thompson, E.: Digital servo calibration and modeling. Technical Report CMU-RI-TR-09-41, Robotics Institute, Pittsburgh, PA (March 2009)
16. Emami, S.: Dynamixel servo motors (2009), http://www.shervinemami.info/dynamixels.html (accessed: November 20, 2011)
17. Anderson, D.: Model Based Inference in the Life Sciences: A Primer on Evidence. Springer Science + Business Media (2008)
18. Futureshop: Wacom bamboo create tablet, cth670m (2013), http://www.futureshop.ca/en-CA/product/wacom-wacom-bamboo-create-tablet-cth670m-cth670m/10180531.aspx (accessed: October 20, 2013)

Evolving Honeycomb Pneumatic Finger in Bullet Physics Engine

Bin Cheng, Hao Sun, and Xiaoping Chen

Dept. of Computer Science, University of Science and Technology of China
Hefei, Anhui, 230027, PRC
{cheng11,hhsun}@mail.ustc.edu.cn, xpchen@ustc.edu.cn

Abstract. Soft robot is becoming a current focus for its inherently compliance and human-friendly interacting with the real world. But most soft robots are designed and fabricated with intuition and empiricism only, which lack of systematic assessment such as force and deformable analysis before fabricating. Before choosing proper soft materials and processing the craft, the experiments of the soft robots can hardly be set up. In this paper, we construct a model of Honeycomb Pneumatic Finger (HPF) with honeycomb pneumatic network embedded which overcomes the shortcoming of embedded rectangular unit. In the meantime, a pressure analysis model is built for the purpose of physical simulation. Based on the model, without choosing any real materials and fabricating, we focus on exploring the correlations between the pressure and the geometrical shapes in physics simulation. Furthermore, we construct a virtual hand consisting of one rigid palm and four HPFs which is simulated with Bullet Physics Engine. By changing the pressure of the corresponding honeycomb units of each finger, the hand can grasp a ball smoothly and lift it up. At last, the deviation between the mathematical analysis and the physical simulation is discussed.

Keywords: Soft robot, Pneumatic finger, Honeycomb structure, Bullet.

1   Introduction

Over the last decade, researchers have developed soft robots with new capabilities in contrast with traditional rigid robots. Actuation mechanisms of soft robots can be divided into the several categories, electromagnetic actuation, thermal actuation, Shape-Memory-Alloys (SMA) and pneumatic actuation. Pneumatic actuation has gained its ground currently because of its inertness to the surrounding environment [1]. It uses a different mechanism to enable dexterous mobility. Through controlling the pressure of each inflatable unit, the robots have distributed deformation with theoretically an infinite number of degrees of freedom [2]. These robots can be used in unpredictable environment in contrast to the rigid robots which are often used in well-defined environment.

At present the main application of pneumatic actuator is the robot hand which needs much flexibility and compliance to grasp and manipulate irregular objects such

© Springer International Publishing Switzerland 2015   411
J.-H. Kim et al. (eds.), *Robot Intelligence Technology and Applications 3,*
Advances in Intelligent Systems and Computing 345, DOI: 10.1007/978-3-319-16841-8_38

as balls and soft materials. However, most of these soft hands are designed and fabricated with intuition and empiricism. Though there are some mathematical models for analysis [19], it cannot describe the dynamic features of the robot when interacting with grasped objects and environment. The dynamic shapes and deformation of the soft hands cannot be predicted before fabrication. This has proven to be quite challenging to validate the controlling algorithm. The creation of pneumatic soft robot requires a simulation platform to help solve the problems above and help fast modeling and fabricating. The main potential features and advantages of simulation platform for soft robot are

- Simulation platform for soft robot provides development and test environment when lack of proper materials or satisfying processing craft.
- Numerous repeated or parallel experiments on the virtual soft robots can be launched quickly without any wastage of the real material.
- The experiment processes which focus on the dynamic features of the soft robots can be easily suspended or interrupted for debugging and recording the relative data.
- With the data of the soft robots in physical simulation tracked as feedback, the controlling and planning algorithm can be validated and revised.

Recently there has been a marked increase in the number of free publicly available physics engines such as PhysX and Bullet which are mainly used to develop video games. They are considered to be helpful applied to implement a soft robot simulation platform described above.

In this paper, we give an approach to evolve HPF in Bullet Physics Engine which is an open source software based on position constraint. It serves to analyze the innovative application model of Honeycomb Pneumatic Finger which overcomes the implicit shortness of rectangle embedded units such that we can see the virtual physical effect that verifies the mathematical model.

2   Background

2.1   Pneumatic Soft Robots

There are two main types of pneumatic soft robots which are based on pneumatic artificial muscle (PAM) and embedded pneumatic networks (EPN). McKibben-type actuator [3] is the outstanding representative of PAM which has high and low-density fibre mesh.   However, we realize that the existing PAM robots are difficult to be built without rigid components which are needed to maintain its original shapes. EPN actuators use chambers embedded in elastomer as repeating components. By changing the pressure of the chambers, it can provide complex motion [4]. EPN is light and totally soft without any rigid components. However, it was pointed out that the embedded unit was small rectangular. Only very soft materials can achieve such big deformation without breaking, so the EPN soft robots made of very soft materials are not suitable for manipulating heavy objects. A new type embedded structure of honeycomb unit has been proposed to create HPN robot [5] (as shown in Fig.1) by nature inspiration and a mathematical model was built to analyze the advantage of it.

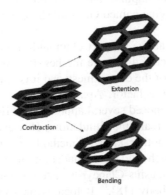

Fig. 1. The deformation modes of HPN robot [5]

The most critical step of manufacturing EPN is to select the suitable material with appropriate deformation and flexibility which supports complex motion and maintains compliance in any forms. Polymers (especially elastomer) are usually used to fabricate EPN. These soft materials allow for continuous deformation enabling ranges of motion limited only by the properties of them [6]. While for robot researchers, understanding the properties of soft materials belongs to the subject of organic materials science and makes the task much more complicate. We need a simulation platform to model the soft robots and simulate different properties of various materials in order to find out potential applications before fabrication. The methods to design automation will open the door to leveraging the full potential of the free form multilateral design space to generate novel deformable robots [7].

2.2   Real-Time Physics Simulation System

In computer graphics and computer games, the physics based animation is widely used. Traditionally in each simulation step the force of each object is accumulated. Then the position and velocity are calculated according to the Newton's second law. But during the simulation it is often desirable to have direct control over positions of objects of vertices of a mesh. A new approach of position based simulation is proposed that gives control over explicit integration and allows manipulating directly the positions of vertices in objects [8]. The commercial physics engine PhysX and the open source physics engine Bullet are both based on this approach. Though they aim at developing video games, researchers still consider them helpful to simulate soft robots.

PhysX which supports cloth and pressure model has been used to simulate artificial muscle-based robot [9]. The muscle-shaped structures are simulated with the cloth feature and employed as actuators. A soft robot which was actuated by muscle wires attached between adjacent segments was simulated in PhysX [10]. Stiffness values were chosen to most closely model the features of the silicon elastomer which was used to create the real soft robot. Bullet is another open source physics simulation library [11] which is a hybrid impulse and constraint-based engine [12] and it's also integrated in the physical engine named Gazebo which is an open source robot simulation platform.

Both simulation systems above support cloth and pressure model where balloons can be simulated as a set of closed triangle meshes. It adds an equality constraint and computes the actual volume of the closed mesh and it is compared against the original volume times the overpressure factor. Apart from the existing methods in physics engines, other researchers proposed several approaches. FEM, FEV [13, 14] and LEM [15, 16] methods are used in real time simulations. However the complexity of given algorithms are still high. Another kind of method of simulating soft body is the mass-spring model [17] which laid the foundation of current approaches. Simple thermodynamics laws and Clausius-Clapeyron state equation has been used for pressure calculation of the soft body [18]. It focuses on soft body deformation rather than inflatable soft chambers. Most of the methods mentioned above cannot be used to simulate soft chambers except the balloon simulation method. But considering that the pneumatic robots are not simple balloons and the walls of each embedded unit were designed thick to maintain their shapes and resist external force, we cannot directly use the cloth and pressure model to simulate pneumatic robots such as our HPF. Currently the lack of physical simulation method for the structure of HPF offers us this opportunity to explore the way of building virtual soft chambers inspired by existing methods.

In this paper, we simplify the mathematical model of HPN in [5] to adapt to the simulation method in Bullet. We also modify the constraint solver in Bullet which uses similar way simulating balloons to study the relationship between the shape of HPN and the pressure of each embedded unit. At last, we use the HPN to construct our virtual HPF which can grip and lift sphere.

3   Honeycomb Pneumatic Finger Analysis

3.1   Pressure-Volume model

The correlation among the air pressure, external force and the original length of chamber walls of HPN has been studied in [5]. The elongations of chamber walls were neglected. The shape of HPN only depends on the angle between the adjacent walls. Actually when inflating or deflating the chamber the length of elastic walls will definitely change. In order to put this model into physical simulation, we reconstruct the mathematical model and give a physics-based approach to simulate this model.

In our model, we study the relationship between pressure and volume of the chamber ignoring the external force. We expect to find out how our HPF based on HPN deforms by only changing the chamber pressure.

HPF is constructed with two columns of hexagonal cylinders as described in Fig.2.

Fig. 2. HPF structure

For every comb unit, the length of the side is L and the height of the cylinder is H. We can easily get the volume of each chamber.

$$V = \frac{3\sqrt{3}}{2}L^2H \tag{1}$$

Let P_i be the pressure of the chamber, P is the outer air pressure. The chamber keeps balance only if $P_i = P$. According to the equation of state of ideal gas the volume of the chamber tends to gets balance and with the pressure P_i we can get the updated volume:

$$P_iV = PV' \tag{2}$$

From (1) (2) we can get the updated L′ and H′:

$$L' = \sqrt[3]{\frac{P_i}{P}}L, H' = \sqrt[3]{\frac{P_i}{P}}H \tag{3}$$

Let I be the axial length of a chamber:

$$I = \sqrt{3}L \tag{4}$$

The analysis model above ignores the tensions of walls and external force which will be solved in the physical simulation.

3.2   Kinematics

In our previous work [5] we consider the HPN as a parallel hyper redundant linkage mechanism. In this paper, we still use this model (shown in Fig.3) but with different definition of I which is described in section 3.1.

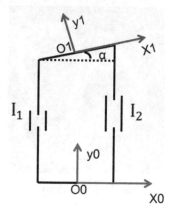

Fig. 3. Simplified Kinematics model

The degrees of bending of the robot depend on the lengths of the left and right chambers, I1, I2 and the distance between them, W. In a similar way, we get a homogeneous transformation matrix:

$$H_1^0 = \begin{bmatrix} \cos(\alpha) & -\sin(\alpha) & 0 \\ \sin(\alpha) & \cos(\alpha) & \frac{l_1+l_2}{2} \\ 0 & 0 & 1 \end{bmatrix} \tag{5}$$

$$\alpha = \arctan\left(\frac{l_2-l_1}{W}\right) \tag{6}$$

Using this matrix the position of one point can be transferred to another coordinate system. If C_i is the point in coordinate i, then $C_0 = H_1^0 C_1$. So we get the position in coordinate i based on the position in original coordinate system:

$$C_0 = H_n^0 C_n \tag{7}$$

$$H_n^0 = H_1^0 H_2^1 \dots H_n^{n-1} \tag{8}$$

4   Physics-Based Dynamic Simulation

In Section 3, we talk about the mathematical model of HPF which makes preparations for the physical simulation. In this section, we use Bullet Physics Engine to simulate our HPF. Bullet is a constraint-based dynamic simulation system. The soft bodies are represented by a set of vertexes and links between them. The points are projected according to a constraint solver to satisfy the constraint. We translate the pressure force into the position variations instead of directly applying the force to vertexes. That is because the external force is out of the HPF system and it may cause so called *ghost forces* making the simulation unstable.

In Bullet the constraints are solved before integrating the positions. And for every point the correction Δp_i has been yielded [8] by constraint function C=0 and the inverse masses $w_i = 1/m_i$:

$$\Delta p_i = -sw_i \nabla_{p_i} C(p_1, \ldots, p_n) \tag{9}$$

$$s = \frac{C(p_1, \ldots, p_n)}{\sum_j w_j \left| \nabla_{p_j} C(p_1, \ldots, p_n) \right|^2} \tag{10}$$

Assuming that the chamber walls will not bend when inflating or deflating, the volume only depends on the length of the chamber wall and height of the hexagonal cylinders as described in (3). When the pressure in the chamber is smaller than outer air pressure, the length of wall and height tend to shrink (as shown in Fig.4).

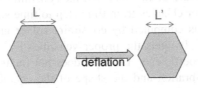

Fig. 4. Chamber walls shrink when deflation

Based on the above assumption we can define the constraint function. For each link connecting two vertexes there is:

$$C(p_1, p_2) = |p_1 - p_2| - \sqrt[3]{\frac{P_i}{P}} R \tag{11}$$

R is the rest length of the link when the inner pressure is P which is the same as the outer air pressure. From (9) (10) (11) we can get:

$$\Delta p_1 = -\frac{w_1}{w_1 + w_2} \left(|p_1 - p_2| - \sqrt[3]{\frac{P_i}{P}} R \right) \frac{p_1 - p_2}{|p_1 - p_2|} \tag{12}$$

$$\Delta p_2 = +\frac{w_1}{w_1 + w_2} \left(|p_1 - p_2| - \sqrt[3]{\frac{P_i}{P}} R \right) \frac{p_1 - p_2}{|p_1 - p_2|} \tag{13}$$

The stiffness k can be incorporated by simply multiply the corrections. In each simulation step we can solve every link according to the above. Given a time step Δt, the whole dynamic HPF is simulated below:

```
(1) for all vertexes i:
(2)     Initialize xi = x0, vi = v0, wi = 1/m0
(3) endfor
(4) loop timestep
(5)     forall vertices i:
(6)         applyGravity(xi)
```

```
(7)              applyExternalForces(x_i)
(8)              v_i = v_i + Δtw_if(x_i)
(9)         endfor
(10)        dampvelocities(v_1,...v_n)
(11)        forall vertices I do p_i = x_i + Δtv_i
(12)        loop Iterations times
(13)             solveConstraint(C_1,...C_{M+M_{coll}+M_P},p_1,...,p_n)
(14)        endloop
(15)        integrateMotion(x_1,...,x_n,v_1,...v_n)
(16)        updateVelocities(v_1,...v_n)
(17)   endloop
```

The main idea of our work is shown in line (13), which projects the distance constraint related to the pressure of chamber. All the constraint function should be called if the pressure of chamber changes from time step to time step. Line (15) obtains the physically valid positions computed by constraint solver and the velocities are updated accordingly. Line (16) gets the proper velocities by other parameters such as friction and restitution coefficients. Line (7) does not include the air pressure force because it may cause vibration and the shape of the chamber tends to distort. The effect of the pressure force is translated into the proper distance between each pair of vertexes.

5   Experiments

Based on the mathematical model in Section 3 and the physical dynamic model in Section 4 we set up two experiments to verify our approach and observe the effects of physical simulation. Section 5.1 studies the relationship between chamber pressure and the length of chamber walls. It also explores how the shape of HPF changes while the chamber deforms. Section 5.2 builds up a 3D model of HPF represented by a set of vertexes and links. By fixing four HPFs to a rigid board, we construct a virtual hand which can grip and lift up a ball in simulation using Bullet. At last, the deviation between the mathematical model and the physical model is discussed.

5.1   Mathematical Analysis Simulation

First we calculate the axial length of chamber with different inner pressure. Then we observe the HPF's bending performance giving the left and right chambers different pressures. Before we start the simulation, we need to amend the equation (3) considering the situation in physical simulation model. If we generate constraints for every link when deflation, it may cause over crumple since Bullet generates much more links for bending effect. For smoothing the changing rate of the length, we simply square it. Thus we get new equations:

$$L' = \sqrt[6]{\frac{P_i}{P}}L, H' = \sqrt[6]{\frac{P_i}{P}}H \tag{14}$$

In the next section, the variables L' and H' are also replaced in the corresponding equations. We set both L and H as 1. The pressure of outer air is also 1. The curve of the relationship between chamber length and inner pressure is shown in Fig.5 according to (4) and (14). We can see that the chamber length increases when the pressure increases and the increasing rate become slower. That makes sense because when inflation the chamber will expand anyway until the material cannot bear the extension force and the chamber explodes.

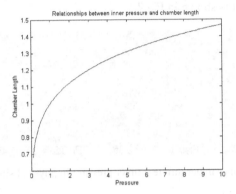

Fig. 5. Chamber length with different inner pressure

Later, the mathematical model mentioned in section 3.2 is used to observe how the shapes change under different inner pressure of the two column chambers. We set w=1, n=10. Let Pi be the inner pressure of the left column, the inner pressure of the right one is set as constant 1. The various displacements are shown in Fig.6. HPF is bending further when the inner pressure decreases. Excellent flexibility is gained on the basis of the experiment result.

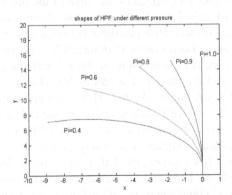

Fig. 6. The various shapes under different air pressure of left chamber

5.2   Physics-Based Simulation

To perform the simulation, a three-dimensional design was created using 3D Studio Max and then the face and node files were generated which are required by Bullet. Our HPF contains 4236 nodes and 2852 faces and each column has four chambers. We keep the pressure of right column chambers as constant 1. All the nodes belong to the walls are obtained for projecting constraint between their links. We set the mass of the total HPF as 50 and the stiffness value as 0.5. Then we change the pressure of the left column. The effect is shown in Fig.7.

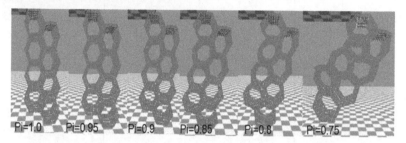

Fig. 7. Shapes under different inner pressure in physical simulation

The inner pressure of top chamber of the left column stays constant 1 which is the same as the outer pressure. Both of the top chambers of the two columns are fixed in experiment. When one chamber is deflating, its volume decreases and it drags the chambers next to it. Hence we get the curling deformation. From Fig.7, we can see that HPF bends further when the pressure of the three chambers in the left columns decreases. If we link more honeycombs together, HFP will tend to curl up which behaves similarly like the displacement curve in section 5.1.

With the bending actuator above, we construct a virtual hand consisting of one rigid board as a palm and four HPFs fixed on the edge of the board. In order to use this structure to grasp and lift a ball stably, the detection of collision must be taken into consideration. We simply divided HPF into eight collision clusters that every chamber is approximate to a cluster. Each cluster is deformable and automatically generated using KNN algorithm. We set the pressure of the left column chambers as 0.8, the radius of the ball as 8 and the mass of the ball as 50. The above box is a reference object. When starting the simulation, the four HPFs curl up at the same time in order to enclasp the whole ball. Each HPF deforms as described in Fig.7. Four HPFs behave like a claw. At first the fingertips touch the ball. As the HPFs are soft, the contact area is adaptive and thus the ball is squeezed smoothly to the top of the hand against the palm (shown in Fig.8). After grasping the ball firmly, then a force is applied to the center of the palm to lift the hand and the ball up. In comparison to our HPF which does not include any rigid components, traditional dexterous robot hands require several rigid joints to perform such deformation. Complex controlling algorithm need also to be designed to achieve the grasp task. We can see that our HPF is self-adaptive when grasping because of its soft and compliant features.

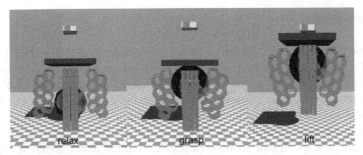

Fig. 8. Virtual hand with soft HPNs performs grasping and lifting

5.3   Deviation Analysis

With the experiments of the mathematical model and physical model above, it's not proper to avoid talking about the deviation between them. After unifying the parameters the displacements of both two models under different inner pressure are shown in Fig.9. The solid lines denote the displacement of mathematical model and the dotted lines denote the displacement of the physical model. The deviation between each pair of displacements can be easily identified.

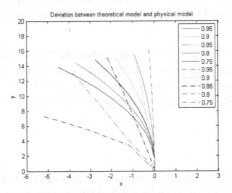

Fig. 9. Deviation statistics

From Fig.9 we can see that the deviation is mainly reflected in Y-axis. That is because in physical simulation HPN is affected by gravity which we ignore in our mathematical model and it tends to extend in Y-axis. When the pressure decreases HPF is mainly affected by its own contractility and the gravity can be ignored relatively. The main reason of the deviation is that L' in equation (14) still declines quickly as the pressure decreases in physical simulation.

6 Conclusion

In this paper, we proposed a design of honeycomb pneumatic finger (HPF) based on the model of HPN. The structure overcomes the shortcoming of embedded rectangular unit. Both mathematical model and physical model are constructed for analysis. We focus on the correlations between the lengths of chamber walls and the inner pressure neglecting external force and internal tensions.

The experiments show that novel HPF has good flexibility and stability without any rigid components. It can achieve various curvatures under different inner pressure. Based on this structure, we construct a virtual hand with one rigid palm and four HPFs which can grasp a ball and lift it up. Our design is confirmed with good application and gives a novel way of simulating soft chambers. With this real-time physical simulation system Bullet, control and planning algorithm of HPF can be easily implemented and tested without fabricating any real objects.

In the future, we still need to reduce the deviation between mathematical model and physical model. Based on our simulation method, more complex deformable shapes can be achieved by linking up more chambers together. With the combinations of different pressure of each chamber leading to different deformation, plenty of controlling and planning algorithm can be implemented for specialized application. Besides, for the purpose of fabricating real HPF, the relationship between parameters in the physical simulation and the properties of real materials need also to be studied.

Acknowledgements. This research is supported by the National Natural Science Foundation of China under grant 61175057 and the Research Fund for the Doctoral Program of Higher Education of China under grant 20133402110026, as well as the USTC Key-Direction Research Fund under grant WK0110000028.

References

1. Lu, Y., Kim, C.J.: Micro-finger articulation by pneumatic parylene balloons. In: 12th International Conference on Transducers, Solid-State Sensors, Actuators and Microsystems, vol. 1, pp. 276–279. IEEE (2003)
2. Trivedi, D., Rahn, C.D., Kier, W.M., et al.: Soft robotics: Biological inspiration, state of the art, and future research. Applied Bionics and Biomechanics 5(3), 99–117 (2008)
3. Kang, B.S., Kothera, C.S., Woods, B.K.S., et al.: Dynamic modeling of Mckibben pneumatic artificial muscles for antagonistic actuation. In: IEEE International Conference on Robotics and Automation, ICRA 2009, pp. 182–187. IEEE (2009)
4. Ilievski, F., Mazzeo, A.D., Shepherd, R.F., et al.: Soft robotics for chemists. Angewandte Chemie 123(8), 1930–1935 (2011)
5. Sun, H., Chen, X.-P.: Towards Honeycomb PneuNets Robots. In: Kim, J.-H., Matson, E., Myung, H., Xu, P. (eds.) Robot Intelligence Technology and Applications 2. AISC, vol. 274, pp. 331–340. Springer, Heidelberg (2014)
6. Morin, S.A., Shepherd, R.F., Kwok, S.W., et al.: Camouflage and display for soft machines. Science 337(6096), 828–832 (2012)

7. Hiller, J., Lipson, H.: Automatic design and manufacture of soft robots. IEEE Transactions on Robotics 28(2), 457–466 (2012)
8. Müller, M., Heidelberger, B., Hennix, M., et al.: Position based dynamics. Journal of Visual Communication and Image Representation 18(2), 109–118 (2007)
9. Glette, K., Hovin, M.: Evolution of artificial muscle-based robotic locomotion in PhysX. In: 2010 IEEE/RSJ International Conference on Intelligent Robots and Systems (IROS), pp. 1114–1119. IEEE (2010)
10. Rieffel, J., Saunders, F., Nadimpalli, S., et al.: Evolving soft robotic locomotion in PhysX. In: Proceedings of the 11th Annual Conference Companion on Genetic and Evolutionary Computation Conference: Late Breaking Papers, pp. 2499–2504. ACM (2009)
11. Coumans, E.: Bullet Physics Engine (2010)
12. Boeing, A., Bräunl, T.: Evaluation of real-time physics simulation systems. In: Proceedings of the 5th International Conference on Computer Graphics and Interactive Techniques in Australia and Southeast Asia, pp. 281–288. ACM (2007)
13. Teran, J., Blemker, S., Hing, V., et al.: Finite volume methods for the simulation of skeletal muscle. In: Proceedings of the 2003 ACM SIGGRAPH/Eurographics Symposium on Computer Animation. Eurographics Association, pp. 68–74 (2003)
14. Rabaetje, R.: Real-time simulation of deformable objects for assembly simulations. In: Proceedings of the Fourth Australasian User Interface Conference on User Interfaces 2003, vol. 18, pp. 57–64. Australian Computer Society, Inc. (2003)
15. Balaniuk, R., Salisbury, K.: Dynamic simulation of deformable objects using the long elements method. In: Proceedings of the 10th Symposium on Haptic Interfaces for Virtual Environment and Teleoperator Systems, HAPTICS 2002, pp. 58–65. IEEE (2002)
16. Costa, I.F., Balaniuk, R.: LEM-An approach for real time physically based soft tissue simulation. In: Proceedings of the IEEE International Conference on Robotics and Automation, ICRA 2001, vol. 3, pp. 2337–2343. IEEE (2001)
17. Provot, X.: Deformation constraints in a mass-spring model to describe rigid cloth behaviour. In: Graphics Interface, pp. 147–147. Canadian Information Processing Society (1995)
18. Maciej, M., Mark, O.: Pressure model of soft body simulation. arXiv preprint physics/0407003 (2004)
19. Lipson, H.: Challenges and Opportunities for Design, Simulation, and Fabrication of Soft Robots. Soft Robotics 1(1), 21–27 (2014)

Towards Coexistence of Human and Robot: How Ubiquitous Computing Can Contribute?

Jingyuan Cheng[1], Xiaoping Chen[2], and Paul Lukowicz[1]

[1] German Research Center for Artificial Intelligence (DFKI),
Trippstadter Strae 122, D-67663, Kaiserslautern, Germany
[2] Multi-Agent Systems Lab, University of Science and Technology of China,
Jinzhai Road 96, Hefei, 230026, China

Abstract. After the ISO 10218-1/2 in 2011, safety factors for industry robot are standardized. As robotics expands its area from industry further into service, educational, healthcare and etc., both human and robot are exposed to a space with more openness and less certainty. Because there is no common safety specification, we raise in this paper our own hypotheses on the safety requirements in dense human-robot co-existing scenarios and focus more on demonstrating the possibilities provided by the research field named Ubiquitous Computing.

1 Introduction

The number of robots over the world keeps on growing. According to the World Robotics studies [1], 159,346 units of industry robots were sold in 2013, 16,067 professional service robots and about 3,000,000 personal and domestic use robots were sold in 2012. As robots' population grows, the physical even emotional contacts between robots and human are also growing, freeing human from certain labor work and meanwhile bringing potential risks. As in early ages robots were implemented mainly in industry, safety specifications have been developed mainly for industrial robots, e.g. ISO 10218-1/2:2011 sets the rule on both robot itself, the robot system and integration [2] [3]; in US the under revision ANSI/RIA R.15.06-2012 re-opens the allowance of human and robot working in a loop. In industry, the general trend shifts from strict isolation of robot from human to detailed specifications on reducing hazards. Out side of industry, however, robots are already in close contact with human, especially in the case of service robot. Since there is no fixed global specification till now, researchers follow their own ideas on whether and how to separate robot from public audience at Expo's and demonstrations. Domestic-use robot providers take care of human by hiding rigid components inside, reducing the robot's weight and speed, and implementing obstacle detecting sensors.

Ubiquitous computing as a fast developing field focuses on pushing the one central powerful computer to multiple little computing units into the environment and onto human beings. This indicates tiny sensing and processing units, and naturally, the application field of environment and human activity monitoring, which are also important factors a robot might need, when comes to dense contact with human beings.

© Springer International Publishing Switzerland 2015
J.-H. Kim et al. (eds.), *Robot Intelligence Technology and Applications 3*,
Advances in Intelligent Systems and Computing 345, DOI: 10.1007/978-3-319-16841-8_39

We perform an initial survey on what exists in Ubiquitous computing and could be used by robotics in this paper. The contribution lies in:

1) we analyze and raise our own hypotheses on the safety requirements in dense human-robot co-existing scenarios;

2) we perform a survey on possibilities provided by Ubiquitous Computing, the merging of which and robotics could potentially support dense human-robot co-existence.

2   Hazards in Dense Human-Robot Co-existing Scenarios

Vasic and etc. gave a detailed survey on safety issues in human-robot interactions [4]. Starting from industry, the danger comes when human gets trapped between robot and an object (e.g. a wall) or when human comes into collision with a robot [5]. A detailed list of significant hazards can be found in ISO 10218-1 as annex, including: Mechanical, electrical, thermal, noise, vibration, radiation, material/substance, ergonomic, the hazards associated with environment and combined hazards. The hazard should be analyzed and minimized from technical points of view, however, in the real applications, there are still unexpected errors and failures which can not be exactly predicted:

- mechanics failure: aging of motors, connectors;
- electronics failure: aging of components and isolation material, out of power half the way of operation;
- program failure: program bugs, untested scenarios;
- operational error: untrained engineers, operators, and users;

Besides regular maintenance, the above listed hazards are minimized in industry applications by:

- strictly pre-defined environment and space (robot cell);
- strictly pre-defined operation routine;
- authorization of properly trained operators, maintenance workers and programers;
- speed limitation when human is present;
- protective stop function and an independent emergency stop function.

While robot go out to factory and into family or other social areas, the above conventional rules become invalid. The situation is similar to that of computer going from military use to civil and then personal use. The difference is however the actuation, the capability of active physical movement brings more potential hazards. Moreover, the robot enters an open environment where changes may happen anytime and anyhow, the users are most often non-professional and unexperienced people. Animals (e.g. pets) might come into close contact with the robot, which might even bring damage to robot. (e.g. a child might see a home service robot and pour water onto it just out of curiosity.) Due to these obstacles, the most sold service robot now is still household robots, which are small in size, carry out comparatively simple and fixed tasks.

In general, to avoid potential damage to human and to itself, a robot in dense coexistence and frequent contact with human needs to know:

Who I am discover identification, including itself, the human being(s) and possible other robot nearby;

Where I am discover context, viz. gather by itself or from environment useful information;

How to work adapt to environment and be able to find the balance between performance and potential hazard level;

How to survive protect first human then itself from damage.

3   Key Factors and Ubiquitous Computing Solutions

Ubiquitous Computing (or in other names: Pervasive Computing, Internet of Things, Ambient Intelligence) is a concept raised by Mark Weiser first in the late 80's [6], "where computing is made to appear everywhere and anywhere." The questions raised above can be mapped into several key factors in Ubiquitous Computing accordingly and listed below.

3.1   Identification

Identification is like a key or a pointer which is linked to further information: parameters of the owner, history and trace, allowance to access databank or use certain resources. The available solution includes:

- **Barcode**: Printed or displayed 1D or 2D machine readable image, with numbers and letters embedded. A camera plus corresponding algorithm can read the information quickly and very reliably. Thanks to the almost neglectable cost (printed on a paper or displayed on a screen), it is widely used in registration system like tickets, good tags in supermarket, book numbering. The code however, must be put to the surface and facing the reader, viz. the tagged object can not be read when it is inside a container or blocked by other objects.
- **Radio-frequency identification (RFID)**: Information transferred wirelessly through electromagnetic fields, the system is composed of a Tag and a reader. The RFID tag where information is stored can be flat, small and bendable. RFID system over-performs barcode in that it doesn't have to be exposed, because electromagnetic field can transmit through most of the material. Active tags with battery enable a higher communication range and passive tags offer a lower cost. RFID tags are used in shipping and on production lines to track goods, or embedded in ID cards for an unique identification of the owner. In research, it is used for indoor-localization [7], to identify which object is being used [8] or to track where the the object is [9].
- **Face recognition and optical character recognition (OCR)**: Barcode and RFID tag need to be attached additionally to object or human, who is willing to show his/her/its identity. In most of the dynamic scenarios (e.g.

in a classroom or restaurant), the up-to-date default setting still doesn't assure most of the people and objects provide his/her/its own identification. Via facial or voice recognition, human can be recognized. With OCR, text printed on an object can be recognized. There are mass amount of research done in all the three directions, even real-time face recognition on wearable device is now possible [10].

3.2   Context

Context, according to Dey's definition [11], is:

"any information that can be used to characterize the situation of an entity. An entity is a person, place, or object that is considered relevant to the interaction between a user and an application, including the user and applications themselves."

Human is able to feel the environment through sensations (vision, sound, balance, touch, smell, taste, temperature, pain), analyze the situation with the brain, store the abstract information, and use the information real-timely or later to improve performance or avoid hazardous situation. For computer or robot, this is not completely straightforward. Depending on the robot's task, some sensation might be unnecessary.

However, one basic requirement to enable coexistence of robot and human is to avoid collision. This is valid for all the service robots, whatever task it has, and can be grouped into three levels:

- a) **Avoid damage when collides**: the robot should slow down or stop when it is already in direct contact with a person.
- b) **Avoid collision**: The robot should avoid entering the void range of a person when planning its movement path, and slow down already before possible collision with human. A broad research on pre-collision safety strategies can be found in [12].
- c) **Avoid secondary damage**: The robot should detect challenging situation and try to avoid it (e.g. rugged floor or running pets), which might results in the robot's falling then unexpected and uncontrolled collision, either from the falling robot or flying away item the robot is carrying (e.g. a cup of hot coffee) d. This also includes decision at critical point, to not collide into a second person when retreating from the collision with the first person.

The key parameters involved here include:

- **Force**: Force can be detected either independently through torque sensing in the joint [13] or through pressure sensitive artificial skin [14]. These techniques stands as the last guard when collision is already happening.
- **Distance and Localization**: It is safer however to stop or change direction already from a distance. Some sensors are already used in distance measuring on robot, e.g. laser and infrared [15], time of flight of sound [16]. Visual information has already been used in the last century for building an environment map [17].

There are many other localization methods developed in Ubiquitous computing: GPS is already a common means for outdoor localization. Indoor localization using time of arrival of ultrawideband (UWB) signal [18] enjoys a very high precision but is limited to simple room setup, because reflections from furnitures and people mess up the original signal. WIFI based indoor localization locates the user by matching the local signal strength to a pre-built signal strength map [19]. Inertial unit (accelerometer and gyroscope) combined with WIFI signals and GPS (for a concrete coordinate when entering and leaving the building) can be used to further improve in-door map and localization precision [20] [21]. Magnetic coupling sensor replaces the map of WIFI signal with a field generated by the coils at a certain frequency, which is hardly influenced by normal environment, thus is very robust [22]. Zhou and etc. proposed a large area high spacial precision pressure sensing matrix, which may provide a real-time obstacle map [23]. Whereas capacitive sensing have been used as touchless input interface [24], it can be used as an emergent trigger when the robot enters into void range of human (viz. cm level). 3-D motion input device based on multiple infrared projector(s) and camera(s) are used for gesture control with limbs and fingerssong2008vision, they can be used for distance measuring between robot and human, too. Also robot itself is used to build up indoor/outdoor maps [25].

– **Warning**: If the robot is not able to avoid collision by itself, it should warn human beings which might be involved in or influenced by the collision, either through vibration, sound of a wearable device, or through audio or visual warning from the robot itself or in the environment.

Beside the key techniques listed above, in ubiquitous and wearable computing, there are already plenty of sensor systems and data mining algorithms designed for environment monitoring and human activity recognition. When multiple sensors are in use, fusion could be implemented to achieve higher precision and to avoid system level fail due to fail of single sensor(s) [26]. There is also research in wireless sensor network [27], to enable connecting distinct sensors into a net with low-power, small-size solutions. The question left here is how to transfer the information to a robot.

4   Conclusion

We analyzed in this paper the hazards in dense human-robot co-existing scenario, which, with the fast development and population growth of service robots, will come in the future sooner or later. We gave our hypotheses on the hazards and perform a survey on existing techniques in the research field named Ubiquitous Computing, which could help minimize the hazards. This paper is supposed to serve as a starting-point for supporting robotics development with ubiquitous computing.

Acknowledgement. This work was partially supported by the collaborative project SimpleSkin under contract with the European Commission (#323849) in the FP7 FET Open framework. The support is gratefully acknowledged.

References

1. World robotics industrial robots 2013 - summary; service robots 2013 - summary. Technical report, IFR Statistical Department, VDMA Robotics and Automation association (2013), http://www.worldrobotics.org/uploads/media/Executive_Summary_WR_2013.pdf
2. ISO10218-1:2011: Robots and robotic devices – safety requirements for industrial robots, part 1: Robots (2011)
3. ISO10218-2:2011: Robots and robotic devices – safety requirements for industrial robots, part 2: Robot systems and integration (2011)
4. Vasic, M., Billard, A.: Safety issues in human-robot interactions. In: 2013 IEEE International Conference on Robotics and Automation (ICRA), pp. 197–204. IEEE (2013)
5. Jiang, B.C., Gainer Jr., C.A.: A cause-and-effect analysis of robot accidents. Journal of Occupational Accidents 9(1), 27–45 (1987)
6. Weiser, M.: The computer for the 21st century. Scientific American 265(3), 94–104 (1991)
7. Sanpechuda, T., Kovavisaruch, L.: A review of rfid localization: Applications and techniques. In: 5th International Conference on Electrical Engineering/Electronics, Computer, Telecommunications and Information Technology, ECTI-CON 2008, vol. 2, pp. 769–772. IEEE (2008)
8. Mantyjarvi, J., Paternò, F., Salvador, Z., Santoro, C.: Scan and tilt: towards natural interaction for mobile museum guides. In: Proceedings of the 8th Conference on Human-Computer Interaction with Mobile Devices and Services, pp. 191–194. ACM (2006)
9. Mamei, M., Zambonelli, F.: Pervasive pheromone-based interaction with rfid tags. ACM Transactions on Autonomous and Adaptive Systems (TAAS) 2(2), 4 (2007)
10. Utsumi, Y., Kato, Y., Kunze, K., Iwamura, M., Kise, K.: Who are you?: A wearable face recognition system to support human memory. In: Proceedings of the 4th Augmented Human International Conference, pp. 150–153. ACM (2013)
11. Dey, A.K.: Understanding and using context. Personal and Ubiquitous Computing 5(1), 4–7 (2001)
12. Arkin, R.C.: Homeostatic control for a mobile robot: Dynamic replanning in hazardous environments. Journal of Robotic Systems 9(2), 197–214 (1992)
13. Haddadin, S., Albu-Schäffer, A., Hirzinger, G.: Safety evaluation of physical human-robot interaction via crash-testing. In: Robotics: Science and Systems, vol. 3, pp. 217–224 (2007)
14. Cannata, G., Maggiali, M., Metta, G., Sandini, G.: An embedded artificial skin for humanoid robots. In: IEEE International Conference on Multisensor Fusion and Integration for Intelligent Systems, MFI 2008, pp. 434–438. IEEE (2008)
15. Graf, B., Hägele, M.: Dependable interaction with an intelligent home care robot. In: Proceedings of ICRA-Workshop on Technical Challenge for Dependable Robots in Human Environments, pp. 21–26 (2001)
16. Schmitz, N., Spranger, C., Berns, K.: 3d audio perception system for humanoid robots. In: Second International Conferences on Advances in Computer-Human Interactions, ACHI 2009, pp. 181–186. IEEE (2009)
17. DeSouza, G.N., Kak, A.C.: Vision for mobile robot navigation: A survey. IEEE Transactions on Pattern Analysis and Machine Intelligence 24(2), 237–267 (2002)
18. Alavi, B., Pahlavan, K.: Modeling of the toa-based distance measurement error using uwb indoor radio measurements. IEEE Communications Letters 10(4), 275–277 (2006)

19. Chintalapudi, K., Padmanabha Iyer, A., Padmanabhan, V.N.: Indoor localization without the pain. In: Proceedings of the Sixteenth Annual International Conference on Mobile Computing and Networking, pp. 173–184. ACM (2010)
20. Evennou, F., Marx, F.: Advanced integration of wifi and inertial navigation systems for indoor mobile positioning. Eurasip Journal on Applied Signal Processing 2006, 164–164 (2006)
21. Leppäkoski, H., Collin, J., Takala, J.: Pedestrian navigation based on inertial sensors, indoor map, and wlan signals. Journal of Signal Processing Systems 71(3), 287–296 (2013)
22. Pirkl, G., Lukowicz, P.: Robust, low cost indoor positioning using magnetic resonant coupling. In: Proceedings of the 2012 ACM Conference on Ubiquitous Computing, pp. 431–440. ACM (2012)
23. Zhou, B., Cheng, J., Sundholm, M., Lukowicz, P.: From smart clothing to smart table cloth: Design and implementation of a large scale, textile pressure matrix sensor. In: Maehle, E., Römer, K., Karl, W., Tovar, E. (eds.) ARCS 2014. LNCS, vol. 8350, pp. 159–170. Springer, Heidelberg (2014)
24. Cheng, J., Bannach, D., Adamer, K., Bernreiter, T., Lukowicz, P.: A wearable, conductive textile based user interface for hospital ward rounds document access. In: Roggen, D., Lombriser, C., Tröster, G., Kortuem, G., Havinga, P. (eds.) EuroSSC 2008. LNCS, vol. 5279, pp. 182–191. Springer, Heidelberg (2008)
25. Thrun, S., Burgard, W., Fox, D.: A real-time algorithm for mobile robot mapping with applications to multi-robot and 3d mapping. In: Proceedings of the IEEE International Conference on Robotics and Automation, ICRA 2000, vol. 1, pp. 321–328. IEEE (2000)
26. Brooks, R.R., Iyengar, S.S.: Multi-sensor fusion: fundamentals and applications with software. Prentice-Hall, Inc. (1998)
27. Yick, J., Mukherjee, B., Ghosal, D.: Wireless sensor network survey. Computer Networks 52(12), 2292–2330 (2008)

Electronic Artificial Skin
for Application in Pressure Sensor

Yumao Gu [1], Yuanzhen Dai[1], Yang Liu[2], and Xiaoping Chen[2]

[1] School of Physical Sciences, University of Science and Technology of China,
No.96, JinZhai Road Baohe District, Hefei, Anhui, 230026,P.R.China
[2] School of Computer Science and Technology, University of Science and Technology of China,
443 Huangshan Road, Hefei, Anhui, 230027, The People's Republic of China

Abstract. Combining with the cognition of human skin, a kind of flexible, stretchable large-area electronic device network (electronic skin) have been demonstrated for the detection of deformation and pressure, which allow robots to maneuver within environment safely and effectively. Through the application of soft technology, we potentially reduce the mechanical and algorithmic complexity involved in robot design. In our experiments we used soft materials coating AgNWs and the FFT algorithms to measure pixel pressure. Polydimethylsiloxane (PDMS) has overwhelmingly been used as substrates of E-skin and other stretchable electronics. Arduino is also used in our experiment as a single-board microcontroller, intended to make the application of interactive objects or environments more accessible.

1 Introduction

With the development of robot technology, we have increasingly found the underlying mechanical properties (shape, flexibility, stretchability, etc.) are particularly important in some applications. The robot system currently uses flat and rigid wafer-based electronics, which are intrinsically incompatible with curvilinear and deformable organisms. Therefore, the flexible material and mechanical strategies used in soft robotics are urgently needed. [1-3]As a biological blueprint of flexible material, human skin are widespread concerned by scientists. Future robots wearing such a flexible pressure sensitive skin can "feel" a touch or pressure. Being stretchable and able to detect the pressure or deformation at the same time would allow for feedback control. It can help realize a plethora of user-friendly applications.

To date, the different types of electronic skin have been manufactured. Through the use of low cost chemical and biosensor technology and the integration of chemical and biological sensors, gas and liquid phase analytes[4-10] have been demonstrated on rigid substrates using CNTs, graphene, inorganic NWs, and organic materials as the active sensor element. There are many electronic skin can sense temperature [11-15]. Moreover, some others have additional features, such as self-healing [16-19],and energy storage[20-28].

Here, we will focus on detecting pressure. Efficacious tactile sensing arrays could be used in designing robots that can collect information about their surroundings to make right responses. The robots can do more complicated work.

© Springer International Publishing Switzerland 2015

J.-H. Kim et al. (eds.), *Robot Intelligence Technology and Applications 3*,

Advances in Intelligent Systems and Computing 345, DOI: 10.1007/978-3-319-16841-8_40

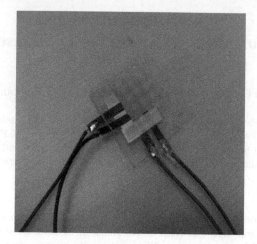

Fig. 1. A simple example of the proposed model

2   The Model

Flexible transparent capacitive sensors can be demonstrated for the detection of deformation pressure. Electric capacity would be affected by several variables, such as area, dielectric and thickness. But a fixed value of the capacity means fixed value impedance. So we can use square waves which are easy to get in digital circuits, to measure the capacity, influenced by different pressure.

2.1   Capacity

Capacitance is the ability to accommodate the electric field, namely the charge given to the reserves under potential difference. As the thickness of our sensor is low combined with the area of each element, it can be seen as parallel plate capacitor, and the edge effects can be ignored. So the approximate formula of the capacity is $C=\varepsilon S/4\pi kd$. With pressure on the sensor, εd is obviously changed, for the dielectric in it is stretchable and flexible. When the pressure is increased, d is lower, and ε changes too. Normally S won't change. But in some linear materials, S may also change when the pressure is on especial aspects.

2.2   Impedance and Square Waves

Impedance of capacitive sensors can be calculated by $Xc=1/\omega c$. So sinusoidal AC's passing through will be influenced by capacity.

A square wave is composted of several sine waves. And we all know that square waves in digital circuits are readily available. Therefore we end of the square wave signal to the sensor, the measurement signal at the other end, and the decomposition of a particular frequency from a sine wave. By observing the peak value of the sine wave, we can confirm its corresponding pressure. Decomposition of signals is approached by Fourier analysis. So we could get the spectrum of the signal.

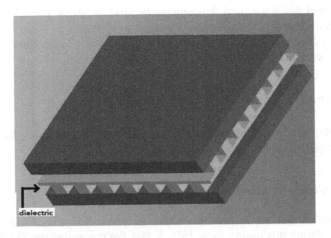

Fig. 2. Structure of the Flexible transparent capacitive sensors

2.3   Material

As materials for constituting E-skin, one defining characteristic is mechanically compliant. That is, the materials can be allowing flex and stretch without physic damage. Polydimethylsiloxane (PDMS) has overwhelmingly been used as substrates of E-skin and other stretchable electronics, for its commercial availability and well-researched properties.[29] For dielectrics, we should choose materials of different purpose. CNTs, grapheme [30] and nanowires [31] (NWs) are good choices for conductor because they have well chemical stability as well as high charge carrier mobility (μ) .

Many groups have made efforts to simultaneously achieve high stretchability, high conductivity, high transparency, and a low-cost large-area fabrication.[32]In our experiments, we use a high performance transparent electrode by embedding an ultra thin silver nanowire (AgNW) network in the surface layer of PDMS.

We first prepared banding AgNWs coating on glass. AgNWs are dissolved in water with a concentration of 1mg/ml.An average diameter is 60 nm. We use a spin-coating method with the speed of 600rpm/min. AgNWs in the network are uniformly distributed and randomly oriented. We can control the times of spin-coating to control the thickness of the nanowires. We have found that the conductivity of the tape increases when the thickness increases, but the transparency weak. For the fabrication of capacitive sensor arrays, AgNW coatings were deposited through a contact mask.

Then liquid PDMS were degassed, and drop cast over the AgNW coatings on glass substrate. A second sheet of glass was pressed onto substrate to control the thickness of the PDMS layer. The PDMS can infiltrate the pores in the network. After curing, the AgNW network is transferred from the original glass substrate and embedded in the surface of the PDMS. Thus, we got flexible electrode with high conductivity.

We used several kinds of materials as the dielectric spacer to increase sensitivity toward small pressures, such as thin PDMS film,3M Scotch 924 ATG Tape and even pan paper. Different characteristic responses of a sensor pixel in capacitive sensor arrays occurred when we used those dielectric spacers. Areas coated with nanowires are connected to the signal source via copper tapes.

When the electrodes in good contact with the dielectric layer, the device is able to measure the pressure like touching.

3   Method

To reduce the cost as well as to improve accuracy, we use modules those are easy to get to accomplish signal's generation and acquisition.

3.1   Fast Fourier Transform

Signal waves we get are record by discrete numbers. So we can use DFT (discrete Fourier transform) processing data. FFT A fast Fourier transform (FFT) is an algorithm to compute the discrete Fourier transform (DFT) and it's inverse. Fourier analysis converts time (or space) to frequency and vice versa; an FFT rapidly computes such transformations by factorizing the DFT matrix into a product of sparse (mostly zero) factors. [33] It has been described as "the most important numerical algorithm of our lifetime". [34]

Each square wave, we read 8 data. And a set of data includes 512 values to reduce deviation. And we just need the amplitude of a special frequency.

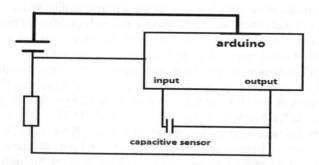

Fig. 3. A simple model of signal acquisition circuit

3.2   Arduino and Signal Acquisition

Arduino is a single-board microcontroller, intended to make the application of interactive objects or environments more accessible. [35] We use it for square waves' generation and signal acquisition. With 10-bit analog to digital converter, it yields a resolution between readings of: 5 volts / 1024 units or, .0049 volts (4.9 mV) per unit. So we use a 1-M-Ω-pull-down resistor to reduce input impedance

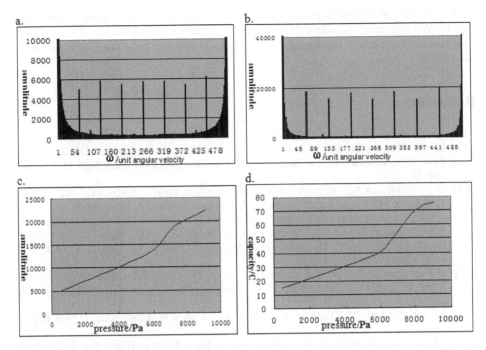

Fig. 4. Data from E-skin with PDMS as dielectric spacer.   a. Result without pressure; b. Result with pressure of 10kpa; c. Our system changes obviously when pressure improves from 1kpa to 10kpa; d. This is a direct result of capacitance meter with high accuracy measurement.

4   Data and Discussion

As the pressure increases from 1kpa 10kpa, we can see that the amplitude of several feature points increased by about 3.7 times, and not decreased in the middle range, which means that we can measure the capacitance sensor is at least as small 1kpa's pressure.

Meanwhile, even more exciting is that with our system can measure at very low cost small capacitance (pF). Compared capacitance table to solve the problem of measuring small capacitance from the process, we have optimized the measurement method, the process of difficult to resolve in a way the computer calculated the same time, also has good tensile properties, can still work in the stretch 50 percent of cases.

5   Conclusion

We fabricated a kind of E-skin by employing composite electrodes and dielectric spacer which exhibits good stretchability and sensitivity. The sheets have a relatively simple, solid-state sandwich structure. The capacitive sensors can be pressed, bent, or stretched and work well.

We use FFT algorithm to process signals so that we can better combine E-skin with robots. We measured impedance rather than capacitance. It is in favor of the acquisition and the feedback of the signals. E-skin must eventually have wide applications in the future.

References

[1] Kim, D.H., Lu, N.S., Ghaffari, R., Rogers, J.A.: NPG Asia Mater. 4, 15 (2012)
[2] Kim, D.H., Lu, N.S., Huang, Y.G., Rogers, J.A.: MRS Bull. 37, 226–235 (2012)
[3] Kim, D.H., Ghaffari, R., Lu, N.S., Rogers, J.A.: Annu. Rev. Biomed. Eng. 14, 113–128 (2012)
[4] a) Sokolov, A.N., Roberts, M.E., Bao, Z.A.: Mater. Today 12, 12 (2009); b) Roberts, M.E., Sokolov, A.N., Bao, Z.N.: J. Mater. Chem. 19, 3351 (2009)
[5] a) Sun, Y., Wang, H.H.: Appl. Phys. Lett. 90, 213107 (2007); b) Su, P.-G., Chuang, Y.-S.: Sens. Actuators, B 145, 521 (2010); c) Sun, Y., Wang, H.H.: Adv. Mater. 19, 2818 (2007)
[6] Fu, D., Lim, H., Shi, Y., Dong, X., Mhaisalkar, S.G., Chen, Y., Moochhala, S., Li, L.-J.: J. Phys. Chem. 112, 650 (2008)
[7] a) Lee, C., Ahn, J., Lee, K.B., Kim, D., Kim, J.: Thin Solid Films 520, 5459 (2012); b) Yang, G., Lee, C., Kim, J., Ren, F., Pearton, S.J.: Phys. Chem. Chem. Phys. 15, 1798 (2013)
[8] Yavari, F., Chen, Z.P., Thomas, A.V., Ren, W.C., Cheng, H.M., Koratkar, N.: Sci. Rep. 1, 1 (2011)
[9] Chung, M.G., Kim, D.-H., Seo, D.K., Kim, T., Im, H.U., Lee, H.M., Yoo, J.-B., Hong, S.-H., Kang, T.J., Kim, Y.H.: Sens. Actuators, B 169, 387 (2012)
[10] Bui, M.P.N., Pham, X.H., Nan, K.N., Li, C.A., Kim, Y.S., Seong, G.H.: Sens. Actuators, B 150, 436 (2010)
[11] Lee, C.-Y., Wu, G.-W., Hsieh, W.-J.: Sens. Actuators, A 147, 173 (2008)
[12] Yu, C., Wang, Z., Yu, H., Jiang, H.: Appl. Phys. Lett. 95, 141912 (2009)
[13] Jeon, J., Lee, H.-B.-R., Bao, Z.: Adv. Mater. 25, 850 (2013)
[14] a) Zhang, C., Ma, C.A., Wang, P., Sumita, M.: Carbon 43, 2544 (2005); b) Martin, J.E., Anderson, R.A., Odinek, J., Adolf, D., Williamson, J.: Phys. Rev. B 67, 094207 (2003)
[15] Zha, J.-W., Li, W.-K., Liao, R.-J., Bai, J., Dang, Z.-M.: J. Mater. Chem. A 1, 843 (2013)
[16] a) Blaiszik, B.J., Kramer, S.L.B., Grady, M.E., McIlroy, D.A., Moore, J.S., Sottos, N.R., White, S.R.: Adv. Mater. 24, 398 (2012); b) Odom, S.A., Chayanupatkul, S., Blaiszik, B.J., Zhao, O., Jackson, A.C., Braun, P.V., Sottos, N.R., White, S.R., Moore, J.S.: Adv. Mater. 24, 2578 (2012)
[17] a) Mynar, J.L., Aida, T.: Nature 451, 895 (2008); b) Cordier, P., Tournilhac, F., Soulie-Ziakovic, C., Leibler, L.: Nature 451, 977 (2008)
[18] Li, Y., Chen, S., Wu, M., Sun, J.: Adv. Mater. 24, 4578 (2012)
[19] a) Tee, B.C.-K., Wang, C., Allen, R., Bao, Z.: Nat. Nanotech. 7, 825 (2012); b) Gong, C., Liang, J., Hu, W., Niu, X., Ma, S., Hahn, H.T., Pei, Q.: Adv. Mater. 2013, 30 (2013)
[20] a) Yu, C., Masarapu, C., Rong, J., Wei, B., Jiang, H.: Adv. Mater. 21, 4793 (2009); b) Li, X., Gu, T., Wei, B.: Nano Lett. 12, 6366 (2012)
[21] Niu, Z., Dong, H., Zhu, B., Li, J., Hng, H.H., Zhou, W., Chen, X., Xie, S.: Adv. Mater. 25, 1058 (2013)
[22] Zhao, Y., Liu, J., Hu, Y., Cheng, H., Hu, C., Jiang, C., Jiang, L., Cao, A., Qu, L.: Adv. Mater. 25, 591 (2013)
[23] Gaikwad, A.M., Zamarayeva, A.M., Rousseau, J., Chu, H., Derin, I., Steingart, D.A.: Adv. Mater. 24, 5071 (2012)

[24] a) Lee, H., Yoo, J.-K., Park, J.-H., Kim, J.H., Kang, K., Jung, Y.S.: Adv. Energy Mater. 2, 976 (2012); b) Wang, C., Zheng, W., Yue, Z., Too, C.O., Wallace, G.G.: Adv. Mater. 23, 3580 (2011)

[25] Xu, S., Zhang, Y., Cho, J., Lee, J., Huang, X., Jia, L., Fan, J.A., Su, Y., Su, J., Zhang, H., Cheng, H., Lu, B., Yu, C., Chuang, C., Kim, T.-I., Song, T., Shigeta, K., Kang, S., Dagdeviren, C., Petrov, I., Braun, P.V., Huang, Y., Paik, U., Rogers, J.A.: Nat. Commun. 4, 1543 (2013)

[26] Kim, D.-H., Lu, N., Ma, R., Kim, Y.-S., Kim, R.-H., Wang, S., Wu, J., Won, S.M., Tao, H., Islam, A., Yu, K.J., Kim, T.-I., Chowdhury, R., Ying, M., Xu, L., Li, M., Chung, H.-J., Keum, H., McCormick, M., Liu, P., Zhang, Y.-W., Omenetto, F.G., Huang, Y., Coleman, T., Rogers, J.A.: Science 333, 838 (2011)

[27] Arriola, A., Sancho, J.I., Brebels, S., Gonzalez, M., De Raedt, W.: IET Microwaves Antennas Prop. 5, 812 (2011)

[28] a) Kaltenbrunner, M., Kettlgruber, G., Siket, C., Schwödiauer, R., Bauer, S.: Adv. Mater. 22, 2065 (2010); b) Kettlgruber, G., Kaltenbrunner, M., Siket, C.M., Moser, R., Graz, I.M., Schwödiauer, R., Bauer, S.: J. Mater. Chem. A 1, 5505 (2013)

[29] Sun, Y., Rogers, J.A.: J. Mater. Chem. 17, 832 (2007)

[30] a) Small, W.R., in het Panhuis, M.: Small 3, 1500 (2007); b) Kordas, K., Mustonen, T., Toth, G., Jantunen, H., Lajunen, M., Soldano, C., Talapatra, S., Kar, S., Vajtai, R., Ajayan, P.M.: Small 2, 1021 (2006)

[31] a) Takei, K., Takahashi, T., Ho, J.C., Ko, H., Gillies, A.G., Leu, P.W., Fearing, R.S., Javey, A.: Nat. Mater. 9, 821 (2010); b) Wu, W., Wen, X., Wang, Z.L.: Science 340, 952 (2013)

[32] Hu, W., Niu, X., Zhao, R., Pei, Q.: Appl. Phys. Lett. 102, 083303 (2013)

[33] Van Loan, C.: Computational Frameworks for the Fast Fourier Transform. SIAM (1992)

[34] Gilbert, S.: Wavelets. American Scientist 82(3), 253 (May-June 1994) (retrieved October 8, 2013)

[35] Official slogan. Arduino Project (retrieved December 31, 2013)

An Intelligent Rover Design Integrated with Humanoid Robot for Alien Planet Exploration

S. Aswath[1], Nitin Ajithkumar[1], Chinmaya Krishna Tilak[1], Nihil Saboo[1],
Amal Suresh[1], Raviteja Kamalapuram[1], Anoop Mattathil[1], H. Anirudh[1],
Arjun B. Krishnan[1], and Ganesha Udupa[2]

[1] Department of Electronics and Communication Engineering,
Amrita Vishwa Vidyapeetham, Kollam- 690525, Kerala, India
[2] Department of Mechanical Engineering, Amrita Vishwa Vidyapeetham,
Kollam- 690525, Kerala, India

Abstract. This paper describes an innovative approach to the design and implementation of a Rover intended for alien environments. The paper enumerates
the various issues faced by the Rover in such an environment and attempts to
solve each of them using innovative design modifications. Although built from
existing designs, the Rover has been modified for better stability, higher order
functioning and improved debugging mechanisms. The Rover features a flexible segmented body with a multipurpose arm, corresponding multi wheel mechanism and *Kinect* module integration for advanced image processing. The
system control, for both the Rover as well the robotic arm integrated with it, is
done using feasible yet extremely efficient microcontrollers and microprocessors such as Arduino, Raspberry pi etc. Inspired from nature's design, a reflex
mechanism has also been integrated into the Rover design, to minimize damage
by automated safety reflexes. The rover also carries a mobile humanoid robot
for more precise human like investigation. The Rover finds applications in the
exploration of other planets, deep sea vents and other hostile environments. It
offers a possibility of integrating numerous features such mineral collection and
sampling, landscape mapping, moisture detection etc. Such an effort may even
prove to be instrumental in detection and study of biological activity in worlds
other than ours.

Keywords: Rover, *Kinect*, Microcontroller, Microprocessor, Humanoid Robot.

1 Introduction

The human body, with its limited form and senses, can only explore a small portion of
the physical world. However this limitation is a handicap to the infinite potential of
the human intellect, which strives to discover the ends of the world. Technology has
more or less removed this limitation by allowing man to explore to the deepest seas
and the farthest galaxies. This paper deals with one such technological advancement.
Technology has come a long way since the first rovers were designed for deep sea
explorations. Today most of the missions to deep sea vents, distant galaxies and

© Springer International Publishing Switzerland 2015   441
J.-H. Kim et al. (eds.), *Robot Intelligence Technology and Applications 3,*
Advances in Intelligent Systems and Computing 345, DOI: 10.1007/978-3-319-16841-8_41

neighboring planets are handles using multifunctional Rover bots. The earliest promi- nent rover, *Mars Pathfinder,* was a U.S. spacecraft that landed on mars with a roving probe on

Mars on July 4, 1997.The mission carried a series of scientific instruments to ana- lyze the Martian atmosphere, climate, geology and the composition of its rocks and soil. Sojourner, the Mars Pathfinder rover, made observation that raise and answer questions about the origins of the rocks and other deposits at the Ares site of Mars. The *Opportunity* successfully investigated soil and rock samples and taken panoramic photos of its landing site. The sampling technology used in this rover allowed NASA scientists to make hypotheses concerning the presence of hematite and past presence of water on the surface of Mars. *Curiosity,* another historic milestone in the history of rover technology, was assigned the role to investigate Martian climate and geology. The assessment of whether the selected field site inside Gale Crater had ever offered environmental conditions favorable for microbial life, investigation of the role of water; and planetary habitability studies in preparation for future human exploration were also achieved in this exploration [1-3]. In these types of rovers integration of Humanoid robot for multifunctional activities are not studied.

This paper proposes a Rover system design which is suitable for all terrains. The design consists of a segmented structure with six employable wheels, addressing the need of flexibility in high risk missions. The rover has a worm like structure which allows it to move across most terrains smoothly and efficiently. One key element that makes this rover design unique is its *Kinect* module integration. The *Kinect* allows a much more accurate and detailed visual data analysis when compared to other devic- es. This is of particular use in this design. To address the various image processing needs of the rover, it is fitted with the state of the art *Kinect* module which acts as eyes of the rover. It processes visual data regarding its immediate environment and using an algorithm to control the suspension of the wheels. This helps to provide both improved protection to the payload as well as efficient tractability across the terrain.

The rover is integrated with a robotic arm which is controlled from the station us- ing a *Kinect* sensor. This enables the handling of objects in the environment greatly increasing the functionality of the rover system. This arm is used to remove small obstacles and collect samples depending on the need of the mission.

To enable the rover to function in a semi-autonomous mode a number of sensors have been used in the system. To avoid catastrophic damage to the rover due to wide- ly varying terrains, an ultra-sonic sensor is used. Much like the effect of a reflex action in the human body, the sensor detects the anomaly and sends the data to the microcontrollers for stopping the rover and averting any danger.

The mechanical system that controls the rover's motor system is quite complex but extremely efficient. The array of sensors combined with system containing a number of commonly available microcontrollers and microprocessors, makes the movement and the working of the rover extremely accurate and comfortably feasible. The motor drivers and the sensors used are interfaced with microcontrollers like the Arduino and is further controlled using state of the art microprocessors like Raspberry pi. The Rover has been design with a semi-autonomous control which allows for fast error correction as well as efficient control. This papers describes both the overall working as well as that of the individual subsystems present in the rover. The humanoid robot on board the rover, adds up to particularly remarkable feature like wider range of exploration, longer life span of operation, greater range and greater accuracy in the experiments and tests conducted during the project.

2   Design Overview

The rover design consists of the following subsystems: - sensor system, humanoid robot system, communication system, robotic arm system, power system, and motor drive system. The diagram below, shows how each subsystem is connected to each other in its functionality. All the subsystems are required for the full functionality of rover on the alien planet. The Raspberry Pi controls all the Arduino, which further controls the other subsystem of the rover like sensory, power, drive etc. The design was optimized in such a way that despite its dependency, the errors in one subsystem will not result in the breakdown of the whole rover system. This allows for a greater probability in data protection and retrieval of rover after the expiry of the missions allotted to it. The figure 1 shows the complete block diagram of the rover system.

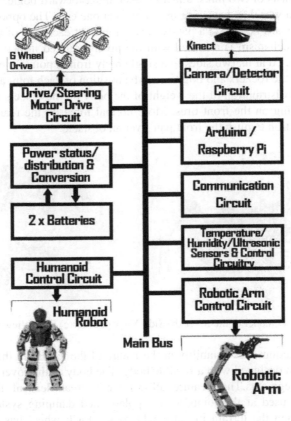

Fig. 1. Rover System Block Diagram

3   Structural Design - Rocker Bogie Mechanism

The degree of mobility of the rover can be judged by the ability of the rover to surmount obstacles whose size is large when compared to the size of the wheels. The rover must be designed in such a way that the rear wheels provide enough power and traction to drive the front wheels against and over the obstacle. Ordinarily a 4 wheeled rover cannot climb over obstacles greater than the tire size simply because the traction required for it cannot be attained. The lack of traction means that there is not enough force required to keep the front tires in contact with the obstacle while forward thrust is given. This necessitates the need for a system that allows such movement. The rocker bogie suspension allows the rover to move over obstacles whose sizes are significantly larger than the wheel diameter because it makes use of an extra set of wheels to provide greater forward thrust. The extra set of wheels also divides the traction force required from each wheels to 1/6th the total value. Moreover reduced forward thrust is required because the front wheels only need to lift 1/3 of the total weight of the rover. Acting together the rear four wheels provide enough traction to keep the rover from slipping.

Each side consists of two links, a main rocker and a forward bogie. The main rocker is attached to a wheel and steering mechanism at one end. The opposite end is connected to the forward bogie via a passive pivot joint. At each end of the forward bogie, a steering mechanism is attached with the pivot mounted in-between. The suspension is connected at its two sides to a single body from a point on each main rocker. The length of the rockers and bogies and the position of each joint are optimized to provide the ideal distribution of the weight of the body on the wheels with the lowest normal force acting on the front tires. More normal force on the rear wheels, translates to more traction to push the front pair over an obstacle.

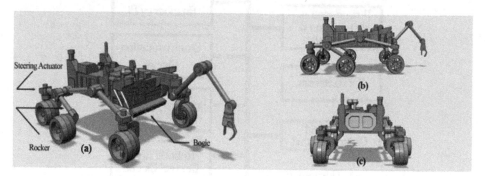

Fig. 2. (a)3D model of the rover (b) Side-View of rover(c) Rear-View of rover

Taking into account the variability in the nature of the terrain in the alien planets, the rover has been designed with a flexible body. The body of the rover is divided into three different segments.This feature allows for a great deal of flexibility. The torsional springs used at the joints as a supplemented damping system. This is of particular use when the terrain in extremely rocky. Each wheel has been given an independent suspension attaching it to the main frame of rover. This acts as the primary damping system of the rover.

The rover body is made from lightweight Aluminum metal. The figure 2(a) show the 3D model of the rover with 2(b) and 2 (c) shows side and rear view. An Al alloy is used in the body. The passive linkage system made up of square Al alloy tubing. The wheel and mounting fixtures are welded at appropriate places. There are no elastic elements in the rover bogie suspension system except for the wheels. A two wheeled rocker arm is used on a passive pivot attached to a main bogie that is connected differentially to the main bogie on the other side. The body of the rover is attached to the differential. This means that it gets suspended at an angle that is the average of the two sides. By passing on a portion of a wheel's displacement to the main bogie the ride is further smoothened by the rocker. Each wheel is driven and steered independently. To eliminate as many dynamic effects as possible, the maximum speed of these rovers are limited. This also allows that the motors to be geared down so that the wheels can individually lift a large portion of the entire vehicle's mass.

The structural design ie. The rocker bogie mechanism, of the rover has been adapted from previous versions of the rover designs. However in this particular design, we introduce a segmentation of the overall body, thus offering it must better robustness and flexibility when compared to previous designs.

4   Motor Control and Drive System

The proposed design of the rover has been equipped with a total of six independently driven wheels. Each wheel will have its own drive motor. The use of six wheel drive will assure that the rover keeps in contact with the ground even in the case of rough terrain. Individual motors will be chosen to simplify the drivetrain design and reduce the number of mechanical parts needed to run the system efficiently. Additionally this approach makes the rover less vulnerable to damage and offers a greater chance of recovery in the event of motor failure. The steering mechanism will be made applicable for the front and rear wheel. They will also work in conjunction with variable speed mechanisms. The figure 3 illustrates the concept of driving the rover motors.

PLANETARY MOTOR   PHOENIX SERIES 24V/30A   ARDUINO MEGA

Fig. 3. The pictorial representation of motor driving

In this particular design, planetary DC motors are to be used because they consume less power, while providing the required torque. This is a great advantage when compared to other three phase AC motors which require high power. The PWM control is achieved using the Arduino Mega microcontrollers. It is interfaced with the Phoenix Series 24V/30A motor driver for variable speed control. The PWM control method makes use of optimization of the width of pulses within a pulse train to control the speed of the motor. The PWM works based on duty cycle - Higher the duty cycle greater will be the motor speed. The speed control mechanism gives the rover the ability to safely maneuver through relatively confined spaces. The motor driver also acts as an H-Bridge, which has the ability to change the direction of the motor in both the forward and reverse direction. The H-Bridge circuitry is shown in figure 4.

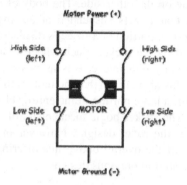

Fig. 4. The H-Bridge circuitry

Using planetry motor datasheet, calculation for the motor requirement were made based on the entire rover system.Adding all the torque acting on the system produce the following equation [4]

$$T_{system} = J_L.\alpha + T_L + T_{BR} + T_{ACC}$$

T_{system} =Total required torque
T_{ACC}= Acceleration torque of the system
T_{BR}=Breakaway torque of the motor
T_L=Torque required by the load
J_L= Moment of intertie of the load
α =Angular acceleration of the system

Taking into account the number of wheels producing the torque, the torque needed from each wheel can be found by dividing T_{system} by six.This shows that the torque per wheel must be 1/6 of T_{system}, which is the reuired torque of the motor.

5 Humanoid Robot System

The humanoid robot on board the rover, adds up to particularly remarkable feature. It essentially makes the exploration more human-like than ever before. The availability of the humanoid bot opens up a lot of opportunities for a wider range of exploration, longer life span of operation, greater range and greater accuracy in the experiments and tests conducted during the project. In this particular rover design, the Arduino board is interfaced with the humanoid robot CM-530 controller via the Zigbee module. Another Zigbee which as acts the receiver is attached to the Humanoid controller. The binary information sent from the Arduino is received at the humanoid controller via the Zigbee wireless interface.

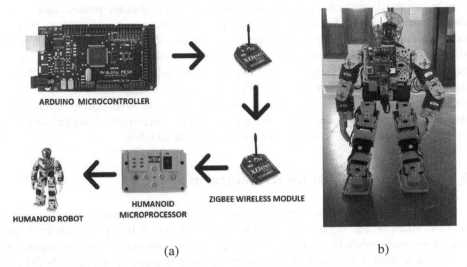

(a)   b)

Fig. 5. (a) The pictorial representation of humanoid robot working (b) The prototype of humanoid robot

The humanoid robot constitutes 18 high torque (15 kgcm) dynamical servo motors, which are positioned at various joints of the humanoid robot. The servo motors are arranged in such a way that the combined motion of the motors resemble human motion. The servo motors are controlled using the CM-530 controller. And based on the binary data received from the humanoid controller, particular program is executed and the humanoid performs specific tasks. The movements of servos are controlled with the help of programming software Roboplus. The RoboPlus software has two components- a) RoboPlus task and b) RoboPlus motion. The RoboPlus motion is GUI supported. This can be used to create and modify data regarding the robot's motion. This data is used in RoboPlus task to write the required program for the Humanoid Robot. The figure 5 (a) shows the pictorial representation of humanoid robot system of the rover. This particular feature of a mobile humanoid robot on board the rover, offers a very significant advantage in the areas where direct interference of human being is quite impossible. Hence it would be a productive area to do further research on. The mobile humanoid robot battery charging is done through the thin film solar panel glued to its body. The intelligent charger to which the battery and solar panel is connected take cares of the auto cut-off when the battery is fully charged. The prototype of humanoid robot used is as shown in figure 5 (b). [5-7]The humanoid robot uses image processing module named HaViMo 2.0. HaViMo is an Integrated Image Processing / Vision Module solution for Robotic and embedded applications. It gives these systems the ability to perform advanced image processing tasks like BLOB

Tracking, Region Growing and Image Girding while being operated in low power MCUs. All the data is processed in HaViMo which then delivers the processed visual results to the MCU. The processing of the received images is also done in the RoboPlus software. The humanoid bot can take the images in a more advanced way with more DOF when compared to the rover itself. And can perform on-spot experiments similar to what humans do, in manned missions, which makes the process more precise than ever before. The humanoid robot also serves the purpose of moisture detection, detection of new minerals and their collection etc. The humanoid robot is linked with the rover wirelessly. An automatic program is executed in humanoid as an interrupt to return to rover's range of detection, when the humanoid is too far away from the rover. Like never before the humanoid robot can be used to repair the rover thus increasing the life span of the rover over existing rovers. The rover also have mechanism to repair the humanoid to some extent. Both interconnected repair services increases the overall life span of both rover and humanoid robot.

6   Robotic Arm and *Kinect* Control in Rover

Our hands have been endowed with the capacity to handle any kind of object no matter how small or big, regular or irregular, light or heavy, it is. This capability is of utmost necessity while handling unexpected and foreign elements. For this purpose, a robotic arm similar to a human arm, with the same degrees of freedom should be used. Controlling such a system is extremely difficult using the ordinary remote control systems. The problem is resolved, in this rover using of the feature of Microsoft *Kinect*, a state of the art image processing device. The gestures captured by the *Kinect*, located at the station, are translated into control signals that are used to control several motors that drive the robotic arm, attached to the rover. This provides an actual replication of the arm movement giving the robotic arm the same robustness and dexterity as that of a human arm.

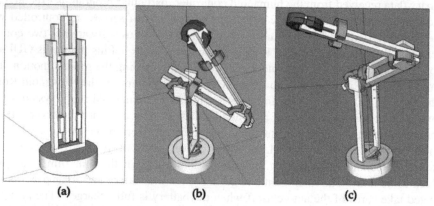

(a)   **(b)**   **(c)**

Fig. 6. (a) Basic model of the Haptic arm at resting position. (b) Haptic Arm implemented with cutter blade replicating normal lifting of arm. (c) Haptic Arm replicating more complex movement of the arm involving all the degree of freedom.

The design of the arm has been made in such a way that it occupies minimum space by folding upon itself, while not in use. On the other hand, when fully extended it can reach out to quite a large spread of space. This design of the robotic arm is such that it allows the rover access to areas like holes, or steep landscapes where the rover may not be able to physically travel. Activities like collecting samples, testing ground etc. in confined spaces can hence be done using the robotic arm. Figure 6 (a) shows the model of the haptic arm whose base could be aligned parallel to the body of the rover to minimize the area while not in use. When the arm is required, it takes the position as shown in the Figure 6.

The *Kinect* control of the arm makes the system more user friendly although the complexity increases. The *Kinect* used to control the robotic arm is placed at the ground station along with the user. To achieve the same, an algorithm involving the use of 3D coordinate geometry is used. For greater precision in gesture control, all the reference points and angles are taken with reference to points within the human body. This means that even when the controller's body is not facing the *Kinect*, a calculated detection of the referance points as decided by the system allows the Robotc arm to replicate the gesture with great precision.

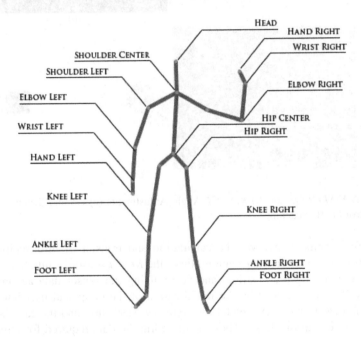

Fig. 7. Skeletal image obtained with Kinect SDK

The above fig. 7. shown image illustrates how a skeletal image is obtained using the Kinect SDK. This skeletal image contains many joints and links. Using the joints mainly focussed in the arm region, we are able to extract the necessary data to be sent to the microcontroller that controls the robotic arm. This data can be in the form of angles.The use of the Skeletal image ensures the exactness of measurement and the proper transfer of data from the earth station to the rover's robotic arm module.

Apart from the *Kinect* used to control the robotic arm, another *Kinect* is mounted in front of the rover whose angle of image capture can be adjusted to different angles so as to face the ground or remain paraller to the body of the rover as the situation demands it. This is useful for two purposes – (i) it could scan the ground for irregularities and make the adjustments in the suspension and braking system. This allows the rover to perform in an autonomous and stable fashion without compromising on its precesion (ii) the *Kinect*'s 3D scanning capability con be used for studying any foreign object by making a 3D image of the selected object which is held in position by the robotic arm as shown in fig. 8.

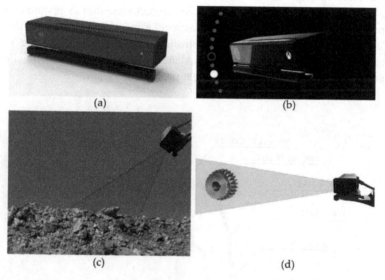

Fig. 8. (a) *KINECT* Sensor (b) *KINECT* Angle Adjustments (c) Ground Scanning for suspension control (d) 3D Scanning

The first segment involves the calculation and transfer of information from the base station to the rover via a satellite. First, the *Kinect* sensor is initialized at the base station by adjusting the position, and customizing the sensor data according to our demands. This is followed by the initialization of the Graphical user interface. This means the detection of the skeletal image of the user's arm and the calculation of the angles between various links. These angles form the data required for mimicking the movement of the Robotic arm. The control signals are generated with a comparative algorithm i.e. the system compares the previous angle and the present angle, and the difference is sent as a signal the motor. The output is then given as a negative feedback so as to minimize the error and improve precision. Two control signals are set in order to control the clockwise and anti-clockwise movement of the motor. This generates about 10 control signals thus giving the robotic arm a space constituting 5 degrees of freedom. After sending these values to the rover, these control signals are channeled to the respective motor control drivers. These motors drivers rotate the motor to attain the precise position. The figure 9 shows the *Kinect* control flow chart.

The Kinect has several advantages when compared to stereoscopic cameras. Stereoscopic Camera provide a three dimensional visual image of the environment. However, this is not enough for automation of the system. The stereoscopic camera output includes inputs from two different cameras kept a two separate positions (corresponding to angles with respect to a point) focusing a certain point. Analyzing these two frames for study of the external environment is highly impractical.

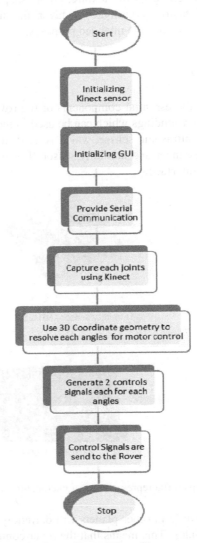

Fig. 9. Kinect Control Flow Chart

The 3D scanning feature of KINECT allows us to directly scan the immediate surrounding or any object thus allowing an easy understanding of the object or the environment. This helps in automation of the rover. The 3D scanning of the ground gives us a precise data regarding the topography of the immediate environment, so that the suspension of the rover's wheel system can be adjusted to maintain the stability of the rover without. The detailed study of an environment includes understanding the texture, shape and size of the objects found in that environment. A 3D model of any object can be made using 3D scanning feature of the KINECT. Reconstruction of the scanned fragments, which saves the texture, color, shape and size can be used for further study. These factors contribute to favor the use of the KINECT when compared to a Stereoscope, for the intended purposes.

7   Sensor System

This system is one of the essential components of the rover. It gathers the necessary information from its surroundings which can be used to monitor the rover. We will be discussing mainly the ultrasonic sensor, which is used in this design for or obstacle detection, temperature sensor and humidity sensor. The figure 10 shows the Arduino microcontroller senor interface.

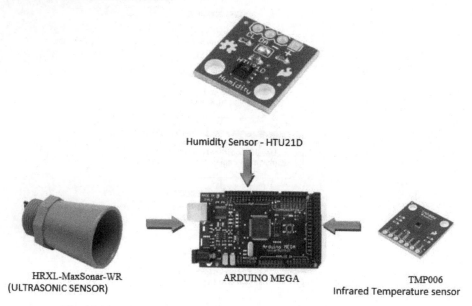

Humidity Sensor - HTU21D

HRXL-MaxSonar-WR
(ULTRASONIC SENSOR)

ARDUINO MEGA

TMP006
Infrared Temperature sensor

Fig. 10. The pictorial representation of microcontroller sensor interface

The Rover upon landing in other planets will definitely face a communication lag which of about 20 minutes. This means that the rover control team at the base station on earth cannot instantly see what is happening to the rover at any given moment. Also they cannot send commands to prevent the rover from running into a rock or falling off a cliff. Therefore an alternate way is required to guide the rover and prevent collisions that could result in its damage, and hence jeopardizing the mission. Like in the human body where a reflex action mechanism is used to reduce the time taken by the brain to process and analyze an external stimulus, an ultrasonic sensor is attached in front of the rover body which can be used to detect an obstacle in front of the rover and hence prevent damage by making it stop. This works by checking if the ultrasonic sensor is not receiving a reflected sonic wave at a particular moment. In this design we are using a HRXL-MaxSonar-WR ultrasonic sensor because it works in low voltages and thus can achieve high precision. The infrared temperature senor TMP006 and Humidity Sensor HTU21D is used to sense the temperature and humidity status of the alien planet at a particular point. All the three sensor are interfaced with the Arduino microcontroller for processing and sending the necessary outputs to the required subsystems.

8   Power Monitoring and Distribution System

The total power system is divided into various blocks: solar cell array, power distribution and monitoring, over voltage protection, high current electronic switching device, voltage regulator, batteries. The system is designed for both power supply and backup power. Two batteries are used for this purpose. While one battery is recharging from the solar array through power monitoring and charging control the other one supplies main power for the rover system. The solar array used is 100W cadmium telluride solar panel array because its 13.5% efficient than other panels and heat resistance. The voltage from the panel is always constant in spite of changing climatic condition. The charging time calculation is done using the luminosity in alien planet.

Power monitoring and distribution system a microprocessor based unit designed to timely control the charging and supply power to the rover system. This system charges the batteries using a technique called trickle charging to keep the battery's charge complete at all times. This is also safer for the battery and provides it with a longer life cycle. Charging the battery whenever power is available or between partial discharges rather than waiting for the battery to be completely discharged. It is used with batteries in cycle service, and in applications when energy is available only intermittently. The figure 11 shows the flow chart of power monitoring and distribution system.

Then the power from the batteries are drawn into high current electronic switching system (basically SMPS). This power is made available to the rover system at various ranges. Lower voltages like 3.3v to 5 v are provided for the integrated sensors and higher voltages are provided for microcontrollers and SBC (single board computers, raspberry pi in this case) and some other high power devices. To minimize wastage of energy like in linear power supply systems, the switching-mode supply continually switches between low-dissipation, full-on and full-off states spending very little time in high dissipation transition. Voltage regulation is done by varying the duty cycle of the switching i.e. ratio of on-to-off time. The advantage of switched-mode power supply is higher power conversion efficiency.

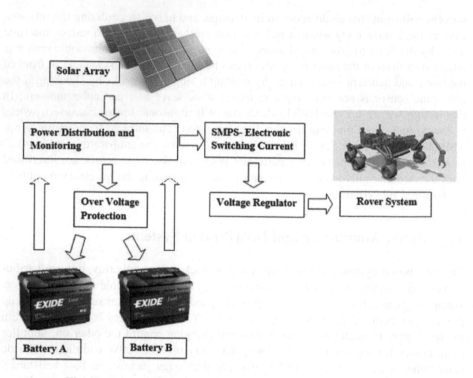

Fig. 11. Shows power monitoring and distribution system flow chart

It is also amounts to a smaller and lighter system when compared to a linear supply due to the smaller transformer size and weight.

8.1   Battery Model

The batteries that best suit rover system are selected on the basis of various parameters like energy density per mass, operating temperature, discharge rate and voltage characteristics. Since these batteries only lag in charging process, special control electronics are needed to protect the battery from overvoltage. By avoiding complete discharge of the battery, cycle life can be increased. An over voltage protection and charge monitoring system are installed for a reliable power supply, protection and maintenance. Since the technique used to recharge is a slow process (using solar array), a technique called trickle charging is used which will allow the battery to remain at its fully charged level.

8.2   Battery Control Unit

The Battery Control Unit contains all the control systems and associated power electronics circuitry required for the battery charging process. It functions by switching the power connections to individual cells. This is control using special control signals sent from the Battery Monitoring Unit

Some of the basic functions of this unit are: controlling voltage and current profile during charging and discharging process, charging individual batteries to equalize the charge on all cells in battery chain, isolating the unit during fault conditions, and responding to rover's operation mode.

9   Communication System

Communication system plays a major role in the rover. Once the communication with earth station is lost, the entire rover mission goes useless. Hence measures are taken that provides reconnection of communication link with the rover. A radio data modem is used for taking over the control of the motor drive and the sensors that indicate the status of the rover, while a secondary communication channel is used for real time communication with the rover to receive continuous video images and data from *Kinect*.

10   Applications

The rover design that has been proposed in this paper has a number of applications particularly in space and deep sea explorations. In this paper we have discussed the application of the rover design in the exploration of alien planets. Undoubtedly the rover can be used for landscape mapping, discovering new landmarks in alien landscapes, examination of mineral content, detection of water etc. This can be done by installing the necessary detection systems and sensors associated with it, on the rover. In dedicated applications such landscape mapping the *Kinect* sensor installed on the rover can be used. Additional sensors can also be installed to increase the accuracy of the mapping. Algorithms can be implemented to detect a particular landscape by a comparison and detection mechanism. This could even be extended to a search to find the most inhabitable space in planets like mars where human exploration is imminent in the near future.

The detection of mineral content can be executed with much better accuracy in the proposed rover design when compared to previous designs. The use of the robotic arm to collect samples is a significant advantage. Water detection which of particular interest in detection of life, can also be done by installing the necessary detection systems on board. The availability of a humanoid bot on board the rover is a particularly useful feature. It essentially makes the exploration more human-based than ever before.

11   Conclusion and Future Works

The rover design as proposed in the paper can be used in alien planet explorations. The different environmental issues that are faced by the rover once touchdown takes place on the alien surface, are taken into account in designing the rover. The rover has been designed for greater stability, improved functionality and better debugging mechanisms. Its flexible segmented body, embedded multipurpose robotic arm, multi wheel mechanism and *Kinect* module integration for advanced image processing, contribute to its improved functionality and flexibility of its overall design. The system control, done using the Arduino microcontroller and the Raspberry pi microprocessor makes the design feasible and easy to implement. The array of sensors used in the rover act to form an efficient sensory information base which is used by the rover to achieve a number function including a reflex mechanism for greater safety. The humanoid bot which is placed on board the rover will be instrumental in many of the advanced functions of the mission designated to the rover. This essentially opens up the possibilities of alien planet explorations and can potentially increase the information we have about these planets manifold.

Considering the many improvements that have been made in this proposed design when compared to other designs the possibilities of this rover are too many to enumerate briefly. To improve its functionality in landscape mapping long range cameras can be installed with image processing technology like the *Kinect*. This could greatly increase the range and depth of the landscape mapping process. Algorithms can be implemented to detect a particular landscape, compare it to existing data stored in a memory and conclude if new sites have been found. Mineral content detection can be improved by using superior equipment and necessary detection mechanisms. The mineral testing can be done by either installing an immobile testing unit, remotely connected to the base station as well the rover, and supplying the collected samples as and when required . Otherwise a more efficient approach can be taken by installing a system on board the rover for preliminary detection and allowing detailed testing to happen elsewhere. The availability of the humanoid bot allows a lot of improvements to be made in the overall design depending on the type of mission. The bot can also be equipped to make repairs and set up bases on alien land which can also help in the other functions that the rover is designed to perform. As advanced future work we have the idea of sending humanoid robot alone with solar panel suit and necessary subsystem for alien planet missions as the humanoid bot opens up a lot of opportunities for a wider range of exploration, longer life span of operation, greater range and greater accuracy in the experiments and tests conducted during the project. Also the humanoid robot can help in building artificial habitation for humans to live on mars in future with the artificial earthen condition on mars. The certain amount of water took to mars can be recycled as done in rocket habitat and oxygen can be recycled with plants inside the habitat, which makes humans life on mars possible.

Acknowledgements. The authors would like to thank Mechatronics and Intelligence Systems Research Lab, Department of Mechanical Engineering, Amrita University, Amritapuri campus for providing support to carry out the research and experiments.

References

1. http://www.wikipedia.org/wiki/Curiosity_(rover)
2. http://www.wikipedia.org/wiki/Mars_Pathfinder
3. http://www.wikipedia.org/wiki/Opportunity_(rover)
4. Ridenour, L.N.: Radar System Engineering, vol. 1. McGraw Hill (1947)
5. Aswath, S., Tilak, C.K., Sengar, A., Udupa, G.: Design and development of Mobile Operated Control System for Humanoid Robot. In: Advances in Computing, 3rd edn., vol. 3, pp. 50–56
6. Aswath, S., Tilak, C.K., Suresh, A., Udupa, G.: Human Gesture Recognition for Real-Time Control of Humanoid Robot. In: Proceedings of International Conference on Advances in Engineering and Technology, Singapore, March 29-30 (2014)
7. Krishnan, A.B., Aswath, S., Udupa, G.: Real Time Vision Based Humanoid Robotic Platform for Robo Soccer Competition. In: Proceedings of International Conference on Interdisciplinary Advances in Applied Computing, India, October 10-11 (2014)

Unified Minimalistic Modeling of Piezoelectric Stack Actuators for Engineering Applications

Ajinkya Jain[1], Rituparna Datta[1,2], and Bishakh Bhattacharya[1]

[1] Smart Materials Structure and Systems Lab,
Department of Mechanical Engineering,
Indian Institute of Technology Kanpur, Pin 208016, INDIA
[2] Department of Electrical Engineering, KAIST
291 Daehak-ro, Yuseong-gu, Daejeon 305-701, Republic of Korea
{jajinkya,rdatta,bishakh}@iitk.ac.in
rdatta@rit.kaist.ac.kr

Abstract. Piezoelectric (PZ) actuator is widely recognized for its high precision and displacement accuracy even at nanometer ranges. A minimalistic model is proposed in the present work, for PZ stack actuators. In the proposed model, various stack assembly arrangements have been assumed. Separate series and parallel assembly arrangements are suggested for both mechanical and electrical parts of the PZ actuators. The linearized constitutive equations formulated by IEEE, is considered to take into account the electromechanical coupling of the PZ actuator.

In the proposed model, stiffness of the connectors in stack assembly have also been taken into account and is modeled as connector spring. To include the effects of connector spring, the relationships of force and voltage with actuator displacement is replaced by the relations between force and voltage with the displacement of point of actuation at the physical system. This leads to a more realistic model of PZ actuator to be used in applications requiring actuator to be modeled as a black-box. With the advent of technology, more and more complex and compact actuating system are emerging into existence. Engineering applications, such as in the field of robotics, that require a black-box modeling of actuators, need simplistic models of the actuators to decrease the computational complexity. The proposed model, being a minimalistic one, qualifies as an ideal candidate for such applications.

Keywords: Piezoelectric (PZ) actuator, Minimalistic model, Series and parallel modeling, Smart System.

1 Introduction

Piezoelectric (PZ) materials fall under the category of smart materials which show coupling of electrical and mechanical properties of materials. PZ actuators are becoming a popular choice amongst both industries and academia for varied engineering applications due to its low power consumption, short response time,

© Springer International Publishing Switzerland 2015   459
J.-H. Kim et al. (eds.), *Robot Intelligence Technology and Applications 3*,
Advances in Intelligent Systems and Computing 345, DOI: 10.1007/978-3-319-16841-8_42

high output force, compact size and good controllability. Apart from these physical benefits, PZ actuators also posses better design characteristics like scalability and ease of control as compared to conventional actuators.

In literature, various models have been proposed to model the voltage response of PZ stack actuator [1–3]. Chee et al., [4] have performed a review on various models developed in relation with PZ actuator for intelligent structures. Further reviews on the usage and modeling of PZ actuators can be found in [5–7].

A widely accepted model for PZ actuator was published by a standards committee of IEEE [8] in 1987. A constitutive model with two linear relationships was formulated, which connect the stress and electric field to strain and electric displacement. However, considering the IEEE model as obsolete, Goldfarb and Celanovic [1] devised a model based on physical principles only. A generalized Maxwell resistive capacitor model as a lumped parameter representation of rate independent hysteresis was proposed in that model.

An extended version of the work done by Goldfarb and Celanovic [1] can be found in [2]. In this work, Adriaens et al., [2] have considered first order differential equation to describe the hysteresis effect and a partial differential equation is used to describe the mechanical behavior. Similar models also have been proposed in [9, 10]. The proposed models used a series connection between a hysteresis operator and linear dynamics to model the characteristics of PZ actuator.

Some recent works have extensively focused on the development of more accurate models of hysteresis [11–14]. In [14], Prandtl-Ishlinskii (PI) operator is used to model the hysteresis, while in [11] use of a modified rate dependent Prandtl-Ishlinskii (PI) operator to account for the hysteresis of PZ actuator, is proposed. A different approach to model hysteresis using least squares support vector machine approach, has been discussed in [13]. In some other works, black -box modeling of the PZ actuators have been explored [15, 16]. In [16], computational methods of fuzzy subtractive clustering and neuro-fuzzy networks were used to model PZ actuators. In yet another work, different approaches of black-box modeling techniques have been also addressed by Mohammadzaheri et al., [15].

In literature, various models have been proposed to account for hysteresis in electrical domain of PZ actuator; which are highly complex and computationally expensive. These models are neither efficient nor suitable to be used in engineering applications, in which actuator is treated as a black-box. If these complex models are used in such applications, special attention is required for actuator analysis itself, which is a tedious task. To address this challenge, a minimalistic model is proposed in the present work. The proposed model is motivated from the works discussed in [1, 2]. However, to further simplify earlier models, the present work considers the linearized relationships of IEEE [8] to explain the electro-mechanical coupling, while physical modeling is solely based on dynamical models proposed in [1]. As the model uses both linearized relationships and simplified physical model, the model qualifies as a minimalistic one.

The main advantages of the proposed model are, it is simple, easy to visualize and have low computational complexity. Due to its simplicity, the proposed model is an ideal choice for various engineering and science applications. Examples of such applications can be found in [17–21].

The proposed modeling of PZ stack is discussed in Section 2, in details. In this section, different combinations of capacitors and springs are modeled separately by decoupling the PZ stack. Section 3 discusses the usage of the model formulation based on specific application, while in Section 4 the conclusion and scope of future work are discussed.

2   Piezoelectric Model Formulation

The constitutive relationship explaining the electromechanical coupling of PZ materials, connect electric displacement density and mechanical strain to the mechanical stress and electric field strength applied across it. The relationship in matrix form is shown below (eq: 1):

$$
\begin{bmatrix} \mathbf{D}_{3\times1} \\ \mathbf{S}_{6\times1} \end{bmatrix} = \begin{bmatrix} \mathbf{d}_{3\times6} & \boldsymbol{\epsilon}_{3\times3} \\ \mathbf{s}^E_{6\times6} & \mathbf{d}^t_{6\times3} \end{bmatrix} \times \begin{bmatrix} \boldsymbol{\sigma}_{6\times1} \\ \mathbf{E}_{3\times1} \end{bmatrix} \tag{1}
$$

where,
D is the electric displacement field
S is the mechanical strain
σ is the mechanical stress
E is the applied electric field strength
d is the piezoelectric charge constant, which indicates the intensity of the piezoelectric effect
ϵ is the dielectric constant of PZ material for a constant σ
s^E is the elastic compliance matrix
d^t is the transpose of the piezoelectric charge constant matrix

Considering the direction of actuation along the direction 3 (Figure 1), and neglecting the changes in other dimensions, the equation (1) reduces to

$$
D_3 = d_{33}\sigma_3 + \epsilon_{33}E_3. \tag{2}
$$

$$
S_3 = \frac{1}{E_p}\sigma_3 - d_{33}E_3. \tag{3}
$$

where considering plane stress assumption, $s^E_{33} = \frac{1}{E_p}$, E_p being the modulus of elasticity of PZ material.

Considering small unidirectional displacement in mechanical strain, stress and electric field strength appearing in Equations (1) - (3) can be written as:

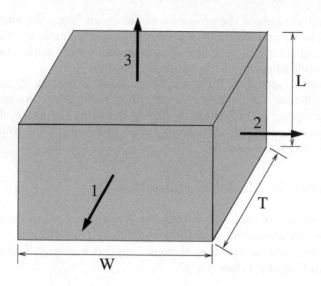

Fig. 1. Illustration of piezoelectric stack in three dimensional space

$$\epsilon = \frac{\Delta L}{L},$$

$$\sigma = \frac{F}{A}, \tag{4}$$

$$E = \frac{V}{L}.$$

where,
L is the length along actuation direction (direction 3)
A is the normal cross-sectional area
F is the Force of actuation
V is the applied voltage

The following relations can be obtained by substituting the values from equation (4) in equation (3):

$$D_3 = d_{33}\left(\frac{F}{A}\right) + \epsilon_{33}\left(\frac{V}{L}\right). \tag{5}$$

$$\Delta L = \left(\frac{1}{E_p}\right)\left(\frac{F}{A}\right)L - d_{33}V. \tag{6}$$

Equation (6) can be rearranged to develop a relationship between the external applied voltage V, actuator displacement ΔL and the force delivered by the PZ stack actuator F. For the stack actuator, the contribution of the connectors of actuator stack assembly should also be taken in account while calculating the stack force and displacements. Connectors can be modeled as solid bars under axial compression which can be represented as a connector spring. In the Figure 2, L is the length of the PZ stack, ΔL is the change in length of the PZ stack, k_{con} is the representation of the stiffness of the connection between PZ stack and physical system and z is the displacement at the point of actuation of physical system. Combining the connector spring modelling with the force-voltage relationship of actuator displacement, an equivalent relationship for the displacement of point of actuation at the physical system z with force F_{st} and external voltage V_{st} can be developed. The force delivered by the PZ stack F_{st} to the physical system can be represented as

$$F_{st} = k_{con}(z - \Delta L_{st}),$$

$$\Delta L_{st} = z - \frac{F_{st}}{k_{con}}. \tag{7}$$

The effective stiffness of the connector spring k_{con} can be found out using

$$k_{con} = \frac{AE}{L_o} \tag{8}$$

where,
A is the cross-sectional area of the actuator assembly
E is the Young's Modulus of the actuator assembly connector material
L_o is the length of the connectors in the direction of actuation

In the present work, electromechanical coupling of PZ stack is decoupled and modeled separately as different combinations of capacitors and springs. According to the requirement, applicable formulation can be used. The modeling details are described below.

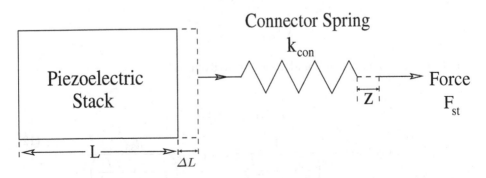

Fig. 2. Piezoelectric stack modeling considering connection with the physical system

2.1   Case A: Series Modeling for Both Electrical and Mechanical Systems

Both the electrical and mechanical part are considered to be made of n_{eff} effective elements connected in series. In electrical part, the external voltage will be divided among n_{eff} elements equally. Hence, the total applied external voltage V_{st} can be written in terms of effective voltage across each element, V, as

$$V_{st} = nV. \tag{9}$$

Similarly, considering series modeling for mechanical system, the net force delivered by the stack F_{st} in terms of effective force delivered by each element, F, can be represented as shown in equation 10. Also, the net displacement of the stack will be summation of individual displacement of each element.

$$F_{st} = F,$$
$$\Delta L_{st} = n\Delta L. \tag{10}$$

Substituting equation (9) and (10) in equation (6) we get

$$\Delta L_{st} = \left(\frac{1}{E_p}\right)\left(\frac{F_{st}}{A}\right) nL - d_{33}V_{st}. \tag{11}$$

Equating equation (7) with equation (11), we get

$$\left(\frac{1}{k_{con}} + n\frac{L}{E_pA}\right) F_{st} = z + d_{33}V_{st}$$

$$or, \quad F_{st} = \frac{z + d_{33}V_{st}}{\left(\dfrac{1}{k_{con}} + \dfrac{nL}{E_pA}\right)}$$

The relation between F_{st} and z can be written as

$$F_{st} = \alpha_1 z + \beta_1 V_{st}$$

$$where \quad \alpha_1 = \frac{1}{\left(\dfrac{1}{k_{con}} + \dfrac{nL}{E_pA}\right)}, \quad \beta_1 = \frac{d_{33}}{\left(\dfrac{1}{k_{con}} + \dfrac{nL}{E_pA}\right)} \tag{12}$$

2.2 Case B: Series Modeling for Electrical and Parallel Modeling for Mechanical Systems

Considering series modeling for electrical part and parallel modeling for mechanical part, the external voltage will be divided among n_{eff} elements equally and the net force delivered by the stack F_{st} will be the sum of effective forces delivered by each element. Furthermore, the total displacement of the stack will be same as that of individual element.

$$V_{st} = nV,$$

$$F_{st} = nF, \tag{13}$$

$$\Delta L_{st} = \Delta L.$$

Substituting equation (13) in equation (6) we get

$$\Delta L_{st} = \left(\frac{1}{E_p}\right)\left(\frac{F_{st}}{nA}\right)L - d_{33}\frac{V_{st}}{n}. \tag{14}$$

Equating equation (7) with equation (13), we get

$$\left(\frac{1}{k_{con}} + \frac{L}{E_p nA}\right)F_{st} = z + \frac{d_{33}V_{st}}{n},$$

$$or, \quad F_{st} = \frac{z + \dfrac{d_{33}V_{st}}{n}}{\left(\dfrac{1}{k_{con}} + \dfrac{L}{E_p nA}\right)} \tag{15}$$

The relation between F_{st} and z can be written as

$$F_{st} = \alpha_2 z + \beta_2 V_{st}$$

$$where \quad \alpha_2 = \frac{1}{\left(\dfrac{1}{k_{con}} + \dfrac{L}{E_p nA}\right)}, \quad \beta_2 = \frac{d_{33}}{\left(\dfrac{n}{k_{con}} + \dfrac{L}{E_p nA}\right)} \tag{16}$$

2.3 Case C: Parallel Modeling for Electrical and Series Modeling for Mechanical Systems

Considering parallel modeling for electrical part and series modeling for mechanical part, the external applied voltage will be same for each element and the net force delivered by the stack F_{st} will be same as that delivered by an individual element. However, the total displacement of the stack will be the summation of displacement in individual element.

$$V_{st} = V,$$

$$F_{st} = F, \tag{17}$$

$$\Delta L_{st} = n\Delta L.$$

Substituting equation (17) in equation (6) we get

$$\Delta L_{st} = \left(\frac{1}{E_p}\right)\left(\frac{F_{st}}{A}\right)nL - nd_{33}V_{st}. \tag{18}$$

Equating equation (7) with equation (18), we get

$$\left(\frac{1}{k_{con}} + \frac{nL}{E_pA}\right)F_{st} = z + d_{33}nV_{st},$$

$$or, \quad F_{st} = \frac{z + d_{33}nV_{st}}{\left(\frac{1}{k_{con}} + \frac{nL}{E_pA}\right)} \tag{19}$$

The relation between F_{st} and z can be written as

$$F_{st} = \alpha_3 z + \beta_3 V_{st}$$

$$where \quad \alpha_3 = \frac{1}{\left(\frac{1}{k_{con}} + \frac{nL}{E_pA}\right)}, \quad \beta_3 = \frac{nd_{33}}{\left(\frac{1}{k_{con}} + \frac{nL}{E_pA}\right)} \tag{20}$$

2.4   Case D: Parallel Modeling for Both Electrical and Mechanical Systems

Fourth type of modeling can be with parallel modeling for both electrical and mechanical part. The total applied voltage will be same for each element and total force delivered by the stack will be the sum of forces delivered by individual element. The total displacement (ΔL_{st}) will be same as that of a single element.

$$V_{st} = V,$$

$$F_{st} = nF, \qquad (21)$$

$$\Delta L_{st} = \Delta L.$$

Substituting equation (21) in equation (6) we obtain

$$\Delta L_{st} = \left(\frac{1}{E_p}\right)\left(\frac{F_{st}}{nA}\right)L - d_{33}V_{st}. \qquad (22)$$

Equating equation (7) with equation (22), we obtain

$$\left(\frac{1}{k_{con}} + \frac{L}{E_p nA}\right)F_{st} = z + d_{33}V_{st},$$

$$or, \quad F_{st} = \frac{z + d_{33}V_{st}}{\left(\frac{1}{k_{con}} + \frac{L}{E_p nA}\right)} \qquad (23)$$

The relation between F_{st} and z can be written as

$$F_{st} = \alpha_4 z + \beta_4 V_{st}$$

$$where \quad \alpha_4 = \frac{1}{\left(\frac{1}{k_{con}} + \frac{L}{E_p nA}\right)}, \quad \beta_4 = \frac{d_{33}}{\left(\frac{1}{k_{con}} + \frac{L}{E_p nA}\right)} \qquad (24)$$

3   Selection of Model: Results and Discussions

In the present work, four different models are proposed to model the PZ actuator behaviour. Based on the requirement, any one out of these four models can be chosen to represent the PZ actuator. The decision of choosing a model is to be based on the experimental data. Through experimental data, force F_{st} is to be measured at some finite external voltages and at corresponding actuator displacements z. To ascertain the appropriate modelling for actuator, relationships given for F_{st} in the different cases will be used as linear approximations for the experimental data. The slopes and intercepts of the linear approximations for each case can be varied by changing the number of effective elements n_{eff}. The best fit approximation will represent the modelling to be used for the actuator.

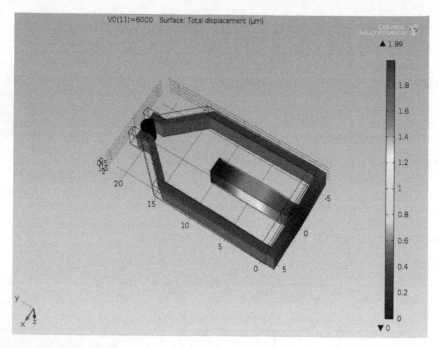

Fig. 3. Simulation results for a Microgripper actuated by piezoelectric actuator as suggested in [22] done on COMSOL Multi-physics Software. *Range of Applied Voltage : 0 − 6000 V.*

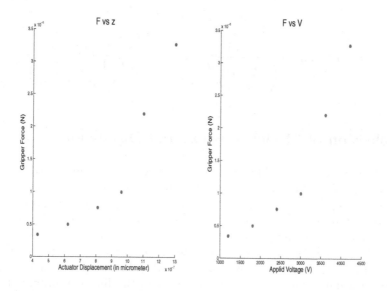

Fig. 4. Simulation data plots: Measured Gripping Force F vs actuator displacement z and External applied Voltage V_{ext} vs displacement z.

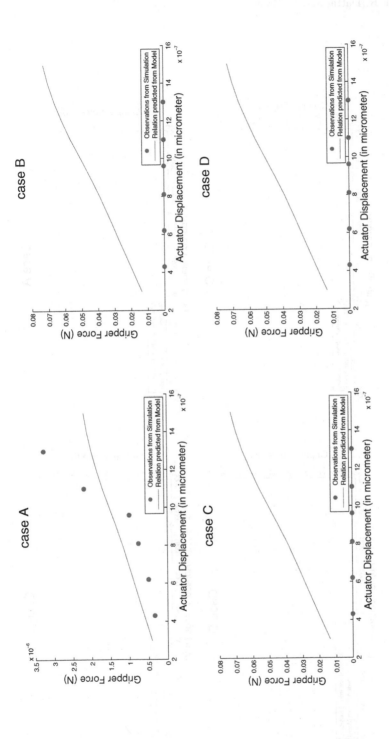

Fig. 5. Plots of gripping force versus actuator displacement, comparing the relation between gripping Force (F) and actuator displacement (z) as predicted by the proposed model with the simulation results for all the four cases, ($n = 35,000$, for case-A, for rest $n = 1$) *Note: Only modeling in Case A provides comparable results with the simulation observations. Only one effective element is considered for Case B, C and D, as it gives the best possible match .*

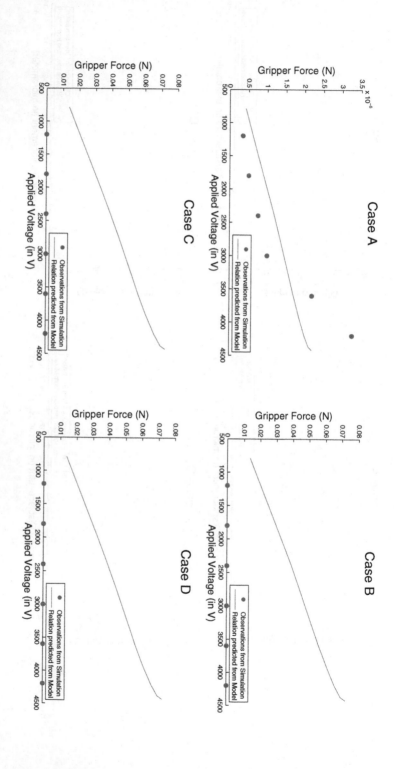

Fig. 6. Plots of gripping versus applied voltage, comparing the relation between gripping force (F) and applied voltage V) as predicted by the proposed model with the simulation results for all the four cases, ($n = 35,000$, for case-A, for rest $n = 1$) *Note: Only modeling in Case A provides comparable results with the simulation observations. Only one effective element is considered for Case B,C and D, as it gives the best possible match.*

The model selection procedure can be better understood from the simulation results provided. Numerical Simulations were done on COMSOL Multiphysics software to validate the effectiveness of the proposed modeling. Considering micro-gripper design as proposed in [22], simulations were performed with piezoelectric actuator made of Lead Zirconate Titanate (PZT-5A), loaded from inbuilt materials library of the software. An external voltage was applied on the actuator in the range $0 - 6000$ Volts (shown in Figure 3). Gripping Force F by the gripper on the cylindrical object to be held, is plotted with the actuator displacement as shown in Figure 4. The material properties used for the simulation are as following:

$$
\begin{aligned}
d_{33} &= 3.74 \times 10^{-10} \ C/N \\
\epsilon_{33} &= 1.505 \times 10^{-8} \ C/m^2 \\
E_p &= 1/1.64 \times 10^{-11} \ Pa \\
L &= 10 \ \mu m \\
A &= 4 \ \mu m^2
\end{aligned}
\tag{25}
$$

For k_{con} calculation,

$$
\begin{aligned}
L_0 &= 2 \ \mu m \\
E &= 200 \ GPa
\end{aligned}
$$

As described previously, the model to be used to describe the behavior of PZ actuator used in the microgripper, can be determined by observing the data obtained from the simulation. By comparing the trends of $Fvsz$ and $FvsV$ plots for the four cases with the observed data, it can be deduced that the simulated actuator can be best modeled using the modeling as proposed in case A *viz. Series modeling for both Electrical and Mechanical domains.* The relations of F with displacement z and V_{st} for the simulated actuator are obtained by considering $n_{eff} = 35,000$. The plots are shown in figures 5 and 6.

4   Conclusions

In the present work, four different combinations of series and parallel assembly for PZ actuator are proposed. The proposed model is an simplified integration of the linearized IEEE model with the physical property based model proposed in [1] and extended in [2]. While the physical model is based on the works of Goldfarb and Celanovic [1], the electro-mechanical coupling of the actuator is governed by IEEE model. Hence produced minimalistic model takes care of the electro-mechanical coupling of PZ actuators at a very low computational cost.

The relationships of force and voltage with point of actuation are considered to derive the constitutive equations of the model. These relationships are derived by modeling the connectors in the stack assembly as a connector spring. In this way, the proposed work is superior to the previous works for more accurate description of force - voltage relationships with out considering hysteresis.

The future scope of this work lies in the field of applications that use black-box modeling for actuators. From the proposed formulation, the black-box for these application is replaced by a minimalistic and precise model of the PZ actuator. The authors wish to apply the derived formulations in various optimization and designing problems related to robotics such as robot gripper design, biped locomotion, surgical robots etc. The formulation can also be extended for minimalistic modeling of other types of actuators.

Acknowledgments. Authors gratefully acknowledge partial funding provided by the Department of Biotechnology, India (Project: DBT /ME /2014172).

References

1. Goldfarb, M., Celanovic, N.: Modeling piezoelectric stack actuators for control of micromanipulation. IEEE Control Systems 17(3), 69–79 (1997)
2. Adriaens, H., De Koning, W., Banning, R.: Modeling piezoelectric actuators. IEEE/ASME Transactions on Mechatronics 5(4), 331–341 (2000)
3. Gu, G.-Y., Zhu, L.-M., Su, C.-Y., Ding, H.: Motion control of piezoelectric positioning stages: modeling, controller design, and experimental evaluation. IEEE/ASME Transactions on Mechatronics 18(5), 1459–1471 (2013)
4. Chee, C.Y., Tong, L., Steven, G.P.: A review on the modelling of piezoelectric sensors and actuators incorporated in intelligent structures. Journal of Intelligent Material Systems and Structures 9(1), 3–19 (1998)
5. Irschik, H.: A review on static and dynamic shape control of structures by piezoelectric actuation. Engineering Structures 24(1), 5–11 (2002)
6. Chopra, I.: Review of state of art of smart structures and integrated systems. AIAA Journal 40(11), 2145–2187 (2002)
7. Anton, S.R., Sodano, H.A.: A review of power harvesting using piezoelectric materials (2003–2006). Smart Materials and Structures 16(3), R1 (2007)
8. I.E.E.E. Standard on Piezoelectricity: An American National Standard, ser. I.E.E.E. Transactions on sonics and ultrasonics. IEEE (1987)
9. Croft, D., Devasia, S.: Hysteresis and vibration compensation for piezoactuators. Journal of Guidance, Control, and Dynamics 21(5), 710–717 (1998)
10. Dimmler, M., Holmberg, U., Longchamp, R.: Hysteresis compensation of piezo actuators. Tech. Rep. (1999)
11. Ang, W.-T., Garmon, F., Khosla, P., Riviere, C.: Modeling rate-dependent hysteresis in piezoelectric actuators. In: Proceedings of the 2003 IEEE/RSJ International Conference on Intelligent Robots and Systems (IROS 2003), vol. 2, pp. 1975–1980 (October 2003)
12. Deng, L., Tan, Y.: Modeling of rate-dependent hysteresis in piezoelectric actuators. In: IEEE International Conference on Control Applications, CCA 2008, pp. 978–982. IEEE (2008)

13. Xu, Q.: Identification and compensation of piezoelectric hysteresis without modeling hysteresis inverse. IEEE Transactions on Industrial Electronics 60(9), 3927–3937 (2013)
14. Seki, K., Ruderman, M., Iwasaki, M.: Modeling and compensation for hysteresis properties in piezoelectric actuators. In: 2014 IEEE 13th International Workshop on Advanced Motion Control (AMC), pp. 687–692 (March 2014)
15. Mohammadzaheri, M., Grainger, S., Bazghaleh, M.: A comparative study on the use of black box modelling for piezoelectric actuatorssome other works. The International Journal of Advanced Manufacturing Technology 63(9-12), 1247–1255 (2012)
16. Mohammadzaheri, M., Grainger, S., Bazghaleh, M.: Fuzzy modeling of a piezoelectric actuator. International Journal of Precision Engineering and Manufacturing 13(5), 663–670 (2012)
17. Canfield, S., Frecker, M.: Topology optimization of compliant mechanical amplifiers for piezoelectric actuators. Structural and Multidisciplinary Optimization 20(4), 269–279 (2000)
18. Datta, R., Deb, K.: Multi-objective design and analysis of robot gripper configurations using an evolutionary-classical approach. In: Proceedings of the 13th Annual Conference on Genetic and Evolutionary Computation, pp. 1843–1850. ACM (1850)
19. Sui, L., Xiong, X., Shi, G.: Piezoelectric actuator design and application on active vibration control. Physics Procedia 25, 1388–1396 (2012)
20. Borodinas, S., Vasiljev, P., Mazeika, D.: The optimization of a symmetrical coplanar trimorph piezoelectric actuator. Sensors and Actuators A: Physical 200, 133–137 (2013)
21. Datta, R., Pradhan, S., Bhattacharya, B.: Analysis of a seven link robot gripper with an integrated piezoelectric actuation system. In: Proceedings of the 13th International Conference on Control, Automation, Robotics and Vision, ICARCV 2014 (2014)
22. Keoschkerjan, R., Wurmus, H.: A novel microgripper with parallel movement of gripping arms. In: Proceedings of 8th International Conference on New Actuators, pp. 321–324 (2002)

Conception of a Tendon-Sheath and Pneumatic System Driven Soft Rescue Robot

Chen Jianfei and Wang Xingsong

School of mechanical engineering, Southeast University (211189), Nanjing, Jiangsu, China
cjfseu@163.com, xswang@seu.edu.cn

Abstract. The rescuing robot can help the workers a lot in searching and rescuing survivors. The design of a tendon-sheath and pneumatic system driven rescue robot is proposed in this dissertation. The robot is designed to solve the insert problem into the ruins and provide some necessary help for the survivor fined by the rescue robot. The introduction of the robot in this paper is divided into two parts: the introduction of mechanical structure and the control system. The robot consists of a searching head, a flexible slender arm, display terminal and a driving unit. The searching head is the most important part of the whole system which is comprised of two pneumatic actuators and a backbone spring. The driving unit is composed of the tendon-sheath, servos, solenoid valves and an air pump, all of these elements are control by MCU. The whole robot system can be packed into a suit case for better carrying.

Keywords: searching and rescuing robot, soft robot, Tendon-Sheath-Driven.

1   Introduction

In human history, natural and man-made disasters such as earthquakes, mining accidents, Terrorist attacks which have caused great loss to people happen every year. So rescue after the disasters is extremely important. However, the circumstance after the disasters is very complicated and many disasters may induce serious secondary disasters which make the rescue even more difficult [2].

Considering the above problems, many schools and scientific institutions have done a lot of work on studying rescue robotics and different kinds of robots have been applied to the search and rescue of the survivors which have dramatically improved the efficiency of rescuing. Compared with conventional rescuing methods, rescuing robots have advantages in finding the survivors, providing waters and oxygen, rescuing etc.

Snake-inspired robot is a typical example of rescuing robot. This kind of robots usually have flexible structures and can adapt to different situations. Cornell Wright etc. from US developed a snake-like robot that includes a motor and gear train, an SMA wire actuated brake, a custom intermodal connector and custom electronics featuring several different sensors for rescuing purpose [4].This robot is able to achieve diverse tasks such as urban search and rescue, mine rescue, industrial inspection, and reconnaissance. In [5], students from South-East University introduces a flexible slender rescue robot. Except for a driving head, they also designed a pushing mechanism which is used to push the robot into crevices in ruins from the back.

© Springer International Publishing Switzerland 2015
J.-H. Kim et al. (eds.), *Robot Intelligence Technology and Applications 3*,
Advances in Intelligent Systems and Computing 345, DOI: 10.1007/978-3-319-16841-8_43

However, the driving head can only provide very small force which is not enough to pull the robot into ruins. Students from Shanghai Jiao Tong University have developed another flexible slender rescue robot equipped with multiple sensors. [6]It consists of two parts: the driving head and a flexible slender arm in the back. This robot can take pictures in ruins, provide water and oxygen for the survivors and communicate with the survivors through an audio system. What's more, 3D reconstruction and positioning under the ruins are achieved in the system which make the rescuing more efficient.

This paper intends to introduce a tendon-sheath driven soft rescue robot that do not have insert problem into the ruins during the movement and can adapt to different size of crevices. The introduction of the robot in this paper is divided into two parts: the introduction of mechanical structure and the control system.

2   Mechanical Design of the Rescue Robot

2.1   Overview of the Tendon-Sheath and Pneumatic System Driven Rescue Robot

The tendon-sheath and pneumatic system driven rescue robot proposed in this paper consists of a searching head and a driving unit. The special feature of the mechanism is the flexibility of the searching head by using a compliant spring backbone and two deformable pneumatic actuator made by silicon rubber, so that the robot can adapt to different circumstances of ruins and crawl forward in the ruins. Figure 1 is the schematic diagram of the whole system.

To conventional flexible soft rescue robot, the robot may probably be stock in the ruins because the circumstance in the ruins is very complicated. When the robot goes into the ruins further and further, the force needed to pull the robot will gradually become larger for there may be more curves and other obstacles in the path. So it is almost impossible for one actuator to drive the robot into the ruins. In this paper, we choose to install several actuators along the flexible soft robot which can provide bigger force and the robot can avoid to be stock in some curves.

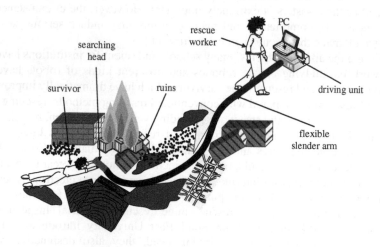

Fig. 1. schematic diagram of the whole system

The robot is driven by Tendon-Sheath and air pressure. The tendon-sheath is comprised of a flexible steel cable and a spring case. The brake cable in bikes is the typical application of the tendon-sheath. It also has been in good application in the field of medical rehabilitation and scientific detection etc. such as endoscopic robot, surgical robot and mechanical hand etc. The tendon-sheath is almost the only way to transfer force and displacement in long distance and the size of the robot can be smaller with the application of tendon-sheath. Figure 2 shows the schematic diagram of tendon-sheath transmission, the single active configuration is adopted in the robot introduced in this paper. Figure 3 depicts the tendon-sheath driving unit, The MG946R servo is chosen to provide the power, each of them can output max torque of 13KG·CM which is enough to drive the tendon-sheath. The servos are installed on the bracket and four pulleys are driven by four independent servos respectively. The sheath is fixed on the aluminum driving unit and the tendon is fixed on the pulley, the servos drive the pulleys and the pulleys drives the tendon at the same time, this is the way that force transfers. The size of this driving unit is $185 \times 70 \times 65 \text{(mm)}$

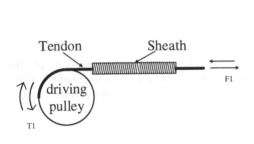

Fig. 2. single active tendon-sheath transmission   **Fig. 3.** Tendon-sheath driving unit

2.2 Design of the Searching Head

The searching head is one of the most important part of the robot which is designed to implement 3 main functions:

1. The ability to bend in four directions (up and down, left and right)
2. The airbags on both sides can expand and contract under the pressure of air, and they can provide support during the movement of the robot.
3. The soft manipulator can hold necessary food to the survivor and remove some obstacles in the path.

The searching head is comprised of a soft manipulator and a pneumatic actuator. The soft manipulator is designed to hold necessary good to the survivor and remove some obstacles in the path. The design of the soft manipulator is not contained in this dissertation. Figure 4 shows the assembled shape of the searching head.

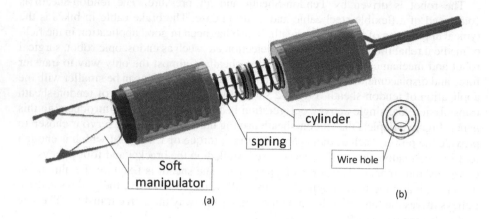

Fig. 4. (a) assembled shape (b) cross-sectional view of the cylinder.

2.3   Design of the Flexible Bending Joint

As is shown in figure 2, the joint consists of a backbone spring and multiple plastic cylinders and it has 2 DOF. To control this module, cylinders are connected by wires through four holes pierced in the cylinders. By pulling and releasing the four wires, 2 DOF motions can be obtained.

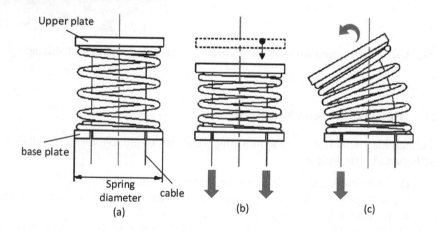

Fig. 5. (a) Original state. (b) Pulling on all four cables yields a translational motion.   (c) Pulling on one of the cables results in a rotational motion.

This flexible bending joint can perform three states of motion, the first state is to keep its original shape without applying any force on the cables. The second state is to flex in one direction by pulling one of the cables which at the same time will cause the upper plate to slightly rotate and translate. With this characteristic, the robot can change its moving direction and search a bigger range. By applying a same force on each cable, the joint can contract along the direction of its own body and the joint will elongate to the original state when the forces are removed because of the spring. The recurring cycles of contraction and elongation makes the robot crawls forward. Figure 5 shows these three states of motion.

2.4   Design of the Pneumatic Actuator

The pneumatic actuator is made of silicon rubber (Eco flex 0030, smooth on.inc). The cured rubber is very soft, very strong and very "stretchy", stretching many times its original size without tearing and will rebound to its original form without distortion. The materials are formed in a mold, and it takes four hours for the silicon rubber to mold. This actuator is controlled by air pressure. When air is supplied in the airbag, the pneumatic actuator contracts in the axial direction and expands in the radial direction. By using the deformable pneumatic actuator made by silicon rubber, the robot can adapt to different circumstances of ruins and crawl forward in the ruins.

Figure 6 is the Schematic diagram and a prototype of this Mechanism. Air tubes are contained between the silicon rubber and the bellows, when air is supplied into the airbag, the pneumatic actuator will contract in the axial direction and expands in the radial direction. The original diameter of the airbag is 40mm, and the airbag can expand to 110mm in maximum when inflated. The original length of the searching head is 180mm and it can contract to 110mm if there is about 10N placed on the tendon-sheath. Two pneumatic actuators are used as support during the movement of the robot.

2.5   Pattern of Movement

Peristaltic crawling is the way that insects such as an earthworm and inchworm etc. moves forward. The movement pattern of our robot is inspired by the peristaltic crawling of earthworm. The rescue robot contracts and expands as following patterns, the expanded airbag is used as support.

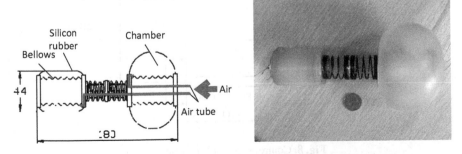

Fig. 6. (left) Schematic diagram of the pneumatic actuator, (right) a prototype

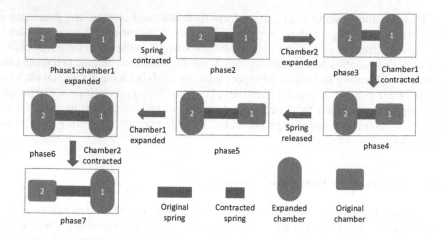

Fig. 7. Flowchart of the movement pattern

3   The Control System

Figure 7 shows the control system of the rescue robot. The robot can be controlled through a gamepad or the computer. The gamepad analog signal output is converted into a digital signal by using an A/D converter. Then the signal is input to the driving circuit board. After calculation and analysis, the driving circuit board will send driving signal to the servos and solenoid valves through RS232 data line. When the rescue robot needs accurate control, the control signal can be sent through the interactive interface in the computer directly.

To save the time and money, an usb gamepad is refitted as the controller. Each rocker on the gamepad is a Two-dimensional potentiometer, so the gamepad can output four analog signals and two digital signals. The four analog signals can control the robot to bend in four directions and the two digital signals can control the robot move or stop.

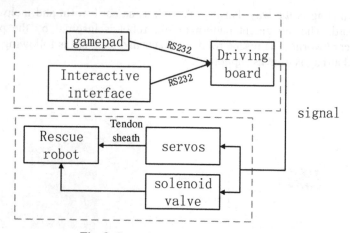

Fig. 8. Control system of the rescue robot

4   Conclusions and Prospection

How to send the robot into complicated crack in ruins is the key problem in the rescuing work, to solve this problem, the initial design of a tendon-sheath driven crawling rescue robot is proposed in this dissertation. It is designed to provide some necessary help for the survivors trapped in the ruins. However, the idea still stays at theoretical level, and a lot of work needs to be done in the future.

1) The prototype need to be improved and experiments to test its effectiveness should be done.
2) More sensors can be installed in the searching head such as camera, temperature sensor and carbon dioxide sensor.
3) Medical functions such as oxygen and water transmission can be added to the robot.
4) A 3D reconstruction algorithm should be developed to estimate the shape of the flexible slender arm
5) An interactive interface with friendly UI need to programmed based on C++Builder

References

1. Casper, J., Murphy, R.R., Micire, M.: Issues in Intelligent Robots for Search and Rescue. In: SPIE Ground Vehicle Technology II, Orlando, FL, pp. 292–302 (2000)
2. Yun, C.: Analysis on the Key Disaster Factors of Seismic Mountain Hazards Chains. In: 2010 International Conference on Remote Sensing, Hangzhou, China, October 5-6, pp. 245–248 (2010)
3. Wolf, A., Choset, H.H., et al.: Design and control of a mobile hyper-redundant urban search and rescue robot. Advanced Robotics 19(3), 221–248 (2005)
4. Wright, C., Buchan, A., Brown, B., et al.: Design and Architecture of The Unified Modular Snake Robot. In: 2012 IEEE International Conference on Robotics and Automation River Centre, Saint Paul, Minnesota, USA, May 14-18, pp. 4347–4354 (2012)
5. Chen, L., Wang, X., Tian, F.: Tendon-Sheath Actuated Robot and Transmission System. In: International conference on Mechatronics and Automation, Changchun, China, pp. 3173–3178 (2009)
6. Jiang, M.: Long-Thin Multisensory Mobile Robot For Interior Debris Searching. Shanghai Jiao Tong University, Shanghai (2012)
7. Yoon, H.-S., Oh, S.M., et al.: Active Bending Endoscope Robot System for Navigation through Sinus Area. In: 2011 IEEE/RSJ International Conference on Intelligent Robots and Systems, San Francisco, CA, USA, September 25-30, pp. 967–972 (2011)
8. Yanagida, T., Adachi, K., et al.: Development of a peristaltic crawling robot attached to a large intestine endoscope using bellows - type artificial rubber muscles. In: 2012 IEEE/RSJ International Conference on Intelligent Robots and Systems, Vilamoura, Algarve, Portugal, October 7-12, pp. 2935–2940 (2012)

AUT-UofM Humanoid TeenSize Joint Team; A New Step Toward 2050's Humanoid League Long Term RoadMap

Taher Abbas Shangari[1], Faraz Shamshirdar[1], Mohammad Hossein Heydari[1], Soroush Sadeghnejad[1], Jacky Baltes[2], and Mohsen Bahrami[1]

[1] Amirkabir University of Technology, No. 424, Hafez Ave., Tehran, Iran,
{taher.abbasi,faraz.shamshirdar,heydari,
s.sadeghnejad,mbahrami}@aut.ac.ir
[2] University of Manitoba, Winnipeg, MB, R3T 2N2, Canada
jacky@cs.umanitoba.ca

Abstract. RoboCup is an important challenge problem in robotics. The goal of RoboCup is to build a team of humanoid robots that can beat the human world champion in 2050. More importantly, RoboCup acts as an important benchmark in developing useful and practical humanoid robots for society. An ongoing concern for RoboCup is the fact that as robots are getting ever more complex and the costs of developing a team of humanoid robots is increasing rapidly. As a result it becomes difficult for teams to participate in the RoboCup competitions. Few teams will be able to build eleven adult-sized humanoid robots to compete in 2050. This paper describes a practical and tested way for developing a team of humanoid robots with full collaboration amongst the team members. This helps in reducing the costs of developing a team. The AUT-UofM team found out that developing a joint team is easier than trying to build one individually. The collaboration between Amirkabir University of Technology, Iran and the University of Manitoba, Canada started in 2014. The joint team project overcame great challenges in fostering collaboration between teams separated by geography, culture, and politics. We achieved the creation of a novel Teen-Size humanoid robot design, generated a new analytical walking engine good balance and push recovery and improved previous monocular vision-based localization. This research is based on experiences of both universities provided from long time participation in RoboCup humanoid league in recent years.

1 Introduction

A humanoid robot, which can perform skillful tasks using its arms, could be considered as a robot, with applications on both on Earth and in Space [1]. Humanoid robotics is a challenging research field, which has received significant attention so far and is going on its way to play a central role in robotics research. Besides, the study of humanoid robots, their stability, mobility, communication, and different tasks in uneven surfaces by humanoid robots have been the focus of many researchers over the last two decades.

© Springer International Publishing Switzerland 2015
J.-H. Kim et al. (eds.), *Robot Intelligence Technology and Applications 3*,
Advances in Intelligent Systems and Computing 345, DOI: 10.1007/978-3-319-16841-8_44

Building a humanoid robot is a difficult systems engineering and project management task. It is an inter-disciplinary problem that requires collaboration with electrical engineering, mechanical engineering, cognitive science, robot control, computer vision, artificial intelligence, and software engineering. At the heart of the intelligent robotics problem is the capability of a robot to sense its environment via various modalities, to form perceptions from the sensed data, to create a partial world model, to reason about the world model, to determine a set of goals, to create plans to achieve those goals, and to control the motors so that it modifies the real world to eventually satisfy its goals.

An ongoing concern in the RoboCup humanoid league in particular, is the fact that as humanoid robots are getting more complex, it becomes more and more difficult for teams to participate in the RoboCup competitions. The costs of building, but also of maintaining and running the robots make participation too expensive for all but the well-funded institutions. The problem will be more pressing in the future when the number of players will be raised to eleven players.

The Teen and Adult sized sub-leagues of the RoboCup humanoid league have almost the most complex and expensive robots and are thus greatly affected by these financial issues. The introduction of a minimum height (45cm) and the increase in the maximum height of (90cm) for the robots in the KidSize sub-league has clearly shown the strong direction of the humanoid league towards larger robots.

Members of RoboCup community including members of the Technical Committee (TC) of the Humanoid League and the trustees are well aware of this problem and there have been many attempts at encouraging joint teams. For example, researchers from the Freie University Berlin received a RoboCup federation Grant in 2012 and developed a nice common communication platform for humanoid robots [2]. However, only a few joint teams have taken part in the RoboCup humanoid league so far. This is partly due to the fact that teams to date perform better when they work singly. Additionally, at the moment forming a joint team results in additional costs for both partners because of the necessary meetings that must be held between two partners and these meetings need communications and travels.

The Amirkabir Robotic Institute and Mechanical Engineering Department of Amirkabir University of Technology have been successfully participating the in RoboCup Humanoid League competitions since 2011 [3]. The AUTMan KidSize team participated with great success during the 2013 RoboCup events all over the world. In particular, it won 2nd place in the Kid size sub-league at the RoboCup 2013 World Championship in Eindhoven, the Netherlands. Team Snobots from the University of Manitoba is a very experienced team and has competed in the RoboCup humanoid league since its very beginning in 2002 [4]. The team performed well at RoboCup 2013 in Eindhoven and won the 3rd place in the Technical Challenge awards. For 2014, team AUTMan and Snobots decided to join forces and formed a joint team AUT-UofM.

In 2014, team AUT-UofM focuses on the following new developments: (a) to develop methods for successful collaboration between teams separated by geography, culture, and politics, (b) to create a novel TeenSize humanoid robot design, (c) generating a new analytical walk engine with better balancing and push recovery, (d) to create improved methods for vision-based localization. There have been several previous attempts at collaboration in the RoboCup humanoid league. However, the collaboration was usually limited to teams that were based in close proximity, or teams that had clearly separated areas of concern (e.g., teams develop either hardware or software only), or to teams that developed robots individually and then formed a team of independent players. The team AUT-UofM faces several challenges. Not only is the team located halfway around the world in Canada and Iran (e.g., scheduling meetings is difficult because of the different time zones), but members also come from very different cultural backgrounds. Furthermore, both teams have previously made great contributions and developed successful designs in hardware and software. Therefore, a close collaboration between team members is expected that will result in a new hardware design which utilizes benefits of designs of both teams and also a close collaboration in the development of the necessary software modules (e.g., ball detection, landmark detection, localization). However, we expected and experienced initial problems as both teams tried to convince the other side to continue using their existing system. In spite of all these problems that need to be overcame, it was an extremely important and valuable lessons for all involved. We believe that such collaboration is vital for the future of RoboCup. Lessons learned from this and similar close collaboration will be crucially important for the future of RoboCup humanoid league. Furthermore, a successful collaboration between team Snobots and team AUTMan will also demonstrate some of the advantages of collaboration for other teams that are interested in participating in the TeenSize humanoid league. This will make the teen and adult sized sub-leagues more attractive and affordable for participants and will hopefully increases the number of participants in those sub-leagues.

2   Related Works

Standardized hardware platforms, such as the DARwIn-OP [5], Nimbro-OP [6] and Nao robot [7], produced by Aldebaran Robotics have had a positive impact on Robo-Cup Humanoid and Standard Platform Leagues. Due to their low cost and also ease of use in RoboCup Humanoid KidSize and TeenSize sub-leagues, now it is convenient to make a team for entering in the RoboCup Humanoid League community.

Successful participation of DARwIn-OP and Nimbro-OP developed as the prototypes of KidSize and TeenSize humanoid robots, foster interest in robotics, and encourage research collaboration through community-based improvement of the platform [6]. However, having larger robots in order to have a more social interaction in human being environments and also participate in the competitions, forced teams and researchers to build their own robots. The most successful examples of commercial large robots are Honda Asimo [8], HRP [9] series, Toyota Partner Robots [10], and Hubo [11]. The maintenance costs, refrain teams from using these robots in a soccer competition.

In spite of having price affordability, ease of manufacturing, assembling and maintenance, low weight, makes AUT-UofM to think about establishing a new team and also making his own robot. In the first prototype of this joint team, configurable actuators, various sensors, and enough computational power have been used. It makes the robot to do various operations, from image processing, path planning and robust stable controlling of dynamic motions.

Technically, AUT-UofM has used ROS as his framework in order to make the robots more reliable and flexible. AUT-UofM also steps toward a more analytical walking than his previous walking algorithm in KidSize class which could reach a high speed but low stability in previous competitions. Using ROS enables this team to use a complicated decision stage based on knowledge base which is also used for localizing robots more accurately.

3   Hardware Design

AUTeen is 105cm in height and 7.5Kg in weight. New robot's kinematic structure is with 20 degree of freedoms (DOF). The design is such that made us to use 6 DOFs for each leg, 3 degree of freedoms for each arm. Video camera of the robot will be hold by 2 servo motors as a Pan-Tilt mechanism. Serial mechanisms have been used in legs to make the robot as simple as possible. The Dynamixel MX and RX series manufactured by Robotis drive the robot joints [11]. MX106 series in leg joints, RX64 series in arms and RX24 series in neck have been used. To reach a higher performance and saving more energy, two motors have used in the knee joints. Configuration of such actuators grants the robot wide range of motion and fast reaction. In this robot, AUT-UofM has focused on robustness, weight reduction and also nimble reaction. Using CM9.04 results in more compatibility and easier sensory information fusion so CM9.04 (OpenCM9.04 is an open-source controller that runs under 32bit ARM Cortex-M3 from ROBOTIS.co [12]) is adopted as a low level controller and Device Communication Manager (DCM) in our robot. USBzDXL is also used as a direct motor controller for Dynamixel's [4] actuators from main processor, this component is needed to communicate with devices like actuators and sensors in different way such as I2C, RS485, Serial TTL, analog and etc. CM9.04 is just used for sensor fusion (low level filtering) and user interface function. The CM9.04 controller working on 72MHz and communicates with upper layer (PC) on serial interface @1Mbps. In low-level computation on this board, we drive three types of different sensors: Motion 9DOF IMU sensor, integrated foot pressure sensors based on FSR sensors, internal actuators load, speed and absolute position sensors for debugging mode. The CM9.04 runs lower-layer algorithms which can provide 3D posture of COM in pitch, roll and yaw at more than 100Hz with high resolution of orientation by filtered and combination of internal gyro, accelerometer and manometer sensors of GY-80 IMU [13]. Figure 2 shows low-level controller and main controller and peripheral connected device. In AUTeen, MAXData QutePC-3001 [15] mini embedded board is used as a main controller. High performance and low power consumption are the main factor for using this kind of main boards as a main processor in humanoid robots. In figure 1 the TeenSize soccer robot of team AUT-UofM is shown.

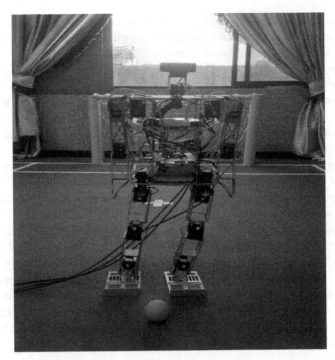

Fig. 1. Humanoid Soccer robot designed and built by team AUT-UofM.

4   Software Development

There is always some requirement, which leads a software development team to keep their work modular, maintainable and reusable. It's clear that these concepts should get noticed in a software life-cycle of a RoboCup team. Besides, it is needless to say it is how much important having a well-designed code at framework level to have a reliable system during development, test and debug. There are some concepts, which are essential for running a "working" robot on the field, like message passing interface for communicating between different parts of your robot code like vision module and motion planner, etc. A level of being real-time and running tasks scheduled, flexibility in being synchronous or asynchronous with the other parts of a software, flexibility in code design while having simple and useful rules, and many other needs are highly important for a robotic framework and as these needs are growing, one cannot start to code his framework from scratch and overcome all the challenges in the best way. So, not to re-invent the wheel, we focused on our needs in different parts and started to take a look at working frameworks and development systems in robotics and figured that ROS is what we need. Some benefits are listed below:
- Anonymous message passing.
- Recording and replay of messages, which ends up the code with highly modular nodes which are publishing/subscribing data isolated from what happens out in relation with others.

- Remote procedure calls, which makes it possible to have services with synchronous request/response interactions between processes, while your system out there between publishers/subscribers are still asynchronous.
- Flexibility in altering your message types, or request/response types will keep your hand open to stay creative in the inter-process communications.
- Being open-source gives you a great support by the community and keeps you up-date.
- Lots of already developed robotic packages and stacks which may reduce your effort to produce your own code from scratch.
- Supporting both C++ and Python gives us flexibility and freedom in code generating for different tasks.
- Easy integration with other systems like fawks, which makes your code easily ex-tendable to integrate modules from other frameworks.
- Fully command-line functionality which leads us to having remote development and 24 hour access to our test bed. Thus, using ROS, we can free our mind from system-level issues and focus a little bit more on software engineering issues like module design, test, documentation and debugging the modules themselves without concern-ing framework-level problems. Now in our team, software related groups are sitting around and discuss their way of communication and message types using ROS and work on their internal system design to produce reliable modules, which will be work-ing together like a charm, in 2014 competitions.

5   Motion

Team AUT-UofM is going to use inverted pendulum method as a main base for de-veloping walking. B-human team could adopt an inverted pendulum walking which diminishes double support phase and can reject some minor disturbance due to keep-ing inverted pendulum walk like [7]. Some optimizations have been done to the me-thod as listed below:

1. There were some gaps between transmissions from straight walking to Omni directional walking in the paper. Since it was not certainly defined how the suffi-cient parameters (s, r, x,) can be found in an Omni directional walking, so a cost function has been used and by optimizing in each parameter selection step, per-fect parameter for the next walking step has been found.
2. Together the disturbance rejection which was explained in the paper, we add a heuristic push recovery (the same as we were doing in the past) to increase the disturbance rejection capability.

In Figure 2 the results of the cost function optimization for finding the best in-verted pendulum parameter for a turning maneuver is shown. It is obvious that the optimization method could find a good-looking walking behavior. In this part in addi-tion to walking engine we also work on whole body controller which designs other motion activity for our robot.

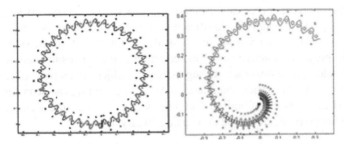

Fig. 2. The generated trajectory using new cost function for finding the next inverted pendulum parameters. The red dots shows origin of each step, the cross signs are the origin of inverted pendulum and the blue lines is the trajectory of COM in a turning maneuver.

6   Cognition

In this section a vision-based approach to self-localization which uses a novel framework to integrate image features with Robust Monte Carlo localization is introduced. The framework consists of five layers. In the first layer, the process of image transmission from camera to memory is described. The second layer discusses mapping and classifying the color space. The third layer focuses on segmentation and also building regions by integrating segments. The fourth layer is about how to extract features from image regions. In the last layer these detected or recognized objects and clues are used to robustly localize the robot.

Image Transmission Layer. To reduce the computation needed for image processing, YUYV is chosen as the video stream format to work on. It is chosen because: 1- It is the format supported by v4l2 Linux library and most of cameras. 2- It allows reduced bandwidth for chrominance components, thereby typically enabling transmission errors or compression artifacts to be more efficiently masked by the human perception than using a "direct" RGB representation. 3- It has a separate channel (Y or luma) to transmit luminance so light varying that is an annoying phenomenon in the RoboCup soccer fields would be under control.

Color Mapping Layer. A modified version of kd-tree algorithm is used to map colors from an arbitrary 3-channel color space C to a set of colors S. A class label $l_i \in$ S is assigned to every pixel $P_i \in$ C.

Each color component is represented by an 8-bit value and k = |C|=5 stands for the number of defined class labels.

$$C \bullet S$$

C = {0, 1 ... 255} and S = {Green, White, Yellow, Blue, Orange}.

Segmentation Layer. It consists of two sub-layers. These two, segments the mapped image using vertical scan lines at first and then build the regions by integrating the segments [15]. The outputs of these layers are some regions.

Considering current processor limitations, processing every pixel of a 0.48-megapixel image will result in significant frame-rate reduction. so only the pixels along a set of vertical scan lines are considered for initial processing. Doing so enables us to prevent camera resolution reduction and as a result increases the performance of later processes such as segmentation, object detection, object tracking and localization. Having robot kinematics data, are generating the scan lines equidistantly on the field plane is an option. The other way is to space the scan lines equidistantly on the image plane. it can be done using Binary Tree.

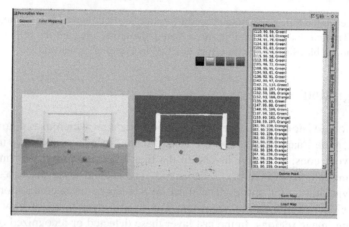

Fig. 3. GUI of the color mapping layer. The left image is original image. The right image shows the mapped colors. This color mapping is done using a look-up table. The look-up table is based on a modified version of kd-tree outputs.

Fig. 4. Segmentation layer. The vertical colored rectangles in the image are actually simple segments created by our segmentation algorithm, each color shows segment label. (yellow, orange, green, etc.)

The pixels along the scan lines are traversed until a certain threshold of consecutive particular pixels e.g. green pixels is exceeded. This process is called color transition. In the following the same process is done to detect next color transition. The information stored in each color transition includes the (x, y) image coordinate of start color class label and end color class label. A segment is built up using these two pixels and the corresponding pixels in the next scan line.

The previous process reduces the segment resolutions. After unifying the related segments to build regions, the region bounds are extended i.e. pixels that both have the same label as the regions and also are attached to the regions [15].

After segmenting the image and building regions, green horizon determination will be done. Determining the green horizon reduces the computation needed for image processing significantly. The robot will be able to look for the ball, the obstacles and junctions beneath the horizon. Besides, most of the goal post area is above the field border. To detect the field border, processing of the image is started at the top of image and if a certain threshold of consecutive green pixels is exceeded, these green pixels are considered as the green horizon.

Object Detection Layer. This layer deals with feature extraction of image regions, rather than straightforwardly extracting features calculated from the whole image. There are four sub-layers: Ball Detection, Line Detection, Goal Post Detection and Obstacle Detection.

The first feature used to classify the regions is color of the regions e.g. orange regions are candidate to be considered as a ball segment. After this level of classification, each region will be processed by one of the mentioned sub-layers.

Fig. 5. Border Detction (Green Horzion Determination). The Blue line surrounding the green field is shown is the calculated convex hull.

Afterward the features are used to detect, track or recognize the objects or clues in the field e.g. ball, goal posts, T junctions, X junctions, Obstacles and so on.

Self-Localization Layer. The last layer is about how to use these detected or recognized objects and clues to robustly localize the robot. Self-localization comes in many different types [17]. Position tracking which is the simplest localization problem, global localization is the second type that is more challenging than position tracking and the third type which is the most difficult is kidnapped robot problem. Because in RoboCup soccer fields, the robot ought to deal with all these three types of localization, Robust Monte Carlo was the approach mostly chosen for robot self-localization in humanoid RoboCup soccer fields.

7   Robot Navigation

A robot working in dynamic environments needs path planning and navigation based on the planned path. Humanoid robots navigation has some more difficulties than other mobile robots due to their bipedal walking.

J. Garimort and A. Hornung have solved the challenges by implementing footstep planning and navigation algorithms for humanoid robots [13-14].

Fig. 6. The left image shows the output of Ball Detection sub layer. Violet circles emphasize the ball candidates. The right image shows the output of Goal-Post detection in the case that only one of the bars is seen. A red rectangle around the right bar of the goal post is drawn in the image.

Footstep planner package is used for navigating and finding the optimal path with the map providing the obstacles positions. The map is provided by the localization module. The planner is based on SBPL and capable of dynamic re-planning. The supported planners algorithms are: ARA*, AD*, R* as the foot parameters differ for each robot, the parameters are changed so that fit to our new built robots' configuration. So, they plans path based on new parameters.

8   Conclusion

In this paper problems of developing a humanoid robot team e.g. necessity of high budget, cultural differences, and different time zones are stated. Afterward building a joint team fully based on collaboration was introduced as the best way of overcoming the above mentioned problems. Then a ROS-based framework was presented for AUT-UofM Humanoid robot. Later building a novel TeenSize humanoid robot design, generating an analytical walking engine, improving methods for monocular vision-based localization are explained. The next plan is developing learning by imitation vision framework that aims to lower the computational costs of learning in comparison to Reinforcement learning that currently is being used. Automatic tuning of the camera parameters such as White Balance Temperature, Gain and Back Light Compensation in order to having a better raw image is another plan that is set to be reached in the future.

References

1. Kaneko, K., Harada, K., Kanehiro, F., Miyamori, G., Akachi, K.: Humanoid robot HRP-3. In: IEEE/RSJ International Conference on Intelligent Robots and Systems, IROS 2008, pp. 2471–2478. IEEE (2008)
2. FUB-KIT team members: FUB-KIT Humanoid TeenSize Robot League Team Description Paper. RoboCup2013, Eindhoven, The Netherlands (2013)
3. AUTMan team members: AUTMan Humanoid Kis-Size Robot League Team Description Paper, RoboCup2013, Eindhoven, The Netherlands (2013)
4. Snobot team members: Snobot Humanoid Kis-Size Robot League Team Description Paper, RoboCup2013, Eindhoven, The Netherlands (2013)
5. Ha, I., Tamura, Y., Asama, H., Han, J., Hong, D.W.: Development of open humanoid platform DARwIn-OP. In: SICE Annual Conference (2011)
6. Allgeuer, P., Schwarz, M., Pastrana, J., Schueller, S., Missura, M., Behnke, S.: A ROS-based Software Framework for the NimbRo-OP Humanoid Open Platform" In Proceedings of 8th Workshop on Humanoid Soccer Robots. In: 13th IEEE-RAS International Conference on Humanoid Robots (Humanoids), Atlanta, GA (2013)
7. Gouaillier, D., Hugel, V., Blazevic, P., Kilner, C., Monceaux, J., Lafourcade, P., Marnier, B., Serre, J., Maisonnier, B.: Mechatronic design of NAO humanoid. In: IEEE International Conference on Robotics and Automation, ICRA (2009)
8. Hirai, K., Hirose, M., Haikawa, Y., Takenaka, T.: The Development of Honda Humanoid Robot. In: Int. Conf. on Robotics and Automation, ICRA (1998)
9. Kaneko, K., Kanehiro, F., Morisawa, M., Miura, K., Nakaoka, S., Kajita, S.: Cybernetic human HRP-4C. In: IEEE-RAS Int. Conf. on Humanoid Robots (2009)
10. Takagi, S.: Toyota partner robots. J. of Robotics Society of Japan 24(2), 62 (2006)
11. Park, I.-W., Kim, J.-Y., Lee, J., Oh, J.-H.: Mechanical design of humanoid robot platform KHR-3 (KAIST Humanoid Robot 3: HUBO). In: IEEE-RAS Int. Conf. on Humanoid Robots (Humanoids), pp. 321–326 (2005)
12. http://www.robotis.com

13. ROBOTIS, OpenCM9.04 (2013), http://www.robotis.com/xe/darwin_en (last accessed January 28, 2014)
14. ROBOTIS. Dynamixel Actuator, http://www.robotis.com/xe/dynamixel_en (last accessed January 25, 2014)
15. Budden, D., Fenn, S., Walker, J., Mendes, A.: A novel approach to ball detection for humanoid robot soccer. In: Thielscher, M., Zhang, D. (eds.) AI 2012. LNCS, vol. 7691, pp. 827–838. Springer, Heidelberg (2012)
16. Härtl, A., Visser, U., Röfer, T.: Robust and efficient object recognition for a humanoid soccer robot. In: Behnke, S., Veloso, M., Visser, A., Xiong, R. (eds.) RoboCup 2013. LNCS, vol. 8371, pp. 396–407. Springer, Heidelberg (2014)
17. Thrun, S., Fox, D., Burgard, W., Dellaert, F.: Robust Monte Carlo localization for mobile robots. Artificial Intelligence 128(1), 99–141 (2001)

ROBO+EDU: Project and Implementation of Educational Robitics in Brazillian Public Schools

Ana C.R. Ribeiro, Dante A.C. Barone, and Lucas E.P. Mizusaki

Informatics Institutes, Federal University of Rio Grande do Sul
Porto Alegre, Rio Grande do Sul, Brasil
{carolribeiro2,dante.barone,lepmizusaki}@gmail.com

Abstract. Educational Robotics is the use of robots and robotic kits in school projects. The students build and program robots to perform a diversity of tasks, sometimes in interdisciplinary activities. That way, some cognitive and group work and project organization skills are trained and exercised, and it also serves as a first with engineering for the students. The ROBO+EDU project aims at developing a Open Source kit for Educational Robotics for Brazilian schools through the Introduction to Educational Robots project from the Mais Educação program. Basing itself in the Arduino hardware, the group is building a kit able to compete with closed platforms already being commercialized, and that can benefit from the already existing development development communities around the platform.

Keywords: Educational Robotics, Open Source, Arduino.

1 Introduction

This article is about the project ROBO+EDU – Educational Robots from the Brazilian government program Mais Educação, it's perspectives, implementation stages and the challenges encountered during it's development. The project aims for the dissemination of Educational Robots and the training of teachers working in public schools of basic education. This objective is pursued through the development of didactic materials and both technologies and educational methodologies, needed in order to insert educational robotics in school context. This project is being developed under the Mais Educação program, from the Brazilian ministry of education, which constitutes as part of a strategy to bring Integral Education to the country public schools, extending the school day to a journey of, at least, seven hours, through workshops in various areas of knowledge.

Thus, this article presents the projects contribution to the use of robots in education, followed by a presentation of the project. Next, the challenges faced in the projects expansion are analyzed and, lastly, the next steps of the project are presented.

© Springer International Publishing Switzerland 2015   495
J.-H. Kim et al. (eds.), *Robot Intelligence Technology and Applications 3,*
Advances in Intelligent Systems and Computing 345, DOI: 10.1007/978-3-319-16841-8_45

2   Educational Robotics

The use of educational robots is a learning activity that derives from the construction of robots by students, being a robot a mechanism composed of a processor using sensors and actuators to interact with an environment through a program that describes it's behavior to perform a specific task. Educational robots might be understand as a "set of resources aiming at learning scientific and technological knowledge linked to the other areas of knowledge, using activities such as design, building and robot programming" (LOPES & FAGUNDES, 2006, p.3).

In LOPES & FAGUNDES (2006), the use of Educational Robots is noted as being usually tied to a technical training, restricted to middle and higher professional education. However, to the authors, there is one use to robots which "accounts to the potential these equipment may achieve in education of schoolchildren: the Educational Robotics- (ER)" (p.2).

The use of robots in education may contribute in a significant way to both teachers and students. To the teachers, it allows the exploration of the most diverse areas, creating a number of experiments with these tools. For the students, it allows to put into practice their theoretical knowledge, making the study more interesting and allowing the construction of knowledge. In this approach, the knowledge is neither in the subject, nor in the environment, but it arises through continuous interaction between both (PIAGET, 1972). This way, the starting point for the knowledge is the subject actions in the real world. In order to know a certain object or situation, it is necessary to act over it, changing it, altering it, learning about it through this transformation. Through these group of different actions that allow the subject to build transformation structures, it is possible to build mental structures which are the basis of knowledge (PIAGET, 1972).

A number of factors may promote the inclusion of educational robotics as an activity related to educational purposes. Robotic kits that are easy to build and use, and easy-to use programming interfaces, directed to child audiences.

> Through educational robotics, the student can develop the capacity to solve problems, to use logic in an effective way and to learn concepts from maths and physics. This way, theoretical concepts that are usually approached without a link to the real world are put to the practical uses. Educational robotics create an environment characterized by technology and creativity, stimulating learning intuitive concepts, kinematics being an example (MORELATO, NASCIMENTO et al, 2010, p.81).

Taking this in consideration, it is important to promote the use of Educational Robotics in school, both in public as in private institutions. It can bring important contributions to the school context, turning the activities more attractive and varied, or providing important discoveries in learning process.

3   The ROBO+EDU Project

The ROBO+EDU project aims to provide initial training and continual education of teachers and other professionals of basic education acting in schools and/or public education systems in Educational Robotics. This through an Open Source kit being developed by the project proponents.

The main motivation behind this research happened because of the robotic kits market in Brazil, affected by heavy prices and low offer, the schools cannot afford them. On top of that, difficulties to import materials and the usual low variety in the kits functions are also limiting factors. This way, it is strategic to develop low cost kits, specially Open Source kits, which may sparkle an autonomic development community. It is also important to plan training and monitoring in order to promulgate the practice of Educational Robotics in the country.

3.1   Kit ROBO+EDU

The ROBO+EDU project was proposed based on the experiences on a previous project called ROBOTEKA (BARONE & MIZUSAKI, 2012), with the objective to organize a group of ten schools in the city of Goiânia, to work with Educational Robotics (using LEGO Mindstorms NXT kits). Although the offer was well received by the schools, there were a number of negative points in the work. Firstly, the purchase of the kits had a number of bureaucratic obstacles to import, and since the offer in national market was limited, the kits delivery was delayed for almost a year. Also, the kits instruction manual was in English, a serious limitation when working with small public schools, which may not have a foreign language teacher. Lastly, there were some concerns about the kits limitations. Sometimes, the schools did not have enough pieces of a certain type to use on their projects, access to spare parts was also not possible. Since the kits were commercial, any need should be met through buying more pieces.

The group decided to use an Open Source development strategy, by building an open project which could be used by any school or independent student. This kit should be a collection of specifications for pieces, sensors, actuators and controllers, which could be consulted in available documentation, and could be built by anyone, or produced on demand to local industries. Since it is Open Source, there are no production rights on those kits, and they would be easier for the schools to buy. According to Cobo (2009, p. 130)

> The support architecture promoted by Open Source promotes an adaptable and innovative creativity, the cooperation and imagination amongst those who do not necessarily know each other, neither find themselves in the same physical environment. The Open Source does not favor age, gender, social status or ethnicity, but wages by adding creativity, knowledge and experience in a robust community, especially after the popularization of Internet

A kit of educational robots should have a controller, sensors, motors and basic mechanical devices. The ROOB+EDU kit consists in a injected plastic lid over the chassis, on which multiple sensors are coupled, ultrasonic, light, temperature, RGB and a pushbutton. It also has two servos with continuous rotation to locomotion, and a ball caster as rear support. It is powered over a USB cable or a couple of AA batteries. It is designed to build a small autonomous car.

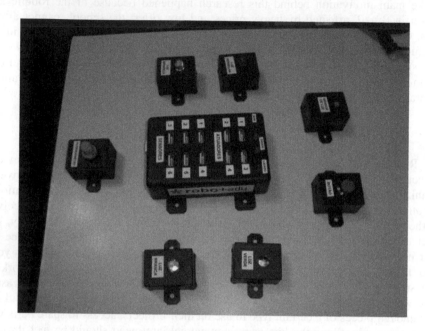

Fig. 1. Open Source kit prototype

The kit is based on the Arduino project, and its controller is a shield, a board which is mounted over the micro controller to extend its functionalities through the drivers needed to read the sensors and power the motors. Being an Open Source project with a standardized input, it is possible for others to propose specific sensors for the robot, with the board resolution of 10 bits. Our infrared and sonar sensors with a resolution of 1:30, so the board is more than enough to handle them. We encapsulated the sensors using USB connectors, which proved to be resilient and available enough for mass production (in contrast with the mini-USB version), while being familiar to most students. The same connectors are used for the actuators, currently only servo motors are included. Every component is being powered by the Arduino supply, limited but enough for the first applications.

Specific libraries were built in order to make reference to the sensors through a single command. This is a necessity in order to be possible to program the kit without detailing the pins of the board, a concept that many students had difficulties to understand and may be, through these commands, bypassed altogether. It is also possible to make future developments with double precision sensors (using two ports, instead of only one).

There are two issues that must be addressed in future versions of the kit. Firstly, the actuators are powered by the board itself. While the Arduino board is rugged, it can only provide 02, Watts of potency from an IO pin, and a max 1 Watt from the combined pins. Future developments will require a separate power supply, in order to power other types of actuators. Secondly, the mechanical structure is still crude and not flexible enough to be used in different assemblies. Commercial kits usually have a set of mechanical parts that can be used in numerous ways, and the ROBO+EDU should have a full list of pieces which may be ordered by the schools. With the popularization of 3D printers it may even be possible for them to print their own projects. A solution is being investigated in the OpenStructures project, using proportional grids to maintain the interconnectivity of different pieces.

It is also being developed, in association with another project called ROBOCETI, a visual programming platform called Visuino (JOST et al., 2012), which is also Open Source and seeks to facilitate the kits use in the interested educational institutions. This platform is multi-language, so it can be configured to other languages other than the standard Arduino, so it's possible to use it with other kits and technologies. As noted above, through the use of specific libraries, it is possible to achieve a high level of abstraction to teaching, making it easy to assimilate the logic that subsides the robot project.

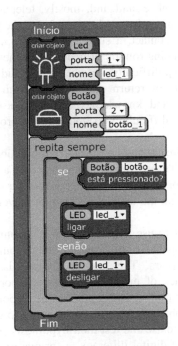

Fig. 2. Visuino design reference

3.2   Kit Implementation

To disseminate this kits to the schools taking part on the project, workshops will be granted to the teachers and monitors responsible to the activities in each school taking part on the project. To the moment, there are 260 schools interested, located all over country. Given this, we opted to provide the workshops in each of the 9 state capitals, and in the federal district, covering every region and allowing participation of most schools.

This project presents its participants with an introduction to Educational Robotics, it's implications and possible applications. So the courses will need to have a brief introduction on robotics, followed by practical examples, both in hardware and in software, with the aforementioned kits, which will be given to the schools after the workshop. This way it should be possible to promote spaces where the teachers might feel motivated and confident enough to pursue the use of robotics on their own schools.

4   Challenges to Implement Educational Robotics

In the beginning of the project, 56 schools decided to take part in these activities. The first step to this was to make contact with these schools and to make the first tutorials and support materials to any kit the schools were already using. The contact with the 56 schools was made through e-mail and, mostly, telephone. In this first moment, there were difficulties to contact these schools and find the people responsible in each institution. Established the contact, a survey was made to find out which of these schools were already developing robotic projects. From these, 25 had bought different types of kits, 11 were having difficulties to use them, 7 did not wish to continue with these projects and 13 did not return our contact. From the variety of situations encountered, multiple kits and knowledge, a new approach was required. These questions reflect a series of difficulties to implement a project like this in a national level.

Even with the schools already having the educational robotics kits, and the teachers and students recognizing it's possibilities, resistance is a recurring factor amongst them. One of the main factors for this resistance is the fear to break the kit, or the lack of extensive knowledge about the technology, which prevents an effective manner to the work.

It is known that the fear to use a specific tool is a common factor for the digital immigrants, or anyone who hasn't followed, from a young age, the technological development of the last years. The fear to "break" is a common one, and is exacerbated through the lack of maintenance in the laboratories and equipment of public schools. Usually, the teacher ends up being the responsible for the maintenance and the solution of any problem, which makes difficult to use certain equipment and does not leave them at will to use and test their hypothesis and new learnings.

This consists in a sort of digital illiteracy, a serious problem as seen in Araújo e Glotz (2009, p.7), since many of the teachers, who are responsible for their students literacy, "also suffer the same problem many of their students also have", that is, many of them does not know how to use the technologies they're working with.

Another factor that is frequently reported by the teachers taking part in the trainings, is the lack of connection between teacher responsible for the Educational Robotics activities and the classroom teachers. If there is only one responsible teacher to maintain these activities, he may be isolated from the rest of the learning activities, being this his only demand. In discussions, many teachers comment that technological resources should be better used and problematized if they don't stay in only one room, or are used in only one class period. This way both teacher and students can have freedom to use these equipments, in a more frequent and attractive way. According to Araújo e Glotz (2009), these teachers show a "rejection regarding the use of those on their activities because they are unaware on the ways to use them in their activities, and of their actual potential" (p.7). Coscarelli e Ribeiro (2011) also stress that

> while not favoring the access to the computer and not turning it into an ally for education, especially for the popular classes, the school will be contributing to another form of exclusion of the students, noting that this will also exclude them from many other instances of contemporary society, which requires its citizens a degree of increasing literacy (p. 32).

This way, it is up to the school to allow the student do assume a central role on his own learning process, not being treated as a passive receiver of information, but acting and being protagonist of his apprenticeship. For this, according to Guimarães (2012), the learning process must become increasingly "personalized, focused on individual needs and interests" (p.127).

Keeping this in view, it is understood that, in order to increase the use of Educational Robotics to be expressive, it is essential to keep track of the activities realized in the schools. This will provide teachers with spaces in which they can count on a qualified professional to give assistance in their educational work. In fact, when it comes to technology, most teachers have great difficulties to adopt them in their daily routines, while , in another extreme, their students easily manipulate electronic devices. If the teacher does not have a "minimal" domain over the resources presented by digital technologies, it is useless to talk on how to develop teaching strategies to exploit their functionalities.

5   Final Considerations

While Educational Robotics is becoming a promising field for educational projects, the necessary technology is still scarce in Brazil, making its application on public schools as part of a larger educational project impossible. On one hand, there are products in the market, with limited supply and a limited flexibility, while also requiring the schools to work with a specific commercial brand. We defend the use of Open Source kits as the best alternative for these activities, but they will need an an active development community in order to grow. To such end, it is necessary to search or support in programs, such as the Mais Educação, and build partnerships between universities, technical schools and independent development groups which can help the schools through technical support. Building both software and hardware on popular Open Source platforms may attract hobbyist developers and help to further spread the technology.

> In light of the ideas here exposed, it seems advisable that the initial student-technology interactions teach children and adolescents to recognize, explore and learn how to program Open Source softwares. As evidenced by certain experiences, learning to program in an early age stimulates logical and analytical thinking, independent learning, collaboration and adaptive and innovative creativity, among other skills (COBO, 2009, p. 130)

The project ROBO+EDU is still developing and testing its prototypes, but the use of Open Source technologies is promising, and points to a new kind of educational technology that can interoperable and more flexible than those already in the market. While the electronic design and software platforms are presented, it is still lacking a flexible mechanical solution to compete with the commercial platforms, and the current prototype can be built in specific ways (namely, as a autonomous car). In developing an Open Source solution, one of the biggest challenges are the construction of supply chains, more than once components were changed for those more available in the market. This is especially difficult when dealing with public founded projects, with specific and inflexible rules for bidding.

Therefore, it is understood that the accompaniment of the development team in the inclusion of this tool is essential to the effectiveness of the project. The development of new teaching skills is a determining factor in pedagogical practices, especially those involving the use of new digital technologies. Although there is a great offer of training in digital technologies, they usually treat them as an end, while technology itself will not be able to change traditional teaching practices. It is proposed that teacher training should explore and develop both technical and pedagogical skills, guided by the sense that knowledge reconstruction can relate information and its practice, since most teachers are not yet very familiar with digital technologies. Teacher training need to include reflections on how these technologies help foster an environment of interaction and development of knowledge amongst those involved in the learning process.

References

Araújo, V.D.L., Glotz, R.E.O.: O Letramento Digital enquanto Instrumento de Inclusão Social e Democratização do Conhecimento: Desafios Atuais. Paidéi@ (Santos) 2, 1–26 (2009)

Cobo, C.: Aprendizaje de Código Aberto. In: Balaguer, R. (org.) Plan Ceibal. Los ojos del mundo en el primer modelo OLPC a escala nacional. Pearson Educación, Montevideo (2010)

Coscarelli, C.V., Ribeiro, A.E. (Orgs.): Letramento Digital: aspectos sociais e possibilidades pedagógicas, 3rd edn. Autêntica, Belo Horizonte (2011)

Guimarães, L.S.R.: O aluno e a sala de aula virtual. In: Litto, F., Formiga, M. (orgs.) Educação a distância: o estado da arte, vol. II. Pearson Education Brazil, São Paulo (2012)

Lopes, D.Q., Fagundes, L.C.: As Construções Microgenéticas e o Design em Robótica Educacional. RENOTE - Revista Novas Tecnologias na Educação 4, 1–10 (2006)

Morelato, L.A., Nascimento, R.A.O., D'abreu, J.V.V., Borges, M.A.F.: Avaliando diferentes possibilidades de uso da robótica na educação. Revista de Ensino de Ciências e Matemática - REnCiMa 2, 80–96 (2011)

Piaget, J.: Desenvolvimento e Aprendizagem. Tradução: Slomp, P.P. In: Lavatelly, C.S., Stendler, F. (eds.) Reading in Child Behavior and Development. Hartcourt Brace Janovich, New York (1972)

Smartphone Controlled Robot Platformfor Robot Soccer and Edutainment

Thomas Tetzlaff, Reza Zandian, Lukas Drüppel, and Ulf Witkowski

South Westphalia University of Applied Science, Lübecker Ring 2, 59494 Soest, Germany
lukas.drueppel@stud.fh-swf.de,
{tetzlaff.thomas,zandian.reza,witkowski.ulf}@fh-swf.de

Abstract. This paper describes a low cost autonomous robot platform with an integrated smartphone. The robot is engineered to play robot soccer in the AMiREsot league as well as be used as an educational platform and for research purposes. The robot's base board includes the electronics required for driving and controlling the motors, accessing the infrared sensors used for object detection and communicating with the smartphone via USB. The smartphone, which has a camera and several integrated sensors, serves as the brain of the robot. The design was focused around low cost and high performance via the integration with a smartphone. Preliminary tests of the platform include object detection and ball following using image processing techniques and the camera of the smartphone. Following testing, it has been concluded that this low cost robot platform with reduced hardware complexity is suitable for education, research and entertainment purposes.

Keywords: AMiREsot robot soccer, mini robot, smartphone, Android, education platform.

1 Introduction

The increasing number of robots in industry and the rising complexity of their tasks, often create challenging issues involving the cooperation and localization between multiple robots as well as the robot's ability to interact with the environment. In order to solve these issues, a simulation and modeling environment is required to further the development process and examine proposed theories. One of the best ways of implementing the simulation environments is by focusing on the most technologically demanding robot soccer activities such as image processing, sensor integration, signal processing, software programming, telecommunications, artificial intelligence, etc.. In addition, the entertainment and enjoyment factors when working with mobile robots ease technological difficulties, attract the younger generation and convey educational concepts to students.

The Federation of International Robot-soccer Association (FIRA) has developed several robot soccer leagues with the aim of implementing theortical algorithms in a dynamic, complex and adaptive environment using multi-robot systems. One of these leagues is AMiREsot (Autonomous Mini Robot for Research and Edutainment),

© Springer International Publishing Switzerland 2015

J.-H. Kim et al. (eds.), *Robot Intelligence Technology and Applications 3,*

Advances in Intelligent Systems and Computing 345, DOI: 10.1007/978-3-319-16841-8_46

which focuses on implementing autonomous mini robots supported by on board vision systems in a soccer game [1]. The first guidelines for the AMiRE Soccer Tournament rules were defined at the workshop held at the 4th AMiRE Symposium in 2007. According to the game rules [2], robot teams should compete with one another fully autonomously and without any human interaction or external data processing unit supporting the system from outside of the field; however interaction between multiple robots is allowed. Robots should use their own vision system for processing and monitoring their surroundings, but they can also use other sensors as long as these sensors do not disturb or disrupt similar sensors on the opponent team's robots. The use of wheels is allowed to simplify the design of the robots. The robots should compete together in team sizes of either one, three or five robots inside a pitch with the dimensions of 2.0 m by 1.4 m.

Considering the rules and conditions defined in the AMiREsot league, the Department of Electronics and Circuits Technology of the University of South Westphalia has defined its goal around the designing and developing of a modular robot platform for these soccer games. The first version of this platform was introduced in [3] and presented in TAROS 2012. The features of this platform include its modularity, availability of a wide range of sensors, communication modules, peripherals and processing power, the possibility to simply change the features of the robot based on the application context and finally its interaction with a user via a set of user interface platforms. A structure diagram for this platform along with samples of the actual platform are shown in Fig. 1 and Fig. 2 respectively.

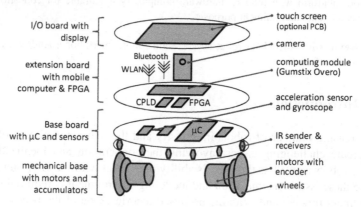

Fig. 1. Modular architecture of the AMiREsot robot soccer platform presented in [3]

According to the assessment present in the works of students from several educational semesters, this platform contains a wide variety of features suitable for applications that range from simple obstacle avoidance to maze exploration and map building [3]. The results of the student's work prove that the robot can serve as an interacting tool which generates a high motivation for students to work in the mentioned areas [3].

F. Mondada et al. [4] pioneered the design and development of mini robots for research and education by introducing the Khepera 1 mini robot platform. Their platform was an out of the box solution which was reasonable in price and was used in education for many years [5]. Similar work which is intended for research and education is presented by S. Herbrechsmeier et al. [5] from the University of Bielefeld. Their system is a modular platform which integrates a sensor processing unit, an actuator unit and performs the cognitive processing required to develop an autonomous mini robot. Other institutes also performed similar projects aimed at bringing a low cost mini robot to market. An example is the Romo robot [6], which is designed around iPhone devices and is used for education and entertainment. This platform uses a smartphone as its processing unit, while its base only contains light indicators and the mechanical parts required for motion. Other similar examples are mentioned in [7][8].

Fig. 2. AMiREsot with touch-screen in the field presented in [3]

Although the previously presented system from the University of South Westphalia was very effective for educational purposes and successful with regards to its implementation for its defined applications, the overall cost of fabrication and maintenance was still relatively high, making it unaffordable for many organizations. Additionally, the complexity of the hardware is a major issue which requires time and effort for manufacturing and assembly. Furthermore, the technological demands of robot soccer (e.g. image processing) require a powerful processing unit with additional memory space, increasing the cost and complexity of the design. Therefore this paper presents a new platform which contains the beneficial features of the previous platform but is designed around a very low price point and minimum hardware complexity. The proposed idea is to use a smartphone as the main processing unit which also contains an internally integrated camera, several internal sensors (gyroscope, accelerometer, temperature sensor, etc.), and communication modules (Bluetooth, Wi-Fi, etc.) in combination with a basic driver circuit which interacts with the smartphone through its USB interface. Due to mass production, smartphones usually offer high processing power for a low price considering their provided features (e.g. sensors, display). These facets make smartphones suitable for being used and integrated in robotics. As many features are internally integrated inside the smartphone, students and researchers can focus on developing techniques, algorithms, software solutions and improving state of the art solutions instead of wasting time on developing hardware which ultimately increases the costs.

In this paper, the robotic platform is explained from both the hardware and software point of view. Section 2 discusses the hardware in detail, including the platform's electronic and mechanical parts. In Section 3 the smartphone programming environment, the required application for communicating with the base and other details relevant to the software are explained. Section 4 provides examples of application areas with some basic results. Section 5 describes the steps for future implementation of the platform. The final section draws conclusions and summarizes the results.

2   Hardware Architecture

The hardware of the developed platform is constructed of a main PCB which also serves as the chassis, motors and wheels, battery packs, infrared sensors and a smartphone.

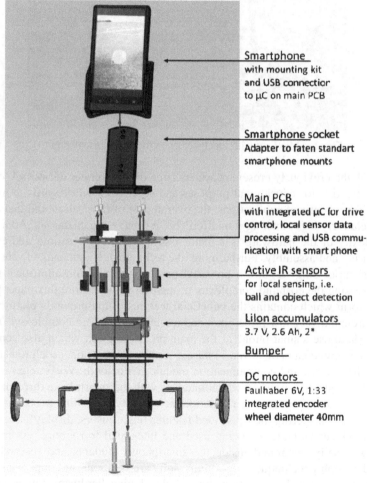

Fig. 3. The assembly plan of the AMiREsot robot platform with integrated smartphone

This section discusses the criteria considered when designing the main PCB, determining its functionality, and choosing the mechanical parts and their features which includes the smartphone holder. Fig. 3 shows the structure diagram of the evolved robot.

2.1 Main PCB

As shown in Fig. 4, the former AMiREsot robot platforms use a metal chassis as the mechanical base which includes different cut outs for the connectors of the motors, batteries and the IR sensors. The connection to the IR sensors is complex and requires a flex-PCB which greatly raises the production costs. To reduce production costs, the mechanical base is replaced by the main PCB board. The motors and motor mounting angles are the remaining needed mechanical parts. The already mentioned flex-PCB used to connect the IR-sensors shown in Fig. 4 (right) is no longer needed due to the fact that the sensors can be directly attached to the main PCB board as shown in Fig. 5.

Fig. 4. Previous version of the mechanical base (top view and bottom view) and equipped robot

To use the main PCB as a chassis, the motors have to be connected to the PCB. The dimensions of the main PCB have not changed with respect to the previous metallic base, allowing the use of the motor holders from the previous design. The robot platform's main PCB has to support several applications necessary for operating a smart robot with complex behaviour. These range from processing IR sensor data to handling communication with the smartphone. The aim is to be able to use the board for different kinds of applications and not just for robot soccer. As a result of this, the robot can be used for different activities defined by particular lab work. Examples of simple individual robot behaviours include obstacle avoidance, e.g. the Braitenberg behaviour [20], wall following or following a moving object. Additionally, the robot should be able to act as part of a team of robots performing multi-robot behaviours, e.g. swarming or cooperative sensing. In this case, the robot will be equipped with an additional smartphone, see section 4. This platform is well suited to be used for educational purposes. Students can easily increase their knowledge and experience in microcontroller programming, the implementation of sensors, robotic behaviour design and different wired communication systems.

To achieve success in the aforementioned application areas, the robot platform has to be able to drive its motors, read the IR distance sensors and communicate with the smartphone. The interfaces utilized for these purposes are the I²C bus and USB. The primary purpose of the IR sensor system is for the detection of objects nearby and at a wide angle (i.e. ball detection), the identification of other robots and the detection of the pitch border. Obstacles can be detected via eight infrared sensors, four being located on the front and four on the back of the robot. The robot system is propelled by two DC motors with a 33:1 gear, thus providing the robots with the appropriate acceleration and driving speed necessary for moving quickly around objects or following other robots or objects like a ball [10]. The motors are internally equipped with incremental encoders, which enable measuring the driven distance and therefore controlling the speed of the wheels.

Fig. 5. Main PCB serving as the chassis with mounted motors, wheels and sensors

In order to accommodate the required processing tasks, the robot contains a 32 bit ARM cortex M3 microcontroller from STMicroelectronics (STM32F103RBT6) [7]. This controller supports multiple peripherals including 2 integrated A/D-converters with up to 16 channels for processing the analogue signals of the 8 infrared sensors. Furthermore, the required bus interfaces needed for communication with the IR sensors, motors and smartphone are available. The microcontroller supports up to 7 timers which clock several time controlled processes, including a PWM signal for each motor. In total, up to 50 I/O-pins are available for interfacing external devices [8]. For near range obstacle detection, the already mentioned 8 infrared sensors (type TCRT5000 from Vishay Semiconductors [9]) are integrated. They are compact in size and can be mounted on the main PCB using an adaptive extension PCB to position them at different angles. A driver IC (Texas Instruments TLC5927IDBQR) handles current management for the infrared sensors. The sensing range can be adjusted via an external resistor which is set through the GPIOs of the microcontroller.

2.2 Function of the Standalone Main PCB

Robots need ambient monitoring sensors to behave autonomously. Since the sensors on the main PCB are limited to IR distance sensors and incremental encoders in the motors, the standalone main PCB has reduced functionality. The current functions of the microcontroller on the main PCB include obstacle avoidance, wall following, following or tracking a moving object and guiding a soccer ball (shown in Fig. 6).

Fig. 6. Standalone main PCB guiding (dribbling) a soccer ball using IR-sensors

2.3   Smartphone Socket

The aim of this robot platform is to fuse a microcontroller and smartphone. The sensors in the smartphone, i.e. gyroscope, accelerometer and camera, are used to navigate the robot. Hence the smartphone must be mechanically fixed to the robot. To preserve flexibility, a specific type of smartphone is not required and a flexible adapter is implemented as shown in the Fig. 7 (left). This adapter, which has been fabricated using 3D printing, enables the fastening of standard smartphone mounts. Fig. 7 shows two connected mounts: the mount in the middle is a purchased Google Nexus 4 mount from Brodit, and the one on the right is a self-developed mount for a SONY XPERIA.

Fig. 7. Smartphone socket without and including smartphone mounts for different models

3   Smartphone Integration

The previous version of the AMiREsot-robot was equipped with a FPGA, Gumstix Overo board, and a microcontroller which serve as central processing units. In order to add extra peripherals, e.g. different types of sensors, a camera, a display or user input buttons to the system, auxiliary PCBs needed to be attached to the main PCB.

In most smartphones, a high quality digital camera is integrated which enables the device to capture images of its environment. To enable communication with external devices, modern smartphones are equipped with various interfaces (e.g. WiFi, Bluetooth, NFC, USB etc.). The integrated touch display and operating system provide a convenient operating interface to the device. One of the operating systems is Android, which is an open source OS released by Google. The "Android SDK" provides a huge number of tools to help ensure a quick and easy start when developing applications for Android phones. These characteristics all combine to form a single powerful processing unit on which basic robot soccer applications can be run.

3.1   Software Running on the Main PCB for USB Communication with the Smartphone

To enable communication with the smartphone, the integrated USB interfaces of the microcontroller and smartphone are utilized. In order to support communication over the Universal Serial Bus, the main PCB needs to be programmed to act as a USB-device.

The microcontroller on the main PCB is programmed in the C programming language using the Keil µVision 4 [13] software development environment shown

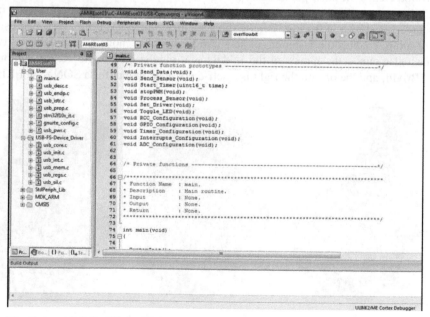

Fig. 8. A snapshot of Keil µVision 4 Integrated Development Environment (IDE) [13]

in Fig. 8 and the Keil ULINK-ME [14] programming device. STMicroelectronics provides a USB-device library for STM32-microcontrollers to help facilitate the development of applications for the microcontroller's USB-interface. The general structure of this library is shown in Fig. 9.

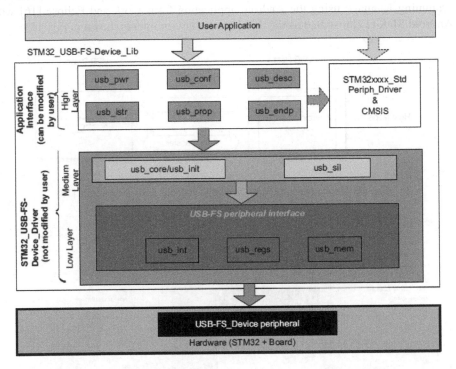

Fig. 9. STMicroelectronics USB-FS-Device-Library structure [15]

Based on this library, the main PCB is programmed to act as a USB-device with two bulk-endpoints and unique vendor and product IDs. For transmissions from the main PCB to the smartphone, "Endpoint 1" is set up as an IN-endpoint with a maximum packet size of 32 bytes, while "Endpoint 2" is setup as an OUT-endpoint with a maximum packet size of 8 bytes. As soon as data packets are received, an interrupt is triggered, and the data is processed.

The data transmitted by the smartphone consists mainly of driving commands, which need to be executed by the main PCB. In order to perform this, a four channel timer enclosed in the microcontroller is setup in pulse-width modulation mode to control the motor drivers based on the desired speed values.

Once the robot is moving, the incremental encoders implemented in the motors provide a signal containing information about the motor's rotation speed and direction. In order to be able to process this information, another timer is setup in encoder mode and increments or decrements the timer's counter based on the encoder signal. As soon as the smartphone makes a request for the sensor values, the buffer's contents including the sensor values will be transmitted.

3.2   Software Modules Prepared for the Smartphone to Enable Communication with the Main PCB Base

The application running on the smartphone uses the USB Host mode (Fig. 10) which has been supported since Android 3.1. The software is developed in the Java programming language using the Android application framework and Eclipse [16] with Android SDK [17] installed as the development environment (shown in Fig. 11).

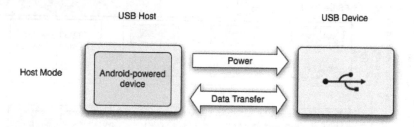

Fig. 10. Hardware configuration of the Android USB Host mode [18]

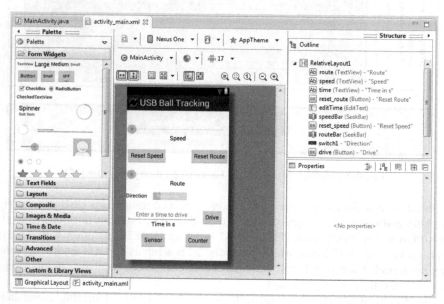

Fig. 11. A snapshot of Eclipse development environment with Android SDK installed [16][17]

An intent filter is implemented in the software running on the smartphone. Once a device with the specified vendor and product IDs is attached to the phone, the intent filter recognizes the action and starts the application. The intent contains the information needed to setup a connection to the USB-device. As soon as the connection is established, the Android application retrieves information from the endpoints specified on the microcontroller of the main PCB, and sorts them by direction of data flow and type of transmission. From this point on, the application is able to either write data to an OUT-endpoint or to request data from an IN-endpoint.

In order to ensure consistency in the data packets sent to the main PCB, a communication protocol is defined, where the first byte sent is a protocol byte. This protocol byte indicates the transmission of driving commands; 0xFF in this case. The second byte contains information regarding the desired rotating direction of each motor. The four following bytes contain the desired speed values that the motors should adopt. The transmission protocol is shown in Fig. 12.

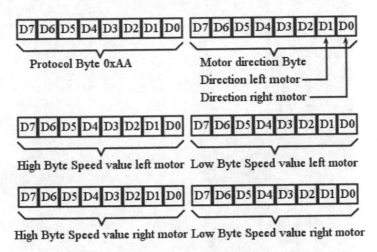

Fig. 12. Motor Control Transmission Protocol

In order to be able to process both the incremental encoder data and the infrared sensor values, the application periodically requests the transmission of this data. The application can be adapted to respond to data transmitted from the main PCB (e.g. to change the rotational speed of the motors).

4   Current Robot Skills

One of the main tasks in robot soccer is determining the location of the ball on the pitch. For the initial implementation of the robot soccer algorithms, an Android application has been developed that tracks and follows a colored ball using the integrated camera in the smartphone. In order to be able to visualize the environment, the smartphone's camera is used to capture image frames which are directly displayed on the screen. The OpenCV [19] image processing toolbox is used to process the camera frames by running standard image processing algorithms.

Before the image processing can start, the Android application needs to be notified of the ball's color. This is done by the user touching the ball area on the display. The application builds a 5x5 matrix around the touched pixel and assumes the average value to be the ball's color. Once the color is stored, image processing algorithms are used to filter the color of the ball from the camera frame. To detect the position of the ball in the image, the circumference of the ball is found using a contour finding algorithm and its center coordinates are extracted. A minimum bounding rectangle is fitted to the ball's circumference and displayed on the screen, along with a point indicating the center of the ball. Fig. 13 shows a robot running the ball-tracking application.

Fig. 13. Smartphone based ball tracking via image processing performed on a phone

Based on the information gathered during the image processing (e.g. center and radius of the ball), the robot's motors are controlled to follow the ball. Once the ball is in range of the robot's infrared sensors, the motors are stopped and the output voltage of the infrared sensors can be captured. For the image processing to work, the color of the ball needs to clearly separate from the colors appearing in the robot's environment.

5   Future Implementations

Implementation of soccer algorithms and improvements to the main PCB, based on the current state of AMiREsot, are planned and in the developmental stage. Right now priority is given to algorithms which search for the ball on the pitch, arrive at the ball using the ball tracking algorithm explained above, and that push the ball in a certain direction using data gathered from the infrared sensors.

Another important task is the ability to orient the robot on the pitch. It is planned to achieve this task by combining line detection with the incremental encoder values. A future version of the main PCB will be additionally equipped with an accelerometer and a gyroscope. These will help facilitate the robot's ability to orientate itself on the pitch by providing useful data regarding the robot's movement.

To be able to implement simple soccer algorithms, such as moving the ball towards the goal, the robot needs to gain knowledge of the location of the opposing goal. This is to be done through image processing and using the data gathered from the motion sensors.

Furthermore, feedback control for the DC-motors is needed to ensure that the desired rotational speed is reached and stable. To implement feedback control, a reasonably accurate model of the robot needs to be created. Based on this model, simulations in MATLAB/Simulink can be run to get insight on the robot's behavior (e.g. step response, bode plot), and a controller can be designed using an appropriate design method. In addition to the modeling and simulation of the system, it still needs to be determined whether to implement the feedback control on the main PCB or on the smartphone. Due to latencies in the Android application caused by the process scheduler of the Android OS, adaptive control may be required when executing the feedback control on the smartphone.

6   Conclusion

This work's intention was to reduce the production costs of a mobile robot for use in research and education. Firstly, the hardware complexity was reduced by using the main PCB as the chassis. This also allows for the mounting of distance sensors directly to the main PCB. Thus the flex PCB used to connect IR sensors in previous versions is no longer needed. The main PCB with its integrated microcontroller can perform simple tasks like obstacle avoidance or run a wall following algorithm. The main PCB provides a platform which allows for research and education projects involving microcontroller based programming and advanced control, e.g. motor control. Furthermore, the evolved robot platform allows for a cost effective design by fusing a microcontroller and smartphone together using USB. To serve this purpose, a flexible smartphone adapter was developed which enables the use of common smartphone mounts to fasten a smartphone mechanically to the robot. By combining a microcontroller and smartphone, tasks can be split by complexity. The microcontroller can handle simple tasks like controlling the motors and measuring distance values. An application has been completed which involves the following of a ball by utilizing the camera integrated within the smartphone. The complete objective of developing a cost effective robot platform with a reduction in hardware complexity was achieved. The new platform will serve as a reference design to groups interested in robot soccer programming as well as smartphone enabled robotics edutainment. The future objective is to compete at the 2015 FIRA world cup robot soccer tournament in the AMiREsot league.

References

1. The Amiresot game, http://www.fira.net/contents/sub03/sub03_2.asp (accessed June 2014)
2. The AMiREsot Robot Soccer Tournament, http://www.fira.net/contents/data/AmireSotDescription.htm (accessed June 2014)

3. Tetzlaff, T., Wagner, F., Witkowski, U.: Modular Mobile Robot Platform for Research and Academic Applications in Embedded Systems. In: Advances in Autonomous Robotics - Joint Proceedings of the 13th Annual TAROS Conference and the 15th Annual FIRA RoboWorld Congress, Bristol, U.K. (2012)

4. Mondada, F., Franzi, E., Ienne, P.: Mobile Robot Miniaturisation: A Tool for Investigation in Control Algorithms. In: Yoshikawa, T.Y., Miyazaki, F. (eds.) Experimental Robotics III. LNCIS, vol. 200, pp. 501–513. Springer, Heidelberg (1994)

5. Herbrechtsmeier, S., Rückert, U., Sitte, J.: AMiRo – Autonomous Mini Robot for Research and Education. In: Advances in Autonomous Mini Robots. Proceedings of the 6th AMiRE Symposium (2011)

6. The Romo robot official website, http://www.romotive.com (accessed September 2014)

7. Caprari, G., Arras, K.O., Siegwart, R.: Robot Navigation in Centimeter Range Labyrinths. In: 1st International Symposium on Autonomous Minirobots for Research and Edutainment (AMIRE 2001), Paderborn, Germany, pp. 83–92 (2001)

8. Caprari, G., Siegwart, R.: Design and Control of the Mobile Micro Robot Alice. In: 2nd International Symposium on Autonomous Minirobots for Research and Edutainment, Brisbane, Australia, pp. 23–32 (2003)

9. Faulhaber: Datasheet DC-Getriebemotoren, http://www.faulhaber.com/uploadpk/DE_2619_SR_DFF.pdf (accessed January 2012)

10. STMicroelectronics. Overview about the STM32F103ZG-Microcontroller, http://www.st.com/web/catalog/mmc/FM141/SC1169/SS1031/LN1565/PF164487?sc=internet/mcu/product/164487.jsp (accessed April 2014)

11. STMicroelectronics. Overview about the STM32F103PG-Microcontroller, http://www.st.com/st-web-ui/static/active/en/resource/technical/document/datasheet/CD00161566.pdf (accessed June 2014)

12. Vishay: Reflective Optical Sensor with Transistor Output, http://www.vishay.com/docs/83760/tcrt5000.pdf (accessed January 2012)

13. ARM Ltd.: Keil μVision 4, http://www.keil.com/uvision/uv4.asp (accessed June 2014)

14. ARM Ltd.: Keil ULINK-ME, http://www.keil.com/ulinkme/ (accessed June 2014)

15. STMicroelectronics: STM32 USB-FS-Device development kit. rev 12, http://www.st.com/st-web-ui/static/active/en/resource/technical/document/user_manual/CD00158241.pdf (accessed June 2014)

16. The Eclipse Foundation: Eclipse, https://www.eclipse.org (accessed June 2014)

17. Google Inc. & Open Handset Alliance: Android SDK, http://developer.android.com/sdk/index.html (accessed June 2014)

18. Google Inc. & Open Handset Alliance: USB Host and Accessory, http://developer.android.com/guide/topics/connectivity/usb/index.html (accessed June 2014)

19. Itseez: OpenCV4Android, http://opencv.org/platforms/android.html (accessed June 2014)

20. Braitenberg, V.: Vehicles: Experiments in synthetic psychology. MIT Press, Cambridge (1984)

AMiRoSoT: An Autonomous, Vision Based, Low Cost Robot Soccer League

Joaquin Sitte[1] and Ulf Witkowski[2]

[1] Queensland University of Technology
Brisbane QLD 4000, Australia
j.sitte@qut.edu.au
[2] South Westphalia University of Applied Science
59494 Soest, Germany
witkowski.ulf@fh-swf.de

Abstract. The Autonomous Mini-Robot Soccer Tournament (AMiRoSoT) league that we describe in this article uses small size robots and a small size field to make the league widely accessible for education and research by keeping the entry barrier to the competition low. We describe the motivation for small size soccer robots and the design criteria applied to the formulation of the league's rules. We examine the requirements for the robots capable of playing in the league and discuss possible low cost robot realisations. An analysis of the goal kicking behaviour illustrates the breath and depth, and the educational and research opportunities, of the league.

1 Introduction

The idea of using the soccer game played by robots as a *Grand Challenge* problem in Artificial Intelligence arose in 1992 [8] , at par with other landmark problems such as computer chess. Robot Soccer was to provide a platform for research into design principles of autonomous agents, multi-agent collaboration, strategy acquisition, real time reasoning and sensor fusion [4]. Later, in 1998, Kitano and Asada [5] formulated the grand challenge as follows:

> By mid-21st century, a team of fully autonomous humanoid robot soccer players shall win the soccer game, comply with the official rule of the FIFA, against the winner of the most recent World Cup.
> *H.Kitano and M. Asada IROS 1998*

The robot soccer grand challenge poses an ambitious goal. In contrast to the abstract computer chess challenge, where the IBM Deep Blue team defeated world champion Garry Kasparov in 1997, the robot soccer challenge is grounded in the real world by sensing the environment and acting on it to fulfil a purpose. Winning a soccer game is a collaborative task requiring fast decisions and dexterous motion by the players in an environment made highly uncertain by the actions of the other players on the field. Robot soccer in its various leagues provides a benchmark for fast moving multiple robots that collaborate on a collective task. Meeting the robot soccer grand challenge requires solving most of the problems in the way of building useful autonomous machines for a wide

© Springer International Publishing Switzerland 2015
J.-H. Kim et al. (eds.), *Robot Intelligence Technology and Applications 3*,
Advances in Intelligent Systems and Computing 345, DOI: 10.1007/978-3-319-16841-8_47

range of task. Most of the engineering and scientific challenges in building mobile factory robots with the versatility of human workers, personal robot assistants, care robots, rescue robots and many others, are also present in the robot soccer game. We can only imagine what a machine that has the dexterity and cognitive abilities of a human soccer player could do.

The robot soccer grand challenge has motivated yearly international robot soccer championships in various leagues since 1996. For a history of early robot soccer see [8] and [2]. These competitions have stimulated many lines of robotics research and have contributed to significant technical advances. Despite the advances, a long road is still ahead. The surface has barely been scratched and there is the potential for technical spin-offs along the way. The robot soccer game, or any similar physical team game, provides a well defined, yet evolving benchmark problem, that by lacking any direct commercial value is on neutral ground for private and industrial players to collaborate.

As by 2014, the two largest robot soccer championships are the RoboCup and the FIRA (Federation of International Soccer Associations) championships. RoboCup aligns closely with the robot soccer Grand Challenge. The participating teams typically come from tertiary institutions and are run by research students. The FIRA league has a strong educational focus.

Robot soccer benefits from the world-wide popularity of soccer as a spectator sport. Thereby, robot soccer has been a very successful motivator for students to learn about science and engineering. Robot soccer is open for wide participation at all levels of technical competence from primary to graduate school. Regrettably the educational and entertainment virtues of robot soccer tend to overshadow its unique features as a technological grand challenge and benchmark for autonomous systems engineering.

The structure of this paper is as follows: In section 2 we examine the contributions and limitations of small size soccer leagues; this leads us to specify the new autonomous small size AMiRoSOT league in Section 3. The new league preserves the advantages of a small league while overcoming many of its limitations. In Section 4 we analyse the characteristics of the robots for the league and briefly explore some options for building the robots. One of these options is building a robot using a mobile phone as the main processing unit. This is described in detail in a companion paper. Section 5 addresses some practical issues related to keeping a low entry barrier for competing in the game. Section 6 relates the demands of an autonomous goal kicking behaviour to the hardware and software characteristics of the AMiRoSoT robots. Section 7 presents the conclusions.

2   Small Size Leagues

Within each of the RoboCup and FIRA robot soccer championships there are several leagues that differ in their resource requirements and educational and research values. Larger size robot leagues always received the most attention. It is easier to make larger robots behave autonomously by equipping them with fast computers for handling on-board vision. In recent years the humanoid robot soccer leagues are drawing the most attention.

The small size robot leagues (180 mm diameter or less) use *global* vision. This means the robots receive movement commands from an off-field computer that processes the video images from a camera that looks at the play field from above. These robots are just small remotely controlled vehicles without on-board sensors. Without their own sensors they cannot be autonomous. Games of up to eleven robots per team are feasible because of the low cost and the small size of the players. The games in the small size leagues can be quite fast and the leagues are well suited for developing game strategies and studying multi-robot cooperation [1]

In the past the only small size robots capable of an autonomous, yet simple, soccer game where the Khepera mini-robots designed by Francesco Mondada at EPFL and produced by the K-Team company [7]. The Khepera robots were small (60 mm diameter) two-wheeled differentially steered minirobots. The accessory for the Khepera that made an autonomous soccer game possible was a camera using a single line of 256 monochrome pixels. With this camera the goal, the ball and other players could be detected in a simplified environment. These single line images could be processed on the robot. Together with infra-red proximity sensors and wheel position encoders the robot could be programmed to execute simple soccer player behaviours such as ball and goal location, ball dribbling, goal kicking, goal keeping, evading the opponent and more. The first Khepera robot soccer competition was the Danish Championship in Robot Soccer held in 1997 [6]. The game was one robot against one robot played in a box about 1m long and about 0.70 m wide that would easily fit on a desk. The technology available at the time limited the Khepera robot soccer to a slow game.

The great virtue of the Khepera robot soccer was its accessibility by students. At around 3000.- Swiss Francs in 1997 the Khepera robots where not cheap, but they were affordable for university computer science and electrical engineering departments. With the Khepera robots, key ideas of autonomous behaviour in the real world could be demonstrated. In the soccer game students could learn about autonomous localisation and navigation, motion control, sensor signal processing, signal feature extraction, how to deal with sensor signal noise and uncertainty, real time computing, and much more. We used Khepera robot soccer for 8 years as a student term project in teaching Computational Intelligence. Starting from scratch, in a single semester, teams of two students could program basic autonomous soccer playing behaviours that had to be evaluated in end of semester competitions.

After more than a decade the Khepera league was superseded by advances in technology. The need arose for a successor league that providing a quantum leap in capabilities by taking advantage of new low cost technology, such as fast 32-bit processors and cheap cameras, while preserving the accessibility of the league.

3   The AMiRESot League

It was clear that a successor to the Khepera robot soccer game had to be played by autonomous robots that use vision for perceiving what happens around them. Small low cost robots with on board vision processing would enable a robot soccer game with technical and research level challenges that can keep the game interesting over many years to come. A first draft for set of rules defining the AMiRoSoT league was formulated in a workshop at the 5-th AMiRE (Autonomous Mini-Robots for Research and Edutainment) Symposium in Buenos Aires in 2007. In 2008 we published the rules for the AMiRESot league [9]

To keep the hardware costs low the robot players must fit into a vertical cylinder of 100 mm diameter. With this diameter the cross section area of the robot is nearly four times that of the Khepera robot and is sufficient to house the required computation, sensing and communication resources [3].

The other cost and accessibility factor is the play field. The larger the field the more expensive it is to build, to handle and to obtain the floor space for playing the game. The play field must be small enough to be stored in a student's or hobbyist's home and that it can be set up and taken away with little effort. Scaling down a typical FIFA compliant soccer field 50:1 gives a play field of 2.00 m × 1.4 m. This is not quite desktop size but it will fit into a small school or university laboratory, a living room or a garage. With the specified size of the robots there is enough room on this play field for teams of 5 robots each. The only alteration is the width of the goal. The FIFA goal width cannot be scaled down by 1:50 because a 100 mm wide robot blocks almost the full goal width. Therefore we specify a goal width of 400 mm. At this width the body the robot goal keeper can only obstruct 25% of the goal width, leaving 75% open for scoring a goal.

Despite the recent trend towards humanoid soccer robots we decided to stick to wheeled robots. Low cost wheeled robots are already capable of fast movement on the play field. The same cannot be said about humanoid robots, even though in the future biped robot soccer players will be capable of a much more human-like ball kicking and dribbling than wheeled robots ever can. At the current state of humanoid robot loco-motion the cost and effort required in building, or even just programming, a humanoid robot capable of agile walking is out of proportion. This is because, a humanoid con-figuration is not necessary for any of the other key capabilities needed for a successful robot soccer game, such as fast visual localisation and navigation, and fast execution of cooperative game strategies.

Fast locomotion is essential for a game that is interesting to watch as well as for a game that serves as a performance benchmark for autonomous behaviours. Any ad-vances in fast, resource efficient vision and cooperative game strategies will be trans-ferable to larger humanoid robots. The AMiRESot league has all the essential elements of autonomous behaviour. It is goal driven, there is uncertainty arising from the other players action, it relies on vision as human players do and the time available for sensor information processing is strictly limited by the physical dynamics of the game.

4   Robots for AMiRoSoT

The cost of the soccer robots for the AMiRoSOT league must be such that students and hobbyist can own one. Friends may come together to form impromptu teams, unpack a field and let their robots play. Although it is possible to construct such robots with current technology, there are not any available for purchase.

The simplest choice for wheeled locomotion of the AMiRoSoT robots is to have a two-wheel differentially steered drive train such as that widely use for mobile robots. The two wheels are on an axis that is a diameter of the cross section of the vertical cylindrical body. Each wheel is driven by its own motor. The difference in speed of the two motors determines the turning rate of the body. With such a drive train robots can turn on the spot, which gives them great agility. In addition a vertical cylinder shape avoids that the robots get stuck in corners. A two-wheeled, differentially steered mobile base is cheap and can be built by students and hobbyists. All that is required are a pair of wheels, a pair of good quality DC motors with gear box and motor axle position encoders, a motor power control circuit, a base plate and some brackets to attach the motors, and of course, a battery. The cylindrical body case can be made of PVC pipe or by 3D printing.

Even if mechanical construction of a soccer robot is relatively simple, the electronic hardware requires advanced electronic design capabilities. At least a 32 bit processor in the class of a high end ARM Cortex, a camera, half a dozen infra-red proximity sensors, a microphone, a power management and an USB interface are the minimum required. Putting all the hardware together, even from off-the-shelf parts is a major undertaking beyond the capacity of most students and hobbyist. And, it has to fit all into that 100 mm diameter cylinder. Finally to make it all work the right software has to be installed and configured.

To remain true to our goal of a league with a low entry barrier, a supplier for at least the ready made computation part had to be found. The computing part must be in the form of a few compatible circuit boards that fit into the available robot body space. To meet that goal turned out more difficult than expected. A suitable high performance and low cost robot could not be found on the market. The successor model of the Khepera II robot, the Khepera III, produced by K-Team company was too big for the league's size limit and too costly. Combining off-the-shelves components into a package for self assembly by prospective participants would not meet the accessibility demand either. This only left the option of a custom design to be manufactured by a third party and offered for purchase. In 2011 Tetzlaf and Witkowski [10], presented a prototype mini-robot for the AMiRESot league and Herbrechtsmeier et al. [3] described a full featured design, still under construction, of an autonomous mini-robot for research and education suitable for the AMiRESot league. Both designs required the fabrication of custom designed printed circuit boards (PCB). Thus the barrier for entering the AMiRESot soccer league remained still too high for the majority of interested teams.

A custom designed board set manufactured in small quantities will never be able to come near or beat the cost advantage of the computing power of a mass produced mid-range mobile phone. Therefore in early 2013 we decided to try an alternative approach that would take advantage of the extraordinary performance/cost ratio of the new generation of smart phones. Even if smart phones are not designed for educational robotics, it is well worth the effort to adapt them to our application.

The prototype mini-robot for the AMiRoSoT league build by Ulf Witkowski's group at the South Westfalia University of Applied Sciences and shown in Figure 1 is described in a companion paper presented at this conference. An Android smart phone provides computing power, communication, camera and inertial sensors. For Android phones, Google provides the Android Software Development Kit (SDK) that has all the necessary tools for writing application software and is free of charge. The motor controller and infrared sensor board for the mobile base is available for purchase.

Fig. 1. Prototype for the mobile phone soccer robot

5 Practical Aspects of the Game Setup

The price to pay for the simple differential steering is the need for rear and front castors that prevent the robots from falling over. Ready made castor wheels on the market are too big for our size of robot. Spherical rollers have been tried, like those in perfume and deodorant flasks, but they easily clog with dirt and end up sliding, instead of rolling. The simple alternative is to use small support posts that have rounded low friction ends that slide on the ground. We will refer to these supports as castors as well.

The problem with castors is that ground contact is on four points, which restricts to moving on flat surfaces. Whenever the front castor moves over an elevation one or both drive wheels are lifted off the ground and the robot becomes uncontrollable or simply gets stuck. An elastic suspension of the castors or the drive wheels can overcome this problem. However, the mechanical complexity of these improvements is not worth the effort because soccer fields are supposed to be flat. To alleviate the problem in a simple way the sliding supports are made such that, when the robot is level they do not quite reach the ground. In this way three points will touch the ground and small irregularities of the ground will not lift off the drive wheels. The result is that the robot will either be slightly tilted forward or backward. Upon acceleration or deceleration the robot will flip from one inclination to the other. For a camera rigidly attached to the robot body the image will shift on each flip, complicating the video processing.

The surface of the play field must be flat without irregularities and it also must be such that the two-wheeled robot can move with good grip, without excessive friction. It is unlikely that these conditions are met by simply marking a soccer field and using the ground as the field surface. A portable play field with a suitable surface has to be provided. Again a simple solution could be a sanded and painted panel of plywood. A 2 m by 1.4 m wooden panel is awkward to carry and to store. To facilitate transport and storage the panel might be cut into two or more smaller panels. It may be difficult to lay out the panels for a game so that no uneven joints appear that hinder the movement of the robots. Rolling out a carpet sounds like the ideal solution. A 1.4 m long roll of carpet is easy to carry and store, and has no joints. The problem here is that ordinary carpet, looped or otherwise, causes too much friction for the small wheels. Instead of carpet a vinyl floor covering offers a smooth surface that can be rolled up and stored away quite easily.

Another issue is the field boundary. It would be ideal to delimit the play field with line markings as on a real soccer field. To make the robot stay inside the field by using visual recognition of the field markings is not too difficult. It is much more difficult to make the robots keep the ball inside the field. In human soccer there will almost always be a player to intercept the ball before it leaves the field and the game is not disrupted too often by the ball being kicked out of the field. This is not the case for the AMiRoSoT league. Therefore a low height frame around the field to contain the ball is needed. The frame should only be slightly higher than the radius of the ball to make sure the ball rebounds back onto the field when hitting the field border. To stand out from the surroundings, the frame is painted white, the same as the line markings on the field. The frame also needs to be easily assembled and disassembled for transportation and storage. The goals are openings in the frame to which goal boxes are attached. For easy visual distinction the inside of the goal boxes are painted black. Vertical cylindrical goal post are fitted onto the frame at each end of the goal. The goal posts are round so that they have the same apparent thickness when viewed from any angle. This is a concession to vision algorithms because the goal posts are the main landmarks for the robot localisation and navigation on the field.

Finally, it is also useful to gain a rough estimate of the dynamics of the game to gauge the time available for the various actions of the robots. The mass of the base of the smart phone robot is around 320 g and the mass of the phone is 110 g, for a total mass of 430 g. The stall torque of the DC motors is around 5 mNm which results in an acceleration of around 1 m/s^2 for 40 mm diameter wheels. With these parameters robot speeds of up to 0.5 m/s are possible. At 0.5 m/s the robot need 4 s to cross the field lengthwise. A squash ball has a mass of 25 g. For the ratio of ball mass to robot mass the final speed of a stationary ball when kicked by the robot is nearly twice the speed of the robot. Therefore ball speeds may reach 1 m/s. At a video frame rate of 30 frames/s the ball can travel up to 33 mm between frames.

6   Ball Kicking with a Cylindrical Robot

To illustrate the variety of tasks to be carried out in a game let us investigate the ball kicking behaviour in its simplest form: when the ball is at rest and there is no goal keeper nor any other robots on the field. In the ball kicking behaviour we would like to see the robot moving swiftly to a position somewhat behind the ball, in line with the ball and the centre of the goal. From there it should gather some more momentum in direction to the goal until it collides with the ball. The momentum transferred to the ball in the collision must be in the direction to the centre of the goal and the ball should speed off in that direction. As soon as the robot collides with the ball the robot should stop. The trajectory to be followed by the robot might look like trace A in Figure 2.

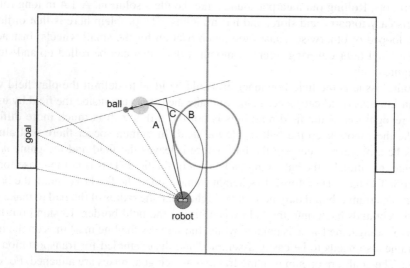

Fig. 2. The most desirable robot trajectory for kicking the ball into the goal is trace A

The sensors available to the soccer robot for the ball kicking behaviour are the video camera, wheel encoders and infra-red proximity sensors. With the camera the robot can perceive all the objects relevant to the soccer game across the whole play field. The robot also has accelerometers, gyroscope and a compass, but for the time being we shall rely mainly on the camera and the wheel encoders to assess the situation on the play field and to find the ball and shoot it into the goal.

There are always two perspectives for designing a behaviour: the perspective of the external observer and the perspective of the robot. What we just did was looking at the task from external observer's perspective, which is for us human designers the most intuitive approach. An external observer sees the robot on the play field in its position relative to the other objects in and around the play area. A person watching could easily steer the robot by remote control to kick the ball. This is also how the task appears in the leagues that use *global vision* from an overhead camera.

An autonomous robot player has to carry out the task from its own perspective, which makes the task very different from that of the external observer. First, the viewpoint of the robot is only slightly elevated above the play field surface. This view is almost a sidewise projection of the objects on the field. Contrary to an overhead viewpoint, which projects displacements of objects on the field surface as proportional displacements on the camera's image plane, in the sidewise projection only the lateral displacement component appears as horizontal displacement in the image plane. The radial displacements component appears only as a change of size (depth). Second, the limited field of view of the robot's camera, typically around 40°, only shows a fraction of the play field. Third, when the robot is placed on the play field its position relative to the goal and the ball is unknown. The 2D positions of the objects of relevance to the goal kicking task have to be calculated from the camera images. The objects of relevance are the ball and the goal opening. The latter being identified by the black and white transition at the goal edges or alternatively by the goal posts.

The first action the robot might do when waking up on the play field is to turn on the spot until it finds the ball in the field of view. Finding the ball could mean to use a colour blob detection algorithm on the camera image. Once the ball is in view the robot could turn until the vertical middle line of the field of view passes through the centre of the ball. The ball would then be straight ahead of the robot and the robot could move towards it. However to kick the ball the robot has to reach a position some distance from the ball, leaving the ball between the goal and the robot. For this the robot has to move to a position somewhere to the side of the ball, instead of heading straight to the ball. But, to which side? To decide this the robot needs to know whether the goal is to the left of the ball or to the right. There is no guarantee the goal is visible to the robot while it approaches the ball and therefore the robot cannot decide to which side of the ball it should steer. Without this information the robot could decide to move towards ball but stop at a distance that leaves room to manoeuvre. If the goal is not in the field of view of the camera the robot could now start turning on the spot to find the goal. Once the goal has been found the robot has enough information to move to a position behind the ball, relative to the goal, suitable for kicking the ball into the goal. In this scenario it is clear that there is no need for the robot to move to the ball before searching for the goal. The robot can do this from where it was first placed on the field. By determining its position relative to the goal and the ball, before making any movement, leaves more room for deciding on the best way to approach the ball. Whatever the position of the robot is, it needs to measure its position relative to the ball and the goal for a successful kick.

Following a trajectory like *A*, in Figure 2 is likely to be quite difficult. Trajectory *C* is an approximation to *A* easier to achieve. Once the robot has localised itself, calculating a suitable position behind the ball for kicking only requires a simple geometric calculation and the robot can move towards it using odometry. However all locations computed from image data are likely have a substantial error, therefore the robot needs to relocalise visually once it has moved the calculated distance. Errors in odometry will also add to the uncertainty because the actual position may differ from the calculated position, which already was affected by visual measurement errors. It may require several repositions before the robot has aligned itself with the ball and the centre of the goal for carrying out the kick. Each reposition is an iteration of the first computation because it requires the computation of the next movement and the estimation of the kicking position.

6.1 Robot Localisation on the Play Field

For the robot localisation means knowing it's relative position to some chosen landmarks (reference objects). When the robot is first placed on the play field it does not know what objects there are in the world around it and where they are. The robot motion on the play field is in two dimensions and therefore we only need two stationary landmarks to uniquely find the position of the robot relative to the landmarks by triangulation. The required measurements are the distances to the landmarks and the observed angle between them.

On the play field the goals and the markings on the field are at known relative positions. The edges of the goals are suitable landmarks and they should be made easy to detect in the camera images. In the AMiRESot field the inside of the boundary frame of the goals is painted black, while the rest of the play field frame is white. From any point of the play field the edges of the goal are visible as a black-white transition on the play field border frame.

When the robot is placed on the play field it can begin its visual exploration of the environment by turning on the spot in a predefined direction, say, anti clock wise, searching for the ball and goal edges in the video images. Whenever one of the searched objects is detected and centred in the image the wheel encoder readings are stored and the distance of the object is estimated from the image.

The distances to the goal edges can be estimated (with a calibrated camera) from the apparent height of goal edges in the image because the true height of the vertical transition edge is the height field frame. The angles are obtained from the readings of the wheel encoders while the robot turns on the spot. The reference direction, that is the direction where the angle is zero, is arbitrary because only differences in direction readings are relevant. Before the robot starts to turn it can read the wheel encoder counts and take this value as the reference.

7   Conclusion

After almost twenty years the robot soccer game continues to appeal to hobbyist, students and robotics researchers. It appeals to the wider public because of its competitive nature and its similarity with the very popular human soccer game. It appeals to teachers and students because by participating in the game through building and programming, robot soccer offers hands on learning opportunities on widespread topics of high relevance across our technological civilisation. It appeals to researchers for the variety of scientific and technical problems posed by the robot soccer grand challenge.

Advances in technology are quickly incorporated in the game to give the robot players ever more dexterity and cognitive abilities. The game is played in various leagues that have different educational, research and entertainment values. We have described a new league for small sized robots that has a lower entry barrier than the other mayor leagues while maintaining a high degree of relevance to the aims of the robot soccer games. The league sticks to the use of wheeled robots which contributes to keep the cost low. At the current state of technology it is much easier to play a fast game with wheeled robots where fast visual localisation and navigation, and cooperative game strategies can be put to the test.

The educational benefits of robot soccer can only be realised if the game is accessible to a wide student population. The high cost and the large amount of work required to derive an intellectual reward are the main deterrents to participating. By attending to practical accessibility aspects the AMiRoSoT stimulates maximum exploitation of cost reducing technology. An example of this is the use of the sensing and computing capabilities of mobile phones. Furthermore robotic technology can only play a role in society if it is affordable. This was convincingly shown by market penetration gained by autonomous vacuum cleaners once their cost benefit ratio became attractive. The small size autonomous robot soccer league stimulates advances in this direction by widening the pool of contributors.

References

1. Biswas, J., Mendoza, J.P., Zhu, D., Choi, B., Klee, S., Veloso, M.: Opponent-driven planning and execution for pass, attack, and defense in a multi-robot soccer team. In: Proceedings of the Thirteenth International Joint Conference on Autonomous Agents and Multi-Agent Systems, AAMAS 2014, Paris, France (May 2014)
2. FIRA.org: About fira/brief history, http://www.fira.net/contents/sub01/sub01_1.asp (accessed July 30, 2014)
3. Herbrechtsmeier, S., Rueckert, U., Sitte, J.: Amiro - autonomous minirobots for research and education. In: Rückert, U., et al. (eds.) Proceedings of the 6th International Symposium on Autonomous Minirobots for Research and Edutainment (AMiRE 2011), pp. 101–112. Springer (2012)
4. Kitano, H., Asada, M., Kuniyoshi, Y., Noda, I., Osawa, E.: Robocup: The robot world cup initiative. In: Proc. of IJCAI 1995 Workshop on Entertainment and AI/Alife, Montreal (1995)
5. Kitano, H., Asada, M., Noda, I., Matsubara, H.: Robocup: Robot world cup. IEEE Robotics & Automation Magazine 5(3), 30–36 (1998)
6. Legolab: Khepera robot football, http://legolab.daimi.au.dk/AdaptiveRobots/robot_football.html (accessed July 30, 2014)
7. Mondada, F., Franzi, E.: Biologically inspired mobile robot control algorithms. Monograph (1993); photocopy without reference
8. RoboCup.org: A brief history of robocup, http://www.robocup.org/about-robocup/a-brief-history-of-robocup/ (downloaded May 24, 2012)
9. Sitte, J.: Amiresot 2008 robot soccer rules 2008, version 1.0 (2008), http://www.amiresymposia.org/AmireSotRules2008.pdf
10. Tetzlaff, T., Witkowski, U.: Modular robot platform for teaching digital hardware engineering and for playing robot soccer in the amiresot league. In: Rückert, U., et al. (eds.) Proceedings of the 6th International Symposium on Autonomous Minirobots for Research and Edutainment (AMiRE2011), pp. 113–122. Springer (2012)

Model Checking of a Training System Using NuSMV for Humanoid Robot Soccer

Yongho Kim[1], Mauricio Gomez[1], James Goppert[2], and Eric T. Matson[1]

[1] M2M Lab, Purdue University, West Lafayette, Indiana, USA
{kim1681,mgomezmo,ematson}@purdue.edu,
[2] FDCHS Lab, Purdue University West Lafayette, Indiana, USA
jgoppert@purdue.edu

Abstract. Model checking is a technique to perform a formal verification process that allows a system to have robustness and correctness. In a given system model as a Finite State Machine (FSM), model checker explores all possible states in brute-force manner. In this paper, we apply this technique to a training system, which teaches a humanoid soccer robot how to intercept a ball that is passed from other players, to verify that the system is failure-safe in a given requirements. Several Computation Tree Logic (CTL) properties to define a critical or potential situation are specified based on the functionality of the system. We show the results of the given properties using NuSMV, a symbolic model checker introduced by Carnegie Mellon University.

Keywords: Model Checking, Formal Verification, Training System for Humanoid Robot Soccer, NuSMV.

1 Introduction

Model checking is an automated technique that identifies logical errors in a specified system. This technique has been used in various fields that require a verification process at system level [1,2] due to the high complexity of integrated systems. A system model in model checking is usually structured as a Finite State Machine (FSM) consisting of states and transitions that represent behaviors of the system, while properties to verify the system are formalized from the system requirements, and are explained in unambiguous manner. The technique explores all possible states through available transitions at the time to validate a given property. Even though this technique has been generally used to check system at design phase, it also can be utilized to maintain validity of system at operation time [3].

In the field of humanoid robot soccer, their systems generally have a well-organized strategy playing an important role in a match to win. The strategy not only contains if-then rules to make a proper decision at a certain time, but it also has to cover as many exceptional cases as possible to occupy better position in the match. However, they cannot cover all the cases because there exists a lot of uncertainties that may be occurred in a match. In a situation of [4], for example, when a robot falls down by a collision with an opponent robot or by

© Springer International Publishing Switzerland 2015
J.-H. Kim et al. (eds.), *Robot Intelligence Technology and Applications 3*,
Advances in Intelligent Systems and Computing 345, DOI: 10.1007/978-3-319-16841-8_48

losing its balance while the robot was going to kick the ball, the robot will try to stand up to kick the ball. If the opponent robot has already taken the ball or kicked it while the robot is standing up, the robot may try to kick the ball that is not detected in its sight anymore. Thus, the robot may not go back to search for ball or ready state because of the unexpected situation. Once a system model is established, model checking technique can check these kinds of potential errors or situations by specifying a property.

In this research, we apply the model checking technique to a training system to discover any potential logical error that interrupt operation of the system. The training system consists of two robots: a robot soccer trainer and a robot soccer player. The goal of the system is to teach the robot soccer player to have some knowledge to successfully intercept a ball when the ball is passed from team players. A FSM and some properties for the system can be specified by defining steps that are needed to complete the training process, and by restating the system requirements. Thus, model checking technique can verify whether the given properties are satisfied in the given state space.

In the rest of the paper, related works of model checking technique are explained in Chapter 2. In Chapter 3, we introduce the training system with detail steps of the training process and a system model. Explanation of how verification process works using a model checker is given in Chapter 4. Finally, we conclude our research and present future works.

2   Related Works

Needless to say, model checking has been widely used in the fields of software and hardware engineering in order to take its powerful functionality for verifying systems or designs of systems. However, model checking technique has being recently utilized to the field of robotics. In [5], a Flapping-Wings Micro Air Vehicle (FWMAV) has been modeled as a hybrid system and validated using Polyhedral Invariant Hybrid Automation (PIHA) [6] to account for bounded disturbances that affect to FWMAV. In particular, Konur, et al. [7] used probabilistic model checking for analyzing behaviors of swarm robots. In their model, several states that represent actions of a foraging robot are defined with probabilities that affect to the successful foraging. The authors used both a statistical simulation tool such as Monte Carlo and a model checking tool named PRISM [8] to compare the results given from the tools. They emphasized that model checking can guarantee the safety properties under all possible situations, which cannot be covered by the statistical simulation approach.

In [9], the authors have developed a model checking program of a control system for an autonomous rotorcraft. They implemented the model checking program using Java PathFinder [10], one of the model checking tools, to check any logical errors in the system that violate the system requirements. According to their research, finding these logical errors by performing simulations is difficult because the control system has continuous states that make the system model to have infinite states. For example, a massive number of interactions with shared variables may cause concurrency of multiple tasks that results in logical errors in the system. However, due to the fact that model checking approach explores all possible situations, it consumes massive computational resources as the number of states increases. Thus, the authors pruned away some paths that are unnecessary to reduce the usage of resources. This research shows the fact that model checking also can be used in real time system with some resource-efficient path exploration algorithms such as depth first search (DFS) algorithm or Dijkstra algorithm.

3   System Modeling

In order to verify a system, the system should be modeled as FSM with some properties that represent the requirements of the system. In this research, we consider a training system that allows a humanoid soccer robot to have a knowledge that is used to intercept the ball successfully. Moreover, CTL properties are defined to verify that the training system satisfies given system requirements.

3.1   Training System for Humanoid Robot Soccer Player

The training system consists of two robots: a trainer robot and a trainee robot. The system aims at teaching the trainee to be able to get a ball passed by the trainer robot. The trainee has a knowledge-based ability to memorize its movement that was used to intercept the ball for the previous trials. Thus, the trainee is able to find out the best movement to catch the ball as many trials as the trainer passes the ball with the same or similar trajectory.

At the initial stage of a trial, the two robots face each other from a given distance and the trainer has a ball. We assume that the initial stage is given by operator (i.e. a human) for every trial. Fig. 1 shows the sequential flow of the system. A ready and kick signal are used to trigger opponent's next action (e.g. ready signal makes the trainer kick a ball). Detail explanation of each sequential step in a trial are presented as follows,

1. The trainee checks its state (e.g. balance, fall down state). When the trainee is ready, a ready signal is fired.
2. When the trainer receives the ready signal, it starts training by showing up a kick signal that means *I am going to pass the ball.*
3. As soon as the trainee recognizes the kick signal, it begins tracking the ball using vision sensor to estimate the trajectory of the ball.
4. After the trainee gets five complete sets of the balls position, a required number of side steps to intercept the ball is calculated based on the estimation of the trajectory. The trainee sidesteps for the calculated number (See Fig. 2).
5. An evaluation to check whether or not the trainee successfully intercept the ball is proceeded. If the trainee can see the ball after the movement, the result of the trial is a success. Otherwise, the trial is a failure.
6. The trainee memorizes the steps that were used to the trial with a result of the evaluation. The memorized steps will affect the calculation of steps for next trial that has the same trajectory of the ball.

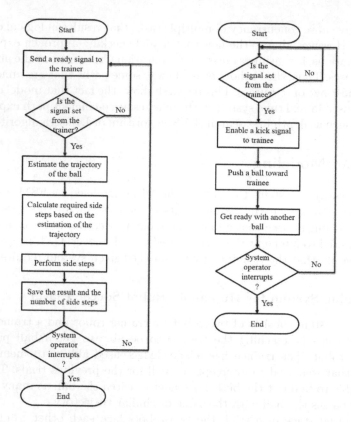

Fig. 1. (Left) a sequential flow of the trainee robot, (Right) a sequential flow of the trainer robot

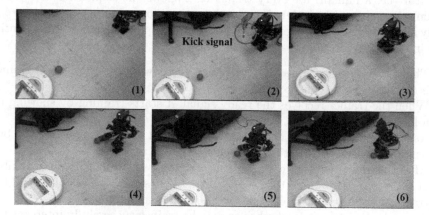

Fig. 2. A demonstration of a trial. (1) Initial position of the trainer and the trainee. (2) We use a green stick to make the trainee see the kick signal. (3) The trainer pushes the ball toward the trainee. (4), (5) The trainee recognizes the trajectory of the ball and start sidestepping. (6) The trainee evaluates whether or not the calculated sidesteps were sufficient to intercept the ball.

In a systematic view of the training process, the system works properly and infinitely without any errors as long as ball is given to the trainer every trial. System in real environment, however, does not always go in a right way. Even though system designer builds some procedures in order to deal with some of expected failure situations, this does not guarantee that the system is failure-safe. This gives us why a formal verification process is useful.

3.2 Symbolic Model of The Training System

We define a symbolic transition system for the training system introduced in Chapter 3.1. A *Soccer Training System* is a tuple $STS=\{S,Act,I,\rightarrow,AP,V\}$, where,

- $S=\{s_0,s_1,...,s_{16}\}$ is a set of states,
- Act is a set of actions,
- $I=\{s_0\} \in S$ is an initial state,
- $V=\{v_{ready\_signal},\ v_{kick\_signal},v_{trials},v_{data},v_{step}\}$ is a set of variables,
- $\rightarrow\subseteq S\times Act\times S$ is a transition relations,
- AP is a set of atomic propositions.

The training system has repeatable sets of states and transitions. Each step in the training process illustrates a state, while each action that fires to go to next step represents a transition. If a state s_i has more than one successor, next state is selected non-deterministically. Let $\pi=s_0,s_1,...,s_n$ be an *infinite path fragment* that consists of infinite number of states. Let $\widehat{\pi}=s_0,s_1,...,s_n$ be a *finite path fragment* such that $s_0=I$, and there exists a transition $s_i \Rightarrow s_{i+1}$ *for every* $0 \leq i \leq n$.

We define three symbolic modules as shown in Fig. 3. The three modules run individually such that each module has an initial state. The boxes and the arrows respectively indicate states and transitions. The trainee module has two failure

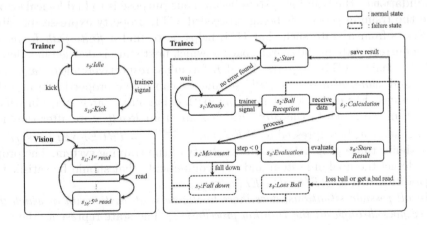

Fig. 3. A Finite State Machine for the training system

states that are expected to be occurred during the training process. The trainer module is simply described with only two states. The transition from Idle to kick in the trainer module is triggered by the signal from the trainee $v_{ready\_signal} \in V$. The vision module collects data of the ball every period and sends the data to the trainee.

In this research, we consider the trainee as a main actor of the system so that the symbolic model is designed in terms of a view of the trainee. We assume all the time that takes for each action is too short to be ignored, i.e. receiving ball data from vision, calculation, and storing data. Additionally, even though an *infinite path fragment* may exist in *STS* because no terminal state is given, we only consider a *finite path fragment* for the verification (i.e. 100 trials). It is the fact that our purpose of this research is to find any logical errors that are represented as a *finite path fragment*, and can be occurred in system level at operation time.

The system starts from *Start* state, the initial state *I*, and moves through the arrows. In *Ready* state, the trainee waits until the signal $v_{kick\_signal} \in V$ is fired by the trainer. There are two types of states: normal and failure state. One trial ends successfully if all the normal states were passed sequentially. If a given *finite path fragment* has at least one or more failure states, at least one or more failure trials exist among all trials.

Atomic propositions can be defined to depict actions in the system. A proposition of *the trainer signal makes the trainee to begin tracking the ball*, for example, can be denoted by a transition of $s_1 \xrightarrow{kick\_signal} s_2$, in which the signal changes the current state from *Ready* to *Ball reception*.

3.3   Computational Tree Logic (CTL) Properties for Specification

Since the system is modeled as a FSM, some properties can be defined for verification of the system. A property is an expression that clearly declares requirements of the system, and composed of formulas and atomic propositions. We use CTL for validation of the training system because our purpose is to find logical errors while the training process is being proceeded. CTL property expresses possible successors from a state such that CTL helps us to find a *finite path fragment* from a current state in the search space in which next state is decided in a deterministic manner. CTL supports *A* and *E* formulas, respectively mean '*along all paths*' and '*along at least one path*' so that we can define properties in branch-time manner. For example, At least one path exists in a search space in which the system can reach to the *Store Result* state from *Ready* state after the kick signal is set. can be expressed as '$EF((s=s_1 \xrightarrow{v_{kick\_signal}} =TRUE) \rightarrow EF(s=s_6))$'.

We define three properties to check validity of the training system. The properties depict the goal of system and the requirements that should be satisfied.

Specification 1. $AF((s=I) \rightarrow EF(s=s_6))$

"*In all possible situations, does there finally exist at least a path in which the trainee goes through all the training procedures?*" The state represents the goal of the training system and it must be satisfied. If model checker finds any of *finite path fragments* that satisfies the property, the state is verified.

Specification 2. $AG(v_{trials} \geq 1 \land (s=s_0) \land \neg v_{ready\_signal})$

"*In all possible situations, does the trainee always continue the training after it falls down?*" In humanoid robot soccer, one of challenges during match is to continue playing soccer when player falls down by a collision with an opponent or losing balance itself. This phenomenon is also applicable in the training system. The trainee may fall down if it loses the balance when it moves.

Specification 3. $AG(v_{kick\_signal} \land v_{ready\_signal})$

"*In all possible situations, is there always a situation in which the signals of the trainer and the trainee are triggered at the same time?*" Normally, the trainee sets a signal to say *I am ready* whereas the trainer sets a signal to say *I am passing the ball*. It would be ambiguous if the two signals are set at the same time.

In the following chapter, we put the symbolic model of the training system and the properties defined in this chapter to model checker. Running model checker outputs results of the specifications, and provides counterexamples if any of the specifications are false.

4   Evaluation

We use a symbolic model checking tool named as NuSMV[11] to verify the training system designed in the previous chapter. NuSMV provides user commands that allow a user to execute a verification process with a symbolic model.

In the training system, the two robots are physically separated, and run on different system environments. Moreover, the vision system that is running on the trainee's system grabs an image frame from a front camera, and calculates position of the ball every period. Thus, the models should be treated as an individual processor in NuSMV. We use the *process* keyword, used to model interleaving concurrency, to emulate the systems' behaviors in NuSMV environment. The *process* keyword enables the modules running in parallel. However, the *process* keyword is deprecated and no longer supported in the version of 2.5 or higher. We used NuSMV with the version of 2.5.4.

NuSMV generates 1 cluster for 367 nodes with 14 initial sets of states for the training system. Due to the fact that the training system consists of repeatable sets of states, the size of the cluster is small enough. Thus, running time for the validation with specifications is too short to measure. NuSMV ran with the specifications given in Chapter 3.3 and gave us the following results as shown in Table 1 with counterexample if it exists.

The specification 1. is true in the given symbolic model of the training system, and has no counterexample. The result proves that the system is able to run the training process for at least one trial.

The specification 2. is false and NuSMV provides a counterexample as shown in Fig. 4 meaning that at least one situation in which the property is not true was found. According to the given counterexample, the trainee is not able to continue after it losses the ball. The trainee generates a signal to go to the next trial after it stores a result in *store_result* state. In *f_lossball* state, the signal is

Table 1. Results of the verification with the given specifications

Item	CTL property	Result
Spec. 1	$AF((s{=}I){\rightarrow}EF(s{=}s_6))$	TRUE
Spec. 2	$AG(v_{trials} \geq 1 \land (s{=}s_0) \land \neg v_{ready\_signal})$	FALSE
Spec. 3	$AG(v_{kick\_signal} \land v_{ready\_signal})$	FALSE
Spec. 3-1	$AG(v_{trials} > 1 \land v_{kick\_signal} \land v_{ready\_signal})$	TRUE

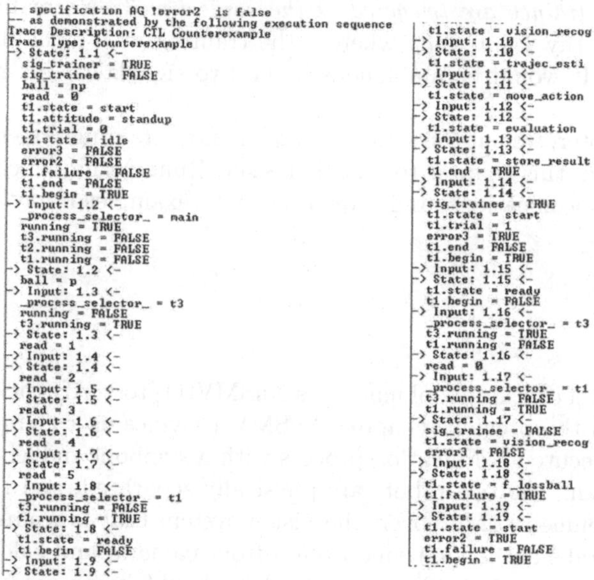

```
-- specification AG !error2  is false
-- as demonstrated by the following execution sequence
Trace Description: CTL Counterexample
Trace Type: Counterexample
-> State: 1.1 <-
  sig_trainer = TRUE
  sig_trainee = FALSE
  ball = np
  read = 0
  t1.state = start
  t1.attitude = standup
  t1.trial = 0
  t2.state = idle
  error3 = FALSE
  error2 = FALSE
  t1.failure = FALSE
  t1.end = FALSE
  t1.begin = TRUE
-> Input: 1.2 <-
  _process_selector_ = main
  running = TRUE
  t3.running = FALSE
  t2.running = FALSE
  t1.running = FALSE
-> State: 1.2 <-
  ball = p
-> Input: 1.3 <-
  _process_selector_ = t3
  running = FALSE
  t3.running = TRUE
-> State: 1.3 <-
  read = 1
-> Input: 1.4 <-
-> State: 1.4 <-
  read = 2
-> Input: 1.5 <-
-> State: 1.5 <-
  read = 3
-> Input: 1.6 <-
-> State: 1.6 <-
  read = 4
-> Input: 1.7 <-
-> State: 1.7 <-
  read = 5
-> Input: 1.8 <-
  _process_selector_ = t1
  t3.running = FALSE
  t1.running = TRUE
-> State: 1.8 <-
  t1.state = ready
  t1.begin = FALSE
-> Input: 1.9 <-
-> State: 1.9 <-

  t1.state = vision_recog
-> Input: 1.10 <-
-> State: 1.10 <-
  t1.state = trajec_esti
-> Input: 1.11 <-
-> State: 1.11 <-
  t1.state = move_action
-> Input: 1.12 <-
-> State: 1.12 <-
  t1.state = evaluation
-> Input: 1.13 <-
-> State: 1.13 <-
  t1.state = store_result
  t1.end = TRUE
-> Input: 1.14 <-
-> State: 1.14 <-
  sig_trainee = TRUE
  t1.state = start
  t1.trial = 1
  error3 = TRUE
  t1.end = FALSE
  t1.begin = TRUE
-> Input: 1.15 <-
-> State: 1.15 <-
  t1.state = ready
  t1.begin = FALSE
-> Input: 1.16 <-
  _process_selector_ = t3
  t3.running = TRUE
  t1.running = FALSE
-> State: 1.16 <-
  read = 0
-> Input: 1.17 <-
  _process_selector_ = t1
  t3.running = FALSE
  t1.running = TRUE
-> State: 1.17 <-
  sig_trainee = FALSE
  t1.state = vision_recog
  error3 = FALSE
-> Input: 1.18 <-
-> State: 1.18 <-
  t1.state = f_lossball
  t1.failure = TRUE
-> Input: 1.19 <-
-> State: 1.19 <-
  t1.state = start
  error2 = TRUE
  t1.failure = FALSE
  t1.begin = TRUE
```

Fig. 4. A counterexample of the specification 2. Reading order is from top-left to bottom-right.

not generated by the trainee. Therefore, we found one of logical errors we did not expect at system design phase.

The specification 3. is false for the verification of the system. As shown the counterexample in Fig. 5, v_{kick_signal} is triggered at the beginning of the trial; and v_{ready_signal} is fired at *store_result* state. Just after that, the two signals are 'TRUE' at the same time in the state step 1.14. However, Fig. 6 shows the expected state flow for the two signals.

According to the NuSMV manual [12], the reason of the phenomenon is that *process* keyword only affects values that are assigned by the process. In the training system, initial value of v_{kick_signal} is *TRUE*, and is not set by the process at the beginning. This leads to putting current state into undesirable state. After the first trial, v_{kick_signal} is changed by the process. Therefore, we changed the property 3 to property 3-1 shown in the Table 1, and the result of the specification becomes true, which means that the training system satisfies the property 3-1.

```
NuSMV > check_ctlspec
-- specification AG !error3  is false
-- as demonstrated by the following execution sequence
Trace Description: CTL Counterexample
Trace Type: Counterexample
-> State: 1.1 <-
  sig_trainer = TRUE
  sig_trainee = FALSE
  ball = np
  read = 0
  t1.state = start
  t1.attitude = standup
  t1.trial = 0
  t2.state = idle
  error3 = FALSE
  error2 = FALSE
  t1.failure = FALSE
  t1.end = FALSE
  t1.begin = TRUE
-> Input: 1.2 <-
  _process_selector_ = main
  running = TRUE
  t3.running = FALSE
  t2.running = FALSE
  t1.running = FALSE
-> State: 1.2 <-
  ball = p
-> Input: 1.3 <-
  _process_selector_ = t3
  running = FALSE
  t3.running = TRUE
-> State: 1.3 <-
  read = 1
-> Input: 1.4 <-
-> State: 1.4 <-
  read = 2
-> Input: 1.5 <-
-> State: 1.5 <-
  read = 3

-> Input: 1.6 <-
-> State: 1.6 <-
  read = 4
-> Input: 1.7 <-
-> State: 1.7 <-
  read = 5
-> Input: 1.8 <-
  _process_selector_ = t1
  t3.running = FALSE
  t1.running = TRUE
-> State: 1.8 <-
  t1.state = ready
  t1.begin = FALSE
-> Input: 1.9 <-
-> State: 1.9 <-
  t1.state = vision_recog
-> Input: 1.10 <-
-> State: 1.10 <-
  t1.state = trajec_esti
-> Input: 1.11 <-
-> State: 1.11 <-
  t1.state = move_action
-> Input: 1.12 <-
-> State: 1.12 <-
  t1.state = evaluation
-> Input: 1.13 <-
-> State: 1.13 <-
  t1.state = store_result
  t1.end = TRUE
-> Input: 1.14 <-
-> State: 1.14 <-
  sig_trainee = TRUE
  t1.state = start
  t1.trial = 1
  error3 = TRUE
  t1.end = FALSE
  t1.begin = TRUE
```

Fig. 5. A counterexample of the specification 3. Reading order is from top-left to bottom-right.

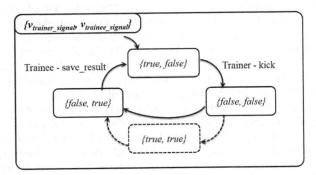

Fig. 6. A simple FSM for the two signals. Dashed are undesirable state and transitions.

5   Conclusion and Future Work

In this research, we have applied a model checking technique to the training system in order to verify that the system runs properly, and to find any logical error in the system. As a result, we have found a logical error and verified that the system is failure-safe in the given specifications for all possible cases.

However, if a symbolic model does not express a system clearly, the verification process is not accurate and may cause logical errors that do not exist in the system. Furthermore, specifications must logically represent system requirements to avoid false-validation of system.

For future work, we will build a model checker that verifies a plain system. Once requirements or system properties are changed in the plain system, the model checker modifies the symbolic model of the system based on the change in real time, and performs a verification to determine availability of a specific function (i.e. if one of motors has an error, the model checker checks whether or not the humanoid robot continues playing soccer). One of the challenges would be how to automatically define symbolic model and properties according to the given system model (i.e. FSM parser, auto-generated specifications).

Acknowledgement. This material is based upon work supported by the National Science Foundation under the Grant Number CNS-1239171.

References

1. Chiappini, A., Cimatti, A., Macchi, L., Rebollo, O., Roveri, M., Susi, A., Tonetta, S., Vittorini, B.: Formalization and validation of a subset of the European Train Control System. In: 2010 ACM/IEEE 32nd International Conference on Software Engineering, vol. 2, pp. 109–118 (2010)
2. Păsăreanu, C.S., Mehlitz, P.C., Bushnell, D.H., Gundy-Burlet, K., Lowry, M., Person, S., Pape, M.: Combining Unit-level Symbolic Execution and System-level Concrete Execution for Testing NASA Software. In: Proc. of the 2008 International Symposium on Software Testing and Analysis, pp. 15–26 (2008)
3. Gallagher, J.C., Humphrey, L.R., Matson, E.: Maintaining Model Consistency during In-Flight Adaptation in a Flapping-Wing Micro Air Vehicle. In: Kim, J.-H., Matson, E., Myung, H., Xu, P. (eds.) Robot Intelligence Technology and Applications 2. AISC, vol. 274, pp. 517–530. Springer, Heidelberg (2014)
4. Ogino, M., Kikuchi, M., Ooga, J., Aono, M., Asada, M.: Optic Flow Based Skill Learning for a Humanoid to Trap, Approach to, and Pass a Ball. In: Nardi, D., Riedmiller, M., Sammut, C., Santos-Victor, J. (eds.) RoboCup 2004. LNCS (LNAI), vol. 3276, pp. 323–334. Springer, Heidelberg (2005)
5. Goppert, J., Gallagher, J.C., Hwang, I., Matson, E.: Model Checking of a Flapping-Wing Mirco-Air-Vehicle Trajectory Tracking Controller Subject to Disturbances. In: Kim, J.-H., Matson, E., Myung, H., Xu, P. (eds.) Robot Intelligence Technology and Applications 2. AISC, vol. 274, pp. 531–543. Springer, Heidelberg (2014)
6. Chutinan, A., Krogh, B.H.: Verification of Polyhedral-Invariant Hybrid Automata Using Polygonal Flow Pipe Approximations. In: Vaandrager, F.W., van Schuppen, J.H. (eds.) HSCC 1999. LNCS, vol. 1569, pp. 76–90. Springer, Heidelberg (1999)
7. Konur, S., Dixon, C., Fisher, M.: Analysing Robot Swarm Behavior via Probabilistic Model Checking. Journal of Robotics and Autonomous Systems 60, 199–213 (2011)
8. Hinton, A., Kwiatkowska, M., Norman, G., Parker, D.: PRISM: A tool for automatic verification of probabilistic systems. In: Hermanns, H., Palsberg, J. (eds.) TACAS 2006. LNCS, vol. 3920, pp. 441–444. Springer, Heidelberg (2006)
9. Lerda, F., Kapinski, J., Maka, H., Clarke, E.M., Krogh, B.H.: Model Checking In-The-Loop. In: 27th American Control Conference, pp. 2734–2740 (2008)
10. Visser, W., Havelund, K., Brat, G., Park, S., Lerda, F.: Model Checking Programs. Automated Software Engineering 10, 203–232 (2003)
11. Cimatti, A., Clarke, E.M., Giunchiglia, E., Giunchiglia, F., Pistore, M., Roveri, M., Sebastiani, R., Tacchella, A.: NuSMV 2: An OpenSource Tool for Symbolic Model Checking. In: Proc. of the 14th International Conference on Computer Aided Verification, pp. 359–364 (2002)
12. User manual of NuSMV with the version of 2.5 (October 2014), http://nusmv.fbk.eu/NuSMV/userman/v25/nusmv.pdf

Some Effects of Culture, Gender and Time on Task of Student Teams Participating in the Botball Educational Robotics Program

David P. Miller[1], Steve Goodgame[2],
Gottfried Koppensteiner[3], and Mao Yong[4]

[1] University of Oklahoma, USA
dpmiller@ou.edu
[2] KIPR , Norman OK, USA
sgoodgame@kipr.org
[3] PRIA, Vienna Austria
koppensteiner@pria.at
[4] ITCCC, Beijing, P.R. China
maoyong@itccc.org.cn

Abstract. Botball is a middle and high school robotics education program with a focus on coding and software development skills along with the application of the engineering design process. Botball is used in hundreds of schools throughout the world. This paper discusses some of the differences and similarities of student teams based on their culture, robotics and software experience, gender and time on task. Some of these factors correlate to student interest in STEM careers while other correlate to the performance of the team.

1 Introduction

The impact of rapid technological innovations on modern societies has been amplified by the globalization of the economy[1]. Several reports, e.g., [2], have been published that document the need to encourage innovation in engineering education, and grow a strong, talented, and innovative STEM workforce [3].

In the US, college enrollment rose from 15.3 million in 2000 to 20.4 million in 2009 [4]. However, percentages of students enrolled in STEM fields decreased from 12.9% in 2000-01 to 10.7% in 2008-09 [5], with no STEM discipline represented in the four most popular majors [4]. New approaches are needed that will attract students not previously inclined towards STEM to become STEM majors. The underrepresentation of women in STEM fields, while slowly improving [6], is a situation that affects many countries [7], [8], [9].

Robotics contests and robotics themed education programs and camps have become a popular method for both increasing interest in engineering disciplines [10], [11] and reducing the gender gap in those disciplines [12], [13]. One of these programs is the Botball Educational Robotics Program [14] which is used by thousands of middle and high school students in a number of countries.

© Springer International Publishing Switzerland 2015
J.-H. Kim et al. (eds.), *Robot Intelligence Technology and Applications 3,*
Advances in Intelligent Systems and Computing 345, DOI: 10.1007/978-3-319-16841-8_49

2 The Botball Robotics Education Program

2.1 Program Overview

The Botball Educational Robotics Program engages middle and high school aged students in a team-oriented robotics competition. By exposing students to an inquiry-based, learn-by-doing activity that appeals to their hearts as well as their minds, Botball addresses industry needs for a well-prepared, creative, yet disciplined workforce with leadership and teamwork experience.

In January, February, and March, the Botball Educator Workshops provide team leaders and mentors with technology training and an introduction to the details of that year's game. Then, after a build period of about 7 weeks, students bring their robots to their regional tournament to compete against others in the current season's game challenge.

Students use science, engineering, technology, math, and writing skills to design, build, program, and document robots. Each team creates one to three autonomous robots that will work together to score points in that year's challenge. Botball gives students the tools to develop sophisticated strategies using artificial intelligence with embedded systems. Students learn to program their robots using C, C++, or Java. The robots need to be able to start when the starting light comes on, stop themselves by the end of the round, and accomplish the game tasks using only the code provided by the students and the input of the robot's sensors. The design of the robot mechanics, sensors, and code are all done by the students.

The finale of the Botball season is the GCER[1] Conference [15] which contains the International Botball Tournament [16]. All teams are eligible to participate in this contest. The conference includes technical sessions where most of the talks are given by the student attendees and detail technical discoveries/innovations or best practices from their robotics experience. Botball and GCER events are created and produced by KIPR[2], and presented in Europe through PRIA[3] and in China through ITCCC[4].

2.2 The 2014 Game Challenges

The 2014 game was centered around physical therapy. The team's robots are supposed to act as a physical therapist, setting up and demonstrating various therapeutic exercises, and putting away some of the equipment. The game board and the various scoring areas are shown in Figure 1. For example, to demonstrate gross motor skills the robots can move the poms into the upper or lower storage areas. Raising and placing tasks would include putting hangers on the rack. Picking and lowering is demonstrated by removing the cubes from the shelves. Fine motor tasks would include placing the cubes in the smaller tubes.

---

[1] Global Conference on Educational Robotics.
[2] KISS Institute for Practical Robotics.
[3] Practical Robotics Institute Austria.
[4] International Teenage Competition and Communication Center.

Fig. 1. Game table and scoring areas

Table 1. Points assigned for game piece placement

Game Piece	On Your Side	@ Fine Motor Task (FMT) Area (touching tape)	On Exercise Bench	On Exercise Bench @ FMT Area	Lower Storage Area	Sorted in Lower Storage Area	Upper Storage Area	Sorted in Upper Storage Area	Fine Motor Task Area	Orange PT Bin	Yellow PT Bin	Lower Hanger Rack	Upper Hanger Rack
Botguy	5	15	50	100									
Blue Cube*1		10			1	5	3	10					
Green Poms					1	5	3	10					
Pink Poms					1	5	3	10					
Your Hanger												10	25
Blue Hanger													Your Upper Hanger Score x2 (for each blue hanger)
Yellow Cube	2	5			1	5	3	10	6	6	25		
Orange Cube	3	5			1	5	3	10	6	100	6		

*1 without Botguy

Game points are awarded by the position of each of the game objects at the end of the round. Teams run unopposed in three seeding rounds, where the average of the top two scores is their seed score. They then compete in double elimination rounds. Their rank in the double elimination, combined with their seed and documentation scores are used to calculate their overall standing in the tournament. The specific points are summarized in Table 1.

2.3   Botball Kit Technology

The Botball kit contains two *Link* robot controllers; small battery operated Linux controllers with four servo channels, four DC motor channels, serial port, WiFi, eight analog to digital ins and eight DIO channels. The *Link* also has a color touch screen for independent operation and data display. Servos, motors, analog and digital sensors, and iRobot Create platform, color cameras and an Asus Xtion 3D sensor are included in the kit, along with a selection of metal, Lego and other hardware to supply structural components.

There is no machining, soldering or gluing required or allowed. All sensors and motors come with connectors installed, and communication cables are supplied. Only the included tools (wrenches, screwdrivers and *U-Glu*) are required in order to build the robots.

The *KISS Platform* programming environment works under Windows, OS X and Linux operating systems and provides a platform independent programming environment for ANSI C, C++ and Java programming. Botball specific libraries have also been added to allow easy access to motor drivers, sensor ports and simplified access to OpenGL and OpenCV functionality.

2.4   Typical Robotics Team Design and Skills

All team leaders (typically the teacher/mentor and/or a couple of student members) participate in a multi-day training session which goes over the kit materials, the programming environment and teaches basic robot skills such as how to:

- follow a line;
- detect an obstacle with a touch sensor;
- detect an obstacle with a range sensor;
- track a colored object with a camera;
- move in a specific pattern;
- start with a signal light and stop after a specified time.

3   Botball Teams by Global Region

At each regional tournament all participating students are encouraged to fill out a survey about their experience. While not everyone participates in the survey, the response rate would appear to be above 50% for all regions except for China; the low response rate from the Chinese students was probably due to the survey only being available in English.

For the US $n = 2194$; in China $n = 17$; the Middle East had $n = 157$ and Austria had $n = 70$ responses.

3.1 Team Size

The survey data is by individual team members, not by teams. One of the questions was for the respondents to give the size of the team on which they were a member. Figure 2 shows the percentage of respondents who indicated each of the team sizes. Note that there is an unrealistic bias in this graph (and for almost all our data) to reflect the results of the larger teams. If there was a single team with 21 members members and two teams with 5 members each, and the response rate was consistent across teams, then two-thirds of the respondents would indicate team size of 21 or more while one third would indicate teams of five or less, while two-thirds of the teams are small and one-third is large.

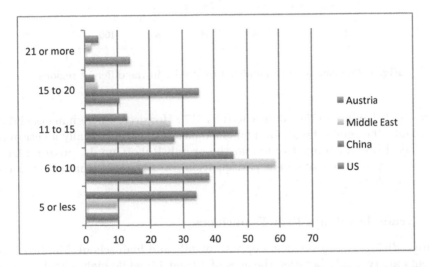

Fig. 2. Percentage of respondents who served on teams of specific size

Taking this bias into account one can see that Austrian teams tended to be the smallest, with majority being five or less. The Middle East and US having most of their teams with less than ten participants (though the US regions had more variation). China tended to have the largest teams, with most teams having more than ten participants.

3.2 Gender Balance

All Botball regions had more boys participating then girls. Figure 3 shows the breakdown. China had just over and the US just under one third female participation. The other regions having slightly more than 15%.

There have been many theories for the gender imbalance in robotics programs (and STEM fields in general). One is that girls are more attracted to artistic disciplines and eschew formal competitions. This is the stated motivation for

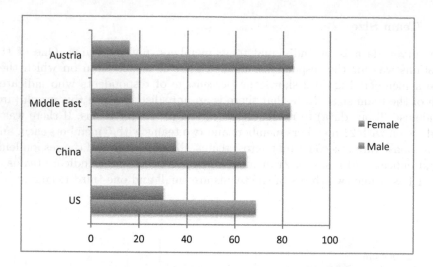

Fig. 3. Percentage of respondents by gender in the different regions

RoboCup Jr having a robot dance activity [17]. However, where numerical data is available, the gender breakdown for the dance activity is virtually identical for the general Botball competition, and the overall RobotCup Jr. program has a much lower percentage of female participation [18] suggesting that other factors may be at play.

3.3   Grade Level and Past Experience

Figure 4 divides the respondents up into middle and high school. Middle school students are typically between the ages of 10 and 13, while high school students are typically 14-19. Different countries have different definitions for these levels, and as in all cases – these are self reported values – and so may be subject to some amount of misinterpretation.

It has been proposed that the earlier girls are exposed to STEM activities, the more likely they are to be interested in those activities [3]. Indeed, the authors' personal experience with elementary school students doing robotics has shown that early intervention (before age 7) leads to a larger female cohort being interested at age 10 then without early intervention [19]. The relatively large percentage of middle school participants in the US could be part of the reason why US Botball has a higher percentage of female students than some of the other regions. However, this does not explain the high female participation in China. Figure 5 again makes China an anomaly by this theory as this is the first year that China has had Botball, and so none of the students have prior Botball experience. The distribution in experience amongst the other regions is quite similar to one another. However, when other robot contest experience is taken into account (Figure 6), the Chinese contradiction is removed (though a Middle East anomaly is introduced).

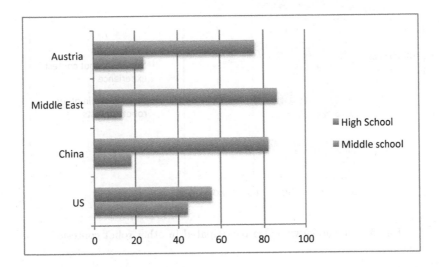

Fig. 4. Percentage of respondents in middle or high school

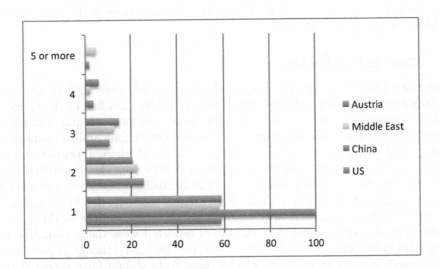

Fig. 5. Percentage of respondents who had years of experience in Botball

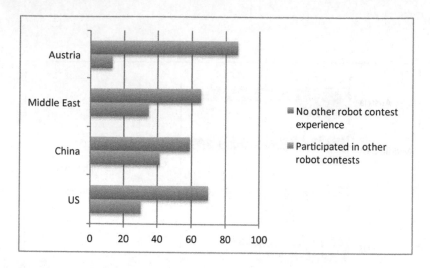

Fig. 6. Percentage who also participated in other robot contests

3.4 Team Roles

Participation in eight roles/assignments were queried in the survey. Figure 7 shows the results. Austrian and Middle East had the largest percentage of respondents in each of the roles. This is consistent with the team size information as these regions tended to have smaller teams then the other two geographic regions and therefore had more of their respondents needing to take on multiple roles within the team. In regions with larger teams more of the participants could focus on one or two roles.

3.5 Time and Technology

More than half of US participating students reported spending less than six hours per week working on their Botball project. More than half of the Chinese students reported spending more than ten hours per week (Figure 8). There appears to be a correspondence between the amount of time and the variety of sensors used on the team's robots (Figure 9). However, there appears to be no correspondence between experience and sensor variety use. There is virtually no difference between US rookie teams (first time Botball at the school) and the US teams in aggregate. It appears that the US teams were the less likely to make high use of sensors as compared to teams from other areas of the world.

One should keep in mind that the hours reported were the personal hours of that team member, while the sensor use was by team (though reported by individuals). Ideally, all members of the same team should have reported the same sensor usage. Some errors in sensor usage reporting are very likely. For example, it very likely that every team used at least one light sensor (as robots start by detecting the starting light). Yet in every region only about 85% of respondents report that one or more light sensors were used by their teams.

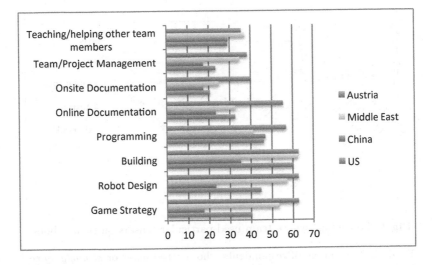

Fig. 7. Percentage who participated in each designated team role/activity

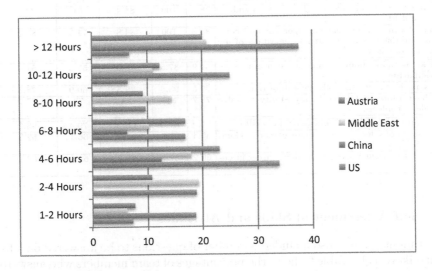

Fig. 8. Hours per week spent by participant on Botball

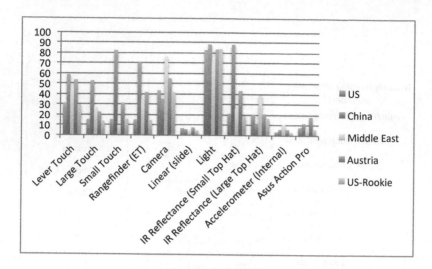

Fig. 9. Percentage whose team used particular sensors on their robots

Table 2. Percentage of respondents who marked *agree* or *strongly agree*

Question	USA	China	Mid-East	Austria	US Rookie
a) I can write a basic program in the C programming language.	62.2	93.8	51.7	71.9	55.9
b) My programming skills have improved after participating in Botball.	68.9	93.8	57.3	39.7	67.2
c) I like to program.	64.6	81.3	48.3	78.1	62.6
d) My communication skills have improved.	69.4	93.8	72.9	53.1	69.3
e) My time management skills have improved.	62.3	93.8	75	42.2	61.9
f) I am more accepting of other team members views and ideas.	78.5	100	85.4	61	79.2
g) The teamwork, project management and communication skills will benefit me regardless of my chosen profession.	85.3	100	84.8	71.9	83.2
h) I know more about engineering concepts after participating.	81.9	87.5	78.5	57.8	81.4
i) My problem solving skills have improved.	79.5	93.8	79.3	60.9	78.3
j) I have more confidence in my ability to solve complex problems.	78.7	100	84	73	77.7
k) I plan on attending a college or university.	94.2	81.3	95.2	79.4	93.2
l) I would consider a degree an/or career in Computer Science.	55.4	87.5	46.9	66.1	52.7
m) I would consider a degree and/or career in engineering.	66.2	87.3	72.3	52.3	63.1

3.6 Self Assessment of Skills and Attitudes

The student surveys also included a number of questions to be answered on a five point Likert scale. Table 2 shows the percentages of team members who answered *agree* or *strongly agree* for that question. The questions have been grouped in the table by topic. In the survey, the question order appears close to random so that the respondents are less likely to be influenced by their answer to the previous question.

Table 3. Percentage of respondents who marked *agree* or *strongly agree* by gender

Question	Male	Female
a) I can write a basic program in the C programming language.	64.0	57.6
b) My programming skills have improved after participating in Botball.	67.3	68.3
c) I like to program.	65.8	60.6
d) My communication skills have improved.	68.0	73.9
e) My time management skills have improved.	60.9	69.1
f) I am more accepting of other team members views and ideas.	76.6	84.9
g) The teamwork, project management and communication skills will benefit me regardless of my chosen profession.	84.8	86.7
h) I know more about engineering concepts after participating.	80.9	81.9
i) My problem solving skills have improved.	79.0	79.9
j) I have more confidence in my ability to solve complex problems.	79.8	77.7
k) I plan on attending a college or university.	93.9	94.8
l) I would consider a degree an/or career in Computer Science.	59.3	46.1
m) I would consider a degree and/or career in engineering.	72.2	52.2

The first three lines in the table refer to student ability and interest in programming. The majority of respondents all feel that they have at least a basic ability to write C programs, and in all regions except Austria, that ability appears to primarily result from their participation in Botball. Most of the Austrian participants are attending technical high schools where programming courses are required. It is possible that a majority of the Austrian participants came to Botball already able to program. Austrian teams also had the highest percentage of their team members working on programming the robots.

Mid-East students appear to have the least confidence in their programming capabilities. Not surprisingly, that cohort also seems to enjoy programming the least. Whether this cause or effect is unknown.

All regions report that the team skills were valuable. Similar results were seen with the general engineering and problem solving skills.

Approximately 95% of respondents from the US and Mid-East regions are planning to go to university. Approximately 80% in the other regions. Respondents were able to indicate consideration of both engineering and computer science majors/careers. The selection for computer science reflected the percentages shown in the *I like to program* question. The relatively low percentage of students indicating consideration of an engineering career in the Austrian region at first appears surprising. As mentioned previously, most of those students are in a technical high school that already has them declaring majors. Many of those students are already enrolled in computer, mechanical or electrical engineering majors. Those students who selected an engineering career have probably already made significant investments in time and coursework in that area and are more likely to follow through than might be true for the US students enrolled in a general education program.

Table 3 presents the same data broken down by gender rather than region. It is interesting to note that all of the questions that dealt with *Did you improve...* during your time in Botball, i.e., questions **b,d,h,i**, girls agreed as strongly or more strongly than boys – they felt they had improved. However, when it comes to programming, fewer felt they had basic capabilities (a), felt they liked to program (c), and (l) were less likely to major in or pursue a career in computer science (though 46%) were interested; which is still many times what is actually seen in universities or industry.

4   Analysis of Survey Results

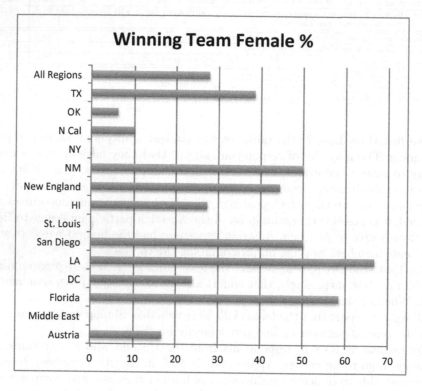

Fig. 10. Percentage of respondents from the regional winners who are female

4.1   The Role of Gender in Tournament Performance

In the US, approximately 30% of Botball students are female (Figure 3). Figure 10 shows the gender breakdown for each of the regional winning teams[5]. Six of the reported regional winners have considerably higher female participation than average for their country, and five have significantly less. Gender makeup of the team does not appear to be an indicator of the team's performance in their regional.

---

[5] The winning teams in the Chicago and China regionals did not provide survey data. In NY, St. Louis and the Middle East Regionals, the survey data for the winning team indicates an all male team.

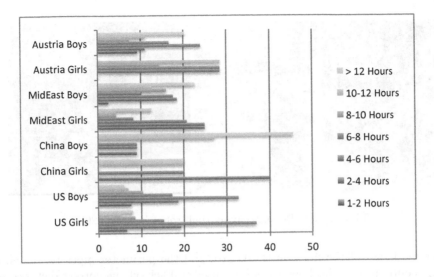

Fig. 11. Regional hours per respondent: boys vs girl

Figure 11 shows the weekly hour commitment in each country divided by gender. Especially for the US (where n for girls is an order of magnitude larger than any other country) the distribution for boys and girls is almost identical. For the other countries the distribution of hours by gender is also very similar.

4.2 Usage of Xtion 3D Sensor

This is the first year in Botball for the Beijing Yuying High School Team. They had only 6 weeks to get familiar with Botball hardware and software, and prepare their strategy for the Chinese regional tournament. The students had no previous programming experience.

Unique features of the team:

- Students had no previous experience with computer programming
- Students had experience with Lego and other mechanical building systems
- All of these students had successfully completed a general technology course, which is mandatory for Chinese high school students. That class included wood working or machining. These students were also some of the organizers/members of a local maker club.
- The team students are classmates and friends.

The team members were very familiar with building mechanical systems and with each other. They had already established collaborative abilities and were familiar with each other's skills and work habits. These factor allowed this team to spend most of their time on strategy and software – which were the parts of the Botball competition new to them.

The team focused on one robot and two primary tasks: placing their hangers on the upper rack and getting an orange cube into the orange marked tube. The

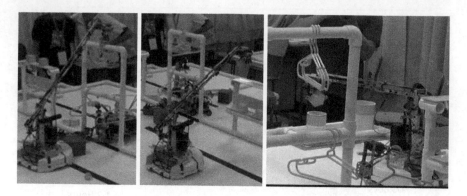

Fig. 12. The Beijing Yuying High robot in action[20]

robot made use of the lines on the board, and the perimeter pipes to navigate to the center area. Wheel encoders were used to back off the correct distance so that the arm would drop the hangers over the pole. The Xtion 3D sensor was used to find the position of the cubes on the top shelf while the robot was in the starting box (at which point it was properly oriented to the shelves). After dropping the hangers, the robot picked up the left cube and drops it in the orange tube (see Figure 12).

While about 6% of teams were using the Xtion sensor at the time of the survey (Figure 9), this was the only team that successfully deployed it, at any region, at their tournament. This team had the sensor integrated into their strategy early on, and it was critical for them to be competitive in scoring. Using this sensor for locating objects to grasp requires more mathematical manipulation then any of the other sensors in the Botball kit.

4.3   Hourly Commitment for Winning Teams

Figure 13 shows the average weekly hourly commitment of the winning regional team minus the regional average commitment. Almost all of the winning teams spend hours more on Botball than the average team member for their particular regional. The major exception is team Hanalani in the Hawaii region. However, the Hanalani team was the largest team in that region being 50% larger than the average Hawaii team size. So the hourly commitment of the team is larger than that of the regional average. Texas and Austria also have winners where the team is larger than the average for their region. Only Florida is the exception having an hourly commitment slightly smaller than the regional average (about 6%) and a team size 3% smaller than the Florida average.

In almost all cases, the winning team spent more team time on the project than average for their region. It should be noted that the US weekly hourly commitment by students varied from a low of 5.2 hrs per week per student in Texas to 8.4 hrs in the Hawaii region. Austria and the Middle East both have weekly student commitments of 8.5 and Chinese students average 9.8 hrs. It

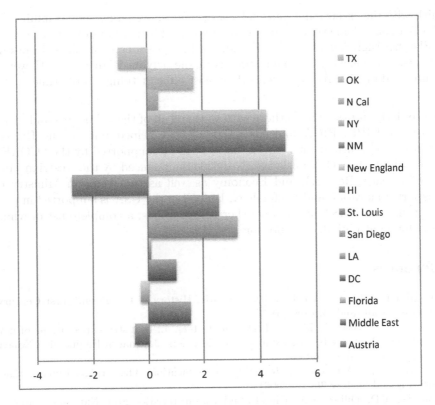

Fig. 13. Difference in weekly hour commitment winners vs regional average

should be remembered that all of this data is self reported and assumes identical distributions within the bins of the survey questions (2 hour increments).

5   Conclusions

In this paper we have reviewed some of the survey and performance data from the 2014 Botball Regionals. The most current data available appears to indicate that Botball does a better job than some other robot programs (e.g., Robocup Jr) in attracting girls. However the gender distribution seems to vary greatly by country – and there is not a good set of data for the same years for the same countries for different programs. Such data is needed in order to do a proper comparison of a program attractiveness to girls. However, the data analyzed from this survey does indicate that girls can contribute as much to robotics teams as boys, and that those that participate have a similar level of involvement to that of the boys. Both genders, but especially girls, found that the program was very beneficial in improving their programming and engineering skills.

The factor that seems to have the best connection to good performance in the program (i.e., successful robots) is the level of time commitment by the team (which with the exception of unusually large teams, is a larger time commitment by individuals). Some time savings can be achieved by previous experience (as in the Fletcher High School team in Florida) or by a previously established working relationship amongst team members (as in the Yuying High team). However, learning and making robot systems, like most difficult things, takes time.

Acknowledgments. The authors with to thank all of the volunteers and members of the KIPR, PRIA and ITCCC staff that helped put on the Botball workshops and tournaments. Botball in Austria was supported by the SCORE!-Project within the COIN initiative (No. 839097), financed by the Austrian Ministry of Science, Research and Economy as well as the Austrian Ministry of Transport, Innovation and Technology. Botball in the USA is supported in part by a grant from NASA and many other organizations, a complete list of which can be found at botball.org/sponsors.

References

1. Friedman, T.L.: The World is Flat: A Brief History of the Twenty-first Century. Farrar, Straus and Giroux (2005)
2. Lang, J.D., Cruse, S., McVey, F.D., McMasters, J.: Industry expectations of new engineers: A survey to assist curriculum designers. Journal of Engineering Education 88(1), 43–51 (1999)
3. Sullivan, J.F.: A call for k-16 engineering education. The Bridge, Linking Engineering and Society 36(2) (2006)
4. Snyder, T.D., Dillow, S.A.: Digest of education statistics 2011. National Center for Education Statistics (2012)
5. Ginder, S., Mason, M.: Postsecondary awards in science, technology, engineering, and mathematics, by state: 2001 and 2009. web tables. nces 2011-226. National Center for Education Statistics (2011)
6. Ramirez, F.O., Wotipka, C.M.: Slowly but surely? the global expansion of women's participation in science and engineering fields of study, 1972-92. Sociology of Education, 231–251 (2001)
7. Hill, C., Corbett, C., St Rose, A.: Why So Few? Women in Science, Technology, Engineering, and Mathematics. ERIC (2010)
8. Dobson, I.R.: It'sa man's world: the academic staff gender disparity in engineering in 21st century australia. Global Journal of Engineering Education 14(3), 213–218 (2012)
9. Roy, R.: Thinking comparative engineering education: India and the rest. Issues and Ideas in Education (March 2013)
10. Barker, B.S., Ansorge, J.: Robotics as means to increase achievement scores in an informal learning environment. Journal of Research on Technology In Education 39(3), 229–243 (2007)
11. Whitehead, S.H.: Relationship of Robotic Implementation on Changes in Middle School Students' Beliefs and Interest toward Science, Technology, Engineering and Mathematics. PhD thesis, Indiana University of Pennsylvania (December 2010)
12. Stein, C., Nickerson, K.: Botball robotics and gender differences in middle school teams. In: Proceedings of ASEE Annual Conference, Salt Lake City, UT (June 2004)

13. Hartmann, S., Wiesner, H., Wiesner-Steiner, A.: Robotics and gender: The use of robotics for the empowerment of girls in the classroom. Gender Designs IT: Construction and Deconstruction of Information Society Technology 13, 175–188 (2007)
14. KIPR: Botball robotics education, http://www.botball.org (2009)
15. KIPR: Global conference on educational robotics (2013), http://kipr.org/gcer
16. Miller, D.P.: Robot contests at GCER 2011. IEEE Robotics and Automation Magazine 18(4), 10–12 (2011)
17. Eguchi, A., Hughes, N., Stocker, M., Shen, J., Chikuma, N.: RoboCupJunior – A decade later. In: Röfer, T., Mayer, N.M., Savage, J., Saranlı, U. (eds.) RoboCup 2011. LNCS, vol. 7416, pp. 63–77. Springer, Heidelberg (2012)
18. Sklar, E., Eguchi, A.: RoboCupJunior — four years later. In: Nardi, D., Riedmiller, M., Sammut, C., Santos-Victor, J. (eds.) RoboCup 2004. LNCS (LNAI), vol. 3276, pp. 172–183. Springer, Heidelberg (2005)
19. Miller, D.P., Stein, C.: So that's what pi is for!" and other educational epiphanies from hands-on robotics. In: Druin, A., Hendler, J. (eds.) Robots for Kids: Exploring New Technologies for Learning, pp. 219–243. Morgan Kaufman (2000)
20. Beijing Yuying Secondary School: IMF-Botball 2014 trailer (2014), http://bit.ly/1wX8PKq

Part III
Applications for Robot Intelligence Technology

Part III
Applications for Robot Intelligence Technology

Coordinated Control of a New Pneumatic Gripper

Jae Chung

Department of Mechanical Engineering
Stevens Institute of Technology
Castle Point on Hudson,
Hoboken, NJ 07030, USA
jae.chung@stevens.edu

Abstract. This paper describes a new pneumatic grasping device, Intelligent Pneumatic Gripper (IPG) that was developed to achieve the primary objective to lift an object to a position without losing contact or slip as fast as possible. Recent developments in pneumatic actuators and valve allow them to be considered for application which previously only electric motors were suitable. Pneumatic system's inherent low stiffness and direct drive capabilities enable smooth compliant geared electric motor systems. Moreover, pneumatic actuators can cost up to 10 times less than electric motors, while offering a higher power to weight ratio. Each joint of the IPG arms is revolute and actuated by a pneumatic actuators consisting of a low friction pneumatic cylinder and a regulator valve with a position sensor. Each valve supplies regulated pressure to a single chamber of pneumatic cylinder. The pneumatic cylinder extends gradually with the applied force-causing side arm to grasp. The grasping motion of IPG is produced by four pneumatic cylinders. As a result of the IPG physical configuration, the force exerted on the arms depends on the position of stroke of the right and left arms. In IPG, grasping motion is initiated by the pneumatic actuators with two touch sensors for ensured grasping of the object. Lifting motion is engaged after grasping motion. Figures 1 and 2 illustrate the grasping and lifting motion of the IPG gripper system. A nonlinear force feedback controller is designed to control both the position and velocity of the end effector and the constraint force between the gripper and the environment. Simulation studies were performed to illustrate the efficacy of the developed control method

1   Introduction

The robots to assist humans in their routine activities must perform many different complex tasks. However the status of robotics research is far from building robots that can independently decide how to grip an object to accomplish every day tasks. At this point, a robot gripping system with enough intelligence to grasp an object with correct force and velocity, as humans do, is considered.

Analytical studies of the grasping and finger by robot hand have been done by many researches [1,2,3,4]. Yoshikawa and Nagai have divided the finger forces into two different forces, manipulation force and internal force, defined the manipulation force, which generates the required external object force. Shimoga and Goldenberg

J.-H. Kim et al. (eds.), *Robot Intelligence Technology and Applications 3,*
Advances in Intelligent Systems and Computing 345, DOI: 10.1007/978-3-319-16841-8_50

[5] have studied modeling and controlling the impedance of a soft finger and showed experimentally how the presence of passive damping helps reduce the peak impact forces that occur as a rigid object is grasped by fingers of a robotic hand from soft materials. Another approach to control robot hands is several force and position control schemes devised for robotic interaction tasks. Chiavervini and Sciavicoo (1993) proposed a parallel approach to force and position control, where position trajectories are sacrificed due to force demands. For physiotherapy, specifying specific position demands would be difficult, as they would be masked by the dominance of the force loop. Force is controlled in constrained directions, while position is controlled in unconstrained directions [6,7,8,9]. Based on the force control current researchers have studied the grasping robot, gripper and finger of the manipulator. However the requirement of the dexterous manipulation of an object with robotic mechanism in sophisticated tasks and the difficulties in realizing such dexterity are stimulating many researchers to tackle the problem regarding the development.

Additionally, the technology has enabled the automation of many process, however, these are mainly focused on grasping without slip. In the grasp of rigid objects, additional considerations have to be made with respect to manipulation, gripping, and sensor required. Such objects may slip during operation of robot for example by increasing mass of object or obtaining external force which may result in drop of the object grasped by the gripper of robot.

2   Human Assist Non-stationary Device for Lifting (HANDL)

The gripper is a critical component of an industrial robot since it interacts with the environments and object, which is grasped and manipulated. Among many problems such as rigidity, lightness, multi-task capability and lack of maintenance, basic requirements for an industrial robot gripper can be recognized in a low-cost and reliable design. Recent developments in pneumatic actuators and valve allow them to be considered for application which previously only electric motors were suitable. Pneumatic system's inherent low stiffness and direct drive capabilities enable smooth compliant geared electric motor systems. Moreover, pneumatic actuators can cost up to 10 times less than electric motors, while offering a higher power to weight ratio.

Each joint of the HANDL arms is revolute and actuated by a pneumatic actuators consisting of a low friction pneumatic cylinder and a regulator valve with a position sensor. Each valve supplies regulated pressure to a single chamber of pneumatic cylinder. The pneumatic cylinder extends gradually with the applied force-causing side arm to grasp. As stated earlier, the grasping motion of HANDL is produced by four pneumatic cylinders as shown in Fig 1.

As a result of the HANDL physical configuration, the force exerted on the arms depends on the position of stroke of the right and left arms. In HANDL, grasping motion is initiated by the pneumatic actuators with two touch sensors for ensured grasping of the object. Lifting motion is engaged after grasping motion. Figures 1 and 2 illustrate the grasping and lifting motion of the HANDL gripper system.

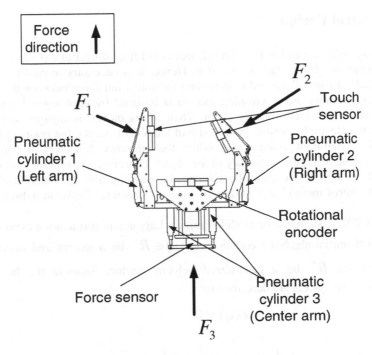

Fig. 1. Grasping direction of HANDL

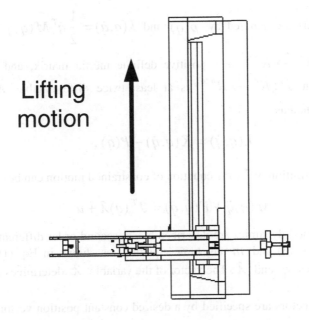

Fig. 2. Lifting motion of HANDL

3   Control Design

The primary objective of the HANDL gripper is to lift an object to a position without losing contact or slip as fast as possible. Hence, it is necessary to control both the position and velocity of the end effector and the constraint force between the gripper and the environment. First, grasping motion is initiated by applying a force to the object based on its weight for secure grip. Then, lifting motion is engaged to move up the object to a desired position. In case slip of the object occurs, the grasping force is controlled to maintain stable grasp while the reference trajectory for lifting is modified on line to improve stability of the object. Therefore, the control system to be developed has two active controllers – one for grasping and the other lifting. The developed control method will have a closed-loop structure similar to hybrid control methods.

The HANDL system can be modeled using a Lagrangian formulation expressed by a set of differential-algebraic equation. Let $q \in R^n$ be a generalized coordinated vector and $\dot{q} \in R^n$ be a generalized velocity vector. Suppose the holonomic constraints of the system are described by

$$\phi(q) = 0 \tag{1}$$

where $\phi^T = [\phi_1, \cdots \phi_m]$ is at least twice differentiable. The potential and kinetic energy functions are denoted by $p(q)$ and $k(q,\dot{q}) = \frac{1}{2}\dot{q}^T M(q)\dot{q}$, respectively, where $M : R^n \to R^{n \times n}$ is a positive definite inertia matrix, and the potential energy function $P : R^n \to R^{n \times n}$ is at least twice differentiable. A Lagrangian function is defined as

$$L(q,\dot{q}) = K(q,\dot{q}) - P(q). \tag{2}$$

Using the definition of L, the equation of constrained motion can be expressed as

$$M(q)\ddot{q} + F(q,\dot{q}) = J^T(q)\lambda + u \tag{3}$$

The constrained dynamics are described by n second order differential equations as shown in Eq. (3) and m algebraic equations as shown in Eq. (1) in terms of $n + m$ variables q and λ. The vector of the variable λ determines the constraint forces.

Regulation vectors are specified by a desired constant position vector $\varphi(q_d) = 0$ and $f_d = J^T(q_d)\lambda_d$ for some constant vector $\lambda_d \in R^m$. To achieve regulation of position and force to the specified position and force vector (q_d, f_d), it is necessary to guarantee the desired values are an equilibrium of the closed loop equations. This can be achieved by the following controller:

$$u = \frac{\partial P(q)}{\partial q} - \frac{\partial P_d(q)}{\partial q} - C\dot{q} \tag{4}$$

where $P_d(q)$ is a desired potential energy function that is chosen to satisfy the following equation:

$$\frac{\partial P_d(q_d)}{\partial q} = f_d \tag{5}$$

Thus, (q_d, λ_d) is an equilibrium of the closed loop system Eq. (6), which requires that the gradient of the desired potential energy function $P_d(q)$ at q_d be parallel to the constant force vector f_d. The $n \times n$ matrix C is assumed to be symmetric and to satisfy $\dot{q}^T C \dot{q} > 0$ for all $\dot{q} \neq 0$ satisfying $J(q_d)\dot{q} = 0$.

A Lyapunov function for the constrained system can be used to guarantee the local stability of the equilibrium (q_d, λ_d). In particular, the modified potential energy is represented as

$$P_{md} = P_d(q) - P_d(q_d) - \phi^T(q)\lambda_d. \tag{6}$$

The modified potential energy function can be used to form a Lyapunov function for the constrained system as

$$V(q, \dot{q}) = \frac{1}{2}\dot{q}^T M(q)\dot{q} + P_{md} \tag{7}$$

Therefore, (q_d, λ_d) is locally asymptotically stable based on the invariance principle. However in case of HANDL, the force feedback is necessary due to applying in the direction normal to the constraint surface at the contact point during the gripping motion. Therefore, this is done by using the Jacobian matrix $J^T(q)$ as a projection. The reason is that $\dot{q}^T J^T(q) = 0$ or the range of $J^T(q)$ is normal to the velocity $\dot{q}$ which is on the tangent plane at the contact point. Therefore, the control with force feedback is

$$u = \frac{\partial P(q)}{\partial q} - \frac{\partial P_d(q)}{\partial q} - C\dot{q} + J^T(q)G_f(\lambda - \lambda_d) \tag{8}$$

where G_f is an $m \times m$ force feedback matrix. This control can be used for position/force control. It is straight forward to show that the closed loop system is asymptotically stable by using Eq. (7).

The controller, Eq. (4) is in general a nonlinear feedback controller. In the following, we choose a particular function $P_d(q)$ which results in a simple affine linear feedback control law. The desired potential energy function is chosen as

$$P_d(q) = P(q) - P(q_d) - \left[\frac{\partial P(q_d)}{\partial q} - J^T(q_d)\lambda_d \right]^T e + \frac{1}{2}e^T W e. \qquad (9)$$

Where $e = q - q_d$ and W is a diagonal matrix. By checking Eq. (5) the modified energy function is

$$P_{cd}(q) = P_d(q) - \lambda_d^T \phi(q) \qquad (10)$$

It is easy to verify that $P_{cd} = 0$ and $\dfrac{\partial P_{cd}}{\partial q} = 0$. Thus $P_{cd}(q)$ has a local minimum at P_d. With this choice, the controller, Eq. (8) takes the following specific form:

$$u = \frac{\partial P(q_d)}{\partial q} - C\dot{q} - J^T(q_d)\lambda_d - W(q - q_d) \qquad (11)$$

Eq. (15) represents an affine feedback control law. The first two terms form a constant bias term, and the third and fourth terms represent the feedback of position and velocity errors. C is a diagonal matrix. In addition to Eq. (11), feedback of the constraint force error can be introduced to tune the constraint force error response. Such a feedback control, including feedback of the constraint force error, is

$$u = \frac{\partial P(q_d)}{\partial q} - C\dot{q} - J^T(q_d)\lambda_d - W(q - q_d) + J^T(q_d)G_f(\lambda - \lambda_d). \qquad (12)$$

Here, Eq. (12) is the control with force feedback. However, when the HANDL slips the object, the desired force and position should be changed to prevent the object from falling down. During the slipping motion, the energy E_p of the object is dissipated. Therefore, the energy to grasp the object should be added to the dissipated energy to prevent the object from falling down while the slip occurs. Also, the lifting velocity should be reduced to facilitate regrasp of the object subject to slip. Therefore, the new desired velocity is changed during slip as

$$\dot{q}_{nd} = \dot{q}_d - \dot{q}_s \qquad (13)$$

where $\dot{q}_s$ is the reduction of the velocity due to slip. It is obtained from the slip sensor mounted on the HANDL. In addition, the original grasping force is modified to a new one as

$$f_{nd} = \frac{E + E_d + f_d}{S} \tag{14}$$

where E is the energy of the HANDL to grasp the object without slip, E_d is the dissipated energy during slip, and S is the position to increase the force of the gripper. Therefore during slip of the object, the lifting velocity of lift and the grasping force are modified on line.

4   Simulation

Simulation study of the HANDL was performed to investigate efficacy of the developed control method. Figure 3 illustrates slip motion of the object employed for simulation study. The object slips to 0.05m between 3 to 4 second during a lifting motion. Simulation parameters were chosen as $M_{11} = 15\ kg$, $M_{22} = 5\ kg$, $M_o = 1.5$kg, $W_{11} = 55$, $W_{22} = 24$, $C_{11} = 4.3$, $C_{11} = 5.3$, and $G_f = 5$ where M_{11} is the mass of the first link for lifting motion of HANDL, M_{22} is the mass of the second link for grasping motion of HANDL, and M_0 is the object mass .

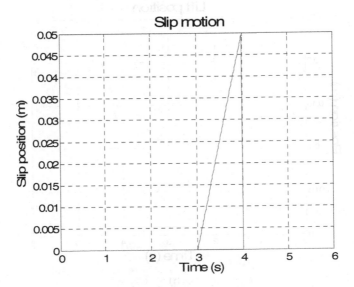

Fig. 3. Slip motion of object

Figure 4 (a) shows a desired trajectory modified according to Eq. (13). The trajectory for lifting motion is modified based on the amount of slip. This means that when the object slips, the lifting velocity is reduced by the slipping velocity of the object. Figure 4 (b) shows the actual position of the lift, which is consistent with Figure 4 (a). Fig 5 (a) shows the desired grasping force by Eq. (14), which prevents further slip of the object. Fig 5 (b) shows the actual grasping force.

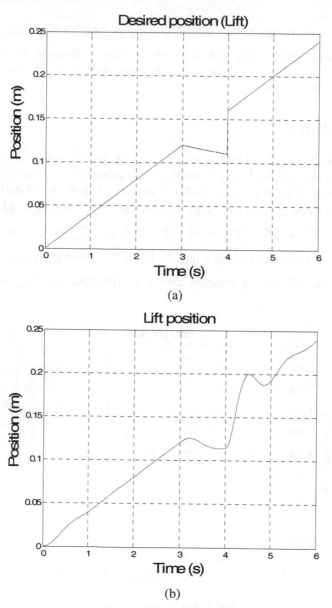

(a)

(b)

Fig. 4. New desired trajectory and output position of Lift of HANDL

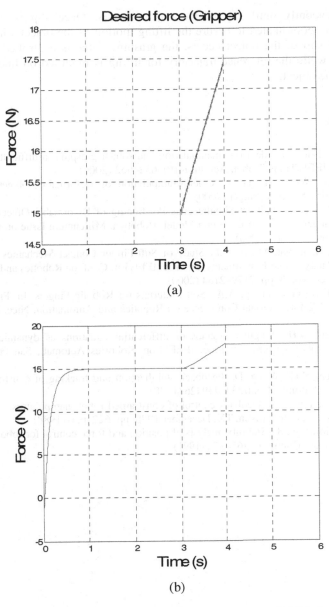

Fig. 5. New desired force and actual force of gripper of HANDL

5   Conclusion

A control method for the HANDL system was developed based on the Lyapunov's direct method to control the lifting position and the grasping force of an object. The developed control system consists of a position and force controllers. The controllers work independently until slip of the object occurs. Once slip is detected, the controllers are coordinated to ensure the lifting motion of the object without further slip. In case slip of the object occurs, the grasping force is controlled to maintain stable grasp while the reference trajectory for lifting is modified on line to improve stability of the object.

References

1. Bicchi, A.: Hands for dextrous manipulation and robust grasping: a difficult road toward simplicity. IEEE Trans. Robot. Autom. 16(6), 652–662 (2000)
2. Henrich, D., Wörn, H. (eds.): Robot Manipulation of Deformable Objects. Advanced Manufacturing Series. Springer (2000)
3. Hirai, S., Wada, T.: Indirect Simultaneous Positioning of Deformable Objects with Multi Pinching Fingers Based on Uncertain Model, Robotica. Millennium Issue on Grasping and Manipulation 18, 3–11 (2000)
4. Xydas, N., Bhagavat, M., Kao, I.: Study of Soft-Finger Contact Mechanics Using Finite Elements Analysis and Experiments. In: Proc. IEEE Int. Conf. on Robotics and Automation, San Francisco, vol. 3, pp. 2179–2184 (2000)
5. Shimoga, K.B., Goldenberg, A.A.: Soft Materials for Robotic Fingers. In: Proceedings of the 1992 IEEE International Conference on Robotics and Automation, Nice, France (May 1992)
6. McClamroch, N.H.: Singular systems of differential equations as dynamic models for constrained robot systems. In: Proc. IEEE Con. Roborics Automat., San Francisco, CA (1986)
7. McClamroch, N.H., Wang, D.: Feedback stabilization and tracking of constrained robots. IEEE Trans. Automat. Contr. 33, 419426 (1988)
8. Yun, X.: Dynamic state feedback control of constrained robot manipulators. In: Proc. 27th Con$ Decision Contr., Austin, TX, December 1988, pp. 622–626 (1988)
9. Wen, J.T., Murphy, S.: Stability analysis of position and force control for robot arms. IEEE Trans. Automat. Contr. 36, 365–371 (1991)

Robust Camera Calibration for the MiroSot and the AndroSot Vision Systems Using Artificial Neural Networks

Awang Hendrianto Pratomo[1,2], Mohamad Shanudin Zakaria[2], Mohammad Faidzul Nasrudin[2], Anton Satria Prabuwono[2], Choong-Yeun Liong[2,3], and Izwan Azmi[2]

[1] Department of Informatics Engineering, Faculty of Industrial Engineering,
UPN "Veteran" Yogyakarta, Indonesia
[2] Pattern Recognition Research Group, Center for Artificial Intelligence Technology,
Faculty of Information Science and Technology, Universiti Kebangsaan Malaysia,
43600 UKM Bangi, Selangor, Malaysia
[3] School of Mathematical Sciences, Faculty of Science and Technology,
Universiti Kebangsaan Malaysia, 43600 UKM Bangi, Selangor, Malaysia
awang@upnyk.ac.id,{msz,mfn,antonsatria}@ftsm.ukm.my,
lg@ukm.my, izwanazmi90@gmail.com

Abstract. The MirosSot and the AndroSot soccer robots have the ability to recognize, and navigate within, their environments without human intervention. An overhead global camera, usually at a fixed position, is used for the robot's vision. Because of the lens distortion, images obtained from the camera do not accurately represent the robot's environment. The distortions affect the coordinates. A technique to calibrate the camera is required to transform the skewed coordinates of the objects in the image to the physical coordinates, which define their real-world position. In this study, a method is proposed for camera calibration using an artificial neural network (ANN) in a two-step process. First, ANN was used to select the camera height and the lens focal lengths for high accuracy. Second, ANN was used to map a coordinate transformation from the camera coordinates to the physical coordinates. During the learning process, the weight of each node in the ANN model changed until the best architecture is reached. The experiments thus resulted in an optimum ANN architecture of 2×4×25×2. The accuracy and efficiency of the camera calibration method were obtained by relearning using the ANN whenever changes to the environmental occurred. Relearning was done using the new input data set for each respective environmental change. Based on our experiments, the average transformation error of the calibration method, using many types of camera, camera positions, camera heights, lens sizes, and focal lengths, was 0.18283 cm.

Keywords: Camera Calibration Technique, Neural Network, Robot Soccer, Global Vision System.

1 Introduction

If the camera is not positioned exactly over the target, the environment will cause the image projection to be distorted and asymmetric. The camera height and position will

© Springer International Publishing Switzerland 2015
J.-H. Kim et al. (eds.), *Robot Intelligence Technology and Applications 3*,
Advances in Intelligent Systems and Computing 345, DOI: 10.1007/978-3-319-16841-8_51

affect the resulting image projection. As the distance of the objects from the center of the camera increases, the differences between the positions of the detected objects from those of the real objects become greater. Objects photographed by a camera in the perpendicular direction will be more accurate projections, whereas if the object is photographed by a camera at an inclination, larger differences in the image projection will occur [1].

The coordinate transformation process is influenced by the type of lens, the height of the camera and the position of the camera. Fig 1 shows the difference in the image projection when the camera is located off-center with respect to the position of the object. The primary issue with respect to lens distortion and camera position is finding an accurate transformation to obtain the actual robot position from the camera projection.

(a)   (b)

Fig. 1. (a) An image object photographed by a camera from the perpendicular direction, (b) an image object photographed by a camera from 100 cm from the center of the field

Camera calibration is used to transform the object position into the actual position. Camera calibration with a view obtains the detailed information from the camera. Camera calibration from the image itself is the primary problem in computer vision [2]. Additionally, camera calibration is the primary process in computer vision used to determine the actual position of the robot from an image [3, 4, 5] Camera calibration in a mobile robot vision system is used to transform the robot position in the camera image to its physical position. The objective of the camera calibration method used in our study is to resolve the non-linear lens distortion using artificial neural networks.

2   State of the Art

2.1   Calibration Grid

A transformation system is calibrated by comparing an object that has a known position in the physical coordinate system to its camera coordinate system position [6, 7, 8]. The transformation process from camera coordinates into physical coordinates requires measurement accuracy and precision [9, 10]. The measurement accuracy of the transformation process is determined from the mass errors that result from the transformation process on the actual object position in the physical coordinates [11]. The camera calibration process also requires a reference point from the image, which is the manifestation of an actual point in the real world [8, 12]. The calibration process is facilitated by the availability of a calibration grid. An example calibration grid is shown in Fig. 2. National Instruments provides limitations that must be met in the construction of a calibration grid as follows [1]:

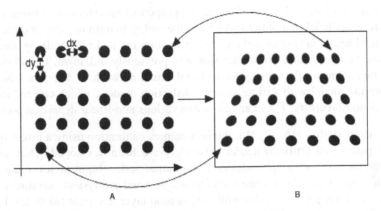

Fig. 2. Calibration process requires a reference point to transform physical coordinates into camera coordinates (http://zone.ni.com/reference/en-XX/help/372916J-01/nivisionconcepts/spatial_calibration/)

1. The distances between points in the x and y directions are the same.
2. The dots must fill the entire workspace.
3. The distances between points are the same.
4. The cutoff distances between the centers of the dots are the same.
5. The dots should be in vertical columns and horizontal rows that are perpendicular to calibrate the grid.

There are many types of grid calibration that can be used, such as grid calibration points, grid calibration chessboard, grid calibration lines, and grid calibration scale curve. From these, the most frequently used are the grid calibration points and the grid calibration chessboard.

2.2   Camera Calibration Using Artificial Neural Networks

A neural network is one of the models that simulate the learning process of the human brain [13]. The human brain consists of millions of neurons; each neuron connects one cell to another cell. An artificial neural network is a collection of nodes with the ability to learn and to store information. Based on the research conducted by [14], an artificial neural network is similar to the human brain in two ways: neural networks have the ability to acquire knowledge through a learning process and, in neural networks, knowledge storage has synaptic weight as a strength connection between the neurons.

Artificial neural networks have been implemented in various fields of science and technology, especially in robotics and image processing. Robotics and image processing are interlinked because, through image processing, a robot can see the world. According to [15], camera calibration has the same features as artificial neural networks in various areas, such as the following:

1. A neural network is developed using the properties of several non-linear neurons; therefore, artificial neural networks have the ability to train non-linear data.
2. Artificial neural networks and camera calibration use a similar training mechanism to find the coefficients from data that was previously unknown. In addition, they calculate unknown data, using an artificial neural network to determine the camera calibration model with various camera distortion models, which can reduce the linear distortion errors. They can also solve various non-linear distortion problems.

In previous studies [16, 17, 18], implicit camera calibration using a broad spectrum of layer perceptron artificial neural networks have been developed. They used two hidden layers with two input nodes and two output nodes. Input nodes in the form of x' and y' camera coordinates are transformed to x and y physical coordinate results. The first layer has ten (10) nodes while the second layer has eight (8) nodes. Learning was implemented using 5000 iterations. This study used a calibration grid point, a point column distance of 25 mm, a point row distance of 20 mm, and a calibration grid size of 11×9 (99-point calibration). The results were compared with the two-step coordinate transformation method developed by Tsai (Tsai 1987). The average error in this method was 0.6559. The neural network method reduced the error level compared with Tsai's two-stage method by as much as 11.45%.

Wang et al. [19] developed an application for the correction of distortion using an artificial neural network with standard grid plates. The standard grid was used to calculate the distortion of the camera. The standard grid was produced using lines spaced 1 mm apart. The calibration grid size in their study was 19×25. The region bounded by the intersection of the center line with the 10th and the 13th column is the center region of the camera with no distortion. The calibration process uses a back propagation artificial neural network with two hidden layers and a feed forward algorithm. Lee and Oh [20] developed a method for camera calibration using an artificial neural network to estimate the global position using a camera mounted on the ceiling. Their method uses input data in the form of pixels per image distortion with an output of pixels per image to be repaired.

3   Methodology

This section describes the approach used in this study. The camera calibration method is based on artificial neural networks, using the camera position, lens focal length, and the current condition.

3.1   Training Set Data Preparation

The provision of the training data set from the images involves several parameters, such as the type of camera, the lens size, the camera position, and the direction of the camera. The resulting image data set was used both for the development and for the experiments used in the camera calibration.

3.2   Development of the Camera Calibration Method

Objects captured by the camera are not in their actual position; therefore, a process is required to transform from camera coordinates into physical coordinates. The purpose of the camera calibration process is to acquire the physical positions of the objects from images. Development of the camera calibration method in this study uses an artificial neural network.

3.3   Camera Calibration Performance Measurements and Experiments

Performance measurements were conducted to obtain the best method for the camera calibration process. For this purpose, a series of experiments to obtain the best performance of every method is proposed. The experimental design of the camera calibration method in our study is divided into two sections: the experimental camera calibration method based on the camera type and lens type and the experimental camera calibration method based on the camera position and camera direction. The camera types that we used in this study are based on CCD-sized sensors, brand cameras, and camera drivers. The types of lenses used in our study are based on the size of lens, brand of lens, and lens focal length.

Camera Calibration Experiments Based on Camera Height and Lens Focal Length

Experimental studies of the camera calibration methods were performed for several camera heights and lens focal lengths. The experiments were conducted to determine the best artificial neural network model architecture to transform the camera image to the robot position based on camera height and lens focal length. This experiment used fifteen images obtained from a Basler Scout SCA 640-70F camera.

Camera Calibration Experiment Based on Camera Position and Orientation

Experiments were performed based on the camera position and the orientation using a Basler Scout SCA 640/120 fc, CCD Sensor Size 1/4 and Fujinon lens (lens size 1/3) camera. The set of experiments for the camera calibration method based on the camera height, the camera position, and the direction the camera was facing is shown in Table 1. The coordinates of the camera position for the set of experiments is also shown in Table 1. The camera position and orientation is shown in Fig. 3.

Table 1. Set of experiments for the camera calibration based on camera height and camera rotation

Experiment set	Camera Position	Camera Coordinates	Camera Height	Camera Rotation
Experiment 1	A1	(0,0)	240	0^0
Experiment 2	B1	(100,0)	240	0^0
Experiment 3	C1	(0,-80)	240	0^0
Experiment 4	D1	(100,-80)	240	0^0
Experiment 5	A2	(0,0)	200	0^0
Experiment 6	B2	(100,0)	200	0^0
Experiment 7	C2	(0,-80)	200	0^0
Experiment 8	D2	(100,-80)	200	0^0
Experiment 9	E	(-40,40)	220	0^0
Experiment 10	F	(40,40)	220	0^0
Experiment 11	G	(40,-40)	220	0^0
Experiment 12	H	(-40,-40)	220	0^0
Experiment 13	A1	(0,0)	220	10^0
Experiment 14	A1	(0,0)	220	-30^0
Experiment 15	A1	(0,0)	220	20^0
Experiment 16	A1	(0,0)	220	30^0
Experiment 17	G	(0,0)	220	-30^0
Experiment 18	E	(0,0)	220	-30^0
Experiment 19	F	(0,0)	220	30^0
Experiment 20	H	(0,0)	220	30^0

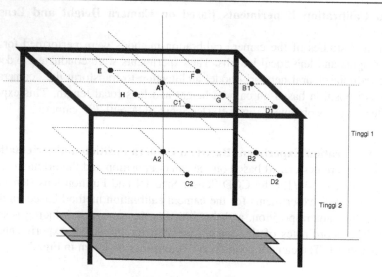

Fig. 3. Camera position

4   Results and Discussion

4.1   Artificial Neural Network Method for Camera Calibration

The development of the camera calibration method using an artificial neural network was divided into several steps as follows:

1. Determine the physical coordinates from the image.
2. Train the physical coordinates and the image coordinates using a neural network to obtain the function that will be used to transform camera coordinates into physical coordinates.
3. Determine the neural network model that is most suitable to correct for lens distortion.
4. Determine the most suitable activation function and learning rate.
5. Simulate data outputs to ensure that the method can be implemented in a mobile robot system.

4.2   Development of the Camera Calibration Method Using an Artificial Neural Network

The following section describes the development process of the camera calibration method based on the current state of global overhead mobile robot vision.

Development of the Camera Calibration Method

The camera calibration is a step in the sight of the three-dimensional (3D) computer to extract the metric information from a two-dimensional (2D) image. Camera calibration methods in our study use two input values and two output values in artificial neural networks. The parameters that are used in our study are the camera coordinates x' and y' for the artificial neural network inputs and the physical coordinates for the outputs. In this study, the results of experiments performed using artificial neural network models are shown in Table 2. The data in Table 2 show that the best ANN model is a 2×4×25×2 model with two inputs to the artificial neural network as camera image coordinates (x', y'), four nodes in the first hidden layer, twenty five hidden nodes in the second layer, and two outputs for the physical coordinates as the result. The resulting artificial neural network model is shown in Fig. 7. The previous study by Woo and Park [17,18] used a 2×10×8×2 artificial neural network model with two inputs, ten hidden nodes in the first layer, eight hidden nodes in the second layer, and two nodes for the output. The study reported in [15] uses a 2×5×2 artificial neural network model, i.e., with two inputs, five hidden nodes and two output nodes. Both of these prior methods used back propagation techniques.

The coordinate transformation process was implemented by calculations performed on the detected position of the robot using weights derived from the ANN learning. Some examples of images used in the development and testing of the camera calibration method based on the current condition are shown in Fig. 4. Results of the coordinate transformation from the image using the ANN method are shown in Fig. 5.

Table 2. Results of the development tests of an ANN architecture using two inputs and two outputs with the Levenberg–Marquardt learning algorithm

Test No	Architecture RNB	Learning Time	Average Error (ΔX, ΔY)		Maximum Error (ΔX, ΔY)	
			X	Y	X	Y
1	2-8-4-2	0:00:55	0.3821	0.2642	3.3068	2.646
2	2-4-8-2	0:06:47	0.6755	0.4295	3.3000	2.1950
3	2-8-16-2	0:08:26	0.2406	0.1912	5.0200	2.2140
4	2-16-8-2	0:21:57	0.6422	0.5623	2.4200	3.2640
5	2-8-24-2	0:27:44	0.3982	0.2555	3.0700	2.2610
6	2-4-25-2	0:05:10	0.1539	0.1346	1.9800	1.1760
7	2-25-4-2	0:48:32	0.5718	0.4249	2.5470	2.6410
8	2-24-8-2	0:37:39	0.5460	0.7868	2.9205	2.3317

Note: ANN Model $2 \times 4 \times 25 \times 2$ using two inputs is the best model from the ANN testing

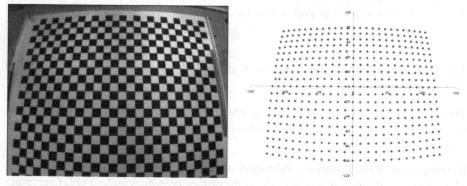

Fig. 4. Sample image and coordinates obtained from a camera height of 200 cm, with lens focal length of 2.8 mm, and camera coordinate position of (3,63)

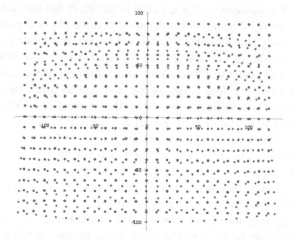

Fig. 5. The result of a coordinate transformation using the ANN method from an image obtained from a camera height of 200 cm, with lens focal length of 2.8 mm, and camera coordinate position (3, 63)

a. Acquisition of the Coordinate References

The coordinate procurement process is conducted by placing a chessboard calibration grid on the field, as shown in Fig 6. The reference coordinates are obtained from the intersecting corners of all the black and white squares. The camera coordinates the acquisition process by setting up the number of the grids to use. In our study, the calibration grid was 21×17 and the sides of all the black and white squares were 10 cm. The next step is to set up the reference coordinates at the upper left corner (X, Y), which in our case was at coordinates (-100, 80). Then, the distance between the vertical and horizontal lines are determined. In our case, that distance was 10, i.e., the distance between the black and white squares in the calibration grid we used was 10 cm.

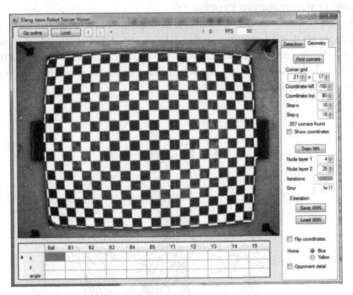

Fig. 6. The image-capturing result using the calibration grid

b. Parameter Preparation for the ANN Learning

To set the parameters for the ANN learning process, the number of hidden nodes in each layer of the ANN model must be determined. In our study, there were four hidden nodes in the first layer and twenty five hidden nodes in the second layer. The maximum number of iterations was set to 10^6 and the minimum error was set to 10^{-12}. A summary of the parameters used in our study are as follows:

— Number of input artificial neural network data = 2
— Output nodes = 2
— Hidden nodes in layer 1 = 4
— Hidden nodes in layer 2 = 25
— Minimum error = 10^{-12}
— Weights initialized using the Nguyen-Widrow method

- Activation function = bipolar sigmoid
- Learning rate = 0.05
- Training method = Levenberg-Marquardt

c. Learning of the ANN model

The learning process was executed to obtain the weight of each node in the hidden layers. The ANN model is a mathematical model of the coordinate transformation process. After the learning is completed, the weight of each node in the hidden layers is obtained.

Coordinate Transformation Process

The coordinate transformation process was implemented by performing calculations on a detected robot with weights derived from the ANN learning. A neural network model from the engine of the robot soccer MiroSot and AndroSot vision systems is shown in Fig. 7. A mathematical model of the coordinate transformation process was derived using the ANN equations 1 and 3.

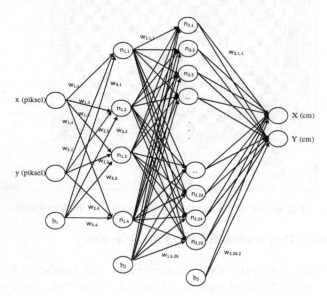

Fig. 7. Development of the ANN model using two inputs and two outputs for the coordinate transformation process from camera coordinates into physical coordinates in the robot soccer MiroSot and AndroSot vision systems

As shown in Fig. 7, the following can be defined:

x and y are the image coordinates

X and Y are the results of the coordinate transformation

$n_{1.k}$ is the node in layer one at line k

$n_{2.j}$ is the node in layer two at line j

f_1 is the transfer function for inputs to layer 1

f_2 is the transfer function for nodes from layer 1 to layer 2
f_3 is the transfer function for nodes from layer 2 to the output
$w_{i.k}$ is the weight of the input node i to the node in layer 1, line k, for $i = 1,2,3$ and $k = 1,2,3,4$
$w_{1.j.k}$ is the weight of the node in layer 1, line k to the node in layer 2, line j, for $k = 1,2,3,4,5$ and $j = 1,2,...,25$
$w_{2.j.l}$ is the weight of the node in layer 2, line j to the output line l, for $j = 1,2,...,26$ and $l = 1,2$
b_1, b_2, and b_3 are the biases of layer 1, layer 2 and layer 3, respectively

The following relations are also true:

$$n_{1.k} = f_1 \left(w_{1.k}x + w_{2.k}y + w_{3.k}b_1 \right), k = 1, 2, 3, 4 \tag{1}$$

$$n_{2.j} = f_2 \left(\sum_{k=1}^{4} w_{2.k.j}n_{1.k} + w_{1.5.j}b_2 \right), j = 1, 2, ... , 25 \tag{2}$$

and the following can be derived:

$$X = f_3 \left(\sum_{j=1}^{25} w_{2.j.1}n_{2.j} + w_{2.26.1}b_3 \right) \tag{3}$$

$$X = f_3 \left(\sum_{j=1}^{25} w_{2.j.1} f_2 \left(\sum_{k=1}^{4} w_{1.k.j} f_1 \left(w_{1.k}x + w_{2.k}y + w_{3.k}b_1 \right) + w_{1.5.j}b_2 \right) + w_{2.26.1}b_3 \right) \tag{4}$$

$$Y = f_3 \left(\sum_{j=1}^{25} w_{2.j.2}n_{2.j} + w_{2.26.2}b_3 \right) \tag{5}$$

$$Y = f_3 \left(\sum_{j=1}^{25} w_{2.j.2} f_2 \left(\sum_{k=1}^{4} w_{1.k.j} f_1 \left(w_{1.k}x + w_{2.k}y + w_{3.k}b_1 \right) + w_{1.5.j}b_2 \right) + w_{2.26.2}b_3 \right) \tag{6}$$

4.3   Experiments and Discussion of the Results of the Camera Calibration Method Based on the Current Camera State-of-the-Art

Coordinate transformation is the process of transforming from the camera coordinates into the physical object coordinates. The coordinate transformation process using the ANN method is as described above. The coordinate references necessary to carry out the transformation process are obtained by a chessboard calibration grid. Several tests were conducted to verify the effectiveness of the coordinate transformation process using artificial neural networks. The tests were conducted to compare with the calibration method by Woo and Park [17, 18] and Xiaobo et al. [15]. Woo and Park employed a calibration method with a 2×10×8×2 artificial neural network [17, 18] and Xiaobo et al. used a 2×5×2 artificial neural network [15]. In this study, the camera calibration method was tested using various images obtained with a camera using various heights and lens focal lengths. The results are shown in Table 3.

Differences in the average error in the resulting coordinates from the transformation process using our ANN model, the ANN model from Woo and Park [17,18], and the ANN model used by Xiaobo et al. [15], were tested using the analysis of variance.

Table 3. Average and maximum error of the result of the coordinate transformation process using our ANN method, the Woo and Park method, and the Xiaobo et al. method

		Our method (ΔX, ΔY)		Woo and Park(ΔX, ΔY)		Xiaobo et al. (ΔX, ΔY)	
		X Axis	Y Axis	X Axis	Y Axis	X Axis	Y Axis
Exp 1	Average	0.1559	0.1651	0.2977	0.3442	1.9047	1.5929
	Maximum	0.8580	1.1270	2.2580	1.9540	7.7300	6.4540
Exp 2	Average	0.2183	0.1552	0.5276	0.3438	1.8448	1.5069
	Maximum	1.0300	0.7390	2.7370	1.6690	7.0600	6.2790
Exp 3	Average	0.1681	0.1680	0.3390	0.1937	1.8614	0.9718
	Maximum	0.7000	1.8080	2.4150	1.3700	7.7800	3.9830
Exp 4	Average	0.0640	0.0676	0.0882	0.0808	1.3321	1.4196
	Maximum	0.3172	0.2896	0.4297	0.3733	7.4602	4.5135
Exp 5	Average	0.1052	0.0780	0.6022	0.2635	1.1559	0.8651
	Maximum	0.4786	0.3983	1.9291	1.2643	7.2643	4.8080
Exp 6	Average	0.2534	0.2153	0.3502	0.2340	1.2183	0.9552
	Maximum	1.3925	1.0076	1.3727	1.1431	7.1431	3.2896
Exp 7	Average	0.2320	0.1979	0.3181	0.2174	1.1681	0.7680
	Maximum	1.1555	0.9486	1.3822	1.3641	5.3641	4.3983
Exp 8	Average	0.2860	0.2468	0.3247	0.1905	1.0640	0.8676
	Maximum	1.0310	0.8043	1.2268	0.9408	6.5276	5.0076
Exp 9	Average	0.2210	0.2124	0.2471	0.1875	1.1052	0.9780
	Maximum	0.9972	1.2560	1.2229	1.0975	7.3390	3.9486
Exp	Average	0.2407	0.1902	0.2435	0.2415	1.2534	0.2153
	Maximum	0.9631	0.7197	1.0636	1.2716	6.0882	4.8043
Exp	Average	0.2518	0.1923	0.3689	0.3711	0.9705	1.1979
	Maximum	0.9532	0.9612	1.6132	1.6195	6.6022	5.2534
Exp	Average	0.2093	0.1914	0.3592	0.2284	0.8292	1.2284
	Maximum	0.8326	0.8117	1.2428	0.9705	7.3502	4.2320
Exp	Average	0.2274	0.2342	0.3899	0.2981	0.8649	1.2981
	Maximum	0.9537	0.7374	1.4425	1.8292	6.3899	4.2860
Exp	Average	0.2586	0.2433	0.3110	0.2274	1.0160	1.2274
	Maximum	1.0094	1.2998	1.3066	0.8649	7.3110	3.2210
Exp	Average	0.2834	0.2097	0.2972	0.2926	1.1427	1.2926
	Maximum	1.2356	1.0801	0.9645	1.0160	7.2972	3.2407
Exp	Average	0.2741	0.2509	0.4864	0.2340	1.4450	1.3983
	Maximum	1.5785	0.8800	1.6064	1.1427	5.4864	3.2518
Exp	Average	0.2965	0.3555	0.3169	0.3699	1.1875	1.0076
	Maximum	1.1697	1.1895	1.1665	1.4450	6.3169	5.2093
Exp	Average	0.3074	0.2397	0.2965	0.3555	1.2415	0.9486
	Maximum	1.1558	1.1118	1.1697	1.1895	7.2965	4.2274
Exp	Average	0.3510	0.1868	0.3960	0.4413	1.3711	1.2860
	Maximum	1.2633	0.6934	1.7337	1.5752	7.3960	3.2860
Exp	Average	0.2807	0.2121	0.4053	0.3096	1.2153	1.2210
	Maximum	1.6548	0.7672	2.1121	1.8687	4.4053	4.2210

Notes: Exp is Experiment; our ANN model is $2 \times 4 \times 25 \times 2$; the Woo and Park model [18,19] is $2 \times 10 \times 8 \times 2$; and the Xiaobo et al. [15] model is $2 \times 5 \times 2$.

Descriptives

PURATA

					95% Confidence Interval for Mean			
	N	Mean	Std. Deviation	Std. Error	Lower Bound	Upper Bound	Minimum	Maximum
1	20	.217430	.0611557	.0136748	.188808	.246052	.0658	.3260
2	20	.309760	.0816368	.0182545	.271553	.347967	.0845	.4357
3	20	1.185948	.2461822	.0550480	1.070731	1.301164	.7343	1.7488
Total	60	.571046	.4653696	.0600790	.450828	.691264	.0658	1.7488

Test of Homogeneity of Variances

PURATA

Levene Statistic	df1	df2	Sig.
14.816	2	57	.000

Multiple Comparisons

Dependent Variable: PURATA

	(I) MODEL	(J) MODEL	Mean Difference (I-J)	Std. Error	Sig.	95% Confidence Interval	
						Lower Bound	Upper Bound
Dunnett T3	1	2	-.092330*	.0228085	.001	-.149389	-.035271
		3	-.968517*	.0567211	.000	-1.114881	-.822154
	2	1	.092330*	.0228085	.001	.035271	.149389
		3	-.876187*	.0579958	.000	-1.024925	-.727450
	3	1	.968517*	.0567211	.000	.822154	1.114881
		2	.876187*	.0579958	.000	.727450	1.024925
Games-Howell	1	2	-.092330*	.0228085	.001	-.148134	-.036526
		3	-.968517*	.0567211	.000	-1.111320	-.825715
	2	1	.092330*	.0228085	.001	.036526	.148134
		3	-.876187*	.0579958	.000	-1.021374	-.731001
	3	1	.968517*	.0567211	.000	.825715	1.111320
		2	.876187*	.0579958	.000	.731001	1.021374

*. The mean difference is significant at the .05 level.

Fig. 8. Results of the comparison of the average error differences between our ANN model with current ANN models using post-hoc Dunnet's T3 and Games-Howell for SPSS

With respect to the SPSS output terms, homogeneity of variance was not found (P-value >0.05, where the P-value is the probability value). Therefore, the post-hoc analysis method with the assumption of non-homogeneous variance was used. Examination using post-hoc Dunnet's T3 and Games-Howell showed that the results of the analysis are significant in both tests (P-value <0.05), and therefore it was concluded that there is a significant distinction between the average errors for all three methods. Based on the minimum error values obtained, it was concluded that our method is the best, followed by the ANN model by Woo and Park [17, 18], and then the ANN model by Xiaobo et al. [15]. The SPSS analysis results are shown in Fig 8.

5   Conclusions

In our study, coordinate transformation using an artificial neural network method was used to transform the camera pixel coordinates into physical coordinates. Testing and analysis were conducted to verify the effectiveness of the camera calibration method considering many factors, including the camera height, the direction of the camera, the camera rotation, the type of the camera and the lens size. The following conclusions were derived from the overall results of our experiments with the ANN models for the robot soccer MiroSot and AndroSot camera calibration.

The neural network model used in our study, i.e., the 2×4×25×2 model, successfully achieved an objective camera calibration for the robot soccer MiroSot and AndroSot systems. The development of the camera calibration method was done using the architecture and the learning method of the general ANN model to adapt to continuous environmental changes. The relearning method was used to adapt to necessary changes, using new data input for changes in the environments. Relearning was originally implemented to improve performance and efficiency, and to reduce the errors for changing or new complex environments. Relearning on the artificial neural network originally used to adjust the weight was able to be implemented instantly and continuously using new data sets for the new conditions. The accuracy of the experiments using ANN has maximum and average errors that are smaller than those of the ANN model developed by Woo and Park [16, 17, 18] and Xiaobo et al. [15]. This is evidenced by a series of tests conducted using the image database. The maximum errors from the coordinate transformation process using the artificial neural network method based on the current condition are 1.4428 cm. All of the test results from the camera calibration process support the conclusion that the camera calibration method using artificial neural networks is the best solution to resolve the non-linear lens distortion problems, other than the current camera calibration method. The accuracy and efficiency of the camera calibration method using the artificial neural network depends only on the relearning process when the robot environments change.

References

1. Relf, C.G.: Image Acquisition and Processing with LabVIEW. CRC Press (2004)
2. Fraga, L.G., Schultze, O.: Direct Calibration by Fitting of Cuboids to a Single Image Using Diferential Evolution. International Journals of Computer Vision, 119–127 (2008)
3. Chen, L., Zheng, X., Hong, J., Qiao, Y., Wang, Y.: A Novel Method for Adjusting CCD Camera in Geometrical Calibration Based on a Two-dimensional Turntable. International Journal of Optic for Light and Electron Optics 121(5), 8–11 (2008)
4. Mendonca, M., Silva, I.N., Castanho, J.E.C.: Camera Calibration Using Neural Network. In: Proc of Int Conf in Central Europe on Computer Graphics, Visualization and Computer Vision, pp. 61–64 (2002)
5. Peng, E., Li, L.: Camera Calibration Using One-Dimensional Information and Its Applications in Both Controlled and Uncontrolled Environments. International Journal on Pattern Recognition 43(3), 1188–1198 (2010)
6. Claus, D., Fritzgibbon, A.W.: A Rational Lens Distortion Model for General Cameras. In: Proceedings of International Conference on Computer Vision and Pattern Recognition, CVPR 2005, pp. 213–219. IEEE Computer Society (2005)
7. Sugawa, R., Takatsuji, M., Echigo, T., Yagi, Y.: Calibration of Lens Distortion by Structured-Light Scanning. In: Proceedings of International Conference on Intelligent Robots and Systems, pp. 832–837 (2005)
8. Hartley, R., Kang, S.B.: Parameter-Free Radial Distortion Correction with Center of Distortion Estimation. International Journal of IEEE Transaction on Pattern Analysis and Machine Intelligence 29(8), 1309–1321 (2007)

9. Lee, D.J., Cha, S.S., Park, J.H.: Stereoscopic Vision Calibration for Three-dimensional Tracking Velocimetry Based on Artificial Neural Networks. In: Proceedings of SPIE International Symposium, Optical Science and Technology, pp. 1–13 (2003)

10. Kim, J.K., Kweon, I.S.: Camera Calibration Based in Arbitrary Parallelograms. International Journal of Computer Vision and Image Understanding, 1–10 (2009)

11. Miks, A., Novak, J.: Estimation of Accuracy of Optical Measuring Systems with Respect to Object Distance. International Journal of Optics Express 19(15), 14300–14314 (2011)

12. Hartley, R., Kang, S.B.: Parameter-Free Radial Distortion Correction with Center of Distortion Estimation. International Journal of IEEE Transaction on Pattern Analysis and Machine Intelligence 29(8), 1309–1321 (2007)

13. Kusumadewi, S.: Artificial Intellegence: Teori dan Aplikasinya. Penerbit Graha Ilmu, Yogyakarta (2003)

14. Haykin, S.: Neural Network: A Comprehensive Foundation. New York MacMilan, Penerbit (1994)

15. Xiaobo, C., Haifeng, G., Yinghua, Y., Shukai, Q.: A New Method for Coplanar Camera Calibration Based on Neural Network. In: Proc. of Int. Conf. on Computer Application and System Modeling, pp. 617–621 (2010)

16. Woo, D.-M., Park, D.-C.: Implicit camera calibration using an artificial neural network. In: King, I., Wang, J., Chan, L.-W., Wang, D. (eds.) ICONIP 2006. LNCS, vol. 4233, pp. 641–650. Springer, Heidelberg (2006)

17. Woo, D.M., Park, D.C.: Implicit Camera Calibration using Multilayer Perceptron Type Neural network. In: Proceedings of the First Asian Conference on Intelligent Information and Database System, pp. 313–317 (2009a)

18. Woo, D.-M., Park, D.-C.: Implicit camera calibration based on a nonlinear modeling function of an artificial neural network. In: Yu, W., He, H., Zhang, N. (eds.) ISNN 2009, Part I. LNCS, vol. 5551, pp. 967–975. Springer, Heidelberg (2009)

19. Wang, H., Cao, G., Xu, H., Wang, P.: Application of Neural Network on Distortion Correction Based of Standard Grid. In: Proceedings of International Conference on Mechatronics and Application, pp. 2717–2722 (2009)

20. Lee, J.H., Oh, S.Y.: An Absolute Robot Pose Estimation System Based on a Ceiling Camera Image Using a Neural Network. In: Proceedings of International Conference on System, Man, and Cybernetics, pp. 3839–3843 (2006)

Image Classification Using Convolutional Neural Networks With Multi-stage Feature

Junho Yim, Jeongwoo Ju, Heechul Jung, and Junmo Kim

Department of Electrical Engineering
KAIST 291 Daehak-ro, Yuseong-gu, Daejeon, Republic of Korea
{creationi,veryju,heechul,junmo.kim}@kaist.ac.kr
https://sites.google.com/site/siitkaist

Abstract. Convolutional neural networks (CNN) have been widely used in automatic image classification systems. In most cases, features from the top layer of the CNN are utilized for classification; however, those features may not contain enough useful information to predict an image correctly. In some cases, features from the lower layer carry more discriminative power than those from the top. Therefore, applying features from a specific layer only to classification seems to be a process that does not utilize learned CNN's potential discriminant power to its full extent. This inherent property leads to the need for fusion of features from multiple layers. To address this problem, we propose a method of combining features from multiple layers in given CNN models. Moreover, already learned CNN models with training images are reused to extract features from multiple layers. The proposed fusion method is evaluated according to image classification benchmark data sets, CIFAR-10, NORB, and SVHN. In all cases, we show that the proposed method improves the reported performances of the existing models by 0.38%, 3.22% and 0.13%, respectively.

1 Introduction

Image classification is an important topic in artificial vision systems, and has drawn a significant amount of interest over the last decades. This field aims to classify an input image based on visual content. Currently, most researchers have relied on hand-crafted features, HoG [1] or SIFT [2] to describe an image in a discriminative way. After that, learnable classifiers, such as SVM, random forest and decision tree are applied to extracted features to make a final decision. However, when a lot of images are given, it is too difficult problem to find features from those. This is the one of reasons that deep neural network model is coming. A few years ago, Hinton et al. [3] revealed the fascinating performance of deep belief nets, which use an effective deep learning algorithm, contrastive divergence (CD), in which each layer is trained layer by layer. Owing to deep learning, it becomes feasible to represent the hierarchical nature of features using many layers and corresponding weights. However, when the input dimension is too large to use, the deep belief network takes a long time to train. At that time, CNN [4],

© Springer International Publishing Switzerland 2015
J.-H. Kim et al. (eds.), *Robot Intelligence Technology and Applications 3*,
Advances in Intelligent Systems and Computing 345, DOI: 10.1007/978-3-319-16841-8_52

sharing weights by convolution method, solved this problem and improved the classification performance for various datasets. It should be noted that all those studies mentioned above have only used features from the top layer to train the following fully-connected layers. In contrast to this approach, Pierre et al. [5] bridged between the lower layer's output and the classifier to take the global shape and local details into account. This use of multi-stage features improved the accuracy over systems that use single stage features on a number of tasks, such as in pedestrian detection and certain sorts of classification. Motivated by many advantages of the multi-layers features, we propose an alternative multi-stage strategy that can be applied to a standard one track CNN whose weight parameter is fixed after the training has been finished without the multi-stage strategy in mind. The experiment results show that our approach can further improve performance of a standard one track CNN. Note that the proposed approach is different from the one in [5] in that the work in [5] trains the multi-stage architecture from the beginning, whereas the proposed method can be applied to a standard CNN whose training has been already finished. This paper includes our approach's motivation in Section 2 to easily help to understand why this model provides good result. The following, Section 3, describes our proposed model and explains how it works. We report experiment result on a various image classification data sets in Section 4, and conclude our research in Section 5.

2   Motivation

Since Matthew et al. [6] invented a probe to look inside a feature map, if one carefully observes the visualized features at each layer, one can obtain intuition as to why multi-stage features could enable further improvement of image classification. When comparing the visualization of features and the corresponding image patches, the latter has the greater variation since CNN mainly focuses on a discriminant structure. For other discoveries in [6], lower layer features are usually simpler than those of higher layers. The meaning of this discovery is that simple images are well activated at lower layers and complex images have high activation value at higher layers. Also, the lower layer features are focused on a smaller area in an image and the higher layer features are focused on a larger area in an image. For these reasons, the deep neural network model that uses the last layer features only finds it hard to classify the dataset which contains both simple and complex objects. This forces us to bind features from multi-stages in an effective way.

3   Model Discription

We propose a novel architecture using a two track deep neural network model. The first track is the deep convolutional neural network model, in which we want to enhance the ability; the second track is the assistance model which can raise the ability of the first track model by using multi-layer features. Any deep

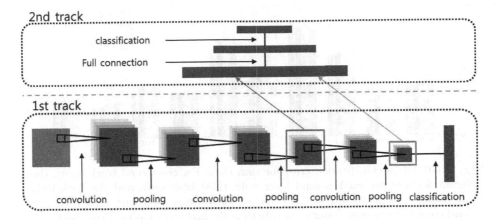

Fig. 1. Architecture of our proposed two track deep neural networks (DNN) model, composed of the first track, a learned CNN model, and the second track, assists the first model to enhance the ability. The particular layers features of the first track go to the input node of the second track model. For the purpose of mixing each layers feature information coming from the first track, the second track model is composed by fully connected layer.

convolutional neural network model that is composed of at least two convolutional layers with a pooling layer can be suitable for the first track model. As mentioned above, by using a pooling layer for the CNN, lower layer features and higher layer features are focused on different ranges. This is the reason that the CNN model with the pooling layer is suitable for the first track. For the second track, we use the restricted Boltzmann machine, RBM for the fully connected layer. Utilizing unsupervised training for the RBM, this model produces good initial weights for the following back propagation system which is supervised learning using label information. We append one more fully connected layer on the top layer with a classifier function as a softmax classifier. The second track operates after the learning of the first track has completed. As illustrated in Fig 3, the visible node of the second track comes from the particular layers feature of the first track. The number of nodes of the second layer in the second track is affected by the number of input nodes.

4 Experiments

In our experiments, we took the first track model for the simple convolutional neural network feature extractor described in Fig 3; this model is composed of three convolutional pooling layers, a fully connected layer, and the softmax classification layer. For the inputs of the second track model, the second and third pooling layers outputs of the first track model were used. The reason that we did not take the first layers output is that this output had much lower level features like edge levels and dimensions that were too large to use for the input.

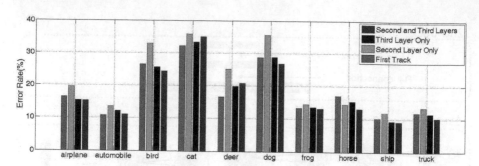

Fig. 2. CIFAR-10 classification error for each class. For the second track input, three different features are used; second layer only, third layer only, and the both layers. In addition, the first track result is attached for a comparison. For horse case, using second layer features show lower error rate than the first track model. Using both layers features improves performance in most cases.

For the unsupervised learning of fully connected layers of the second track, we used a learning rate of 0.001 with 50 epochs. Once the fully connected layers were trained we used them for the initial weights of supervised learning using label information. We trained 100 epochs for the supervised training. The whole set of experiments had the same setup with above.

4.1   CIFAR-10

CIFAR-10 is a dataset of natural RGB images of 32×32 pixels [7]. It contains 10 classes with 50,000 training images and 10,000 test images. All of these images have different backgrounds with different light sources. Objects in the image are not restricted to the one at center, and these objects have different sizes that range in orders of magnitude. For the first track model, we used the convolutional neural network model described in [8] (layers-18pct.cfg); this model is composed of three convolutional layers with 5×5 filters and 32 feature maps per layer, with a fully connected layer at the top of the layer. This model is shown in Fig 1. Before the training of the first track, we subtracted the mean values of the training set from each image. We trained for 120-10-10 epochs with an initial learning rate of 0.001 and a weight decay factor of 10. After the learning of the first track model, we extracted the second and third pooling layer features for each image and used them as the input for the second track model[1], described above. As shown in Table 1, using the second track model with the first track model is better than using only the first track model. Due to the fact that we used the features from the first model and not from the other model, it can be suggested that we enhanced the first model. To demonstrate this insight, we performed an additional experiment. For the training of the second track model, we used three different features, which are from the second pooling layer, the

---

[1] The number of nodes of second layer : 2000

Fig. 3. Ten classes of CIFAR-10 dataset images. Images have a different background with various light sources. Object in the image is not located on a center with various sizes.

third pooling layer, and both layers. We compared the results by class. Using the third layer feature was found to be better than using the second layer feature for most classes. However, what we have to focus on is that the result of using second layer feature only is pretty well and gets a better performance than the result of first track model for some case, horse class. This means that useful features are existing in the second layer feature. Furthermore, for most cases, by using both layers features, we were able to obtain a better performance than was possible when using the first track only.

Table 1. CIFAR-10 classification error (%) using our model and the first track model

Task	Proposed Method	First Track	Improvement
CIFAR-10	18.0 ± 0.11	18.38	0.38

4.2 NORB

We evaluated our two track model on the small NORB dataset (normalized-uniform), which is intended for 3D object recognition systems [9]. This dataset contains images of 50 toys belonging to 5 generic categories with 6 sets of lighting conditions, 9 elevations, and 18 azimuths. Each image consists of the binocular pair of 96×96 gray images with a normalized object size and a uniform background. We trained and tested the system on 24,300 images. Before using the images, the images were down-sampled to 48×48 and subtracted by the per-pixel mean. We used the same setup as that used in the CIFAR-10 experiment for the first track model. However, the first and third convolutional layers had 64 feature maps; the second convolutional layer had 32 feature maps. We trained

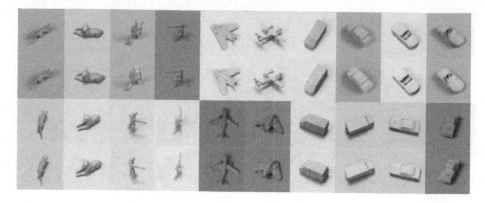

Fig. 4. Five classes of small NORB dataset images. Each two columns represent one class. Images have a same background with various light sources and object is on the center of image.

this model for 150-10-10 epochs with an initial learning rate of 0.001 and a weight decay factor of 10. For second track learning, we used the second and third pooling layer features of the first track. Because the input had large dimensions, we used the 3,500 unit fully connected layer for the second track. In this experiment, we saw a surprising improvement. As can be seen in Table 2, the proposed method reduces the error rate by 3.22%

Table 2. Small NORB classification error (%) using our model and the first track model.

Task	Proposed Method	First Track	Improvement
small-NORB	7.69 ± 0.13	10.91	3.22

4.3   SVHN

The Street View House Numbers dataset (SVHN) contains 10 digits [10], similar to the MNIST dataset [11]. Each image represents one digit. The challenging point of the SVHN dataset is that each image may contain multiple digits with different colors and various light sources. The training set contains 73,257 images; the testing set consists of 26,032 images. All images are cropped to 32 × 32 size and subtracted by per-pixel means. Our experiment for the SVHN dataset was set up in the same way as the CIFAR-10 experiment for the first track model. We trained the CNN model which contains three convolutional layer with 64, 64, and 128 feature maps with 5 × 5 filter size in layers 1, 2, and 3, respectively. This model was trained for 500-30-30 epochs with an initial learning rate of 0.001 and weight decay factor of 10. When the first track model was finished,

Fig. 5. Ten classes of SVHN dataset images. Each image represents the multiple digit in the real-world house number images. Images are cropped to 32 × 32 color images.

the second and third pooling layer features were input into the input node of the second track model, which contained 3,000 units with fully connected layers. Table 3 shows the classification performance of the proposed training model. Our proposed two track model enhances the performance of the first track model by 0.13%.

Table 3. SVHN classification error (%) using our model and the first track model.

Task	Proposed Method	First Track	Improvement
SVHN	5.95 ± 0.04	6.08	0.13

5   Conclusion

We propose a two track deep neural network model that is composed of an already learned CNN model and a fully connected layer model. Our model improves the learned CNN model's performance by using intermediate layer features. Via experiments in which we used our model on various datasets, we were able to demonstrate the needs of not only the top layers features but also those of the other layers features. The improved performance that resulted from the use of our model is due to the characteristic in which each layer's features focus on a different range of images. In the future, we will deal with a fine-grained dataset that requires a system to consider both global and local shape features.

Acknowledgement. This research was supported by the MOTIE (The Ministry of Trade, Industry and Energy), Korea, under the Technology Innovation Program supervised by KEIT (Korea Evaluation Institute of Industrial Technology), 10045252, Development of robot task intelligence technology.

References

1. Dalal, N., Triggs, B.: Histograms of oriented gradients for human detection. In: IEEE Computer Society Conference on Computer Vision and Pattern Recognition, CVPR 2005, vol. 1, pp. 886–893. IEEE (2005)
2. Lowe, D.G.: Distinctive image features from scale-invariant keypoints. International Journal of Computer Vision 60(2), 91–110 (2004)
3. Hinton, G.E., Osindero, S., Teh, Y.W.: A fast learning algorithm for deep belief nets. Neural Computation 18(7), 1527–1554 (2006)
4. Jarrett, K., Kavukcuoglu, K., Ranzato, M., LeCun, Y.: What is the best multi-stage architecture for object recognition? In: 2009 IEEE 12th International Conference on Computer Vision, pp. 2146–2153. IEEE (2009)
5. Sermanet, P., Kavukcuoglu, K., Chintala, S., LeCun, Y.: Pedestrian detection with unsupervised multi-stage feature learning. In: 2013 IEEE Conference on Computer Vision and Pattern Recognition (CVPR), pp. 3626–3633. IEEE (2013)
6. Zeiler, M.D., Fergus, R.: Visualizing and understanding convolutional neural networks. arXiv preprint arXiv:1311.2901 (2013)
7. Krizhevsky, A., Hinton, G.: Learning multiple layers of features from tiny images. Computer Science Department, University of Toronto, Tech. Rep. (2009)
8. Krizhevsky, A., Sutskever, I., Hinton, G.: Imagenet classification with deep convolutional neural networks. Advances in Neural Information Processing Systems 25, 1106–1114 (2012)
9. LeCun, Y., Huang, F.J., Bottou, L.: Learning methods for generic object recognition with invariance to pose and lighting. In: Proceedings of the 2004 IEEE Computer Society Conference on Computer Vision and Pattern Recognition, CVPR 2004, vol. 2, pp. II–97. IEEE (2004)
10. Netzer, Y., Wang, T., Coates, A., Bissacco, A., Wu, B., Ng, A.Y.: Reading digits in natural images with unsupervised feature learning. In: NIPS Workshop on Deep Learning and Unsupervised Feature Learning, vol. 2011 (2011)
11. LeCun, Y., Bottou, L., Bengio, Y., Haffner, P.: Gradient-based learning applied to document recognition. Proceedings of the IEEE 86(11), 2278–2324 (1998)

Traversability Classification Using Super-voxel Method in Unstructured Terrain

Soohwan Song and Sungho Jo

Dept. of Computer Science, KAIST
291 Daehak-ro, Yuseong-gu, Daejeon 305-701, Korea
dramanet30@kaist.ac.kr, shjo@kaist.ac.kr

Abstract. Estimating the traversability of terrain in an unstructured outdoor environment is one of the challenging issues in autonomous vehicles. When dealing with a large 3D point cloud, the computational cost of processing all of the individual points is very high. Thus voxelization methods are used extensively. In this paper, we propose a more fine-grained voxelization algorithm in the context of unstructured terrain classification. While the current shape of a voxel is a fixed-length cubic, we construct a flexible shape voxel which has spatial and geometrical properties. Furthermore, we propose a new shape histogram feature that represents the statistical characteristics of 3D points. The proposed method was tested using data obtained from unstructured outdoor environments for performance evaluation.

Keywords: unmanned vehicle, unstructured terrain, traversability classification, point cloud, voxel.

1 Introduction

Correctly classifying an outdoor environment into traversable and non-traversable regions is still a challenging task in autonomous vehicles. In particular, recognition in environments with unstructured terrain remains a difficult problem since the information obtained from sensors is highly inaccurate. In this paper we explore the traversability classification problem for sparse and unstructured terrain data. There are two important issues of concern.

The first is to efficiently process the large amount of 3D point cloud data. For problems of traversability classification in 3D point clouds, *voxelization* methods are generally employed for efficiently processing large amounts of data [1] [2] [3]. Voxelization is a method to divide the 3D terrain into fixed-length cubes and extract their features (Fig. 1(a)). When a point cloud is converted into voxels, it is possible to lose the geometric information of the point cloud because of its fixed-length cubic shape. In order to overcome this problem, we employ a flexibly shaped *supervoxel* (Fig. 1(b)) which includes spatial and geometrical properties. The Voxel Cloud Connectivity Segmentation algorithm was recently proposed to generate supervoxels [4]. The algorithm clustered voxel-clouds which were generated from RGB-D images, and

© Springer International Publishing Switzerland 2015
J.-H. Kim et al. (eds.), *Robot Intelligence Technology and Applications 3,*
Advances in Intelligent Systems and Computing 345, DOI: 10.1007/978-3-319-16841-8_53

segmented only structured indoor scenes. Since this algorithm is unsuitable for sparse point clouds in outdoor environments, we modify it to be usable for sparse point clouds.

The second issue with respect to traversability classification is to define an appropriate feature which represents a local area well. There have been many studies on feature extraction for traversability classification. For point cloud classification, the *saliency features* [3] [5] have most commonly been used. The saliency features were used to capture the surface-ness, linear-ness, and scatter-ness of the local area. However these features could give inaccurate results when neighborhoods are not representative of the local geometry such as within sparse regions. In this paper, we propose a new histogram-type shape feature which works well in any environment. The new shape feature represents the statistical characteristics of 3D points with respect to shape and can be computed relatively fast. Furthermore, for more accurate classification, we extract a histogram-type color feature and combine the visual feature with the new shape feature.

The proposed supervoxel method and the new histogram-type features were tested using data obtained from unstructured outdoor environments for performance evaluation. Furthermore, this paper investigates whether the proposed methods can improve the performance of traversability classification.

(a)   (b)

Fig. 1. Examples of (a) voxel and (b) supervoxel over-segmentation

2   Supervoxel

2.1   Point Features and Distance Measure

This section describes how to extract point features and measure the distance between them. Common approaches to extract a point feature is to directly use a surface normal [6] or compute a histogram feature which captures the variations of surface normals in a local patch [7] [8]. Since the surface normal is estimated by fitting a plane to some neighboring points, it will be greatly affected by density of neighboring points. So the normal is unsuitable for a point feature in sparse point clouds. Moreover, plane fitting methods become very computationally expensive when applied to millions of points. Thus we need to find other point features which do not consider the surface normal for quickly extracting the point features in sparse point clouds.

This paper extracts geometric information by analyzing local distributions of consecutive points. In [9], the authors used the *consecutive point information* (CPI) that can be obtained from a 2D LIDAR system. It used the angles between the y-axis and the line passing through two consecutive vertical points. In our work, we convert the point feature into a form that can be used with a 3D LIDAR system by adding information about consecutive horizontal points.

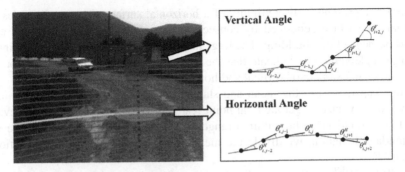

Fig. 2. Example of vertical and horizontal angels

In order to extract the CPI feature, we should define the order of consecutive points. The Cartesian coordinates of a 3D point cloud are converted to spherical coordinates (azimuth Θ and elevation ϕ) and they are discretized to predefined intervals. All points are sorted by azimuth and elevation, and represented as follows:

$$
\begin{bmatrix}
 & \Theta_{1st} & \Theta_{2nd} & \Theta_{3rd} & \\
\phi_{1st} & p_{1,1} & p_{1,2} & p_{1,3} & \\
\phi_{2nd} & p_{2,1} & p_{2,2} & p_{2,3} & \cdots \\
\phi_{3rd} & p_{3,1} & p_{3,2} & p_{3,3} & \\
 & & \cdots & &
\end{bmatrix}
$$

Each $p_{i,j}$ is a vector whose elements include the 3D Cartesian coordinates of ith elevation and j^{th} azimuth point. The consecutive vertical point of a point $p_{i,j}$ is $p_{i+1,j}$ and consecutive horizontal point is $p_{i,j+1}$. Let $v_{i,j}^V$ be the vector passing through two consecutive vertical points, it can be computed as $v_{i,j}^V = p_{i+1,j} - p_{i,j}$. Thus the angle $\theta_{i,j}^V$ between the x-y plain and $v_{i,j}^V$ is computed as follows:

$$
\theta_{i,j}^V = \sin^{-1}\left(\frac{v_{i,j}^V \cdot z}{|v_{i,j}^V| \times |z|}\right)
\tag{1}
$$

where z is a z-axis unit vector. The θ^V represents the inclination angle of the surface generated by consecutive vertical points. Small angles indicate a flat surface such as the ground and large angles indicate vertically oriented object surfaces. Furthermore, the surface of trees or bushes has a large angle variation.

Let $v_{i,j}^H$ be the vector passing through two consecutive horizontal points, it can be computed as $v_{i,j}^H = p_{i,j+1} - p_{i,j}$. The angle $\theta_{i,j}^H$ between two consecutive vectors $v_{i,j-1}^H$ and $v_{i,j}^H$, can be computed as follows:

$$\theta'^H_{i,j} = \cos^{-1}\left(\frac{v_{i,j}^H \cdot v_{i,j-1}^H}{|v_{i,j}^H| \times |v_{i,j-1}^H|}\right), \; \theta_{i,j}^H = \min\left(\theta'^H_{i,j}, \pi - \theta'^H_{i,j}\right) \tag{2}$$

The θ^H represents a discontinuity of a horizontal surface. The small angles indicate that the surface generated by consecutive horizontal points is flat such as the ground or the wall of a building. The large angles represent a discontinuous point, and the large angle variations indicate that the surface is scattered such as a tree or bush. Fig. 2 shows an example of vertical and horizontal angles.

In this paper, we use the mean and standard deviation of the neighboring angles within a window size as geometrical point features. The mean and standard deviation should be normalized to be within a range of [0, 1]. In order to normalize the mean and standard deviation, we divide the mean by $\pi/2$ and the standard deviation by $\mathbf{std}(0, \pi/2)$.

Let $\hat{\theta}^V$ and $\hat{\theta}^H$ be the normalized means of vertical and horizontal angles, and $\hat{\sigma}^V$ and $\hat{\sigma}^H$ be the normalized standard deviations of each angle, the point feature f_k used for the supervoxel over-segmentation task is defined like this:

$$f_k = [x_k, y_k, z_k, \hat{\theta}_k^V, \hat{\theta}_k^H, \hat{\sigma}_k^V, \hat{\sigma}_k^H] \tag{3}$$

where x, y, and z are the Cartesian coordinates of a point.

In order to compute the point feature distances, it is necessary to normalize spatial distances by their respective maximum distances. We define the normalization constant as the distance between the center and maximally distant point in a cluster. Let R_{seed} be a seed resolution of initial cluster, we can normalize our spatial distance d^c by dividing by the normalization constant $\sqrt{3}R_{seed}$. Since point shape features are already normalized, it is not necessary to normalize the shape distance d^s. By applying a weight of point distance w_{point}, the distance of two point features f_i and f_j is computed as follows:

$$d^c = \sqrt{(x_i - x_j)^2 + (y_i - y_j)^2 + (z_i - z_j)^2}$$

$$d^s = \sqrt{(\hat{\theta}_i^V - \hat{\theta}_j^V)^2 + (\hat{\theta}_i^H - \hat{\theta}_j^H)^2 + (\hat{\sigma}_i^V - \hat{\sigma}_j^V)^2 + (\hat{\sigma}_i^H - \hat{\sigma}_j^H)^2}$$

$$d = \sqrt{w_{point}\left(\frac{d^c}{\sqrt{3}R_{seed}}\right)^2 + (1 - w_{point})(d^s)^2} \tag{4}$$

2.2   Supervoxel Segmentation

The supervoxel segmentation algorithm is summarized in Algorithm 1. The algorithm was originally motivated by previous work in [4]. The supervoxel segmentation algorithm in [4] used a dense depth image and it clustered the voxel-clouds. Therefore we modified the algorithm to be usable in sparse point clouds by clustering cloud points instead of voxel-clouds.

Algorithm 1 Supervoxel segmentation

---

1. Construct neighborhood graph G.
2. Generate voxels V_k with resolution R_{seed}.
3. Initialize cluster center points P_k^{seed} to center of V_k.
4. Change the center points P_k^{seed} to the lowest gradient in radius R_{search} of each center.
5. Assign each P_i in V_k to each cluster C_k and calculate distances D_i^{min} between $P_i \in C_k$ and P_k^{seed}.
6. **for** each cluster C_k **do**
7.   Extract the neighbor points P^* in radius $2 \times R_{seed}$ of P_k^{seed}.
8.   Calculate distances D^* between P^* and P_k^{seed}.
9.   Remove points whose distance D^* are higher than minimum distance D^{min} from P^*.
10.   Find the connected points $\hat{P}^*$ of P^* from P_k^{seed} by using **BFS** of G.
11.   Assign the each points of $\hat{P}^*$ to cluster C_k and update D^{min} to its distance D^*.
12. **end for**

We begin by constructing a neighborhood graph G, which ensures that disconnected points are not clustered in the same cluster. We obtained the neighborhood graph by connecting the six nearest neighbors of each point. Then, we divide the whole point cloud into a voxelized grid V_k with seed resolution R_{seed}. The seed points are initialized to the nearest point to the center of each voxel grid cell. The initial points are changed to the lowest gradient position in the search range R_{search} of each point. The ith point gradient g_i is computed as:

$$g_i = \Sigma_{k \in p_{adj}} \frac{|f_i - f_k|}{N_{adj}} \tag{5}$$

where p_{adj} and N_{adj} represent the neighboring points of the neighborhood graph and number of neighbor points, respectively.

We generate supervoxels by clustering points based on their distance between each seed point, P^{seed}. Each point is assigned to the nearest cluster center. First, the points within a same V_k are initialized to the same cluster C_k, and their distances to P_k^{seed} are stored in variable D^{min}. Then, the method iterates the following procedures for each cluster C_k. First we extract the neighbor points P^* in radius $2 \times R_{seed}$ of P_k^{seed} and calculate their distances, D^*, to P_k^{seed}. Then, the points whose distances D^* are lower than their D^{min} are assigned to cluster C_k and update D^{min}. In order to ensure the connectivity of the points in C_k, only points connected to P_k^{seed} are selected. We find the connected points by traversing G using *breadth-first search* (BFS).

2.3   Features of Supervoxel

In this section, we define the features of each supervoxel cluster. The feature space consists of two different histogram feature sets, one with color characteristics and the other with geometric characteristics.

The first feature set F^c is a *LAB* color histogram derived from the color components in a supervoxel. F^c is produced by discretizing the colors in the supervoxel into a number of bins, and counting the number of points in each bin. Let F_b^c be a color histogram value of b index in a supervoxel cluster, it is computed as:

$$F_b^c = \frac{1}{N}\sum_{i=1}^{N} \mathbf{I}(f_i^c \in bin(b)) \tag{6}$$

where f_i^c is a LAB color value of ith point, N is the number of points in a cluster and $\mathbf{I}$ is the indicator function. The color histogram F^c represents the probability mass function of the point color feature.

The second feature set F^s is a shape histogram of a supervoxel, which is derived from the point shape feature $f^s = [\hat{\theta}^V, \hat{\theta}^H, \hat{\sigma}^V, \hat{\sigma}^H]$. Let F_b^s be a shape histogram value defined as:

$$F_b^s = \frac{1}{N}\sum_{i=1}^{N} \mathbf{I}(f_i^s \in bin(b)) \tag{7}$$

Similar to F^c, the shape histogram F^s represents the probability mass function of the point shape feature. As mentioned earlier, the performance of saliency features [3] [5] or other shape histogram features [7] [8] derived from surface normals are strongly influenced by the density of local points. However, our shape histogram feature is derived from CPI and works well in any environment. Furthermore, it can be computed relatively fast. We empirically obtained good results by dividing the feature values into three bins for angle means and two bins for angle standard deviations. Therefore, the number of shape histogram bins is $3^2 \times 2^2 = 36$.

3   Traversability Classification

In order to train the classification model, we clustered all of the supervoxels in the training dataset using *k-means clustering*, and modeled each cluster M_k by its mean value F_{M_k}. Separate sets of cluster models were maintained for positive and negative examples. We adopted the χ^2 distance function as a similarity measure of clusters, defined as:

$$D(F, F') = w_{voxel} \sum_{i=1}^{B^S} \frac{\left(F'^S_i - F^S_i\right)^2}{F'^S_i + F^S_i} + (1 - w_{voxel}) \sum_{j=1}^{B^C} \frac{\left(F'^C_j - F^C_j\right)^2}{F'^C_j + F^C_j} \tag{8}$$

where B^S and B^C represent the number of histogram bins, and w_{voxel} is a weight factor. The weight factor w_{voxel} allows us to control the relative contributions of the two components.

In order to find the traversable regions, the learned positive and negative models are compared to new input regions. If the similarity ratio of a new region is greater than a threshold λ, the region is classified as a traversable region. The classifier is defined as:

$$H(F_k) = \mathbf{I} \left[\frac{D(F_k, F_{M_t})}{D(F_k, F_{M_{nt}})} \geq \lambda \right] \tag{9}$$

where M_{nt} is the non-traversable model closest to F_k, and M_t is the closest traversable model [10] .

4   Experimental Results

In order to classify drivable regions, we obtained the point cloud data from a HDK-32E LiDAR sensor and 640x480 pixel images from a forward-facing camera. The environmental data captured was unstructured terrain which included foliage and dense vegetation over 40cm high. A relatively dense data frame was constructed by combining six consecutive overlapping data frames, and we extracted twenty consecutive dense data frames. We then manually labeled the dense frames for traversability classification. The last ten dense frames were used as a testing dataset and the preceding ten frames were used as the training dataset. For the supervoxel segmentation, we set the supervoxel parameters R_{search} and w_{point} to $0.5 \times R_{seed}$ and 0.5 respectively. Two hundred traversable and non-traversable models were trained for the classification task.

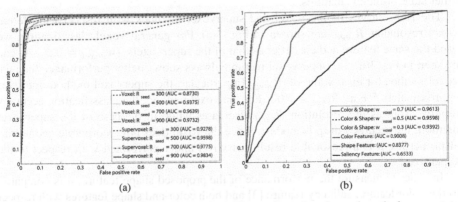

Fig. 3. ROC curves showing the results of (a) the supervoxel and voxel methods respect to different R_{seed} and (b) supervoxel method under different feature sets.

(a) Voxel: color & shape (ACC = 0.8914)

(b) Supervoxel: color & shape (ACC = 0.9418)

(c) Supervoxel: color feature (ACC = 0.9181)

(d) Supervoxel: shape feature (ACC = 0.7976)

Fig. 4. Traversability classification results under different settings. The green and red points are correctly classified traversable and non-traversable regions respectively, and the blue points are incorrectly classified regions. ($R_{seed} = 500$, $w_{voxel} = 0.5$, and $\lambda = 1$).

To evaluate the performance of our proposed method, we divided the experiment into two subsections. First, the performance of the supervoxel method was compared to a general voxel method. Second, we compared the proposed shape feature against other state-of-the-art features.

The ROC curves of the voxel and supervoxel methods with respect to different voxel resolution R_{seed} are shown in Fig. 3(a). For general voxel classification, we used the same feature and classifier as that of the supervoxels ($w_{voxel} = 0.5$). As can be seen in Fig. 3(a), the supervoxel method always shows better performance than the voxel method for each value of R_{seed}. In particular, the supervoxel method improves approximately 5% at $R_{seed} = 300$. For the voxel method, the classification accuracy rapidly drops as the resolution R_{seed} decreases while in the case of the supervoxel method, the accuracy drop is modest. Since the voxel does not incorporate geometrical properties, it is a reasonable result that voxel under-performed with respect to the supervoxel.

In order to measure the performance of the proposed shape feature, it is compared to the color feature, saliency feature [3] and both color and shape features with respect to different weight factors w_{voxel}. Fig. 3(b) shows the ROC curves of the supervoxel method ($R_{seed} = 500$) with respect to different features. In this figure, accuracy of the proposed shape feature is lower than that of the color feature. It indicates that the color is the most powerful feature for traversability classification. However, the major outcome of the experiment is that the performance of the color feature can be improved by more than 6% with the help of the shape feature. Since w_{voxel} represents degree of contribution of shape feature, the best performance at $w_{voxel} = 0.7$ indicates that the shape feature contributes more to the classification task than color feature. Furthermore, the shape feature performs better than the saliency feature. The saliency feature is unusable in sparse unstructured environments since it is influenced by the local point density, while our proposed feature works well in sparse environments.

Fig. 4 shows examples of traversability classification under different settings. As mentioned above, the performance of the voxelization method (Fig. 4(a)) is lower than the supervoxel method (Fig. 4(b)). Furthermore, we can verify that the combination of the shape feature (Fig. 4(d)) and color feature (Fig. 4(c)) offers excellent performance improvements. The results clearly indicate that the supervoxel method and its histogram-type features could improve both the performance and accuracy of traversability classification in unstructured terrains.

5   Conclusion

In this paper, the performance and accuracy of traversability classification in unstructured terrains is improved in a number of ways. The supervoxel method, a new, fine-grained voxelization method in sparse point clouds, is proposed for efficiently processing large amounts of point cloud data. Unlike traditional voxelization methods, it has geometrical properties and is usable for sparse point clouds. This paper proposes a new histogram-type shape feature that incorporates consecutive point information. The experimental results show that both the supervoxel method and the shape histogram features are successful and show better performance in most cases than other state-of-art methods.

Acknowledgements. This work was supported by the Technology Innovation Program, 10045252, Development of robot task intelligence technology that can perform task more than 80% in inexperience situation through autonomous knowledge acquisition and adaptational knowledge application, funded By the Ministry of Trade, industry & Energy (MOTIE, Korea).

References

1. Heckman, N., et al.: Potential negative obstacle detection by occlusion labeling. IEEE/RSJ. Int. Conf. Intell. Rob. & Syst (2007)
2. Stoyanov, T., et al.: Path planning in 3D environments using the normal distributions transform. IEEE/RSJ Int. Conf. Intell. Rob. & Syst (2010)

3. Bogil, S., Myungjin, C.: Traversable ground detection based on geometric-featured voxel map. In: Bogil, S., Myungjin, C. (eds.) IEEE. Korea-Japan Joint Workshop on Frontiers of Computer Vision, pp. 31–35 (2013)
4. Papon, J., et al.: Voxel cloud connectivity segmentation-supervoxels for point clouds. In: IEEE Conf. Computer Vision and Pattern Recognition, pp. 2027–2034 (2013)
5. Lalonde, J.F., Vandapel, N., Huber, D.F., Hebert, M.: Natural terrain classification using three-dimensional ladar data for ground robot mobility. Journal of Field Robotics 23(10), 839–861 (2006)
6. Rabbani, T., van den Heuvel, F., Vosselmann, G.: Segmentation of point clouds using smoothness constraint. In: Int. Archives of Photogrammetry, Remote Sensing and Spatial Information Sciences, pp. 248–253 (2006)
7. Rusu, R., et al.: Learning informative point classes for the acquisition of object model maps. In: IEEE. Int. Conf. Control, Automation, Robotics and Vision, pp. 643–650 (2008), 2008
8. Rusu, R., Blodow, N., Beetz, M.: Fast point feature histograms (FPFH) for 3D registration. In: IEEE Int. Conf. on Rob. & Autom (2009)
9. Yungeun, C., Seunguk, A., MyungJin, C.: Online urban object recognition in point clouds using consecutive point information for urban robotic missions. Robotics and Autonomous Systems (2014)
10. Dongshin, K., et al.: Traversability classification using unsupervised on-line visual learning for outdoor robot navigation. In: IEEE Int. Conf. on Rob. & Autom. (2006)

Soft Peristaltic Actuation for the Harvesting of Ovine Offal

M. Stommel[1], W.L. Xu[2], P.P.K. Lim[3], and B. Kadmiry[3]

[1] Department of Electrical and Electronics Engineering,
Auckland University of Technology, New Zealand
mstommel@aut.ac.nz
[2] Department of Mechanical Engineering, The University of Auckland, New Zealand
p.xu@auckland.ac.nz
[3] Callaghan Innovation, 24 Balfour Rd, Auckland 1052, New Zealand
{patrick.lim,bourhane.kadmiry}@callaghaninnovation.govt.nz

Abstract. Many tasks in lamb meat processing have been automated by mechatronic systems during the past years. However, the extraction of edible organs from the unordered organ package has remained a challenge. Traditional sensing methods and hard robotic effectors are not suitable for the slippery and deformable tissue in varying geometric constellations. In this paper, we propose a soft peristaltic method to bring the organ package into the optimal configuration for the removal of single organs. We give a system overview, discuss its viability, and point out the challenges in its implementation.

A deformable xy-sorting table is proposed to order the organ package. By producing moving wave shapes on its surface, the table changes the geometric configuration of the organs as perceived and controlled by a machine vision module. When an organ is in the optimal position, it is picked up and removed by traditional robotic solutions.

Keywords: Soft robotics, peristalsis, meat processing.

1 Introduction

Although significant to meat processing [1, 2], the harvesting of edible ovine offal, such as liver, heart, and kidneys, has not been automated yet [3]. Where skilled labour is not available, ovine offal is left unprocessed. Given the full organ package of heart, lung, liver, kidneys, and intestines, which has been separated from the carcass, the problem is to separate single organs from the package, thereby cutting connecting tissue such as arteries, and feed single organs to subsequent stages of processing according to their classification.

Automatic approaches used for organ harvesting in small animals (poultry) [4] cannot be used in sheep due to differences in size and mechanical properties. Moreover, the purely mechanical solutions used in poultry are not compliant and damage the organs frequently. Sensor feedback could improve compliant operation, in particular visual feedback. Machine vision for the recognition of

© Springer International Publishing Switzerland 2015
J.-H. Kim et al. (eds.), *Robot Intelligence Technology and Applications 3*,
Advances in Intelligent Systems and Computing 345, DOI: 10.1007/978-3-319-16841-8_54

internal organs has been solved for some medical applications [5–7], but not applied to ex vitro recognition. Rigid, visually servoed industrial robots seem however unsuitable for several reasons: First, the interaction of soft tissue with a hard effector is not well understood and can not be modelled realistically or taken into account for control. Secondly, the removal of single organs from the unordered package requires a complex repositioning of single/multiple organs, which in turn requiring a detailed and precise recognition of the object shape and the identification of manipulable/occluding parts. This level of detail does not seem to be provided by the above mentioned medical vision algorithms.

In this paper, we propose a solution that (1) achieves a compliant actuation, (2) has lower requirements on sensor interpretation, and (3) avoids the need for a complex handling/manipulation of organs during removal. We outline the technical approach and discuss its viability based on the existing literature on the topic, as well as experience from our past studies.

2   Soft Robotics

Soft-bodied robots are designed to simulate the morphology and mechanical function of animals and natural structures. Examples include a robotic elephant trunk, octopus arm, tongue. soft-bodied robotics is mostly based on the principle of muscular hydrostat, which focusses on the interaction of compressable/expandable materials and pressure exerting materials and drives [8,9].

Soft-bodied robots are attractive for the handling of internal organs because of the inherent compliance to the environment. Under strain, they adapt to the surface geometry, which reduces the danger of pinching or piercing an object. Compared to rigid robots, they are also simple to build and low-cost. Integrated rigid fibers and components can improve the mechanical properties by trading robustness against compliance [10, 11].

The modelling of soft-bodied robots is challenging due to the nonlinear behaviour of shape memory alloy (SMA) or pneumatic actuators [12].

The use of soft-bodied robots has been demonstrated for medical simulation, locomotion, and gripping. A soft robot simulating a human esophagus has been used to study swallowing and peristaltic food transport without risk for a human subject [13]. A caterpillar-like movement has been demonstrated for a soft-bodied robot actuated by a resistive heating of SMA elements working as longitudinal muscles [14]. A soft silicone body with inflatable cells and pneumatic drive has been used to design a four legged walking robot [15]. Explosions have been used to trigger jumps [16]. While the silicone rubber used to form the soft body withstood a series of short gas explosion, failures where caused by charred gas input lines. Robotic grippers are inspired by the octopus or starfish. They have been demonstrated for SMA [17] and pneumatic drives [11]. An alternately hard/soft vacuum gripper has been developed where vacuum is used to compress the loose particle filling of a flexible vacuum suction cap in order to turn it rigid and let it keep its shape [18].

The nonlinear elasticity [12] of the rubber material used in most soft robots can be modelled by empirical models (e.g. Mooney-Rivlin, Ogden, Yeoh model).

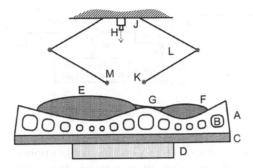

Fig. 1. The conceptual robotic animal offal sorting system. The peristaltic table consists of a flexible silicone layer (A) with inflatable cells (B) on a rigid ground (C). Inflation is controlled in a way that a single organ (F) is moved away from the organ package (E). Peristalsis is supported by shaking the table (D). A camera (H) gives optical feedback on the organ configuration. A flexible vacuum gripper (K) mounted on a manipulator arm (L) is attached to a dual-arm robot (J). The gripper picks up the isolated organ and lifts it up in order to expose connecting tissue (G). The connecting tissue is cut by a visually controlled cutting end-effector (M) that is attached to the other manipulator arm.

Due to a lack of theoretical solutions for the continuous case, the kinematic and dynamic behaviour is approximated within a discrete pattern of interconnected modular elements [19].

Pneumatic drives can be used to realise complex movements with many degrees of freedom [15]. Applied to a soft silicone body, they are clean, safe, low-cost, and achieve a compliant actuation. The periphery consists of air source and pressure regulators. Electroactive polymers (EAP), and most notably dielectric elastomers (DE) are another important actuation technique [20]. The technique is based on the change of mechanical properties of the material when applying a high voltage (>1 KV). Multi-layer or helical structures amplify mechanical movements to the scale needed for the actuation of a robot. They combine fast response, noise-free operation, high resilience and light weight [21]. However, there is a risk of dielectrical breakdown, and the relation between voltage and deformation is nonlinear.

In spite of recent developments in creating deformable surface pressure sensors using soft capacitors [22] or resistors [23], reliable measurements of the shape of the actuated robot by integrated sensors are currently not available. For our proposed system, we therefore outline a camera-based shape calibration.

3   Proposed System and Its Operation

3.1   Principle of Operation

The proposed system consists of a soft-bodied xy-sorting table, a machine vision system, and an industrial robot. The industrial robot lifts single organs and cuts

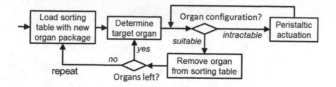

Fig. 2. Principle of operation

them from the organ package. The xy-sorting table applies peristaltic waves to the organ package in order to bring single organs in the optimum position and orientation for pickup. Starting from optimum conditions, the pickup and removal of single organs is relatively straightforward using industrial robotics. Figure 1 illustrates the setup. The peristaltic ordering of the organ package avoids the very unpromising approach of using industrial robots and visual servoing for repositioning, reorienting, and manipulation of the organ package.

Figure 2 illustrates the principle of operation: The unordered organ package (heart-lung package, liver, kidneys, intestines) is loaded on the sorting table. Pneumatic control is used to produce moving wave shapes on the surface of the table. Like in peristaltic food transport [13], these waves drive the organ package. Under the force of their weight, the organs will move towards energy minima as defined by the shape of the surface. This allows for a purposeful repositioning, reorientation, and local separation of organs. A machine vision module determines the state of the current organ positions. If a suitable state has been reached, then a single organ is lifted, cut, and removed from the organ package. If the state is unsuitable for organ removal, peristaltic control continues. A graph model defines transitions between states by patterns of peristaltic movement, which allows to plan a trajectory from the current state to a desired state.

3.2 Soft XY-Sorting Table

In order to realise moving wave patterns, we propose a two-dimensional array of inflatable cells embedded in a compliant, soft upper layer of the sorting table. A related design has already been implemented for the 1D case [13]. The geometry of the cell structure had been optimised for peristaltic actuation by finite element analysis. Ecoflex silicone rubber and a pneumatic drive with digital interface had been used. To support peristalsis, we propose an additional drive for shaking the table. In order to remove single organs, a bi-arm robot is necessary that is equipped with multiple vacuum suckers on a soft structure on one arm, and a self-aligning cutting device on the other arm. The soft vacuum gripper is designed to adapt to the soft surfaces of the organs without damaging them. Top-mounted and side cameras sense the current organ state.

3.3 Recognition of Organs and Organ Configurations

Organs have already been segmented and classified from image material in medical applications [5–7]. Although many approaches are based on MR and CT images, often conventional local feature extractors like SIFT [24] or LBP [25] are used. The challenges in transferring these methods to meat processing consist in the use of conventional cameras and 3D sensors, and the reassessment of the boundary conditions. In particular, it is not possible to make assumptions about the relative positions of organs.

A precise segmentation of the organs will not be necessary. Instead, we propose to encode the relative layout of the organs on the table into a feature descriptor, similar to those used in full human body tracking, where the relative positions of the limbs are of interest [26]. The feature space can then be sub-divided into discrete regions, each representing a class of similar organ configurations. These organ configurations form the basis of the graphical model of states and actuations.

3.4 Peristaltic Positioning of Organs for Removal

The goal of peristaltic actuation and shaking is to bring single organs of the organ package into the optimal position for pickup and removal. The organ in question must be unoccluded, on top of the package, and properly positioned and oriented. In order to create wave shapes that translate, rotate and separate organs, the inflatable chambers of the sorting table must be pressurised independently. The pneumatic drive consists primarily of a D/A converter, solenoid valves to control air pressure, and an air compressor. Actuation patterns can be created by a central pattern generator [27, 28], a biologically inspired technique used in robotics to simulate the repetitive signal patterns generated by neural networks. For calibration, the 3D shape of the (empty) sorting table can be measured by a vision sensor (e.g. stereo or time-of-flight). This allows for mapping the dependency between the deformation of the table surface and the control signals. The dependency between the control signals and the actual movement of the organs has not been researched yet and must be studied using simulations and experiments on real organs and dummy objects.

Feedback from the machine vision module might not be detailed enough to perform fine corrections to the wave patterns to compensate for small disturbances. Instead, machine vision provides a general assessment of the organ configuration in real-time. The main purpose is to compare the current situation to a list of known states in order to determine a sequence of actuation patterns that lead to a desired state. The connection between states and actuation patterns can be modelled by a probabilistic nondeterministic state machine. The success rates for initiating a state change by applying an actuation pattern must be trained from experiments. The machine vision module can give feedback on the duration of a certain actuation pattern by signalling the successful or unsuccessful state change. In the case of an unsuccessful state change (caused by noise or a disturbance), a new path to the desired state must be determined.

3.5   Removal of Single Organs

The removal of single organs proceeds in two steps. First, a vacuum gripper lifts the organ, then a dedicated tool cuts the exposed tissue connections between to the organ package. The vaccum gripper should consist of multiple small suction cups that are compatible with the soft tissue. The cups should be mounted on a flexible structure [29] in order to self-adapt to the uneven and varying surface geometry. Since peristalsis is used to prepare optimal starting conditions, it is not necessary to use the gripper to realign or uncover the organ. This is a substantial simplification of the task, possibly to an open-loop control problem.

Complex visual servoing of the cutting device might be avoided by using a self-aligning tool, e.g. curved scissors [3] or knifes [30]. Again, the preparation of the organ configuration by peristaltic sorting reduces the complexity of the problem. A detailed scene understanding and classification of organ parts is not necessary. The cutting operation can be planned based on 3D data and geometric features. The tool design requires a study of the mechanical properties of the connecting tissue.

4   Discussion

4.1   Novelty of the Idea

The combination of soft actuation, industrial robotics and machine vision has not been proposed for the sorting and separation of animal offal before. In fact, there is currently no automatic system for sorting lamb organs. Automatic approaches for organ sorting exist for the processing of small animals like poultry or squid. Because of missing sensor feedback, these systems often damage the organs to be harvested, which limits the yield. Our solution improves over the existing ones by using a soft, compliant robot that prevents such damages. Machine vision systems have not been developed yet for the purpose of offal classification. We therefore propose an innovative transfer of techniques from medical applications taking into account the differing boundary conditions. We propose a combination of the latest techniques (soft bodied robots, 3D printing, machine learning) from multiple fields. The design of the xy-sorting table is a novel two-dimensional extension of the one-dimensional peristaltic food transport in an artificial aesophagus. We extend the idea to complex peristaltic planar movement and rotation. The machine vision part is based on the latest spatially sensitive techniques from the recognition of deformable objects. These techniques have not been applied to the recognition of ex vitro organ configurations, a problem unknown to machine vision before the proposal of peristaltic sorting. The same holds for the proposed optical calibration of the soft robot by 3D image recording.

4.2   Design of the Soft Actuator

Because of its advantageous mechanical properties, silicone rubber is a popular choice in soft robots [14, 15]. In a study using Ecoflex 0030, Chen et al. [12]

found tensile strains at break above 600% for five of specimens. The hardness is below the Shore A scale and it can be cured at room temperature using 3D printed moulds [15, 31]. The material is also easy to combine with reinforcing parts and fibers that alter the mechanical properties [10, 11]. A pneumatic drive is proposed because it is compliant, low cost, and does not need high voltages. An electric shaker is a simple and effective support for the peristaltic actuation of the soft structure. A bi-arm robot is preferred over two separate robots to reduce the costs. A soft gripper with multiple vacuum suckers complies with the different textures and shapes of the organs to manipulate.

4.3   Recognition of Organs and Organ Configurations

Humans recognise organs from images by colour, shape, and textural features. However, shapes vary strongly and are difficult to extract automatically due to occlusion, deformation, and structures in the organs themselves. Colour is an important feature which is relatively easy to extract, although a calibration is needed (or assumptions, e.g. of a 'grey world'). Most evidence for a successful automatic recognition in the literature comes from medical applications [32–34]. Due to differences in the sensor setup (CT, MR, X-ray instead of colour), these algorithms are not directly applicable. Medical sensors seem inappropriate because of their price, size, and implications on robot design. In terms of algorithms, good results have been achieved using general purpose feature extractors (SIFT [24], LBP [25]) which have originally been proposed for colour/grey-level images. A transfer of these methods seems therefore realistic. The use of 3D data from time-of-flight or comparable sensors simplifies the detection of object centres and object outlines. A shape sensitive descriptor for organ configurations is a transfer from human-body-tracking [26], where configurations of limbs are modelled. The descriptor is based on histograms in a grid partitioning of the image, a technique which is well studied in many other descriptors [24, 35, 36]. The vector form is compatible with the mathematical notation used in many statistical and machine learning methods.

4.4   Peristaltic Positioning of Organs for Removal

In order to reposition organs in the plane, it is at least necessary to produce elongated wave fronts in two directions. A rotation, spreading, or compaction of the organ package requires radial wave patterns with small directional error. Dynamic actuations are thinkable that perform manipulations not possible by slow planar movements. It is therefore necessary to control each inflatable cell individually. The optical calibration is necessary because earlier studies have found a nonlinear dependency between pressure and shape, which does not only result from the cell design, but from the rubber material itself [12].

The proposed state machine has several advantages: First, it guides the choice of control patterns in a way that the organ package is brought into the optimal starting condition for organ removal. Secondly, its probabilistic nature allows for the estimation of success rates for a certain chain of actuations. This allows for the choice of more successful states for organ removal as well as the optimisation of the control patterns. The method is able to model the dynamic behaviour of the organ package by introducing more states.

4.5   Removal of Single Organs

A vacuum gripper is the preferred solution because the technique is well developed, has low requirements on machine vision, and is easier to control than a mechanical gripper. Since the vacuum gripper is placed in the middle of the object, the computer vision module must only provide the object position but no detailed shape description. Whereas a hard gripper might damage the organs due to punctual force peaks resulting from the slippery material, the vacuum gripper uses the compliance of the material to function. However, in order to compensate for deformations of the organs during lift-up, the gripper should be flexible, too. An example of a deformable vacuum gripper is the 'universal gripper' [18] which can be alternatingly soft and hard by controlling the amount of air in a loose particle filling. The use of peristalsis in preparation of the removal of single organs is a major simplification of the whole task. It is therefore not necessary to identify organ movements and deformations caused by mechanical grippers, a largely unsolved problem that also requires a much higher level of detail than the peristaltic control. With peristalsis, the removal of single organs can also be reset easily to handle error conditions.

5   Conclusion

We proposed a robotic approach to the sorting of ovine offal that is based on a soft peristaltic xy-sorting table. In particular, our approach solves the problem of handling the soft internal organs and bring them into a state that is tractable using current industrial technologies. The alternative of using industrial robots to deal with an unordered organ package would be unpromising. Compared to approaches used in other animals, our method minimises the stress for the organs. Our method is robust because it exploits the natural tendency of organs to assume energy minima. This reduces the variance in vision and control variables. The method is able to recover from disturbances and error states because of a probabilistic state modelling and easy restarts. The peristaltic sorting table is cheap, has many options for material design, and can be manufactured easily using 3D printed moulds. Nonlinearities in the material can be calibrated from optical measurements.

However, the proposed approach is challenging in many ways. Inspite of many prototypes of soft robots, there is no theory that guides the design of the peristaltic table. Constrains from the area of application have not been quantified yet. For example the mechanical properties of organ packages are largely unknown at the moment. Effective actuation patterns have not been designed yet. Secondly, a computer vision system for the ex-vitro recognition of organs must be designed. This requires a study of sensors and feature extractors and has an effect on the space of organ configurations. Algorithms need to be developed to train the graph model from observed situations, to identify desired states, and to plan paths in the graph model in order to reach the optimal configurations for organ removal. Then tools for gripping and cutting must be developed. Scissor-like tools might only be sufficient for heart, liver, and kidneys. The removal of the spleen and the unfolding of the intestines are much more complex.

References

1. "MIA, Annual Report," Meat Industry Association MIA (Trade Association representing New Zealand meat processors, exporters and marketers) (2013)
2. Meat Technology Update, CSIRO Food and Nutritional Sciences: Meat Industry Services (June 2008)
3. Choi, S., Zhang, G., Fuhlbrigge, T., Watson, T., Tallian, R.: Applications and Requirements of Industrial Robots in Meat Processing. In: International Conference on Automation Science and Engineering (CASE), vol. 2, pp. 1107–1112. IEEE (2013)
4. Jansen, T.C., Spijker, R.: Method and apparatus for mechanically processing an organ or organs taken out from slaughtered poultry. U.S. Patent No. 20110237171 A1 (2011)
5. Furst, J.D., Susomboom, R., Raicu, D.S.: Single organ segmentation filters for multiple organ segmentation. In: Annual International IEEE Conference on Engineering in Medicine and Biology Society (EMBS), pp. 3033–3036 (2006)
6. Pauly, O., Glocker, B., Criminisi, A., Mateus, D., Möller, A.M., Nekolla, S., Navab, N.: Fast Multiple Organ Detection and Localization in Whole-Body MR Dixon Sequences. In: Fichtinger, G., Martel, A., Peters, T. (eds.) MICCAI 2011, Part III. LNCS, vol. 6893, pp. 239–247. Springer, Heidelberg (2011)
7. Venkatraghavan, V., Ranjan, S.: Generic Framework for Organ Localization in CT and MR Images. In: Applied Imagery Pattern Recognition Workshop (AIPR). IEEE (2011)
8. Kim, S., Laschi, C., Trimmer, B.A.: Soft Robotics: a Bioinspired Evolution in Robotics. Trends in Biotechnology 31, 287–294 (2013)
9. Trivedi, D., Rahn, C.D., Kier, W.M., Walker, I.D.: Soft robotics: Biological inspiration, state of the art, and future research. Applied Bionics and Biomechanics 5, 99–117 (2008)
10. Shepherd, R.F., Stokes, A.A., Nunes, R.M.D., Whitesides, G.M.: Soft Machines That are Resistant to Puncture, and That Self Seal. Advanced Materials 25(46), 6709–6713 (1998)
11. Stokes, A.A., Shepherd, R.F., Morin, S.A., Ilievski, F., Whitesides, G.M.: A Hybrid Combining Hard and Soft Robots. Soft Robotics 1 (2013)
12. Chen, F.J., Dirven, S., Xu, W.L., Li, X.N.: Soft Actuator Mimicking Human Esophageal Peristalsis for a Swallowing Robot. IEEE/ASME Transactions on Mechatronics (2013)
13. Dirven, S., Chen, F.J., Xu, W.L., Bronlund, J.E., Allen, J., Cheng, L.K.: Design and Characterization of a Peristaltic Actuator Inspired by Esophageal Swallowing. IEEE/ASME Transactions on Mechatronics (2013)
14. Lin, H.-T., Leisk, G.G., Trimmer, B.: GoQBot: A Caterpillar-inspired Soft-bodied Rolling Robot. Bioinspiration & Biomimetics 6, 026007 (2011)

15. Shepherd, R.F., Ilievski, F., Choi, W., Morin, S.A., Stokes, A.A., Mazzeo, A.D., Chen, X., Wang, M., Whitesides, G.M.: Multigait soft robot. Proceedings of the National Academy of Sciences, PNAS (2011)
16. Shepherd, R.F., Stokes, A.A., Freake, J., Barber, J., Snyder, P.W., Mazzeo, A.D., Cademartiri, L., Morin, S.A., Whitesides, G.M.: Using Explosions to Power a Soft Robot. Angewandte Chemie International Edition 52, 2892–2896 (2013)
17. Laschi, C., Cianchetti, M., Mazzolai, B., Margheri, L., Follador, M., Dario, P.: Soft Robot Arm Inspired by the Octopus. Advanced Robotics 26, 709–727 (2012)
18. Brown, E., Rodenberg, N., Amend, J., Mozeika, A., Steltz, E., Zakin, M., Lipson, H., Jaeger, H.: Universal robotic gripper based on the jamming of granular material. Proceedings of the National Academy of Sciences (PNAS) 107(44), 18809–18814 (2010)
19. Kang, R., Branson, D.T., Caldwell, E.G.D.G.: Dynamic Modeling and Control of an Octopus Inspired Multiple Continuum Arm Robot. Computers & Mathematics with Applications 64, 1004–1016 (2012)
20. Brochu, P., Pei, Q.: Advances in Dielectric Elastomers for Actuators and Artificial Muscles. Macromolecular Rapid Communications 31, 10–36 (2010)
21. Carpi, F., Frediani, G., Turco, S., Rossi, D.D.: Bioinspired Tunable Lens with Muscle-Like Electroactive Elastomers. Advanced Functional Materials 21, 4152–4158 (2011)
22. Wong, R.D.P., Posnerb, J.D., Santos, V.J.: Flexible microfluidic normal force sensor skin for tactile feedback. Sensors and Actuators A: Physical 179, 62–69 (2012)
23. Park, Y.L., Chen, B.R., Wood, R.J.: Design and Fabrication of Soft Artificial Skin Using Embedded Microchannels and Liquid Conductors. IEEE Sensors Journal 12, 2711–2718 (2012)
24. Lowe, D.G.: Object Recognition from Local Scale-Invariant Features. In: International Converence on Computer Vision (ICCV), pp. 1150–1157 (1999)
25. Ojala, T., Pietikäinen, M., Harwood, D.: Performance evaluation of texture measures with classification based on kullback discrimination of distributions. In: IAPR International Conference on Pattern Recognition (ICPR), pp. 582–585 (1994)
26. Stommel, M., Beetz, M.: Sampling and Clustering of the Space of Human Poses from Tracked, Skeletonised Colour+Depth Images. Technical Report 70, Center for Computing and Communication Technologies, University of Bremen, Germany (2013)
27. Zhu, M.Z., Xu, W.L., Bronlund, J.: CPG-based control of a swallowing robot. International Journal of Computer Applications in Technology (2013)
28. Matsuoka, K.: Sustained Oscillations Generated by Mutually Inhibiting Neurons with Adaptation. Biological Cybernetics 52, 367–376 (1985)
29. Martinez, R.V., Branch, J.L., Fish, C.R., Lin, L., Suo, Z., Whitesides, G.M.: Robotic Tentacles with Three-Dimensional Mobility Based on Flexible Elastomers. Advanced Materials 25, 205–212 (2013)
30. Singh, J., Potgieter, J., Xu, W.L.: Ovine automation: robotic brisket cutting. Industrial Robot: An International Journal 39(2), 191–196 (2012)
31. Xia, Y.N., Whitesides, G.M.: Soft lithography. Angewandte Chemie International Edition 37(5), 550–575 (1998)
32. Cuadra, M.B.: Atlas-based segmentation and classification of magnetic resonance brain images. PhD thesis, Ecole Polytechnique Federale De Lausanne (2003)
33. Glocker, B., Pauly, O., Konukoglu, E., Criminisi, A.: Joint Classification-Regression Forests for Spatially Structured Multi-Object Segmentation. In: Fitzgibbon, A., Lazebnik, S., Perona, P., Sato, Y., Schmid, C. (eds.) ECCV 2012, Part IV. LNCS, vol. 7575, pp. 870–881. Springer, Heidelberg (2012)

34. Liu, X., Song, Q., Mendonca, P., Tao, X., Bhotika, R.: Organ Labeling Using Anatomical Model-Driven Global Optimization. In: IEEE International Conference on Healthcare Informatics, Imaging and Systems Biology (HISB), pp. 338–345 (2011)
35. Bay, H., Ess, A., Tuytelaars, T., van Gool, L.: SURF: Speeded Up Robust Features. Computer Vision and Image Understanding (CVIU) 110(3), 346–359 (2006)
36. Ke, Y., Sukthankar, R.: PCA-SIFT: A More Distinctive Representation for Local Image Descriptors. In: Computer Vision and Pattern Recognition (CVPR), vol. 2, pp. 506–513. IEEE (2004)

The Design of ARM-Based Control System
of Unmanned Research Catamaran

Zhang Yan-na, Shan Ti-kun, Ding Fei, and Yang Wei-min[*]

College of Electromechanical Engineering, Qingdao University of Science and Technology,
Qingdao 266061, China
{qustzyn,dingfei1215}@126.com, stkun@163.com,
yangwm@mail.buct.edu.cn

Abstract. The control system of the small unmanned research catamaran is divided into two parts, boat-borne system and onshore control system, which connect through ZigBee. The design of control system and the selection of electronic module adopt the modular, systematic ideas for the two structures of the unmanned research catamaran respectively. The writing and debugging of the main programs of whole system and the device drivers of each module have been completed, based on the IDE of KEIL, and the experiment of motion control, transmission distance and automatic navigation test have been carried out, which has confirmed the feasibility of the whole design.

Keywords: unmanned research catamaran, ZigBee, control system.

1   Introduction

Most of the foreign unmanned boats have been mainly used in the military field and resource exploration, and less for the scientific research and marine environmental monitoring [1]. The study of unmanned boat was started relatively late in China. The way of connection between the console and the hull of small unmanned research catamaran are mainly radio waves, GPRS, GSM, Bluetooth and so on, both in China and abroad [2]. The hull often uses mono-hull, and both the key technology and the movement of the hull have some limitations. With the application of unmanned boats expanded in marine range and the diversification of sensors, the catamaran relying on its advantages of high stability and high effective loading capacity, becomes the main type of unmanned boat in the field of marine science research [3]. On this occasion, the authors have used a catamaran as the hull of unmanned research boat, which was designed in this paper with the 500 meters-control distance, and for which the authors designed a set of control system based on ARM.

2   Control System Design and Electronic Module Selection
for the Hull

In order to improve the stability of the hull and the scalability of the system, a catamaran has been chosen as a carrier for boat-borne system. The basic parameters are shown in Table 1. A three-dimensional modeling is showed in figure 1.

---

[*] Corresponding author.

© Springer International Publishing Switzerland 2015
J.-H. Kim et al. (eds.), *Robot Intelligence Technology and Applications 3,*
Advances in Intelligent Systems and Computing 345, DOI: 10.1007/978-3-319-16841-8_55

Table 1. The basic parameters of the USV hull

Project	Working range	Length [m]	Width [m]	Height [m]	Hull material	Battery	Camera
Parameter	Within 500m	1.8	1.2	0.8	FRP	Lithium-ion battery	Speed Dome Camera

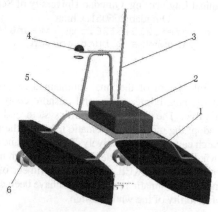

1.Main hull 2.Antiwater tank placed with circuit and module 3. Transmitting antenna 4. Speed dome camera 5.Stent 6. Propeller

Fig. 1. 3D structure of the boat

The control system of the small unmanned research catamaran is divided into two parts, boat-borne system and onshore control system, which connect through ZigBee. The control frame is showed in figure 2.

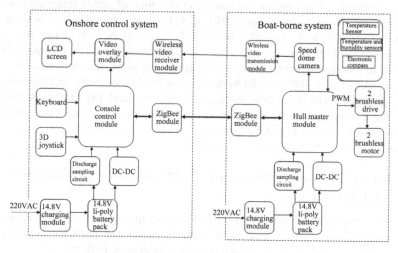

Fig. 2. Control frame of USV

2.1   Master Chip Selection

While the main control module of unmanned research catamaran is related to the normal operation of the whole system. STM32F107VCT6 of ST's STM32 series have been used as the master chip on the basis of stability, development cycle, cost and other aspects, which has a maximum operating frequency of 72MHz, integrating 512KB Flash memory and 64KB SRAM memory as well as a variety of DMA channels, a A/D converter of 12-bit, 16-channel scale with double-sampling and hold capability.-There are 5 USART interfaces (ISO7816 interfaces, LIN, IrDA compatible, debug control). Motion controller need to control a variety of movements of unmanned research catamaran and to realize the capability of automatic navigation, thus richer peripheral interfaces and stronger data processing capabilities are required.

2.2   Electronic Module Design and Selection

The hull's minimum system design of ARM [4] is the main control board of STM32F107 taking main clock crystal oscillator, RTC clock crystal oscillator, reset circuit, power-on display LED, download circuit, and the rest I/O ports are led out through two rows of pins for later use.

The STM32F107 download module selects SW mode for it needs only four lines, and can complete online debugging and it also has the function of download process.

DS18B20 digital temperature sensor is adopted, featuring small, low spending hardware, strong anti-interference ability, high accuracy, and is packaged with 316L stainless steel, featuring waterproof ability and seawater corrosion resistance properties.

The control system, electronic circuits, batteries, etc. on the hull are placed in a waterproof safety box, inside which the temperature and humidity have a crucial effect on the service life and safety of the various electronic components. Therefore, a temperature and humidity sensor can be added in the circuit design, then the sensor can real-time monitor the temperature and humidity inside the box, timely warning when temperature or humidity exceeded once. DHT11 temperature and humidity sensor is used in this design [5].

A 3-axis digital compass IC of Honeywell HMC5883L is adopted, with a 12-bit ADC and low interference AMR sensor, whose precision exacts to $1° \sim 2°$, and it also has obtained compass heading, hard magnetic, soft magnetic, and automatic calibration library.

3   Control System Design and Electronic Module Selection for Onshore Console

The master chip on shore should have a serial connection with ZigBee modules, which can collect two-dimension joystick data through two-way AD and signals of twelve buttons. Therefore, STM32F107VCT6 is adopted as the main control chip for the onshore console, considering the match of data from MCU between upper and lower machine-position.

Three-dimensional potentiometer control circuit diagram is shown in figure 3, PCO, PC2 are outgoing line slot of potentiometer. Being a total of five lines, they are +5V,

GND and three signal lines. In the figure, 1 is to +5V, 5 to GND. The capacitance between +5V power and GND is for filtering, making potentiometer signals better.

STM32F107 has a 16-chip 10-bit ADC and can apply three-dimensional potentiometer directly.

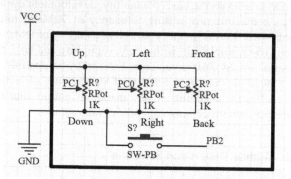

Fig. 3. Three-dimensional joystick circuit diagram

4   Drivers Design for Control System

The modular programming ideas [6] are used to design the SCM main program. The main program uses the typical foreground and background framework based on embedded system, namely the main program being composed of the infinite loop routine running in the background and the interrupting service routine running in the foreground. Background program mainly completes the operation that is not strict about time, like reading, processing and displaying of data. In contrast, foreground program mainly handles the operation that is strict about time. The main program's framework of the SCM is showed in figure 4.

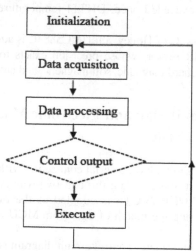

Fig. 4. Main program framework

PID algorithm is used to realize the function of automatic navigation. As being showed in figure 5, the given direction value does difference with the currently detected direction value, getting a difference value. After the PID calculation, the two propellers move. Then if compared the two direction value again, the result has been found that it gradually making less error than the given difference value, which means the heading of hull is gradually stabilized and the purpose of automatic navigation can be achieved.

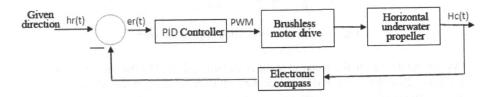

Fig. 5. Automatic navigation system framework

5   Experiment

In order to verify whether the device can operate successfully, the electronic module are assembled and debugged in the laboratory, until successful operation. The device is showed in Figure 6.

Fig. 6. System hardware building as a whole

As putting the hull in the water at point A in figure 7, the authors use joystick potentiometer to control the hull motion. The experiment shows that the potentiometer can control the hull's forward, backward and its turning movement. When the joystick is pushed evenly, the hull would accelerate smoothly as well. The hull remains stable in the water, and back and forth, side to side phenomenon does not emerge.

As putting the hull in the water at point B, the authors force the hull to drive along route BC. When it reaches point F, the information, sent by the sensor on the hull, cannot feed back to the onshore console in real time, which indicates the connection breaks off. The distance between point B and F is 133m after measuring, which means the maximum connection distance of the system is 517.4 meters. As being showed in Figure 7.

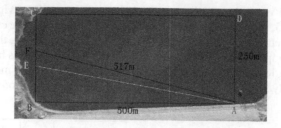

Fig. 7. Maximum connection distance diagram

As putting the hull in the water at point A and towards E (The distance between point B and E is 120 meters), the authors force the hull to drive along route AE automatically at highest speed after the automatic navigation button is pressed, then the driving route of the hull would have an intersection point in G-spot with route BC. After measuring, the distance between the EG is 7 meters. After calculating, the angular deviation is 1.84° while the hull running 514 meters in the still water. As being showed in Figure 8.

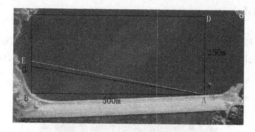

Fig.8 Experiment diagram of automatic navigation

6   Conclusion

The article have designed a set of control system, based on ARM and embedded system, for the unmanned research catamaran within 500 meters-control distance. And it can run successfully after being confirmed by experiments which show that the movement controlling system runs well, the connection distance can be up to 517.4 meters as to meet the design requirements, and the deviation of automatic navigation is smaller as to meet the daily requirements. In summary, the system as well as the connection is operating successfully and normally, which confirms the feasibility of the whole design.

References

1. Manley, J., Vaneck, T.: High Fidelity Hydrographic Surveys Using an Autonomous Surface Craft. In: Proceedings of Oceans Community Conference 1998, MTS, Baltimore (September 1998)

2. Caccia, M., Bibuli, M., Bono, R., et al.: Basic navigation, guidance and control of an unmanned surface vehicle. Autonomous Robots 25(4), 349–365 (2008)
3. Wang, X.-W., Zhang, W., Zheng, M.: Intuitionistic fuzzy hybrid operator catamaran boat main scale decision-making method. Boat 24(4), 1–6 (2013)
4. Huang, Z.-W., Deng, Y.-M., Wang, Y.: ARM9 embedded system design fundamental tutorial, vol. 8, pp. 1–7. Beijing University of Aeronautics and Astronautics Press, Beijing (2008)
5. Ni, T.-L.: The application of single bus sensor DHT11 in the measurement and control of temperature and humidity. Microcontroller and Embedded System Application 6, 60–62 (2010)
6. Baker, B.: A combined finite-time speed and yaw controller for an under actuated unmanned surface catamaran using way-point navigation. The University of Texas at San Antonio (2013)

The Design and Experiment of Unmanned Surveyed Catamaran

Ding Fei, Shan Ti-kun, Zhang Yan-na, and Yang Wei-min[*]

College of Electromechanical Engineering, Qingdao University of Science and Technology,
Qingdao 266061, China
{dingfei1215,qustzyn}@126.com, stkun@163.com,
yangwm@mail.buct.edu.cn

Abstract. This topic has designed a remote control distance within 500 meters of small unmanned surveyed catamaran (USC). The USC is divided two sections, hull parts and shore console system. Each section has its own energy. The communication between the two systems is according to ZigBee., control units can complete hull motion control and real-time image acquisition work.

Keywords: unmanned surveyed catamaran, ZigBee, launching experiments.

1   Introduction

Currently ocean unmanned intelligent monitoring systems are divided into three main categories: Remotely Operated Vehicle (ROV), Autonomous Underwater Vehicle (AUV), Unmanned Surface Vessel (USV) [1]. Most foreign unmanned vessel is mainly used in the military field and resource exploration, less for scientific research, marine environmental monitoring. America's first unmanned ship was used for scientific research in 1997 by the Massachusetts Maritime Academy made "ARTEMIS" [2], while China's first unmanned ocean surveillance ship "Sky One" developed by the China Aerospace Science and Industry Corporation and the China Meteorological Administration Atmospheric Sounding Technology Center in 2008 [3]. This topic attempts to design a remote control unmanned surveyed catamaran use for investigation shallow areas that ordinary research vessel can't reach.. It must be able to independently move to the specified location under control, record the relevant data and real-time image to the shore console system. The remote distance is about 500m.

2   System Components

Unmanned surveyed catamaran design is a complicated systematic project. The system consists of two parts of the shore stations and control onboard systems, designed with modular, systematic design ideas [4]. We choose a catamaran as unmanned research vessel hull, hull control system uses ZigBee wireless communication technology [5],

---

[*] Corresponding author.

© Springer International Publishing Switzerland 2015
J.-H. Kim et al. (eds.), *Robot Intelligence Technology and Applications 3,*
Advances in Intelligent Systems and Computing 345, DOI: 10.1007/978-3-319-16841-8_56

wireless video transmission using a separate stereo power 2.4G wireless video transmit&receive modules, and show the data on the video by On-Screen Display module. As shown in Figure 1.

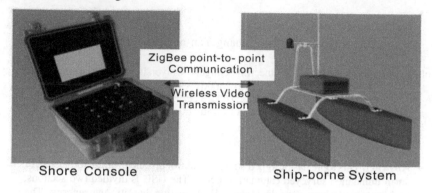

Fig. 1. System components of USC

3   Main Function and Parameters

Main function: 1. The real-time video shooting by high-speed dome camera can display on the shore console screen. By manipulating the shore on the keypad can dominate camera's zooms, manual / automatic zoom switch, manual focus adjustment etc.. 2. Control the hull forward, backward, turn and so on by console. 3. The battery power, battery power hull, water temperature, the internal temperature and humidity, electronic compass, date, time and status information can be superposed on the console screen in real-time. Parameters are as Table 1~2.

Table 1. Main characteristics of the USC' control console

Project	Length	Width	Height	Screen	Buttons&Joystick	Battery
Parameter	406mm	330mm	174mm	10.1 Inch	13 buttons 1 joystick	Lithium-ion battery

Table 2. Main characteristics of the USC' hull

Project	Length	Width	Height	Air Weight	Hull Material
Parameter	1.8*m*	1.2*m*	0.8*m*	25Kg	FRP
Project	Propeller Structures	Front Thrust	Maximum Speed	Battery	Camera

Parameter	2 brushless DC thrusters	16kgf	5knots	Lithium-ion battery	Speed Dome Camera

4   Functional Unit Design

Size design of the hull: Because catamaran has large deck area, small wave resistance, good stability, short cycle roll and many other advantages, so choose a catamaran as unmanned research vessel hull [6]. Its parameter: maximum speed 5 knots, within the scope of the low-speed boat [7].

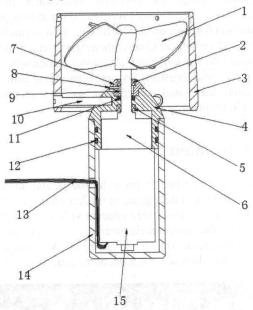

1. Propeller 2. Upper bearing 3. Fairwater sleeve 4. Conical head 5. Lower bearing 6. Reduction gearbox 7. Header 8. Oil inlet port 9. Oil chamber 10. Bracing piece 11. Motive seal "X" seal ring 12. Static seal "O" seal ring 13. Eectric wire 14. Scroll 15. Brushless Direct Current Motor

Fig. 2. Construction of underwater propulsor

For the low-speed catamaran: sheet body aspect ratio $\dfrac{L}{B}$ are the main factors affecting the catamaran sheet resistance of the body, it has the same rules and mono-hull for high speed catamaran, the coefficients used to express the length of the drainage volume. Wave resistance is a small proportion of the total resistance, then select catamaran focus on reducing the cost and reduce the wet surface, and generally

$\dfrac{L}{B} = 6 \sim 8$. Spacing ratio is one of the main factors affecting the inter-chip interference resistance of the body. There is little effect on the catamaran interference resistance, without obvious clustering rule $\dfrac{K}{B} > 2$, and generally, generally $\dfrac{K}{B} = 2$ close to the real ship.

The semi-hull length L = 1.8 m, width B = 0.3 m, spacing K=0.9m, length-width ratio $\dfrac{L}{B} = 6$, spacing ratio $\dfrac{K}{B} = 3$.

Underwater propeller design: the common underwater propulsion there are propeller, Cote catheter propeller, channel thruster, flat rotating propeller, water-jets, serial propulsion, imitation fish propulsion, MHD etc.. Which can be used in unmanned survey boat propeller has propeller, Cote catheter thrusters, water-jets, MHD. Cote pusher catheter has small dimensions, high propulsion efficiency, which can be either used as a master propeller and thrusters. So we choose Cote catheter propeller as our unmanned surveyed catamaran's underwater propeller. The construction of underwater propulsor is as shown in Figure 2.

5   Prototype Experiment

Hardware Structures and Experiment: prior to the actual launching, the various components were assembling and debugging in the laboratory. When the entire system is connected, power on, 1s later the screen appears video and characters, temperature and humidity parameters, the direction parameter, electricity information of console and hull is update complete, as shown in Figure 3. All components are working properly. The whole process does not crash phenomenon.

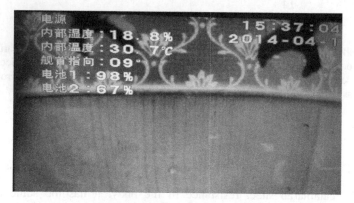

Fig. 3. OSD on screen

Thruster Control Linear Optimization Experiments: to test the linearity of propulsion motor speed, using a tachometer to measure motor speed under different duty cycles. The measured data generate PWM Output Signals-Motor Speed ratio curve, shown in Figure 4. The figure shows, when the duty cycle of the PWM output signal at 10%-65%, the motor speed is linear change. When the duty cycle of the PWM output signal is 0-10% and above 65% the motor speed is non-linear change. So need to do linear calibration for the duty cycle of the PWM output signal between 0-10% and above 65%.

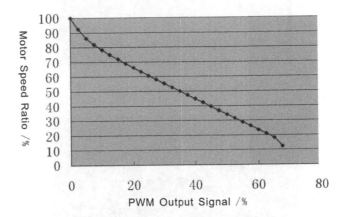

Fig. 4. PWM output signal - motor speed ratio curve

Anti-electromagnetic Interference Experiment: in order to verify the anti-electromagnetic interference capability of communication and other functions of the system, we bring the USC to the Electrical Testing Department of Quality Supervision & Inspection to do anti-electromagnetic interference experiment. After testing, the communication is all right, video occasional slight fluctuation, normal use is ok. As shown in Figure 5. The design of the radio communication scheme is feasible.

Fig. 5. Picture of testing result

Fig. 6. Motion of hull

Fig. 7. Real-time image

Motion Control Experiments: put the hull in the water at A point, and using shore console control the hull motion. The hull's control flexible, sports attitude smooth. Figure 6 shows the hull in the water sports gesture. The hull drive about 500m at maximum speed to point B, spend time 181s, calculated maximum hull speed in still water is about 5.37 knots. It's control distance reached 500m, and able to reach the designated place.

Image Transmission Experiment: when the hull is still at point B, we record the real-time image as shown in Figure 7.

6   Conclusion

Through the practical validation, the unmanned surveyed catamaran designed in this topic has high anti-interference ability wireless communication system, good communication system, good performance of motion control system, communication distance and image transmission achieved the reach needs, reach the basic needs of the investigation. Therefore, small boats with this control method can be applied to unattended operation in related fields.

References

1. Zhao, J.-H., Gao, Y.-B., Zhu, G.-W., et al.: Marine Observation Technology Progress. Marine Technology 27(4), 1–2 (2008)
2. Manley, J., Vaneck, T.: High Fidelity Hydrographic Surveys Using an Autonomous Surface Craft. In: Proceedings of Oceans Community Conference 1998. MTS, Baltimore (September 1998)
3. Manley, J.E.: Unmanned surface vehicles, 15 years of development. In: OCEANS 2008, pp. 1–4. IEEE (2008)
4. Baker, B.: A combined finite-time speed and yaw controller for an underactuated unmanned surface vessel using way-point navigation. The University of Texas at San Antonio (2013)
5. Li, W.-Z., Duan, C.-Y.: ZigBee Introduction to the Wireless Network Technology and Practice. Beijing University of Aeronautics and Astronautics Press, Beijing (2007)
6. Hussain, N.A.A., Sathyamoorthy, D., Nasuddin, N.M., et al.: Development of a Prototype Unmanned Surface Vessel (USV) Platform. Defence S&T Technical Bulletin 6(1) (2013)
7. Wang, X.-W., Zhang, W., Zheng, M.: Decision-making of catamaran principal dimensions based on intuitionistic fuzzy hybrid geometric operator. Ships 24(4), 1–6 (2013)

Techniques for Designing an FPGA-Based Intelligent Camera for Robots

Miguel Contreras, Donald G. Bailey, and Gourab Sen Gupta

School of Engineering and Advanced Technology
Massey University, Palmerston North, New Zealand
{M.Contreras,D.G.Bailey,G.SenGupta}@massey.ac.nz

Abstract. This paper outlines useful techniques to design and develop an intelligent camera on a Field Programmable Gate Array (FPGA). Some of the development and testing issues of porting a software algorithm onto a hardware platform are discussed, as well as ways to avoid the corresponding problems. To demonstrate the importance of these techniques, an intelligent camera designed to calculate the position, orientation and identification of soccer playing robots is used as a case study.

1   Introduction

Intelligent cameras can be an integral part of many robotic or automated systems. Bailey et al. [1] describes an intelligent camera as an extension of the smart camera by directly processing the pixels as they are streamed from the camera. As such they are well suited for applications where limited space, limited energy, and fast and reliable image processing is required in a self-contained unit. Field Programmable Gate Arrays (FPGAs) are ideal for this as they offer parallel processing, compared to serial processing from a conventional computer CPU. Combined with streamed processing it is possible to begin processing each frame directly as the pixels are read from the sensor, without the need for large latency memory units. In some instances, it is possible to calculate all of the useful information even before the frame is fully captured.

There is a lot of research outlining the use of FPGAs utilising image processing algorithms to create smart cameras [2-4]. However there is very little literature on design techniques for the development of FPGA-based intelligent cameras. Furthermore, there is even less literature which explicitly identifies the development issues and pitfalls. Designing an intelligent camera is a complex task, and there are numerous pitfalls. During the design of our own intelligent camera we have encountered many of them. This paper attempts to address this lack of literature by explicitly identifying these pitfalls, in an effort to reduce the learning curve so that others may learn from our mistakes. This paper is primarily aimed at early developers of intelligent cameras. Nevertheless, some of the discussions will benefit even those who have experience with image processing and FPGA algorithm development.

© Springer International Publishing Switzerland 2015
J.-H. Kim et al. (eds.), *Robot Intelligence Technology and Applications 3*,
Advances in Intelligent Systems and Computing 345, DOI: 10.1007/978-3-319-16841-8_57

The techniques discussed will help simplify the design of the intelligent camera and decrease the development time.

Section 2 identifies many of the pitfalls and techniques used to develop an intelligent camera. These techniques have been divided into three categories developing an algorithm testing scheme, developing the initial software algorithm, and developing the resulting hardware algorithm. Each of these sections will outline techniques to overcome the pitfalls, decrease the development time and simplify the development.

The techniques discussed in this paper will be explained using the FPGA-based intelligent camera currently being developed at Massey University. This intelligent camera is designed for robot soccer to identify individual robots using colour patches and calculate their positions and orientation on the field. As described by Contreras et al. [5] the algorithm follows a modular design of blocks or components. Each block is responsible for applying a filter or a process from the overall algorithm, and passing the results onto the subsequent blocks. Further information on the functionality and design on the algorithm can be found in [1, 5]. The software algorithm was developed using Matlab and the hardware description language (HDL) used to program the FPGA was Handel-C. However the techniques discussed in this paper are universal and can be used with any other software package or HDL.

2   Issues with Intelligent Camera Design

Despite all of the advantages that an intelligent camera can offer to projects requiring image processing applications they are not commonly employed. This is due to the greater complexity and longer development time of an FPGA-based camera compared to more conventional software approaches. This paper will look at a few of the more common issues that can hinder the development process.

- Design of parallel systems is complex
- Timing constraints must be explicitly handled due to parallelism
- Limited hardware debugging resources at run-time
- Long compile times for hardware algorithms

Designing a hardware algorithm is not straightforward. Unlike software algorithms which follow a serial command structure FPGAs allow for parallel algorithms to be executed. Parallel algorithms allow for faster processing as many functions can be calculated simultaneously. Because of this the design can become very complex, especially as the algorithm becomes more intricate.

Timing in particular becomes an issue as different operations have different latencies. This is not as much of an issue in software as timing is implicit in the sequence of commands. However, hardware must handle the timing explicitly the ensure information is processed correctly. In many cases errors can occur when implementing the correct logic but at the wrong instance.

Debugging is not as simple for hardware algorithm development as it is for software algorithms. Although compilers check for syntax errors, finding logic and timing errors is much more difficult in hardware. There are two main difficulties here. First, in software, only one thing happens at a time. So the changes resulting from a single clock cycle are small. In hardware, many things happen simultaneously. A lot can change in a single clock cycle, making it significantly more difficult to identify errors and track down there root causes. Many development environments provide a software simulation package meant to emulate the functionality of the FPGAs resources. These provide a systematic overview of the functionality as the parallel image processing algorithm is executed. Even so, finding the causes of unwanted or unanticipated changes in data can be time-consuming. Second, when interfacing the algorithm with real-time hardware, it is necessary for the algorithm to process data at real-time rates. For simulation, it is also necessary to simulate the operation of hardware external to the FPGA. Therefore, additional simulation models must be developed, debugged, and tested before even simulating the algorithm. For an intelligent camera the simulation model for the camera will need to stream pixel data from a file. Any errors in this model (for example timing errors) can invalidate perfectly working algorithm code. Due to the complexity of most image processing algorithms, simulation can be a very laborious and time consuming method of debugging. Furthermore once the algorithm is encoded onto the FPGA there are limited mechanisms for debugging errors.

Code compilation is an inevitable part of any programming and image processing project. The more complex the algorithm is the longer it takes to compile. This is true for both software and hardware compilers. However it takes longer to compile using HDL compilers as more optimization and timing passes must be processed. Hardware has an added place and route phase where logic is mapped to specific resources on the FPGA. The very large number of possible mappings makes the process poorly defined and difficult to optimise, especially when approaching the capacity of the FPGA. This drastically increases the development time. This is especially true when parts of the algorithm are dependent on the surrounding environment, and must be adjusted frequently.

During the development of our own intelligent camera many techniques were explored in an attempt to minimize these issues. This paper will outline some of the more important and successful principles.

- It is faster to test theories in a software environment than in hardware.
- A modular algorithm design can make programming and testing simpler.
- It is important to clearly define the interfaces between modules.
- Built-in functions in an image processing suite (such as Matlab) can both help and slow down algorithm development
- Some form of communication with the intelligent camera is critical.
- A parameterized hardware design can help make it easier and faster to make changes.
- There are many resources available to help debug during run-time in hardware.

Designing an image processing algorithm in hardware is very difficult. These compilers are not designed to easily show the effects of different filters or techniques compared to a software development environment such as Matlab. Even if the filter is relatively trivial it is better to first test its behaviour in a software environment to investigate how it affects the algorithm as a whole. Using a software environment can save a lot of development time as it allows for easy interaction and fast prototyping of different algorithms.

Numerous image processing environments also come with built-in functions and filters that can further decrease development time. Although great care should be taken when utilising these built-in functions and filters. Even though they may return quick and desirable results, they may be too complex to run efficiently on an FPGA. Once an algorithm has been developed it is possible to use the software environment to modify the design to closer resemble the functionality of the FPGA. Even though the parallelism will not be duplicated, the algorithm can be adapted to loosely mimic its functionality. This makes converting the algorithm to a hardware language much simpler and helps reduce its complexity as a whole.

Both a modular and parameterised algorithm design can be very useful when developing and testing in both software and hardware. A modular design allows each filter or block to be tested individually and simplifies the algorithm into smaller blocks. With each module compartmentalised it is important to clearly define and minimize the transfer of information between each module. This makes design, debugging, and testing much simpler as each module becomes localised. Similarly making each block parameterised can simplify changes to the interaction of the blocks without changing their functionality. However this can add some complexity to the design of each block. Even with the added complexity it still benefits the development process by eliminating the possible of accidentally introducing errors.

Communication between the FPGA and a computer is important. Many off-the-shelf FPGA development boards come with a variety of communication ports, such as USB, RS232, and general purpose IO. These interfaces are essential for transferring captured images from the sensor to begin the development of an algorithm. They can also be used to send commands to the intelligent camera to execute different functions, such as adjusting thresholds. This allows for settings to be changed during run-time without the need to recompile, thus reducing the total time spent compiling the algorithm. Results and processed images can also be transferred from the FPGA in order to test the functionality of the algorithm, in order to measure performance and debug errors. A solid form of communication can be one of the most useful ways to test and develop the intelligent camera during run-time.

There are several other resources available to help debug an FPGA at run-time other than a solid form of communication. Many off-the-shelf FPGAs come with resources such as LEDs, switches and buttons that allow for interaction with the intelligent camera. These can be used in many different ways, from displaying information to enabling and disabling filters. This can help identify faults without the need for extensive simulation examinations.

3   Developing an Algorithm Testing Scheme

When developing any kind of image processing solution, it is very important to follow a procedural development plan. In a very basic form this includes gathering test images, algorithm design, and algorithm testing. This is usually done using a software image processing suite, such as Matlab. This allows for quick prototyping and testing of algorithms with visual and numeric results. Since the image processing algorithm development process is largely heuristic, it involves a lot of trial and error. The long compilation times of HDLs means that making even small changes to an algorithm is no longer interactive, hampering the design exploration process. Therefore it is best to first design the algorithm in a software environment, and then optimize the resulting working algorithm into an HDL. This allows us to fully test the algorithm to see how it works and behaves under different circumstances before any difficult hardware programming begins. Fig. 1 illustrates the process used to fully develop an intelligent camera on an FPGA.

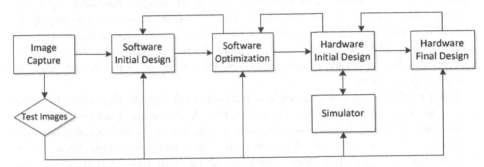

Fig. 1. Development and testing process for an FPGA-based intelligent camera

The first step in the design of any image processing algorithm is to capture a set of test images from the camera. A basic skeleton code will need to be developed on the FPGA to allow images to be taken from the camera and transferred to a computer, shown in Fig. 2.

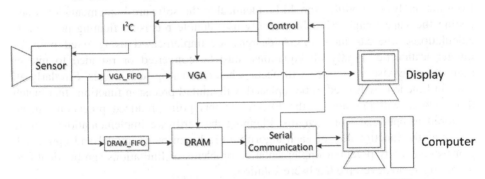

Fig. 2. Block diagram for a basic image capture algorithm on an FPGA

During initialization the control unit initialises the camera resolution, pixel skipping, gain and other camera parameters via the I²C controller. Under normal operation the captured pixels are loaded into the VGA_FIFO only. The VGA_FIFO acts as a buffer to load the data into the VGA module which controls displaying the pixels onto a monitor. The serial communication module receives the commands coming from the computer and controls sending the test images to the computer. Basic serial commands include camera adjustment via I^2C and commands to initiate a test image capture. When an image capture command is received the controller waits until the start of the next frame, to ensure a full frame is captured. At the start of the next frame the pixel data is loaded into the DRAM via the DRAM_FIFO. Once the image has been saved into memory it can then be sent to the serial module where it will be sent to a computer. Using an RS232 connection to transfer an image is very slow, taking over one and a half minutes to transfer a single 640x480 image at 57600 baud rate. This is why the image must first be saved to memory so that no information is lost.

The next stage in the development is to start designing the algorithm in a software environment. The desired characteristics of a good image processing algorithm development package are for it to be interactive and to allow fast prototyping of algorithms. Matlab is one such package. This allows various design decisions to be effectively explored to get the desired results from processing the image. Fig. 1 incorporates two software design processes, the initial software design, and the software optimization.

The aim of the first software implementation is to design the algorithm and make sure it works as intended, as quickly as possible. At this stage it may be easier and quicker to use built-in image processing functions to get the basic functionality of the algorithm. An example of this could be implementing standardized Bayer interpolation or YUV transforms which would be built into Matlab. This saves development time and allows these filters to be tested without needing to program them from scratch.

Once the initial design is completed, the algorithm needs to be optimized to more closely follow the hardware functionality. This is used to help test the effectiveness of the algorithm using the advantages and limitations of the hardware environment. It also provides a standard to test the hardware implementation against, as the results from the hardware modules should be identical to the software implementation when testing the same image. Matlab generally uses double precision floating point in its calculations, whereas most FPGA designs are implemented using fixed point or integer arithmetic. Firstly all equations must be converted or rounded to integer numbers otherwise deviations will occur when comparing test images. Similarly all built-in functions will need to be replaced with limited precision functions that match those which will operate on the FPGA. At this point, reduced precision can be explored to optimize word lengths. Planning the hardware implementation is made easier as the functionality of each algorithm block can be examined and optimized. Consequently any potential implementation problems or limitations can be identified before trying to develop a hardware solution.

Once the algorithm is designed from a software aspect it can be adapted for a hardware implementation. Parallel algorithm design is not straightforward. Because there are multiple processes running simultaneously, not only does the logic and functionality need to be correct but they must also be synchronised. Hardware adds a temporal aspect that must be explicitly handled, unlike serial coding which handles this implicitly. This can make the debugging process very intricate with multiple modules running in parallel. In software debugging each step executes one instruction. This makes error identification relatively easy as only one thing changes each step. The debugging process for hardware becomes more complicated as the parallelism allows for multiple processes to occur each clock cycle, generating numerous changes at once. This can make isolating an error very difficult.

The first step to debug a hardware algorithm is to create a test bench within the simulator to emulate the functionality of the camera. It must emulate capturing pixels, streamed from a text file, and pass them to each block for testing. Using this test bench it is possible to test each algorithm module individually or the complete algorithm using the test images captured earlier. To speed up the simulation, smaller sub-images can be used if the algorithm has been designed appropriately. This simulation is implemented on a computer so it operates serially, however it will simulate parallel operations when intended. This will accurately simulate the operation of the algorithm though it will take much longer to do so. Another advantage of the simulation is that it allows a step by step procedural overview of each block, such as when a register is read or written to, or what values are present in a FIFO etc. This can be very helpful to identify timing or synchronisation errors. The main purpose is to test the modules and compare the results to the software algorithm. The aim is to make the results of the simulation identically match the software results. Any differences present at this time could mean an error in the adaptation of the algorithm. Subsequent testing and debugging will need to be done to find the error and correct it.

A disadvantage of the test bench is that it must be programmed from scratch. This can introduce unintended errors or limitations that may not be applicable in the actual hardware implementation. Special care should be taken when creating this test bench to ensure that any faults that occur are not being caused by the simulator itself.

We made this mistake ourselves in the first revision of our test bench. The camera requires a blanking period at the end of each row before transmitting the next one. In an attempt to compress the simulation time smaller case-specific test images were used with shorter blanking periods. Normally the blanking period would last for several hundred clock cycles, but was shortened to a few dozen. During the testing of the connected component module several errors were reported and it was assumed that the module was not programmed correctly. After much investigation into the logic and timing no problems could be found that could explain the incorrect results. It was not discovered until later that the errors were being caused by the test bench itself and not the module. The connected component module required several clock cycles in the blanking period to finish calculating and assembling each blob. Normally there would be plenty of time to complete them on the FPGA; however the blanking period was too small in the simulator. Although this limitation was relatively easy to correct, it shows that not all errors are necessarily caused by the modules.

Finally once the algorithm is adapted and tested in simulation it is time to run some real-world tests on the hardware itself. This is done by comparing test images processed on the FPGA with identical images processed by software. There are two ways to test the algorithm on the FPGA. The first is to upload a previously captured test image onto the FPGA, processing the image and then downloading the results back to the computer for comparison with the ones calculated by the software implementation. The second method captures a new test image, stores it in memory, and then processes it, avoiding the need to upload a test image. Then both the test image and results are downloaded to the computer and compared with the software implementation. The first method allows for faster overall processing as the software algorithm has already calculated the results. Another advantage is that the camera is not actually required. Hardware emulates the camera by providing the pixel stream from memory. This can be helpful if the camera will be operating in areas where it may not be possible to have a computer for a long period of time, such as outdoors. However, an advantage of the second method is it can help find limitations in the algorithm that were not initially considered by the initial test images.

4   Techniques for Designing in Software

Before the algorithm can be implemented onto an FPGA it must first be designed. HDL compilers are not designed to allow for easy image processing algorithm development. So it is necessary to first design and test the algorithm in software. Even when the algorithm has been developed complications can occur with converting into an HDL. This is where a software application can be used to quickly test changes to the design or operation of the algorithm.

Software Algorithm Development

The initial design of the algorithm in software is a very important step that should not be overlooked. Even filters that may seem trivial should be tested and designed in software to see how they react with the rest of the algorithm. A software package, such as Matlab, is interactive and allows for quick prototyping of concepts to get the functionality correct. An example of this from the development of our own intelligent camera, was trying to identify each robot. It was theorized that the colour patches could be identified using shape recognition, using compactness [6].

$$Compactness = Perimeter^2/Area \qquad (1)$$

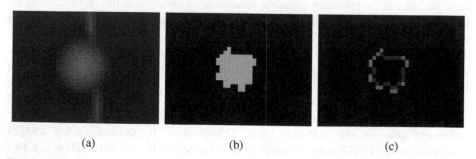

| (a) | (b) | (c) |

Fig. 3. Images from case study. (a) Image of the ball after the Bayer interpolation. (b) Image after colour thresholding. (c) Perimeter outline of the ball.

This is a unitless value that can distinguish between circular, square, and rectangular patches on the basis of shape. However, dealing with discrete images means that many measurements are only approximations of their continuous counterparts. It is well known that measuring a perimeter can be difficult [7, 8]. Therefore it is important to first test the theory in software to identify any possible limitations or inaccuracies before developing the hardware algorithm and discovering it does not work.

Fig. 3 shows the ball from one of the test images. The ball looks circular after the Bayer interpolation. However, the thresholding modules removed a lot of the edge pixels. This left the circle very misshaped. Using the compactness equation it was very hard to distinguish the difference between the circles and squares.

In this case several different methods of estimating the perimeter were compared before deciding that none were sufficiently reliable, and a different shape factor was used. By using the software development platform we were able to quickly prototype this theory and visually observe its effects. This allowed us to make quick decisions and explore other options.

Another advantage of creating a software algorithm is it creates a gold standard for which to test the hardware implementation against. If the software algorithm is capable of calculating correct results, then the goal is to make the FPGA return identical results. The software algorithm can therefore be used to create case-specific test images by partially processing test images captured from the camera.

When to Use Built-in Image Processing Functions

Matlab like many other image processing suites come with built-in functions and filters used to support image processing. Examples are Bayer pattern demosaicing and various morphological noise reduction filters. In the first stage of algorithm design an argument can be made whether or not these built-in functions should be used. On one hand the aim of the first algorithm design is to devise a solution to the problem as quickly as possible. On the other hand these filters exploit techniques and algorithms that may not function on an FPGA, therefore making the results less useful. Though, this may not be a problem when applying simpler, standardised filters such as an RGB to YUV conversion. Special care should be taken when applying filters such as noise reduction or Bayer interpolation, as there are many different types. A lot of papers have been written comparing the difference of image quality between different Bayer interpolations[9, 10]. For example, a gradient-corrected linear interpolation as described by Malvar et al. [11], similar to the algorithm used in Matlab, gives better results than a simple nearest-neighbour interpolation. Nevertheless the nearest-neighbour interpolation would be computationally simpler to implement on an FPGA. It is good practice, especially in this case, to try the simplest algorithms first. A complex Bayer interpolation could take a long time to develop and test compared to a simpler solution. Why spend weeks or months developing a complex solution when a simple one will return adequate results? Also, if the nearest-neighbour interpolation is likely to be implemented on the FPGA then it should be used in the software development; otherwise any results calculated would be less applicable.

5   Techniques for Designing and Testing in Hardware

There are many difficulties with programming an HDL. One of the most common, and at times hardest to debug, is timing and synchronisation errors [4, 12]. Commonly these errors will cause a register to be overwritten before the data has been read, causing the algorithm to fail. These errors become increasingly hard to find as the algorithm becomes larger and more complex. In a software implementation this is not a problem as there is plenty of memory to assign each variable an individual register. In a hardware solution this can be impossible or certainly impractical to do especially as the size and complexity of the algorithm increases. With pixel processing, for example, the same hardware processes all the pixels, just at different times. This is because memory and resources are limited on FPGAs. There are some techniques that can help simplify algorithms, and thus make it easier to debug and resolve timing errors.

Modular Algorithms

A modular algorithm breaks up the entire algorithm into smaller interconnected blocks. Each block is responsible for applying a specific filter or process and passes the results to the subsequent block. Each block operates independently from each other but is kept synchronised by signals passed between them. By reducing the algorithm into smaller blocks each module becomes less complex and can be tested individually.

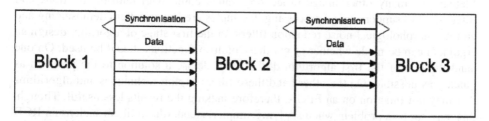

Fig. 4. Block diagram showing a basic modular design

The modular design utilises the hardware's parallel properties to run each block concurrently on streamed data. Unlike software image processing where entire frames are captured and processed, this design processes each pixel individually. This greatly decreases the latency as each frame is processed as it is captured. Therefore there is no need to store each frame into memory. However, since each block requires a different amount of time to process (latency) a data synchronous control signal is used to keep the algorithm synchronised. The processed data is passed between each block though signal carriers, and there is no limit to how many there can be. In fact in more complex situations a single block can pass multiple synchronisation and data signals to various different blocks. For example, Fig. 5 shows the output of the connected component module sending data to various other blocks.

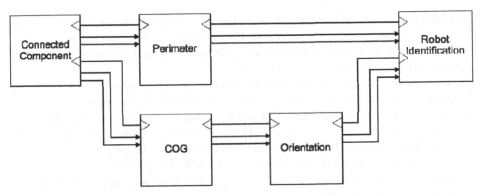

Fig. 5. Block diagram illustrating the functionality of multi input/output modules

The connected component module is responsible for detecting blobs from thresholded pixels and gathering data about them. This information is needed by multiple blocks to calculate the robot position, orientation, and identification. Therefore the blob detection block outputs separate data and synchronisation signals to the various blocks down stream.

Parameterised Algorithms

When programming in any HDL, register widths and data paths must be statically defined. This means that an 8 bit value can only fit in an 8 bit register, unless specific concatenation or splitting is used to change the width of the value. The natural instinct when programming an HDL is to hard-code the lengths of these registers. Most of the time, this creates a simple and easy to follow design that works perfectly. The problem comes if the design changes and several of the data paths change width. It can be a tedious and error-prone process going through the design making all the changes. However, a technique that can help simplify the development, especially when used with a modular design, is to parameterise each block. This allows adjustments to be made quickly without changing the functionality of each block. Example parameters may be to adjust the width of data pixels, change the position of words within concatenated strings, or changing the latency of synchronisation signals. Each change on its own could be responsible for dozens of register changes, and could introduce errors.

There are a number of ways to parameterise an algorithm, but the simplest is to have all register widths point back to global variables instead of hard-coding them. By adjusting the global variables the changes will propagate throughout each block without changing their functionality. The disadvantage of this method is that it may take longer to encode each block. Nevertheless it can make the development of the algorithm simpler and faster should any changes be necessary.

A variation of the first method requires utilising control variables inside each block as well as global variables. Utilizing all of the global variables in one position can create a very large and complex group. Particularly if some of the values are only used by a small number of blocks. By removing some of the less common global variables and only implementing them within their required blocks it is possible to reduce the number of global variables. The drawback to this method is that each control variable will need to be changed individually, although this is still simpler than changing dozens of individual registers.

An example of where a parameterised algorithm would be useful can be examined from the case study. The original algorithm for the intelligent camera was designed to use an 8-bit data stream from the image sensor. It was later discovered that there were problems distinguishing some colours. This was resolved by increasing the camera pixel resolution to the full 12-bits per pixel available from the camera. This required massive changes to the hardware algorithm to allow for the extra data to be processed. This took considerable time to manually change each block and debug any errors that were inadvertently introduced. However, with a parameterised design only a few parameters would need to be changed and the blocks would not need to be retested as they would still function as before, only processing wider data words.

Techniques to Help Debug at Run-time on an FPGA

One of the hardest parts of developing and testing the algorithm on an FPGA is the lack of a run-time debugger. Once the FPGA is running the design will either work or not, and when it fails often there will be no indication as to the cause. There are some ways to troubleshoot this however. Many off-the-shelf FPGA development boards have features, such as LEDs, LCD displays, 7-segment number displays, switches, and buttons. These can all be used to help debug the code or relay what process is being executed. For example the 7-segment number display can be utilised to act as a frame counter, or to display threshold values. LEDs can be used to indicate whether a communication command has been sent or received, or when a command finishes executing. Switches can be used to activate or deactivate specific filters to see the effect they have. Similarly, buttons can be used to make adjustments to filter characteristics and thresholds.

An example from the case study shows how many of these features where used together to help capture test images. When a serial command is sent from the computer to the FPGA to initiate an image capture command an LED will light up. This is to indicate that the command has been received and the process will begin. As the image is transferred from DRAM to the serial module the 7-segment counter is used to keep track of which row is being processed. This helped identify issues with data loss, whether it was caused due to FPGA or connection error. Finally when the test image was fully sent another LED would light up to signal the end of the process.

Some FPGA manufacturers will make basic code available to operate these hardware features. However it is usually only available for a limited number of HDLs. This means that drivers will first need to be created before these features can be used. This is not too much of a problem as these are relatively simple to write.

6   Summary and Conclusion

Using an FPGA to implement image processing is not a new idea. FPGA-based intelligent cameras can be very powerful, allowing for fast and accurate processing from a smaller and more power efficient platform, compared with conventional computers. Even though intelligent cameras offer many advantages for different robotic applications, they are difficult to develop. With this in mind, it is surprising to find a lack of literature detailing the development process for intelligent camera design. The techniques discussed in this paper are meant to as a guide to help others new to this field to quickly and efficiently develop their own projects. To an expert some of these techniques may seem obvious; however it is likely that at some point, like us, they have made the same mistakes.

Because the development of any image processing algorithm can be quite complex it is best to first design it using a software package, such as Matlab. Once an optimized algorithm has been developed it needs to be redeveloped to operate efficiently on an FPGA. The design and testing of the hardware implementation can be simplified by making the algorithm modular. Also, by parameterising each module, changes can be effected easily without changing the functionality of each block. Overall, a solid testing scheme must be present to ensure both software and hardware algorithms operate as designed, thus aiding in the development of the intelligent camera.

Acknowledgements. This research has been supported in part by a grant from the Massey University Research Fund (11/0191).

References

[1] Bailey, D.G., Gupta, G.S., Contreras, M.: Intelligent Camera for Object Identification and Tracking. In: Kim, J.-H., Matson, E., Myung, H., Xu, P. (eds.) Robot Intelligence Technology and Applications. AISC, vol. 208, pp. 1003–1013. Springer, Heidelberg (2013)

[2] Johnston, C., Gribbon, K., Bailey, D.: Implementing image processing algorithms on FPGAs. In: Eleventh Electronics New Zealand Conference (ENZCon 2004), Palmerston North, New Zealand, pp. 118–123 (2004)

[3] Dias, F., Berry, F., Serot, J., Marmoiton, F.: Hardware, Design and Implementation Issues on a FPGA-Based Smart Camera. In: First ACM/IEEE International Conference on Distributed Smart Cameras (ICDSC 2007), Vienna, Austria, pp. 20–26 (2007)

[4] Lim, Y., Kleeman, L., Drummond, T.: Algorithmic Methodologies for FPGA-Based Vision. Machine Vision and Applications 24, 1197–1211 (2013)

[5] Contreras, M., Bailey, D.G., Gupta, G.S.: FPGA Implementation of Global Vision for Robot Soccer as a Smart Camera. In: Kim, J.-H., Matson, E., Myung, H., Xu, P. (eds.) Robot Intelligence Technology and Applications 2. AISC, vol. 274, pp. 657–666. Springer, Heidelberg (2014)

[6] Davies, E.: Machine Vision: Theory, Algorithms, Practicalities, 3rd edn., pp. 193–194. Morgan Kauffmann, San Francisco (2005)

[7] Kulpa, Z.: Area and Perimeter Measurement of Blobs in Discrete Binary Pictures. Computer Graphics and Image Processing 6, 434–451 (1977)

[8] Ellis, T., Proffitt, D., Rosen, D., Rutkowski, W.: Measurement of the Lengths of Digitized Curved Lines. Computer Graphics and Image Processing 10, 333–347 (1979)

[9] Ramanath, R., Snyder, W., Bilbro, G., Sander, W.: Demosaicking methods for Bayer color arrays. Journal of Electronic Imaging 11, 306–315 (2002)

[10] Jean, R.: Demosaicing with The Bayer Pattern. Department of Computer Science, University of North Carolina (2010)

[11] Malvar, H., Li-Wei, H., Cutler, R.: High-Quality Linear Interpolation for Demosaicing of Bayer-Patterned Color Images. In: IEEE International Conference on Acoustics, Speech, and Signal Processing (ICASSP 2004), Montreal, Canada, pp. 485–488 (2004)

[12] Gribbon, K., Bailey, D., Johnston, C.: Design Patterns for Image Processing Algorithm Development on FPGAs. In: IEEE TENCON of Region 10, Melbourne, Australia, pp. 1–6 (2005)

Robust Object Recognition Under Partial Occlusions Using an RGB-D Camera

Yong-Ho Yoo and Jong-Hwan Kim

Department of Electrical Engineering, KAIST
291 Daehak-ro, Yuseong-gu, Daejeon, Republic of Korea
{yhyoo,johkim}@rit.kaist.ac.kr

Abstract. For a robot to execute a specific task, the robot firstly has to understand what objects are in robot's view. To complete a specific task in a given time, the computation time for recognition is also important. There are much research for increasing recognition accuracy, but the recognition speed is not enough to be applied in real environment. On the other hand, there are also much research for reducing the computation time for recognition, but the recognition accuracy needs to be further improved. Nowadays, deep network has come into the spotlight due to its speed and accuracy. Deep network doesn't need to find hand-tuned features. This paper proposes a deep network-based object recognition algorithm. The main contribution is that objects could be recognized under occlusion, as objects are often laid to overlap each other. The occlusion makes object recognition accuracy worse. To overcome this problem, the dataset for training consists of not full images but partial information of images and corresponding ground truths. The object region could be found very quickly by using an RGB-D camera. By assuming that most objects are on the stable plane, object regions are taken easily. Experimental results demonstrate such consideration of contextual information (e.g. objects are on the table) makes the performance of recognition better.

Keywords: object recognition, occlusion, deep learning, deep belief network, RANSAC.

1 Introduction

Recently, the progress of an RGB-D camera that provides both color and dense depth information makes a paradigm shift in computer vision. There are many applications using this RGB-D camera, e.g. object recognition [1], people tracking visual odometry and Simultaneous Localization and Mapping (SLAM) [2,3]. In addition to computer vision, the usage of the RGB-D camera is extended to robotics [4,5]. When a specific situation is given to a robot, this robot has to understand the situation. This situation is mainly judged by camera images. Existing cameras such as an RGB camera and a stereo camera [6] have limit to estimating the distance between camera and some objects. But, this limitation

© Springer International Publishing Switzerland 2015
J.-H. Kim et al. (eds.), *Robot Intelligence Technology and Applications 3,*
Advances in Intelligent Systems and Computing 345, DOI: 10.1007/978-3-319-16841-8_58

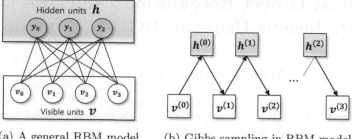

(a) A general RBM model.   (b) Gibbs sampling in RBM model.

Fig. 1. A general RBM model and Gibbs sampling process in RBM

has been overcome by the development of depth camera that makes it possible to estimate the exact distance.

For a robot to carry out a given task, the robot has to understand a current place and situation. Objects located in front of the robot's view have to be detected and distinguished. As a scene given to the robot is changed from moment to moment, noticing this change of scenes is essential to increase recognition accuracy and improve recognition speed.

In this paper, we propose robust object recognition under partial occlusions using the RGB-D camera. A plane is easily detected using depth information that are obtained by the RGB-D camera. By assuming objects are located on a plane such as a table, object recognition accuracy could be increased and recognition speed could be faster by taking only pixels located on the plane. In particular, a database is made up of image patches and corresponding ground truths. By making the database as patch-based images and labels, robust object recognition is possible even if there are partial occlusions. A learning algorithm is based on Deep Belief Network (DBN) [7,8]. In a multi-layer neural network, weights are pre-trained by Restricted Boltzmann Machine (RBM) [12]. Then, weights are fine-tuned by back-propagation. The advantage of DBN is that features are extracted by not hand-crafted method like SIFT [9], SURF [10], but learning. It accelerates feature extraction so that computation cost is considerably reduced.

The remainder of the paper is organized as follows: Section 2 describes some preliminaries such as RANdom SAmple Consensus (RANSAC) [11], RBM and DBN. The procedure of pre-processing and the proposed learning algorithm are detailed in Section 3, and experimental results are presented in Section 4. Finally, concluding remarks follow in Section 5.

2   Preliminaries

2.1   Restricted Boltzmann Machine

As RBM [12] is a special kind of stochastic neural network, it consists of visible units and hidden units. In RBM, there are no connections between units in the same layer. There are only connections between units in the different layer as shown in Fig. 1(a). RBM is a kind of Markov Random Field so that it can be represented as probabilistic function as

$$p(v, h) = \frac{1}{Z} e^{-E(v,h)} \tag{1}$$

where Z is the partition function and v and h are visible units and hidden units, respectively. And energy function E in Eqn. 1 is defined as

$$E(v, h) = b'v - c'h - h'Wv \tag{2}$$

where W represents the weights between hidden and visible units and b, c are off-sets of visible and hidden units respectively. Parameters in RBM are obtained by a stochastic gradients of log-likelihood. A log-likelihood gradients are as follows:

$$\frac{logp(v)}{w_{ij}} = <v_i h_j>_{data} - <v_i h_j>_{model} \tag{3}$$

In Eqn. 3, the first term could be calculated directly from given data. But, the second term is computationally impossible to obtain exactly. So, the second term is approximated by the Gibbs sampling method to sample v and h. Using $p(h|v)$ given visible units v, hidden units h are sampled. Then, visible units v could be derived by using $p(v|h)$ given hidden units h. By iterating this procedure as shown in Fig. 1(b), sampled hidden units and visible units are approached to accurate samples of $p(v, h)$. To speed up this iterative sampling process in practice, samples are obtained by only one step of Gibbs sampling [13].

2.2 Deep Belief Network

Deep Belief Network model consists of stacked RBM models as shown in Fig. 2. The first step is to evaluate hidden units in RBM in bottom of DBN. After that, evaluated hidden units are new input data in upper RBM model. By doing this

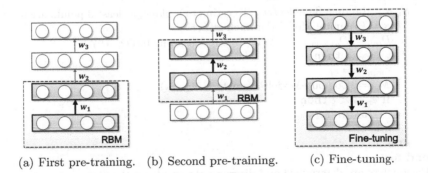

(a) First pre-training.   (b) Second pre-training.   (c) Fine-tuning.

Fig. 2. A general RBM model and Gibbs sampling process in RBM

procedure iteratively, the uppermost hidden units could be evaluated. These units are linked to output units and all weights are updated by back-propagation and this procedure is called fine-tuning.

3   Algorithms

3.1   RANSAC-Based Plane Detection

As a specific task is given to a person, he or she finds some objects that are needed to execute a given task. In an indoor environment, there are many objects on the stable plane such as table or desk. This fact could be applied to robots as well. In images given to a robot, a searching space can be limited to pixels on the plane so that recognition speed could be increased. A 3D point cloud is needed to estimate a plane in a image. The 3D point cloud is easily derived by the RGB-D camera.

In the 3D point cloud obtained by the RGB-D camera, there are not only pixels that lie in a plane, but also many outliers. These outliers means pixels far from the estimated plane. Sometimes, depth camera might not give exact distance information so that pixels in a plane could be outliers. To release this noisy depth information, the RANSAC algorithm that is robust to outliers is used. After finding parameters for plane equation about inliers, the distance between each point cloud and derived plane is assigned to all pixel in the image. Pixels in inlier set are derived as

$$P_{inlier} = \{p_i | |D(p_i)| < D_{th}\} \tag{4}$$

Algorithm 1 Object region estimation using RANSAC

1: **INPUT:** A 3D point clouds D
2: **OUTPUT:** Binary image P
3: Binary image means whether each pixel belong to object region or not.
4: Get 3D point clouds D using depth image
5: **for** $i = 0 \rightarrow$ max iterations **do**
6:     $S_i \Longleftarrow$ Random Sample(D)   ▷ To estimate a plane, at least 3 points are needed.
7:     $M_i \Longleftarrow$ Compute Model Parameters(S_i)
8:     $\{e, D_{inlier}, D_{outlier}\} \Longleftarrow$ VerifyingModelParameter(D, M_i)
9:     **if** $e_i < e_{min}$ **then**
10:         $e_{min} \Longleftarrow e_i$
11:         $M \Longleftarrow$ ComputeModelParameters(D_{inlier})
12:         **if** $e_{min} < \epsilon$ **then**
13:             Return M_i
14:         **end if**
15:     **end if**
16: **end for**
17: Parameters are determined and discriminate points on the plane.
18: $P^* \Longleftarrow$ AbovePlane(D)   ▷ Object region is restricted on the plane.

Algorithm 2 Learning Algorithm

1: **INPUT:** Patch-based RGB images X and ground truth Y in a training set
2: **OUTPUT:** Parameters W in deep neural network
3: Initialize all parameters in Deep Network.
4: Derive hidden units by training RBM.
5: **for** $i = 0 \rightarrow$ the number of hidden layer-1 **do**
6:     **for** $j = 0 \rightarrow$ max iteration **do**
7:         $v_i^1, v_i^1 \Longleftarrow \text{GibbSampling}(p(v_i^0, h_i^0))$ ▷ Iterate Gibbs sampling only 1 time.
8:         $W_{i,i+1}^{new} \Longleftarrow W_{i,i+1}^{old} = W_{i,i+1}^{old} + \mu(< v_i^0 h_i^0 > - < v_i^1 h_i^1 >)$ ▷ Update weights
9:         $v_{i+1} \Longleftarrow h_i$ ▷ Let hidden units h_i be visible units v_{i+1}
10:     **end for**
11: **end for**
12: **for** $i = 0 \rightarrow$ max iteration **do**
13:     **for** $j =$ the number of layer-1 $\rightarrow 1$ **do**
14:         $W_{j,j+1}^{new} = W_{j,j+1}^{old} + \mu \delta_{j+1} v_i$ ▷ v_i Back-propagate error to lower layer.
15:   ▷ v_i means current layer's unit values.
16:                         ▷ δ_{i+1} means error back-propagated from upper layer.
17:     **end for**
18: **end for**

---

where D_i is the distance between p_i and the estimated plane and D_{th} is threshold to discriminate between inliers and outliers. Pixels on the plane are calculated as

$$P_{object} = \{p_i | D(p_i) \geq D_{th}\}. \tag{5}$$

3.2 Learning

To recognize objects, input images and corresponding ground truth are necessary. A dataset for both object recognition and scene parsing consists of RGB images and ground truth images labeled for all pixels. In this paper, patch-based object recognition is executed; therefore the pixel-based dataset has to be converted to the patch-based dataset. An arbitrary patch's ground truth may have many kinds of labels. In this case, patch level's ground truth is assigned to the most frequent label. After patch-based database is constructed, parameters could be obtained by training. Neural network has two hidden layer of size H and a softmax output layer of dimension C that is total number of classes. Parameters are pre-trained by RBM and these are trained using a back-propagation and a cross-entropy loss function.

4 Experiment Results

Primesense Carmine 1.09 that is a kind of RGB-D camera was used to capture real indoor scene and objects. This camera's depth ranges from 0.35m to 1.4m. Objects in a dataset were four kinds of bottles that contain pepper, mustard, vinegar and sesame. The reason why we chose these objects was that robot may

(a) Recognition result in training dataset. (b) Recognition result in test dataset.

Fig. 3. Object recognition result in training set and test set

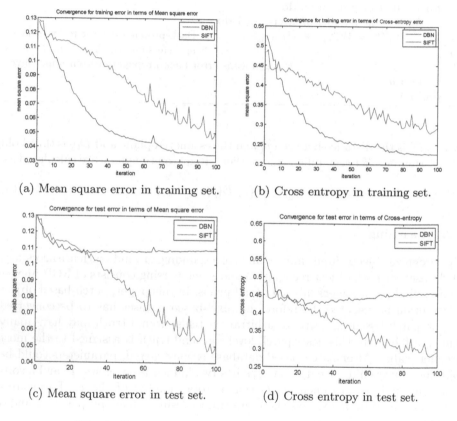

(a) Mean square error in training set.

(b) Cross entropy in training set.

(c) Mean square error in test set.

(d) Cross entropy in test set.

Fig. 4. Convergence of error in training set and test set

work in a kitchen. There were total 80 labeled pairs of RGB and depth images in the dataset. We used the 40 odd-numbered images as training set, and the remaining 40 even-numbered images as the test set. For learning, two models were used. These models were similar except output units' activation function and error criteria. The first model used a sigmoid function with mean square error. The second model used soft-max function with cross entropy error. The results of object recognition are described in Fig. 3. In the figure, images in the first row are original RGB images. Images in the second row and third row are the the results of DBN and neural network with SIFT descriptor, respectively.

Figs. 4(a) and 4(b) compare training images' convergence in terms of mean square error and cross entropy error, respectively. In these graphs, the model learned by DBN is better than the model that used the SIFT descriptor in terms of convergence error. As shown in Fig. 4(c) and 4(d) that show the results of test images, the model that uses SIFT descriptor is better than the proposed model.

5   Conclusion

In this paper, robust object recognition under occlusion was proposed using RANSAC with the RGB-D camera. Deep Belief Network are used for training and this model was compared with neural network with the SIFT descriptor. Using RANSAC for finding a plane in 3D point cloud captured from depth information, it was possible to recognize objects much faster. In addition, our algorithm could be applied to object recognition under occlusion. Our future work includes the object recognition for the increased number of objects. Although the performance of the proposed method is better than neural network with the SIFT descriptor in training set, the performance of the proposed method in test set is not better. This overfitting problem in DBN has to be solved in the near future.

Acknowledgements. This work was supported by the Technology Innovation Program, 10045252, Development of robot task intelligence technology, funded by the Ministry of Trade, Industry & Energy (MOTIE, Korea).

References

1. Lai, K., et al.: A large-scale hierarchical multi-view rgb-d object dataset. In: 2011 IEEE International Conference on Robotics and Automation (ICRA). IEEE (2011)
2. Huang, A.S., et al.: Visual odometry and mapping for autonomous flight using an RGB-D camera. In: International Symposium on Robotics Research, ISRR (2011)
3. Kim, D.-H., Kim, J.-H.: Image-Based ICP Algorithm for Visual Odometry Using a RGB-D Sensor in a Dynamic Environment. In: Kim, J.-H., Matson, E., Myung, H., Xu, P. (eds.) Robot Intelligence Technology and Applications. AISC, vol. 208, pp. 423–430. Springer, Heidelberg (2013)
4. Henry, P., et al.: RGB-D mapping: Using depth cameras for dense 3D modeling of indoor environments. In: The 12th International Symposium on Experimental Robotics, ISER (2010)
5. Lenz, I., Lee, H., Saxena, A.: Deep learning for detecting robotic grasps. arXiv preprint arXiv:1301.3592 (2013)
6. Helmer, S., Lowe, D.: Using stereo for object recognition. In: 2010 IEEE International Conference on Robotics and Automation (ICRA). IEEE (2010)

654   Y.-H. Yoo and J.-H. Kim

7. Bengio, Y.: Learning deep architectures for AI. Foundations and Trends® in Machine Learning 2(1), 1–127 (2009)
8. Hinton, G.E.: Deep belief networks. Scholarpedia 4(5), 5947 (2009)
9. Lowe, D.G.: Object recognition from local scale-invariant features. In: The Proceedings of the Seventh IEEE International Conference on Computer Vision, vol. 2. IEEE (1999)
10. Bay, H., Tuytelaars, T., Van Gool, L.: SURF: Speeded up robust features. In: Leonardis, A., Bischof, H., Pinz, A. (eds.) ECCV 2006, Part I. LNCS, vol. 3951, pp. 404–417. Springer, Heidelberg (2006)
11. Fischler, M.A., Bolles, R.C.: Random sample consensus: a paradigm for model fitting with applications to image analysis and automated cartography. Communications of the ACM 24(6), 381–395 (1981)
12. Salakhutdinov, R., Hinton, G.E.: Deep boltzmann machines. In: International Conference on Artificial Intelligence and Statistics (2009)
13. Bengio, Y., Delalleau, O.: Justifying and generalizing contrastive divergence. Neural Computation 21(6), 1601–1621 (2009)

Face Verification Across Pose
via Look-Alike Ranked List Comparison

Sojung Yun and Junmo Kim

Dept. of Electrical Engineering, Korea Advanced Institute of Science and Technology
291 Daehak-ro, Yuseong-gu, Daejeon 305-701, Republic of Korea
ysjtweety@kaist.ac.kr, junmo@ee.kaist.ac.kr

Abstract. Face verification is a process to determine whether the two input faces are of the same person or not. Traditional face verification approaches have focused mainly on frontal faces, but for diverse applications, face verification should also work on faces across different pose variation. Therefore, this paper presents a method that is robust to pose changes from -90 to 90 degrees via ranked list of look-alikes. Proposed method works in two steps. First, we measure the similarity between the probe image and all the images in the library and get two ranked lists of look-alikes. Second, we measure the similarity between the two ranked lists, which is considered as the final similarity between the two images. We suggest a way to re-rank the look-alike lists emphasizing the duplicates in the list. Our experimental results on the CMU Multi-PIE database, which is one of the most extensive database in terms of pose variation, show improved performance over the other methods.

Keywords: face verification, ranked list, re-ranking, Jarvis-Patrick clustering, Kendall's tau, Gabor filter, principal component analysis, linear discriminant analysis.

1 Introduction

Face of the same identity can look different when presented in different conditions such as pose, illumination, expression change. In practical applications of face recognition such as surveillance cameras, people do not always keep their faces looking in to the camera with neutral expression. Therefore, the importance of face recognition in unconstrained condition is notable, and it has been studied extensively. There have been some successful methods for face recognition under uncontrolled illumination variations by using a hybrid Fourier-based facial feature extraction and a score fusion scheme [1], and by local tenary patterns(LTP) descriptor [2]. Also, exteneded curvature gabor(ECG) classifier-based approach was proposed for low-resolution face image recognition [3].

This paper will mainly focus on the pose problem. Pose variation in face recognition has always been considered as an important problem, and many promising methods have been proposed such as 3D morphable model(3DMM) [4], generic elastic model(GEM) [5], correspondence latent space discriminant analysis(CLS) [6], and so on. Recently a pose adaptive filter(PAF) [7] was proposed

© Springer International Publishing Switzerland 2015
J.-H. Kim et al. (eds.), *Robot Intelligence Technology and Applications 3*,
Advances in Intelligent Systems and Computing 345, DOI: 10.1007/978-3-319-16841-8_59

using a deformable model for pose estimation. All these works perform fine on moderate pose variations where the two eyes are still visible, but perform very poorly on recognizing faces with extreme pose change(e.g. horizontal rotation of 90°). A recent survey on face recognition across pose [8] pointed out that the protocols for testing face recognition across pose are not even unified. This indicates that we still have a lot to do in developing a pose robust face recognition system.

There is an interesting approach for solving pose variance problem through the Doppelgänger list comparison [9]. Based on the observation of [10] that the variations between the images of the same face due to illumination and viewing direction are almost always larger than image variations due to change in the face identity, [9] suggests a data-driven approach for face recognition. They describe an image by an list of identities from a library covering samples of diverse variation, where the order is determined by the similarity between the image and the individual elements of the library. There are a few papers utilizing ranked list comparison for face recognition. There has been a very similar approach measuring the dissimilarity between two faces using the rank-order distance for clustering a face dataset [11]. Liu et al. [12] proposed a ranked order list fusion method for face recognition after plastic surgery. Hwang et al. [13] used the rank order list for merging multiple classifiers.

In this paper, we adapt this ranked list-based method for pose-robust face verification across horizontal pose change spanning 180°. We use Gabor feature dimension reduced by PCA and LDA to represent each image, and compare the cosine similarity between each probe and the library images. We point out some problems with the previous distance measures for measuring similarity between ranked lists, and suggest a new method to efficiently re-rank the ranked list. The simple framework of our work is illustrated in Fig. 1.

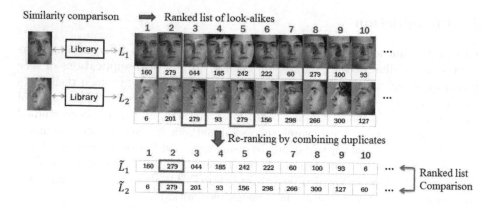

Fig. 1. The pipeline of our method

2   Distances between Rankings

The metrics for information retrieval have been studied extensively, and its main concern is to measure the distance between permutations. The method of measuring distance between permutations (or ranked lists) has a rich and long history. The two most commonly used and classical measures are Spearman's footrule distance and Kendall's tau. Let $\tau_i(j)$ denote the rank of element j in the list L_i. The Spearman's footrule distance, which is given by (1), measures the l_1 distance between ranks, and the Kendall's tau given by (2) measures the number of pairwise inversions. Let $\mathcal{P}$ be the set of unordered pairs of distinct elements. If element j and k are in the same order in τ_1 and τ_2, $\overline{K}_{j,k}(\tau_1, \tau_2) = 0$; and if j and k are in the opposite order in τ_1 and τ_2, $\overline{K}_{j,k}(\tau_1, \tau_2) = 1$. Recently, these methods were extended to accommodate element and position weights [14].

$$F(L_1, L_2) = \sum_j |\tau_1(j) - \tau_2(j)| \tag{1}$$

$$K(L_1, L_2) = \sum_{(j,k)\in\mathcal{P}} \overline{K}_{j,k}(\tau_1, \tau_2) \tag{2}$$

Another classical similarity measure was proposed by Jarvis and Patrick [15]. Jarvis-Patrick clustering is a method based on similarity between neighbors. One or more neighbors in common are used to judge the similarity of the objects under study. In this method, the number of neighbors to examine is used as a parameter. Previous work on face verification based on ranked list [9] emphasizes the significance of the top p people in the ranked list on measuring similarity between the probe and library. Applying Jarvis-Patrick's measure to calculate the similarity between ranked list of look-alikes has been proved to be efficient from the work of [9]. If we only consider the upper p rank in the ranked list including N identities, the measure is given by (3) where $[a]_+ = \max(a,0)$.

$$R(L_1, L_2) = \sum_{j=1}^{N} [p + 1 - \tau_1(j)]_+ \cdot [p + 1 - \tau_2(j)]_+ \tag{3}$$

However, the listed methods do not consider the duplicates in the ranked list but only consider the *first* of each individual subject. In our application of the ranked list, we use the permutation of the identity of each image, and there can be more than one image of a single identity in the library since the library would consist of images with different poses for each subject. Multiple images of the same person with small pose differences may appear in the upper rank of the ranked list. These repeating elements can be meaningful in measuring the similarity between the two permutations. Particularly, in Jarvis-Patrick clustering, which gives much weight to the upper ranked elements, considering only the *first* for each identity in the ranked list may lose a large amount of significant information. Therefore, we suggest a process to re-rank the ranked list combining all repetitions in the ranked list to a single element. For example, in Fig. 1, subject 279 appears twice in the top 10 of the ranked list. With the proposed re-ranking procedure, the rank of subject 279 switches to the second in L_2. For subject k, the summation of $f(rank) = 1/\sqrt{rank}$ for all elements of subject k in the ranked list, which is simply the reciprocal of the square root of the element rank, is the new score for re-ranking. A new ranked list is made by descending order of $S(k)$. The scoring function (4) was made for the purpose of emphasizing the upper rank elements while giving less attention to the remaining elements. Fig. 2 is the plot of $f(rank)$ when the total length of the list N is 100. In Fig. 2, we can observe that $1/rank$ emphasizes only approximately the upper 5% rank of the list and almost ignores the remaining ones. On the other hand, $f(rank) = 1/\sqrt{rank}$ emphasizes approximately the upper 10% rank of the list, and considers the remaining rank with less weight.

$$S(k) = \sum_{\forall i \in \{i \mid ID(i)=k\}} \frac{1}{\sqrt{rank(i)}}, \qquad i = 1, \ldots, N \qquad (4)$$

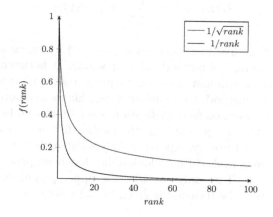

Fig. 2. Weighting function for computing the new score for re-ranking

3   Face Verification via Look-Alike Lists

3.1   Finding Look-Alikes from the Library

The framework of the proposed method is shown in Fig. 1. The first procedure is to make a ranked list of look-alikes. Every image in the face library and the two input probe images are aligned with eye coordinates and are normalized to 60 x 80 pixels. Gabor features [16] are extracted at five scales and eight orientations at each face image pixel. The extracted Gabor features are projected into a low-dimensional subspace by PCA [17] and fisher LDA [18]. Given a pair of probe images, we find the look-alikes of each probe in the library by calculating cosine similarity between the extracted low-dimensional features. The comparison results are the two ranked list of look-alikes, where the first is the most similar to the probe. We denote the look-alike lists for probe 1 and probe 2 as L_1 and L_2.

3.2   Comparing Look-Alike Lists

The look-alike list is a permutation of the identity labels in the library set. The library consist of n distinct identities and can have several images for each identity. If there are k images for each identity equally, the length of the look-alike list will be $N = k \times n$. We can compare the similarity between L_1 and L_2, which is eventually regarded as the similarity between probe 1 and probe 2. For better performance, we apply the re-ranking method (4) resulting in $\widetilde{L_1}$ and $\widetilde{L_2}$. The methods for measuring the distance between ranked lists are described in section 2. Our experimental results shows that using Jarvis-Patrick clustering (3) after re-ranking by proposed method gives the best result. The threshold p $(10 < p < n)$ in (3) is decided adaptively.

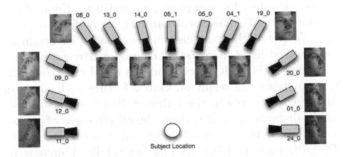

Fig. 3. Multi-PIE database

4   Experimental Results

Fig. 3 shows the Multi-PIE database [19] used for the experiment. Multi-PIE is the most extensive face database in terms of pose variation containing images of 337 people recorded in up to four sessions over the span of five months. Thirteen cameras were located at head height, spaced in 15° intervals spanning 180° horizontally, which makes 13 pose differences. Although Multi-PIE also provides 18 illumination and 6 expression changes, we fixed the illumination and expression for our experiment to focus on solving the pose variation problem. In Multi-PIE, we used 249 subjects as the library and the training data for PCA and LDA, and the remaining 88 subjects as the probe. For each subject, we used 13 images spanning each pose change, which makes 1144 test images and 3237 training images (size of the library). The probes were randomly selected as 3000 positive and negative pairs each among the test images, and each pair always consisted of the two images with pose difference.

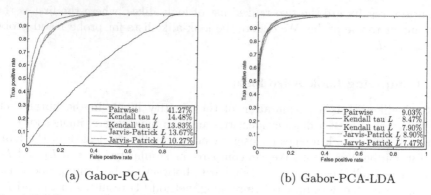

(a) Gabor-PCA   (b) Gabor-PCA-LDA

Fig. 4. ROC curves obtained by using pairwise distance and the two similarity measures between rankings for face verification across all poses. L denotes the original ranked list obtained by similarity comparison with the probe and the library. $\widetilde{L}$ denotes the new ranked list after re-ranking by the proposed method.

We tested two measures for comparing ranked lists, Kendall's tau (2) and Jarvik-Patrik's measure (3). The Spearman footrule perform comparable to the Kendall's tau. We tested the two measures on both the original ranked list L and the new ranked list $\widetilde{L}$ as a result of proposed re-ranking method. The result in Fig. 4 shows the ROC curve and the equal error rate(EER) of each method. Fig. 4(a) shows the result on Gabor feature with dimension reduced by PCA. The large gap between the pairwise distance measure and the other ranked list based distance measures show the effectiveness of ranked list based method. Fig. 4(b) shows the result on Gabor feature with dimension reduced by PCA and LDA. Although the LDA itself has a good discriminative performance, improved peformance using $\widetilde{L}$ on both (2) and (3) verifies the effectiveness of the proposed re-ranking method. The length of L is 3237, which is the total size of the library, and the length of $\widetilde{L}$ is 249, which is the number of subjects in the library. Threshold of (3) on L was set as $p = 100$, and threshold on $\widetilde{L}$ was set as $p = 30$ on the result of Gabor-PCA, and $p = 74$ on the result of Gabor-PCA-LDA, which was the threshold showing the best result.

5   Conclusion

In this paper, we proposed a method using look-alike ranked list comparison based on Gabor feature with dimension reduced by PCA and LDA for pose robust face verification. Since the face recognition method utilizing ranked list comparison has been first proposed, a few follow-up research has been done. We showed potential of this ranked list-based method as a useful solution for the challenging pose change problem in face recognition by suggesting a novel re-ranking method. The state-of-the-art face recognition performance on the database *in the wild* with small pose variation has shown to be comparable to the human peformance. We expect that combining these state-of-the-art methods with a good ranked list comparison method can promote the improvement in pose-robust face recognition performance comparable to the human performance.

Acknowledgments. This research was supported by the MOTIE (The Ministry of Trade, Industry and Energy), Korea, under the Technology Innovation Program supervised by KEIT (Korea Evaluation Institute of Industrial Technology),10045252, Development of robot task intelligence technology. This work was also supported by the ICT, broadcasting R&D program, 2014-044-025-003, Technology Development of Virtual Creatures with Digital Emotional DNA of Users, funded by MSIP(Ministry of Science, ICT and Future Planning), Korea.

References

1. Hwang, W., Wang, H., Kim, H., Kee, S.C., Kim, J.: Face recognition system using multiple face model of hybrid fourier feature under uncontrolled illumination variation. Trans. Img. Proc. 20(4), 1152–1165 (2011)
2. Tan, X., Triggs, B.: Enhanced local texture feature sets for face recognition under difficult lighting conditions. Trans. Img. Proc. 19(6), 1635–1650 (2010)
3. Hwang, W., Huang, X., Noh, K., Kim, J.: Face recognition system using extended curvature gabor classifier bunch for low-resolution face image. In: 2011 IEEE Computer Society Conference on Computer Vision and Pattern Recognition Workshops (CVPRW), pp. 15–22 (2011)
4. Blanz, V., Vetter, T.: Face recognition based on fitting a 3d morphable model. IEEE Trans. Pattern Anal. Mach. Intell. 25(9), 1063–1074 (2003)
5. Prabhu, U., Heo, J., Savvides, M.: Unconstrained pose-invariant face recognition using 3d generic elastic models. IEEE Trans. Pattern Anal. Mach. Intell. 33(10), 1952–1961 (2011)
6. Sharma, A., Haj, M.A., Choi, J., Davis, L.S., Jacobs, D.W.: Robust pose invariant face recognition using coupled latent space discriminant analysis. Comput. Vis. Image Underst. 116(11), 1095–1110 (2012)
7. Yi, D., Lei, Z., Li, S.Z.: Towards pose robust face recognition. In: Proceedings of the 2013 IEEE Conference on Computer Vision and Pattern Recognition, CVPR 2013, pp. 3539–3545. IEEE Computer Society, Washington, DC (2013)
8. Zhang, X., Gao, Y.: Face recognition across pose: A review. Pattern Recogn. 42(11), 2876–2896 (2009)
9. Schroff, F., Treibitz, T., Kriegman, D., Belongie, S.: Pose, illumination and expression invariant pairwise face-similarity measure via doppelgänger list comparison. In: Proceedings of the 2011 International Conference on Computer Vision, ICCV 2011, pp. 2494–2501. IEEE Computer Society, Washington, DC (2011)
10. Adini, Y., Moses, Y., Ullman, S.: Face recognition: The problem of compensating for changes in illumination direction. IEEE Trans. Pattern Anal. Mach. Intell. 19(7), 721–732 (1997)
11. Zhu, C., Wen, F., Sun, J.: A rank-order distance based clustering algorithm for face tagging. In: Proceedings of the 2011 IEEE Conference on Computer Vision and Pattern Recognition, CVPR 2011, pp. 481–488. IEEE Computer Society, Washington, DC (2011)
12. Liu, X., Shan, S., Chen, X.: Face recognition after plastic surgery: A comprehensive study. In: Lee, K.M., Matsushita, Y., Rehg, J.M., Hu, Z. (eds.) ACCV 2012, Part II. LNCS, vol. 7725, pp. 565–576. Springer, Heidelberg (2013)

13. Hwang, W., Roh, K., Kim, J.: Markov network-based unified classifier for face identification. In: Proceedings of the 2013 IEEE International Conference on Computer Vision, ICCV 2013, pp. 1952–1959. IEEE Computer Society, Washington, DC (2013)

14. Kumar, R., Vassilvitskii, S.: Generalized distances between rankings. In: Proceedings of the 19th International Conference on World Wide Web, WWW 2010, pp. 571–580. ACM, New York (2010)

15. Jarvis, R.A., Patrick, E.A.: Clustering using a similarity measure based on shared near neighbors. IEEE Trans. Comput. 22(11), 1025–1034 (1973)

16. Liu, C., Wechsler, H.: Gabor feature based classification using the enhanced fisher linear discriminant model for face recognition. Trans. Img. Proc. 11(4), 467–476 (2002)

17. Turk, M., Pentland, A.: Eigenfaces for recognition. J. Cognitive Neuroscience 3(1), 71–86 (1991)

18. Belhumeur, P.N., Hespanha, J.A.P., Kriegman, D.J.: Eigenfaces vs. fisherfaces: Recognition using class specific linear projection. IEEE Trans. Pattern Anal. Mach. Intell. 19(7), 711–720 (1997)

19. Gross, R., Matthews, I., Cohn, J., Kanade, T., Baker, S.: Multipie. Image Vision Comput. 28(5), 807–813 (2010)

Landmark Tracking Using Unrectified Omniirectional Image for an Automated Guided Vehicle

Zahari Taha and Jessnor Arif Mat-Jizat

Innovative Manufacturing, Mechatronics, and Sports Laboratory,
Faculty of Manufacturing Engineering, Universiti Malaysia Pahang, 26600,
Pekan, Pahang, Malaysia
zaharitaha@ump.edu.my, jessnor@imamslab.com,

Abstract. In this paper, a study on landmark tracking using unrectified omnidirectional image for automated guided vehicle is presented. Omnidirectional image from a catadioptric camera may appear distorted against the height of an object. However, for a flat object on the floor, the distortion is negligible thus can be advantageous for on-the-ground landmark; Landmark used in this study was Code-128 standard barcode. The barcode is modified to suit the detection process where the barcode adopted cyan instead of white background and bears a red strip on top for orientation. The image processing can directly begin tracking landmarks when no distortion rectification in the image was required. We adopted a topological map approach where the automated guided vehicle moves from landmark to landmark. Experiments were conducted on a small four wheel drive, four wheel steering automated guided vehicle. The results were measured through number of successful consequent tracking of the landmark.

1   Introduction

Automated Guided Vehicle (AGV) is an autonomous transport system typically used in manufacturing plants to transport material from one station to another. Traditionally, AGV is made to travel along guided paths such as wire in the ground or dedicated fixed line on the floor in the manufacturing plants. This type of AGVs is called fixed path AGVs In recent years, many researches concentrate on technologies allowing the AGVs to operate without a physical path referred to as free ranging AGVs. Free ranging AGVs must be able to perform, position and orientation estimation, and goal seeking based only on their onboard sensor and perception systems.

The adoption of camera on to a mobile robot for visual navigation was started by Moravec with "Stanford Cart". [1]The cart was used to drive through cluttered environment autonomously using images broadcasted by its onboard camera system. The cart moved from initial position to a specified final position while avoiding any object it deemed as obstacle. The path planning was solely based on based on the images it acquired. The Stanford cart then becomes pioneer on autonomous guided

© Springer International Publishing Switzerland 2015
J.-H. Kim et al. (eds.), *Robot Intelligence Technology and Applications 3,*
Advances in Intelligent Systems and Computing 345, DOI: 10.1007/978-3-319-16841-8_60

vehicle research. Since then the research on the vision autonomous vehicle continues gathering interest even in the recent time.

Recently, the usage of omnidirectional camera onto mobile robot had been studied. Omnidirectional camera offers an extended field of view of 360° along the horizontal line with one camera rather than monocular camera. This extended field of view may obtain much more information from it images such as bearing information, AGV localization, or artificial landmarks recognition [2]. The omnidirectional camera wide field of view allow the vision system to recognize landmarks more faster as no tilting, panning or AGV reorientation needed [3]. Although the image range may be shorter than perspective camera, but omnidirectional camera can captures image range that is more relevant to robotics navigation.

For researches in robotics navigation using landmarks, the term natural landmark or artificial landmark is usually discussed. The term natural landmarks often refers to certain features in the detected environment without placing it purposely in the scene. These features may be door frames, window frames, ceiling lamps or air conditioner vent [4-6]. After detecting and recognizing the features, the AGV main computer then runs an algorithm to localize itself and then update the AGV position and orientation. This type of landmarks navigation may be most useful for map based navigation. On the other hands artificial landmarks may be introduced to the environment in order to assist navigation. The artificial landmarks usually consist of patterns design specifically to assist detection and recognition. The design may include encoded symbols, readable by landmarks detections and recognition algorithms [6-8].

2   Modelling of the Automated Guided Vehicle

2.1   Automated Guided Vehicle Model

In this paper, a bicycle model is used to model the AGV shown in figure 1 where v is the AGV's velocity with reference to its centre of gravity, (COG) and δ is the angle

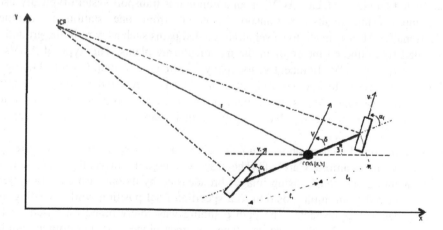

Fig. 1. Bicycle model of the AGV

from the AGV longitudinal axis to vehicle's velocity. The angle δ is known as side-slip angle. Side-slip angle allows a vehicle translational move to its either side without having to rotate. Meanwhile, β is the heading angle of the AGV. It is measured from the X-axis to the vehicle longitudinal axis. The velocity, side-slip angle, and heading angle are derivative from the motors driving the wheels at each axle. Since the motors run independently from each other, the front wheel velocity, v_f, and the rear wheel velocity, v_r, have to be defined. the steering angle for both, the front wheel, α_f, and the rear wheel, α_r, are also defined at same location where the angle is the rotational deviation of the wheels from the AGV's longitudinal axis. l_f and l_r then is defined as the length between the centre of gravity of the AGV to the front and rear axle respectively.

As the AGV moves in the global coordinates system, its movement can be calculated by using rate of change horizontal distance ($\dot{X}$) or the horizontal velocity, rate of change vertical distance ($\dot{Y}$) or the vertical velocity and the rate of change heading angle ($\dot{\beta}$) or the angular velocity thus bring us the following equations:

$$\dot{X} = v \cos(\beta + \delta) \tag{1}$$

$$\dot{Y} = v \sin(\beta + \delta) \tag{2}$$

$$\dot{\beta} = \frac{v \cos \delta \, (\tan \alpha_f - \tan \alpha_r)}{l_f + l_r} \tag{3}$$

where

$$\delta = \arctan \frac{l_r \tan \alpha_f + l_f \tan \alpha_r}{l_f + l_r} \tag{4}$$

$$v^2 = \frac{v_f{}^2 + v_r{}^2 + 2 v_f v_r \cos(\alpha_f - \alpha_r)}{4} \tag{5}$$

2.2 Position and Orientation

In the global coordinates, the AGV whereabouts can be tracked by locating the position with respect to the origin [x, y] and the orientation of the AGV from the X-axis to its longitudinal axis [β] as shown in figure 2. The coordinate $[X1,1,\beta1]$ is the current pose of the AGV and the coordinate $[X2,Y2,\beta2]$ is the desired pose of the AGV. The difference between current and final poses of the AGV can also be expressed in polar coordinates where ρ is the shortest distance between the AGV and the landmark and τ is the angle between the landmark, AGV and the x-coordinate.

$$\rho = \sqrt{(X_2 - X_1)^2 + (Y_2 - Y_1)^2} \tag{6}$$

$$\tau = \arctan \left(\frac{Y_2 - Y_1}{X_2 - X_1} \right) \tag{7}$$

As the AGV moves, the camera also moves as the camera is installed on board the AGV. Therefore, the camera have to locate and detect the nearest landmark relative to its current position and must determine the relative position and orientation of the AGV to the global coordinates. Since the landmark's pose is known in the global position, the landmark is a key component in the frame transformation from the global coordinate system to the camera frame coordinate system. The orientation of the landmark is given by the perpendicular axis of the landmarks red strip in terms of the global orientation β.

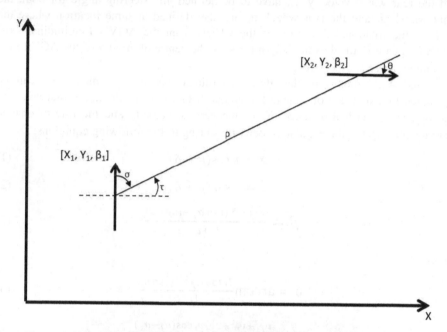

Fig. 2. Position and orientation of the AGV and the landmark in global coordinates

Fig. 3. On-board omnidirectional camera view

The camera coordinate system is fixed to the AGV's local coordinate system where as the AGV moves the camera local coordinate system remain the same as the AGV local coordinate system. The camera system can measure the relative distance, ρ_c, from AGV to the landmark and also the relative angle, τ_c, from the AGV to the landmark. The ρ_c and the τ_c are the measurement of ρ and τ in camera local coordinate system shown in figure 3.

Nevertheless, in order to transform the AGV local orientation to the known landmark's global orientation and to make use of equation 1 to 7, an angle between the x-axis of the AGV local coordinate and the red strip of the landmark, γ, is measured. The relationship between the global coordinates system and the AGV local system is given by the equation 8 and 9 where β_1 is the orientation of the AGV in global coordinates system and β_2 is the orientation of the landmark in global coordinates system.

$$\beta_1 = \beta_2 - \gamma \; if \; \beta_2 - \beta_1 > 0 \tag{8}$$

$$\beta_1 = 180 - \gamma + \beta_2 \; if \; \beta_2 - \beta_1 < 0 \tag{9}$$

In order to calculate τ from τ_c, a transformation angle (Ta) between the local coordinates system and the global coordinates system must be determined. The AGV will only rotate about the z-axis in relation to the global coordinates system and the x-axis and the y-axis of the global coordinate system, as well as the x-axis and the y-axis of the image remain at 90° to each other throughout the movement. The transformation angle is given by equation 10. Thus from the calculated transformation angle, τ can be calculated from equation 11.

$$T_a = 90 - \beta_1 \tag{10}$$

$$\tau = \tau_c - T_a \tag{11}$$

3   Experimental Setup

3.1   Selecting Landmark

The landmark chosen for this study is enlarged Code-128 standard barcode. It can be changed to any Code-128 barcode standard that suits the operation of the company as the standard barcode is widely used as machines tag for asset inventory. As the barcode is unique to the particular machine where it was tagged, the AGV can localize itself about the machine and perform the same task on the machine station even if the machine were moved due to layout changes. However, a little modification is required to the standard barcode to make it more salient to the background. The barcode used in this study adopting cyan instead of white background and bears a red strip on top for orientation as shown in figure 4. These landmarks were placed in a specific known location in global coordinate perspective and thus creating interest point for the AGV to move toward.

Fig. 4. Barcode used in this study

3.2   The Camera

The camera used in this study is catadioptric camera where the object is reflected on a smooth circular conical shape mirror into a perspective camera. This reflected image is highly distorted against the height of the object due to the surface contour of the mirror. However, a flat-on-the-floor object experience negligible distortion. The distortion is uniform throughout the tall object image and thus allowing the barcode to be read even if the barcode is placed vertically as shown in figure 5.

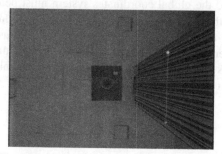

Fig. 5. Distortion of vertically placed barcode

3.3   The Experiments

The experiments are conducted indoor with two cameras. The main camera is the omnidirectional camera on-board the AGV. Image processed from this camera governs the movement of the AGV. However, this camera can only track the position of the AGV in relation to the local coordinates system. Thus, a second camera is needed to track the AGV's position in global coordinates system. The second camera was a perspective camera placed on the ceiling about 3500mm height from the floor. The figure 6 below shows a picture of the experimental setup.

Three experiments were conducted where the landmark were placed, first in front of the AGV, second on the right of the AGV, and third on the left of the AGV. The AGV then approaches the landmark and align itself to the landmark. The trajectory of the AGV was simulated using mathematical computer software prior to the actual run. The result from simulation and the actual run is compared.

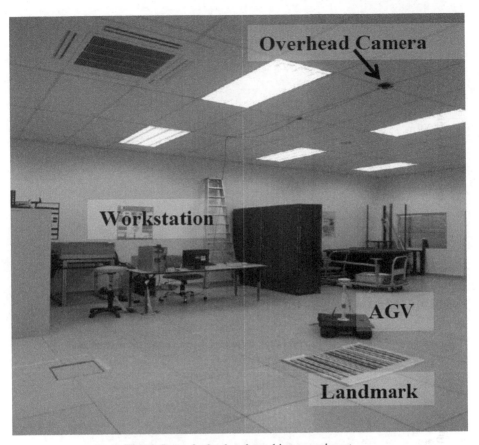

Fig. 6. Setup for landmark tracking experiment

4   Result and Discussion

In the first trajectory experiment, the starting pose of the AGV was (76, 315, 0) and the landmark was positioned at (500, 350, 0) in pixel coordinates with reference to the global coordinates system. The landmark used in this experiment was an A1 size paper, which is mapped to 211 x 149 pixels area. Figure 7 shows the simulated path and the actual path of the AGV.

The actual trajectory shows a deviation from the simulated trajectory. The final pose of the AGV in the experiment was (500, 311, -1) while the final pose of in simulation was (500, 350, 0). The AGV had deviated 39 pixels which is an actual distance of 15.6 cm from the simulated final position. A comparison between the experimented data and the simulation data showed a variance of 650.2 and the standard deviation of 25.5 pixels. This standard deviation value is 5.3% deviation from the whole tracked vertical position which consists of 480 pixels. The deviation was also less than quarter of the AGV's width of 118pixels (47cm). The orientation of the final position only deviate 1° from the centre of the landmark.

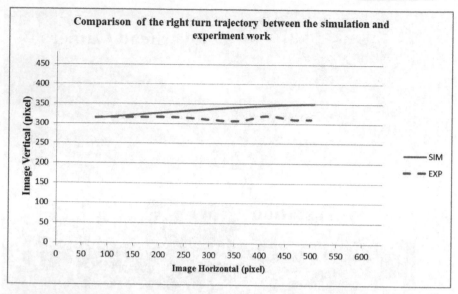

Fig. 7. Comparison of straight trajectory between the simulation and experimental work

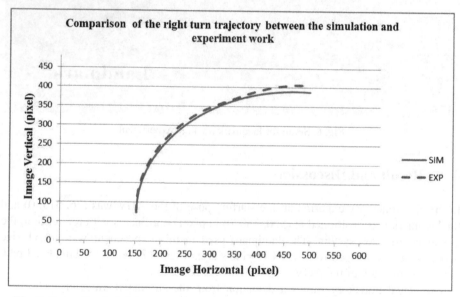

Fig. 8. Comparison of right-turn trajectory between the simulation and experimental work

When the landmark was placed to the right of the AGV, the start position of the AGV was placed at the (153, 74, 91) pixel coordinate and position of the centre of the landmark was placed at (500, 350, 0) pixel coordinate. Figure 8 shows the simulated and actual path of the AGV. The simulation and the experiment works showed a similar progression as the AGV approach the landmarks. However, at 500 horizontal pixel points, the AGV stopped at 405 vertical pixels, while the simulation stopped at 388 vertical pixels.

The deviation between the simulation and the experimental result was 17 pixels, which represent a 6.8 cm in the measurement. However, the progression showed a variance of 115 and the standard deviation of 10.72 pixels when the actual and simulated data is compared. This standard deviation value represented 2.2% of deviation from full 480 vertical image pixels. The absolute data also showed tendency of larger deviation at the end of tracked position compared to the initial position. On the other hand, the orientation of the final position also must be considered. The final orientation of simulated result was -5.8° while the final position of the experiment was 0°.

When the landmark was placed to the left of the AGV, the AGV was placed at the starting pose (610, 74, 86) and the landmark was positioned at (130, 350, 180). Figure 9 shows the comparison between the simulated and the experimental work. The simulation and the experiment works showed a similar progression as the AGV approach the landmarks. At 130 horizontal pixel points, the simulation showed the vertical pixels at 364 while the experimental work showed that the AGV stopped at 372 vertical pixels.

The progression showed a variance of 332.75 and the standard deviation of 18.72 pixels when the actual and simulated data is compared. This standard deviation value represented 3.8% of deviation from full 480 vertical image pixels. However, the errors value seems to be consistent at all phases of the experiment; initial middle or final phase. The final orientation of simulation work was 211° while the final position of the experiment was 203°.

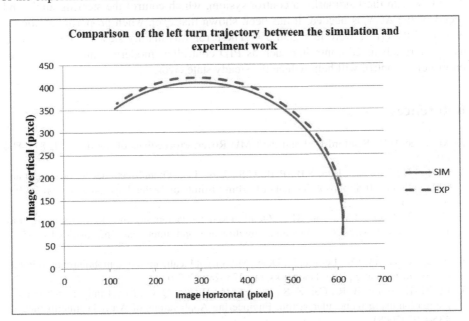

Fig. 9. Comparison of the left turn trajectory between the simulation and experiment

When all data from the three trajectory were compared, it was evident that a steady state error exist. This error may have been resulted from the proportional control used to control steering angle and the speed. Proportional control feedback did not take into account of the previous error nor the mean error over time in order to control the parameters. A classical control of proportional-integral-derivative control may significantly reduce the steady state error caused by proportional control.

However, implementation of intelligent control to the system such as fuzzy controller may reduce the steady state error further. Chul-Goo Kang and Hee-Sung Kwak had demonstrated the elimination of the steady state position errors in a robotic manipulator by using fuzzy control algorithm [9]. Their work successfully reduces steady state error up to 90% from the two input; position error and velocity error, fuzzy controller. In this research, only proportional control is used because the primary concern was to generate the trajectory for the AGV to align and position itself with a landmark.

5   Conclusion

This work had showed that a landmark can be tracked successfully using unrectified image from omnidirectional camera and generate the trajectory for the AGV to align and position itself with the landmark. This enables more work to be concentrated on recognizing the landmark instead of rectifying image from camera. In order to align the AGV with the landmarks, a control system, which control the steering angle and speed of the AGV is needed. It has been shown that even when proportional control was used the errors were less than 5%. The error was mainly steady state error thus using more advanced controller such as PID controller, modern control principal or intelligence control will help reduce the steady state error.

References

1. Moravec, H.P.: Stanford Cart and the CMU Rover. Proceedings of the IEEE 71, 872–884 (1983)
2. Lee, D.H., Sun, J.S., Han, S.B., Park, C.S., Kim, J.H.: Omnidirectional robot system and localization for fira robosot. Journal of Harbin Institute of Technology (New Series) 15, 96–99 (2008)
3. Taha, Z., Chew, J.Y., Yap, H.J.: Omnidirectional vision for mobile robot navigation. Journal of Advanced Computational Intelligence and Intelligent Informatics 14, 55–62 (2010)
4. Ayala, V., Hayet, J.B., Lerasle, F., Devy, M.: Visual localization of a mobile robot in indoor environments using planar landmarks, pp. 275–280 (2000)
5. Li, M.-H., Hong, B.-R., Cai, Z.-S., Piao, S.-H., Huang, Q.-C.: Novel indoor mobile robot navigation using monocular vision. Engineering Applications of Artificial Intelligence 21, 485–497 (2008)
6. Zhang, H., Zhang, L., Dai, J.: Landmark-based localization for indoor mobile robots with stereo vision, pp. 700–702 (2012)

7. Wu, C.-J., Tsai, W.-H.: Location estimation for indoor autonomous vehicle navigation by omni-directional vision using circular landmarks on ceilings. Robotics and Autonomous Systems 57, 546–555 (2009)
8. Briggs, A.J., Scharstein, D., Braziunas, D., Dima, C., Wall, P.: Mobile robot navigation using self-similar landmarks, pp. 1428–1434 (2000)
9. Chul-Goo, K., Hee-Sung, K.: A fuzzy control algorithm reducing steady-state position errors of robotic manipulators. In: Proceedings of the 1996 4th International Workshop on Advanced Motion Control, AMC 1996-MIE, vol. 1, pp. 121–126 (1996)

Human Pose Estimation Algorithm
for Low-Cost Computing Platform
Using Depth Information Only

Hanguen Kim[1], Sangwon Lee[1], Youngjae Kim[1], Dongsung Lee[2],
Jinsun Ju[2], and Hyun Myung[1]

[1] Urban Robotics Laboratory (URL), Korea Advanced Institute of Science and
Technology (KAIST), 291 Daehak-ro (373-1 Guseong-dong), Yuseong-gu,
Daejeon 305-701, Republic of Korea
[2] Image & Video Research Group, Samsung S1 Cooperation, 168 S1 Building,
Soonhwa-dong, Joong-gu, Seoul, Republic of Korea

Abstract. In this paper, we present human pose estimation algorithm
that use depth information only. To estimate human poses on a low
cost computing platform, we propose a human pose estimation algorithm
that mixes a geodesic graph and a support vector machine (SVM). The
proposed algorithm can work for any human without calibration and thus
anyone can use the system immediately. The SVM-based human pose
estimator uses randomly selected human features to reduce computation.
The human pose estimation is evaluated through several experiments and
the results showed that our approaches perform fairly well.

1 Introduction

In recent years, human pose estimation that enable more natural communication methods for human beings have become an important topic in the areas
of computer vision and multimedia [1]. Using human pose for interaction with
computer-assisted systems can provide many important benefits [2]. Also recent
advances in 3D depth sensors such as Microsoft Kinect have created many opportunities for games, security, surveillance, and entertainment [3]. Kinect is a
low cost RGB-D (Red, Green, Blue, and Depth information, respectively) sensor
based on structured light technology but is limited to indoor use [4]. The structured light sensor infers the depth at any image location by projecting a known
infrared light pattern on to a scene and evaluating the distortion of the projected
pattern. Kinect combines a structured light sensor with a regular RGB sensor
that can be calibrated to the same image frame. Despite their favorable properties, depth sensors provide data that contains noise. Estimating human full-body
poses from such noisy data remains a challenging problem [5]. In spite of these
weaknesses, there remains strong interest in improving the current methods of
human pose estimation [6], [7], [8].

In the case of human pose estimation algorithms, traditional methods can be
classified into model-based and model-free algorithms, depending upon whether

© Springer International Publishing Switzerland 2015   675
J.-H. Kim et al. (eds.), *Robot Intelligence Technology and Applications 3*,
Advances in Intelligent Systems and Computing 345, DOI: 10.1007/978-3-319-16841-8_61

a priori information about the object shape is employed [9] or not. Following the work reported in [10], which proposed a human pose estimation algorithm on a GPU (Graphic Processing Unit), several studies have extended their work or focused on the use of parallel processing efficiently. Their results represent the best performance to date, but it is currently difficult to run their algorithms on a mobile platform.

To address these problems, this paper presents human pose estimation algorithm. The aim of the algorithm is operation on a low cost platform without calibration for any human so that anyone can use the system immediately. To this end, we propose a method that mixes a geodesic graph and an SVM-based human pose estimator from depth information only. The proposed method uses a small amount of randomized human feature points on a geodesic graph to estimate body parts. Body parts that typically involve a lot of movement are then estimated by the value of the geodesic distance.

The remaining sections are organized as follows. In Sections 2, the proposed human pose estimation algorithm is described in detail. In Section 3, experimental evaluations are provided. Finally, we present concluding remarks in Section 4.

2   Human Pose Estimation Algorithm

Our human pose estimation algorithm utilizes geodesic distance and an SVM, and works on only depth images without RGB information. For comparison with Open NI, we divide the human body into 15 parts: head, neck, torso, L/R (left/right) shoulders, L/R elbows, L/R hands, L/R hips, L/R knees, and L/R feet. The proposed algorithm locates these 15 body parts in depth images.

A flow diagram of the proposed human pose estimation algorithm is shown in Fig. 1. In pre-operation, background subtraction is performed on the input depth image, and the human body ROIs (region of interests) are extracted ensuring that each extracted ROI contains a human body without occlusion. For a human body in each ROI, a geodesic distance iso graph is generated, and it is used for feature points generation. A definition of the geodesic distance iso graph is given in Section 2.1. Instead of training and classifying all pixels on the human body, a small number of feature points are randomly generated on the geodesic distance iso graph. For coordinate consistency of body parts, the origins of the feature point coordinates are then translated to the central moment of a human body, and the x, y, z (depth) coordinates of feature points are normalized to fit into a pre-defined human body size. The processed feature points are classified into one of body parts by the nonlinear multiclass SVM classifier using x, y and depth information of the feature points.

2.1   Geodesic Distance Iso Graph

Geodesic distance between two points in a graph is the sum of edge costs in the shortest path connecting the two points. When a human body ROI is extracted, a graph for the human body is generated from the human body ROI image.

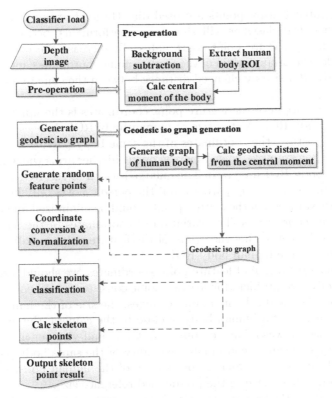

Fig. 1. Human pose estimation flow diagram

For each node in the human body ROI graph, geodesic distance from central moment of the human body ROI is calculated using Dijkstra's algorithm. The cost in the Dijkstra's algorithm is computed using x, y, z - position and depth - information. After calculating the geodesic distances, a geodesic iso graph is generated by uniformly dividing the interval between the minimum and maximum geodesic values into multiple levels. As the geodesic distance value of the nodes corresponds to the distance from the central moment of the human body, the highest geodesic distance iso levels are shown at the head, hands and feet points. Using these characteristics of geodesic distance iso graph, head, hands and feet can be found easily.

2.2   Human Feature Generation

From the geodesic iso graph, we generate feature points for training and prediction of a classifier. The feature points are generated randomly on the nodes of geodesic iso graph except the zero-level nodes. As the geodesic iso level of the nodes around the central moment of the human body ROI is zero, the feature points are not generated near the central moment. As a result, the feature points are evenly distributed on the human body, and are not concentrated on the torso part that has largest area.

If the generated feature points are used directly for learning and prediction of the classifier, the classifier will show poor performance because of lack of consistency among feature point coordinates. To increase the consistency, we should consider the proper origin of the feature point coordinates and the change of human body ROI size by changed distance between the camera and a human body.

Originally the origin of the feature point coordinates is the upper left vertex of the human body ROI. As the size of human body ROI changes with respect to the human posture, the coordinates of the same body part vary significantly with postures. For example, when a person opens his/her arms widely, the width of the human body ROI increases. The enlarged width of the ROI alters the y-coordinate values of the body parts even if the person moves only his/her arms. The coordinate variation of the feature point should be minimized for consistency of feature point coordinates. The variation of coordinates of the feature point can be reduced by converting the origin of the feature point coordinates to the central moment of the human body ROI.

In addition to the origin of feature point coordinates, we should also consider the human body size problem for feature point consistency. When the distance between the camera and a human body changes, the size of the human body in images also changes. If a human body is close to the camera, the human body size and distance between feature points are large, and vice versa. The change of human body size causes coordinate inequality of the same body part for the same poses. To ensure coordinates consistency of the human body, we need to adjust the human body size to the pre-defined reference size.

In order to adjust the human body size, we use depth information. It is obvious that the human body size in images is proportional to distance (depth) with negative slope. From this fact, we can define the following equation:

$$D = k_a \times W_s + k_b, \tag{1}$$

where D, W_s, k_a, k_b represent an average depth of the human body, a shoulder width, and proportional coefficients, respectively. If the reference shoulder width is denoted by W_{ref}, a scale factor S_D for a certain average depth of the human body can be calculated as follows:

$$S_D = \frac{W_{ref}}{W_s} = \frac{W_{ref} \times k_a}{D - k_b}. \tag{2}$$

The feature point coordinates are multiplied by the scale factor given by Equation (2) to normalize the human body size to the pre-defined size. We conducted an experiment to obtain the coefficients in Equation (2). The details are given in Section 3.

2.3   Feature Point Classification

For classification of feature points, an SVM is employed [11], [12]. An SVM is a supervised machine learning algorithm that is used for classification, regression

Fig. 2. Examples of training data

and other learning tasks. The SVM constructs optimal hyperplanes in high-dimensional space in the training process. Optimal hyperplanes are the those that have the largest margin between the hyperplanes and the nearest training data point which is called support vector. The SVM was originally designed as a binary linear classifier, but it can be modified as a nonlinear multiclass classifier. For multiclass classification, the commonly used methods include one-against-all, one-against-one, and directed acyclic graph SVM (DAGSVM) [13]. In this paper, a nonlinear multiclass SVM is employed for body parts classification of feature points. By applying kernel trick, the SVM can be transformed to a nonlinear classifier [14]. The idea of kernel trick is that by applying the nonlinear mapping through the kernel function, the linearly non-separable data can be separated by a linear SVM. We choose the one-against-one method as a multiclass classification method. This method constructs $n \times (n - 1)/2$ classifiers, where n is the number of classes. Each constructed classifier is trained for two classes with training data. In the classification, class of an input data is determined by votes from classifiers.

To train the SVM classifier, we generate a training data set using joints information obtained from OpenNI. The joints information from OpenNI includes locations of 15 joints: head, neck, torso, L/R (left/right) shoulders, L/R elbows, L/R hands, L/R hips, L/R knees, and L/R feet. The feature points generated randomly from the geodesic distance iso graph are labeled as one of 15 joints through the nearest neighbor algorithm. For each feature point, the distance from all joint points is calculated and the joint with smallest distance is chosen as its label. Fig. 2 shows some examples of the generated training data. The generated ground truth data files is saved to a file containing x, y, z coordinates of feature points and body part label information. The origin of x, y coordinates in ground truth data is converted to the central moment of human body, and normalized when generated.

3   Experiment

In this section, the proposed human pose estimation algorithm is evaluated. We conducted the experiment on a PC with Intel i5 3.0 GHz quad-core CPU. A depth camera ASUS Xtion pro is used and installed at about 2 m height. The input depth image size from the depth camera is originally 640×480, but it is reduced to half size 320×240 to reduce processing time. We use radial basis

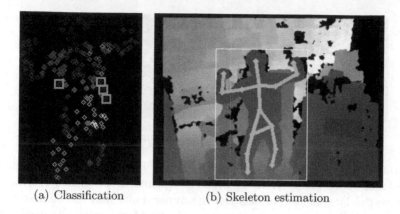

| (a) Classification | (b) Skeleton estimation |

Fig. 3. An example of feature points classification (a) and pose estimation (b)

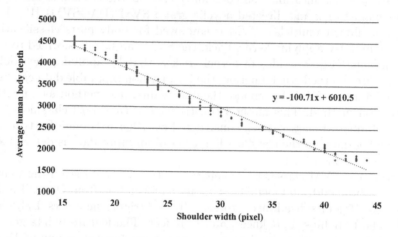

Fig. 4. Regression result of shoulder width and depth

kernel-based C-SVM for classification [11], [12], [15]. The gamma value of the kernel function is about 0.03, and the C value of C-SVM is about 0.5.

We investigate the performance of the proposed pose estimation algorithm using OpenNI human joints information as ground truth data, and the error metric is calculated for 1,000 frames of various poses.

Scale Factor for Normalization. We conducted an experiment to determine the relationship between average depth of the human body and shoulder width. Fig. 4 shows the experiment results. The linear regression from the experiment indicates that shoulder width is proportional to depth with negative slope as we assumed in Section 2.2. Equations (1) and (2) can be rewritten as follows:

$$D = -100.71 \times W_s + 6010.5, \tag{3}$$

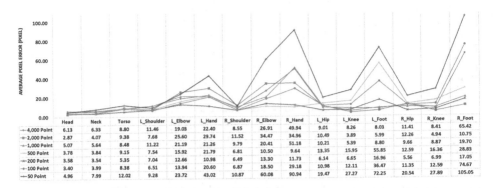

	Head	Neck	Torso	L_Shoulder	L_Elbow	L_Hand	R_Shoulder	R_Elbow	R_Hand	L_Hip	L_Knee	L_Foot	R_Hip	R_Knee	R_Foot
4,000 Point	6.13	6.33	8.80	11.46	19.03	22.40	8.55	26.91	49.94	9.01	8.26	8.03	11.41	8.41	65.42
2,000 Point	2.87	4.07	9.38	7.68	25.60	29.74	11.52	34.47	34.96	10.49	3.89	5.99	12.26	4.94	10.75
1,000 Point	5.07	5.64	8.48	11.22	21.19	21.26	9.79	20.41	51.18	10.21	5.39	8.80	9.66	8.87	19.70
500 Point	3.78	3.84	9.15	7.54	15.92	21.79	6.81	10.50	9.64	13.35	15.95	55.85	12.59	16.36	28.83
200 Point	3.58	3.54	5.35	7.04	12.66	10.98	6.49	13.30	11.73	6.14	6.65	16.96	5.56	6.99	17.05
100 Point	3.40	3.99	8.38	6.51	13.94	20.60	6.87	18.50	29.18	10.98	12.11	36.47	11.35	12.59	74.67
50 Point	4.96	7.99	12.02	9.28	23.72	43.02	10.87	60.08	90.94	19.47	27.27	72.25	20.54	27.89	105.05

Fig. 5. Average pixel error of pose estimation with respect to the number of feature points: 50, 100, 200, 500, 1,000, 2,000, and 4,000

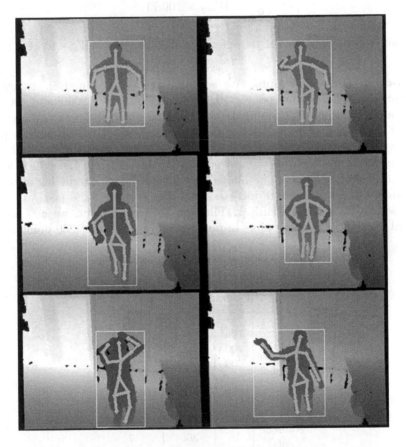

Fig. 6. Examples of human pose estimation experiment results

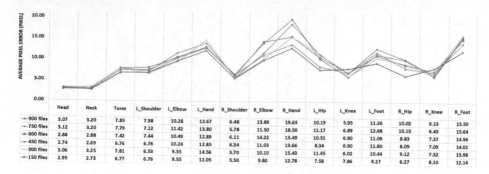

Fig. 7. Average pixel error of pose estimation with respect to training data size: 150, 300, 450, 600, 750, and 900

$$S_D = \frac{W_{ref} \times 100.71}{6010.5 - D}. \tag{4}$$

In our experiments, we set the reference shoulder width W_{ref} to 12 pixels.

Number of Feature Points. We compare the pose estimation accuracy of different number of feature points: 50, 100, 200, 500, 1000, 2000, and 4000. For every case the number of training data is 150 files. Fig. 5 shows the average pixel error of different feature points number for each body parts. 50 feature points shows the highest error, whereas 200 feature points shows the smallest error. These results are attributed to underfitting and overfitting. Below 200 feature points, underfitting occurs, and above 200 feature points, overfitting occurs. As the number of feature points increases, the frame rate of the pose estimation decreases, as shown in TABLE 1. Through this experiment, we can see that optimal number of feature points in pose estimation is about 200 points.

Table 1. Frame rates of pose estimation

	Points	Training data size	Frame rate (fps)
Number of feature points	50		10
	100		9
	200	150	8
	500		3
	1000		2
	2000		1
	4000		less than 1
Number of training data	200	150	8
		300	7
		450	7
		600	6
		750	6
		900	6

Number of Training Images. We further investigate the pose estimation error with respect to training data size: 150, 300, 450, 600, 750, and 900 files. The number of feature points is 200 points in all cases. As shown in Fig. 7, there is not a significant difference in accuracy between them. This is due to similar poses in large training data sets. As the number of training data increases, similar poses in the data set can exist. The frame rate of pose estimation decreases as the number of training data increase, but the difference between them is quite small. For fast pose estimation, the number of training data should be small as long as training data includes various poses that do not overlap with each other.

4   Conclusion

In this paper, we have presented human pose estimation and gesture recognition algorithms using the depth information only. In the case of the human pose estimation algorithm, we proposed a method that mixes a geodesic graph and an SVM-based human pose estimator. The SVM-based human pose estimator uses a small amount of randomized human features. The proposed algorithm can be executed on a low cost platform and anyone can use the system immediately without prior calibration. We have evaluated our method by several experiments, and the human pose estimation results demonstrated the acceptable performance of our approaches.

Our future work will include dealing with the above problems and increasing the robustness of human pose estimation. Moreover, we plan to further develop the advanced monitoring system for facilities such as production lines, building lobbies, etc. Finally, a future version of our method will allow application in outdoor environments.

Acknowledgment. This research was financially supported by Samsung S1 Cooperation. This work is also financially supported by Korea Minister of Ministry of Land, Infrastructure and Transport (MOLIT) as U-City Master and Doctor Course Grant Program.

References

1. Correa, P., Marques, F., Marichal, X., Macq, B.: 3d human posture estimation using geodesic distance maps. In: Conference Speech and Computer (SPECOM), pp. 31–34 (2004)
2. Bellucci, A., Malizia, A., Diaz, P., Aedo, I.: Human-display interaction technology: Emerging remote interfaces for pervasive display environments. IEEE Pervasive Computing 9(2), 72–76 (2010)
3. Zhang, Z.: Microsoft kinect sensor and its effect. IEEE Multimedia 19(2), 4–10 (2012)
4. Freedman, B., Shpunt, A., Machline, M., Arieli, Y., et al.: Depth mapping using projected patterns. WO Patent 2,008,120,217 (October 10, 2008)

5. Ganapathi, V., Plagemann, C., Koller, D., Thrun, S.: Real time motion capture using a single time-of-flight camera. In: IEEE Conference on Computer Vision and Pattern Recognition (CVPR), pp. 755–762 (2010)
6. Shotton, J., Sharp, T., Kipman, A., Fitzgibbon, A., Finocchio, M., Blake, A., Cook, M., Moore, R.: Real-time human pose recognition in parts from single depth images. Communications of the ACM Magazine 56(1)
7. Hernández-Vela, A., Zlateva, N., Marinov, A., Reyes, M., Radeva, P., Dimov, D., Escalera, S.: Graph cuts optimization for multi-limb human segmentation in depth maps. In: IEEE Conference on Computer Vision and Pattern Recognition (CVPR), pp. 726–732 (2012)
8. Schwarz, L.A., Mkhitaryan, A., Mateus, D., Navab, N.: Human skeleton tracking from depth data using geodesic distances and optical flow. Image and Vision Computing 30(3), 217–226 (2012)
9. Poppe, R.: Vision-based human motion analysis: An overview. Computer Vision and Image Understanding 108(1), 4–18 (2007)
10. Shotton, J., Fitzgibbon, A., Cook, M., Sharp, T., Finocchio, M., Moore, R., Kipman, A., Blake, A.: Real-time human pose recognition in parts from single depth images. In: IEEE Conference on Computer Vision and Pattern Recognition (CVPR), pp. 1–8 (2011)
11. Cortes, C., Vapnik, V.: Support-vector networks. Machine Learning 20(3), 273–297 (1995)
12. Burges, C.J.: A tutorial on support vector machines for pattern recognition. Data Mining and Knowledge Discovery 2(2), 121–167 (1998)
13. Hsu, C.W., Lin, C.J.: A comparison of methods for multiclass support vector machines. IEEE Transactions on Neural Networks 13(2), 415–425 (2002)
14. Aizerman, A., Braverman, E.M., Rozoner, L.: Theoretical foundations of the potential function method in pattern recognition learning. Automation and Remote Control 25, 821–837 (1964)
15. Amari, S.I., Wu, S.: Improving support vector machine classifiers by modifying kernel functions. Neural Networks 12(6), 783–789 (1999)

Dense Optical Flow Estimation
with 3D Structure Tensor Models

Tongwei Lu[1], Ying Ren[2], Wenting Liu[2], and Anyuan Chen[2]

[1] School of Computer Science and Engineering, Wuhan Institute of Technology, China
lutongwei@gmail.com
[2] Hubei Key Laboratory of Intelligent Robot (Wuhan Institute of Technology), China
{593523182,liuwengting123,1041319944}@qq.com

Abstract. Since the 3D structure tensor at each pixel can interpret the local between frames well, it can be used to estimate dense flow. According to the assumptions of brightness constancy, the optical flow estimation can be converted to the calculation the eigenvector of the structure tensor, rather than the complex calculation of linear system. Iterative coarse-to-fine refinement is used to improve the performance. Experimental results show that the proposed algorithm is robust and effective for computing the dense flow.

Keywords: optical flow, 3D structure tensor.

1   Introduction

As the rapid expansion of surveillance applications, more and more researchers focus on studying optical flow, which is used to get the foregrounds' motion velocity. Studies of optical flow mainly include dense flow and sparse flow. In this paper, we focus on the study of dense flow.

Researchers have challenged to improve both the efficiency and accuracy of the dense flow. Some dense optical flow estimation methods based on the original Horn's idea calculate velocity by point [1]. Brox [2] proposed a coarse-to-fine strategy using warping technique [3,4] based on the brightness constancy. Bruhn [5] proposed a combination of local methods(namely Lucas–Kanade) and global methods(namely Horn and Schunck), Liu [6] proposed optical flow can be calculated using generalized eigenvalue analysis without solving a linear system,which is demonstrated by affine motion model.

Calculating velocity by-point would be time consuming and tedious, however, the results comes to be more stable. A few prior works have proposed using 3D structure tensor as motion model. Pless [7] described a first real-time method for factoring background motions into multiple flow fields, which can represent global patterns of local motions effectively. Thanks to the optical flow benchmark dataset at Middlebury pages, a significant improvement has already been achieved by researchers [8].

© Springer International Publishing Switzerland 2015
J.-H. Kim et al. (eds.), *Robot Intelligence Technology and Applications 3,*
Advances in Intelligent Systems and Computing 345, DOI: 10.1007/978-3-319-16841-8_62

Our work is organized as follows. Section 2 gives the algorithm of estimation of optical flow, and section 3 shows the experimental results. Finally, section 4 gives the conclusion.

2   Estimation of Optical Flow

The structure tensor has been widely used in the image analysis field for optic flow estimation and segmentation [7].It is created based on the spatio-temporal gradients in frames.

2.1   The Mixture Global Motion Models

$\nabla I = (\nabla I_x, \nabla I_y, \nabla I_t)^T$ is used to interpret the spatio-temporal gradients where, are the partial derivatives of I with respect to x, y and t. The structure tensor $\sum$ at each pixel is defined as:

$$\sum = \nabla I \nabla I^T \tag{1}$$

The motion appears to happen in place where ‖>0. The images are blurred using Gaussian smoothing to decrease the influence of noise,

$\sum$ defines a Gaussian distribution $N(0, \sum)$. Like previous work does [9], mahalanobis distance is used to detect anomalous measurements, comparing to a preselected threshold Tb.

$$dist = \nabla I^T \sum \nabla I \tag{2}$$

If the dist between the new measurement and the Gaussian distribution is less than Tb, we update the corresponding Gaussian distribution.

$$\sum_{i,t} = (1-\beta)\sum_{i,t-1} + \beta \nabla I \nabla I^T \tag{3}$$

$\beta = \alpha P(N_i \mid \nabla I), \alpha \in [0,1]$, $P(N_i \mid \nabla I)$ is the likelyhood, α is the preselected adaptive rate of Gaussian distribution [7].

The weight of the Gaussian distribution is updated as:

$$\omega_{i,t} = (1-\beta_i)\omega_{i,t-1} + \beta_i \tag{4}$$

2.2   Relationship Between Structure Tensor and Optical Flow

According to the brightness constancy [3]

$$I(x, y, t) = I(x+u, y+v, t+1) \tag{5}$$

Linearizing Equation (5) by Taylor expansion, it is changed to:

$$I(x, y, t) = I(x, y, t) + u\frac{\partial I}{\partial x} + v\frac{\partial I}{\partial y} + 1\frac{\partial I}{\partial t} \tag{6}$$

From equation (6), we can get the optical flow constraint equation:

$$uI_x + vI_y + I_t = 0 \tag{7}$$

Where, $I_x = \dfrac{\partial I}{\partial x}$, $I_y = \dfrac{\partial I}{\partial y}$, $I_t = \dfrac{\partial I}{\partial t}$

It is also equal to $\nabla I^T \begin{pmatrix} u \\ v \\ 1 \end{pmatrix} = 0$. The optical flow vector $(u, v, 1)^T$ must be

orthogonal to vector ∇I^T. Multipling ∇I to each side of the equation, it can be changed to:

$$\nabla I \nabla I^T \begin{pmatrix} u \\ v \\ 1 \end{pmatrix} = 0 \tag{8}$$

Consider that, if the smallest eigenvalue of the structure tensor $\sum$(It must be symmetric matrix) closes to be zero, the corresponding eigenvector can be considered as the solution of the equation. So the confidence of the solution can be judged by the equation:

$$conf = 1 - \frac{\lambda_1}{\lambda_2} \tag{9}$$

Here, λ_1, λ_2, λ_3 is the eigenvalue of the structure tensor and $\lambda_1 < \lambda_2 < \lambda_3$. In this case, we can estimate the 3D optical flow by estimation of eigenvector.

2.3   Coarse-to-fine Process

The end condition of the coarse-to-fine iterative process is the confidence, but it is a litter different to Liu's work [6]. Their idea comes from the common sense that appropriate smoothing of the image frames improves the estimation of optical flow [10]. So it's iterative process use bigger Gaussian filter to smooth the image. Our coarse-to-fine process will take adjacent pixels into account. The model is shown in Fig.1. Pixels whose confidence measures are lower than the threshold should be reanalyzed in the next iteration by calculating a bigger patch based structure tensor, where ∇I_{mean} is the weighted mean of all pixels' ∇I in the patch.

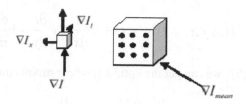

Fig. 1. ∇I_{mean} is the weighted mean of all pixels' ∇I in the patch

3   Experiment

Since it need to build motion patterns at the beginning, the computational cost is somewhat larger than simply solving equations by-point. As time goes, the motion patterns become stable and solving equations is replaced by patterns matching.

One group images of all-frames dataset(http://vision.middlebury.edu/flow/)is tested. Compared to   Simple Flow [11], the result is shown in Fig.2. It showes only the right-top part for showing here.

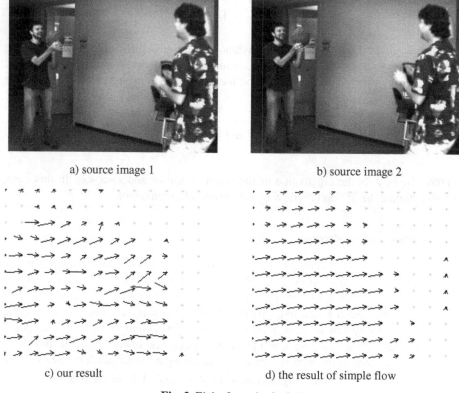

a) source image 1   b) source image 2

c) our result   d) the result of simple flow

Fig. 2. Eight-frame basketball

The result of interpolation error and running time are shown in Tab.1, compared to Simple Flow.

Table 1. The comparison of interpolation error(SD) and running time

	interpolation error(SD)	average running time(ms)
Our method	19.5	732
Simple Flow	22.6	24383

As the experiment shows, the proposed algorithm can get good result neither need calculating equations by-point, nor building multiple pyramids. At the beginning, without any models built, it should be time consuming to build models. The initial models are not stable and there may be some error patterns to match, so the results of the initial frames may be relatively poor. After some frames, modeling process becomes stable, so does the running speed. Since there is neither Gaussian pyramids building nor interpolation process in this algorithm, it takes less time and space. After the motion patterns are built, the noise pixels can be matched into right patterns.

Second, we test our method on our traffic dataset. It is a rainy scene, so the quality of the dataset is really bad. The result is shown in Fig.3.

a) frame 5: the running time closes to be six seconds.

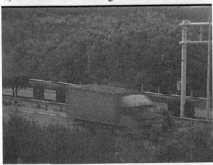

b) frame 11:the running time closes to be four seconds.

Fig. 3. 28-frame flow in traffic scene. At the beginning, without any models built, it should be time consuming to build models. The initial models are not stable and there may be some error patterns to match, so the results of the initial few frames may be relatively poor. After some frames, modeling process becomes stable, so does the running speed.

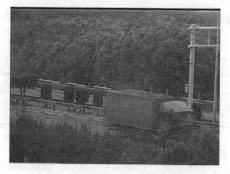

c) frame 17:the running time closes to be two seconds

d) frame 20:the running time closes to be one second

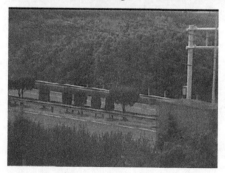

e) frame 28:the running time closes to be one second.

Fig. 3. *(Continued)*

4   Conclusion

The 3D structure tensor is used to compute dense flow. The initial models are not stable and there may be some error patterns to match. As the frames go on, modeling process becomes stable. Once the motion patterns are built, the noise pixels can be matched into right patterns. Experiments show that the algorithm is robust and effective. The algorithm can be applied to the analysis of motion effectively such as behavior recognition and matching moving target.

Acknowledgements. This work is supported by the Science Research Funds of Wuhan Institute of Technology (12126031), Open Research Funds of Hubei Key Laboratory of Intelligent Robot (Wuhan Institute of Technology) and Science Research Funds of Education Department of Hubei Province (Q20131507, 2012FFA134).

References

1. Horn, B., Schunck, B.: Determining optical flow. Artificial Intelligence 16, 185–203 (1981)
2. Brox, T., Bruhn, A., Papenberg, N., Weickert, J.: High accuracy optical flow estimation based on a theory for warping. In: Pajdla, T., Matas, J.(G.) (eds.) ECCV 2004. LNCS, vol. 3024, pp. 25–36. Springer, Heidelberg (2004)
3. Black, M.J., Anandan, P.: The robust estimation of multiple motions: parametric and piecewise smooth flow fields. Computer Vision and Image Understanding 63(1), 75–104 (1996)
4. Mémin, E., Pérez, P.: A multigrid approach for hierarchical motion estimation. In: Proc. Sixth International Conference on Computer Vision, pp. 933–938. Narosa Publishing House, Bombay (1998)
5. Bruhn, A., Weickert, J., Schnörr, C.: Lucas/Kanade Meets Horn/Schunck: Combining Local and Global Optic Flow Methods 61(3), 211–231 (2005)
6. Liu, H., Chellappa, R., Rosenfeld, A.: Accurate dense optical flow estimation using adaptive structure tensors and a parametric model. IEEE Trans. IP(12), 1170–1180 (2003)
7. Wright, J., Pless, R.: Analysis of Persistent Motion Patterns Using the 3D Structure Tensor. In: IEEE Workshop on Motion and Video Computing, pp. 14–19 (2005)
8. Baker, S., Scharstein, D., Lewis, J.P., Roth, S., Black, M.J., Szeliski, R.: A database and evaluation methodology for optical flow. IJCV (92), 1–31 (2011)
9. Pless, R., Larson, J., Siebers, S., Westover, B.: Evaluation of Local Models of Dynamic Backgrounds. In: Computer Vision and Pattern Recognnition, pp. 73–78 (2003)
10. Barron, J., Fleet, D., Beauchemin, S.: Performance of optical flow techniques. IJCV 12, 43–77 (1994)
11. Tao, M.W., Bai, J., Kohli, P., Paris, S.: SimpleFlow: A Non-iterative, Sublinear Optical Flow Algorithm. Computer Graphics Forum (Eurographics 2012) 31(2) (May 2012)

Improvement of Dust Detection System Using Infra-red Sensors

Donghoe Kim[1], Wonse Jo[1], Bumjoo Lee[2], Jinung An[3], and Donghan Kim[1]

[1] Department of Electronics and Radio Engineering, Kyung Hee University, Yongin, 446-701, Republic of Korea
{genesis130,wonsu0513,donghani}@khu.ac.kr
[2] Department of Electrical Engineering, Myongji University, Yongin, Korea
leebumjoo@gmail.com
[3] Pragmatic Applied Robot Institute, DGIST, Daegu, 771-873, Republic of Korea

Abstract. This paper proposes a sensor, which is capable of determining the type of dust through the dust detection sensor system that uses infrared sensor. Recently, cleaning robot is highly favored in the service robot market, which has been lead to development in various fields. Researches including navigation, localization, and developing the key components are underway to improve the performance, where the researches related to sensor are actively progressed accordingly. The dust detection device is developed to improve the performance of cleaning robot by determining the type of dust. The developed dust detection device is consists up of 1 transmitter and 24 receivers. The amount of transmitted and reflected light emanated from light sensor varies according to the characteristics of various types of dust, which passes through the middle of the dust detection device. The changes in the amount of light from 24 receivers are collected as the experiment data using the test bed of proposed dust detection device. Lastly, this paper confirms determining the type of dust by analyzing the pattern using the pixel-based dust similarity method.

1   Introduction

Recently, cleaning robot is leading the service robot market, which becomes as new home appliance that can lead the changes in the pattern of human life. The size of service robot market is consistently expanding by increasing the 'technical accessibility' of robot to the public. By escaping from non-practicality and non-realistic robot technology studied in research level, the cleaning robot increases the familiarity to other service robots by providing valuable services in real life [1,2]. Thus, worldwide robotics and electronics companies and research institutes are accelerating the development of key technologies in order to dominate the cleaning robot market in advance. In Korea, a variety of cleaning robot products are being launched with remarkable technological growth from the major company and the small businesses related to the robotics. In addition, the research is actively progressed in both university and research institutes to enhance the performance of cleaning robot [3].

© Springer International Publishing Switzerland 2015
J.-H. Kim et al. (eds.), *Robot Intelligence Technology and Applications 3*,
Advances in Intelligent Systems and Computing 345, DOI: 10.1007/978-3-319-16841-8_63

A method for improving the performance of cleaning robot can be largely divided into three parts: robot navigation, developing intelligence i.e. localization, developing key components of robot [4-6]. The key components include camera, actuation, battery, control devices, etc., and in recent years, research development and application of dust detection sensor technology is being discussed [7,8]. As an example of dust sensor in practical application, iRobot Roomba has a feature that can determine the presence or absence of dust. However, a significant improvement to cleaning performance cannot be expected because it simply determines the presence or absence of dust.

This paper proposes a dust detection system that is advanced than existing technology by using an infrared sensor. Since the proposed sensor system is capable of identifying the type of dust, better performance can be expected than existing technologies that only has the degree of simplicity [9].

Kim et al. [10] initially introduced the dust detection device. However, the existing dust detection device takes too long time to receive data. In addition, there is a drawback, which reduces the performance when detecting at least two types of dust. Thus, the proposed paper develops a new dust detection device to cope with these problems and to enhance the dust distinguishing performance simultaneously.

This paper is organized as follows: Section II proposes the data detection method using infrared sensor. Section III describes the conventional dust detection device. Section IV presents the newly developed and upgraded dust detection device. Section V presents the data filtering method and the data similarity analysis method with experiment results. Concluding remarks follow in Section VI.

2   DATA Detection Using Infrared Sensor

2.1   Principle of Dust Sensor System

The proposed dust detection device is made up of infrared sensor with two components: transmitter that generates a light and receiver that detects light reflected or transmitted

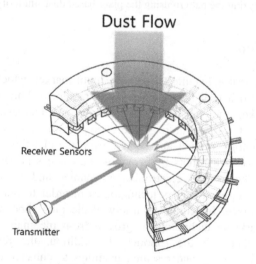

Fig. 1. Concept of dust detection sensor

from dust. The device uses one transmitter and many receivers. When dust passes through the central part of sensor, the proposed sensor system determines the type of dust by analyzing the change in the amount of light collected in the receiver.

Fig. 1 shows the concept of an optical dust detection sensor. According to the presence of dust, it shows that an emitted light from the transmitter gets received as a reflected light by the receiver.

2.2   Algorithm

Fig. 2 shows the flowchart of dust detection sensor system. Initially, a dust is inserted into the system, and then sensor values are extracted by using the device (shown in Fig. 6). Note that this extracted data from sensor is called 'raw data' in this paper. The amount of transmitted and reflected light varies based on the type of inserted dust, which is recorded as the 'sensor value'.

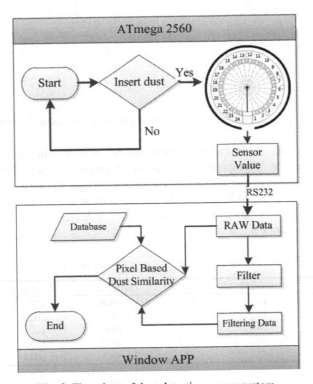

Fig. 2. Flowchart of dust detection sensor system

The sensor data go through filtering process and pixel-based dust similarity algorithm. The type of dust is distinguished in the final step. In the filtering process, the average (AVR) method and the moving average (M/A) method are used to compare the performance. The detailed procedure of filtering process is described in the next section. Pixel-based dust similarity method is used as the dust type determination method, where this method is described in Section V.

3   Conventional Dust Detection Device

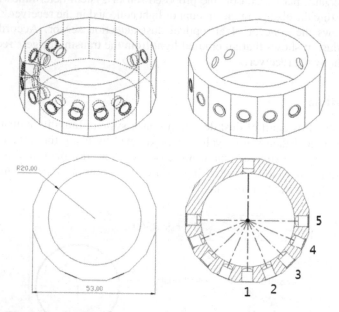

Fig. 3. Configuration of conventional sensor transmitter and receiver

Fig. 3 shows the configuration of conventional sensor transmitter and receiver. Sensor '1' to sensor '5' represents the order of receiver, where sensors ('1', '2') facing the transmitter detect the light that passes through the dust and sensors ('3', '4', '5') detect the reflected light. In addition, when there is no dust in the dust sensor, a large amount of light enters to sensors ('1', '2'), whereas a very small amount of light enters to sensors ('3', '4', '5').

Table 1. Sensor data received from conventional receiver unit

Time	Receiver 1	Receiver 2	Receiver 3	Receiver 4	Receiver 5
1	42	773	1008	1019	1017
2	41	659	1014	1020	1018
3	44	822	1012	1020	1017
4	56	792	987	1011	1014
5	43	641	993	1014	1012

Table 1 represents the amount of light received in the conventional receivers. Note that the value decreases as more light comes, and vice versa. As shown in Table 1, the value is 41 when there is the maximum amount of light, whereas the value increases to 1021 when there is the minimum amount of light. The amount of light collected in the receiver changes due to time. Note that 8 data are collected within 1 sec at every 0.125 sec in the conventional dust detection device.

Fig. 4. Snapshot of conventional dust detection sensor test bed

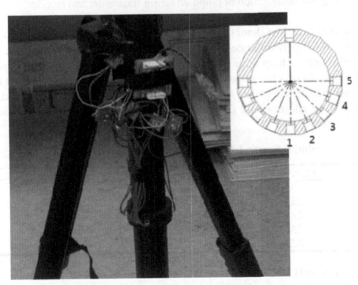

Fig. 5. Snapshot of conventional dust detection sensors

Fig. 4 and Fig. 5 show the snapshot of conventional dust detection sensor test bed. In order to use this device, the dust must be dropped from top to bottom. The dust falls in the central part of sensor (shown in Fig. 5), which is then converted into measured data.

4   Proposed Dust Detection Device

Fig. 6 shows the configuration of proposed sensor transmitter and receiver. The number of transmitter is kept same as 1, but the number of receiver is increased to 24. Note that the size of proposed sensor is identical to the conventional sensor. Similarly, Sensor '1' to sensor '24' represents the order of receiver. Sensors ('11', '12', '13', '14') facing the transmitter detect the light that passes through the dust, whereas the rest of receivers detect the reflected light.

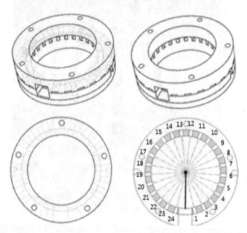

Fig. 6. Configuration of proposed sensor transmitter and receiver

In the conventional dust detection device, it is possible to distinguish two sample types of dust, but the performance is decreased when the number of samples is greater than two. It also takes a long time to gather enough data using the system that receives 8 data per second. In order to cope with these problems in the proposed dust detection device, the number of receiver is increased to obtain more accurate data and to allow various analysis. In addition, the sensor system is improved such that 100 data can be collected per second.

Table 2. Sensor data received from proposed receiver unit

Time	Receiver 8	Receiver 9	Receiver 10	Receiver 11	Receiver 12
1	256	247	212	240	158
2	249	282	278	711	951
3	265	233	260	610	982
4	259	285	305	746	988
5	264	264	268	717	291

Table 2 represents the amount of light received in the proposed receivers. Assuming that the data are in the symmetric, 'Receiver 8' to 'Receiver 12' represents the best seen data being transmitted and reflected among 24 sensors. Compared to the conventional dust detection device, the value increases as more light comes to the receiver. The value is 990 when there is the maximum amount of light, whereas the value drops to 128 when there is the minimum amount of light.

Fig. 7. Snapshot of proposed dust detection sensor test bed

Fig. 8. Snapshot of proposed dust detection sensor

Fig. 7 and Fig. 8 show the snapshot of proposed dust detection sensor test bed. As shown in Fig. 7, the proposed device is properly aligned horizontally, which is difficult to realize in the conventional device (shown in Fig. 4). Significantly, the experimental platform is improved to progress the experiment more accurately and rapidly. Similar to the conventional device, the dust must be dropped from top to bottom. The dust falls in the central part of sensor (shown in Fig. 8), which is then converted into measured data. The measured data are then sent to PC and distinguished its type through the dust detection algorithm.

5   Experimental Results

5.1   Filtering

In Table 2, the same type of dust is used to progress the experiment. However, the measured values in 'Receiver 11' and 'Receiver 12' show the significant differences. This occurrence is due to measurement noise or non-constant distribution when the dust passes the sensor. This difference can cause an error to the dust similarity algorithm explained in the next section. Thus, it is essential to implement a filter to eliminate or to minimize the difference.

Two filtering methods – the average method and the moving average method – are used to filter the measured values from the sensor.

$$\bar{x}^i = \frac{\sum_{j=1}^n x_j^i}{n}, i = 1, \dots 5 \tag{1}$$

In the average method (1), $\bar{x}^i$ represents the average value of ith receiver and x_j represents the measured sensor value according to the time sequence. Lastly, n represents the number of total data used in the average method.

The average moving average method is known as repetitive data processing, which is represented as follows:

$$\bar{x}^i[t] = \frac{1}{n}(x^i[t] + x^i[t-1] + \cdots + x^i[t-n+1]) \qquad i = 1, \dots 5 \tag{2}$$

As shown in (2), the filtered value of ith receiver $\bar{x}^i[t]$ is consists of the average of n number of data from the current raw data $\bar{x}^i[t]$ to the $x^i[t-n+1]$ number of previous data. The measurement noise in raw data can be decreased by using the average method and the moving average method.

$$d(f,g) = \sqrt{\sum_{j=1}^m \sum_{i=1}^n [f(i,j) - g(i,j)]^2} \tag{3}$$

5.2   Pixel-based Dust Similarity

The amount of light received in the receiver gets filtered and then determines the similarity by comparing to the previously measured values. This paper uses the pixel-based image similarity method [11] as the technique for determining the similarity. This method compares the similarity between two sets of data as follows:

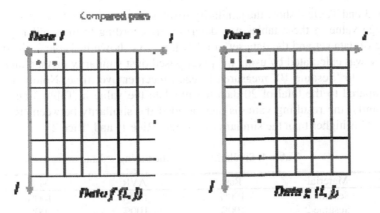

Fig. 9. Two array of data set for checking the similarity

Fig. 9 shows the two array of data set for checking the similarity. $f(i, j)$ and $g(i, j)$ represent the   function of first data set (previously measured values) and the function of second data set (currently measured values), respectively. i represents the order of receiver, j represents the data order, n and m represent the number of total data and the number of receivers, respectively. Thus, two data sets become similar as the calculated value in (3) becomes smaller. Significantly, the type of dust will be distinguished based on this similarity value.

5.3   Results

To verify the effectiveness of the proposed dust detection device, two types of dust – rice and sesame – were used. A certain amount of rice and sesame were inserted into the proposed dust detection device for 10 seconds, and then, 10 data sets of each dust were generated. The final data sets were composed of 'raw' data, average (AVR) data, and moving average (M/A) data, which were calculated with the rate of 100 data per second. Lastly, the similarity was determined by alternately picking two data sets of other using (3).

Table 3. Similarity result when using rice

Material	Raw	AVR	M/A
Rice1	100%	100%	100%
Rice 2	100%	100%	100%
Rice 3	100%	100%	100%
Rice 4	100%	100%	100%
Rice 5	100%	100%	100%
Rice 6	100%	100%	100%
Rice 7	80%	100%	100%
Rice 8	80%	100%	100%
Rice 9	90%	100%	100%
Rice 10	100%	100%	100%

Table 3 and Table 4 show the similarity result of rice and sesame, respectively. The percentage value in these tables was determined according to the similarity between one type of data set and the data set of other kind. As shown in Table 2, the degree of similarity was calculated by using the pixel-based dust similarity method and the data of 'Rice 1' to 'Sesame 10' measured in each receiver over time. Note that 'Rice 1' was compared to the total of 20 data to calculate the value in Table 3 and Table 4. Significantly, the resulting value is decreased if the similarity between 'Rice 1' and 'Sesame 1' is higher than the similarity between 'Rice 1' and 'Rice 2'.

Table 4. Similarity result when using sesame

Material	Raw	AVR	M/A
Sesame 1	100%	100%	100%
Sesame 2	100%	100%	100%
Sesame 3	100%	100%	100%
Sesame 4	100%	100%	100%
Sesame 5	100%	100%	100%
Sesame 6	90%	80%	80%
Sesame 7	90%	80%	80%
Sesame 8	100%	100%	90%
Sesame 9	100%	100%	100%
Sesame 10	100%	100%	100%

The experiment results showed that the similarity in 10 data sets of rice was 95 percent when using the raw data, 100 percent when using both the average and the moving average data. In contrast, in 10 data sets of sesame, the similarity was 98 percent when using the raw data, 96 percent when using the average data, and 95 percent when using the moving average data. The experiment results indicated that a more accurate dust determination result was shown with rice than sesame. This was due to the differences in the amount of infrared transmittance and reflectance. In addition, the similarity results of using the average and the moving average data for 'Rice 7', 'Rice 8', and 'Rice 9' were better than using the raw data. In contrast, the similarity results of using the average and the moving average data for 'Sesame 6', 'Sesame 7', and 'Sesame 8' were poor than using the raw data. The proper rice data was obtained after applying the filter, whereas applying filter in the sesame data removed some characteristics of sesame to decrease the similarity value.

When the conventional dust detection device shown in Section III was used, the similarity in the rice data was 82 percent when using the raw data, 81 percent when using the average data, and 82 percent when using the moving average data. In the sesame data, the similarity was 42 percent in the raw data, 65 percent in the average data, and 64 percent in the moving average data. Compared to the conventional dust detection device, the proposed dust detection device showed the significant performance improvement in the result of determining two types of dust.

6 Conclusion

This paper described the infrared sensor-based dust detection device to enhance the performance. The proposed device was able to differentiate the type of dust using the pixel-based dust similarity method. The detection performance was greatly improved compared to the conventional dust detection device. Applying the filter showed to improve the data performance, but it was confirmed that the performance was poor in some cases.

Testing the proposed dust detection device with various types of dust and with various filters is left as a further work to verify the robustness of the system and to increase the system performance, respectively. Lastly, developing a small-sized detection device that can be used to actual cleaning robot system is left as a further work.

Acknowledgment. This research was supported by Technology Innovation Program of the Knowledge economy (No. 10041834, 10045351) funded by the Ministry of Knowledge Economy (MKE, Korea) and Basic Science Research Program through the National Research Foundation of Korea(NRF) funded by the Ministry of Education, Science and Technology(No. 2012R1A1A2043822) and DGIST R&D Program of the Ministry of Science, ICT and Future Planning of Korea(14-BD-01).

References

1. Fink, J., et al.: Living with a Vacuum Cleaning Robot. International Journal of Social Robotics 2, 1–20 (2013); Bruce, K.B., Cardelli, L., Pierce, B.C.: Comparing Object Encodings. In: Abadi, M., Ito, T. (eds.) TACS 1997. LNCS, vol. 1281, pp. 415–438. Springer, Heidelberg (1997)
2. Jodi, F., DiSalvo, C.: Service Robots in the Domestic Environment: A Study of the Roomba Vacuum in the Home. In: Proceedings of the 1st ACM SIGCHI/SIGART Conference on Human-Robot Interaction, pp. 258–265. ACM (2006)
3. Nam, M.-K.: A Study on present Condition of Development and Market of Artificial Intelligence Robot In and Out of Country. Journal of the Korean Society of Design Culture 16(2), 198–207 (2010)
4. Batalin, M.A., Sukhatme, G.S., Hattig, M.: Mobile Robot Navigation Using a Sensor Network. In: Proceedings of the 2004 IEEE International Conference on Robotics and Automation, ICRA 2004, vol. 1. IEEE (2004)
5. Doh, N.L., Kim, C., Chung, W.K.: A practical path planner for the robotic vacuum cleaner in rectilinear environments. IEEE Transactions on Consumer Electronics 53(2), 519–527 (2007)
6. Kim, H., et al.: User-Centered Approach to Path Planning of Cleaning Robots: Analyzing User's Cleaning Behavior. In: Proceedings of the ACM/IEEE International Conference on Human-Robot Interaction, pp. 273–380. ACM (2007)
7. Chen, Z., Birchfield, S.T.: Qualitative Vision-based Mobile Robot Navigation. In: Proceedings of the 2006 IEEE International Conference on Robotics and Automation, ICRA 2006, pp. 2686–2692. IEEE (2006)

8. Freire, E., et al.: A New Mobile Robot Control Approach via Fusion of Control Signals. IEEE Transaction on Systems, Man, and Cybernetics, Part B: Cybernetics 34(1), 419–429 (2004)

9. Kawakami, H., et al.: Dust detector for vacuum cleaner. U.S. Patent No. 5,163, 202 (November 17, 1992)

10. Kim, D.-H., Min, B.-C., Kim, D.-H.: A Dust Detection Sensor System for Improvement of a Robot Vacuum Cleaner. Journal of Institute of Control, Robotics and Systems 10, 896-500(5 pages) (2013)

11. Ascenso, J., Brites, C., Pereira, F.: Motion compensated refinement for low complexity pixel based distributed video coding. In: IEEE Conference on Advanced Video and Signal Based Surveillance, AVSS 2005. IEEE (2005)

Assessment of the Effectiveness of Acupuncture on Facial Paralysis Based on sEMG Decomposition

Anbin Xiong[1,2], Xingang Zhao[1], Jianda Han[1], and Guangjun Liu[1,3]

[1] State Key Laboratory of Robotics, Shenyang Institute of Automation (SIA),
Chinese Academy of Sciences (CAS), Shenyang, Liaoning, 110016, China,
{xiongab,zhaoxingang,jdhan}@sia.cn
[2] University of Chinese Academy of Sciences (CAS), Beijing, 100049, China
[3] Department of Aerospace Engineering, Ryerson University, Toronto, Canada
gjliu@ryerson.ca

Abstract. Although acupuncture has been extensively and routinely applied in clinic around the world, its use is conventionally based on empirical knowledge rather than scientific evidence. In this paper, we investigate the surface electromyographic (sEMG) activity on the face of patients with facial paralysis to validate the effectiveness of acupuncture. sEMG decomposition method is employed to quantify of the differences between the healthy and diseased sides, which includes 2 parts: to decompose sEMG into motor units action potential trains (MUAPTs) with Gaussian Mixture Model (GMM); to assess the recovery of patients with facial paralysis according to the differences between the MUAPTs of healthy and diseased sides. Finally, an auto-regression model is used to predict the recovery trends. Results indicate that the proposed method can assess the effectiveness of acupuncture on facial paralysis and achieve a high accuracy of 90.94% to predict the recovery trends.

1 Introduction

Acupuncture is a traditional Chinese therapeutic method with more than a history of more than 2000 years. A great deal of valuable experience has been accumulated in the long time clinical practice. Its effectiveness has recently been accepted and recognized by western medicine as an important complementary therapy [1].

Acupuncture has been normally used in clinic to treat the disease of nervous system such as facial paralysis. Facial paralysis is a common disease that involves the paralysis of the structures innervated by the facial nerves and may be caused due to stroke, brain tumour, birth trauma (in newborns), infection or Lyme disease. Although good effects on facial paralysis have been reported in randomized controlled trials [2, 3], the therapeutic effect of acupuncture on the disease of facial paralysis is still under debate [4, 5].

On the other hand, surface electromyography (sEMG) is the electrical activity that can be measured non-invasively to investigate the aspects of motor control and muscle contraction [6]. Researchers utilize sEMG to verify the efficacy of acupuncture during the past decades.

© Springer International Publishing Switzerland 2015
J.-H. Kim et al. (eds.), *Robot Intelligence Technology and Applications 3*,
Advances in Intelligent Systems and Computing 345, DOI: 10.1007/978-3-319-16841-8_64

Rancan *et al.* [7] use sEMG and maximal molar bite force to investigate the activity of masseter before and after a 3-month acupuncture therapy in individuals with temporomandibular disorder. Results indicate the decrease in EMG activity as well as the increase in the values of maximal bite force after acupuncture treatment. Costa *et al.* [8] study the immediate effects of acupuncture with sEMG on the tibialis anterior muscle and find that acupuncture on both Zusanli (ST36) and Yinlingquan (SP9) can reduce the EMG activity to fibialis anterior. Hübscher *et al.* [9] investigate the immediate efficacy of acupuncture compared to sham acupuncture and placebo laser acupuncture. EMG was used to measure the EMG activity of the rectus femoris muscle and result of *t*-test shows that acupuncture treatment was efficacious for improving isometric quadriceps strength in recreational athletes.

The aforementioned researches demonstrate that sEMG can distinguish the differences of muscles activities before and after the acupuncture treatment with simple statistic analysis. However, these differences have not been quantified precisely.

Furthermore, sEMG is a non-linear summation of the electrical activity of the motor units in a muscle. Each action potential (AP) from each active unit contributes a small and constant increase in the total electrical signal recorded at the skin [10]. sEMG decomposition is a technique that resolves a composite sEMG signal into its constituent motor unit action potential trains (MUAPTs) [11] . Various sEMG decomposition methods have been proposed including Adaptive Resonance Theory networks [12], modified k-means [13], Convolution Kernel Compensation [14] and IPUS based on Gaussian Mixture Model (GMM) [15]. Hence, the recruitment and firing of motor units (MUs) can be obtained, which are significant indicators to detect the disorder and disease of the neuromuscular system.

In this study, we propose to quantify the effectiveness of acupuncture on facial paralysis based on sEMG decomposition. sEMG is decomposed using GMM according to the shape information of MUAP; MUAPTs of healthy and diseased sides are obtained respectively; Features are extracted and dimension-reduced with Principal Component Analysis (PCA); then, the healthy and diseased categories are clustered with k-means method; finally, the center-to-center distance between the two clusters are obtained and used to predict the recovery trend with an auto-regression model.

The rest of the paper is organized as follows. Section 2 describes the sEMG processing algorithms. Section 3 presents the experiment settings. The experimental results are analyzed and reported in Section 4. Section 5 draws the conclusions.

2   Method

The quantification of effectiveness of acupuncture on facial paralysis is constituted with two parts: 1) to decompose sEMG into MUAPTs; 2) assess the effectiveness and predict the recovery trend according to MUAPTs. The flowchart of the signal processing procedure is illustrated in Fig. 1.

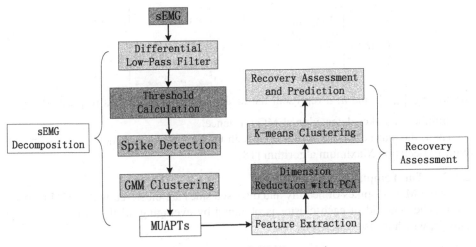

Fig. 1. The flowchart of sEMG precessing

2.1 sEMG Decomposition

2.1.1 sEMG Prepocess and MUAP Detection

According to [16], sEMG is filtered with a 2-order differential filter, that is,

$$x_t = y_{t+2} - y_{t+1} - y_t + y_{t-1} \tag{1}$$

where y_t is the sampled raw signal and x_t is the sampled filtered signal, t is the sample time. Then, the threshold α is calculated according to (2)

$$\alpha = 1.5 \times \sqrt{\frac{1}{N} \sum_{t=1}^{N} x_t^2} \tag{2}$$

where 1.5 is an empirical constant, N is the length of sEMG and x_t is the filtered signal, t is the sample time.

The filtered signal is scanned for locations where the detection threshold is exceeded. Once a MUAP spike is detected, its firing time is defined by the location of the maximum value found within the next 1 ms [17]; and a MUAP spike consists of the 8 neighbouring samples of the peak locations.

$$muap_i = \{x_{peak_i-3}, ..., x_{peak_i}, ..., x_{peak_i+4}\} \tag{3}$$

where subscript $peak_i$ denotes the location of the ith peak of the sEMG.

2.1.2 Clustering with GMM

After the MUAP spikes are detected, a spike matrix can be obtained for the sEMG on healthy and diseased sides:

$$MUAP^{\xi} = \begin{bmatrix} x^{\xi}_{peak_i-3}, x^{\xi}_{peak_i-2}, \dots, x^{\xi}_{peak_i}, x^{\xi}_{peak_i+1}, \dots, x^{\xi}_{peak_i+4} \\ x^{\xi}_{peak_j-3}, x^{\xi}_{peak_j-2}, \dots, x^{\xi}_{peak_j}, x^{\xi}_{peak_j+1}, \dots, x^{\xi}_{peak_j+4} \\ \dots \\ x^{\xi}_{peak_k-3}, x^{\xi}_{peak_k-2}, \dots, x^{\xi}_{peak_k}, x^{\xi}_{peak_k+1}, \dots, x^{\xi}_{peak_k+4} \end{bmatrix}_{q \times 8} = \begin{bmatrix} \mathbf{x}^{\xi}_i \\ \mathbf{x}^{\xi}_j \\ \dots \\ \mathbf{x}^{\xi}_k \end{bmatrix}_{q \times 8} \quad (4)$$

where $\xi \in \{h, d\}$ and h, d represent the healthy and diseased sides respectively; q is the number of spikes detected in sEMG sequence.

Then, the matrix MUAP is clustered with GMM, whose parameters are estimated with Expectation Maximum algorithm [18].

2.1.3   The Template of each MUAPT

After the MUAP spikes of healthy and diseased sides are clustered with GMM respectively, the centroid of each cluster is obtained by averaging all the samples in each cluster, which is also called template.

2.2   Recovery Assessment Based on MUAPT

2.2.1   Feature Extraction

In this paper, we propose to extract features including integral of absolute value (*IAV*), maximum value (*MAX*), median value of non-zero value (*NonZeroMed*), and semi-window energy (*SemiEny*) with a time window of 400ms and a sliding window of 100ms. The calculations of these features are as follows:

a) Integral of Absolute Value (*IAV*)

$$IAV = \frac{1}{N} \sum_{i=1}^{N} |x_i| \quad (5)$$

where x_i is the ith sampling point, and N is the length of time windows.

b) Maximum Value (*MAX*)

$$MAX = \max_{i=1,\dots,N} |x_i| \quad (6)$$

where x_i is the ith sampling point, and N is the length of time windows.

c) Median of Non-zero Value (*NonZeroMed*)

$$S = find(x_i \neq 0)$$
$$NonZeroMed = median(S) \quad (7)$$

where x_i is the ith sampling point, $i = 1, 2, 3, \dots, N$, and N is the length of time windows.

d) Semi-window Energy (*SemiEny1* and *SemiEny2*)

$$SemiEny1 = \sum_{i=1}^{N/2} x_i^2 \quad and \quad SemiEny2 = \sum_{i=(N/2)+1}^{N} x_i^2 \quad (8)$$

where x_i is the ith sampling point, and N is the length of time windows.

After the features are extracted, feature matrice $\mathbf{FM}^{\xi}$ of healthy and diseased sides are formed respectively.

2.2.2   Dimension Reduction with PCA Method

Principal component analysis (PCA) is a statistical procedure that uses an orthogonal transformation to convert a set of observations of possibly correlated variables into a set of values of linearly uncorrelated variables called principal components, which is also used for dimension reduction of matrix [19]. The orthogonal transformation W is given by

$$\mathbf{Y}_{p\times\eta}^{\xi} = \mathbf{FM}_{p\times(5M)}^{\xi} \cdot \mathbf{W}_{5M\times\eta} = \left[PC_1, PC_2, ..., PC_\eta \right]_{p\times\eta} \tag{9}$$

where $\mathbf{Y}_{p\times\eta}^{\xi}$ is the dimension-reduced matrix and W is the orthogonal transformation matrix obtained with PCA method.

2.2.3   k-Means Algorithm

In this paper, the dimension-reduced matrice $\mathbf{Y}_{p\times\eta}^{\xi}$, $\xi \in \{h,d\}$ are clustered with k-means algorithm into 2 subsets while minimizing the within-cluster sum of squares (WCSS) [20].

After that, the sEMG samples have been divides into two subsets, the healthy cluster and diseased cluster. The distance between the centroids of the two clusters is then used to assess the effectiveness of acupuncture on facial paralysis and predict the recovery trends of patients.

3   Experiment and Results

3.1   Participants

Twenty participants, including thirteen male and seven female, randomly selected from a pool of subjects treated in Shenyang Hospital of Traditional Chinese Medicine (SHTCM), Shenyang, China, aged between 22 and 64 years old (44.2±4.84 years), with an average weight of 60±9.45 kg and height of 1.65±0.7m, participated in this test. They were clinically examined with regard to inability to control facial muscles to some extent and received no other treatments before this test.

3.2   Procedure

All participants completed a review of his/her date of onset, data of first consultation, cardinal symptoms, accompanying symptoms etc. in advance to help clinician to determine the acupoints for needling before the test. In this study, the acupoints including Yingxiang(LI20), Sibai (ST2), Quanliao(SI18), Dicang(ST4) were selected by an experienced acupuncturist and will be punctured for 30 minutes each day. The needles were stainless, with a diameter of 0.2mm and length of 30mm. The depth, angle and manipulation of needles insertion depended on the participants' symptoms and the discretion of an experienced acupuncturist.

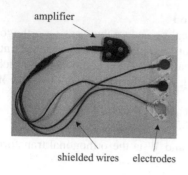

amplifier

shielded wires   electrodes

Fig. 2. The pre-amplifier and the electrodes

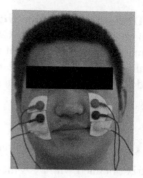

Fig. 3. The position to place the electrodes on the muscle of levator labii superioris

sEMG signals acquired on the bilateral levator labii superioris (LLS) were utilized to assess the muscles activity difference between the healthy and the diseased parts. Results before and after the acupuncture were compared. These analyses were conducted with FlexComp (Thought Technology Co. Ltd.®, Canada), consisting of 6-channel pre-amplifiers (MyoScan-Flex, with a gain of 1000 and a common rejection mode ratio >100 dB, Fig.2) and self-adhesive disposable surface electrodes (Ag/AgCl) with a diameter of 8 mm and a center to center distance of 20 mm, which adhered to the skin that was cleaned previously with alcohol. The positions to affix the electrodes can be seen in Fig.3. All data was acquired and sampled with a frequency of 2kHz and then digitally filtered by a bandpass filter of 2-500Hz and a notch filter of 50Hz for the further process with MATLAB.

The participants were requested to complete cheek-bulging (CB), pouting (PT), grining (GN), nose-wrinkling (NW) etc. before and after the acupuncture. Each movement last for 2s approximately and was repeated 3 or 4 times with an interval of 3s in each session. The participants received therapy 5 days a week and the sEMG recording was conducted with a fixed 7 days interval.

After the sEMG signals are filtered with the 2-order differential low-pass filter given in Eq. (1). The threshold value is calculated to be 49.9 according to Eq. (2). And the MUAP spikes are marked with red circles locating at their peaks' location, as illustrated in Fig.5. The numbers of MUAP spikes, which is shown in legends, on healthy side is greater than that on diseased side because of the larger amplitude of healthy side and the same threshold.

Then, the detected MUAP spikes on healthy and diseased sides are clustered with GMM, respectively. The healthy side is clustered into 5 subsets while the diseased side is into 3 subsets according to Akaike information criterion (AIC) [21]. Hence, the feature matrix of healthy side consists of 25 columns and diseased side 15 columns as described in Eq. (10).

Moreover, the templates of healthy side are compared with those of diseased side; the pairs of templates with the smallest Euclidean distances are plotted in Fig. 6. The numbers in Fig. 6(a-c) are the Euclidean distances between the pairs of the healthy and diseased templates, respectively.

The MUAPTs corresponding to the templates which have similar tracks are selected; and the differences between the healthy and diseased sides are quantified using the feature vectors of the selected MUAPTs, as shown in Fig.7. The feature vectors in red square are selected to quantify the differences between the healthy and diseased side.

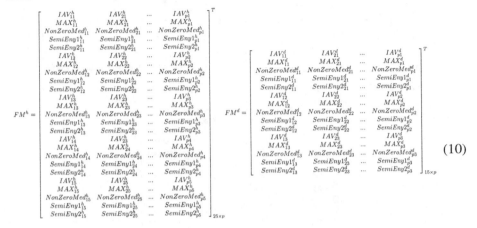

$$(10)$$

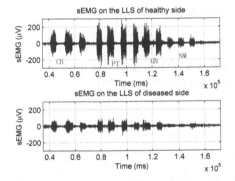

Fig. 4. The raw sEMG acquired on the bilateral levator labii superioris (LLS)

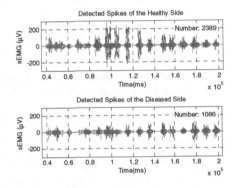

Fig. 5. The detected MUAP spikes of healthy and diseased sides. The number in the legend is the number of MUAP spikes detected in the sEMG sequence.

As given in Eq. (11), PCA method is used to reduce the dimension of the selected 15 columns of the feature vectors into 5, namely $\eta = 5$ in Eq. (9), because 5 components have accounted for 90.1% of variance of the matrix. The selection of components number is an empirical procedure based on the idea that the selected components should account for more than 90% of the variance normally. More details can be found in [22].

$$\mathbf{Y}_{p\times5} = \mathbf{FM}_{p\times15} \cdot \mathbf{W}_{15\times5} = \left[PC_1, PC_2, PC_3, PC_4, PC_5\right]_{p\times5} \qquad (11)$$

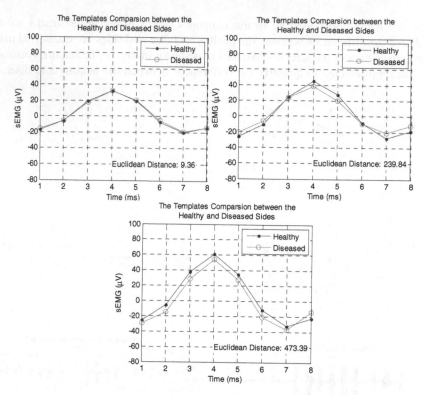

Fig. 6. The templates comparison between the healthy side and diseased side. The number in each plot is the Euclidean distance between the healthy and diseased templates.

$$[FM^h; FM^d] = \begin{bmatrix} IAV^h_{11} & IAV^h_{21} & \dots & IAV^h_{p1} & IAV^d_{11} & IAV^d_{21} & \dots & IAV^d_{p1} \\ MAX^h_{11} & MAX^h_{21} & \dots & MAX^h_{p1} & MAX^d_{11} & MAX^d_{21} & \dots & MAX^d_{p1} \\ NonZeroMed^h_{11} & NonZeroMed^h_{21} & \dots & NonZeroMed^h_{p1} & NonZeroMed^d_{11} & NonZeroMed^d_{21} & \dots & NonZeroMed^d_{p1} \\ SemiEny1^h_{11} & SemiEny1^h_{21} & \dots & SemiEny1^h_{p1} & SemiEny1^d_{11} & SemiEny1^d_{21} & \dots & SemiEny1^d_{p1} \\ SemiEny2^h_{11} & SemiEny2^h_{21} & \dots & SemiEny2^h_{p1} & SemiEny2^d_{11} & SemiEny2^d_{21} & \dots & SemiEny2^d_{p1} \\ IAV^h_{12} & IAV^h_{22} & \dots & IAV^h_{p2} & IAV^d_{12} & IAV^d_{22} & \dots & IAV^d_{p2} \\ MAX^h_{12} & MAX^h_{22} & \dots & MAX^h_{p2} & MAX^d_{12} & MAX^d_{22} & \dots & MAX^d_{p2} \\ NonZeroMed^h_{12} & NonZeroMed^h_{22} & \dots & NonZeroMed^d_{p2} & NonZeroMed^d_{12} & NonZeroMed^d_{22} & \dots & NonZeroMed^d_{p2} \\ SemiEny1^h_{12} & SemiEny1^h_{22} & \dots & SemiEny1^h_{p2} & SemiEny1^d_{12} & SemiEny1^d_{22} & \dots & SemiEny1^d_{p2} \\ SemiEny2^h_{12} & SemiEny2^h_{22} & \dots & SemiEny2^h_{p2} & SemiEny2^d_{12} & SemiEny2^d_{22} & \dots & SemiEny2^d_{p2} \\ IAV^h_{13} & IAV^h_{23} & \dots & IAV^h_{p3} & IAV^d_{13} & IAV^d_{23} & \dots & IAV^d_{p3} \\ MAX^h_{13} & MAX^h_{23} & \dots & MAX^h_{p3} & MAX^d_{13} & MAX^d_{23} & \dots & MAX^d_{p3} \\ NonZeroMed^h_{13} & NonZeroMed^h_{23} & \dots & NonZeroMed^d_{p3} & NonZeroMed^d_{13} & NonZeroMed^d_{23} & \dots & NonZeroMed^d_{p3} \\ SemiEny1^h_{13} & SemiEny1^h_{23} & \dots & SemiEny1^h_{p3} & SemiEny1^d_{13} & SemiEny1^d_{23} & \dots & SemiEny1^d_{p3} \\ SemiEny2^h_{13} & SemiEny2^h_{23} & \dots & SemiEny2^h_{p3} & SemiEny2^d_{13} & SemiEny2^d_{23} & \dots & SemiEny2^d_{p3} \\ IAV^h_{14} & IAV^h_{23} & \dots & IAV^h_{p4} & \dots & \dots & \dots & \dots \\ MAX^h_{14} & MAX^h_{24} & \dots & MAX^h_{p4} & \dots & \dots & \dots & \dots \\ NonZeroMed^h_{14} & NonZeroMed^h_{24} & \dots & NonZeroMed^h_{p4} & \dots & \dots & \dots & \dots \\ SemiEny1^h_{14} & SemiEny1^h_{24} & \dots & SemiEny1^h_{p4} & \dots & \dots & \dots & \dots \\ SemiEny2^h_{14} & SemiEny2^h_{24} & \dots & SemiEny2^h_{p4} & \dots & \dots & \dots & \dots \\ IAV^h_{15} & IAV^h_{25} & \dots & IAV^h_{p5} & \dots & \dots & \dots & \dots \\ MAX^h_{15} & MAX^h_{25} & \dots & MAX^h_{p5} & \dots & \dots & \dots & \dots \\ NonZeroMed^h_{15} & NonZeroMed^h_{25} & \dots & NonZeroMed^h_{p5} & \dots & \dots & \dots & \dots \\ SemiEny1^h_{15} & SemiEny1^h_{25} & \dots & SemiEny1^h_{p5} & \dots & \dots & \dots & \dots \\ SemiEny2^h_{15} & SemiEny2^h_{25} & \dots & SemiEny2^h_{p5} & \dots & \dots & \dots & \dots \end{bmatrix}_{25 \times p}$$

Fig. 7. The feature vectors in red square corresponding to the selected MUAPTs with similar templates are used to quantify the differences between the healthy and diseased side.

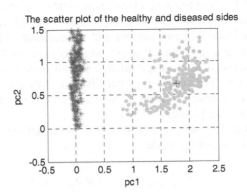

Fig. 8. PC2 with respect to PC1 for a patient with facial paralysis. The samples of healthy side are clearly separated with those of diseased side.

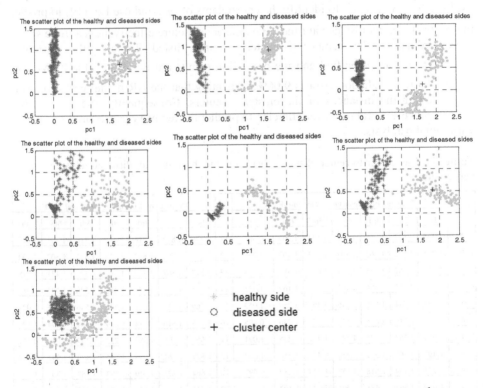

Fig. 9. The clusters and center-to-center distances of patient-1 during acupuncture therapy

The first two components PC1 and PC2 are plotted in Fig. 8, from which we can see that the diseased and healthy sides are clearly separated. The center-to-center distances of patient-1 during the acupuncture treatment are plotted in Fig. 9.

Furthermore, the k-means clustering algorithm is employed to clustering the healthy and diseased side into two subsets and the centroid of each cluster is calculated during a complete acupuncture therapy, which consists of 7 times treatments.

20 patients participated in our experiments, and all their sEMG signals were analyzed in the same way. The center-to-center distances of each patient during his/her acupuncture therapy are listed in Table 1. Moreover, the center-to-center distances of the twenty participants are plotted in Fig. 10.

Finally, we can build an auto-regressive model to predict the recovery trend for the patients. In this study, a 4-order model is employed, i.e.

$$\hat{x}_k = a_1^k x_{k-1} + a_2^k x_{k-2} + a_3^k x_{k-3} + a_4^k x_{k-4} \tag{12}$$

where x_k represents the center-to-center distance at time k; $\hat{x}_k$ is the predicted value, and a_i, $i = 1,......,4$, are the coefficients. Then, for any $k \geq 5$, we can use $[x_{k-1}, x_{k-2}, x_{k-3}, x_{k-4}]$ to identify the coefficients of a_i^k, and use Eq. (12) to predict the center-to-center distance at time k, i.e., the acupuncture effectiveness at time k.

According to the center to center distance listed Table-1, the variables a_1, a_2, a_3, a_4 are identified and the prediction results are listed in Table-2. We can see the largest prediction error is 9.06%, indicating that the AR model performs well in predicting the distances of the clusters' centers. Consequently, the 4-order AR model can be a useful index for clinician to predict the recovery procedure of a patient with facial paralysis.

Table 1. The center-to-center distance between the healthy and diseased clusters of 20 participants

No.	The times of acupuncture / sampling sEMG							No.	The times of acupuncture / sampling sEMG						
	1st	2nd	3rd	4th	5th	6th	7th		1st	2nd	3rd	4th	5th	6th	7th
1	1.855	1.867	1.963	1.722	1.544	1.269	0.985	11	1.778	1.877	1.716	1.632	1.518	1.491	1.315
2	1.785	1.842	1.863	1.703	1.623	1.469	1.256	12	1.691	1.709	1.682	1.549	1.496	1.307	1.134
3	1.753	1.867	1.735	1.718	1.615	1.385	1.164	13	1.813	1.867	1.763	1.682	1.549	1.431	1.218
4	1.555	1.673	1.596	1.463	1.215	1.162	1.038	14	1.581	1.649	1.523	1.418	1.328	1.193	0.983
5	1.655	1.784	1.695	1.546	1.457	1.259	1.053	15	1.769	1.817	1.719	1.623	1.546	1.361	1.027
6	1.795	1.716	1.636	1.549	1.504	1.236	0.885	16	1.775	1.816	1.714	1.629	1.564	1.394	1.189
7	1.653	1.759	1.784	1.598	1.497	1.285	1.091	17	1.653	1.49	1.383	1.222	1.147	1.062	0.967
8	1.498	1.567	1.692	1.538	1.429	1.317	1.067	18	1.829	1.761	1.643	1.586	1.423	1.392	1.162
9	1.731	1.709	1.618	1.537	1.429	1.289	0.997	19	1.864	1.721	1.659	1.593	1.482	1.307	1.125
10	1.639	1.596	1.481	1.391	1.283	1.194	0.979	20	1.745	1.736	1.646	1.509	1.437	1.319	1.164

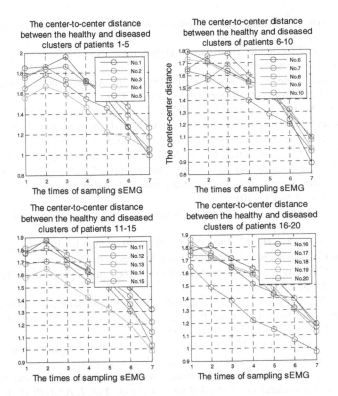

Fig. 10. The center-to-center distances between the healthy and diseased clusters of 20 patients during acupuncture therapy

4   Conclusion

In this paper, we propose a sEMG decomposition based quantification method of the effectiveness of acupuncture on facial paralysis. sEMG on healthy and diseased sides are decomposed into MUAPTs using GMM respectively. Features including *IAV*, *MAX*, *NonZeroMed*, *SemiEny1* and *SemiEny2* are extracted and dimension-reduced with PCA method; then, *k*-means algorithm is employed to cluster the healthy and diseased sides to obtain the center-to-center distance between the two clusters, which are used to predict the recovery trend with an auto-regression model finally. Results indicate the proposed method can achieve an accuracy of 90.94%.

References

1. Baldonado, M., Chang, C.-C.K., Gravano, L., Paepcke, A.: The Stanford Digital Library Metadata Architecture. Int. J. Digit. Libr. 1, 108–121 (1997)
2. Bruce, K.B., Cardelli, L., Pierce, B.C.: Comparing Object Encodings. In: Ito, T., Abadi, M. (eds.) TACS 1997. LNCS, vol. 1281, pp. 415–438. Springer, Heidelberg (1997)

716   A. Xiong et al.

3. van Leeuwen, J. (ed.): Computer Science Today. LNCS, vol. 1000. Springer, Heidelberg (1995)
4. Michalewicz, Z.: Genetic Algorithms + Data Structures = Evolution Programs, 3rd edn. Springer, New York (1996)
5. World Health Organization. Acupuncture: Review and Analysis Of Reports On Controlled Clinical Trials. World Health Organization (June 2002)
6. Wu, B., Li, N., Liu, Y., Huang, C.Q., Zhang, Y.L.: Study on clinical effectiveness of acupuncture and moxibustion on acute Bell's facial paralysis: randomized controlled clinical observation. Chinese Acupuncture & Moxibustion 26(3), 157–160 (2006) (in Chinese)
7. Li, Y., Li, Y., Liu, L.A., Zhao, L., Hu, K.M., Wu, X., Chen, X.Q., Li, G.P., Mang, L.L., Qi, Q.H.: Acupuncture and moxibustion for peripheral facial palsy at different stages:multi-central large-sample randomized controlled trial. Chinese Acupuncture & Moxibustion 31(4), 289–293 (2011)
8. Hsieh, C.L.: Acupuncture as treatment for nervous system diseases. BioMedicine 2, 51–57 (2012)
9. Chen, N., Zhou, M., He, L., Zhou, D., Li, N.: Acupuncture for Bell'spalsy (review). The Cochrane Collaboration, vol. (8). John Wiley & Sons, Ltd. (2010)
10. De Luca, C.J.: Physiology and mathematics of myoelectric signals. IEEE Trans. Biomed. Eng. 26, 313–325 (1979)
11. Rancan, S.V., Bataglion, C., Bataglion, S.A., et al.: Acupuncture and Temporomandibular Disorders: A 3-Month Follow-up EMG Study. The Journal of Alternative and Complementary Medicine 15(12), 1307–1310 (2009)
12. Costa, L.A., Araujo, J.E.: The immediate effects of local and adjacent acupuncture on the tibialis anterior muscle: a human study. Chin. Med. 3, 17 (2008)
13. Hübscher, M., Vogt, L., Ziebart, T., et al.: Immediate effects of acupuncture on strength performance: a randomized, controlled crossover trial. Eur. J. Appl. Physiol. 110(2), 353–358 (2010)
14. Wang, M., Loo, W.T., Chou, J.W.: Electromyographic responses from the stimulation of the temporalis muscle through facial acupuncture points. Journal of Chiropractic Medicine 6(4), 146–152 (2007)
15. Gerdle, B., Ostlund, N., Grönlund, C., Roeleveld, K., Karlsson, J.S.: Firing rate and conduction velocity of single motor units in the trapezius muscle in fibromyalgia patients and healthy controls. J. Electromyogr. Kinesiol. 18(5), 707–716 (2008) (Epub April 24, 2007)
16. Gazzoni, M., Farina, D., Merletti, R.: A new method for the extraction and classification of single motor unit action potentials from surface EMG signals. J. Neurosci. Methods 136(2), 165–177 (2004)
17. Stashuk, D.W.: Decomposition and quantitative analysis of clinical electromyographic signals. Medical Engineering & Physics 21(6), 389–404 (1999)
18. Holobar, A., Zazula, D.: Multichannel blind source separation using convolution kernel compensation. IEEE Trans. Signal Process. 55(9), 4487–4496 (2007)
19. Nawab, S.H., Chang, S.S., De Luca, C.J.: High-yield decomposition of surface EMG signals. Clinical Neurophysiology 121(10), 1602–1615 (2010)
20. McGill, K.C., Cummins, K.L., Dorfman, L.J.: Automatic decomposition of the clinical electromyogram. IEEE Trans. Biomed. Eng. BME-32, 470–477 (1985)
21. Stashuk, D.W.: Decomposition and quantitative analysis of clinical electromyographic signals. Medical Engineering & Physics 21(6), 389–404 (1999)
22. Bailey, T.L., Charles, E.: Fitting a mixture model by expectation maximization to discover motifs in biopolymers, pp. 28–36 (1994)

23. Richard, O.D., Peter, E.H., David, G.S.: Pattern Classification, 2nd edn., pp. 113–117. Wiley-Interscience (October 2000)
24. David, M.: An Example Inference Task: Clustering. In: Information Theory, Inference and Learning Algorithms, August 25, ch. 20, pp. 284–292. Cambridge University Press. Version 7.0 (2004)
25. Akaike, H.: A new look at the statistical model identification. IEEE Transactions on Automatic Control 19(6), 716–723 (1974)
26. Cattell, R.B.: The scree test for the number of factors. Multivariate Behav. Res., 245–276 (April 1966)

Human Detection Algorithm Based on Edge Symmetry

Hao Wang, Jinjin Chen, Baofu Fang, and Shuanglu Dai

Hefei University of Technology,
Hefei, China

Abstract. One of the most important abilities that personal robots need when interacting with humans is the ability to detecte human timely. Due to the scan of the input image without any disparity in the traditional method for human detection, processing speed can not meet the demand of a real-time system appropriately. Under such a circumstance, an edge symmetry based human detection algorithm is proposed. With mechanism of scan lines, the symmetrical value of each pixel is calculated along scan line and candidate regions are picked out. Then candidate regions are verified by using Histograms of Oriented Gradients (HOG) feature and Support Vector Machine(SVM) classifier. Experiment shows that the algorithm has a good command of keeping the precise of the recognition as well as elevating the speed of calculation.

Keywords: machine vision, pedestrian detection, edge symmetry, histograms of oriented gradients, support vector machine (SVM).

1 Introduction

In the next few years, the individual service robots will become a part of our daily life, play important role as our assistant and elder-care companions. Personal service robots need to be able to understand the outside environment, and be aware with of various complex situations. In particular, robots have to complete the human identification and tracking with high quality and precision. However, due to the various outer surroundings and fluctuant of the pedestrian, the human detection is also a challenging problem in the field of computer vision.

Traditional pedestrian detection method usually adopts the sliding window mechanism, a fixed size detection window scanning the whole input image line by line, extracting the features for the judgment condition of pedestrian existence. Dalal et al. [1] proposed the classic pedestrian detection method based on the sliding window mechanism. Dalal focus on the description of character operators with the basic thinking that different kind of the target emerged in the window has obvious difference. Another classic method [2][3] is to simplify the calculation by introducing a cascade of multi-classifier which is also based on sliding window mechanism.

However, the speed of sliding windows mechanism will be retarded by scanning the whole image, which can't be used in a real-time system. For example, even dealing with a 640×480 image with low resolution, the times of scanning will be above 10 million. Considering of such situation, implementations to cut down the

© Springer International Publishing Switzerland 2015
J.-H. Kim et al. (eds.), *Robot Intelligence Technology and Applications 3,*
Advances in Intelligent Systems and Computing 345, DOI: 10.1007/978-3-319-16841-8_65

719

number of the scanning regions has become popular. Combining with stereo vision, Arie et al[4] calculate distance information focusing on only foreground regions in an input image. The size of detection window is determined appropriately and dynamically based on the distance information and paraperspective projection model, which can avoid multiple tests of window size and leads to reduce the number of detection windows. Bertozzi et al[5] pick out the possible rectangle list contain pedestrian considering the use of two stereo camera systems simultaneously: far infrared cameras and daylight cameras. Xia at el[6] confirm the possible pedestrian regions according to the depth information which taken by kinect directly. Such approach above considered the specific equipment which will have some prohibit in the practical application. Therefore, this paper concern on how to decrease the regions scanned with image pictured only by ordinary monocular camera for an accelerated pedestrian detection algorithm.

Relying on the hypothesis that a pedestrian is featured by mainly vertical edges with a strong symmetry respect to a vertical axis; size and aspect ratio satisfying specific constraints; always generally placed in a specific region. Schauland et al [7] make use of Sobel operator to extract boundaries and SVM to deal with the classification while Cosma et al [8] extract all the pedestrian hypotheses using stereo cameras at first, and then refine the pedestrian hypotheses by computing the symmetry axis for every hypothesis and generate new hypothesis.

In this paper, considering the character of boundaries symmetry for pre-dealing with the input image, candidate regions is picked out for SVM classification, which simplified the process of calculation. Experiment shows that the algorithm can speed up the processing effectively as well as decrease the computing expenses at scan stage.

2   Symmetry-Based Pedestrian Detection Method and Coordinate Definition

2.1   Symmetry-Based Pedestrian Detection Method

In vision area, for such a symmetric object, there is at least one symmetry axis that splits the object into two identical but mirror-inverted halves. As Fig.1 shows, a pedestrian can be seemed as a symmetrical object with a vertical symmetrical axis.

Fig. 1. Pedestrian samples

So, as soon as the vertical symmetrical axis is determined, the position of a pedestrian can be calculated. An effective method to calculate such axis is to find the high symmetrical value of the pixel. In general, the operators calculate the symmetry value of a point in the image by summing up the number of the pair pixels with same character that are equidistant but in opposite directions from a symmetry center.

Calculating symmetry for each pixel will be a time consuming process. By introducing mechanism of scan lines to carry out several scan lines that cover predefined regions, standard character of pedestrian is kept and the computing processing will be accelerated without any impact on the detection result. Such regions can be predicted based on the road positions and the geometric constraints.

Several pixel features can be used to calculate the symmetry value such as gray scale value, binary contour and horizontal edges. Gray value can quickly isolate the pedestrian from background, yet easily affected by illumination. Contour is insensitive to illumination variations, but the detection can be affected by symmetric or partially symmetric background objects such as simple billboards or construction symmetry edges. Horizontal line can reduce the influences from background symmetric objects, but the result depends on how well the horizontal lines can be extracted.

This paper uses the camera parameters both inside and outside to determine scanning area, uniformly distributed several horizontal scan lines in the scanning area, and then calculates the size of scanning window for each scan line. By taking its advantage of less sensation to noise and intensity changes in the image, contour-based symmetry detection is adopted. Symmetrical values calculated for each pixel along scan line within the scope of the scan window, determines number of candidate regions based on the distribution of peak symmetry value. Feature vector of each candidate region to classification verification is extracted, eventually determining areas for pedestrians in original image.

2.2   Definition of Relative Coordinate

At first, define the image coordinate system, camera coordinate system, world coordinate system and relative parameters [11].

Fig.2 shows the image coordinate system, whereis the original point;is the coordinate of the system with pixel as its unit.is the intersection between the camera optical axis and image plat; x-coordinate and y-coordinate is paralleled to u-coordinate and v-coordinate correspondingly.is the coordinate of the system with millimeter as its unit .

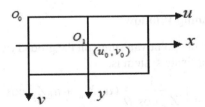

Fig. 2. Image coordinate system

Fig.3 shows the geometry relationship of the camera imaging, whereis the optical center;andis paralleled to x-coordinate and y-coordinate; -coordinate is the axis of the camera which is vertical to the image plat. The camera coordinate system is constructed by origin point and coordinates with as the focal distance. Meanwhile, define a world coordinate system where Rotation matrix and shift matrix describe the position relationship between the camera coordinate system and world coordinate system.

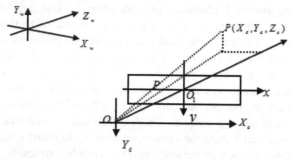

Fig. 3. Camera and world coordinate system

Assume the inner parameter matrix of the camera K and the altitude of the camera is h , the original point of the world coordinate system is at the projection of the camera.

Assume $K = \begin{bmatrix} a_x & 0 & u_0 & 0 \\ 0 & a_y & v_0 & 0 \\ 0 & 0 & 1 & 0 \end{bmatrix}$, outer parameter matrix $M = \begin{bmatrix} R & t \\ 0^T & 1 \end{bmatrix}$, where

$R = \begin{bmatrix} 1 & 0 & 0 \\ 0 & \cos\theta & -\sin\theta \\ 0 & \sin\theta & \cos\theta \end{bmatrix}$ as a rotation matrix around axis X_c , $t = \begin{bmatrix} 0 \\ h \\ 0 \end{bmatrix}$ is a shift

matrix between the camera and the world coordinate system where h is the altitude of the camera.

3   Edge Symmetry Based Pedestrian Detection

3.1   Scale of the Scanning Window

The relationship between the point vanishing on the ground (world coordinate system) and that in the image coordinate system is:

$$\begin{bmatrix} u \\ v \end{bmatrix} = \begin{bmatrix} \dfrac{1}{Z_w \cos\theta}(a_x X_w + u_0 Z_w \cos\theta) \\ v_0 - a_y \tan\theta \end{bmatrix} .$$

As point P in the world coordinate system is known, its projection in the image coordinate system is written as:

$$\begin{bmatrix} u \\ v \\ 1 \end{bmatrix} = \lambda KM \begin{bmatrix} X_w \\ Y_w \\ Z_w \\ 1 \end{bmatrix} \tag{1}$$

When point P is on plat $x-z$ in the world coordinate system, $Y_w = 0$, Eq.(1)can be simplified as:

$$\begin{bmatrix} u \\ v \\ 1 \end{bmatrix} = \lambda K \begin{bmatrix} r_1 & r_3 & t \end{bmatrix} \cdot \begin{bmatrix} X_w \\ Z_w \\ 1 \end{bmatrix} \tag{2}$$

Where $\begin{bmatrix} r_1 & r_2 & r_3 & t \end{bmatrix} = \begin{bmatrix} R & t \\ 0^T & 1 \end{bmatrix}$,

Future more, point   is on an infinite line in   plane,

$$\begin{bmatrix} u \\ v \\ 1 \end{bmatrix} = \lambda K \begin{bmatrix} r_1 & r_3 & t \end{bmatrix} \cdot \begin{bmatrix} X_w \\ Z_w \\ 0 \end{bmatrix} \tag{3}$$

Solve Eq.(3),

$$u = \frac{1}{Z_w \cos\theta}(a_x X_w + u_0 Z_w \cos\theta) , \tag{4}$$

$$v = v_0 - a_y \tan\theta$$

With the induction from Eq.(1) to Eq.(4), v-coordinate of the point vanishing on the ground is only relative to the pitch angle and inner parameter of the camera. As soon as the inner and outer parameter has been determined, the regions for scanning will be determined. For a more simplified calculation, assume the 12 scan lines distribute with same interval in the scanning regions as Fig.4 shows:

Fig. 4. Scan line distribution

The relationship between the point in the image coordinate system and that in the world coordinate system $\begin{bmatrix} X_w \\ Z_w \\ 1 \end{bmatrix} = \begin{bmatrix} \dfrac{(u - u_0)a_y h}{(v - v_0)a_x} \\ \dfrac{a_y h}{(v - v_0)} \\ 1 \end{bmatrix}$:

According to Eq.(1) , the equation above can be written as:

$$\begin{bmatrix} u \\ v \\ 1 \end{bmatrix} = \lambda \begin{bmatrix} a_x & u_0 \cos\theta & 0 \\ 0 & -a_y \sin\theta + v_0 \cos\theta & a_y h \\ 0 & \cos\theta & 0 \end{bmatrix} \begin{bmatrix} X_w \\ Z_w \\ 1 \end{bmatrix} \tag{5}$$

When $\cos\theta \neq 0$, $N = \begin{bmatrix} a_x & u_0 \cos\theta & 0 \\ 0 & -a_y \sin\theta + v_0 \cos\theta & a_y h \\ 0 & \cos\theta & 0 \end{bmatrix}$ is an invertible matrix.

Specialliy, $\theta = 0$, which means the camera plane is parallel to the ground, respectively,

$$Z_w = \frac{a_y h}{(v - v_0)}, \quad X_w = \frac{(u - u_0)a_y h}{(v - v_0)a_x} \tag{6}$$

Let H_{img} be the heights of the pedestrian, then $H_{img} = \dfrac{f}{Z_w} H_{obj}$, where f is the focal distance fixed by a camera, H_{obj} represents the real heights of the pedestrian, according to Eq.(7):

$$H_{img} = \frac{f \cdot H_{obj}}{a_y h}(v - v_0) \tag{7}$$

The height H_{img} of the scanning window at image position v will be calculated out as well as the scale of the window.

3.2   Matrix of the Symmetrical Value

After the scale of the scanning window calculated, a process of border detection need to be done with method of Canny operator so that the symmetry character matrix will be determined. In the bottom center of the scanning window, the symmetrical value can be written as follow according to the regular character of pedestrian as Eq.(8):

$$SymVal\,(x,y)=\frac{100}{H}\cdot\sum_{x'=1}^{W/2}\sum_{y'=0}^{H}I(x,x',y',c) \tag{8}$$

Where W, H represent the width and height of the scanning window; x, y indicate the position of the pixel in the input image. Let c represent the group of axis:

$$c=\begin{cases}\dfrac{2}{3}W\,, & if \quad y'\le\dfrac{3}{7}H \\[2mm] 0\,, & if \quad \dfrac{3}{7}H<y'\le\dfrac{6}{7}H \\[2mm] \dfrac{2}{3}W\,, & if \quad y'>\dfrac{6}{7}H\end{cases} \tag{9}$$

Then symmetrical value $I(x,x',y',c)$ can be calculated as follow:

$$I(x,x',y',c)=\begin{cases}1-3\times\dfrac{|x'-c|}{W}, & if\ |x'-c|\le\dfrac{W}{6}\ and\ I(x-x',y-y')=I(x+x',y-y') \\[2mm] 0.5-1.5\times\dfrac{|x'-c|}{W}, & if\ \dfrac{W}{6}<|x'-c|\le\dfrac{W}{3}\ and\ I(x-x',y-y')=I(x+x',y-y') \\[2mm] 0.25-0.75\times\dfrac{|x'-c|}{W}, & if\ \dfrac{W}{3}<|x'-c|\le\dfrac{W}{2}\ and\ I(x-x',y-y')=I(x+x',y-y')\end{cases} \tag{10}$$

Comparing with several other methods, pedestrian regions are predicted more exactly in this paper than that of others. There are more possible pedestrian regions predicted by our method than that of others.

3.3 Candidate Regions Confirmation

After filtering the symmetrical value matrix with eliminating the point which either small symmetrical value or isolation character, accumulate the matrix according to its column and extract the peak of the result by using non-maximum suppression algorithm by two times. The first time is to determine local maximum point in each region and the second is to find the peak value point in all regions.

The aim of candidate regions is to pick out rectangle regions with possibility including pedestrian. Based on the basic result of peak value point extraction and the statistics of average proportion of pedestrian's heights and width, the only parameter need to get is the bottom of the pedestrian. By searching according to the order of the column, the candidate region will be induced. One of the results is shown right.

Fig. 5. Candidate regions

3.4   Candidate Regions Verification

The method to verify the candidate region is to extract the HOG feature vector and verify it by SVM classifier. A HOG featur vector calculates the gradient direction of each pixel in the target region, quantifies them to a predefined range and histograms are calculated and used as the feature vector for classification. The detailed steps are listed below:

a)Scale the candidate regions to 64×128

b) Divide the image to be 8×8 pixels cell, where each 2×2 cells is constructed to be a block(As Fig.6 shown)

c)Calculate the gradient of each pixel in the image

d)Synthesize all the gradient vector to be a gradient orientation histogram.

e)Connect all the cells in a block in series to form a HOG character vector

f)Connect all HOG character vector of blocks in series to form the feature vector for classification

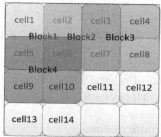

Fig. 6. Block and cell distribution

Support Vector Machine(SVM) is a machine learning method established on the theory of statics learning and VC dimension and principle of minimized structural risk which takes an great advantage in samples with small amount, nonlinear and high dimensional pattern recognition. Its basic principle is to find an optimal separating hyperplane meeting the requirement of the classification and keeping the precise of the recognition as well as maximize the nearest minimum interval. In dealing with high dimensional feature vector, SVM propose to map the feature vector into high-dimensional space by kernel function, avoid complicated calculation in the high-dimensional space directly.

4   Experiments and Results

In the experiment, 1000 positive sample images with pedestrian only and 1000 background images without pedestrian are chosen after united processing (64*128 pixels) from INRIA database established by Dalal et al.[1] With the C_SVC classification model and a fastest linear kernel function of SVM provided by LIBSVM. Optimize the parameter with libsvm. Experiment tests a whole video, part of the results are shown as Fig.7:

Fig. 7. results of experiment

In order to analyze the effect of the detection, regular standard for pedestrian detection algorithm is introduced. The basic evaluation parameter is described in Table 1.

Table 1. Basic evaluation parameters

	ActualPositive	ActualNegative
PredictPositive	TruePos	FalsePos
PredictNegative	FalseNeg	TrueNeg

The main standard include:

$$Pecision = \frac{TruePos}{TruePos + FalsePos} \tag{11}$$

$$\mathrm{Re}\,call = \frac{TruePos}{TruePos + FalseNeg} \tag{12}$$

The result experiments under different scene are listed in Fig.8:

Fig. 8. Different situation results

Where the gray images label the candidate regions out and blow each of them is the corresponding detection results. Fig.10 shows our algorithm has a basic command of completing the detection, however, since the method of classification test is a bit rough for a complete pedestrian situation which misses some human body occasionally. The statistics data is shown in Table 2:

Table 2. Statistics result

	regions confirmation	regions verification
Precision	41%	98%
Recall	95%	85%

Table 3 shows the average CPU times of our algorithm, which is experimented under standard platform(Intel Core2 Quad 2.5GHz) without any hardware acceleration.

Table 3. Average CPU Times

	Average cost (ms)
Candidate region	28
test	32
CPU times	67

Table 4 shows a comparison of our algorithm with paper[1][3][4]:

Table 4. Comparision result

	Precision (%)	Average cost (ms)
paper[1]	85	6559
paper[3]	87	3601
paper[4]	96.1	58
Our algorithm	98	67

The comparison experiment shows our algorithm has accelerated the detection to a large extent under just monocular camera as well as kept the precise of detection. The average cost of CPU times is approach to the previous paper[4] with method of stereo vision as prior knowledge.

5   Conclusions

This paper proposed an edge symmetry based pedestrian detection algorithm with monocular camera. According to the parameter of the camera, the scan lines are settled and the scales of the corresponding windows are calculated. After the processing of extracting edge information, calculating the symmetry value and picking out the peak value in the scanning area, the candidate regions can be determined by outer and inner parameter of camera. Candidate pedestrians are validated by using HOG feature and linear SVM. Experiment shows our algorithm has basically completed the task of pedestrian detection. However, due to the simplified algorithm for test processing, our algorithm may still miss some human body occasionally which makes implementation be our main direction for our future optimization.

References

1. Dalal, N., Triggs, B.: Histograms of oriented gradients for human detection. In: IEEE Computer Vision and Pattern Recognition (CVPR), vol. 1, pp. 886–893 (2005)
2. Viola, P., Jones, M.: Rapid object detection using a boosted cascade of simple features. In: IEEE Computer Vision and Pattern Recognition, vol. 1, pp. I-511–I-518 (2001)
3. Paisitkriangkrai, S., Shen, C., Zhang, J.: Fast pedestrian detection using a cascade of boosted covariance features. IEEE Circuits and Systems for Video Technology 18(8), 1140–1151 (2008)
4. Arie, M., Moro, A., Hoshikawa, Y.: Fast and stable human detection using multiple classifiers based on subtraction stereo with HOG features. In: IEEE Robotics and Automation (ICRA), pp. 868–873 (2011)
5. Bertozzi, M., Broggi, A., Rose, M.D.: A symmetry-based validator and refinement system for pedestrian detection in far infrared images. In: IEEE Intelligent Transportation Systems Conference, pp. 155–160 (2007)
6. Xia, L., Chen, C.C., Aggarwal, J.K.: Human detection using depth information by kinect. In: IEEE Computer Vision and Pattern Recognition Workshops (CVPRW), pp. 15–22 (2011)
7. Schauland, S., Kummert, A., Park, S.B., Urgel, U.I., Zhang, Y.: Vision-based pedestrian detection-improvement and verification offeature extraction methods and svm-based classification. In: IEEE Intelligent Transportation Systems Conference, pp. 97–102 (2006)
8. Cosma, A.C., Brehar, R., Nedevschi, S.: Part-based pedestrian detection using HoG features and vertical symmetry. In: IEEE Intelligent Computer Communication and Processing (ICCP), pp. 229–236 (2012)
9. Treder, M.S.: Behind the looking-glass: A review on human symmetry perception. Symmetry 2(3), 1510–1543 (2010)
10. Teoh, S.S., Bräunl, T.: Symmetry-based monocular vehicle detection system. Machine Vision and Applications 23(5), 831–842 (2012)
11. Ma, S.D., Zhang, Z.Z.: Computer vision: the computational theory and algorithm foundation, p. 53. Science Press (1998)
12. Canny, J.: A computational approach to edge detection. IEEE Pattern Analysis and Machine Intelligence (6), 679–698 (1986)
13. Neubeck, A., Van, L.G.: Efficient non-maximum suppression. In: IEEE International Conference on Pattern Recognition, vol. 3, pp. 850–855 (2006)
14. Vapnik, V.: The nature of statistical learning theory. Springer (2000)
15. Chang, C.C., Lin, C.J.: LIBSVM: a library for support vector machines. ACM Transactions on Intelligent Systems and Technology (TIST) 2(3), 27 (2011)

Pose-Sequence-Based Graph Optimization Using Indoor Magnetic Field Measurements

Jongdae Jung and Hyun Myung

Dept. of Civil and Environ. Engg, KAIST,
291 Daehak-ro, Yuseong-gu, Daejeon 305-701, Korea
hmyung@kaist.ac.kr

Abstract. In this paper we provide a method of handling loop closings in a simultaneous localization and mapping (SLAM) problem by employing indoor magnetic measurements and pose graph optimization. Since the magnetic field in indoor environments has unique spatial features, we can exploit these characteristics to generate the constraints for the pose graph-based SLAM. Specifically, whenever certain motion conditions are satisfied, a series of robot poses along with their magnetic measurements can be grouped into a sequence. A loop closing algorithm is then proposed based on the sequence and applied to the pose graph optimization. Experimental results show that the proposed SLAM system with only wheel encoders and a single magnetometer obtains comparable results with a reference-level SLAM system in terms of robot trajectory, by correctly detecting the loop closings.

Keywords: SLAM, magnetic field, pose graph optimization.

1 Introduction

The problem of concurrently addressing localization and mapping is well defined in the robotics community as a simultaneous localization and mapping (SLAM) problem [1,2]. Current state of the art SLAM algorithms express the problem as a probabilistic constraint graph and solve it with sparse optimization techniques [3,4,5]. Pose graph SLAM is one of the variants where only the robot trajectory is estimated from relative pose measurements. Relative pose measurements are typically obtained from self-motion and loop closures. The self-motion is usually estimated by wheel odometry, and loop closures are obtained by scan matching, place recognition, etc.

Recently, many works have demonstrated the feasibility of applying indoor magnetic measurements to localization problems both for pedestrians [6,7,8] and mobile robots [9,10,11,12]. Since the magnetic field in indoor environments is stable in the temporal domain and sufficiently varying in the spatial domain [13,14], we can also exploit these characteristics to generate constraints of the pose graphs. In terms of the loop closing constraints, magnetic measurements can be helpful in recognizing the previously visited site. A single measurement, however, may not be sufficient to discriminate the ambiguities caused by similar

© Springer International Publishing Switzerland 2015

J.-H. Kim et al. (eds.), *Robot Intelligence Technology and Applications 3*,
Advances in Intelligent Systems and Computing 345, DOI: 10.1007/978-3-319-16841-8_66

areas of the magnetic field. In this case, employing a sequence of measurements can significantly reduce the ambiguities. A loop closing algorithm then can be designed based on the sequence of magnetic measurement and applied to the optimization.

The rest of this paper is organized as follows. Section II introduces the conventional pose graph SLAM algorithm. The design of novel magnetic constraints is then described in Section III and its validation is provided in Section IV. Finally, conclusions are drawn in Section V.

2   Pose Graph Optimization

2.1   Basic Formulation

Let us define a pose graph $\mathbf{x}$ as the collection of $\mathbf{SE}(2)$ robot poses as follows:

$$\mathbf{x} = [\mathbf{x}_1^{\mathrm{T}}, \ldots, \mathbf{x}_n^{\mathrm{T}}]^{\mathrm{T}} \tag{1}$$

where the i-th $\mathbf{SE}(2)$ robot pose is defined as $\mathbf{x}_i = [x_i, y_i, \theta_i]^{\mathrm{T}}$ and n is the number of poses. Given k-th measurement $\mathbf{z}_k$, the residual $\mathbf{r}_k$ is calculated as the difference between the actual and predicted observations as follows (we assume Gaussian noise for all measurements):

$$\mathbf{r}_k(\mathbf{x}) = \mathbf{z}_k - h_k(\mathbf{x}) \tag{2}$$

where h_k is an observation model for the k-th measurement. With an assumption of independence between observations, we can define a target cost function $E(\mathbf{x})$ as a weighted sum of the residuals:

$$E(\mathbf{x}) = \sum_k \mathbf{r}_k(\mathbf{x})^{\mathrm{T}} \Lambda_k \mathbf{r}_k(\mathbf{x}) \tag{3}$$

where Λ_k is an information matrix for the k-th measurement. The optimal configuration of the pose graph $\mathbf{x}^*$ is then obtained by minimizing Eq. 3 and this can be represented as follows:

$$\mathbf{x}^* = \arg\min_{\mathbf{x}} \sum_k \mathbf{r}_k(\mathbf{x})^{\mathrm{T}} \Lambda_k \mathbf{r}_k(\mathbf{x}) \tag{4}$$

In terms of probabilistic inference, the solution $\mathbf{x}^*$ is a realization of the maximum a posteriori estimation of the robot poses given all observations.

With proper linearization, Eq. 4 is deduced into the form of a linear system

$$\mathbf{H}\Delta\mathbf{x} = -\mathbf{g} \tag{5}$$

where $\mathbf{H}$ and $\mathbf{g}$ are the Hessian and gradient of $E(\mathbf{x})$, respectively, and $\Delta\mathbf{x}$ is the increment of the graph. We can calculate the Hessian $\mathbf{H}$ and gradient $\mathbf{g}$ as

$$\mathbf{H} = \sum_{\{i,j\}\in\Upsilon} \mathbf{H}_{ij} \tag{6}$$

$$= \sum_{\{i,j\}\in\Upsilon} \mathbf{J}_{ij}^{\mathrm{T}} \Lambda_{ij} \mathbf{J}_{ij} \tag{7}$$

$$\mathbf{g} = \sum_{\{i,j\}\in\varUpsilon} \mathbf{g}_{ij} \tag{8}$$

$$= \sum_{\{i,j\}\in\varUpsilon} \mathbf{J}_{ij}^{\mathrm{T}}\boldsymbol{\Lambda}_{ij}\mathbf{r}_{ij} \tag{9}$$

where $\varUpsilon$ is a set of constrained pose pairs $\{i,j\}$ and $\mathbf{J}_{ij} = \partial h_{ij}(\mathbf{x})/\partial\mathbf{x}$ is a Jacobian of the observation model $h_{ij}(\mathbf{x})$.

The updated pose graph $\mathbf{x}^*$ is then obtained by adding the increment to the previous poses.

$$\mathbf{x}^* = \mathbf{x} + \varDelta\mathbf{x}. \tag{10}$$

2.2 Design of Constraints

Since the calculation of the residuals and Jacobians is dependent on the constraint types, we can divide the cost function into the corresponding constraint terms. Assuming ground vehicles, pose graphs generally have two types of constraints – odometric and loop closing constraints. The cost function $E(\mathbf{x})$ then can be described as

$$E(\mathbf{x}) = E_{\mathrm{odm}}(\mathbf{x}) + E_{\mathrm{LC}}(\mathbf{x}) \tag{11}$$

where $E_{\mathrm{odm}}(\mathbf{x})$ and $E_{\mathrm{LC}}(\mathbf{x})$ are the cost caused by the odometric and loop closing constraints, respectively. Each cost term can be further described as

$$E_{\mathrm{odm}}(\mathbf{x}) = \sum_{\{i,j\}\in\mathcal{P}} \mathbf{r}_{ij}^{\mathrm{odm}}(\mathbf{x})^{\mathrm{T}}\boldsymbol{\Lambda}_{ij}^{\mathrm{odm}}\mathbf{r}_{ij}^{\mathrm{odm}}(\mathbf{x}) \tag{12}$$

$$E_{\mathrm{LC}}(\mathbf{x}) = \sum_{\{i,j\}\in\mathcal{Q}} \mathbf{r}_{ij}^{\mathrm{LC}}(\mathbf{x})^{\mathrm{T}}\boldsymbol{\Lambda}_{ij}^{\mathrm{LC}}\mathbf{r}_{ij}^{\mathrm{LC}}(\mathbf{x}) \tag{13}$$

where $\mathcal{P}$ and $\mathcal{Q}$ are the sets of constrained pose pairs by odometric and loop closing constraints, respectively, and $\boldsymbol{\Lambda}_{ij}^{\mathrm{odm}}$ and $\boldsymbol{\Lambda}_{ij}^{\mathrm{LC}}$ are information matrices for odometric and loop closing measurements, respectively. The residuals $\mathbf{r}_{ij}^{\mathrm{odm}}$ and $\mathbf{r}_{ij}^{\mathrm{LC}}$ can be defined as

$$\mathbf{r}_{ij}^{\mathrm{odm}}(\mathbf{x}) = \mathbf{z}_{ij}^{\mathrm{odm}} - h_{ij}^{\mathrm{odm}}(\mathbf{x}) \tag{14}$$

$$\mathbf{r}_{ij}^{\mathrm{LC}}(\mathbf{x}) = \mathbf{z}_{ij}^{\mathrm{LC}} - h_{ij}^{\mathrm{LC}}(\mathbf{x}) \tag{15}$$

where $\mathbf{z}_{ij}^{\mathrm{odm}}$ and $\mathbf{z}_{ij}^{\mathrm{LC}}$ are the relative pose measurements from odometry and loop closures, respectively, and $h_{ij}^{\mathrm{odm}}(\mathbf{x})$ and $h_{ij}^{\mathrm{LC}}(\mathbf{x})$ are the corresponding observation models.

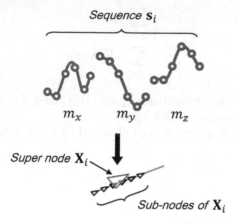

Fig. 1. Magnetic sequence generation. As the robot moves along a linear path, a sequence of three-component magnetic measurements are grouped into a single vector s_i. The corresponding pose nodes are also grouped to generate a super node $\mathbf{X}_i$.

As for the observation models, standard inverse composition of two poses [2] is used to describe both odometric and loop closing measurements:

$$h_{ij}^{\mathrm{odm}}(\mathbf{x}) = \mathbf{x}_j \ominus \mathbf{x}_i \quad (i < j) \tag{16}$$

$$= \begin{bmatrix} (x_j - x_i)\cos\theta_i + (y_j - y_i)\sin\theta_i \\ -(x_j - x_i)\sin\theta_i + (y_j - y_i)\cos\theta_i \\ \theta_j - \theta_i \end{bmatrix} \tag{17}$$

where $\ominus$ represents an operator for the inverse composition of two $\mathbf{SE}(2)$ poses. $h_{ij}^{\mathrm{LC}}(\mathbf{x})$ can be calculated in the same manner.

3   Sequence-Based Loop Closing

Fig. 1 illustrates the procedure of sequence generation. We assume that the robot's path includes linear segments due to the structured indoor environments. Whenever a pre-specified number of robot poses are made along a linear path, all the measurements of the three-component magnetic field are then gathered and grouped into a single sequence vector s_i. The corresponding pose nodes are also grouped as a super node and their indices are stored and managed. The reason for requiring a linear motion is that in this way we can restrain the magnetic fluctuation occurring by the robot's orientation change, which enhances the matching performance. We define the linear motion with the following conditions:

$$\max\{\mathrm{std}(x_i, \ldots, x_{i+N_s-1}), \mathrm{std}(y_i, \ldots, y_{i+N_s-1})\} \leq T_{\mathrm{t}}^{\mathrm{s}} \tag{18}$$

$$\mathrm{std}(\theta_i, \ldots, \theta_{i+N_s-1}) \leq T_{\mathrm{r}}^{\mathrm{s}}$$

where std$(\cdot)$ is a function calculating the standard deviation, N_s is the number of poses in a sequence, and T_t^s and T_r^s are thresholds for the translational and rotational motions, respectively, for sequence generation.

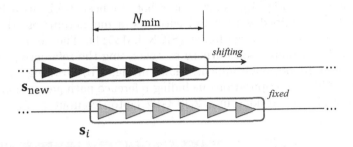

Fig. 2. Graphical representation of sequence-based matching. s_{new} are the newly generated sequence and s_i is an existing sequence to be matched. Shifting of the sequence is started from the point defined by N_{min}. During the matching procedure, only the overlapped sub-nodes are used for the evaluation.

Whenever a new sequence s_{new} is generated, a sequence-based matching is performed to detect a loop closing. Basically, we perform a line by line matching, shifting one line segment while the other is fixed (see Fig.2 for illustration). To prevent a false positive loop closing, we require at least N_{min} sub-nodes to be included during the shifting.

During the matching procedure, the matching score is evaluated using the Euclidean distance D_E between two sequences [15]. When D_E is lower than the threshold T^{LC}, we decide that a loop closing has occurred. The corresponding sub-nodes are then added into a set of LC-constrained pose pairs $\mathcal{Q}$ with $z_{ij}^{LC} = 0$ and are optimized. Algorithm 1 describes the overall procedure for the proposed magnetic sequence-based loop closing algorithm.

Algorithm 1. checkMagneticLC

Input: $\{X_i\}$ (existing super nodes), X^{new} (newly added super node)
Output: $\mathcal{Q}$ (set of LC pose pairs)
1. $\mathcal{Q} \leftarrow \{\}$
2. $s_{new} \leftarrow X^{new}.sequence$
3. **for** $i = 1$ to num. of existing super nodes **do**
4. $s_i \leftarrow X_i.sequence$
5.   **if** $D_E(s_{new}, s_i) \leq T^{LC}$ **then**
6. $\mathcal{Q} \leftarrow$ add $posepairs$
7.   **end if**
8. **end for**

---

4   Experiments

4.1   Experimental Setup

In order to verify the proposed method, we conducted an experiment in a real indoor environment. Fig. 3(a) shows the robot (Pioneer 3-AT) used in the experiments. It is equipped with wheel encoders, a magnetometer (Honeywell's HMR2300), and a laser range finder (SICK LMS511). The laser range finder is merely used for comparison purpose by generating the reference paths[1]. The robot runs in an indoor environment, shown in Fig. 3(b), drawing a 3 m × 3 m square path. All the computations including reference path generation and sensor data acquisition are done in a 2.40 GHz laptop computer. Table 1 shows

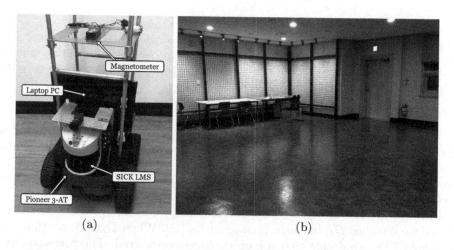

(a)   (b)

Fig. 3. Experimental setup. (a) The robot used in the experiments. (b) Test environments.

all the parameters used in the analysis. The gathered sensor data are processed offline and in an incremental fashion to generate the pose graph SLAM results.

Table 1. Summary of the parameters used in the optimization

T_t^s	0.2 m
T_r^s	10 deg
T^{LC}	10 μT
N_s	7
N_{min}	5

---

[1] http://wiki.ros.org/gmapping

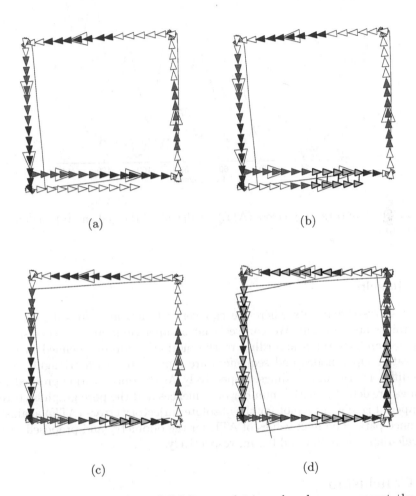

Fig. 4. Optimization procedure. Solid lines and triangular shapes represent the trajectories from the odometry and pose graph SLAM, respectively. The super node is represented with a larger triangular shape. (a) No loop closings are happened yet. (b) As the optimization further proceeds, loop closings (shown in grey nodes) are detected. (c) The pose graph is optimized. (d) Final pose graph.

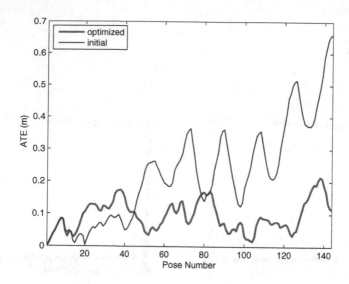

Fig. 5. Absolute trajectory errors (ATE) for the initial (i.e., odometry) and optimized pose graph

4.2   Results

Fig. 4 shows the results where we can verify that the proposed loop closing constraints are working. We can see that a super node and its corresponding sequence are generated according to the motion conditions defined in Eq. 18. Generated super nodes and sequences are represented with triangular shapes with different size and shadings, respectively. As the optimization proceeds, loop closings are detected with a number of sequences and the pose graph is correctly reshaped. Fig. 5 gives a plot for the absolute trajectory errors (ATE) against the reference path. The max values of ATE for the odometry and proposed method are calculated as 0.7 m and 0.2 m, respectively.

5   Conclusion

In this paper, we proposed a method of solving the indoor SLAM problem by employing magnetic field measurements. The proposed SLAM was realized in a pose graph optimization-based framework, where a novel sequence-based loop closing constraint was designed. The experimental results showed that the proposed SLAM system with only wheel encoders and a single magnetometer obtains comparable results with a reference-level laser-based SLAM system in terms of robot trajectory, thereby validating the feasibility of applying magnetic constraints to the SLAM.

References

1. Thrun, S., Burgard, W., Fox, D.: Probabilistic Robotics. The MIT Press, Cambridge (2005)
2. Fernandez-Madrigeal, J., Claraco, J.L.: Simultaneous Localization and Mapping for Mobile Robots: Introduction and Methods. Information Science Reference, Hershey (2013)
3. Kuemmerle, R., Grisetti, G., Strasdat, H., Konolige, K., Burgard, W.: g2o: A general framework for graph optimization. In: Proc. IEEE International Conference on Robotics and Automation (ICRA2011), Shanghai, China, pp. 3607–3613 (May 2011)
4. Kaess, M., Johannsson, H., Roberts, R., Ila, V., Leonard, J., Dellaert, F.: iSAM2: Incremental smoothing and mapping using the Bayes tree. The International Journal of Robotics Research 31, 217–236 (2012)
5. Lee, D., Myung, H.: Solution to the SLAM problem in low dynamic environments using a pose graph and an RGB-D sensor. Sensors 14(7), 12467–12496 (2014)
6. Afzal, M., Renaudin, V., Lachapelle, G.: Magnetic field based heading estimation for pedestrian navigation environments. In: Proc. IEEE International Conference on Indoor Positioning and Indoor Navigation (IPIN2011), Guimaraes, Portugal, pp. 1–10 (September 2011)
7. Bird, J., Arden, D.: Indoor navigation with foot-mounted strapdown inertial navigation and magnetic sensors [emerging opportunities for localization and tracking]. IEEE Wireless Commun. Mag. 18(2), 28–35 (2011)
8. Haverinen, J., Kemppainen, A.: A geomagnetic field based positioning technique for underground mines. In: Proc. IEEE International Symposium on Robotic and Sensors Environments (ROSE2011), Montreal, QC, pp. 7–12 (September 2011)
9. Zhang, H., Martin, F.: Robotic mapping assisted by local magnetic field anomalies. In: Proc. IEEE International Conference on Technologies for Practical Robot Applications (TePRA 2011), Woburn, MA, pp. 25–30 (April 2011)
10. Vallivaara, I., Haverinen, J., Kemppainen, A., Roning, J.: Magenetic field-based SLAM method for solving the localization problem in mobile robot floor-cleaning task. In: Proc. IEEE International Conference on Advanced Robotics (ICAR2011), Tallinn, Estonia, pp. 198–203 (June 2011)
11. Frassl, M., Angermann, M., Lichtenstern, M., Robertson, P., Julian, B., Doniec, M.: Magnetic maps of indoor environments for precise localization of legged and non-legged locomotion. In: Proc. IEEE/RSJ International Conference on Intelligent Robots and Systems (IROS 2013), Tokyo, Japan, pp. 913–920 (November 2013)
12. Jung, J., Lee, S., Myung, H.: Indoor mobile robot localization using ambient magnetic fields and radio sources. In: Proc. International Conference on Robot Intelligence Technology (RiTA 2013), Denver, USA (December 2013)
13. Gozick, B., Subbu, K., Dantu, R., Maeshiro, T.: Magnetic maps for indoor navigation. IEEE Trans. Instrum. Meas. 60(12), 3883–3891 (2011)
14. Angermann, M., Frassl, M., Doniecy, M., Julianyz, B., Robertson, P.: Characterization of the indoor magnetic field for applications in localization and mapping. In: Proc. IEEE International Conference on Indoor Positioning and Indoor Navigation (IPIN2012), Sydney, NSW, pp. 1–9 (November 2012)
15. Cha, S.: Comprehensive survey on distance/similarity measures between probability density functions. International Journal of Mathematical Models and Methods in Applied Sciences 1(4), 300–307 (2007)

An Efficient Ego-Lane Detection Model to Avoid False-Positives Detection of Guardrails

Vitor S. Bottazzi[1], Jun Jo[1], Bela Stantic[1], and Paulo V.K. Borges[2]

[1] School of Information and Communication Technology
Griffith University, Gold Coast, Australia
vitor.bottazzi@griffithuni.edu.au, {j.jo,b.stantic}@griffith.edu.au
[2] Autonomous Systems Laboratory, CSIRO ICT Centre
Brisbane, Australia
vini@ieee.org

Abstract. Detecting lane markings is a challenging task for vision-based systems due to uncontrolled lighting environments present on the roads. Road infrastructures surrounding the painted markings such as guardrails and curbs often reduce the accuracy of existing solutions. The mentioned infrastructure frequently behaves like lane markings increasing the occurrence of candidate features to be selected by lane detectors. Most of the lane detectors use machine learning techniques with long training phases and inflexible models to achieve some level of robustness, therefore an efficient approach capable of performing unsupervised learning is required. The adoption of an efficient model, which can monitor the ego-lane boundaries while identifying false positive references, is discussed in this paper. The proposed architecture allows the combination of multiple image-processing cues to improve accuracy and robustness on vision-based methods. Our method performed with high accuracy in lane marking detection under highly dynamic lighting, including in presence of guardrails.

1 Introduction

Ego-lane position estimation is an important feature that can be used to support both human drivers and self-driving vehicles. Lane detection algorithms are regularly used to identify lane departure propensity while estimating the upcoming geometry of the road. This process can be subdivided into a sequence of steps, including feature extraction, model fitting and lane markings tracking, to perform efficiently. A flexible architecture where different image-processing cues are combined becomes useful to enhance algorithm's performance while monitoring the vehicle's behaviour over time. Several car manufacturers have recently released advanced driver assistance system (ADAS) technologies [1] to enhance their customers' safety and comfort. This tendency allied to the recent phenomena of self-driving cars' prototypes deployed in US roads has promoted the interest of the research community in driver assistance technology for the application on both ADAS and self-driving systems.

© Springer International Publishing Switzerland 2015
J.-H. Kim et al. (eds.), *Robot Intelligence Technology and Applications 3,*
Advances in Intelligent Systems and Computing 345, DOI: 10.1007/978-3-319-16841-8_67

The proposed approach reduces the propensity for detecting false positives that are often introduced by dynamic lighting and linear structures close to the lane markings such as curbs and guardrails. The strategy used in this work is to extract multiple references of the road using different methods in parallel. The references are then compared using a simplified model capable of representing a drivable corridor in the camera's perspective. The principle of using different image-processing techniques leads to detecting errors with different characteristics. We used a variety of datasets to reproduce road conditions with sunlight in different angles, including situations where the road is surrounded by metallic structures such as guardrails and bridges. We show in our experiments that the proposed method is robust and performs well under challenging conditions with average ego-lane detection of 99.15%.

2   Related Work

A variety of lane detection approaches has been presented in literature to support drivers and self-driving vehicles.

The rapidly adapting lateral position handler (RALPH) [2], which is capable of determining vehicle's location, can automatically adapt itself to road feature changes without men's supervision. The RALPH approach processes a trapezoid region-of-interest between 20 and 70 meters ahead of the vehicle, moving forward according to the car speedup. The road sampling creates a low-resolution image (30×32 pixels) where important features such as lane-markings are approximated to the ground truth. The system uses supervised machine learning to control the car's steering by training a neural network. This approach demands correct steering behaviour demonstrated by a human before it starts operating autonomously.

The generic obstacle and lane detection (GOLD) [3] system applies the inverse perspective mapping (IPM) technique [4] to remove the perspective effect of the road and to process the image at the ground plane domain. Lane detection is based on a line-wise determination searching for black-white-black transitions. Nevertheless stereo images can provide extra information regarding non-flat surfaces on the road, the paper constantly assumes that the road is flat. A custom parallel SIMD computer was used to support the image processing effort.

The lane-finding in another domain [5] is a lane extraction algorithm based on frequency domain features. Lane edges are the object of interest of this work and they are captured using multi-resolution Fourier transform (MFT). The first step is the detection of the edge-like features regions, then the regions are subdivided using quad tree while a similar detection is performed at the lower nodes of the tree until every pixel of the image is classified. The deformable template shape model is used to locate diagonally dominant edges. Real-time response was not achieved at that time, limiting the application of the approach.

Labayrade et al. [6] used multiple road model instances to estimate lane markings position. First, a low-level detection computes the intensity gradients, searching for a pair of high magnitude gradients (positive and negative). Then, the centre of the road painting contour is selected as a feature. After that, the width of the lane is used as chief criterion to select valid candidates. Tracking is also performed to compute the past and current position of the borders of the lane, handling non-continuous lane-markings. Additionally, a second longitudinal consistency algorithm is applied using M-estimators. In case two estimated points are close, the average position is used to update the lane model. A confidence value indicating the operating state of the system is based on the percentage of lane border points found. The global reliability is computed as a sum of left and right lane marks confidence.

Recent work [7] still applies IPM with the assumption of approximately parallel lane markings provided by the bird eye's view. Then, a second order Sobel filter following horizontal direction is applied on the IPM image. Furthermore, a quadratic parabola model is chosen to represent the lane marking shape and a region-of-interest is selected to reduce the amount of information to be processed. A multi kernel density based method is used to setup the curve parameters comparing the similarity of a random pixel on the image and the model. The algorithm performs fast at 68.18ms per frame on average. However, the article does not mention about the impact of dynamic lighting in the algorithm's accuracy.

2.1 Limitations of Current Approaches

The robustness of image-processing algorithms is strongly influenced by the lighting conditions. An algorithm that performs faultless at mid-day sunlight might show big discrepancies as the sun start to move towards sunset. First, the sun motion shall change the amount of available light. Second, linear structures with similar behaviour to lane markings (i.e. curbs, guardrails, bridges, etc) contribute to scatter even more the scenario. Nonetheless, the presence of high reflective road surfaces is a factor that still challenges algorithms' accuracy.

Passive sensors such as single cameras have been used in lane mark detection for decades. Cameras are very flexible allowing lane detection, traffic sign identification, obstacle avoidance and traffic monitoring beyond other applications. The intrinsic limitations of dealing with this sensor are:

- The computation power demanded to analyse the huge amount of data provided by digital cameras; and
- High dependability on lighting conditions.

The intrinsic computation effort attached to each operation in the pipeline must be monitored carefully to do not affect the overall system performance. As image analysis is a very computational expensive process, the amount of operations applied to a single frame (i.e. segmentation, contour extraction, object tracking, etc.), may not enable real-time response.

The challenge of detecting and tracking objects in presence of dynamic light conditioned current solutions to adopt supervised machine learning to achieve some degree of robustness. However, road situations out of the training set often get neglected [8, 9] plus long training phases are mandatory. There are mainly three shortcomings identified in literature:

- Efficiency related to the image-processing stage and the complex mathematical models adopted by current solutions [3, 5, 10–12];
- Long training phase and limited flexibility provided by robust solutions [8, 9, 13]; and
- Robustness and accuracy retainment according to dynamic lighting projected on the road surface.

The literature confirmed that there is a high demand for supportive methods able to take precautions under those situations. Many different approaches have been employed to detect potential lane marking references but even complex feature extraction approaches are not capable of providing accuracy retainment under the influence of dynamic lighting without machine intelligence support. Hence, uninteresting contour references (i.e. marking ridges, curbs and guardrails) continue to be mismatched as lane painting features. As far as we know, there is no automatic tuning approach capable of adapting current algorithms to different lighting conditions. Our research proposes a pipeline architecture (see Fig. 1) supported by real-time lighting evaluation, multiple image-processing cues and an efficient mathematical model, which are used in combination to improve ego-lane marking estimation. The next section explains the methodology used to address the ego-lane detection problem.

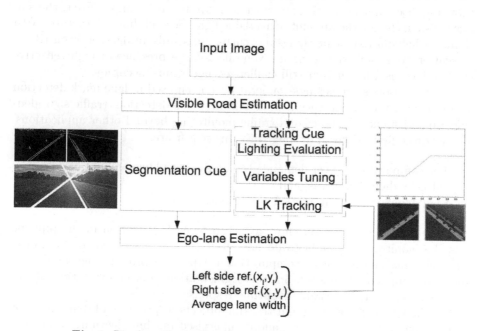

Fig. 1. Pipeline illustrating the parallel image-processing cues

3   The Ego-Lane Detection System

Our method uses a pre-detection stage called *Visible Road Estimation* (Fig. 2a) followed by the two main image-processing cues and the ego-lane calibration (Fig. 2b) before enabling the *Ego-lane Estimation*. The pre-detection stage aims to estimate the vertical position of the road horizon in the image, concentrating the computational effort to the portion of the image where road is visible. In the first main cue, the segmentation process is applied to extract the painted marks and to detect the relative position of the lane markings in the frame. The second main cue performs the ego-lane tracking by monitoring the painted markings dynamics based on its optical flow. The *Ego-lane Estimation* stage performs the model fitting, which aggregates the information from both cues abstracting a drivable corridor, towards robust estimation of the lateral position of the painted markings.

(a) Road horizon estimation   (b) Ego-lane calibration in progress

Fig. 2. Pre-detection stage

The average distance between painted markings is updated every frame and further used for false positives identification. The process of discarding new features that are not coherent with the average distance between painted markings increases the the ego-lane prior reliability. The appearance segmentation and tracking methods are used in parallel to validate the correct ego-lane boundary references.

The first match between the two image-processing cues is defined as our *ego-lane calibration*. From this point, both segmentation and salient points tracking cues interact to maintain an updated ego-lane representation defined by our triangular model, see Fig. 3. Then, left and right ego-lane references are monitored, reinforcing the prior model accuracy with the most recent information about the ego-lane boundaries. The interaction between the parallel cues is depicted in the next sections.

3.1   Appearance-Based Segmentation Cue

The appearance-based cue detects the painted markings on the road surface by using static segmentation regions-of-interest (SSROIs) [14]. The SSROIs are pre-

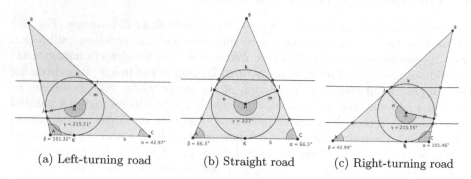

(a) Left-turning road   (b) Straight road   (c) Right-turning road

Fig. 3. Trigonometry applied to lane tracking

positioned according to the visible road area, in order to prioritise portions of the image where the painted markings are likely to appear. It aims to detected lane boundaries using the intersections created by the extended lines extracted from painted markings present on the road surface. The points A, B and C depict the intersections used to setup our prior, see Fig. 3.

The algorithm selects both left and right sides of the ego-lane separately using independent SSROIs. Then, Gaussian smoothing technique is applied to each selected area, in order to reduce noise. Subsequently, the SSROIs are transformed to grayscale preparing for the contour extraction process and Canny algorithm is applied due to it thin and precise contour results. Finally, the Hough transform is applied on each side of the road to detect the linear segments created from ego-lane marking contours, see Fig. 1.

The references are considered valid when the intersections of the extended painted marks intersect each other and the triangle's base, according to an expected distance variation approximating previous frames references. Those past references are used by the tracking parallel cue repositioning the dynamic tracking windows every frame [14]. The feature tracking process increases the probability of lane marking detection by searching at portions of the image with high probability of road painting occurrence. This process is explained in detail in the next section.

3.2   Salient Point Tracking Cue

Painted lane markings often become discontinuous, disappear or change colour (i.e. white-yellow-white) depending on traffic rules. The tracking of the painted markings allows the algorithm to rely on past measurement for a short period of time (i.e. milliseconds). Tracking also enables lane markings reconstruction based on mathematical models. According to the past references of the ego-lane, the trajectory of the vehicle can be calculated and the search for lane marks can be optimised.

The tracking cue uses the Lucas-Kanade [15] algorithm running in parallel with the segmentation cue. Both cues exchange information with our triangular model about past lane marking references. Strong light sources pointed to the road often cause large variations on pixel intensity disrupting tracking. Abnormal painting, road imperfections and shadows are additional examples of external variables that affect the lane tracking accuracy. The optical flow tracking was combined to the segmentation approach to improve accuracy under dynamic lighting scenarios and it can still perform efficiently due to our reduced DTROI [14].

The dynamic tracking regions-of-interest (DTROIs) were idealised to reduce sensibility to noise when strong light sources in movement are detected. The size of the DTROI is reduced proportionally to the intensity of the external lighting variance, reducing mismatches introduced by highly reflective surfaces. It is also designed to focus at the portions of the image where lane marks were previously identified. The painted marks are tracked across consecutive frames using the DTROI (see the green square at Fig. 4a) [16]. A sanity check [14] based on the Laplace expansion theorem is applied to validate candidates (see Fig. 4b). The sanity check only considers features that are close enough to our triangular reference.

(a) The interaction between the SSROI (blue) and the DTROI (green)

(b) Candidates selection

Fig. 4. The left DTROI shows only selected candidates while the right DTROI shows two selected candidates an one outlier

As mentioned previously, painted markings segmentation and tracking can be affected by many different factors. Our algorithm automatically adapts itself according to measured lighting variations in real-time. The input used to evaluate lighting change is the difference between saturation channels [16] extracted from consecutive frames using the HSV colour space. In case saturation values oscillate between 30% and 60% based on minima and maxima variances measured on the run, tuning is performed over variables that have direct influence on the lane markings tracking. Two variables show direct impact over the algorithm sensibility. The first variable is the area of the DTROI. The second variable represents the maximum number of corners to be tracked by the optical flow algorithm. The number of tracked corners increases linearly with the size of the tracking window because a larger portion of the road will be analysed. Finally, the initial threshold used for features segmentation is updated according to the saturation changes. This precaution increased the algorithm's robustness and accuracy when abnormal light variance is detected on the road surface.

A soft-threshold mechanism is applied to the left and right DTROIs while the number of tracked features is adjusted. The motivation for using adaptive windows and soft-thresholds combined is to reduce the influence of noise introduced by strong lighting sources projected on road surface. Additionally, it minimises the overall computational effort as the number of pixels analysed reduces due to dynamic lighting. In the next section, the process of comparing different elements of the road towards accurate ego-lane estimation is explained.

3.3   Ego-Lane Boundaries Estimation

The segmented painted markings are compared with the tracked lane markings to validate the ego-lane boundary. This check enables the maintenance of a prior triangular reference through the analysis of subsequent frames. Additionally, the algorithm considers the lateral motion of the ego-lane markings in respect to the car. On average, the distance between lane paintings is expected to variate between 2.5 m and 4 m. The ego-lane model was idealised to simplify the detection of a drivable corridor, inclusive on non-standardised road datasets.

The *ego-lane calibration* process (see Fig. 2b) consists on the match between the complementary cues references explained previously (see Fig. 4a). It enables the algorithm to initialise the model with a confirmed distance between lane markings, keeping an average distance reference between markings to be used for classification purposes. The prior model represents the drivable corridor using the intersection between current left and right extended painted markings with the prior to represent the ego-lane in perspective. New intersections with the prior are evaluated every frame, based on valid measurements proximity, for an increased model consistency.

After calibration, the segmentation cue updates the model with the lateral position of ego-lane markings represented by the triangle's baseline ($\overline{AC}$) and the boundary of the drivable corridor in perspective ($\overline{AB}$ and $\overline{BC}$), see Fig. 3. The segmented references are updated every frame and used in combination with the average distance between points A and C, see Fig. 1. In case one of the lateral references of the ego-lane disappear, due to lighting influence for example, the temporal information regarding the road boundary can be accessed to optimise the search for the missed painting reference.

4   Results

The first footage (Fig. 5a) used in the experiments contains over 1380 frames and the second footage (Fig. 5b) contains over 1880 frames. The equipment used to record the samples was the GoPro3 camera with resolution of 848×480 running at 30 fps. The images were rectified to reduce the distortion caused by the GoPro3 lens.

(a) Avoid guardrails and curbs   (b) Sunlight reflection

Fig. 5. Robustness in different scenarios with low contrast background and road surface scars including false positives classification

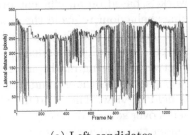
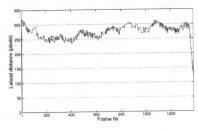

(a) Left candidates   (b) Filtered left reference

Fig. 6. Left painted markings references

The existing methods cited in literature have presented poor efficiency and design due to the expensive computational effort spent analysing large pixel clusters [3, 5, 10, 13] or by the adoption of complex mathematical models [11, 12]. In most of the cases the effort applied to reconstruct accurately the road boundaries doesn't provide a satisfactory reward. The situations listed bellow are ordinary cases where lighting and dynamic environment keeps challenging current solutions and disrupting previous models presented in literature:

- Sunlight projected in angle on the surface of the road (see Fig. 5b);
- Low contrast between the road painting and asphalt (see Fig. 5a); and
- Linear profiles, road scars and objects surrounding the road (see Fig. 5).

Additionally, recent work still leads to false positive detection of guardrails [13, 17] while using expensive computation. In other words, the approaches discussed in section 2 demonstrated to be expensive in runtime, and/or not flexible enough in presence of dynamic lighting.

Our algorithm executed fast calibration in presence of continuous painted markings (approximately $250ms$). The pre-detection stage including the estimation of the horizon position (see Fig. 2a) and the ego-lane calibration (see Fig. 2b) performed in less than $0.5s$. When applied to the first footage, the ego-lane detection algorithm shown 98.3% accuracy (see Fig. 5). The 1.7% false positives detection was caused by curbs within the acceptable variance of the ego-lane width. Fig. 6 shows the left marking reference estimation over time.

Fig. 6a illustrate the disturbances introduced by guardrails and curbs over time. Fig. 6b shows the filtered position of the lane marking avoiding guardrail references. The algorithm performed optimally in the second footage (see Fig. 5b) achieving 100% detection under highly dynamic lighting were the sun's relative position is moving from right to left, including highly reflective road surface during sunset time.

5   Conclusion

In this article an ego-lane detection algorithm is proposed to increase robustness and efficiency under dynamic lighting. Our work is based on an efficient prior model, which approximates the ego-lane markings behaviour to avoid the detection of linear structures surrounding the ego-lane markings. The footage used in the experiments has predominantly continuous painted markings under multiple lighting conditions including low contrast between painting marks and road surface. The camera is positioned *approximately* parallel to the ground aiming to capture a similar perspective of a driver's field of view. The SSROIs were pre-positioned according to extensive experiments over several datasets used for development and automatically re-positioned under the road horizon by the pre-detection stage, with no use of training phase.

The experiments demonstrated that it is possible to achieve highly accurate detection and false candidates classification using and efficient triangular model, which abstract the drivable corridor. The experiments were conduced over 3000 frames, which were separated in two samples, achieving 99.15% ego-lane detection. The algorithm performs fast as well running at $50ms$ *per frame on average.* Our appearance-based approach performs with high accuracy for continuous lane markings and high contrast curbs. However, environmental variables such as missing lane marks, abnormal road painting and wet surfaces can still affect the performance of the system. In the future, the algorithm will be extended to analyse the angles of the optical flow vectors enhancing lane departure and lane merging estimation.

References

1. Hur, J., Kang, S.N., Seo, S.W.: Multi-lane detection in urban driving environments using conditional random fields. In: 2013 IEEE Intelligent Vehicles Symposium (IV), pp. 1297–1302. IEEE (2013)
2. Pomerleau, D., Jochem, T.: Rapidly adapting machine vision for automated vehicle steering. IEEE Expert 11, 19–27 (1996)
3. Bertozzi, M., Broggi, A.: Real-time lane and obstacle detection on the gold system. In: Proceedings of the 1996 IEEE Intelligent Vehicles Symposium, pp. 213–218. IEEE (1996)

4. Borkar, A., Hayes, M., Smith, M.: An efficient method to generate ground truth for evaluating lane detection systems. In: 2010 IEEE International Conference on Acoustics Speech and Signal Processing (ICASSP), pp. 1090–1093. IEEE (2010)
5. Kreucher, C., Lakshmanan, S.: Lana: a lane extraction algorithm that uses frequency domain features. IEEE Transactions on Robotics and Automation 15, 343–350 (1999)
6. Labayrade, R., Douret, J., Aubert, D.: A multi-model lane detector that handles road singularities. In: IEEE Intelligent Transportation Systems Conference, ITSC 2006, pp. 1143–1148. IEEE (2006)
7. Lu, W., Florez, S.A.R., Seignez, E., Reynaud, R., et al.: An improved approach for vision-based lane marking detection and tracking. In: International Conference on Electrical, Control and Automation Engineering, pp. 382–386 (2014)
8. Wang, Y., Dahnoun, N., Achim, A.: A novel system for robust lane detection and tracking. Signal Processing 92, 319–334 (2012)
9. Yoo, H., Yang, U., Sohn, K.: Gradient-enhancing conversion for illumination-robust lane detection. IEEE Transactions on Intelligent Transportation Systems 14, 1083–1094 (2013)
10. Sehestedt, S., Kodagoda, S., Alempijevic, A., Dissanayake, G.: Efficient lane detection and tracking in urban environments. In: Proc. European Conf. Mobile Robots, pp. 126–131 (2007)
11. Wang, Y., Shen, D.G., Teoh, E.K.: Lane detection using spline model. Pattern Recognition Letters 21, 677–689 (2000)
12. Zhou, S., Jiang, Y., Xi, J., Gong, J., Xiong, G., Chen, H.: A novel lane detection based on geometrical model and gabor filter. In: IEEE Intelligent Vehicles Symposium, pp. 59–64 (2010)
13. Felisa, M., Zani, P.: Robust monocular lane detection in urban environments. In: 2010 IEEE Intelligent Vehicles Symposium (IV), pp. 591–596. IEEE (2010)
14. Bottazzi, V.S., Borges, P.V., Jo, J.: A vision-based lane detection system combining appearance segmentation and tracking of salient points. In: 2013 IEEE Intelligent Vehicles Symposium (IV), pp. 443–448. IEEE (2013)
15. Baker, S., Matthews, I.: Lucas-kanade 20 years on: A unifying framework. International Journal of Computer Vision 56, 221–255 (2004)
16. Bottazzi, V.S., Borges, P.V.K., Stantic, B., Jo, J.: Adaptive regions of interest based on HSV histograms for lane marks detection. In: Kim, J.-H., Matson, E., Myung, H., Xu, P. (eds.) Robot Intelligence Technology and Applications 2. AISC, vol. 274, pp. 677–687. Springer, Heidelberg (2014)
17. Deusch, H., Wiest, J., Reuter, S., Szczot, M., Konrad, M., Dietmayer, K.: A random finite set approach to multiple lane detection. In: 2012 15th International IEEE Conference on Intelligent Transportation Systems (ITSC), pp. 270–275. IEEE (2012)

Peak Detection with Pile-Up Rejection Using Multiple-Template Cross-Correlation for MWD (Measurement While Drilling)

Sangwon Lee[1], Byeolteo Park[2], Youngjai Kim[1], and Hyun Myung[2]

[1] Robotics Program, KAIST, Daejeon, 305-701, Korea
[2] Department of Civil and Environmental Engineering, KAIST, Daejeon, 305-701, Korea
{lsw618,starteo,david-kim,hmyung}@kaist.ac.kr

Abstract. In this paper, we propose a novel pile-up rejection algorithm using multiple-template for MWD (Measurement While Drilling) which is very useful in underground localization of drilling devices. Conventional peak detection algorithms that are based on cross-correlation use only a template signal in the peak detection. However, if two peak pulses pile-up too closely, the conventional methods will recognize them as a single peak pulse. If the pile-up signals are recognized as a single peak signal, the false peak information will be given. To solve this problem, we additionally use a template signal that consists of two overlapping peak pulses to detect peak pulses that pile-up closely. We confirmed our algorithm by using actual output data from amplifier connected to a scintillator.

Keywords: Peak detection, Pile-up, Cross-correlation.

1 Introduction

Nowadays, directional drilling is widely used for gathering unconventional resources such as tight gas and shale gas. The purpose of directional drilling is to bore toward a desired direction in underground. The directional drilling requires robotics technologies such as localization and steering control. While drilling, MWD (measurement while drilling) provides geological information of underground. The geological information such as gamma ray spectrum profile is very important especially for the localization of drilling devices in directional drilling.

One of the important techniques for generating a gamma spectrum is detecting peak signals from an amplifier connected to a scintillator. When a time interval between successive gamma rays is shorter than the time width of the amplifier output signal, the signals from the amplifier can be overlapped. This phenomenon is called "pile-up." If false peak information is accumulated in the spectrum by pile-up signals, the gamma spectrum will be distorted. The distorted spectrum makes it difficult to identify the gamma ray source. In order to reduce the distortion of the spectrum, the peak information from pile-up signals should be eliminated. Pile-up signals can be detected by inspecting the time interval between consecutive peak pulses or using deconvolution method [1].

© Springer International Publishing Switzerland 2015
J.-H. Kim et al. (eds.), *Robot Intelligence Technology and Applications 3,*
Advances in Intelligent Systems and Computing 345, DOI: 10.1007/978-3-319-16841-8_68

In detection of peak pulses, our proposed method employs a peak detection method based on cross-correlation [2-7]. The cross-correlation has been used in many applications such as image processing [4-5], speech recognition [8], fingerprint matching [9], and so on. Conventional cross-correlation-based peak detection algorithms use only one template signal. However, our proposed algorithm uses two different kind of template signals to reject pile-up peak pulses.

The remainder of this paper is organized as follows. Section 2 explains our method on how to detect peak pulses and reject pile-up peak pulses using multiple-template-based cross-correlation algorithm. In Section 3, we describe the experimental results with our approach to confirm our method. Conclusion and future work are discussed in Section 4.

2   Pile-Up Rejection

In this section, we explain the multiple-template cross-correlation-based peak detection algorithm to reject pile-up signals. The overview of the proposed algorithm is shown in Fig. 1.

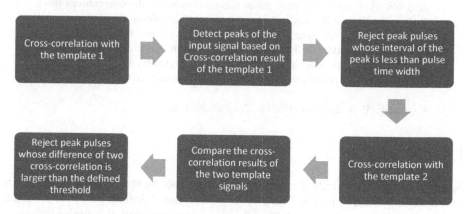

Fig. 1. Overview of the proposed algorithm

2.1   Peak Detection Using Cross-correlation

Cross-correlation methods [2-7] are used to measure the similarity between two signals. The cross-correlation of two continuous functions $f(t)$ and $g(t)$ is defined as follows:

$$(f \star g)(t) = \int_{-\infty}^{\infty} f^*(\tau)g(t + \tau)d\tau \tag{1}$$

where f^* denotes the complex conjugate of f. If we define $f(t)$ as a template signal and $g(t)$ as an input signal, the peak pulse time location in the input signal $g(t)$ is the local maximum of equation (1). As a result, we can find the time location of $g(t)$ where peak pulses exist.

2.2   Multiple-Template Cross-correlation

If all the peak pulses in the input signal are detected using cross-correlation, the pile-up signal can be easily identified by calculating the time difference between consecutive peak pulses. If the time difference between the consecutive pulses is smaller than the time width of the peak pulse, those peak pulses are pile-up signals.

However, when the consecutive peak pulses are located too closely, these pulses are identified as a single peak pulse. To detect these kinds of pile-up peak pulses, we use two template signals as shown in Fig. 2. Template signal 1 is a single Gaussian pulse whose time width of signal is same as an input peak pulse and template signal 2 is two overlapped Gaussian pulses.

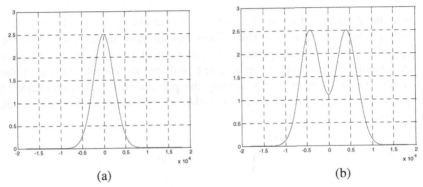

(a)   (b)

Fig. 2. Pulse templates (a) template signal 1: single Gaussian pulse (b) template signal2: overlapped Gaussian pulses

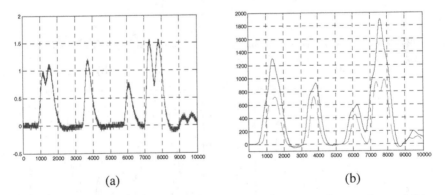

(a)   (b)

Fig. 3. Example of pile-up peak pulses (a) input signal containing pile-up pulses and (b) cross-correlation results with template 1(red dotted line) and template 2(blue solid line). In case of pile-up peak pulses, the cross-correlation with template 2 is much larger than the cross-correlation with template 1.

Because the pile-up pulses and template 2 have high similarity and larger area than normal pulses or template 1, the ratio of the cross-correlation with template 2 to the cross-correlation with template 1 at pile-up pulses is larger than the ratio at a normal pulse. In other words, at a certain time location in the input signal, if the ratio of the cross-correlation with template 2 to the cross-correlation with template 1 is above the predefined threshold, pile-up peak pulses exist in that time location. By detecting the pile-up peak pulses, only normal pulses can be extracted.

3   Experiments

In this section, we present the experiment results of our method. The signal data used in the experiment is actual data from an amplifier connected to a scintillator. All pulses in the input signal have same time width and same shape except the amplitude of the peak.

Fig. 4 is the peak detection results using cross-correlation with only template 1. Red triangles indicate the detected peaks. These results show that in case of too closely pile-up pulses, only one peak pulse of the pile-up peak pulses is detected. In these cases, we cannot decide whether the detected peak pulses are pile-up peak pulses. However, the proposed method can identify the pile-up peak pulses and reject the pile-up peak pulses as shown in Figs. 5 (a), (b), (c), (d) where each signal is same signal in Figs. 4 (a), (b), (c), (d).

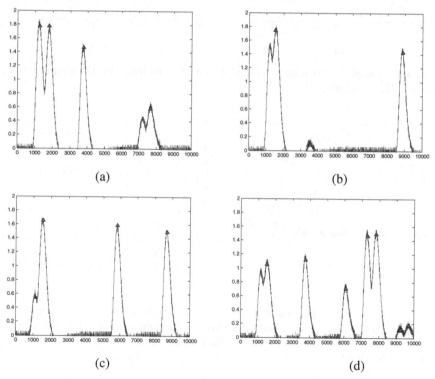

(a)

(b)

(c)

(d)

Fig. 4. Peak detection results indicated by triangular marks using cross-correlation with only template 1

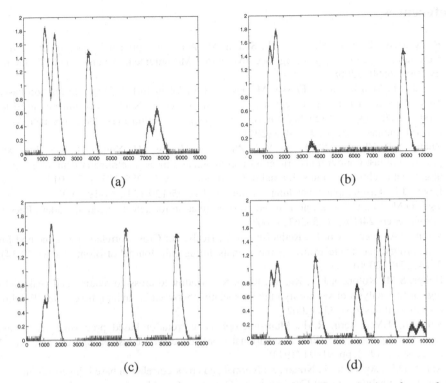

Fig. 5. Peak detection results indicated by triangular marks with pile-up rejection after applying the proposed method where each signal is same signal in. Figs. 4 (a), (b), (c), (d).

4 Conclusion and Future Work

We proposed a new method for pile-up rejection by using multiple-template cross-correlation. Peak detection based on only one template cross-correlation has limitation if the pile-up peak pulses are overlapped too closely. The proposed algorithm can reject pile-up pulses by utilizing multiple templates and can detect normal peak pulses. For a future work, we will separate the pile-up peak pulses to recover the information of each pulse.

Acknowledgement. This research was supported by Korea Institute of Geoscience and Mineral Resources (KIGAM) (Project title: Gamma spectroscopy well logging for physical properties measurement in unconventional reservoirs) and by grant No. 2012T00201725 from by the R&D program of the Korea Ministry of Trade, Industry and Energy (MOTIE). The students are financially supported by Korea Ministry of Land, Infra-structure and Transport (MOLIT) as 「 U-City Master and Doctor Course Grant Program.

References

1. Raad, M.W., Deriche, M., Noras, J., Shafiq, M.: A novel approach for pileup detection in gamma-ray spectroscopy using deconvolution. Measurement Science and Technology 19(6), 65601 (2008)
2. Faisal, M., Schiffer, R.T., Flaska, M., Pozzi, S.A., Wentzloff, D.D.: A correlation-based pulse detection technique for gammaray/neutron detectors. Nuclear Instruments and Methods in Physics Research Section A: Accelerators, Spectrometers, Detectors and Associated Equipment 652(1), 479–482 (2011)
3. Zheng, Y.B., Zhang, Z.M., Liang, Y.Z., Zhan, D.J., Huang, J.H., Yun, Y.H., Xie, H.L.: Application of fast Fourier transform cross-correlation and mass spectrometry data for accurate alignment of chromatograms. Journal of Chromatography A 1286, 175–182 (2013)
4. Lewis, J.P.: Fast template matching. Vision Interface 95(120-123), 15–19 (1995)
5. Tsai, D.M., Lin, C.T.: Fast normalized cross correlation for defect detection. Pattern Recognition Letters 24(15), 2625–2631 (2003)
6. Wren, T.A., Patrick Do, K., Rethlefsen, S.A., Healy, B.: Cross-correlation as a method for comparing dynamic electromyography signals during gait. Journal of Biomechanics 39(14), 2714–2718 (2006)
7. Boker, S.M., Rotondo, J.L., Xu, M., King, K.: Windowed cross-correlation and peak picking for the analysis of variability in the association between behavioral time series. Psychological Methods 7(3), 338 (2002)
8. Sullivan, T.M., Stern, R.M.: Multi-microphone correlation-based processing for robust speech recognition. In: IEEE International Conference on Acoustics, Speech, and Signal Processing, vol. 2, pp. 91–94 (1993)
9. Karna, D.K., Agarwal, S., Nikam, S.: Normalized cross-correlation based fingerprint matching. In: Fifth International Conference on Computer Graphics, Imaging and Visualisation, CGIV 2008, pp. 229–232 (2008)

Wireless Remote Control of Robot Dual Arms and Hands Using Inertial Measurement Units for Learning from Demonstration

Woo-Young Go and Jong-Hwan Kim

Department of Electrical Engineering, KAIST
291 Daehak-ro, Yuseong-gu, Daejeon, Republic of Korea
{wygo,johkim}@rit.kaist.ac.kr

Abstract. This paper proposes real time wireless remote control of robot dual arms and hands using IMUs (Inertial measurement unit). The remote control system is developed for robots to learn various complex tasks from operator's demonstration. The control system helps robots to make human-like motion planning directly through operator's demonstration rather than the RRT (Rapidly-Exploring Random Tree) algorithm or the vector field method. To demonstrate the effectiveness of the developed system, experiments are carried out for real time wireless remote control of robot dual arms and hands. The experiment results show that the remote control system operates well enough to teach a robot various human-like complex tasks.

Keywords: Humanoid robot, robot dual arms, learning from demonstration, teleoperation, remote control, IMU.

1 Introduction

This paper deals with teaching robots to perform complex manipulation tasks such as watering the plants, arranging various toys and toasting the bread by operator's demonstration. A human operator performs a task as a demonstration and then the robot does the same task by using the data from the IMUs attached on operator's arms and hands.

Learning from demonstration or by imitation has been investigated for robots as a strategy for human-robot interaction, in particular, as a way of teaching motions to the robots without serious programming of the RRT (Rapidly-Exploring Random Tree) algorithm or the vector field method [1,2,3,4,5]. A robot observes multiple demonstrations of a specific skill such as using RGB-D sensor with OpenNI skeleton tracking, motion tracking, teleoperation and directly guiding the robot's end-effector or is taught over teaching sessions and then, the robot extracts a generalized motion out of the demonstrations or teachings [6,7,8,9].

This paper focuses on imitation teaching as a teaching strategy for a robot to imitate the teacher's motions for a specific task by encoding the joint angles of the teacher [10]. As the dual arms and hands have multiple degrees of freedom,

© Springer International Publishing Switzerland 2015
J.-H. Kim et al. (eds.), *Robot Intelligence Technology and Applications 3*,
Advances in Intelligent Systems and Computing 345, DOI: 10.1007/978-3-319-16841-8_69

(a)   (b)

Fig. 1. Configuration of IMU. (a)Inertial measurement unit. (b)IMU with LIPB rechargeable battery.

it is not easy to control all of the joints through the mathematical modelling. The developed remote control system is simply designed using IMUs enough to control humanoid robot dual arms and hands without their mathematical models complicated computation.

This paper is organized as follows. Section 2 describes the details of IMU. Sections 3 and 4 presents remote control systems for robot dual arms and hands, respectively. Experiment results are presented in Section 5. Finally, conclusion and further work follow in Section 6.

2   Inertial Measurement Unit

The IMU in this system is EBIMU24GVS which is 2.4GHz, 9 DOF and get roll, pitch, yaw data as shown in Fig. 1. It does not need a power cable because it uses LIPB rechargeable battery. Moreover, it can be used wirelessly if we set up the receiver in Fig. 2 2(a). Fig. 2(b) shows a screen shot of the IMU test program.

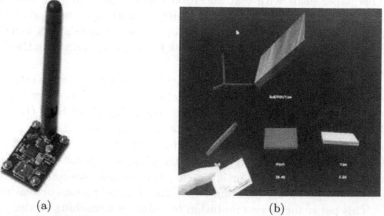

(a)   (b)

Fig. 2. IMU receiver and IMU test program. (a)IMU receiver. (b)Screen shot of the IMU test program.

3 Remote Dual Arms Control System Design

The real time wireless remote control system is developed for Mybot-KSR humanoid manipulators each with a 7 DOFs, which was developed in the Robot Intelligence Technology Lab. at KAIST. The developed system is tested using a Webots simulators of Mybot-KSR as shown in Fig. 3. Mybot-KSR's dual arms are to be controlled to the operator's demonstration using IMUs attached on operator's arms. Although each arm of Mybot-KSR has 7 DOFs, we practically use 4 DOFs of each arm except wrist joints.

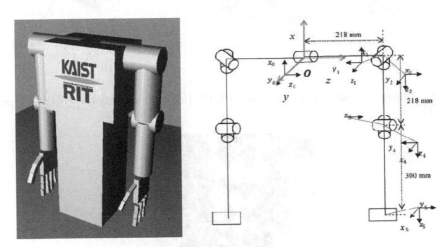

Fig. 3. Mybot-KSR upper body

The data that can be obtained from an IMU are roll, pitch and yaw signal data as shown in Fig. 4. Fig. 5 shows two IMUs attached on operator's arm. Thus, totally 4 IMUs send roll, pitch, yaw signals to the receiver, which are used to control Mybot-KSR's manipulators. Fig. 6 shows the overall process with a Webots simulator for learning from demonstration.

The overall process starts with the operator's demonstration. Second, the IMU receiver receives the signal using radio frequency communication every control period wirelessly.

Third, a learning from demonstration program connects to the IMU receiver port to get the signals from the four IMUs and stores initial pose joint angles to calculate the relative angle from the initial position from demonstration. Based on the initial angles, in every control period, the program obtains the current angles from the IMUs and calculates the differences by subtracting current angles from initial angles and then send the difference of each angle to the Webots simulator.

Finally, the Webots simulator gets the difference of each angle from the learning from demonstration program and operates Mybot-KSR's dual arms' each joint.

In case of the left arm, the first IMU's yaw, roll and pitch signals make motions lift arm forward, lift arm sideways and rotates arm respectively. The second

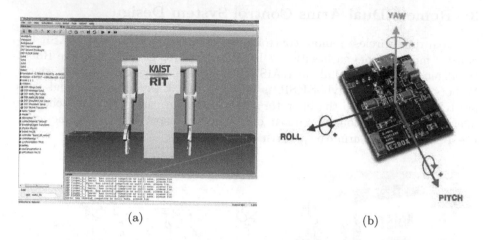

(a)   (b)

Fig. 4. Webots simulator and data type of IMU. (a)Webots simulator. (b)Data type of IMU.

Fig. 5. IMU on operator's arms

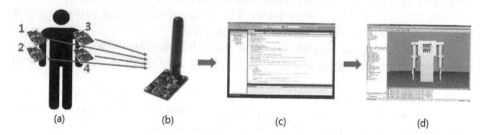

(a)   (b)   (c)   (d)

Fig. 6. Overall process with a Webots simulator. (a)Operator's demonstration with IMUs attached on his arms. (b)The IMU receiver. (c)Learning from demonstration program. (d)Webots simulator.

IMU's roll signal makes the motion bend elbow. In case of the right arm, the third IMU's yaw, roll and pitch signals make motion lift arm forward, lift arm sideways and rotates arm respectively. The forth IMU's roll signal makes the motion bend elbow. With the 4 IMUs, dual arms with total 8 DOFs are controlled.

4 Remote Hands Control System Design

Mybot-KSR's hand has 5 fingers of which specifications are summarized in Table. 1. The robot hand models a real human hand and each finger has one DOF except the ring and little fingers both sharing 1 DOF. Thus, each hand has

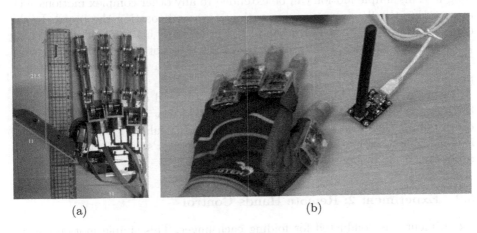

(a)   (b)

Fig. 7. Hand remote control system design. (a)Mybot-KSR's hand. (b)Glove with IMUs and receiver.

Table 1. Specification of Mybot-KSR's hand

Developed humanoid robot hand	
Length	21.5 cm
Finger length (orderly)	11.0 cm, 9.0 cm, 9.0 cm, 9.0 cm, 7.0 cm
Number of fingers	5
Number of joints	15
Number of controlled DOF	5
Fingertip force	4.5 N
Weight	564.0 g
Speed (fully close)	0.28 s
Faulhaber DC-micro motors	012 SR
Motor weight	190.0 g
Motor torque	0.8 mNm
Encoder	IE2-256
Gear head	141:1

5 DOFs such that 4 IMUs are used to control each hand of the Mybot-KSR. Using the IMU hand control system, the four IMUs' roll values make a motion bend fingers and the first IMU's pitch value makes a motion move sideways. Fig. 7 shows a real time wireless wearable glove with IMUs to control the Mybot-KSR's hand.

5   Experiment

5.1   Experiment 1: Remote Dual Arms Control

Experiment was conducted for lifting and throwing a cylindrical box after grabbing it. This simple motion can be extended to any other complex motions such as watering the flowers, arranging various toys and toasting the bread. In the experiment, an operator did a complex motion with his arms while watching the robot arms in the Webots simulator to see if or not they follow his motion. This makes the operator adjust the postures of robot arms to operator's desired postures.

To verify the remote control, Fig. 8 shows the end effect's trajectory using the forward kinematics. The points denote the trajectory of and end effect. From these points, the trajectory of dual arms can be calculated. With this system, it is possible to teach the robot to learn from operator's demonstration and the robot can make human-like motion as shown in Fig. 9.

5.2   Experiment 2: Remote Hands Control

Experiment was conducted for folding each finger. This simple motion can be extended to any other complex motions such as grabbing a cup, mouse and toy. It is really hard to learn how to grab objects because there are multiple degrees of freedom of hand, i.e. 15 DOFs. However, using this system, the robot can learn easily how to grab objects, as the developed robot hand was designed with 4 DOFs. In the experiment, an operator did a complex motion with his hands

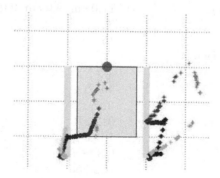

Fig. 8. Trajectory of an end effect of Mybot-KSR from the forward kinematics

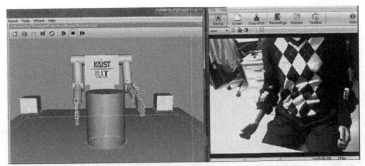

(a) Grabbing the cylindrical box

(b) Lifting the cylindrical box

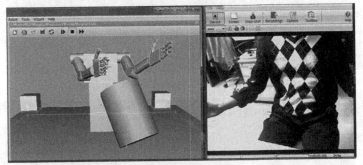

(c) Throwing the cylindrical box

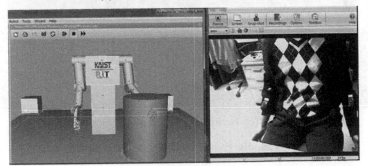

(d) Lowering Mybot-KSR's arms

Fig. 9. Remote dual arms control system experiment

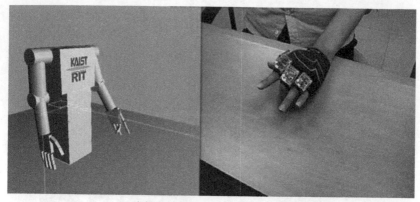

(a) Bending a middle finger

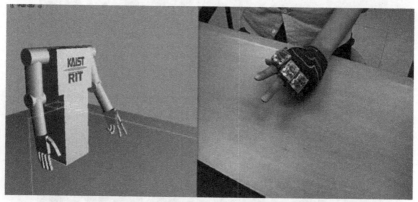

(b) Bending a ring finger and a little finger

(c) Bending all of fingers

Fig. 10. Remote Hands control system experiment

while watching the robot hands in the Webots simulator to see if or not they follow his hand motion as shown in Fig. 10. In this way, the operator could adjust the postures of robot hands to his desired postures.

6   Conclusion and Further Work

In this paper, we developed the system that wireless remote control system of robot dual arms and hands using inertial measurement units to teach various complex tasks through operator's demonstration. A human operator performed a task of grabbing, lifting and throwing a cylindrical box as a demonstration and then the robot did the same task by using the data from the IMUs attached on operator's arms and hands. The effectiveness of the proposed system was demonstrated through experiments carried out for wireless remote control of robot dual arms and hands. The experiment results showed that the remote control system worked well enough to teach a robot various human-like complex tasks.

The drawback of this system is that it is not possible to use more than 4 IMUs at the same time. In case of using more than 4 IMUs at the same time, the delay time of 2 seconds is expected such that real time control is not possible. For this reason, to make a whole motion of Mybot-KSR's upper body at the same time is not possible using IMUs. To make the whole motion of dual arms and hands, we must teach dual arms and each hand separately and then combine all of the upper body's joint angular trajectories.

Acknowledgements. This research was supported by the MOTIE (The Ministry of Trade, Industry and Energy), Korea, under the Technology Innovation Program supervised by KEIT (Korea Evaluation Institute of Industrial Technology), 10045252, Development of robot task intelligence technology.

References

1. Calinon, S., Guenter, F., Billard, A.: On learning, representing and generalizing a task in a humanoid robot. IEEE Trans. Systems, Man and Cybernetics, Part B 37(2), 286–298 (2007)
2. Shon, A.P., Storz, J.J., Rao, R.P.N.: Towards a real-time Bayesian imitation system for a humanoid robot. In: Proc. Int Conf Robotics Automation, Roma, Italy (2007)
3. Ijspeert, A.J., Nakanishi, J., Schaal, S.: Trajectory formation for imitation with nonlinear dynamical systems. In: Proc. IEEE Int. Conf. Intelligent Robots Systems, pp. 752–757 (2001)
4. Atkeson, C.G., Schaal, S.: Robot learning from demonstration. In: Proc. 14th Int. Conf Machine Learning (ICML), pp. 12–20 (July 1997)
5. Pastor, P., Hoffmann, H., Asfour, T., Schaal, S.: Learning and generalization of motor skills by learning from demonstration. In: Proc. Int. Conf. Robotics Automation, Kobe, Japan (2009)
6. Breazeal, C., Scassellati, B.: Robots that imitate humans. Trends in Cognitive Sciences 6(11), 481–487 (2002)

7. Billard, A., Calinon, S., Dillmann, R., Schaal, S.: Robot Programming by Demonstration. In: Siciliano, B., Khatib, O. (eds.) Handbook of Robotics, pp. 1371–1394. Springer (2008)
8. Argall, B., Chernova, S., Veloso, M., Browning, B.: A survey of robot learning from demonstration. Robotics and Autonomous Systems 57(5), 469–483 (2009)
9. Schaal, S.: Is imitation learning the route to humanoid robots. Trends in Cognitive Sciences 3(6), 233–242 (1999)
10. Hersch, M., Guenter, F., Calinon, S., Billard, A.: Dynamic system modulation for robot learning via kinesthetic demonstrations. IEEE Trans. Robotics 24(6), 1463–1467 (2008)
11. Calinon, S., Billard, A.: Incremental learning of gestures by imitation in a humanoid robot. In: Proc. IEEE Int. Conf. Human Robot Interaction, Arlington, Virginia, USA, pp. 255–262 (2007)
12. Go, W.-Y., Kim, J.-H.: Flexible and Wearable Hand Exoskeleton and Its Application to Computer Mouse. In: Proc. International Conference on Robot Intelligence Technology and Applications (RiTA), Denver, U.S.A (December 2013)

Online Learning-Prediction Based Diagnosis Decision Support System Towards Swallowing Dysfunction in Rehabilitation Medicine

Chen Jie[1,2], Li Ping[2], Su Chong[3,*], Fu Dapeng[1], and Chen Yan[1]

[1] Beijing Zhongguancun Hospital, Department of Rehabilitation medicine,
Beijing 100190, China
[2] Tianjin University of Traditional Chinese Medicine, Tianjin 300029, China
[3] College of Information Science and Technology, Beijing University of Chemical Technology,
Beijing 100029, China
suchong@mail.buct.edu.cn

Abstract. Medical diagnosis is a complex and fuzzy cognitive process of learning, such as neural networks of artificial intelligence methodologies, showing great potential can be applied to the development of medical decision support systems (MDSS). In this paper, online learning- prediction based neural networks are developed to support the diagnosis of swallowing dysfunction in Rehabilitation Medicine, along with the increasing accuracy of systematic study when the cases are input. The input layer of the system includes 28 input variables, categorized into five groups and then encoded using the proposed coding schemes. The RBF (Radical Basis Function) algorithms are employed to train the online learning- prediction system, the number of nodes in the hidden layer is determined by the online nodes updating process. Each of the 15 nodes in the output layer corresponds to one swallowing dysfunction disease of interest. A total of 120 medical records collected from the patients suffering from fifteen swallowing dysfunction have been used to train the system, where, 20 cases are used to test the system. Particularly, '5-fold' cross validation is applied to assess the performance of the decision support system. The results show that the proposed online learning- prediction based decision support system can achieve very high diagnosis accuracy (>90%), giving rise to satisfied results and showing validity of the contributions.

Keywords: Medical Decision Support system, On-line Learning- prediction, RBF (Radical Basis Function), Rehabilitation Medicine, swallowing dysfunction.

1 Introduction

It is acknowledged that decision support systems have played increasingly important roles in medicine diagnosis nowadays. However, as diagnosed cases increasing, it's important to enhance the on-line learning and prediction capabilities of the medicine decision support system.

---

* Corresponding author.

© Springer International Publishing Switzerland 2015
J.-H. Kim et al. (eds.), *Robot Intelligence Technology and Applications 3,*
Advances in Intelligent Systems and Computing 345, DOI: 10.1007/978-3-319-16841-8_70

In the 1990's, many experts and scholars introduced new ideas and methods to build decision support systems for medical diagnosis and treatment with AI methodologies. Robinso & Thomson (2001) [1] deeply discussed the functions of patients' age, social status changes, and other factors which led to changes in medical decision-making and treatment process; Smith et al (2003) [2] for medical decision support The system gave the design criteria and performance evaluation for the medicine decision support system; Yana et al (2006) [3] introduced a multiplayer prediction based decision support system to support the diagnosis of heart diseases and collected 352 patients' cases to verify the accuracy of system; Montani et al (2008)[4] proposed to design a modular architecture with case-based Reasoning (CBR) methodology to provide decision support. There're medical decision support strategies in a lot of medicine areas, however, it is conceivable that the problem of how describe on-line learning-prediction based medicine decision support system still represents a challenge.

Nerual and motor dysfunction caused by stroke is a major killer of human. Currently, majority of traditional stroke rehabilitation medicine diagnosis researches are concerned with theoretical relevance and method issues: Andersen (1997)[5] extended the original stroke depression diagnosis and treatment evaluation methods; Leonard et al (1998)[6] assessed the functional outcomes of first-time cerebrovascular accident survivors I to 5 years after rehabilitation in their cases of rehabilitation centers; Ska et al (2003)[7] presented a patient suffering from acquired deep dyslexia with using pictures and vocabulary voice test method to improve the patient's language function; Garcia et al (2009) [8] proposed a high-accuracy algorithm for the detection of the periodicity cycles of both oesophageal and laryngeal voices with low quality which allows the accurate and automatic estimation of pitch, jitter and shimmer; Wadea et al (2010) [9] provided an overview of methodological issues specifically related to the evaluation of rehabilitation interventions studies; Cataldo et al (2012) [10] detected positive and negative prognostic factors associated with functional outcomes in first-time stroke patients admitted to an integrated home care rehabilitative program with 141 patients with a first-time stroke who were admitted to a home care rehabilitation program. However, the aims of this kind of research mostly certain the symptoms or a single disease site considerations of patients, which is lack of comprehensive and integrated patients' body indicators, suffering strong limitations for the rehabilitation medical diagnosis work .

In addition, many studies of the implementation of rehabilitation medicine decision support system based on soft computing theory are deeply discussed: Teodorescua et al (2001) [11] presented the use of fuzzy methods in applications for rehabilitation, namely in tremor diagnosing and control, as well as developed software system; Yardimci (2009) [12] reviewed the application of soft computing methods(such as fuzzy mathematics, neural networks, genetic algorithms and other intelligent methods) in biomedical, neurological diseases; Jamwal et al (2010) [13] deal with forward kinematics (FK) mapping of a parallel robot, especially designed for ankle joint rehabilitation treatments; Shamsuddin et al (2012) [14] presents the findings from study on the initial behavior of autistic children of moderately impaired intelligence when exposed to simple human-robot interaction (HRI) modules executed by a humanoid robot; Cheng et al (2012)[15] reviewed techniques for rehabilitation treatment of Parkinson disease, Alzheimer disease, stroke, and vestibular diseases. However, the decision support systems are biased in favor of feature extraction and treatment program from a local condition of the patients, suffering strong limitations of a macroscopic perspective. Especially, the construction methods of rehabilitation medicine swallowing dysfunction decision support system haven't been discussed.

Online learning-prediction are playing an important role in decision support system. In a number of online learning-prediction algorithms, RBF neural network technology is mostly popular. Han et al (2011)[16] proposed a flexible structure Radial Basis Function (RBF) neural network (FS-RBFNN) and its application to water quality prediction. Hsu et al (2012) [17] proposed an indirect adaptive self-organizing RBF neural control (IASRNC) system which is composed of a feedback controller, a neural identifier and a smooth compensator. Wang et al (2012) [18] proposed new self-correcting scheme of forecasting method based on single model structure which can avoid inherent defects of double model structure and improve the network's online self-correcting ability. Yin et al (2013)[19] proposed an on-line prediction model of ship roll motion via a variable structure radial basis function neural network (RBFNN), whose structure and parameters are tuned in real time based on a sliding data window observer.

Motivated by the discussion above, this paper proposes an approach to online learning-prediction adapted to medicine diagnosis mechanism. Successively, a kind of online learning-prediction RBF neural network models is presented, along with enabling algorithms, aiming at grasping essentials in doctors' diagnosis experiences towards rehabilitation medicine swallowing dysfunction diagnosis. The remainder of this paper is organized as follows. Section 2 introduces the online learning-prediction framework of RBF neural network. In Section 3, the methods of building the decision support system based on an online learning-prediction RBF neural network are presented. The experimental results are reported in Section 4. Section 5 concludes the article and assesses the future perspectives.

2   Online Learning-Prediction Algorithm Based on RBF Neural Network

2.1   RBF Neural Networks' Mathematical Description

Based on a MISO(multi-input/single-output) RBF network, assuming the RBF network has K hidden nodes, Gaussian radial basis function is used in the network. The RBF network model is described as:

$$f(X_i) = \omega_0 + \sum_{k=1}^{K} \omega_k \phi_k (X_i) \tag{1}$$

Where, X_i is the input of the RBF network; $f(X_i)$ is the output of the RBF network; ω_k is the connection weight of kth hidden layer's output; ω_0 is a constant of output offset; $\phi_k(x_i) = \exp(-\dfrac{1}{\sigma_k^2}\|X_i - u_n\|^2)$ is Gaussian radial basis function,

$u_k \in R^n$ is the data center , and, σ_k is the expansion constant of RBF function.

2.2   Hidden Layer Nodes Updating Methods of RBF Network

2.2.1   Judgment of ' Novelty' Node

Traditional 'Novelty' hidden nodes' condition are described as:

$$|e_i| = |y_i - f(X_i)| > e_{\min} \tag{2}$$

$$\min_k \|X_i - u_k\| > \xi \tag{3}$$

Where, the formula (2) represents that if a hidden layer neurons will be added in the network, the error between the network chrome output and the target should be large enough, as well as, $e_{\min}$ is the desired approximation accuracy. The formula (3) indicates the neurons' activity with the minimum distance between the samples and selected center.

New hidden layer nodes' parameters are described as:

$$\begin{aligned} \omega_{k+1} &= e_i \\ u_{k+1} &= X_i \\ \sigma_{k+1} &= v\|X_i - u_n\| \end{aligned} \tag{4}$$

Where: v is the iteration factor; u_n is the most close center of samples X_i.

2.2.2   'Pruning' Principle

In order to obtain the 'parsimony' network structure, we should 'prune' the existing nodes. The hidden layer nodes which make little contributions for the network output should be removed, meanwhile, it does not affect the network performance. To achieve this purpose, 'pruning' criteria is shown as the following definitions:

(1) To remove the network hidden layer nodes which make little, we calculate the hidden layer nodes output for each sample :

$$\sigma_k^i = \omega_k \exp(-\dfrac{1}{\sigma_k^2}\|X_i - u_n\|^2) \tag{5}$$

Get the maximum hidden layer node output $|\sigma_{\max}^i| = \max(\sigma_k^i)$, the output is normalized to

$$\lambda_k^i = \left|\dfrac{\sigma_k^i}{\sigma_{\max}^i}\right| \quad k = 1, \cdots, K \tag{6}$$

If $\lambda_k^i < \delta$, the kth node will be removed.

(2) The hidden nodes with small-scale function will be removed. If $\sigma_k < \chi$, the kth node will be removed.

2.2.3   Online Learning - Prediction

The online learning – prediction algorithm based RBF neural network is shown in fig.1.

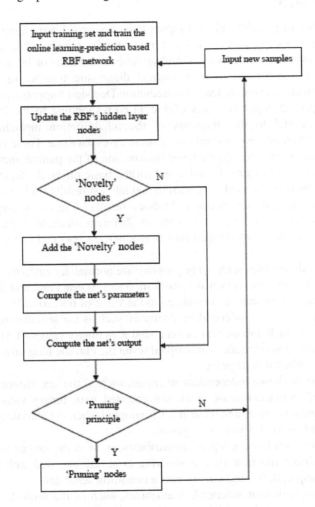

Fig. 1. Online learning – prediction algorithm based RBF neural network

3   System Structure

The online learning-prediction RBF neural network used in this study consists of three layers including an input layer, a hidden layer and an output layer. In this section, we will present the details regarding each of these layers.

3.1   Input Layer

A large number of possibly relevant inputs must be considered during the diagnosis. A physician reaches a correct diagnosis or treatment decision based upon observations, the patient's answers to questions, and physical examinations or lab results. Patients' information is transformed to mathematical diagnostic process variables through rehabilitation medicine online learning-prediction Decision Support System.

In this paper, 28 input variables (Table 1) corresponding to the 28 input nodes, which are essential to the diagnosis of the rehabilitation medicine swallowing dysfunction of interest are extracted from patient cases database. These variables can be divided into five categories: (i) the basic information of the patient, including age and sex, (ii) the history (4 factors in total), and (iii)the basic physical function(11 factors in total), (iv) the first physical examination and lab test results (6 factors in total) and (v) the second physical examination and lab test results (5 factors in total).

The medical diagnose scales consist of different discrete variables. Therein, the 28 input variables (categorized into 5 groups) are encoded using the following schemes:

(a) Numerical variables such as age pressure are normalized on to the interval [0, 1]. For example, the patients' ages may span from 0 to 120 years old, and thereby the age of a 60-year-old patient can be normalized to the value of 60/120=0.50.

(b) Variables with two independent attributes, such as the sex, chronic lung disease and history of stomach disease et al. are encoded with binary values (0, 1). For instance, 1 represents male and 0 female; 1 is adopted when the chronic lung disease is positive, while 0 is used when it is negative.

(c) Variables with two independent attributes, such as the sex, chronic lung disease and history of stomach disease et al. are encoded with binary values (0, 1). For example, 1 represents male and 0 female; 1 is adopted when the chronic lung disease is positive, while 0 is used when it is negative.

(d) Variables with three independent attributes, such as the tongue function, palate function and throat function after swallowing et al. are encoded with binary values (0, 1). For example, 0, 0.5 and 1 represents a condition separately.

(e) Variables with four independent attributes, such as the sense level, control of head and trunk are encoded with binary values (0, 1). For example, 0, 0.33, 0.67 and 1 represents a condition separately.

Table 1. Input variables essential to rehabilitation medicine swallowing dysfunction diagnosis and their corresponding encoding values or ranges

	Number	Index	Description	Scale value	Encode value
Basic information	1	Age	0-120 are normalized on to the interval [0, 1]	[0, 1]	[0,1]
	2	Sex	male=1, female=2	1/2	(0,1)
History of diseases	3	Chronic lung disease	negative =1, positive =2	1/2	(0,1)
	4	Stomach illness	negative =1, positive =2	1/2	(0,1)
	5	Drug	negative =1, positive =2	1/2	(0,1)
	6	Fever pneumonia by infection	negative =1, positive =2	1/2	(0,1)
Basic physical function	7	Consciousness level	clear-headed =1, Drowsiness, but can be woke up =2, Patients have response, but can not open eyes and speck =3, Patients have response to pain =4	1/2/3/4	(0,0.33,0.67,1)
	8	Control of head and trunk	Sit steadily=1, Sit unsteadily =2, Head can be controlled =3, Head cannot be controlled =4	1/2/3/4	(0,0.33,0.67,1)
	9	Breathing's pattern	Normal = 1, abnormal = 2	1/2	(0,1)
	10	Lip's pattern	Normal = 1, abnormal = 2	1/2	(0,1)
	11	Tongue's function	Normal = 1, abnormal = 2, Diminished or lack of =3	1/2/3	(0,0.5,1)
	12	Salivation	negative =1, positive =2	1/2	(0,1)
	13	Movement of the soft palate	Normal = 1, abnormal = 2, Diminished or lack of =3	1/2/3	(0,0.5,1)
	14	Nose Reflux	negative =1, positive =2	1/2	(0,1)
	15	Chew	negative =1, positive =2	1/2	(0,1)
	16	Laryngeal function	Normal = 1, abnormal = 2, lack of =3	1/2/3	(0,0.5,1)
	17	Gastric reflux after eating and drinking	negative =1, positive =2	1/2	(0,1)

Table 1. *(continued)*

	Number	Index	Description	Scale value	Encode value
Stage 1 of the physical examination and lab testing: give 1 tablespoon water (5 ml) with three times	18	Water flows out	None or once=1, more than once=2	1/2	(0,1)
	19	Invalid laryngeal movement	Existing=1, No existing =2	1/2	(0,1)
	20	Repeated swallow	None or once =1, more than once =2	1/2	(0,1)
	21	Coughing when swallowing	None or once =1, more than once =2	1/2	(0,1)
	22	Wheezing when swallowing	Existing=1, No existing =2	1/2	(0,1)
	23	Throat function after swallowing	Normal = 1, weakening or hoarseness = 2, Can not speak = 3	1/2/3	(0,0.5,1)
Stage 2 of the physical examination and lab testing: If phase 1 is normal , repeated the experiment for 3 times. If there're more than 2 experiments whose results are normal, then give a cup of 60ml water	24	Ability to complete phase 1	Can complete stage 1=1, Can not complete stage 1 =2 (If the first phase can not be completed, the following evaluation scale will take the most pessimistic value to operate directly)	1/2	(0,1)
	25	Coughing after swallowing	Existing=1, No existing =2	1/2	(0,1)
	26	wheezing after swallowing	Existing=1, No existing =2	1/2	(0,1)
	27	Throat function after swallowing	Normal = 1, abnormal = 2, Diminished or lack of =3	1/2/3	(0,0.5,1)
	28	Existing aspiration	None=1, possible=2, Existing =3	1/2/3	(0,0.5,1)

3.2 Hidden Layer

Online update "node" learning strategies have been proposed by Figure 1, number of hidden layer nodes are set by prior experience. Number of hidden layer nodes are updating with the new training set data input to determine the hidden layer structure. Therefore, the accuracy and convergence speed of the network are guaranteed.

3.3 Output Layer

Traditional rehabilitation medicine swallowing dysfunction medical diagnostic categories are gave in table 2:

Table 2. Traditional rehabilitation medicine swallowing dysfunction medical diagnostic categories

Traditional basic medical diagnostic categories	
Oral phase	I
Tongue phase	II
Pharynx phase	III
Throat phase	IV

The output layer comprises of 15 nodes and each node corresponds to one rehabilitation medicine swallowing dysfunction of interest. The architecture of the overall decision support system is illustrated in Table 3, as well as the system network framework is shown in Figure 2

Table 3. Output layer nodes (Patient's diagnosis results' categories)

Output layer nodes corresponding to the Patient's diagnosis results' categories					
1 pattern	I (0.067)	II (0.133)	III (0.200)	IV (0.267)		
2 patterns combination	I/II (0.333)	I/III (0.400)	I/IV (0.467)	II/III (0.533)	II/IV (0.600)	III/IV (0.667)
3 patterns combination	I/II/III (0.733)	I/II/IV (0.800)	I/III/IV (0.867)	II/III/IV (0.933)		
4 patterns combination	I/II/III/IV (1.000)					

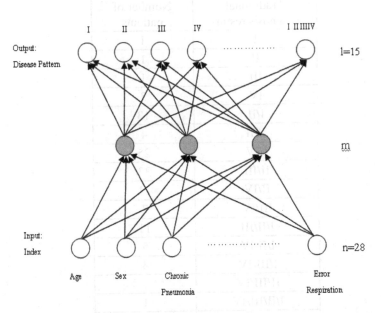

Fig. 2. Network framework of rehabilitation medicine swallowing dysfunction

3.4   System Summary

Based on the discussion above, rehabilitation medicine swallowing dysfunction decision support system consists of 28 input layer nodes, m hidden layer nodes which is to be optimized (based on our experience, m's initial value is 3) and 15 output layer nodes. Where, the accuracy threshold of the network is 0.001, each training step is 500. The network frame is shown in Figure 2.

4   Case Studies and Performance Evaluation

The rehabilitation medicine swallowing dysfunction database used for testing and tuning the system in this study consists of 120 cases gathered from the Beijing Zhongguancun Hospital and Tianjin Nankai Hospital, which are located in Beijing and Tianjin separately, P. R. China. We have trained and tested the system using these medical records and the results are reported in this section.

4.1   Experimental Data and Performance Evaluation

It is acknowleged that the number of the collected cases is large which is more than 30 in medical diagnose research, which is different from the industrial process. The rehabilitation medicine swallowing dysfunction cases are described in detail in table 4.

Table 4. Experimental data

Traditional diagnosis results	Number of patients
I	1
II	1
III	1
IV	3
I/II	13
I/III	4
I/IV	25
II/III	33
II/IV	11
III/IV	7
I/II/III	11
I/II/IV	5
I/III/IV	3
II/III/IV	1
I/II/III/IV	1
total	120

4.2   Decision Support System Performance Evaluation [20]

4.2.1   Cross-Validation Test

It's very important to estimate the accuracy of the classification methods. Cross-validation is a kind of methods which can be used to estimate the predictive model fitting performance. Cross-validation test method divides the original data into two parts, one is the training set, the other is the validation set. In order to evaluate the performance of the classifier, the training set is used to train the classifier, as well as, the validation set is used to Verify the classifier.

K-fold cross-validation divides the data into K disjoint "fold"($S_1, S_2, \cdots S_k$) with K iterations operation. In the ith iteration, S_i is described as the test set, the rest is the training set ($S_1, \cdots, S_{i-1}, S_{i+1}, \cdots S_k$). The accuracy is described as:

$$Acc_{cv} = \frac{1}{n} \sum_{(x_i, y_i) \in D} \delta(I(D_i, x_i), y_i) \tag{7}$$

where n is the size of the dataset D, x_i is the instance of D, y_i is the label of xi, and D_i is the possible label of x_i by the classifier. Here,

$$\delta(i, j) = \begin{cases} 1 & if\ i = j \\ 0 & otherwise \end{cases} \tag{8}$$

The method can overcome the small sample size and get better generalization ability.

In this study, we apply the '5-fold' cross-validation method to verify the performance of RBF network model.

Patient's diagnostic data in table 4 is divided into five subset (F_1, F_2, F_3, F_4, F_5):

(I) Training: $F_1 + F_2 + F_3 + F_4$; Testing: F_5

(Ii) Training: $F_1 + F_2 + F_3 + F_5$; Testing: F_4

(Iii) Training: $F_1 + F_2 + F_4 + F_5$; Testing: F_3

(Iv) Training: $F_1 + F_3 + F_4 + F_5$; Testing: F_2

(V) Training: $F_2 + F_3 + F_4 + F_5$; Testing: F_1

4.2.2   Performance Verification

Based on samples selected randomly from the 120 medicine samples above, the '5-fold' cross-validation estimated accuracy of each test run, as well as, the interval of estimated accuracy and the mean accuracy are listed in Table 5. Here the first column CV1 refers to the first test based on the cross validation method.

Table 5. Accuracy result: '5-fold' cross-validation

	CV_1	CV_2	CV_3	CV_4	CV_5	Interval
ACC_n (%)	91.3%	98.2%	94.1%	94.3%	95.7%	[91.3%,98.2%]
Mean ACC_n (%)	94.8%					

We use the network whose accuracy is 98.2% to show the rehabilitation medicine swallowing dysfunction diagnosis decision support process. The training's cut-off condition is the training error is less than 10^{-3}. Figure 3 presents the training accuracy of the network. Figure 4 shows the diagnosis approximating offsets of original 120 cases. Figure 5 shows the diagnosis approximating offsets of the new 20 cases.

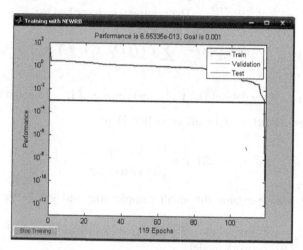

Fig. 3. Training accuracy of the network

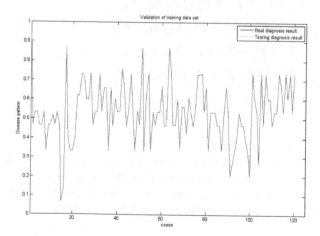

Fig. 4. Diagnosis approximating offsets of original 120 cases

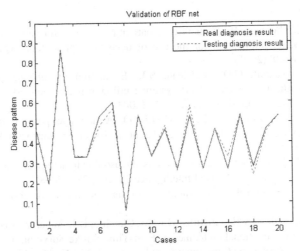

Fig. 5. Diagnosis approximating offsets of the new 20 cases

4.3   Summary

The 'k-fold' cross-validation evaluation methods have been adopted to assess the performance of the proposed online learning-prediction based rehabilitation medicine swallowing dysfunction decision support system. From the results reported in Table 5 and Figures 3-5, it can be found that the system has a strong capability to accurately recognize all the fifteen swallowing dysfunction diseases (>90%).

5   Conclusions

Based on a large number of swallowing dysfunction cases which relevant medical units collected, a class of RBF neural network based online learning-predictive decision support system is proposed in this paper. Towards the continuous input of new cases, the accuracy of the decision support system model is enhanced to assist clinicians and community health workers to make diagnosis decision. Five-fold cross-validation method is overcoming the small-scale cases to verify the accuracy and validity of the proposed   model.

Nonetheless, it should be pointed out that this research remains rather fundamental currently, which is   in desperate need of further investigations on some key   issues, such as how to adapt more comprehensive diagnosis indexes towards resolving specific problems and how to identify the relationship of the indexes.

References

1. Robinson, A., Thomson, R.: Variability in patient preferences for participating in medical decision making: implication for the use of decision support tools. Quality in Health Care 10(1), 34–38 (2001)
2. Smith, A.E., Nugent, C.D., McClean, S.I.: Evaluation of inherent performance of intelligent medical decision support systems: utilizing neural networks as an example. Artificial Intelligence in Medicine 27, 1–27 (2003)
3. Yana, H., Jiang, Y., Zhenge, J., Pengc, C., Lid, Q.: A multilayer perceptron-based medical decision support system for heart disease diagnosis. Expert Systems with Applications 30, 272–281 (2006)
4. Montani, S.: Exploring new roles for case-based reasoning in heterogeneous AI systems for medical decision support. Applied Intelligence 28, 275–285 (2008)
5. Andersen, G.: Post-stroke depression: diagnosis and incidence. Eur. Psychiatry 12, 255–260 (1997)
6. Leonard, C.T., Miller, K.E., Griffiths, H.I., McClatchie, B.J., Wherry, A.B.: A Sequential Study Assessing Functional Outcomes of First-Time Stroke Survivors I to 5 Years After Rehabilitation. Stroke and Cerebrovascular Diseases 7(2), 145–153 (1998)
7. Ska, B., Garneau-Beaumont, D., Chesneau, S., Damien, B.: Diagnosis and rehabilitation attempt of a patient with acquired deep dyslexia. Brain and Cognition 53, 359–363 (2003)
8. García, B., Ruiz, I., Méndez, A., Mendezona, M.: Objective characterization of oesophageal voice supporting medical diagnosis, rehabilitation and monitoring. Computers in Biology and Medicine 39, 97–105 (2009)
9. Wadea, D.T., Smeetsa, R.J.E.M., Verbunt, J.A.: Research in rehabilitation medicine: Methodological challenges. Journal of Clinical Epidemiology 63, 699–704 (2010)
10. Cataldo, M.C., Calcara, M.L., Caputo, G., Mammina, C.: Association of total serum cholesterol with functional outcome following home care rehabilitation in Italian patients with stroke. Disability and Health 5, 111–116 (2012)
11. Teodorescua, H.-N.L., Chelaru, M., Kandela, A., Tofan, I., Irimia, M.: Fuzzy methods in tremor assessment, prediction, and rehabilitation. Artificial Intelligence in Medicine 21, 107–130 (2001)
12. Yardimci, A.: Soft computing in medicine. Applied Soft Computing 9, 1029–1043 (2009)
13. Jamwal, P.K., Xie, S.Q., Tsoi, Y.H., Aw, K.C.: Forward kinematics modelling of a parallel ankle rehabilitation robot using modified fuzzy inference. Mechanism and Machine Theory 45, 1537–1554 (2010)
14. Shamsuddin, S., Yussof, H., Ismail, L.I., Mohamed, S., Hanapiahc, F.A., Zaharid, N.I.: Humanoid Robot NAO Interacting with Autistic Children of Moderately Impaired Intelligence to Augment Communication Skills. Procedia Engineering 41, 1533–1538 (2012)
15. Cheng, Y.-Y., Hsieh, W.-L., Kao, C.-L., Chan, R.-C.: Principles of rehabilitation for common chronic neurologic diseases in the elderly. Journal of Clinical Gerontology & Geriatrics 3, 5–13 (2012)
16. Han, H.-G.: An efficient self-organizing RBF neural network for water quality prediction. Neural Networks 24, 717–725 (2011)

17. Hsu, C.-F., Chiu, C.-J., Tsai, J.-Z.: Indirect adaptive self-organizing RBF neural controller design with a dynamical training approach. Expert Systems with Applications 39, 564–573 (2012)
18. Wang, Y., Liu, G.: A Forecasting Method Based on Online Self-Correcting Single Model RBF Neural Network. Procedia Engineering 29, 2516–2520 (2012)
19. Yin, J.-C., Zou, Z.-J., Xu, F.: On-line prediction of ship roll motion during maneuvering using sequential learning RBF neural networks. Ocean Engineering 61, 139–147 (2013)
20. Kohavi, R.: A study of cross-validation and bootstrap for accuracy estimation and model selection. In: Proceedings of International Joint Conference on Artificial Intelligence (1995)

Cloud-Based Image Recognition for Robots

Daniel Lorencik[1], Jaroslav Ondo[1], Peter Sincak[1], and Hiroaki Wagatsuma[2]

[1] Department of Cybernetics and Artificial Intelligence, Technical University of Kosice
{daniel.lorencik,jaroslav.ondo,peter.sincak}@tuke.sk
[2] Department of Human Intelligence Systems, Kyushu Institute of Technology,
Kitakyushu, Japan
waga@brain.kyutech.ac.jp

Abstract. Paper deals with Cloud-based Robotics approach which seems to be very supported by new technologies in the area of Cloud computing. In this paper, we will present and early implementation of a system for cloud-based object recognition. The primary use of the system is to provide an object recognition as a service for a wide range of devices. The main benefit of using the cloud as a platform are easy scalability in the future and mainly the sharing of already collected knowledge between all devices using this system. The system consist of feature extraction part and the classification part. For feature extraction, SIFT and SURF are used, and for the classification, the MF ArtMap has been used. In this paper, the implementation of both parts will be presented in more detail, as well as preliminary results. We do assume that Cloud Robotics and Brain research for Robots will emerge into a functional system able to share and utilize common knowledge and also personalization in close future.

Keywords: cloud computing, cloud robotics, SIFT, SURF, MF ArtMap, Brain like systems.

1 Introduction

Cloud Computing was introduced for IT domain many years ago. The impact to Intelligent Robotics came only recently when a concept of Cloud Robotics came into the domain of Intelligent Robotics [1], [2]. We do believe that Cloud Robotics should include implementation of Artificial Intelligence on the Cloud and also this technology can bring some major changes in core Artificial Intelligence like pattern Recognition towards continuously changing representation set for learning. Learning approach seems to be incremental and also some brain like inspirations can play important role of the resulting system. Crowdsourcing and also multisource information about brain functioning can bring effect in resulting accuracy of Robotic Intelligence.

2 Cloud Based Framework for Cloud Robotics

We have discussed the system proposal in greater detail in [3]. The proposed system is based on the notion of AI Brick [4] – to provide well-defined system suited for one task – in this case the object recognition. Since the system uses Microsoft Azure as a

© Springer International Publishing Switzerland 2015
J.-H. Kim et al. (eds.), *Robot Intelligence Technology and Applications 3*,
Advances in Intelligent Systems and Computing 345, DOI: 10.1007/978-3-319-16841-8_71

cloud platform, the inherited cloud capabilities will allow for easy scalability in case of increased demand on service, will allow for easy deployment of new versions (as the main logic will be provided as a cloud service) and most importantly, it will allow for knowledge acquisition and sharing from all of the connected clients.

The important feature of the system is that it places no special requirements for the devices that would use it. The only requirements are ability to capture images and to send them over the internet connection to the service.

2.1   Cloud Computing Platform and Technological Aspects

As was already mentioned, the system is based on the PaaS [5] (Platform as a Service) provided by Microsoft Azure. Since our system is intended to be a cloud service, we adopted the modular architecture of Azure cloud services, where user interfaces are created as web roles hosted on virtual computers of variable computing power with the use of ASP.NET, and the background jobs are created as worker roles hosted on dedicated virtual servers of variable computing power. These are interacting with the use of Message Bus and Queues.

The image data are stored as a blob storage, as well as the descriptors extracted from them.

To create a truly cloud based service, we use the No-SQL Azure Tables [6] for cross-referencing the image data, extracted descriptors and the classification data instead of SQL-like databases.

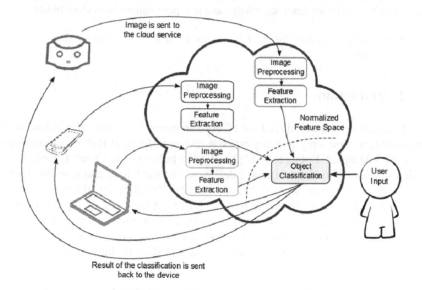

Fig. 1. High level architecture of the proposed system [3]

From the high level architecture proposal on the Fig. 1 it can be seen that only the image is sent over the Internet to the cloud service as an input data. The required preprocessing and feature extraction is done on the cloud. This approach certainly creates a problem in the terms of the speed, as the upload of the image is a time consuming operation.

However, it is necessary to achieve the normalized feature space required for the object classification and also makes the resultant service more widely available, as we do not require any special software of the device for the communication. In the final stage of the service development, we will implement REST-like API for the use in other scenarios (in line with the AI-brick notion).

Such a service can then be utilized in many applications, most notable are the applications of the cloud robotics. An example can be the RoboEarth project [7] which is able to use existing cloud image recognition services like Google Goggles [8].

2.2   Research   Approaches used in Proposal

The image processing is important part of information acquisition for Robots. In image processing a feature space can be used in many forms. We have chosen spectral and also derived descriptors as features for pattern recognition procedure. We are using the SIFT (Scale invariant feature transform) [9] ad SURF (Speeded-Up Robust Features) [10] for features extraction and Membership function ArtMap [11]–[13] and Gaussian classifier for the classification of objects. The main research goal of our work is to adapt these approaches to the cloud environment and to find out which combination of extractor-classifier provides the best results.

We had chosen these two classifiers as the MF ArtMap represents the model-free classifier, whereas the Gaussian represents the model-dependent classifier. One of our research goal is to compare these two methods.

We anticipate the challenge with the adaptation of the classifier methods to the cloud environment. The goal of the proposed system is to provide stable service for all devices connected without regard to the actual number of connected devices. In other words, the service has to be scalable. In the terms of cloud computing that means the virtual machines which are the underlying infrastructure of the service can be at any time rebooted, shut down or started. Therefore the system itself has to be built in a way that reflects these conditions.

We also compare these classification methods to the simple matching, which can prove faster in certain conditions (up to certain number of entries in the table storage).

Another anticipated challenge is how to work effectively with the large sets of data we assume we will amass during the course of experiments and eventual publication of the service for the public use.

As one of the classifier, and the one to be used in the proof of concept experiment, we had considered the use of one from the group of ART neural networks due to the previous experience, more precisely ArtMap [14], [15] neural network subgroup. These networks are able to be trained using supervised learning. Finally, MF (membership function) ArtMap ([13], [16]) neural network was chosen as a classifier. This type of neural network combines theory of fuzzy sets and ART theory. The consequence of this combination is structured output consisting of computed values of the membership function of every found fuzzy cluster of every known class for the input.

This way, it is possible to compute how much the input belongs into every class. The input is classified into the class represented by winner fuzzy cluster. Winner fuzzy cluster is cluster belonging into the output vector, however the value of its membership function is maximal in the output vector.

3   Cloud-Based Image Classification – Software as a Service

We had divided the service into two parts, one called Cloud-based Feature Extraction (CFE) and the second the Cloud-based Classification - CCL. In this section we will talk about the feature extraction part which we had already implemented as a service. In the following text, we will use the abbreviation CFE instead of the Cloud-based feature extractor for describing the service.

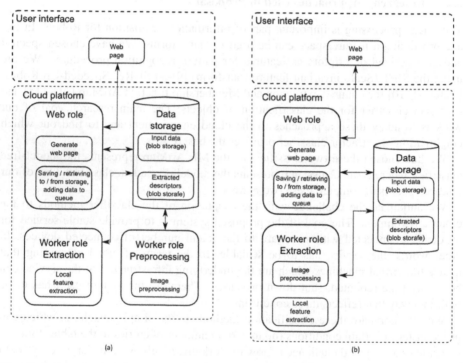

Fig. 2. CFE architectures overview. On the left (a) is architecture version 1, on the right (b) architecture version 2.

3.1   CFE Architecture Version 1

Our first architecture design was to use dedicated roles for extraction and for image preprocessing. The idea was that the image preprocessing was the same for both of the extractors and seemed only fitting to have it scaled automatically based on the actual load. For each of the extractors, separate worker roles were created – for the same reason. The communication with the user was done through another web role.

The inter-roles communication was implemented with the Azure Queues, as compared to the Service Bus Queue have less overhead and are faster. In the queue message, we are sending the unique identification of the image.

The workflow in this architecture was as follows:

1. The user uploads the image via the web page, and chooses the extractor (SIFT, SURF or both of them)
2. The image is stored to the blob storage, and the unique Id of the image is put into the queue for image preprocessing
3. The image preprocessing role accesses the image, and rewrites it with normalized image (scaled down if too big, and set to the shades of gray). Also, the Id of the preprocessed image is put into the queues for selected extractor services
4. The extractor role access the image in the storage by unique Id, and extract local features, which are then stored in blob storage with extractor prefix and image Id. The image and its extracted features are also written to the Azure Table, in which the relations between objects are kept
5. The web page with result table is updated and shows the uploaded preprocessed image along with the extracted features (available as an XML formatted document).

The schema of the architecture can be seen on the left side of the Fig. 2.

This architecture had a drawback in the terms of speed, as can be seen in the Table 1 and Table 2.

3.2 CFE Architecture Version 2

In the second architecture design, we made changes to speed up the process of feature extraction. As can be seen from the **Table 1** and **Table 2** , there is a significant time when the service is literally doing nothing, it just waits for sleep cycle to complete to check the queue for new messages. Since the architecture 1 used 3 queues (with one feeding the other two through the image preprocessing role), we decided to add the image processing to the extraction roles, thereby eliminating the first queue and one worker role. This idea was supported also because the image preprocessing was the least time consuming operation in the cycle.

By the elimination of one worker role, the workflow in architecture 2 changed:

1. The user uploads the image via the web page, and chooses the extractor (SIFT, SURF or both of them)
2. The image is stored to the blob storage, and the unique Id of the image is put into the queue for selected extractor
3. The extractor role access the image in the storage by unique Id, and extract local features, which are then stored in blob storage with extractor prefix and image Id. The image and its extracted features are also written to the Azure Table, in which the relations between objects are kept
4. The web page with result table is updated and shows the uploaded preprocessed image along with the extracted features (available as an XML formatted document).

The schema of the architecture can be seen on the right side of the Fig. 2.

This architecture was quicker than the first. The results of measurements can be seen in the tables Table 3 and Table 4. The speed-up is between 18 and 32%. Currently we are optimizing the code to further speed-up the extraction process.

3.3   Measured Speed Results for the CFE Architectures

For testing, we used 20 images of varying size and complexity, smaller one with resolution 0.16 MPX (mega pixels) and the biggest one had 10.84 MPx. Five of the images were above FullHD resolution. The batch of images can be considered small, but at this stage, we use it only for validation of the design and the rough speed tweaking of the service. After deployment, the testing will be more rigorous with bigger sample size.

We also measured the cloud service run in local emulator, so we can compare these two environments. But even in local emulator, we were using live cloud storage (unemulated), therefore only the roles were run locally.

The infrastructure we used were Small compute instances for all roles, and the sleep cycle for worker roles was set to 2 seconds. We will also experiment with these settings in later stages of research.

In the following tables, the measured values of time taken by the service are shown. The "Time for the user" column shows the time between clicking the upload button and showing the result on the page. The "Sum of time taken by tasks" column shows the sum of time actually consumed by the roles to compute result. The last two rows shows the time for extracting the local features and storing them in storage.

Table 1. Measurements of the CFE architecture 1 - speed on the local emulator

	Time for user	Sum of time taken by tasks [s]	SIFT extraction [ms]	SURF extraction [ms]
min	0:00:02	2.0450	435.1242	710.4513
max	0:04:39	21.4839	8196.7643	12731.7996
Average	0:00:20	5.0472	1860.4650	2287.6808
Median	0:00:05	3.3764	896.0543	1229.4251

Table 2. Measurements of the CFE architecture 1 - speed on cloud environment

	Time for user	Sum of time taken by tasks [s]	SIFT extraction [ms]	SURF extraction [ms]
min	0:00:01	0.9759	197.9281	354.9850
max	0:00:15	11.9751	5967.6276	8374.6194
Average	0:00:04	2.6114	1007.3908	1690.6716
Median	0:00:03	1.7334	473.1524	1074.7474

Table 3. Measurements of the CFE architecture 2 - speed on the local emulator

	Time for user	Sum of time taken by tasks [s]	SIFT extraction [ms]	SURF extraction [ms]
min	0:00:00	1.3733	170.0301	211.0119
max	0:00:10	12.2926	3121.1854	5058.4078
Average	0:00:03	3.0056	632.8390	942.6149
Median	0:00:02	2.3446	369.7710	586.2928

Table 4. Measurements of the CFE architecture 2 - speed on cloud environment

	Time for user	Sum of time taken by tasks [s]	SIFT extraction [ms]	SURF extraction [ms]
min	0:00:00	0.7403	156.2778	169.7089
max	0:00:11	10.0929	3578.1217	7811.9474
Average	0:00:03	1.9533	686.7257	1358.1596
Median	0:00:02	1.3111	349.4731	788.2077

4   Cloud-Based MF ArtMap Classifier

The second part of proposed system is classifier CL – implemented as Software as a service. . Once the image's descriptors are extracted, they are propagated into classifier. The classifier classifies the object on the picture into one of known classes or create new one if the object does not fit to none of known classes.

From the group of ART neural networks we chose ArtMap [14], [15] neural network subgroup. These networks are able to be trained using supervised learning. Finally, MF (membership function) ArtMap [13], [16] neural network was chosen as a classifier. This type of neural network combines theory of fuzzy sets and ART theory. The consequence of this combination is structured output consisting of computed values of the membership function of every found fuzzy cluster of every known class for the input. This way, it is possible to compute how much the input belongs into every class.

We implemented MF ArtMap neural network classifier as separated cloud service. That makes proposed system more modular and allows combination of any classifier and any image descriptor extractors to reach the best results. MF ArtMap neural network is implemented like a data structure. All values of the MF ArtMap classifier and also trained classes, relevant clusters and their settings are stored in cloud table in cloud data store.

During the experiments, we encountered a problem with training new images (**Fig. 3**). Extractor service (CFE) extract different number of descriptors for every input image. This number depends on different factors like a size of the input image, number of detected key points in the image etc. Simultaneously, the MF ArtMap neural network expects constant dimension vector as an input. Therefore, we decided to train MF ArtMap network sequentially - every descriptor as separate input. Once all descriptors of input image are propagated through the MF ArtMap neural network, we obtained vector of values of all membership function of input descriptors to all clusters and all classes. At this point we were able to statistically classify input image into one of known class or create new class of no match was found.

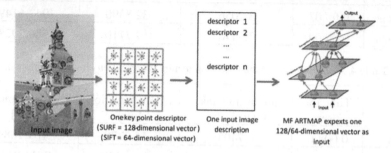

Fig. 3. Graphical description of training problem

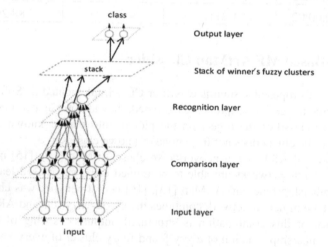

Fig. 4. Modification of MF ArtMap topology for sequential input

Described solution to the problem required the modification of the MF ArtMap topology. On the **Fig. 4** the modified topology is presented. The layer called *stack of winner's fuzzy clusters* has been added. The consequence of this modification is that the output from neural network is not just one winning fuzzy cluster determining the input class, but the output is the set of winner fuzzy clusters. After all descriptors are propagated through the first three layers, the content of the stack is propagated to the output layer, where the winner fuzzy clusters are statistically   evaluated and the class of the input set of descriptors is determined.

4.1   Proof of Concept Experiment

In our experiments with MF ArtMap on the Cloud, we created architecture shown on **Fig. 5**. The robot Nao is capturing the image and send it to the control application on the computer. The Windows Form application relays this image to the cloud service for processing. The image is then processed on the cloud, the features extracted by the CFE service and passed to the MF ArtMap classifier. The result of the classification is then send back to the control application on the computer, which relays the data to the robot. After successful classification, the robot says the result class of the object on the captured image.

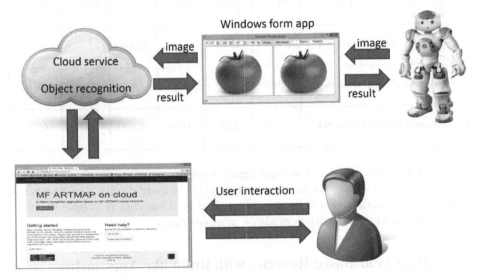

Fig. 5. High-level architecture of the system used as a proof of concept

The experiments were done on two sets of images – set 1 and set 2. First set consisted of logos and simple objects, set number 2 contained images of more complex objects. Both sets were divided 60/40 for learning and testing phase. For comparison, we have used different type of features, SIFT, SURF and spectral RBG features of the image. The results of the experiments are shown in the **Table 5**. The basic intention was to observe a behavior of the CFE on different types of images and there classification accuracy could be influenced by number of features identified on those different type of images. The number of clusters and generalization ability of the MF ArtMap classifier was also observed and taken into consideration. The incrementally of the MF ArtMap classification approach is very good advantage since additional classes will not require the retraining of the neural network but are just incrementally processed in the feature space.

Table 4. Proof of concept – results of the classification using two sets of data

		SET 1			SET 2		
		SIFT	SURF	RGB	SIFT	SURF	RGB
Classification precision	Training set	100,0%	100,0%	90,0%	100,0%	100,0%	91,2%
	Testing set	70,0%	65,0%	70,0%	65,2%	65,2%	56,5%
	Representative set	85,0%	82,5%	80,0%	82,6%	82,6%	73,9%
Number of found clusters		2161	3075	798	2165	2895	681
Generalization of Neural Net		0,223	0,491	0,999	0,149	0,423	0,998

The above classification results are representing average classification rate which was previously evaluated on the contingency tables in more details. The Number of Clusters and Generalization is in correlation since the ideal case is to have few clusters in classification but it also depends on processing data.

5   The Cloud-Based Robotics with Brain like Approaches

Our intermediate goal is to use the gained knowledge to implement an MF ArtMap as a service. The Proof of concept presented in this paper was using the basic structure with MF ArtMap, which was not modified for cloud infrastructure. Regarding this, the synaptic weights of cloud version of MF ArtMap will have to be stored separately. This will allow for easy duplication of trained neural network, or moving the application to more powerful cloud server if there was demand for it. The scaling will then be done by the platform independently of human intervention, thereby providing robustness to the object recognition service. The MF ArtMap will need to be adapted further for the task of object recognition using feature descriptors, as the number of descriptors varies with object. The proof of concept used batch learning, which provided rather unsatisfying results. Therefore, we are working on MF ArtMap input layer modification to allow for inputting all the descriptors at once.

In close future we do believe to add to this framework some brain like approaches mainly from repository maintained by PhysioDesigner project [17]. We believe that implementation of hybrid approaches using selected methods of Artificial or Computational Intelligence and Brain like and more biologically inspirared systems can lead to more accorate results in the Cloud-based framework for Robots. The current testing platform is NAO humanoid robot and we do expect the extend this activity to Pepper humanoid platform next year.

6   Conclusion

We have presented some results of Cloud-based system for Object Recognition useable for Humanoid Robot NAO. We do believe that further work on the approach can be useful for multi robotic platform and also we do expect hybridization of the classical AI approaches with brain like approaches for the benefit if Cloud-based robotic intelligence. We expect problems with the standardization of databases for intelligence including fact that domain oriented knowledge will be preferable and easy to implement versus universal knowledge. Also the learning procedure is expected to be incremental and domain oriented and we do not think that universal learning approach will succeed in the close future.

Acknowledgment. This paper is the result of the Project implementation: University Science Park TECHNICOM for Innovation Applications Supported by Knowledge Technology, ITMS: 26220220182, supported by the Research & Development Operational Programme funded by the ERDF.

References

[1] Kuffner, J.J.: Cloud-enabled robots. In: IEEE-RAS International Conference on Humanoid Robotics (2010)

[2] Mohanarajah, G., Hunziker, D., D'Andrea, R., Waibel, M.: Rapyuta: A Cloud Robotics Platform. IEEE Trans. Autom. Sci. Eng., 1–13 (2014)

[3] Lorencik, D., Tarhanicova, M., Sincak, P.: Cloud-based object recognition: A system proposal. In: Kim, J.-H., Matson, E., Myung, H., Xu, P. (eds.) Robot Intelligence Technology and Applications 2. AISC, vol. 274, pp. 707–715. Springer, Heidelberg (2014)

[4] Ferraté, T.: Cloud Robotics - new paradigm is near. Robotica Educativa y Personal (January 20, 2013)

[5] Mell, P., Grance, T.: The NIST Definition of Cloud Computing Recommendations of the National Institute of Standards and Technology. Nist Spec. Publ. 145, 7 (2011)

[6] Giardino, J., Haridas, J., Calder, B.: How to get most out of Windows Azure Tables, http://blogs.msdn.com/b/windowsazurestorage/archive/2010/11/06/how-to-get-most-out-of-windows-azure-tables.aspx

[7] RoboEarth Project, http://www.roboearth.org/ (accessed: March 20, 2014)

[8] Google Goggles, http://www.google.com/mobile/goggles/#text (accessed: March 20, 2014)

[9] Lowe, D.G.: Object recognition from local scale-invariant features. In: Proceedings of the Seventh IEEE International Conference on Computer Vision, vol. 2, pp. 1150–1157 (1999)

[10] Bay, H., Tuytelaars, T., Van Gool, L.: SURF: Speeded Up Robust Features. In: Leonardis, A., Bischof, H., Pinz, A. (eds.) ECCV 2006, Part I. LNCS, vol. 3951, pp. 404–417. Springer, Heidelberg (2006)

[11] Bodnárová, A.: The MF-ARTMAP neural network. Latest Trends in Applied informatics and Computing, 264–269 (2012)

[12] Smolár, P.: Object Categorization using ART Neural Networks. Technical University of Kosice (2012)

[13] Sinčák, P., Hric, M., Vaščák, J.: Membership Function-ARTMAP Neural Networks. TASK Q 7(1), 43–52 (2003)

[14] Carpenter, G.A.: Default ARTMAP. Boston (2003)

[15] Kopco, N., Sincak, P., Kaleta, S.: ARTMAP Neural Networks for Multispectral Image Classification. J. Adv. Comput. Intell. 4(4), 240–245 (2000)

[16] Sincak, P., Hric, M., Vascak, J.: Neural Network Classifiers based on Membership Function ARTMAP. In: Melo-Pinto, P., Teodorescu, H.-N., Fukuda, T. (eds.) Systematic organisation of information in fuzzy systems, pp. 321–333. IOS Press (2003)

[17] PhysioDesigner, http://physiodesigner.org/

Rubio Simulator

A PBL Based Project of a Rubik Cube Solving Robot Simulator

Bruno Oliveira Cattelan, Ruan Leitõ Nunes, and Dante A.C. Barone

UFRGS, Porto Alegre, Brazil
{bocattelan,rlnunes}@inf.ufrgs.br,
dante.barone@gmail.com

Abstract. The Rubio Simulator came from a project from the Federal University of Rio Grande do Sul, which develop a robot called Robo Rubik[1].The robot was entirely designed by engineering and computer science students at the Federal University of Rio Grande do Sul, and consists of mechanical, electrical and software parts integrating a computing system that was full customized to solve autonomously Rubik cubes at any initial position. However, as the project advanced and started to be more complex, it became impractical to run the robot every time that a test needed to be done, because it took almost two minutes to solve each cube. Also, it was a great risk since the robot's parts were quite expensive, and an accident could cause irreparable damages. Thus the need of a simulator, so that tests could be made without further risks or costs. Both the robot and its simulator have employed extensively Problem Based Learning(PBL)[2].

Keywords: Rubik Cube, Simulator, Software Engineering, Problem Based Learning.

1 Introduction

The *PET Computação* Group[3] from the Federal University of Rio Grande do Sul(UFRGS) is a group of twelve students working on many projects, from research to teaching and outreach activities, which are coordinated by a tutor professor. It is part of a specific Brazilian Ministry of Education program, and its main goal is to collaborate to the intellectual growth and algorithm experience, giving to their integrants better conditions to get a high qualified education, through interdisciplinary and cooperative tasks to be developed. It was created in 1998 and employs extensively PBL methodology for the accomplishment of tasks done by the its constituents, which are undergraduate students of Bachelor in Computer Engineering and Bachelor of Computer Science, undergraduate courses both belonging to UFRGS. The Robo Rubik robot project exists since 2011. It has been only quite recently that the idea of a simulator appeared. More precisely, this idea appeared spontaneously in the year of 2014. It was due to the

© Springer International Publishing Switzerland 2015
J.-H. Kim et al. (eds.), *Robot Intelligence Technology and Applications 3*,
Advances in Intelligent Systems and Computing 345, DOI: 10.1007/978-3-319-16841-8_72

risk of breaking the claws of the robot while testing a new algorithm, which are not available at the Brazilian market of electronic parts, since they are imported and need a long time to be replaced when they break. Also, there were many complaints about the noise coming from the laboratory where the robot was physically located, since it is situated quite close to many lecturer's offices. Furthermore, the Robo Rubik usually takes almost two minutes to complete each cube. As the need to test new possible algorithms grew, it became impractical to use the robot extensively. Also, to testthe speed in which the claws moved would be a risk to the engineers and students that worked near the robot. This problem does not exist in the simulator. After some discussions, a few students gathered and started this new project employing since its beginning the PBL methodology. It was specially well accepted in the group due to its algorithm simplicity. Furthermore, it showed to be a very good way for new students to learn the basics of programming and group cooperation, and also for the possibility to try possible different aspects of robot's real implementations without the necessity of mounting all new possible architectures.

2   The Project

2.1   Problem Based Learning

The PBL method has begun in medicine and then came to be tested in other study areas. It is based in a more liberal way of teaching. Instead of having the students confined in a classroom hearing theoretical subjects, it defends that those students would learn much more if they had instead a problem to solve, and then research and learn by themselves whatever they judged necessary to solve it. This method has already had some interesting results in engineering[3], and this simulator appeared as a good chance to test whether PBL could be efficiently used also in computer science learning. The proposed problem is the construction of a software to simulate the behavior a robot made in other project, the Robo Rubik[figure 1]. This simulator must be able to receive a sequence of movements that the robot should execute, and assess whether they are valid, ie, there was no clash of claws, nor the cube dropped, besides analyzing the sequence of movements, in fact, solve the current configuration rubik's cube.

2.2   The Robo Rubik

The project Robo Rubik was specified in 2011, to be showed at the National Science and Technology Week, SNCT, promoted by the Ministry of Science Technology and Innovation (MCTI), which is a collection of events happening in more than 2000 cities simultaneously in Brazil, in a specific week of October of each year since 2003. In 2014, SNCT will be at its eleventh yearly edition. This project has come to the University as a result of a specific demand from the Brazilian Ministry of Science , Technology and Education to UFRGS's Laboratory of Intelligent Robotics, coordinated also by the Tutor professor of the Computing

PET Group, since the Ministry was already aware about the potentialities of UFRGS's lecturers and students in research and in the promotion of high quality outreach activities aiming the divulgation and popularization of Science. Due to the short time given to accomplish the project, it was not completed for the 2011 event to be held in Brasília, Brazil's capital city. The main parts which were initially completed were: the specification of the project, the purchase of the mechanical and electrical parts and the programming of some subsystems. At half of 2013, with the arrival of new and very motivated skilled students to the Computing PET Group, the idea to relaunch the robot came back strongly.

Fig. 1. The Rubik Robot

The robot is composed of two perpendicular mechanic claws, that act over the cube faces. Each claw is rotated by a servo motor and contains a pneumatic actuator for the opening and closing of the claw. They are controlled by a computer equipped with a webcam, that recognizes the color distribution over the cube and identifies the sequence of movements needed for the resolution. This sequence is then translated to the rotations, opening and closing of the robot's claws, which is transmitted sequentially to the control's subsystem (fig. 3) responsible for commanding the movements of the motors and the pneumatics driving. Since the group had deployed the PBL methodology since the beginning of the project's new phase of development, the set of components were organized in a modular way, where the development of each part was independent of the development of the others. This has permitted that the replacement of some modules by equivalent ones could be provided, in order to comply with the communication protocols established to the sending and receiving data to and from other modules. A flowchart representing the project components, and their intercommunication can be seen at figure 2. The remarkable simple organization of the project had a didactic aspect as goal. In this sense, it was simpler to

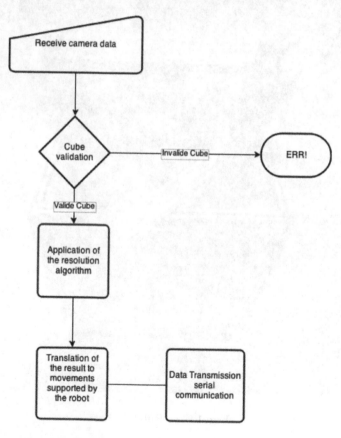

Fig. 2. A Flowchart of the Robo Rubik Behavior

explain the project behavior to new incoming students to the group, abstracting technical details, making it easier to effectively introduce them in the team's work.

2.3   The Simulator

For didactic purposes, it was decided that the computer language to be used would be C++ but limited by mainly using the C functions. The simulator was created in steps; first, it was implemented the cube itself, a series of matrices each one representing a face of the cube[figure 3]. Since this was not aesthetic interesting, a graphic interface[figure 4] was created using the library Allegro[5]. This version was presented in a large event promoted by the University yearly, which is called University Open Doors. Its main objective is to present adequate information for high school students previously to their future undergraduate courses definition. Brazil has a specific entrance to the University exam, called Vestibular, which obligates very young people to decide their future careers at ages of 16 to 17 years old. To complicate things, just UFRGS offers 75 different undergraduate courses, so this is the huge number of options that each student prior to his/her entrance to the University has to face. Later, it were added a series of improvements, such as archive reading to receive a cube from a text file, and a command line option, possibilitating that designers could manually put how is the initial configuration of the cube.

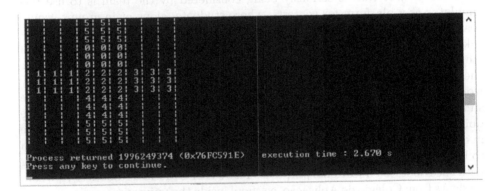

Fig. 3. First graphics of the simulator

Further on, we started adding the claws of the robot, simultaneously creating tests to verify if any accident had occurred. It was then decided that for this first simulator, only some of the possible problems would be taken into account. A list was created, composed of everything that the engineers imagined that could go wrong. From that list, some events were judged more relevant. In this sense the simulator included the possibility to foresee if both claws were open, and most importantly: if the claws crashed into one another, which would cause serious damages to them and a great financial loss to the laboratory.Not only

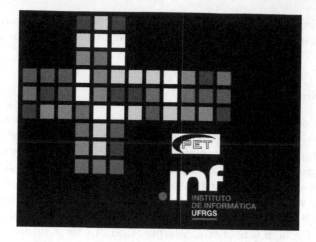

Fig. 4. Graphics of the version presented in UFRGS Portas Abertas

that, but a claw clash would also be a risk to the students around the robot, because small parts could break and hit someone.

2.4   Contests

A future project that is seriously being considered by the team is to use this simulator as a platform for contests, where students, alone or in groups, would search for and create new algorithms for the robot to solve cubes. It would be taken into account the speed in which the cube is solved and the number of steps to get to the final solution. The contestants would also have to consider the opening and closing of the claws and its movements speed, and so they would have to create algorithms accordingly. These would then be put in the simulator, and the group whose solution would present less movements and the quickest solution would win. This would incentivize the students to participate in competitions and to try new approaches to old problems, that sometimes seem to have been already completely solved. Those contests could also be used, along with the robot, to educational purposes. There is in UFRGS a Roboteka[7], where students from Colégio de Aplicação, a school inside the university, can learn using educational robots. This contests would be an interesting way to instigate and encourage those kids into applied sciences, which are not very popular among brazilian students.

3   Conclusion

This is just the first simulator created by UFRGS 's Computing PET group. As the Robo Rubik project grows, so will the Rubio simulator. From all that was learnt in the making of this program, we were able to create not only better algorithms to test the robot, but also ways to improve it. For the next step in our project, we intend to test using the simulator whether it is interesting or not to put more claws into our robot, possibiliting then to speed it and to ameliorate its efficiency. This kind of test would be too expensive to be made without the simulator, and so the computing program, which was initially small, starts to grow and becomes essential the deployment of advanced software engineering techniques and physical new architectures developments as well. As for the contests, depending on the feedback that will be received , the platform could be expanded to an online basis operation where anyone could submit an algorithm to be evaluated. The idea is to base it in some standard and output problems, like the model generally used on programming contests like UVA[6]. The computer science students involved in this project had an incredible opportunity to interact and learn with the computer engineers who worked in the robot. This is a rare opportunity in undergraduate courses, mainly because the study tends to focus only in their specific areas, not comprising any other one. And so, this project and the others which will follow its success will continue to merge both courses, providing a form of studying which the university strongly lacks. Also, the first semester students involved in the simulator claimed to have learned much more about algorithms and programming by working in this project than by their regular classes, because it required knowledge that they would sometimes learn only in the next semesters. This shows a promising future for PBL in computer science as well.

References

1. R.L. Nunes., L.E.P. Mizusaki., F. Ávila.: Soares Dante Augusto Couto Barone Robo Rubik, um projeto para o ensino interdisciplinar de engenharia. In: CAIM 2014 (2014)
2. Savery, J.R., Duffy, T.M.: Problem Based Learning: An Instructional Model and Its Constructivist Framework. Constructivist Learning Environments: Case. Studies in Instructional Design (1996)
3. Martins, I.L.: Educação tutorial no ensino presencial – uma análise sobreo PET. Ministério de Educação. Brasil (2007)
4. Mills, J.E., Treagust, D.F.: Engineering Education Is Problem Based or Project Based Learning the Answer. Australasian Journal of Engineering Education (2004)
5. Official Allegro Web Site, http://alleg.sourceforge.net/
6. UVA Site, http://uva.onlinejudge.org/
7. Mizusaki, L., Barone, D.: Roboteka: Robotic Projects in Brazillian Secondary Schools. In: ICEE: An International Conference on Engineering Education (2011)

Magnetorheological Damper Control in a Leg Prosthesis Mechanical

Cesar H. Valencia[1], Marley Vellasco[1], Ricardo Tanscheit[1], and Karla T. Figueiredo[2]

[1] Electric Engineering Department,
Pontifical Catholic University of Rio de Janeiro (PUC-Rio)
R. Marquês S. Vicente 225 – Gávea, Rio de Janeiro, RJ, Brazil
[2] State University of West District (UEZO)
Av. Manuel Caldeira de Alvarenga 1203 – Campo Grande, Rio de Janeiro, RJ, Brazil
{chvn,marley,ricardo,karla}@ele.puc-rio.br

Abstract. Different models of controllers has proven to be feasible for specific tasks in equipment designed for people with physical disabilities, this paper presents and evaluates the design and simulation of a fuzzy controller for a Magneto-Rheological damper embarked on a Prosthetic Leg Mechanics. As additionally, it was characterized the behavior of the damper in a fuzzy inference system to evaluate its behavior within the proposed control system. The ultimate goal of the control is to decrease the force exerted by the knee; which is used to the full dynamic range of the Magnetorheological Damper (MRD). The extraction of fuzzy rules for the controller used the behavior of the angle of the knee, the knee strength and percentage of gait and for extracting rules were considered MRD damper characteristics of the dynamic response as the strength, the power and the speed of the piston.

1   Introduction

Technological advances offer new challenges in instrumentation and control systems. The case study considered in this work makes use of growing need to develop models that have the ability to adapt and respond appropriately.

The article presents a proposal for fuzzy logic control of a prosthetic mechanical leg used for disabled people. In general, low cost prosthetics have dynamic cushioning systems and those who have them, show a very high market value. Doing the controller design to embark a magnetorheological damper on the prosthesis, would offer a possible alternative with high comfort to the end user.

The controller design was done by modeling human gait, which is recurrent. There have been different ways of subdividing human gait. For the present work, the stages were divided in four, according to the model presented in Bohara (2006).

Usage of a MRD shock absorber is based on its ability to enjoy the properties of the magneto rheological fluid which viscosity is controlled by a magnetic field. The change of state occurs in a short time (*ms*), allowing precise control.

Problem formulation is presented in section 2; modeling the controller that considers the stages of travel and the characteristics of the damper are presented in

© Springer International Publishing Switzerland 2015
J.-H. Kim et al. (eds.), *Robot Intelligence Technology and Applications 3*,
Advances in Intelligent Systems and Computing 345, DOI: 10.1007/978-3-319-16841-8_73

section 3; simulations results are presented in section 4, and finally section 5 reports the findings of the developed work.

2   Problem Formulation

A MRD has the advantage of being a semi active damper, ie, combining the properties of passive dampers for low power required for its control and the advantage of active dampers to have a dynamic response. Its control is a difficult task due to its high nonlinearity. Control algorithms for traditional MRD are highly complex and implementation requires a considerable computational burden, which is why drivers that use methods of computational intelligence as exhibited in Atray (2003), Schurter (2000-2001) and (Jang 1995) are developed obtaining satisfactory results.

Our objectives in this work are generating a Fuzzy controller for the MRD used in mechanical prosthetic legs and validate fuzzy MRD from model to observe its output force to the level of current resulting from the controller.

3   Fuzzy Controller Modeling

To model the controller was established as an objective to reduce the force on the knee in those points that do not modify the maximum force in the knee point where it is equal to 0. For this, there were used angle and force knee values of the 4 modes of human gait according to Bohara (2006).

Next, the used input variables are presented, the characteristics of the MRD, and the controller design.

3.1   Stages of Human Gait

Figure 1 lists the 4 modes of human gait identified Bohara (2006), namely model "Stance Flexion / Extension" mode 2 "PreSwing" mode 3 "Swing Flexion" and mode 4 "Swing Extension".

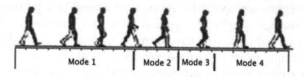

Fig. 1. Stages of Human Gait. Adapted from (Bohara, 2006)

The mathematical model presented in Uyar (2009) was used as a reference. The behavior of human gait, in equation 1 is presented in the model by *Uyar*.

$$
\begin{bmatrix} J_0 + J_1 + 2J_1\cos\theta_2 & -J_1\cos\theta_2 - J_1\cos\theta_2 \\ -J_1\cos\theta_2 - J_3 & J_3 \end{bmatrix} \begin{bmatrix} \overline{\dot{\theta}_1} \\ \dot{\theta}_2 \end{bmatrix} +
$$

$$
\begin{bmatrix} -2J_1\dot{\theta}_1\dot{\theta}_2\sin\theta_2 + J_1\dot{\theta}_2\sin\theta_2 \\ J_1\dot{\theta}_1\dot{\theta}_2\sin\theta_2 \end{bmatrix} +
$$

$$
\begin{bmatrix} G_1\sin\theta_1 + G_2\sin(\theta_1 - \theta_2) \\ -G_2\sin(\theta_1 - \theta_2) \end{bmatrix} = \begin{bmatrix} T_1 \\ T_2 \end{bmatrix} \tag{1}
$$

Where, J_0, J_1, J_2 e J_3 are the parameters of the kinetic energy and G_1 e G_2 are the ones of potential energy in the gait dynamic analysis.

In Figures 2 and 3 are shown the curves of knee angle and force; these being defined in 4 modes of travel.

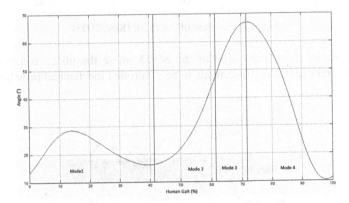

Fig. 2. Knee angle. adapted from (Bohara, 2006)

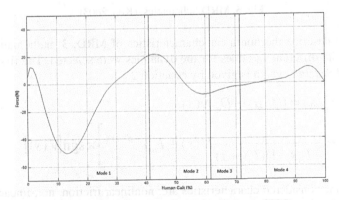

Fig. 3. Force Knee. adapted from (Bohara, 2006)

Experiments for this curves were performed in subjects with 75Kg weight in a complete gait cycle.

3.2   MRD Response

The use of a MRD allows taking advantage of the iron particles content in the fluid. So that, when they are exposed to a magnetic field, they form aligned chains changing the viscosity of the fluid. Figure 4 presents the 3 states of ferrous particles when exposed to a magnetic field.

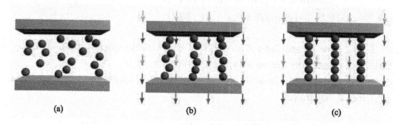

Fig. 4. Activation of the fluid. (Koo, 2003)

Figure 5 shows the composition of the MRD where the piston rod performs the compression and expansion movements to be controlled mechanical prosthesis.

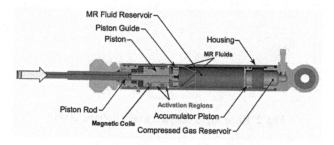

Fig. 5. MRD Schematics. (Koo, 2003)

Trying to describe the nonlinear characteristics of MRD, 3 mathematical models which consider different variables for the dynamics were selected. The first one is the *Milecki and Sedziak*, which is given by equation 2.

$$F_d(s) = ((k_v \mu) + D_{tl})v(s)$$
$$+ \left[\frac{k_h}{(T_m s + 1)(T_e s + 1)} U(s) + F_{tm} e^{-\beta|v(s)|} \right] \times \mathrm{sgn}[V(s)] \tag{2}$$

In that one, the considered characteristics are: nonlinear friction, non-linearities of the magnetic field and the attributes of the magneto-rheological fluid. In the *Bingham* model predicts a Newtonian flow, taking into account the coefficient of viscous damping and the magnitude of the frictional force Coulomb, as shown in Equation 3.

please always cancel any superfluous definitions that are not actually used in your text. If you do not, these may conflict with the definitions of the macro package, causing changes in the structure of the text and leading to numerous mistakes in the proofs.

$$F = f_c \, \mathrm{sgn}(\dot{x}) + c_o \, \dot{x} + f_o \tag{3}$$

Finally, the **Bingham modified** model, or visco elastoplastic, is shown in equations 4 and 5. It takes into account the friction elements over the previous model, the coefficient of rigidity and flow speed.

$$F = k_1(x_2 - x_1) + c_1 \, \dot{x} + f_o \tag{4}$$

$$F = k_2(x_3 - x_2) + f_o \tag{5}$$

In Figure 6 is shown the dynamic damper considering the speed of piston displacement, the applied current, and the generated force.

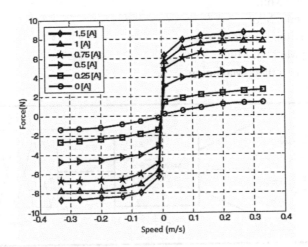

Fig. 6. MRD Dynamic´s. (Tusset, 2008)

In figure 7 is shown an example of an embedded MRD in a mechanical prosthesis "C-Leg from Otto Bock," to absorb knee impacts.

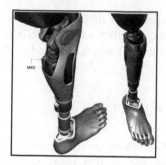

Fig. 7. MRD embeded on a mechanical prosthesis.

3.3   Controller Design

The controller design using fuzzy logic seeks to exercise dominion in the shock absorber variable; in this case said linguistic variable output is current. As input variables were considered the stage of the march, angle, and knee force.

Figure 8 presents the schematic of the designed controller. Said controller has the following settings:

- Type Fuzzy Inference: Mamdani.
- Operator "or": Min.
- Operator "and": Max.
- Defuzzyfication: Center of Mass.
- Implication Method: Min.
- Aggregation Method: Maximum.

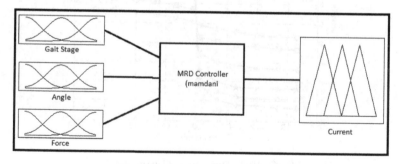

Fig. 8. Controller Block Diagram

In Figure 9, the variable gait stage is shown. This refers to the modes described above, i.e., six words "Esti", "Est1", "Est2", "Est3", "Est4" and "Estf" universe where speech covers a cycle that is a complete gait cycle.

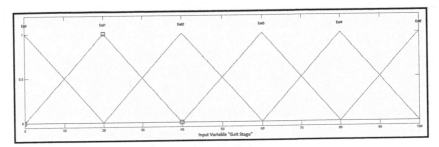

Fig. 9. Sets for Input Variable "Stage Gait"

In figure 10 are presented the terms of the angle variable, which are 6 in total: "Angi" "VLos", "Low", "Medium", "High" and "Angf" with a universe of discourse from 0 to 80 degrees, which are values in the characteristic curve of the human gait

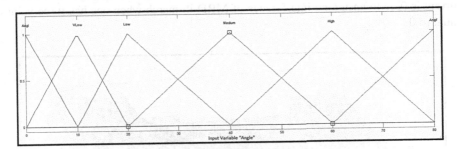

Fig. 10. Sets for Input Variable "Angle"

In figure 11 are presented the 7 terms "Fi", "VLow", "Low", "Medium", "High", "VHigh" and "Ff", these represent the behaviour of the force in Newtons from -70N to 40N. Such values were adopted from the characteristic force curve previously mentioned.

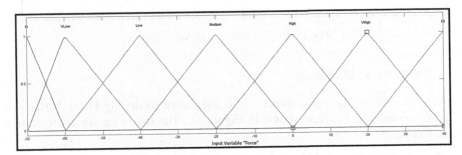

Fig. 11. Sets for Input Variable "Force"

Figure presents 12 the terms used in the output current variable. In total, there are 13 terms that represent typical values that can be used in the MRD.

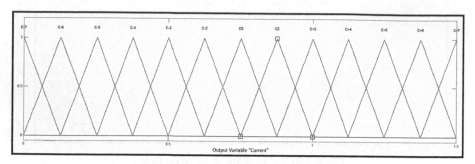

Fig. 12. Sets for Input Variable "Current"

There were 82 used rules in order to reach the initially stated purpose, "decrease the force on the knee at those points where no change is maximum and the force on the knee point where it is equal to 0."

Figure 13 presents the control surface obtained. One can see the regions where the current is 0, ie to ensure no interference of MRD in march when one has force values equal to 0.

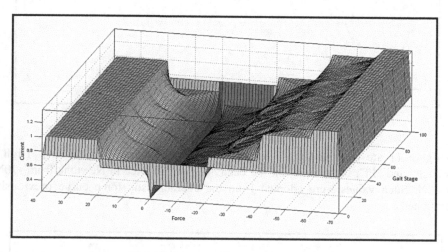

Fig. 13. Current vs Force vs Stage Gait.

3.4   MRD Fuzzy Modelling

In order to evaluate the performance of an alternative modelling Fuzzy MRD was proposed an inference system shown in Figure 14. The following are configuration characteristics.

—   Type Fuzzy Inference: Mamdani.
—   Operator "or": Min.
—   Operator "and": Max.
—   defuzzyfication: Center of Mass.

- Implication Method: Min.
- Aggregation Method: Max.

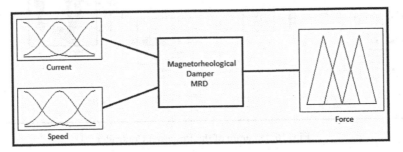

Fig. 14. MRD Block Diagram Modeling

As linguistic input variables were used operating Current with 6 sets, the piston´s speed displacement with 5. As output variable was used the force with 7 sets.

25 rules were generated in full to perform MRD modelling. Figure 15 shows the control obtained surface.

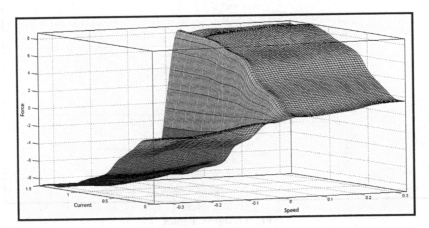

Fig. 15. Force vs Speed vs Current

The presented vales shown in Figure 6 for the control surface are consistent with the MRD dynamic response.

4   Results

Tests were done using the block diagram of Figure 16, where MRD modelling was done with two configurations. The first was with the modified Bingham model while the second used Fuzzy Logic. The best response   was for the first one.

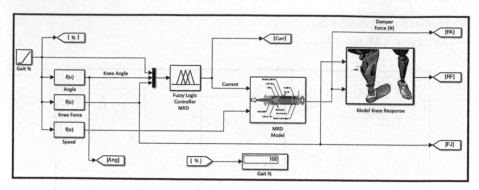

Fig. 16. Diagram of the Proposed Controller

In Figure 17 are shown the signals provided as inputs to the controller. It has the percentage of gait, the angle and strength of the knee.

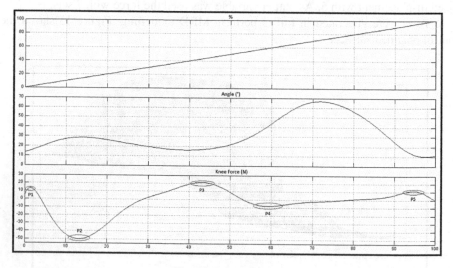

Fig. 17. Inputs Signals

The first result to be displayed is the response using (PID-MRD) PID controller. This driver was developed in earlier work and showed decreased strength in selected points. The response is shown in Figure 18.

In Figure 19 are presented the results using the fuzzy controller and the modified Bingham (FC-MB-MRD) model. It is observed that the results obtained for the selected points were better compared to the previous test.

Finally it is shown in Figure 20 the response using the controller and the Fuzzy MRD modelling (FC-FMRD). It is shown that the results achieved proposed goals, however, with some difficulties.

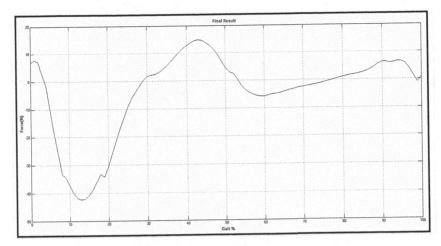

Fig. 18. MRD-PID Result

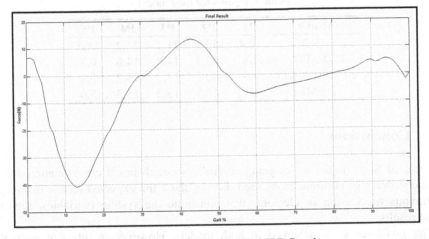

Fig. 19. Fuzzy Controller-MB-MRD Result.

There were selected 5 critical points were the force reached threshold value and it can be observed in Figure 17.

In Table 1 are presented the values of strength at critical points in order to establish the performance of the two sets of models.

The results obtained for knee force after using the MRD fuzzy modeling decreased in selected critical points gait percentage (P1→2%, P2→14%, P3→44%, P4→60% e P5→96%). Besides, those zero points did not have significant changes in magnitude.

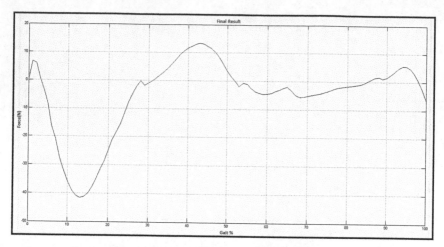

Fig. 20. Fuzzy Controller-FuzzyMRD Result.

Table 1. Results for used model

Model	P1	P2	P3	P4	P5
MRD	12,4	-50,1	21,2	21,2	11,5
PID-MRD	7,8	-42,3	14,8	14,8	6,3
FC-MB-MRD	6,9	-40,7	13,2	13,2	5,1
FC-FMRD	6,7	-41,5	13,3	13,3	5,6

5   Conclusions

The use of fuzzy logic in designing controllers establishes a close relation with a specialist. Various configurations used did not show the expected results inducing a continuous reassessing of elements that constitute the system (variables, sets and rules) results.

The goals set were achieved in both models. However, points that generated unintended slight changes were observed, ie, values resulting force that could disrupt the march.

Settings for different sets of input and output variables are used in the implementation of the controller design. In the case of the input variables of the form sets directly affects the region of influence of the rules being that the triangular sets showed the best response.

The computational cost of the proposed model is similar to PID controller, allowing it to be feasible to use in a real prosthesis, some relevant aspects like a implementation of necessary code in the device hardware are the next research focus.

The ease abstraction of desired behavior for each chosen critical point can justify the decision to work with Fuzzy Logic, always that all characteristics are supervised for the specialist for create the rules bank.

The future research objectives are: proposed model behavior with the real hardware, ease adjusts the rules and model elements for different user configurations and finally implementation cost of proposed model in a prosthesis for default.

References

1. Aphiratsakun, N., Parnichkun, M.: Fuzzy based Gains Tuning of PD controller for joint position control of AIT Leg Exoskeleton-I (ALEX-I). In: IEE International Conference on Robotics and Biomimetics, Bangkok, Thailand, pp. 859–864 (2008)
2. Atray, V.S., Roschke, P.N.: Design, fabrication, testing, and fuzzy modeling of a large magnetorheological damper for vibration control in a railcar. In: IEEE/ASME Joint Railroad Conference, Chicago, IL, USA, pp. 223–229 (2003)
3. Bohara, A.: Finite State Impedance-Based Control of a Powered Transfemoral Prosthesis Tese de Mestrado, Vanderbilt University, EUA, pp. 3016–3021 (2006)
4. Borjian, R., Lim, J., Khemesee, M.B., Melek, W.: The design of an intelligent mechanical Active Prosthetic Knee. In: IEEE Industrial Electronics, Orlando, USA (2008)
5. Wang, D., Liu, M., Zhang, F., Huang, H.: Design of an expert system to automatically calibrate impedance control for powered knee prostheses. In: IEEE International Conference on Rehabilitation Robotics (ICORR), Seattle, USA, pp. 1–5 (2013)
6. Jang, J.-S.R., Sun, C.-T.: Neuro-fuzzy modeling and control. Proceedings of the IEEE 83, 378–406 (1995)
7. Koo, J.-H.: Using Magneto-Rheological Dampers in Semiactive Tuned Vibration Absorbers to Control Structural Vibrations. Doctoral Thesis, Virginia State University, EUA (2003)
8. Nandi, G.C., Ijspeert, A., Nandi, A.: Biologically inspired CPG based above knee active prosthesis. In: IEEE/RSJ International Conference on Intelligente Robots and Systems, Nice, France, pp. 2368–2373 (2008)
9. Schurter, K.C., Roschke, P.N.: Fuzzy modeling of a magnetorheological damper using ANFIS. In: Ninth IEEE International Conference on Fuzzy Systems, FUZZ- IEEE 2000, San Antonio, TX, USA, pp. 122–127 (2000)
10. Schurter, K.C., Roschke, P.N.: Neuro-fuzzy control of structures using magnetorheological dampers. In: Proceedings of the 2001 American Control Conference, Arlington, VA, USA, pp. 1097–1102 (2001)
11. Singh, K., Bhatia, R., Ryait, H.S.: Precise and Accurate Multifunctional Prosthesis Control Based on Fuzzy Logic Techniques. In: International Conference on Communications Systems and Network Technologies (CSNT), Katra, India, pp. 188–193 (2011)
12. Skelly, M.M., Chizeck, H.J.: Real-time gait event detection for paraplegic FES walking. IEEE Transactions on Neural Sustems and Reahabilitation Engineering, 56–98 (2001)
13. Tusset, Â.M.: Controle Ótimo Aplicado em Modelo de Suspensão Veicular Não-Linear Controlada Através de Amortecedor Magneto-Reológico Tese de Doutorado. Universidade Federal do Rio Grande do Sul, Brasil (2008)
14. Uyar, E., Baser, O., Bace, R., Özçivici, E.: Investigation of bipedal human gait dynamics and knee motion control Izmir. Dokuz Eylül University - Faculty of Engineering Department of Mechanical Engineering, Turkey (retrieved August 2009)
15. Weir, R.F., Ajiboye, A.B.: A multifunction prosthesis controller based on fuzzy-logic techniques. In: IEEE Conference on Engineering in Medicine and Biology Society, vol. 2, pp. 1678–1681 (2003)

Appendix: Springer-Author Discount

The appendix should appear directly after the references, and not on a new page

All authors or editors of Springer books, in particular authors contributing to any LNCS or LNAI proceedings volume, are entitled to buy any book published by Springer-Verlag for personal use at the "Springer-author" discount of one third off the list price. Such preferential orders can only be processed through Springer directly (and not through bookstores); reference to a Springer publication has to be given with such orders. Any Springer office may be contacted, particularly those in Heidelberg and New York:

Springer Auslieferungsgesellschaft
Haberstrasse 7
69126 Heidelberg
Germany
Fax: +49 6221 345-229
Phone: +49 6221 345-0

Springer-Verlag New York Inc.
P.O. Box 2485
Secaucus, NJ 07096-2485
USA
Fax: +1 201 348 4505
Phone: +1-800-SPRINGER
(+1 800 777 4643), toll-free in USA

Preferential orders can also be placed by sending an email to orders@springer.de or orders@springer-ny.com.

For information about shipping, please contact one of the above mentioned orders departments. Sales tax is required for residents of: CA, IL, MA, MO, NJ, NY, PA, TX, VA, and VT. Canadian residents please add 7% GST. Payment for the book(s) plus shipping charges can be made by giving a credit card number together with the expiration date (American Express, Eurocard/Mastercard, Discover, and Visa are accepted) or by enclosing a check (mail orders only).

Solar-Hydrogen Energy: An Effective Combination of Two Alternative Energy Sources That Can Meet the Energy Requirements of Mobile Robots[*]

A. Sulaiman[**], F. Inambao, and G. Bright

Mechanical Engineering, University of KwaZulu-Natal,
Howard College Campus, Glenwood, Durban, South Africa
200300000@stu.ukzn.ac.za

Abstract. The continued reliance on fossil fuels for energy, although cheap, is just not sustainable. Fossil fuels have very limited reserves and cannot be expected to last indefinitely. Besides, the process involved in producing energy from fossil fuels releases harmful carbon emissions into the atmosphere resulting in a greenhouse effect and consequently global warming with disastrous effects on climate change. This makes it imperative to seek alternative energy sources which are renewable i.e. in constant supply as well as clean and do not cause harmful effects on climate or cause pollution. Alternative energy sources such as solar and hydrogen, although both clean and renewable, do have their limitations, but this study demonstrates that these challenges can be overcome by effectively combining these two energy sources so that the one complements the other in a way that is workable in a combined system which can meet the energy requirements of modern technologies such as an industrial mobile robot.

1 Introduction

The largest mobilization of people that the world has ever seen on the issue of climate change took place on 21 September 2014, with more than half a million people taking to the streets across the globe in 162 countries to demonstrate their concerns on climate change. The largest demonstration was in the city of New York which was the venue of the UN Summit on Climate Change with about 400,000 people showing up and still more people being turned away by the march organizers because the crowds had swelled beyond the capacity that the route of the march could accommodate.[1]

The number of people that have been affected world-wide and the damages inflicted by extreme weather brought about by climate change has been

---

[*] The author is a candidate for a PhD degree in Engineering in the Mechanical Department, University of KwaZulu-Natal, South Africa.

[**] Corresponding author.

[1] see web-site of Peoples Climate March
http://peoplesclimate.org/media/

© Springer International Publishing Switzerland 2015
J.-H. Kim et al. (eds.), *Robot Intelligence Technology and Applications 3,*
Advances in Intelligent Systems and Computing 345, DOI: 10.1007/978-3-319-16841-8_74

819

unprecedented in recent years.[2] Barely a year ago, the most powerful storm ever recorded in history, namely Typhoon Haiyan, struck the Phillipines on 7 November 2013. This was followed not long thereafter by the Polar Vortex which brought record-breaking cold temperatures from the North Pole to parts of North America in December 2013 and early 2014.

There is now overwhelming evidence that global warming is changing the world's climate.[3] Most notable amongst the causes of global warming is the burning of fossil fuels which is used to produce the world's energy requirements. The world relies largely on fossil fuels for its energy requirements, and it is the combustion of fossil fuels that releases harmful carbons into the atmosphere which contributes to global warming. The current production of energy from fossil fuels causes more damage to the environment than any other single human activity. It contributes to 80% of the air pollution suffered by major cities world-wide and more than 88% of the green-house gas emissions responsible for global warming.[4]

Now more than ever, there is a pressing need on us to reduce our reliance on fossil fuels and develop alternative energy sources. Fossil fuels have very limited reserves and cannot be expected to last indefinitely. We are using up fossil fuels such as coal, petroleum oil and natural gas 100,000 times faster than they are being formed in the earth's layers, with the result that these resources are expected to run out in a matter of a few decades at the very least. To simply go on relying on fossil fuels for energy, although cheap, is just not sustainable. On the other hand, energy derived from alternative sources such as solar, wind, and hydro-power are unlimited, and moreover, very clean in that it does not involve any processes which are harmful to the environment as is the case with fossil fuels which not only affects climate change but also causes pollution.

2   Alternative Energy Sources

2.1   Solar

There is no limit to the amount of sunlight that is generated by the sun. The amount of energy that is received from the sun on the earth's surface is enormous. As a matter of fact, more energy from the sun falls on the earth in one hour than the amount of energy that is used by everyone in the world in one year. And the amount of solar energy reaching the earth in one year is double that of all the energy that can be obtained from all of the other sources such as coal, oil, natural gas and uranium

---

[2] United Nations Office for the Co-ordination of Humanitarian Affairs. see web-site http://www.unocha.org/what-we-do/advocacy/thematic-campaigns/climate-change/humanitarian-impact.

[3] Climate Management : Solving the problem : Casper, Julie Kerr (Global Warming) Facts on File 2010, p. ix.

[4] Ibid. p. 123.

combined. In one square metre of the earth's surface 1,000 watts of energy from the sun reaches the earth.[5]

The only problem with solar energy is that it is dependent on the availability of sunlight. And sunlight is not always available as for instance at night, or even during the day if it is cloudy. This makes sunlight an unreliable source because it cannot provide a constant supply of energy. However, because it is such an abundant source of energy, it could be utilised, during times of availability, to generate whatever energy that is required for consumption at that time, and at the same provide energy to generate additional energy from a secondary source such as Hydrogen [1,2] which could then be made available for consumption during times when solar energy is unavailable.

Solar energy, therefore, on its own cannot provide all of our energy requirements. It is necessary to supplement it with an additional energy source.

2.2   Hydrogen

Hydrogen is another plentiful source of energy. It is the most plentiful element in the universe. Like solar energy, it is clean and also does not require any combustion process that would produce any toxic emissions as in the case of fossil fuels. Hydrogen is the simplest and the lightest element [1-4]. An atom of hydrogen consists of only one proton and one electron, yet it has the highest energy content per unit weight of all fuels. Hydrogen's energy density is 52,000 Btu/lb, which is three times greater than that of gasoline. NASA has used liquid hydrogen since the 1970s to propel the space shuttle and other rockets into orbit.

Despite its simplicity and abundance, hydrogen doesn't occur naturally on the earth like coal, oil or natural gas. It is always combined with other elements. It is found in water (H_2O), which is a combination of hydrogen and oxygen, and it is also found in fossil fuels. So, in essence one has to produce it before you can use it.

From fossil fuels you would need to extract it through a process called steam methane reformation. High temperature and pressure break the hydrocarbon into hydrogen and carbon oxides — including carbon dioxide, which is released into the atmosphere as a greenhouse gas. This obviously is not a clean option as it brings us back to the problem of harmful carbons being released into the atmosphere, thus contributing to global warming and its harmful effects on climate change. This is not a viable option as you are using dirty energy to create clean energy. This doesn't solve the problem – it just moves it around.

A cleaner option would be to extract hydrogen from water. The process by which the hydrogen element can be split from the other constituent element to which it is bound, namely oxygen when it is in the form of water, is by using a process called electrolysis. Electrolysis as the name implies, requires electricity [5].

However, it would be quite a waste to use up one form of energy, namely electricity, just to create another source of energy i.e. hydrogen. Furthermore, the use of electricity which is obtained from the combustion of fossil fuels, would bring us

---

[5] Ibid. p. 130.

back to square one, namely the problem of causing pollution in the process as well as the problem of contributing to green-house gases in the atmosphere which leads to global warming.

But if one is able to use electricity that is produced from a source that is free and renewable then at least one is not wasting one form of energy just to create another form of energy. Furthermore, if the process involved in generating electricity from that source is clean and does not result in harmful waste products such as carbon emissions and toxic gases, which is the case when producing electricity generated from fossil fuels such as coal, then that option would certainly provide a viable alternative. Electricity that is obtained from solar energy provides just the ideal answer to that problem. Solar energy from the sun is both free (apart from the cost of harnessing it), and also a clean source of energy that does not entail any harmful process such as combustion, as is the case for fossil fuels [6].

3   Research

This study demonstrates how a primary source of energy (i.e. solar energy derived directly from the sun), although not in constant supply, can be effectively combined with a secondary source of energy (hydrogen), so as to provide a constant and reliable source of energy to meet the energy requirements of modern technologies such as a mobile robot.

4   The Solar-Hydrogen Energy System Explained

The set-up for a Solar-Hydrogen Energy System is illustrated by the Block Diagram in Figure 1. Thereafter, follows an explanation of each step in the Solar-Hydrogen system, with particular reference to the components that were utilized in this study for the power generation of a mobile robot.

4.1   Photo-Voltaic Cells

The obvious form of energy that we observe from the sun is in the form of heat because it is something that we can all feel and it sustains all life-forms on earth. Normally, when the sun's rays strike an object it makes it hot. This is because in most elements there are electrons which simply vibrate in place when they are struck by solar rays, thereby creating heat. But it has been observed that in certain substances like silicon (commonly used in computer chips), when it is struck by the rays of the sun, it causes electrons to actually move from their resting place. The sunlight knocks electrons loose from their atoms, allowing them to flow freely through in the silicon material. This flow of electrons is a current and is what makes electricity [7].

In order to make the silicon material conduct electricity, it has to consist of two layers like a wafer: a top layer which is doped with Phosphorous to give it a negative charge (-); and the bottom layer is doped with Boron to give it a positive charge (+).

Sunlight consists of little particles of solar energy called photons. When sunlight (or photons) hits the solar cell, the photons are absorbed by the negative layer of the photovoltaic cell and electrons are freed from that layer i.e. the negative layer. These freed electrons will naturally want to migrate to the positive layer. When metal contacts are attached to each layer and a load placed in between, the electrons will flow through the circuit, thereby creating electricity.

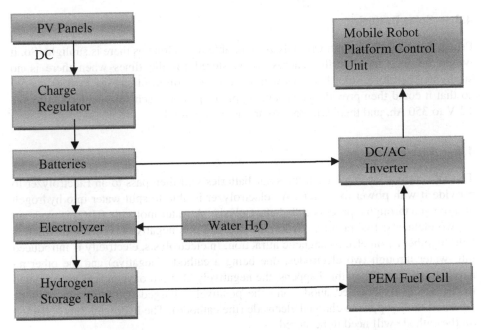

Fig. 1. Block diagram of the proposed power supply for the Mobile Robot

The term "photo-voltaic" gets its name from the process whereby light (photons) is converted to electricity (voltage), hence the term "photo-voltaic".

The basic building block of a PV system is the solar cell. Each individual solar energy cell produces only 1-2 watts. To increase power output, the solar cell is wired together to form a module. These modules (from one to several thousand) are then wired up in serial and/or parallel with one another, into what's called a solar array, to create the desired voltage and amperage output required by the given project.

The electricity that is produced by the PV panels is direct current (DC).

The Photo-voltaic (PV) system used in this study consists of 6 PV modules. The total installed power is 1.2kWp in standard conditions. The specifications of the solar modules are determined as follows: maximum power is 200 W; Open circuit voltage (V_{oc}) is 59.5V; Optimum operating voltage (V_{mp}) is 46.1V; Optimum operating current (I_{mp}) is 4.37A.

4.2   Charge Regulator

The next step in this energy system is a Charge Regulator. As has been noted above, the electrical current that is produced by the PV cells is DC. This current will pass through a Charge Regulator in order to maintain stable current flow which will be regulated at 12V.

4.3   Solar Batteries

Power from the PV system will only be available for as long as there is sunlight. So, it would be necessary for this energy to be stored for the times when there is no sunlight. Therefore, solar batteries will serve as a medium of storage for this energy so that it could then provide a constant supply of power. Each solar battery will have 12 V to 350 Ah, and there will be 4 solar batteries in total.

4.4   Electrolyzer

The DC current available from the solar batteries will then pass to an Electrolyzer to provide it with power to operate. An electrolyzer is able to split water into hydrogen and oxygen through a process called electrolysis. A water molecule (H_2O) is made up of two elements: two positive Hydrogen ions and one negative Oxygen ion which is held together by an electromagnetic attraction. In electrolysis, electricity is introduced into water through two electrodes, one being a cathode (negative) and the other an anode (positive). When this happens, the negatively charged oxygen ions are attracted to the positively charged anode, and the positively charged hydrogen ions will be attracted to the opposite charged electrode (the cathode). The hydrogen thus collected on the cathode will need to be stored.

4.5   Hydrogen Storage Tank

Hydrogen poses very real challenges when it comes to storage. Because it is so light, it occupies a lot of space. As the lightest gas in the atmosphere, it is even lighter than air (14.4 times lighter than air). This means that it would require a large amount of space to store hydrogen and this is what makes it a very cumbersome substance to handle. One, therefore, needs to find a way of squeezing it into a smaller space so that it can be packaged in a form that is more compact and portable enough to be carried on-board a robot in order to use it as a power source. This can be achieved through compression of the gas into a pressurized cylinder.

However, one can achieve higher densities by storing the hydrogen in a liquid form. But this would require extremely low temperatures of -253°C (referred to as cryogenic temperatures) in order to maintain the gas in a liquid state.

Either way, whether as a compressed gas or cryogenic liquid, hydrogen under pressure could quite easily result in a leak. And since hydrogen has one of the widest ranges for the explosive/ignition mixture with air of all gases, this means that whatever the mix proportion between air and hydrogen, a hydrogen leak will most likely result in an explosion with the slightest spark. This is what makes hydrogen so highly flammable.

A promising alternative would be to store hydrogen in a solid state, and this can be done by using a metal alloy. A metal alloy acts like a lattice structure of individual atoms with small spaces in between. When exposed to hydrogen at certain pressures and temperatures, the metal alloy is able to absorb large quantities of the hydrogen gas. The hydrogen gas splits to form two hydride ions. The hydride ion is very small, consisting of only a proton and an electron, and can fit into the small spaces in the metal lattice structure. The metal lattice structure is able to absorb large quantities of the hydrogen and when this happens, the hydrogen gets distributed compactly throughout the metal lattice. When metal alloy combines with hydrogen in this way, it forms simple chemical compounds called metal hydrides. Metal hydrides thus have a higher volumetric hydrogen storage capacity.

Metal hydrides are an efficient method for storing hydrogen. Some hydrides can actually store twice the amount of hydrogen than can be stored in the same volume of liquid hydrogen. When compared with the volume of space that hydrogen occupies in its original form as gas, as a metal hydride the hydrogen is compacted into a solid form which is one thousand times smaller than the original hydrogen gas [8,9].

In this study, therefore, the form in which hydrogen is stored will be a Hydrogen Tank containing a metal alloy in granular form.

4.6   Fuel Cell

We have seen how in electrolysis, when electricity is applied to water (H_2O), the result is that Hydrogen (H_2) and Oxygen (O) is produced. However, since the object is to now get back the electricity for use, one can reverse the process by combining Hydrogen with Oxygen, thereby producing electricity in the process. It is very much like "reverse electrolysis". This process is achieved by means of a Fuel Cell. In simple terms, a Fuel Cell can be considered as an Electrolyzer working in reverse.

In this study, the Fuel Cell is connected to the Hydrogen Storage Tank from which it gets a constant supply of hydrogen. For as long as there is a flow of hydrogen to the Fuel Cell, it will just simply go on producing electricity.

4.7   DC/AC Inverter

The electricity produced by the Fuel Cell is in DC voltage. In order to convert it into the form required for the mobile robot that is used in this study, a DC/AC inverter will be used.

5   The Experiment Set-Up

In the actual experiment set-up, it will not be practical to incorporate all the components of the Solar-Hydrogen Energy system on-board the mobile robot, particularly with regard to the solar panels which require a lot of space. The robot on the other hand needs to be compact so as to be manoeuvrable in the smallest of spaces.

In this model, the setting up of the solar panels as well as the electrolysis process and the production of the hydrogen will be done at a station, or alternatively on-board a second robot which will serve as a Fuel Tanker to supply the main robot which will perform the function designated for that robot.

The Fuel Tanker Robot, therefore, does not need to have size and space constraints of a robot which has been designed for a specific function. The chief function of the Fuel Tanker Robot would be simply to serve as a Fuel Tanker that would be able to go up to the mission robot at a convenient meeting spot for re-fuelling.

The experiment set-up showing the essential components for the power supply on-board the mobile robot, is illustrated in Figure 2.

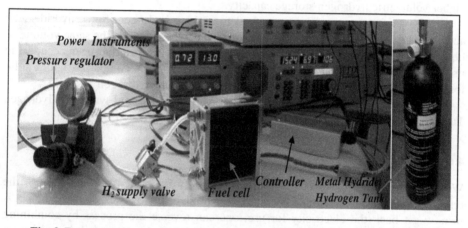

Fig. 2. Experiment set-up showing the Hydrogen Storage System and Power Supply

6   Mechanical Design and Hardware Architecture

The particular type of robot used for this study is designed for an industrial and manufacturing setting, and designated as an Industrial Mobile Robot Platform (IMRP). It serves as a unit load carrier that is capable of carrying a pay-load of up to 40 kg on its deck. The deck is a moving platform, which is powered by a 12V DC Motor. The IMRP has been constructed from square tubing (38mm x 38mm) (see Figures 3 and 4), and its overall gross weight is 120kg.

For improved manoeuvrability, the IMRP has been given omni-directional capabilities i.e. the ability to move not just forward and back-ward, but also side-ways in a crab-like fashion, unlike a conventional vehicle which uses a steering mechanism to steer the two front wheels either to the left or right only, and then too, only whilst the vehicle is moving, which means that the vehicle requires a turning circle and consequently extra room to make the turn. On the other hand, the omni-directional capability for the IMRP is achieved by the use of mecanum wheels (illustrated in Figure 3). A mecanum wheel consists of a hub with a number of individual rollers arranged around that part of its circumference that would ordinarily make contact with the ground but for the fact that these rollers are mounted on its contact surface. Each roller is mounted at a 45° angle to the wheel axis, and is able to rotate about its own axis, whilst at the same time revolving around the axis of the wheel as the wheel itself turns. Using this type of wheel, a vehicle can be propelled directly in a sideway motion without having to make the vehicle move forward at the same time, as you would do for a conventional vehicle. With mecanum wheels, a sideway motion is achieved by controlling the direction of rotation of two selected wheels that would be different from the direction of two other wheels. Similarly, a vehicle can be made to move in various directions by choosing different combinations of wheel rotation direction for the four wheels [10, 11].

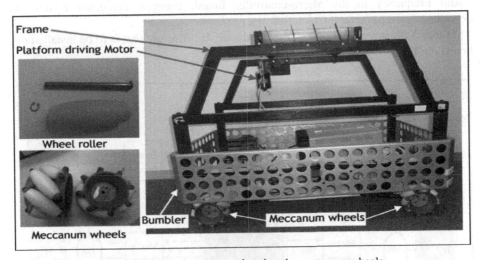

Fig. 3. Mobile robot system showing the mecanum wheels

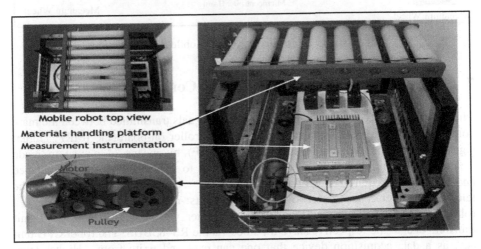

Fig. 4. Mobile robot system showing the front and top view

Each wheel would therefore need to be driven independently and this would require a separate motor for each wheel. Four 12V DC motors are used.

The control of these motors would require a Motor Driver Circuit Board, which in turn will interface with an on-board processor with intelligent functionality, namely a Micro-controller Board (Basic Stamp) that will have the ability to generate four independent PWM signals for each of the four wheels.

The control of movement of each wheel will have to be calculated independently of the other wheels, and the various permutations involved for all four wheels will require higher level computation. This will be carried out by a computer, supported by software specifically written for this purpose. The computer will be controlled by an operator from a remote location, which will transmit digital signals by means of Radio Frequency to the Micro-controller Board, using a Telemetry System as described in the paragraph which follows.

Figure 5 illustrates that part of the Hardware Architecture that is to be installed on-board the IMRP.

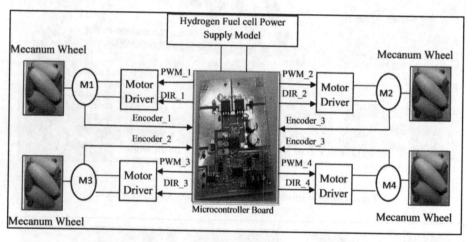

Fig. 5. Hardware architecture of the Mobile Robot system

7 The Telemetry System for Remote Control

The IMRP will be controlled by means of digital signals transmitted between the main computer from a remote spot, and the Micro-controller on-board the IMRP. Digital signals are transmitted via Radio Frequency (RF) by means of a custom made Telemetry System. The system is made according to specifications of the guidance and navigation requirements of a mobile robot. As illustrated in Figure 6, the telemetry system consists of two main components: a USB-Transceiver unit and a robot CPU unit that communicates with each other via RF. The robot CPU unit transfers data to the respective slave modules via the RS485BUS. The robot CPU unit acts as a data acquisition device that one can read and write from. Higher level programming to control the IMRP can be done by the user, as required [10,11].

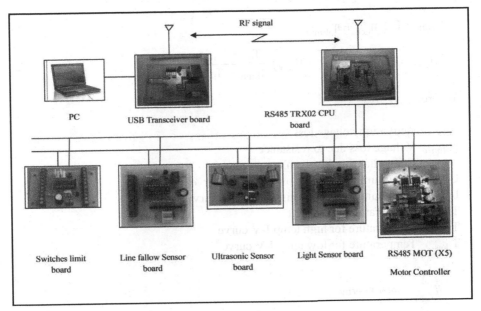

Fig. 6. Mobile Robot Telemetry System Components

8   The Simulation Model

The efficiency of the energy system used in this study is analysed using a software simulation tool. The software chosen for this study is TRNSYS (Transient Energy System Simulation), a very versatile simulation tool with a wide range of use in renewable energy and emerging technologies. The TRNSYS model simulates the performance of the entire energy-system by breaking it down into individual components.

A fuel cell component reads its input such as inlet pressures, physical properties, cell current, number of cells, cooling data and membrane properties, and then runs the sub-routine and calculates the output data such as cell voltage, power and temperature, hydrogen consumption or energy efficiency. By linking the hydrogen consumption output of the fuel cell, and hydrogen production output of the electrolyzer to the hydrogen outflow and hydrogen inflow inputs of the hydrogen tank respectively, the hydrogen tank sub-routine can calculate the hydrogen level in the tank. Upon linking the hydrogen tank output, user power demand and electricity production output of the PV panels to the system controller, the controller can decide how the system should work. The simulation model is shown in Figure 7.

TRNSYS calculates the state of each component at every step. The system consists of several inter-connected components. The components used in the model are: a photovoltaic array module, a fuel cell module, and a hydrogen storage module. The fuel cell converts chemical energy to electricity, in much the same way as a battery. The fuel is pure hydrogen supplied from the hydrogen storage tank. It is possible to use the excess heat from the fuel cell. The cell voltage takes the form:

$$U_{cell} = E + \eta_{act} + \eta_{ohmic} \tag{1}$$

$$U_{cell} = U_{low} + (U_{high} - U_{low}).\frac{T_{fc} - T_{low}}{T_{high} - T_{low}} \tag{2}$$

Where:

η_{act} Activation voltage loss
η_{ohmic} Voltage loss due to resistance
U_{cell} Cell voltage at the given temperature
U_{low} Maximum voltage for low temp I-V curve
U_{high} Maximum voltage for high temp I-V curve
T_{fc} Temperature of fuel cell
T_{high} Temperature for high temp I-V curve
T_{low} Temperature for low temp I-V curve

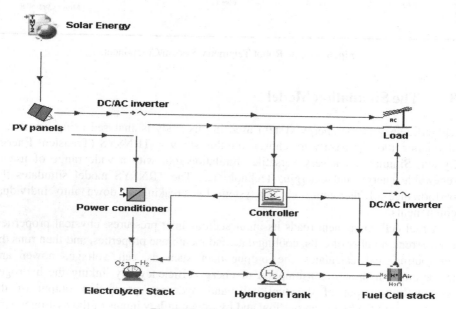

Fig. 7. The Simulation Model

The current for the high (I_{high}) and the low temperature (I_{low}) curve is calculated from this equation 3. To find the current at the working temperature of the fuel cell, the current is calculated by linear interpolation [12-14]:

$$I_{temp} = I_{low} + (I_{high} - I_{low}).\frac{T_{fc} - T_{low}}{T_{high} - T_{low}} \tag{3}$$

Where:

E   Thermodynamic potential
I_{high} Maximum current for high temp I-V curve
I_{low} Maximum current for low temp I-V curve
I_{temp} Maximum current at the given temperature
U_{low} Maximum voltage for low temp I-V curve
U_{high} Maximum voltage for high temp I-V curve
U_{temp} Maximum voltage at the given temperature
T_{fc} Temperature of fuel cell
T_{high} Temperature for high temp I-V curve
T_{low} Temperature for low temp I-V curve

Two main efficiencies are calculated, the electric efficiency, η_{el} and the total efficiency η_{eff}. The reason for calculating two efficiencies is that it is only the electric efficiency that will heat up the cells. The total efficiency also includes the loss of hydrogen that will not heat up the fuel cell (the electric efficiency and the total efficiency will be very close at normal or high production, but will differ at a very low production rate). Thus:

$$\eta_{el} = \frac{V_{fc} \cdot I_{fc}}{V_{fc} \cdot (I_{fc} + k_{kurloss} \cdot V_{fc})}\,_{...} \tag{4}$$

$$\eta_{eff} = \frac{V_{fc} \cdot I_{fc}}{V_{ref} \cdot (I_{fc} + k_{kurloss} \cdot V_{fc} + k_{hydloss} \cdot I_{min})}\,_{...} \tag{5}$$

where:

I_{fc} Current for fuel cell
K Constant for Temperature-Voltage equation
V_{fc} Voltage over fuel cell
V_{ref} Reference voltage
η_{el} electric efficiency
η_{eff} total efficiency

The hydrogen energy storage sub-system, comprising an electrolyzer, hydrogen storage tank, and a fuel cell, is an integral part of a solar-hydrogen power supply system for supplying power to the Mobile Robot. This energy storage is required due to variation of the intermittent and variable primary energy source [14].

9   Conclusion

Studies on alternative energy sources are not a matter of future science any longer, but rather an urgent imperative demanding the attention of present day scientists.

It is somewhat of a fallacy to regard fossil-based fuels as cheap. In reality it carries a huge hidden cost as well as human cost. Already as far back as 2008, the United Nations Office for the Co-ordination of Humanitarian Affairs had already warned that "climate disasters are on the rise. Around 70 percent of disasters are now climate related – up from around 50 percent from two decades ago. These disasters take a heavier human toll and come with a higher price tag. In the last decade, 2.4 billion people were affected by climate related disasters, compared to 1.7 billion in the previous decade. The cost of responding to disasters has risen ten-fold between 1992 and 2008."[6]

"For the last three years, for the first time in history, we're seeing catastrophic disasters that have been more than $2 billion in losses"[7] Major disasters, such as that recently witnessed in the Phillipines when the most powerful storm ever recorded in history struck our planet on 7 November 2013, will serve to mobilise world opinion against the continued reliance on fossil-based fuels, and be an impetus for more studies like this on alternative energy systems.

References

1. Zuttel, A., Borgschulte, A., Schlapbach, L.: Hydrogen as a Future Energy Carrier. WILEY-VCH Verlag GmbH & Co. KGaA (2008)
2. Sherif, S.A., Yogi Goswami, D., (Lee) Stefanakos, E.K., Steinfeld, A.: Handbook of Hydrogen Energy. CRC Press (2014)
3. Pagliaro, M., Konstandopoulo, A.G.: Solar Hydrogen: Fuel of the Future (Society of Chemistry UK) (2012)
4. Luo, F.L., Ye, H.: Renewable energy systems: Advanced Conversion Technologies and Applications. Taylor & Francis Group LLC (2013)
5. Hordeski, M.F.: Alternative Fuels: The Future of Hydrogen. CRC Press (2008)
6. Dincer, I., Rosen, M.A.: Sustainability aspects of hydrogen and fuel cell systems. Energy for Sustainable Development 15, 137–146 (2011)
7. Corbo, P., Migliardini, F., Veneri, O.: Hydrogen Fuel Cells for Road Vehicles. In: Green Energy and Technology. Springer-Verlag London Limited (2011)
8. Broom, D.P.: Hydrogen Storage Materials the Characterization of Their Storage Properties. Springer-Verlag London Limited (2011)
9. Hirscher, M.: Handbook of Hydrogen Storage New Materials for Future Energy Storage. WILEY-VCH Verlag GmbH & Co. KGaA, Weinheim (2010)
10. McComb, G.: Robot Builder's Sourcebook. McGraw-Hill (2003)
11. Xu, W.L., Bright, G., Cooney, J.A.: Visual dead-reckoning for motion control of a Mecanum-wheeled mobile robot. Mechatronics International Journal 14, 623–637 (2003)
12. Lee, S.: Development of a 600 W Proton Exchange Membrane Fuel Cell Power System for the Hazardous Mission Robot. Journal of Fuel Cell Science and Technology (2010)
13. Rubio, M.A., Urquia, A., Dormida, S.: Diagnosis of Performance Degradation Phenomena in PEM Fuel Cells. International Journal of Hydrogen Energy, 1–5 (2009)
14. Ulleberg, O., Morner, S.: TRNSYS Simulation models for solar-hydrogen systems. Solar Energy 59(4-6), 271–279 (1997)

---

[6] United Nations Office for the Co-ordination of Humanitarian Affairs. As reported on its web-site: "Climate Change – Threats and Solutions"

[7] According to Jerry Velasquez from UN Office for Disaster Risk Reduction (UNISDR) as reported by ABC News on its web-site abc.net.au in "Climate change makes super typhoons worse, says UN Meteorological Agency" Thursday, 14 Nov 2013.

Author Index

Printed in the United States
By Bookmasters